FLORA OF IRAQ

VOLUME FIVE

PART TWO

3128
1
2
3
4
W.J.H. delt
Pub by S. Curtis Glazenwood Essex Jany 1 1832.

FLORA OF IRAQ

VOLUME FIVE

PART TWO

LYTHRACEAE TO CAMPANULACEAE

EDITED BY

SHAHINA A. GHAZANFAR

AND

JOHN R. EDMONDSON

With the collaboration of the staff of
the National Herbarium of Iraq of the
Ministry of Agriculture, Baghdad

Published on behalf of the
MINISTRY OF AGRICULTURE
Republic of Iraq
by
ROYAL BOTANIC GARDENS, KEW
2013

First published in 2013 by
Royal Botanic Gardens, Kew,
Richmond, Surrey, TW9 3AB, UK
www.kew.org

Distributed on behalf of the Royal Botanic Gardens, Kew in North America by the University of Chicago Press, 1427 East 60th Street, Chicago, IL 60637, USA

ISBN 978-1-84246-493-9

British Library Cataloguing in Publication Data
A catalogue record for this book is available from the British Library

Design and page layout by Christine Beard
Publishing, Design and Photography
Royal Botanic Gardens, Kew

Frontispiece: *Michauxia laevigata* from Curtis's Botanical Magazine, vol. 59, t. 3128 (1832).

Printed in the UK by Marston Book Services Ltd
Printed in the USA by The University of Chicago Press

CONTENTS

FOREWORD

Work to collect plant samples in Iraq began in the nineteenth century with an increasing number of herbarium specimens preserved since 1920. The first nucleus of an herbarium in Iraq was established in Rustumiya in 1928. Thus, began an activity involving increased cooperation with international institutions, particularly the Royal Botanic Gardens, Kew, and collaboration with foreign experts and Iraqi scientists, greatly assisted by the founding of the National Iraqi Herbarium in Abu Ghraib in 1949.

During this period a range of activities were developed, the most important being the *Flora of Iraq* project which involved many authors and editors from both Iraq and overseas, notably the distinguished Iraqi scientist (late) Professor Dr Ali Al-Rawi, Mr Jan B Gillett and others, under the chief co-editors, Mr CC Townsend and Mr Evan Guest.

Work continued on this project, involving a growing number of Iraqi plant collectors who explored the remotest regions of Iraq and discovered many previously undescribed species. Regrettably it was blocked in the eighties and nineties due to circumstances in Iraq and a shortage of funding. However, after 2003 the new Iraqi Government was determined to rebuild scientific relations between Iraq and the international institutions, particularly Kew Gardens, as well as funding the costs of publication of the *Flora of Iraq* project, and helping to complete the three remaining unpublished volumes.

The book in your hands (part 2 of vol. 5) is the fruit of this cooperation and collaboration. This part describes 16 plant families, from Lythraceae to Campanulaceae, accounts of genera and 322 species many of which are also illustrated.

We can now look forward with confidence to the successful completion of the *Flora of Iraq* project, which provides a means of documenting and managing the remarkable biodiversity of our country.

Mohammed Zain Al-Abdeen Mohammed Raoof
Director General of SBSTS, Iraqi National Herbarium,
Ministry of Agriculture, Abu Ghraib, Iraq

ACKNOWLEDGEMENTS

To pick up an incomplete handwritten manuscript, type, edit and update it into a publishable account for a Flora requires much help and goodwill of many botanists. I (Shahina Ghazanfar) am indebted to John (co-editor) for word-processing all unpublished manuscripts of the *Flora of Iraq* that were written some twenty to thirty years ago (several from longhand, and in some cases from almost illegible handwriting or on paper that was old and faded) to a form where they could be completed, updated and edited.

During the process of completing this Volume, many people from the National Herbarium at Abu Ghraib, Baghdad University, University of Sulaimani, and Nature Iraq have helped me with information on collections and new distribution records from Iraq; to them all we are very grateful. I would like to acknowledge Dr Aqeel Jaber Abbas, Director of the Botany Department, Dr Abdul Latif Rahim, Ikhlas Hussein, Maad Nezar (National Herbarium, Abu Ghraib), Ali Haloob, Saman Abdulrehman, Nabeel Abdulhassan (Nature Iraq and University of Sulaimani), Hadeel Yassen and Maysoon Byati (Baghdad University) for their helpful contributions. I also thank the Bentham-Moxon Trust for funding Ali Haloob to visit Kew and work on the *Flora of Iraq* project. We thank Ihsan Al Shehbaz (Missouri) for his continuous support and help with the many inquiries relating to the Flora, and to Henk Beentje for his generous help with proof reading.

I (John Edmondson) would also like to express my appreciation for the facilities and help afforded by staff at Kew. I am grateful to Michael Pimenov (Moscow) and Mark Watson (Edinburgh) for help and advice with the Umbelliferae, and to Hossain Akhani, Mostafa Assadi and Faroukh Ghahremaninejad (Tehran), Ian Hedge and Sabina Knees (Edinburgh), John Hunnex (Natural History Museum, London), Walter Lack (Berlin) and Karol Marhold (Bratislava) for helping in various ways. I must also thank David Mabberley (Sydney), formerly Keeper at Kew, for taking a personal interest in reviving the project, and the Royal Society of London for funding a study visit by Michael Pimenov to Kew, Liverpool and Edinburgh.

We gratefully acknowledge the considerable legacy of expertise from the original editors, Cliff Townsend and Evan Guest, who defined the format and style of the Flora.

The continuous support of Royal Botanic Gardens Kew Library and Herbarium staff towards this volume is gratefully acknowledged.

INTRODUCTION

to VOLUME 5, PART 2

This latest part of the *Flora of Iraq* covers representative plant families in Iraq of Orders Lythrales to Campanales (families Lythraceae to Campanulaceae), following the sequence set out in Vol. 2 (1965), based on the system of family classification devised by John Hutchinson, formerly Keeper of the Herbarium at Royal Botanic Gardens Kew. Its publication follows a lengthy hiatus during which the systematic revision of families of the *Flora of Iraq* continued, but (due to a lack of funding) the resulting accounts were not edited for publication. All those involved in the project are very pleased that it has proved possible to re-start the process of editing and completing a major floristic project involving close co-operation between the Ministry of Agriculture, The Iraq National Herbarium, Abu Ghraib, Government of Iraq and the Royal Botanic Gardens, Kew.

The process of editing the Flora has involved a number of steps, depending on the degree of completeness of the already existing manuscripts and the availability of taxonomists willing to write new accounts where none previously existed. Happily, most of the accounts for the present Part had already been written, some by staff of the Kew and Baghdad herbaria and others by external contributors. Two people merit special mention here. The late Evan Guest, who co-edited many of the earlier volumes, was a key contributor as well as editor. He collated information on the economic uses and vernacular names (not only in Arabic and Kurdish, but occasionally in other languages) of plants, based on his wide knowledge of the Iraqi flora and his access to relevant information held at Kew. He also applied his vast geographical knowledge to the often problematic question of locating the sites where plants had been collected, standardizing the place-names' nomenclature and publishing a Gazetteer and several Supplements in the Flora itself. It is no exaggeration to state that by helping to launch the project, and bringing the early volumes to completion, he played a key role above and beyond his editorial duties. Cliff Townsend, who worked with Guest throughout the series of early volumes up until the 1980s, was both author and family editor for the largest family in this Part, the Umbelliferae. He also coordinated work by a number of botanists at Kew and elsewhere on the next volume in the series, the soon to be published Vol. 6, covering the large family Asteraceae (Compositae), and supplied many of the accounts of smaller genera himself. His work up until his retirement in 1986 continued even after the publication of parts of the Flora had ceased, with the result that a majority of the accounts of the yet to be published families now exist in draft at various stages of completing and editing.

In preparing these accounts for publication, some improvements have been made in such areas as: the citation of names of authors of scientific names (following the standard list compiled by Brummitt & Powell (1992); family circumscriptions (following the APGIII system); more accurate georeferencing of collection localities, following research into the itineraries of various plant collectors and using the resources of Google Earth and the U.S. Geonames gazetteer; and a wider choice of illustrations, some taken in part from other Floras and others being specially drawn at Kew. The editors have avoided making major changes in the format, wishing to preserve the integrity of the text when viewed as a whole. We have, however, reluctantly ceased citing vernacular names in Arabic script, and have adopted a more selective method of citing herbarium specimens to avoid undue inflation in the size of the volume. Where new, accurately named material has become available subsequent to the preparation of an account, it has been inserted into the account only when it represents a new territory (denoted by a three-letter code), except for very rare species occurring in fewer than five territories, in which case up to three specimens per territory have been cited.

The orthography of scientific names follows the most recent International Code of Botanical Nomenclature, and thereby conforms to the spelling adopted in IPNI (the International Plant Names Index). Minor variations have been acceptable in the citation of book and journal sources in order to match the standard lists published in earlier volumes of the Flora. In a few instances, these have been changed to avoid confusion with other more recently published titles: for example, Karl Heinz Rechinger's *Flora Iranica* was originally cited as *Fl. Iran.*, but with the appearance of a *Flora of Iran* published in Tehran with the abbreviation *Fl. Iran* we have expanded the former Flora's abbreviated title to *Fl. Iranica.*

Collectors' names, too, have been modified somewhat. Certain Iraqi collectors, as cited on specimen labels, spell their names in various different ways, and the preferred standard spelling normally follows the same scheme of transliteration used for geographical names. For example, Al-Kaisi is cited as Qaisi; in every instance the prefix Al- has been omitted for brevity's sake. Collectors are normally cited by surname only, exceptions being necessary to differentiate the Polunin brothers Oleg (O.) and Nicholas (N.) and other similar cases. Only a very few collectors' names have been abbreviated: see list of abbreviations. (Vol. 1, p. 176)

In many instances the accounts appearing in this Flora were written over thirty years ago; knowledge of the taxonomy of the *Flora of Iraq* and adjacent countries has since advanced significantly, allowing a more accurate and complete account to be given of external distribution of species and infraspecific taxa. The most notable sources of floristic data have been *Flora Iranica*, edited by Karl Heinz Rechinger, and *Flora of Turkey and the East Aegean Islands*, edited by Peter Hadland Davis. The former Flora is virtually completed, the latter is not only finished but is also accompanied by several Supplements, and plans are currently being developed for a new *Flora of Turkey* (in Turkish). The *Flora of Iran* (in Persian) is also making good progress, though unlike *Flora Iranica* it does not treat plants occurring in the highlands of NE Iraq. A start has been made on a new *Flora of Arabia*, and *Flora Palaestina* is about to appear in a newly revised edition. It is most gratifying to see the progress in studying the SW Asian flora which has taken place during the period when the *Flora of Iraq* project was dormant, and the compilers of these Floras are about to fill the remaining gaps left by earlier projects such as *Flora of Lowland Iraq* (Rechinger 1956) and the various Floras of Syria, which are now seriously outdated.

A conventional Flora represents only a preliminary stage in developing knowledge of Iraq's plant resources. Many subjects now considered essential for better understanding the ecology and biodiversity of the country are barely touched upon in this Flora, such as the conservation status of rare and endangered flora, the vast changes (and reversals) caused by the drainage (and re-flooding) of the Southern Marshes, and the insidious processes of desertification and deforestation. To a certain extent this Flora is already an historical document, based as it is on records gathered from 180 to 50 years ago for the most part. Its publication offers an opportunity for a new generation of botanists to go out and study the flora in more detail, making use of the greatly improved communications within Iraq and employing modern methods of georeferencing their field collections using GPS and digital photography. Soon, a new type of Flora will emerge which offers new and simpler ways of identifying plants in the field and in the herbarium, as well as providing innovative methods of applying its floristic data for the benefit of Iraq and its people.

Shahina A Ghazanfar
Royal Botanic Gardens, Kew

John R Edmondson
Hon. Research Associate, Royal Botanic Gardens, Kew

LIST OF CONTRIBUTORS

Families treated in Vol 5(2)

Lythraceae – S. A. Ghazanfar, Royal Botanic Gardens, Kew, UK.

Onagraceae – P.H. Raven, Missouri Botanic Garden, St. Louis, MO, USA.

Haloragaceae – C.C. Townsend, Royal Botanic Gardens, Kew, UK.

Gentianaceae – P. Taylor †, formerly Royal Botanic Gardens, Kew

Menyanthaceae – P. Taylor

Primulaceae – P. Taylor

Plumbaginaceae – S. Mobayen, formerly University of Tehran, Iran and University of Montpellier, France

Plantaginaceae – Z. Chalabi-Ka'bi †, formerly Baghdad University, Iraq

Crassulaceae – M. Bywater, Royal Botanic Gardens, Kew

Saxifragaceae – C.C. Townsend

Vahliaceae – C.C. Townsend

Parnassiaceae – J.R. Edmondson, Hon. Research Associate, Royal Botanic Gardens, Kew

Umbelliferae – C.C. Townsend

Valerianaceae – E. Hadač, Institute of Landscape Ecology, Czechoslovak Academy of Sciences

Dipsacaceae – I.K. Ferguson, Royal Botanic Gardens, Kew

Campanulaceae – H.A. Alizzi †, formerly National Herbarium of Iraq, Baghdad

99. LYTHRACEAE

Jaume St.-Hil., Expos. Fam. 2: 175 (1805)

Shahina A. Ghazanfar[1]

Annual or perennial herbs, shrubs and trees. Leaves simple, opposite or verticillate, petiolate to sessile, stipulate with minute stipules or exstipulate, entire, pinnately veined. Inflorescence cymose or racemose or flowers solitary, terminal or axillary, bracteolate. Flowers hermaphrodite, 4 or 6 or 8(–16)-merous, regular or slightly irregular, perigynous or rarely epigynous; androphore present or absent. Sepals united to form a hypanthium, tubular or campanulate, unequal or regular, persistent, ribbed; lobes valvate with small inner triangular appendages in between closing the throat. Petals present or absent, when present often red, purple or orange, free, usually clawed, inserted towards apex of calyx tube, alternating with sepals, folded in bud. Stamens (4–)8–16(–35), adnate to hypanthium below petals; filaments sometimes of different lengths; anthers dorsifixed or versatile, dehiscing longitudinally. Ovary superior (inferior in *Punica*), of 2–4(–6) carpels, 2–6-locular, rarely 1-locular, sessile to stipitate; style 1; stigma ± 1, usually capitate. Fruit capsular or baccate, septicidal or loculicidal or circumscissile, splitting irregularly or indehiscent. Seeds numerous, without endosperm.

A medium-sized family of 625–650 species, cosmopolitan, but widespread in tropical regions, less so in temperate regions. Many species such as *Lagerstroemia speciosa* and *L. indica* (Crepe Myrtle) are cultivated for ornament in warm tropical countries. *Punica granatum* (pomegranate) and *Lawsonia inermis* (henna) are cultivated in many warm and tropical regions of the Old World.

Morphologically the Lythraceae *sensu lato* (including Trapaceae) are varied, without any unique characters that define the family. But, despite the morphological differences and the variable position of the ovary (superior to partly inferior in *Sonneratia*; partly inferior in *Duabanga*; and partly, nearly, or completely inferior in *Punica* and *Trapa*) evidence from molecular data shows *Duabanga* (Indomalaysian), *Sonneratia* (India & Pacific Islands), *Punica* and *Trapa* lie within the Lythraceae.

Four genera and eight species occur naturally in Iraq; one or possibly two species are cultivated.

1. **Punica granatum** *L.* (pomegranate; Arabic: rumman; seeds: chab rumman) is a small tree or shrub with glabrous opposite or fascicled leaves, oblong-lanceolate to obovate; flowers scarlet red or white, showy, 3 cm or more in length; calyx tube adnate to ovary, somewhat fleshy; fruit globose, up to 12 cm in diameter, pale red to scarlet, partitioned by thin leathery yellow septa; seeds red, pink, or white, fleshy, juicy. A native of Iran and NW India, it is cultivated in the central and lower regions of Iraq for its fruit and is also highly regarded for its medicinal properties.

2. **Lawsonia inermis** *L.* (Arabic: henna) is a small glabrous shrub with opposite, elliptic to obovate leaves; flowers 4-merous, white, in terminal panicles, fragrant; capsule ± 5 mm in diameter, dehiscing irregularly. A monotypic genus distributed throughout Africa and Asia. It has been grown in Egypt and the Middle East since ancient times. Its origin is obscure, but it is believed to be indigenous to SW Asia; it is planted and naturalized ± throughout its range. It is cultivated in central and southern Iraq.

A paste made from powdered leaves and water yields an orange-red dye used widely in the eastern world to stain hands and hair during celebrations and otherwise. Also used medicinally as a coolant, against sunburn and itchy skin.

---

[1] Royal Botanic Gardens, Kew, UK.

1. Trees or shrubs 2
   Herbs 3
2. Flowers solitary, large, red, showy with many petals; ovary inferior; fruit 4–12 cm in diameter; seeds covered in juicy pulp (cultivated) . *Punica granatum*
   Flowers in terminal panicles, white with 4 petals; ovary superior; fruit capsular 0.5 cm in diameter; seeds not covered in juicy pulp (cultivated) . . . . *Lawsonia inermis*
3. Flowers solitary or in cymes, axillary; calyx tubular with 6 subulate outer and 6 minute inner appendages . . . . 1. **Lythrum**
   Flowers clustered in leaf axils; calyx campanulate with 4 outer calyx lobes and 4 obsolete inner appendages . . . . 2. **Ammannia**

1. LYTHRUM L.

Sp. Pl.: 446 (1753); Gen. Pl. ed. 5: 205 (1754); Koehne in Engl., Pflanzenr. IV. 216 (1903)

Herbs, annual or perennial rarely shrubby; young branches 4-angled. Leaves opposite and decussate, verticillate or alternate. Flowers solitary in axils, or in cymes or terminal spikes, 6-merous, [mono-, di-, or] trimorphic in relation of filament to style length. Calyx tube (hypanthium) elongate, 6–12 angled or -veined; sepals 4–6; epicalyx present. Petals 4–6 (or absent). Stamens 4–12 (12 or fewer in Iraq plants), in 2 whorls of different lengths, inserted near the base of the tube. Ovary oblong, ± 2-locular; ovules many. Capsule 2-valved, fusiform, included. Seeds 8–many, elongate.

About 35 species, cosmopolitan in distribution; four species found in Iraq.
Lythrum (from the Gr. λύθρον, *luthron*, loosening or atonement).

1. Plant with at least upper parts of stems hairy; flowers in axillary cymes forming terminal spikes; perennial herb/shrub . . . . 1. *L. salicaria*
   Plant glabrous or scabrid; flowers 1 or 2 in axils of leaves; annual herb. 2
2. Calyx tube prominently 12-veined, veins greenish and extending into calyx teeth; calyx teeth distinct, ± 1.5 mm; style exserted . . . . 3. *L. silenoides*
   Calyx tube veins not very prominent and not extending into calyx teeth; calyx teeth minute; style just exserted 3
3. Stems leafy; leaves lanceolate to ovate-lanceolate, 5–15 × 2–4 mm; calyx 5–6 mm . . . . 2. *L. hyssopifolia*
   Stems not very leafy; leaves oblanceolate to oblong-lanceolate to linear-lanceolate, 5–15 × 1–2 mm; calyx tubular, 4–4.5 mm . . . . 4. *L. tribracteatum*

1. **Lythrum salicaria** *L.*, Sp. Pl. 446 (1753); Boiss., Fl. Orient. 2: 738 (1872); Koehne in Engl., Pflanzenr. IV. 216 (1903); Post, Fl. Syria, Palest. Sinai 1: 470 (1932); Polatschek in Fl. Iranica 51: 5 (1968); Dar in Fl. Pak. 78: 7 (1975); Gilbert in Fl. Ethiop. & Eritr. 2(1): 403 (2000); Haining Qin & Graham in Fl. China 13: 276 (2007); Naanaie, Fl. Iran 67: 17 (2010).

Perennial herbs or small shrubs, up to 1.5 m tall; stems erect, branched, 4-angled, upper part hairy. Leaves opposite, sometimes alternate towards the top of stem, sessile, ovate to ovate-lanceolate, 20–85 × 7–35 mm, base rounded, somewhat clasping, apex acute, entire, sparsely to densely pubescent or glabrous or with scabrid margins. Flowers in 1-several flowered cymes forming terminal spikes up to 30 cm; bracts 5–11 × 3–5 mm, broadly lanceolate. Calyx tube 6–6.5 mm, 12-ribbed; sepals ± 1 mm, broadly triangular; epicalyx erect, 1.5–2 mm, exceeding the sepals. Petals purple to pink-purple, oblanceolate, 6–9 mm. Stamens and style exserted. Capsule oblong. Fig. 1: 1–9.

HAB. Damp mountain slopes, gorges, rocky water streams, river banks; alt. 550–1500 m; June–Sept.
DISTRIB. Common in northern Iraq in the mountains, but also found in the foothills and the desert plateau region. **MAM**: 36 km NE Zakho Marga, Qus village, *Rawi* 23529!; Danalok, 16 km E of Amadiya, *Al Dabbagh & Hamid* 45877!; **MRO**: Rowanduz gorge, *Guest* 2993!; 11 km N of Shaglama, *Thamer* 47879!; Berrog mt., on road to Qandil, *Rawi & Serhang* 23946!; **FNI**: Mindan Bridge on Khazir river, *Alizzi & Omar* 35282!; **FNI**: Sa'diya, *Al Kaisi* 47223!; Sangar, *Chakrawarty* 30758!; **LCA**: Zafariniya, *Al Kaisi & Hamad* 47576!; **DWD**: 7 km N of Halabja, *Al Dabbagh* 47009!

Fig. 1. **Lythrum salicaria**. 1, habit flowering branch × $^1/_2$; 2, roots × $^1/_2$; 3, part of flowering branch × 2; 4, part of stem showing leaves × 1; 5, part of stem × 1; 6, flower × 3; 7, 8, sepals and petals × $2^1/_2$; 9, flower with persistent epicalyx segments × 3. Reproduced from Fl. China 13: f. 301, with permission.

Widely distributed worldwide; N Africa west through Arabia to Europe and east to Iran, Afghanistan to India and China; North America; also recorded from southern Africa. $2n$ = 30, 50, 58, 60.

2. **Lythrum hyssopifolia** *L.*, Sp. Pl. 447 (1753); Boiss., Fl. Orient. 2: 739 (1872); Koehne in Engl., Pflanzenr. IV. 65 (1903); Post, Fl. Syria, Palest. Sinai 1: 471 (1932); Rech.f., Fl. Lowland Iraq: 443 (1964); Polatschek in Fl. Iranica 51: 3 (1968); Boulos, Fl. Egypt 2: 145 (2000); Collenette, Wildflow. Saudi Ariabia: 546 (2000); Gilbert in Fl. Ethiop. & Eritr. 2(1): 399 (2000); Naanaie, Fl. of Iran 67: 7 (2010).

Lythrum thymifolium sensu A. Rich. (1847) *non* L. (1753).

Annual herb, up to 50 cm tall; stems ± erect to ascending, branching from base, leafy, woody below, 4-angled above, often reddish-green; the whole plant glabrous. Leaves opposite, sessile, lanceolate to ovate-lanceolate, 5–15 × 2–4 mm, base rounded, somewhat clasping, apex obtuse to acute, entire. Flowers 1 or 2, axillary; bracteoles linear, minute; flowers sessile or pedicel < 1 mm. Calyx tube 5–6 mm, 12-ribbed; teeth ± 1 mm, broadly triangular; epicalyx erect, exceeding the sepals. Petals purple, oblanceolate, 6–9 mm. Style just exserted. Capsule ± 4 mm, just exserted or enclosed in the calyx. Fig. 2: 7–8.

HAB. Ditches, rivers banks, cultivated field, saline swamps; alt. 450–1250 m; Apr.–Sept.
DISTRIB. Common in northern Iraq. **MAM**: Khaira Mt. NE of Zakho, *Rawi* 23252!; Aradin, *Al-Kaisi & K. Hamad* 46043!; **MSU**: Chamchamel, *Rawi* 22854!; Lower Diyala, *Robertson* 1!; **FNI**: Eski Kellek, bank of Zab R., *Gillett* 8181!; **DLJ**: Takrit, *Alizzi & Husain* 33766!; **LCA**: Khalis, saline swamp, *Hunting Tech. Services* 2!; **LEA**: Aziziya, *Khatib & Alizzi* 33480!

Native to southern Europe, now almost cosmopolitan; naturalized in southern Africa, tropical East Africa, N Africa, the Middle East and SW Asia.

3. **Lythrum silenoides** *Boiss. & Noë* in Boiss., Diagn. Pl. Or. Nov. Ser. 2, 2: 55 (1856) & Fl. Or. 2: 739 (1872). — Type: Iraq, Baghdad, *Noë* 293 (G!, holo); Koehne in Engl., Pflanzenr. IV. 68 (1903); Rech.f., Fl. Lowland Iraq: 443 (1964); Polatschek in Fl. Iranica 51: 4 (1968).

Annual herb, up to 40 cm tall; stems ± erect to ascending, branched, woody below, 4-angled or sometimes winged above; the whole plant glabrous. Leaves opposite, sessile, lanceolate to oblong-lanceolate to linear-lanceolate, 5–24 × 1–5 mm, base rounded, somewhat clasping, apex acute, entire. Flowers 1 or 2, axillary; bracteoles minute; pedicel 1–2 mm. Calyx tube 6–7 mm, 12-veined, veins prominent and extending into the calyx teeth; teeth ± 1.5 mm, narrow, acute. Petals purple, oblanceolate, 6–9 mm. Stamens exserted; style exserted beyond calyx teeth. Capsule ± 6 mm, oblong, just exserted or enclosed in calyx. Fig. 2: 1–6.

HAB. River banks, open marshy places and near water; alt. 1000–1200 m; Aug.–Oct.
DISTRIB. Found in northern Iraq. **MAM**: Sarsang, *Wheeler-Haines* W1342!; **MSU**: Ahmad Kolwan, nr. Penjwin, *Al-Radzi* 5327!; Bakraja, Sulaimaniyah, *Al Radzi* 5247!; Sulaimaniyah, *Qardar* 5378!; **FNI**: Mindan Bridge, *Alizzi & Omar* 35294!; **FUJ**: Sheilkh Humus Zimmar, Mosul, *Habib Mashtouf* 5205!

Iran, Afghanistan.

4. **Lythrum tribracteatum** *Salzm. ex Spreng.*, Syst. Veg. 4(2): Cur. Post. 190 (1827); Boiss. Fl. Or. 2: 740 (1872); Koehne in Engl., Pflanzenr. IV. 64 (1903); Post, Fl. Syria, Palest. Sinai 1: 471 (1932); Polatschek in Fl. Iranica 51: 2 (1968); Zohary, Fl. Palaest. 2: 369 (1972).

[incl. *Lythrum tribracteatum* Salzm. ex Spreng. var. *majus* (DC) Maire (1940)]
L. tribracteata Salzm ex Benth., Cat. Pyr. 98 (1826) – nom. nud.
L. bibracteatum Salzm. Ex DC., Prodr. 3: 81 (1828) pro. syn.

Annual herb, up to 35 cm tall; stems ± erect to prostrate to ascending, branched, 4-angled or sometimes winged above; the whole plant glabrous to scabrid. Leaves opposite, sessile, oblanceolate to oblong-lanceolate to linear-lanceolate, 5–15 × 1–2 mm, base rounded, somewhat clasping, apex obtuse, entire, ± glabrous to scabrid on margins. Flowers 1 or 2, axillary; bracteoles linear, ± 1 mm; pedicel ± 1 mm. Calyx tube cylindrical, 4–4.5 mm, 12-veined; teeth triangular, < 1 mm. Petals purple to pale purple, oblanceolate, ± 3 mm. Stamens just exserted; style and stigma included. Capsule ± 4 mm, oblong, glabrous, included in the calyx.

Fig. 2. **Lythrum silenoides**. 1, habit × 2/3; 2, flower in axil × 2; 3, flower, side view × 8; 4, calyx with stamens and styles × 8; 5, capsule with calyx remains × 8; seeds × 24. **Lythrum hyssopifolia**. 7, habit × 1; 8, flower, side view × 8. 1 from *Al-Radzi* 5327; 2, 3, 4 from *Wheeler-Haines* W1342; 5, 6 from *Qardar* 5378; 7, 8 from *Omar et al.* 37295. © Drawn by J.E. Beentje (Feb. 2011).

HAB. Damp places, ditches; alt.?200 m; May.
DISTRIB. **FNI**: Bagsaiya, *Hadač, Harris & Waleed* W1950!
NOTE. A single collection from the Iraq-Iran border area collected from a dry borrow pit beside road in irrigated countryside.

Mediterranean region, N Africa, Palestine and Upper Jordan Valley, Iran, Afghanistan to southern Russia; introduced and becoming weedy in North America.

SPECIES EXPECTED TO OCCUR IN IRAQ

Lythrum thesioides *M.Bieb.*, Fl. Taur.-Cauc. 1: 307 (1808); Boiss., Fl. Orient. 2: 740 (1872); Polatschek in Fl. Iranica 51: 5 (1968).

There is a fragment of a specimen collected from Sarsang (*Wheeler-Haines* W1348) identified as ?*L. thesoides* M.Bieb. which may belong to this species. The fragment cannot be identified with certainty, but the species is found in the southern Mediterranean, southern Russia, Iran, and Afghanistan and is likely to be present in Iraq. It is distinguished by its small calyx tube (1.75–3.5 mm), campanulate in fruit; teeth 0.5 mm, ovate-lanceolate; petals minute or absent; stamens included; capsule just exceeding the hypanthium.

2. **AMMANNIA** L.

Sp. Pl.: 119 (1753); Gen. Pl. ed. 5: 55 (1754); Koehne in Engl., Pflanzenr. IV. 216: 42 (1903); Graham in J. Arn. Arb. 66: 395–420 (1985)

Annual or short-lived perennials of aquatic or wet habitats, glabrous. Leaves decussate, sessile, linear to lanceolate, cordate or auriculate at base. Flowers in sessile or pedunculate axillary cymes, regular, 4-merous (in Iraq); bracteoles 2. Calyx tube campanulate, 8-veined; calyx lobes 4, with appendages equaling or shorter than the lobes or absent. Petals 1–4 or absent, falling soon. Stamens 4(–8), included to exserted. Ovary incompletely 2(–4)-locular; stigma capitate. Capsule membranous, dehiscing irregularly. Seeds obovoid.

About 25 species, cosmopolitan in distribution, but mostly tropical African.
Ammania, commemorating Paul Amman (1634–1691), a German scholar, physician and botanist from Leipzig.
A complex genus with several species showing polymorphic characters and wide distribution ranges, much in need of a revision with molecular analyses.

1. Cymes pedunculate; styles ≥ 1 mm long. 2
   Cymes sessile; styles < 1 mm long 3
2. Cymes ± lax, usually 3-flowered, often the lateral flowers with longer pedicels; peduncle 0.5–2mm; pedicel up to 2 mm; calyx tube pinkish, with prominent veins. . . . . 2. *A. multiflora*
   Cymes ± congested; peduncle up to 1 mm; pedicel ± 1 mm; calyx tube, greenish, veins not prominent . . . . 1. *A. auriculata*
3. Calyx tube glabrous; style ± 0.5 mm . . . . 3. *A. baccifera*
   Calyx tube papillose to scabrid; style absent . . . . 4. *A. verticillata*

1. **Ammannia auriculata** *Willd.*, Hort. Berol. 1: 7, t. 7 (1803); DC., Prodr. 3: 80 (1828); Boiss., Fl. Orient. 2: 743 (1872); Koehne in Engl., Pflanzenr. IV. 216: 45, fig. 5B (1903); Hutch. & Dalz., Fl. W. Trop. Afr. ed. 2, 1: 164, fig. 61 (1954); Rech.f., Fl. Lowland Iraq: 443 (1964); Polatschek in Fl. Iranica 51: 7 (1968); Dar in Fl. Pak. 78: 8 (1975); Graham in J. Arn. Arb. 66: 403 (1985); Verdcourt in Fl. E. Trop. Afr. Lythraceae: 37 (1995); Boulos, Fl. Egypt 2: 147 (2000); Collenette, Wildflow. Saudi Ariabia: 546 (2000); Gilbert in Fl. Ethiop. & Eritr. 2(1): 403 (2000); Ghazanfar, Fl. Oman 2: 67 (2007); Haining Qin & Graham in Fl. China 13: 276 (2007).

Ammannia arenaria Kunth, Nov. Gen. Sp. 6: 190b (1824).
A. auriculata Willd. var. *arenaria* (Kunth) Koehne in E.J. 1: 245 (1880).

Annual herbs, 15–50 cm tall, unbranched or branched with stems ascending, stems 4-angled to almost winged, glabrous or sometimes minutely scabrous. Leaves opposite, narrowly lanceolate or oblong-lanceolate to linear-oblong, 10–65 × 0.5–10 mm, base cordate-auriculate, clasping, apex acute to obtuse, margins entire, 1-veined. Flowers (1–)3–7(–15)

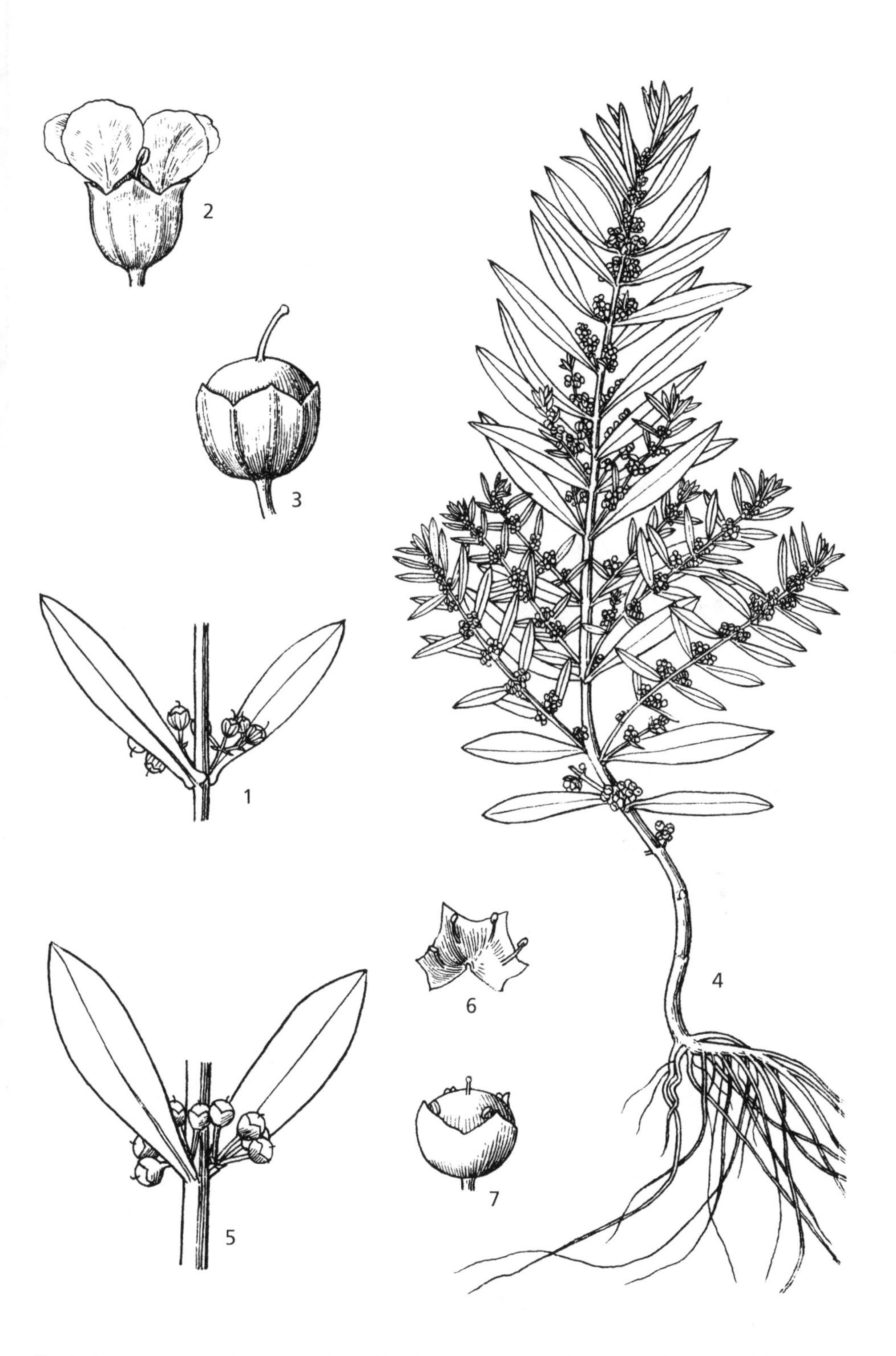

Fig. 3. **Ammannia auriculata.** 1, portion of fruiting branch × 1; 2, flower × 10; 3, fruit × 10. **Ammannia baccifera.** 4, fruiting branch × 2; 5, portion of fruiting branch enlarged; 6, flower enlarged; 7, fruit × 8. Reproduced from Fl. China 13: f. 292, 5,6,7, with permission.

in congested pedunculate axillary cymes; peduncle ± 1 mm; pedicel 1–2 mm; bracteoles linear, ± 1 mm, not reaching the calyx tube. Calyx tube campanulate to urceolate, 1–2 mm, 4 (–8)-ribbed; lobes triangular; appendages minute. Petals reddish purple, suborbicular, ± 1.5 mm, sometimes absent. Stamens exserted. Style as long as or longer than ovary. Capsule 1.5–3 mm in diameter, exceeding the calyx lobes. Fig. 3: 1–3.

HAB. Wet, marshy and irrigated places, rice fields, rivers banks; alt. 10–30 m; fl. Aug.– Nov.
DISTRIB. Common in central and eastern Iraq. **LEA**: 30 km from Aziziya to Kut, *Azizi & Omar* 34626!; 40 km S of Kut, !; **LCA**: Nasiriya, *Rawi* 16594!; Baghdad, *Guest* 13588!; Zafariniya, on banks of Tigris, *Gillett* 5863!; **LSM**: N of Al Azair, Kor Al-Kasser, *Rawi* 16549!
NOTE. A widely distributed, polymorphic species with variations in the number of flowers in cymes, length of peduncle and pedicel size of petals, and length of style. Difficult to tell apart from *A. multiflora* Roxb., *A. senegalensis* Lam. (1791) (distributed in E & W tropical Africa, southern Africa, Egypt) with overlapping characters. In Iraq *A. multiflora* is recorded from higher altitudes.

Pantropical; Africa through Arabia, Iran, Afghanistan to India and China; Australia, the Caribbean Islands, W Indies, N C & S America.

2. **Ammannia multiflora** *Roxb.*, Fl. Ind. 1: 447 (1820); Boiss., Fl. Orient. 2: 743 (1872); C. B. Clark in Fl. Brit. Ind. 2: 570 (1879); Koehne in Engl., Pflanzenr. IV. 216: 48 (1903); Polatschek in Fl. Iranica 51: 7 (1968); Dar in Fl. Pak. 78: 9 (1975); Haining Qin & Graham in Fl. China 13: 275 (2007).

Annual herbs, 15–30 cm tall, unbranched or branched with stems ascending, stems 4-angled to almost winged, glabrous or sometimes minutely scabrous. Leaves opposite, narrowly lanceolate to linear to linear-oblong, 8–25 × 1–3 mm, base cordate, clasping, apex acute to obtuse, margins entire, 1-veined. Flowers (1–)3–7 in lax to congested pedunculate axillary cymes, usually 3-flowered, often the lateral flowers with longer pedicels; peduncle 0.5–2mm; pedicel up to 2 mm; bracteoles linear, ±1 mm, not reaching the calyx tube. Calyx tube reddish green, campanulate to urceolate, ± 1.5 mm, 4–8-ribbed; lobes triangular; appendages minute. Petals reddish purple, suborbicular, ± 1.5 mm. Stamens exserted. Style ± as long as the ovary. Capsule 1.5–3 mm in diameter, exceeding the calyx lobes, reddish brown to reddish wine in colour.

HAB. Wet and marshy places; banks of water channels; rice fields; alt. 20–700 m; fl. Aug.–Oct.
DISTRIB. Common in rice fields in eastern Iraq. **FNI**: Jalaula, *Agnew, Hadač & Waleed* W 1833!; **FAR**: Arbil, near Kani Rash, *Gillett* 9655!

Tropical and north Africa through Arabia, Iran, Afghanistan to India and China; Australia.

3. **Ammannia baccifera** *Linn.*, Sp. Pl. 1: 120 (1753) & ed. 2: 175 (1762); Hiern in Fl. Trop. Afr. 2: 478 (1871); Koehne in Engl., Pflanzenr. IV. 216: 53, fig. 5M (1903); Hutch. & Dalz., Fl. W. Trop. Afr. ed. 2, 1: 165 (1954); Polatschek in Fl. Iranica 51: 8 (1968); Dar in Fl. Pak. 78: 9 (1975); Graham in J. Arn. Arb. 66: 405 (1985); Verdcourt in Fl. E. Trop. Afr. Lythraceae: 45 (1995); Boulos, Fl. Egypt 2: 147 (2000); Collenette, Wildflow. Saudi Ariabia: 546 (2000); Gilbert in Fl. Ethiop. & Eritr. 2(1): 405 (2000); Ghazanfar, Fl. Oman 2: 67 (2007); Haining Qin & Graham in Fl. China 13: 275 (2007).

Ammannia attenuata A.Rich., Tent. Fl. Abyss. 1: 278 (1847).

Annual or perennial herbs, 20–70 cm tall, stems branched, often woody below, ascending, 4-angled to narrowly winged above, glabrous. Leaves opposite, narrowly lanceolate or oblong-lanceolate to oblanceolate, 8–40 × 1.5–8 mm, base cuneate to rounded to cordate, clasping, apex acute to obtuse, margins entire, 1-veined. Flowers 3–15 or more in dense, sessile, axillary cymes; pedicel ± 1 mm; bracteoles linear, ± 1 mm, not reaching the calyx tube. Calyx tube broadly campanulate, tapering at base, 1–2 mm, becoming globose in fruit to 2 mm in diameter; lobes triangular; appendages lacking. Petals absent. Stamens 4, included or barely exserted. Style shorter than ovary, ± 0.5 mm; Capsule ± 2 mm diameter, just exceeding the calyx lobes. $2n = 24, 26$. Fig. 3: 4–7.

HAB. Wet and marshy places, rice fields, rivers banks; alt. 10–20 m; fl. Aug.–Dec.
DISTRIB. Common in central and eastern Iraq. **LEA**: 30 km from Aziziya to Kut, *Alizzi & Omar* 34623!; **LCA**: Baghdad, *Wheeler Harris* W1461!; Abu Ghuraib, *Alizzi & Omar* 34641!; Suwaira, on bank of R. Tigris, *Hashim Amin* 35815!; **LSM**: Abu Chesaf, 25 km S of Al Kahla, *Thamer* 46660!

Tropical and north Africa, Arabia east through Iran to Afghanistan, India, China to Vietnam and Malaysia; Australia and Caribbean Islands.

3. **Ammannia verticillata** (*Ard.*) *Lam.*, Encycl. Meth. Bot. 1: 131 (1783); Boiss., Fl. Orient. 2: 743 (1872); Koehne in Engl., Pflanzenr. IV. 216: 51 (1903); Rech.f., Fl. Lowland Iraq: 442 (1964); Polatschek in Fl. Iranica 51: 8 (1968); Dar in Fl. Pak. 78: 11 (1975).

Cornelia verticillata Ard., Animadv. Bot. Spec. Alt. 2: 9, tab.1 (1764); *Ammannia caspica* M.Bieb., Fl. Taur. Cauc. 2: 457 91803).

Annual herbs, 15–40 cm tall, stems simple or branched, often woody below, ascending, 4-angled to narrowly winged above, glabrous. Leaves opposite, lanceolate to oblong-lanceolate to linear-lanceolate, 7–60 × 2–8 mm, base cuneate to rounded, clasping, apex acute to obtuse, margins entire, 1-veined. Flowers 3–5 in sessile, axillary cymes; pedicel ± 1 mm; bracteoles minute. Calyx tube papillose to scabrid, campanulate to tapering at base, ± 1 mm, becoming globose in fruit; lobes ± 0.5 mm, triangular; appendages prominent. Petals absent. Stamens 4, included or barely exserted. Style absent; stigma sessile to ± sessile. Capsule ± 1.5 mm in diameter, included in the calyx or barely exceeding the lobes.

HAB. Wet and marshy places, ditches, rice fields; alt. 20–900 m; fl. July–Sep.
DISTRIB. Northwestern Iraq. **MAM**: Sarsang, *Haines* W1009!; S of Sharanish village, *Rawi & Nuria* 29059!; **MSU**: Sulaimaniya, *Faddou* 5335!; **FUJ**: Zimmar, *Habib Mashtuf* 5201!; **FPF**: Jalaula, *Haines* W1832!; **LCA**: 5 km WSW of Shamiyia, *Alizzi & Mohammed* 32144!

Iraq to Pakistan and SE Russia.

100. ONAGRACEAE

J.Lindley, Nat. Syst. Bot. ed. 2 (1836)

P.H. Raven[2]

Perennial herbs, rarely annuals, shrubs or small trees. Leaves alternate or opposite, mostly simple; stipules usually absent. Flowers bisexual (very rarely unisexual), mostly actinomorphic, often perigynous, usually with a ± conspicuous hypanthium above the ovary upon which the sepals, petals and stamens are attached. Sepals 2–6, often 4. Petals equal in number to the sepals or wanting. Stamens equal in number to sepals and opposite them or twice as many, in which case the epipetalous ones are usually shorter. Ovary inferior, usually 4 or 5-locular, rarely 1–6-locular, the ovules usually numerous. Fruit usually a loculicidally dehiscing capsule, rarely a berry or indehiscent. Seeds lacking endosperm.

The family is best represented in western North and South America, and consists of 650 species in 17 genera.

A family of little economic importance save for the few general commonly cultivated for their showy flowers, such as *Clarkia, Oenothera, Gaura* and *Fuchsia*. Ba'ali (1946) mentions two species of *Clarkia* (*C. elegans* (= *C. unguiculata*) and *C. pulchella*) and one of *Gaura* (*G. lindheimeri*) among the spring flowers sometimes grown in gardens in Iraq: they are annual herbs with purple, lilac, crimson, rose to white flowers.

The Common (Eng.) or European (Am.) Evening Primrose (*Oenothera biennis* and related species) are also occasionally cultivated in gardens at Baghdad. Though really biennial, these graceful erect herbs flower freely during their first summer, producing a succession of large pale yellow flowers opening in the evening. The fragrantly-scented flowers of these plants are pollinated by nocturnal flying insects, such as hawkmoths (*Sphingideae*), for which they are well adapted. Although originally native of eastern temperate North America, they have been cultivated in Europe for some 350 years and are now widely naturalized there in such habitats as river banks, thickets and sandy places. The root resembling a short red carrot is edible and has been a popular component of salads and soups, or boiled as a vegetable. Eaten as a "mezze", or aperitif "hors d'oeuvres", it enjoyed the reputation of a stimulant to wine-drinking, to which the Latin name of the genus *Oenothera*, derived from an ancient Greek name implying "wine-scenting", is held to be an allusion.

Ba'ali states that one of the species of this group, *Oe. lamarkiana* (= *Oe. erythrosepala*) is sometimes cultivated in our gardens. According to Chittenden (1951) this garden species originated from a mutation in England about a hundred years ago. The peculiar genetic behaviour of the species of this

[2] Missouri Botanic Garden, St. Louis, MO, USA.

alliance, which is attributable to the inclusion of all their chromosomes in a ring at meiotic metaphase I, led de Vries to his classical mutation theory which, although based on some inaccurate premises, has been a cornerstone of the science of genetics. *Oenothera*, as *Epilobium* (q.v.) and other genera of the Onagraceae, has shown itself to be a successful colonizer. Some species readily escape from cultivation and have become naturalized in distant parts of the world far beyond their original areas of natural distribution. For example, Eig, Feinbrun and Zohary (1934) observed that *Oe. drummondii*, a species recommended by Stout (1935) for cultivation in Egyptian flower gardens, and possibly also sometimes grown in gardens in Iraq, had become completely naturalized in Palestine, though strictly limited to the coastal plain of Jaffa. They cite this as an interesting example of rapid invasion, the plant having probably been introduced via the ports of Jaffa or Haifa some 50–60 years ago. The species was first recorded from this locality by Bornmüller and, as they point out, it appears highly improbable that none of the numerous botanists who visited Palestine during the course of the 19th century should have failed to observe this showy plant had it existed there before Bornmüller's visit.

The climate in Iraq is too extreme to suit the genus *Fuchsia*, of which many ornamental species and varieties, nearly all of tropical American origin are grown, and have to a certain extent been naturalized, in Europe, temperate N America and other mild climates.

1. Sepals and petals 4 or 5, stamens 8 or 10 2
   Sepals, petals and stamens 2 .... 3. **Circaea**
2. Petals purplish or white; seeds with a conspicuous chalazal tuft of hairs, free; sepals deciduous in fruit .... 2. **Epilobium**
   Petals yellow; seeds lacking hairs, firmly embedded in endocarp; sepals persistent in fruit .... 1. **Ludwigia**

1. LUDWIGIA L.

Sp. Pl. ed. 1, 118 (1753); Gen. Pl. ed. 5, 153 (1754)
incl. *Jussiaea* L. (*Jussieua* auct.)

Perennial or annual herbs, sometimes woody at base. Leaves opposite or more commonly spirally arranged, simple, mostly entire. Flowers axillary or in definite leafy racemes. Flowers not epigynous. Sepals 4 or 5, less commonly 3 or 6. Petals equal in number to sepals or absent, yellow or white. Stamens equal in number to or twice as many as the sepals. Ovary usually with as many cells as there are sepals, usually many-seeded; stigma hemispherical. Fruit a capsule, indehiscent or apically, terminally or locudicidally dehiscent. Seeds usually numerous, 1- or many-ranked in each cell of capsule.

Ludwigia (commemorating the German physician Christian Gottlieb Ludwig (1709–1773) of Leipzig.)

About 81 species of marsh plants with a predominantly tropical distribution. The genus *Jussiaea*, formerly recognized, comprised those species with the stamens twice as many as the sepals, but the genera as thus delimited are clearly artificial groupings which should be abandoned (see Brenan in Kew Bull. 1953, 163–173; Hara in Journ. Jap. Bot. 238: 289–294, 1953).

1. **Ludwigia stolonifera** (*Guill. & Perr.*) *P.H.Raven* in Reinwardtia 6: 390 (1964); Boulos, Fl. Egypt 2: 150 (2000); Chamberlain & Raven in Fl. Turk. 4: 182 (1972).

Jussiaea diffusa Forssk., Fl. Aegypt.-Arab. 210 (1775), non *Ludwigia diffusa* Buch.-Ham. in Transact. Linn. Soc., 14: 301 (1824).
J. stolonifera Guill. & Perr., Fl. Seneg. Tent. 1: 292 (1832).
J. repens (non L.) Boiss., Fl. Orient. 2: 751 (1872); Zohary, Fl. Palest. ed. 2, 1: 475(1932); Zohary in Dep. Agr. Iraq Bull. 31: 108 (1950).
J. repens L. var. *diffusa* (Forsk.) Brenan in Kew Bull. 8: 171 (1953).
Ludwigia adscendens (L.) Hara subsp. *diffusa* (Forssk.) P.H.Raven in Kew Bull. 15: 476 (1962).

Stems spreading, prostrate or ascending, to at least 1 m long, rooting at the nodes, with tufts of white aerenchyma from submerged stems. Branches densely villous, more rarely glabrous. Leaves narrowly lanceolate to narrowly elliptical, 2–9 × 0.3–2.3 cm, broader on sterile shoots. Flowers axillary. Sepals 5, 5–10 × 2–3.2 mm. Petals 5, bright yellow, 10–18 mm long, 6–10 mm wide, entire. Stamens 10, the episepalous longer. Disc elevated, densely white-hairy; style 4.5–8 mm. Capsule 15–27 mm, dark brown, shining, on a pedicel 0.6–1.7 cm, ribbed, the seed showing clearly as a series of bumps ± 1.5 mm apart between the ribs. Seeds approximately vertical, 1.1–1.2 mm, firmly embedded in coherent cubes of woody endocarp at maturity. Fig. 4: 1–7.

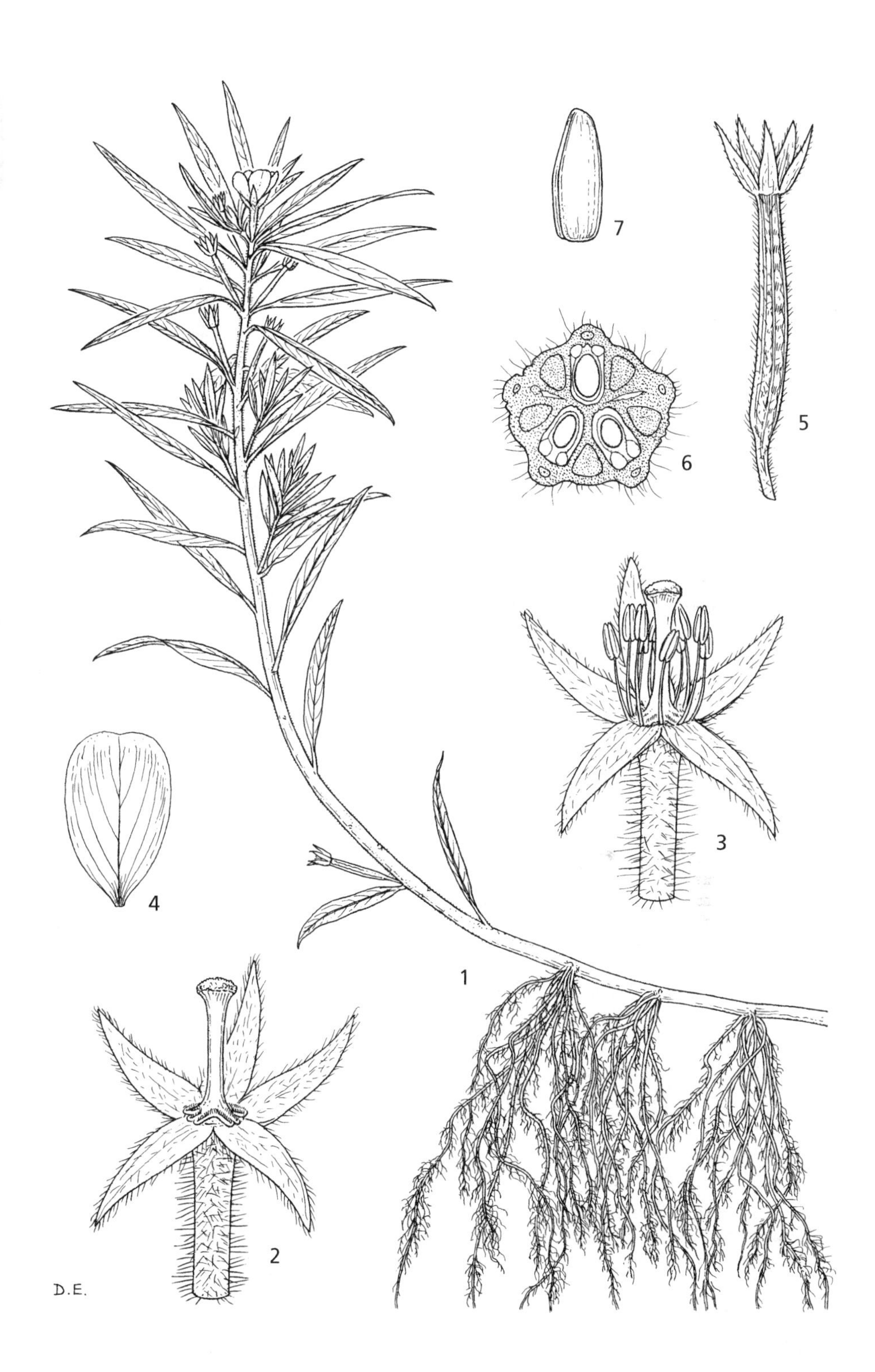

Fig. 4. **Ludwigia stolonifera**. 1, habit × $^{1}/_{3}$; 2, flower showing disc × 3; 3, flowers showing short stamens × 3; 4, petal × 2; 5, capsule × $1^{1}/_{3}$; 6, cross section of capsule × 6; 7, seeds × 6. Drawn by D. Erasmus.

HAB. Shallow water, wet places, moist soil; up to alt. 10 m; fl. & fr. Apr.–Oct.
DISTRIB. Common in southern marshes of Iraq, and also in other marshy localities. **LSM**: Chahala, *Field & Lazar* 46!; Mesaida, *Rawi* 16510!; Halfaya marsh, *Husham (Allizi) & Mahommed* 29606!; Abu Ka'ida & Majarr marshes, *Husham (Alizzi) & Mahommed* 29614!; Garmat Bani Sa'ad, *Rawi* 16562!; Dibn, *Thesiger* 1258!; N of Azair, *Rawi* 16547!; Rabaijish, *Rawi* 12550!; near Nasiriya, *Gamal Abdin* 12!; **DSD**: S. of Faisaliya (nr. Abu Sukhair), *Gillett* 9964!
NOTE. This species, widespread in the western Tropics and Subtropics of the Old World, is tetraploid ($n = 16$), unlike the Asian *L. adscendens* (*Jussiaea repens*) and other closely related taxa, which are diploid ($n = 8$).

Ludwigia stolonifera, KOBANI or GUBANI (Ir., **LSM**) ARMAT (lower Ir., possibly a corruption of Ar. ARMADH for sea-weed, water-weed, green scum on water etc. ?). This plant develops two kinds of roots (cf. *Gillett* 9964): ordinary siliceous anchorate roots and spongy aerating roots ("pneumatophores"), consisting largely of aerenchymatous tissue, which arise from the true roots or from the stems and grow upwards often until they reach the surface of the water. The second kind of root is not developed when the plant grows on a terrestrial habitat. According to Burkhill (1935) the related species *L. adscendens* (*Jussiaea repens*) is used by the Malays of Perak for poulticing in skin complaints and is stocked by Chinese herbalists in the northern part of the Malay peninsula.

Syria, Lebanon, Palestine, Egypt, throughout Africa.

2. **EPILOBIUM** L.

Sp. Pl., ed. 1, 347 (1753); Gen. Pl. ed. 5, 471 (1754); Haussknecht, Monographie der Gattung *Epilobium* (1884)

Perennial herbs, rarely annuals or woody at base. Leaves usually opposite, at least below, and alternate above or in the inflorescence, rarely in whorls of 3. Flowers epigynous or calyx tube absent. Sepals 4, deciduous in fruit. Petals 4, purplish or white, usually deeply notched at the apex. Stamens 8, the epipetalous shorter. Stigma 4-lobed or entire. Ovary 4-locular, with numerous ovules. Capsule slender, elongate, dehiscing loculicidally into 4 valves; seeds with a chalazal tuft of long hairs, wind-borne.

Cosmopolitan in distribution, this genus has about 164 species which occur throughout the temperate and montane regions of the world. It is the largest in the family and contains the great majority of species of Onagraceae found in the Old World. Most of the species are somewhat mesophytic in habitat preference. In this genus, the mode of vegetative propagation is variable, some species having above-ground stolons. Those of Iraq have below ground level, either turions (fleshy overwintering buds), soboles (pale elongate shoots) or leafy rosettes which are ± erect. Sterile hybrids involving *E. hirsutus*, *E. parviflorus* and *E. tetragonum* are known from Europe in all possible combinations, and the hybrid between *E. minutiflorum* and *E. rechingeri* is known from Iran.

Epilobium (from the Greek), Willow-Herb, Willow-weed (Am.). There does not appear to be any general Kurdish name for this common plant, though several unconfirmed local names have been recorded namely QUZALA (Kurd.-Bakrajo, *Radhi* 5238), ?KHARPON (Kurd.-Qala Diza, *Rawi* 9361), both for *E. hirsutum* (Handel-Mazzetti noted ZAHR EL-ASSAL ("honey-flower") for this species in Aleppo; ? BIRZBANK (Kurd.-Amadiya, *Tikriti* 16930) for *E. parviflorum*.

The taller species have sometimes been brought into cultivation as garden ornamentals, particularly the native *E. hirsutum* and the Rosebay, Fireweed (Am.) or French Willow (*E. angustifolium*), but, though beautiful, they are apt to spread and become a nuisance and they are not really suited to the formal garden. The latter species often covers extensive areas of denuded wood after forest fires in N America (hence the name Fireweed). One or two of the smaller species are sometimes grown in English rock gardens. Apparently, there is no local use for the species in Iraq.

1. Stems densely or short villous; stigma deeply 4-lobed. 2
   Stems strigose to glabrescent; stigma capitate or clavate 3
2. Leaves distinctly clasping; flowers up to 2 cm long . . . . 1. *E. hirsutum*
   Leaves short-petiolate, not clasping; flowers 5–10 mm long. . . . . 2. *E. parviflorum*
3. Inflorescence strongly nodding before anthesis; stigma capitate; seeds nearly smooth, obovoid, 1–1.2 mm long . . . . 3. *E. ponticum*
   Inflorescence suberect, not strongly nodding before anthesis; stigma clavate; seeds finely papillate. 4
4. Leaves lanceolate or narrowly ovate, distinctly petiolate, petiole 2–4 mm long; inflorescence strigose with an admixture of glandular hairs. . . . . 4. *E. rechingeri*
   Leaves oblong or oblong-lanceolate, sessile or petiole < 1 mm, blade decurrent; inflorescence lacking glandular hairs . . . . .5. *E. tetragonum*

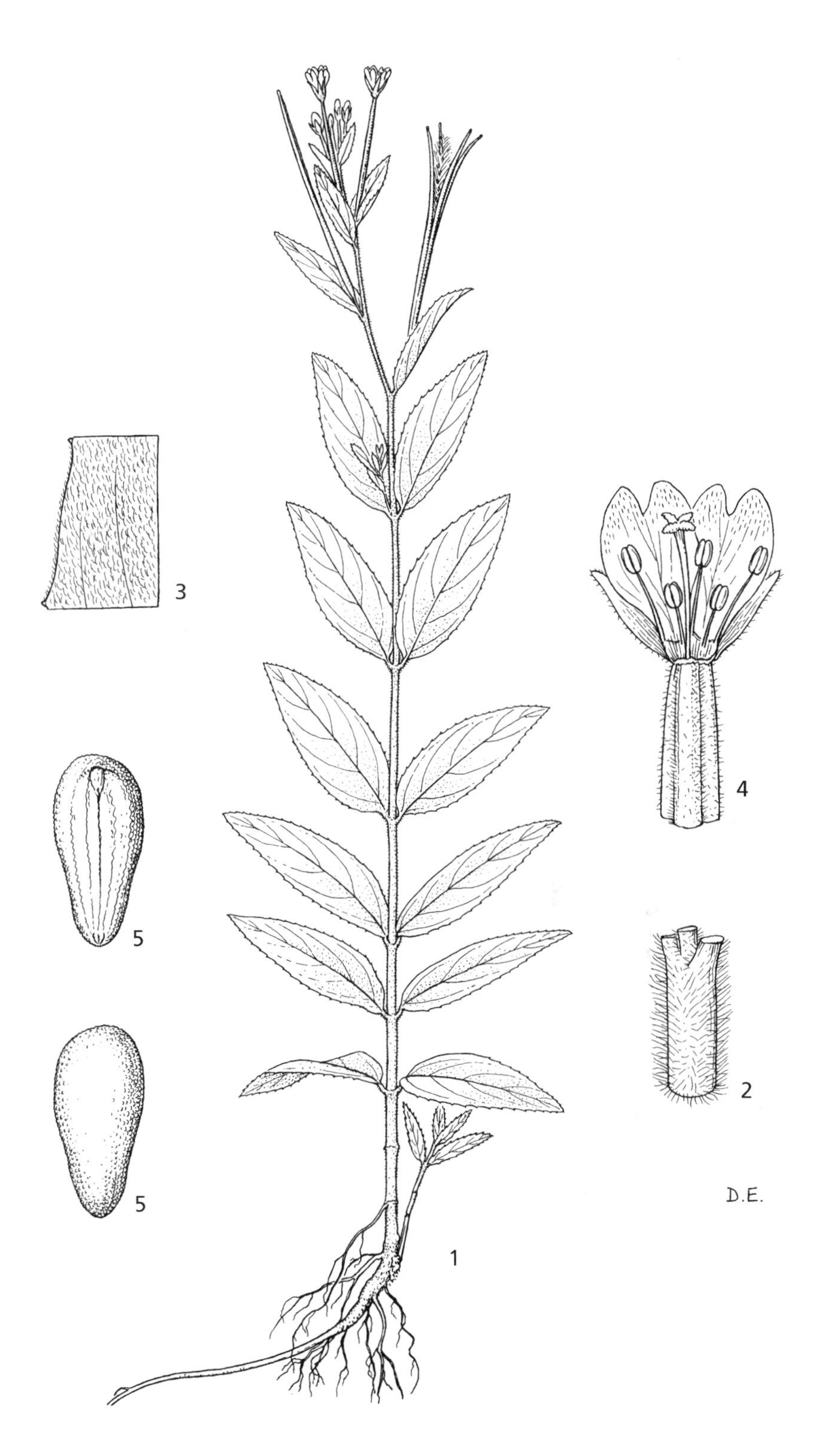

Fig. 5. **Epilobium parviflorum**. 1, habit × 2/3; 2, stem × 4; 3, upper surface of leaf × 4; 4, flower with front sepals and petals removed × 5; 5, seeds × 20. Drawn by D. Erasmus.

1. **Epilobium hirsutum** *L.*, Sp. Pl. ed. 1, 347 (1753); Boiss., Fl. Orient. 2: 746 (1872); Zohary, Fl. Palest. ed. 2, 1: 473 (1932); Guest in Dep. Agr. Iraq Bull. 27: 31 (1933); Zohary in Dep. Agr. Iraq Bull. 31: 108 (1950); Raven in Fl. Iranica 7: 7 (1964); Raven in Fl. Europaea 2: 309 (1968); Chamberlain & Raven in Fl. Turk. 4: 187 (1972); Raven & Hoch in Fl. Pak. 139: 16 (1981); Collenette, Wildflow. Saudi Ariabia: 580 (1999); Chaudhary, Fl. Kingd. Saudi Ariabia: 2(1): 229 (2001).

E. tomentosum Vent., Hort. Cels. 90, t. 90 (1800).
E. hirsutum var. *villosissimum* ("*villosissima*") Koch, Syn. Fl. Germ. Helv. ed. 1, 240 (1835).
E. hirsutum var. *tomentosum* (Vent.) Boiss., Fl. Orient. 2: 746 (1872); Zohary, Fl. Palest. ed. 2, 1: 473 (1932); Blakelock in Kew Bull. 3: 430 (1948).

Rhizome thick, producing white fleshy soboles. Stems to 2.1 m tall, plant robust, branching, entirely covered with ± dense villous pubescence. Leaves oblong to lanceolate, 2–12 × 0.5–3.5 cm, serrate, sessile, clasping. Inflorescence with an admixture of glandular hairs, suberect before anthesis. Flowers large, up to 2 cm; petals bright purplish-rose. Stigma deeply 4-lobed, elevated above the anthers at anthesis. Capsule 4–10 cm. Seeds brown, obovoid, papillate, 1–1.5 mm.

HAB. Usually by streams or springs and in other damp places; sometimes also in hedgerows, old fields etc.; up to alt. 1700 m (usually 700–1700 m); fl. & fr. Jul.-Oct.
DISTRIB. Common throughout forest zone of Iraq, rare on the plains. **MAM**: nr. Zawita, *Guest* 1663!; Sarsang, *Haines* W. 1197!; "ad rivulos montis Gara Kurdist." (Gara Dagh), *Kotschy* 330!; "Chalki on Khabur River" (Chalki Nasara), *Thesiger* 1246!; **MRO**: Baradost Dagh, *Thesiger* 1203!; Bola, *Guest & Alizzi* 15858!; Walash, *Guest* 2667!; Haji Umran, *Gillett* 12439! *Rawi* 24972!; Qandil Range, *Rawi & Serhang* 23945! 25342!; Pushtashan, *Rawi & Serhang* 26487! 26487A!; **MSU**: Bakrajo, *Radhi* 5238! 5245!; Qala Diza, *Rawi* 9361!; Qara Dagh, *Zohary* 16368! (huh); **FNI**: Ain Sifni, *Low* 411!; **LSM**: Halfaya, *Rawi* 16632!
NOTE. *Epilobium hirsutum* is a tall, large-flowered species which is variable in pubescence. Although white-pubescent individuals are common in many parts of its range, particularly in the Near East, this variation does not appear to be geographically correlated and does not, in my opinion, demand taxonomic recognition.
Hairy Willow-Herb, Codlins and Cream (Eng.), Hairy Willow-weed (Am.). A striking plant, by far the most common species in Iraq; suitable for cultivation in a "wild garden" near running water.

Throughout Europe, N E and S Africa, Cyprus, Syria, Lebanon, Palestine, Turkey, Iran, Afghanistan and throughout Central Asia and Himalaya to China and Japan; introduced in east North America.

2. **Epilobium parviflorum** *Schreb.*, Spicil. Fl. Lips. 1456, 155 (1771); Boiss., Fl. Orient., 2: 746 (1872); Zohary, Fl. Palest., ed. 2, 1: 473 (1932); Blakelock in Kew Bull., 3: 430 (1948); Raven in Fl. Iranica 7: 7 (1964); Raven in Fl. Europaea 2: 309 (1968); Chamberlain & Raven in Fl. Turk. 4: 187 (1972); Raven & Hoch in Fl. Pak. 139: 17 (1981).

E. menthoides Boiss. & Heldr., Diagn. ser. 2, 2: 53 (1856).
E. parviflorum var. *menthoides* (Boiss. & Heldr.) Boiss., Fl. Orient. 2: 747 (1872); Zohary, Fl. Palest. ed. 2, 1: 473 (1932).

Rhizome elongate, giving rise to rosettes in the autumn. Stems to 75 cm tall, plant robust, entirely short villous. Leaves oblong to narrowly lanceolate, 2.5–10 × 0.7–3 cm, serrulate, short-petiolate, but not clasping. Inflorescence with an admixture of glandular hairs, ± erect before anthesis. Flowers 5–10 mm; petals purplish-rose. Stigma deeply 4-lobed, surrounded by the anthers at anthesis. Capsule 6–8 cm. Seeds obovoid, brown, papillate, ± 1 mm. Fig. 5: 1–5.

HAB. Damp shady places by streams; alt. 550–2100 m; fl. & fr. Jul.-Aug.
DISTRIB. Occasional in mountain region of Iraq (forest zone, and sometimes in thorn-cushion zone). **MAM**: Amadiya, *Tikriti* 16390!; Sarsang, *Haines* W. 498A!; **MRO**: Shaqlawa, *Gillett* 13413!; Halgurd Dagh, *Guest* 2847!; Haji Umran, *Gillett* 12433!; Pushtashan, *Rawi & Serhang* 23847!; Qurnaqo, *Rawi & Serhang* 26613!; **MSU**: Kanitakht, *Zohary* 16369!; Qara Dagh, NE of Kanitakht, *Zohary & Duvdevani* 16370!
NOTE. *E. parviflorum* is variable in degree of pubescence but not as conspicuously as is *E. hirsutum*. In contrast to that species, it is often self-pollinated.

Throughout Europe, N Africa, Aegean Islands, Cyprus, Syria, Lebanon, Palestine, Turkey, Iran, W Himalaya and southern Central Asia to W China.

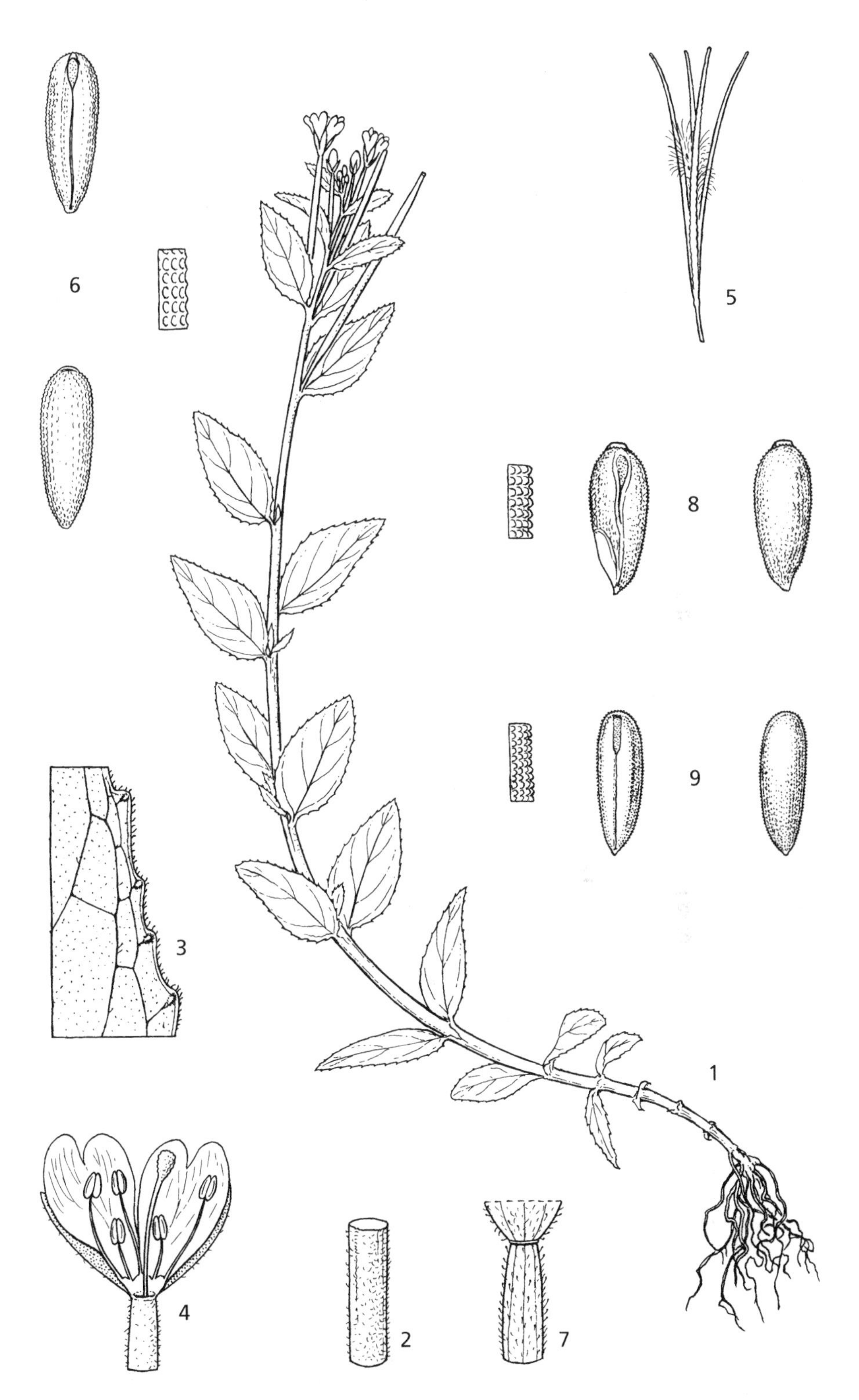

Fig. 6. **Epilobium ponticum**. 1, habit × ⅓; 2, part of stem (detail); 3, section of leaf × 4; 4, flower with front sepals and petals removed × 4; 5, fruit × ⅔; 6, seed details × 14. **Epilobium rechingeri**. 7, part of stem × 4; 8, seed details × 14. **Epilobium tetragonum**. 9, seed details × 14. Drawn by D. Erasmus.

3. **Epilobium ponticum** *Hausskn.*, Mon. Epil. 202 (1884); Raven in Fl. Iranica 7: 13 (1964); Chamberlain & Raven in Fl. Turk. 4: 180 (1972); Hoch & Raven in Fl. Pak. 139: 34 (1981).

[*E. origanifolium* (non Lam.) Boiss., Fl. Orient. 2: 750 (1872), pro parte.]
[*E. frigidum* (non Hausskn.) Blakelock in Kew Bull. 3: 430 (1948).]
[*E.* aff. *warakense* (non Nábělek) Blakelock in Kew Bull. 3: 430 (1948.]

Rhizome soboliferous, scaly. Stems 10–30 cm tall, glabrescent, strigose above, with elevated lines decurrent from the margins of the petioles. Leaves ovate, 1.5–4 × 1–2 cm, weakly serrulate, broadly rounded at the base, short-petiolate. Inflorescence glabrescent or strigose, strongly nodding before anthesis. Flowers 7–10 mm; petals pinkish-purple. Stigmas capitate, surrounded by the anthers at anthesis. Capsule 4.5–6 cm. Seeds obovoid, brown, lacunose, nearly smooth, 1–1.2 mm. Fig. 6: 1–6.

HAB. Among rocks by water; alt. 2000–3000 m; fl. & fr. Jul.-Sept.
DISTRIB. Rare in Iraq; only recorded from the thorn-cushion zone on the Halgurd Range, near the Iranian frontier. **MRO**: Halgurd Dagh, *Guest* 2814!; 2961!; *Gillett* 9558!; Goum Tawera, *Rawi & Serhang* 24723!

Turkey, Iran, Caucasus.

4. **Epilobium rechingeri** *Raven* in Årbok Univ. Bergen Math.-Naturv. Ser. 1962: 26, tab. 7 (1962).

?*E. roseum* (non Schreb.) Zohary in Dep. Agr. Iraq Bull., 31: 108 (1950), specimens not found at Baghdad herbarium.

Underground parts scaly, bearing turions (fleshy overwintering buds). Stems 10–25 cm tall, sparsely strigose above, with weakly elevated lines decurrent from the margins of the petioles. Leaves lanceolate or narrowly ovate, 2–3.5 × 0.9–1.5 cm, weakly serrulate, distinctly short-petiolate (petiole 2–4 mm), broadly cuneate at base. Inflorescence sparsely strigose (rarely glabrous) with a few glandular hairs on the calyx. Flowers 6–7 mm; petals purplish pink. Stigmas clavate, surrounded by anthers at anthesis. Capsule 5–5.5 cm. Seeds obovate, brown, finely papillate, ± 1.5 mm long. Fig. 6: 7.

HAB. On dark rock; alt. 1200–2100 m; fl. & fr. Jul.-Aug.
DISTRIB. Rare, in upper forest and thorn-cushion zones of Iraq, near Turkish and Iranian frontiers. **MAM**: near Sharanish, *Rechinger* 11986!; **MRO**: Rust, *Haley* 170!; Sula Khal, *Rawi & Serhang* 24668!
NOTE. Some individuals of *E. rechingeri* from Iran are entirely glabrous but otherwise indistinguishable from the other plants referred to this species. Hybrids of *E. rechingeri* are known from Iran (l.c., p. 28).

Iran.

5. **Epilobium tetragonum** *Hausskn.* in Österr. Bot. Zeit., 29: 55 (1879); Nábělek in Publ. Fac. Sc. Univ. Masaryk, 35: 115 (1923); Zohary in Dep. Agr. Iraq Bull., 31: 108 (1950); Raven in Fl. Iranica 7: 9. 1964; Raven in Bothalia 9: 318. 1967; Raven in Tutin Fl. Europaea 2: 310. 1968; Chamberlain & Raven in Fl. Turk. 4: 189. 1972; Raven & Hoch in Fl. Pak. 139: 18 (1981).

E. tetragonum var. *minutiflorum* (Hausskn.) Boiss., Fl. Orient. Suppl., 240 (1888); Zohary, Fl. Palest. ed. 2, 1: 174 (1932).

Rhizome stout. Stems 10–75 cm tall, strigose, sometimes densely so, especially above, or glabrescent, with weakly elevated lines decurrent from the margins of the petioles. Leaves oblong-lanceolate, 2–8 × 0.5–2 cm, serrulate; petiole < 1 mm; blade decurrent. Inflorescence usually white-canescent, rarely glabrescent, suberect before anthesis. Flowers 2.5–4 cm; petals very pale pink. Stigma clavate, surrounded by the anthers at anthesis. Capsule 3.5–6 cm. Seeds obovoid, brown, papillate, ± 1 mm, with a very short pellucid beak at the chalazal end. Fig. 6: 8.

HAB. Mountain slopes; alt. 1400–1700 m; fl. & fr. Jun.-Aug.
DISTRIB. Rare, in upper forest zone of Iraq, near Turkish and Iranian frontiers. **MAM**: Ain Nuni, *Nábělek* 268; **MRO**: Haj Umran, *Rechinger* 11323!; Rawi 24944!; **MSU**: Kani Takht, *Zohary* 16366!, 16367!; Hawraman, nr. Tawila, *Rechinger* 12386!
NOTE. *Epilobium tetragonum* is a characteristic forest species of the Orient, most common in Iran and Afghanistan.

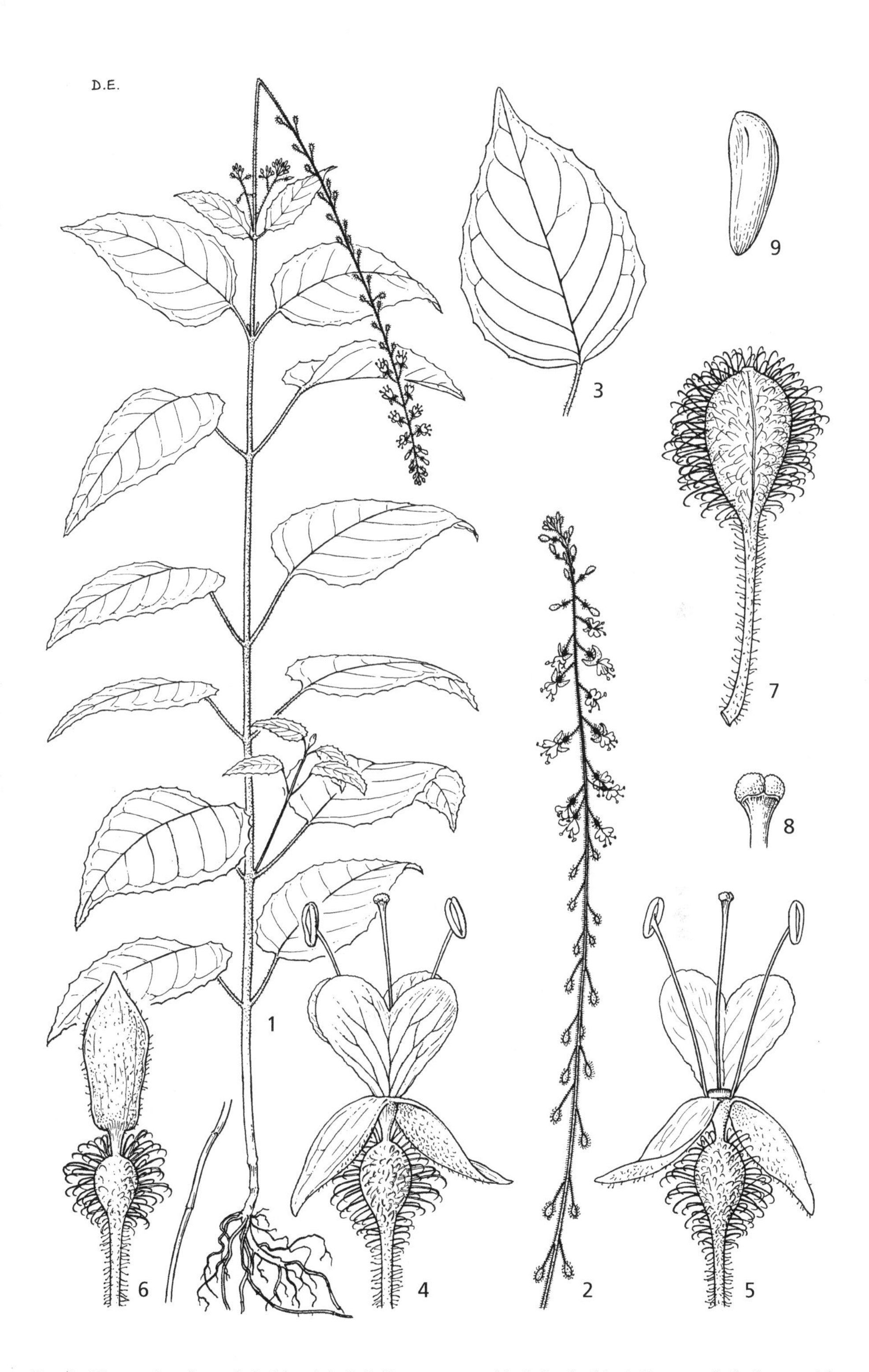

Fig. 7. **Circaea lutetiana**. 1, habit × $^1/_3$; 2, inflorescence × $^2/_3$; 3, leaf × $^2/_3$; 4, flower × 6; 5, flower with front petals removed × 6; 6, flower with fruit × 6; 7, fruit × 6; 8, stigma × 18; 9, seed × 6. Drawn by D. Erasmus.

Lebanon, Turkey, Iran, Caucasus, Central Asia, Afghanistan, Baluchistan, and NW Himalaya to Kumaon.

3. **CIRCAEA** L.

L., Sp. Pl. ed. 1, 12 (1753); Gen. Pl. ed. 5, 24 (1754).

Rhizomatous perennial herbs. Leaves opposite, decussate, ± clustered at the summit of a bare stem, long-petiolate. Inflorescence racemose, terminal, bracteate at the base. Flowers perigynous, the calyx tube sometimes very short. Sepals 2, deciduous in fruit. Petals 2, white or pale rose, usually deeply emarginate. Stamens 2. Stigma 2-lobed or subentire. Fruit indehiscent, 1–2-locular and -seeded, covered with straight or hooked bristles.

About nine species found throughout the North Temperate region.

1. **Circaea lutetiana** *L.*, Sp. Pl. 1: 12 (1753); Boiss., Fl. Orient. 2: 753 (1872); Zohary, Fl. Palest. ed. 2, 1: 476 (1932); Chamberlain & Raven in Fl. Turk. 4: 181 (1972).

Plants 15–60 cm tall, usually unbranched below, with numerous thick, fleshy soboles from the underground rhizome. Stems sparsely strigose or short-villous to nearly glabrous. Leaves ovate, 2.5–11 × 1.5–6.5 cm, truncate to slightly cordate at base, long-acuminate at apex, sparsely and shallowly denticulate; petioles 1–5 cm. Inflorescence glandular-pubescent. Pedicels 4–8 mm in fruit, reflexed at maturity. Hypanthium slender, 1–1.2 mm. Sepals 2–4.5 × 1–3 mm, muriculate. Petals cuneate, 2–4 × 2.2–5 mm, notched for about one-half their length. Filaments 2.5–5.5 mm; anthers 0.4–0.8 mm. Style 4–6.7 mm; stigma deeply 2-lobed, white to reddish, held above and away from the spreading anthers at anthesis; disc present, 0.2–0.4 mm high. Fruit 3–4 × 2–2.5 mm, 2-locular, tapering at base, smooth to slightly rugose, densely covered with white hooked hairs 0.7–1.1 mm long. Seeds 2–3 mm. Fig. 7: 1–9.

HAB. Woodland.
DISTRIB. **MSU**: Hawraman, 1867, *Haussknecht* s.n.!
NOTE. Although this specimen may not be from within the limits of Iraq, the species is to be expected elsewhere along the mountainous regions of the Iranian and Turkish borderland.

Throughout Europe, Syria, Turkey, Iran, the Caucasus and eastward in Siberia.

101. **HALORAGACEAE**[3]

O.G. Petersen, Halorrhagiaceae in Pflanzenfam. 3(7): 226–237 (1889); A.K. Schindler, Halorragaceae in Pflanzenr. 23(IV.225): 1–133 (1905).

C.C. Townsend[4]

Herbs (rarely subshrubs or small trees), usually aquatic or growing in wet habitats. Leaves opposite, alternate or verticillate, the submerged leaves deeply pectinately divided. Stipules usually absent, occasionally represented by intravaginal scales or ochreae. Flowers small, solitary and axillary, or in axillary clusters, or in terminal spikes, corymbs or panicles, unisexual (monoecious or polygamous) or hermaphrodite, actinomorphic, epigynous; perianth uniseriate or biseriate. Calyx (when present) with the tube adnate to the ovary, sepals 2–4, free. Petals 2–4, caducous, imbricate or valvate, or none. Stamens 2, or in 1 or 2 series of 4, short; anthers basifixed, dehiscing by longitudinal slits. Ovary inferior, (1–)4-locular with a single pendulous anatropous ovule in each loculus; styles isomerous with the loculi, short, the stigmas usually plumose or papillose. Fruit a nut or drupe or schizocarp, the carpels separating at maturity into 1-seeded nutlets; seeds with copious endosperm.

Ten genera and about 130 species, cosmopolitan but especially well represented in the southern hemisphere: Two species found in Iraq.

[3] The family name "Haloragidaceae" was previously used, however the spelling "Haloragaceae" is the conserved name.
[4] Royal Botanic Gardens, Kew. Updated by J. Osborne, Royal Botanic Gardens, Kew.

1. MYRIOPHYLLUM L.

Sp. Pl. ed. 1: 992 (1753); Gen. Pl. ed. 5: 429 (1754)

Perennial rhizomatous herbs, aquatic (or occasionally growing on wet mud), only inflorescence usually appearing above the water surface. Leaves verticillate in whorls of 3–6, pinnately divided into simple, narrowly linear or capillary segments, rarely (not in Iraq) opposite or alternate; bracts sometimes simple, entire or toothed, more frequently similar to the leaves but smaller. Flowers commonly sessile and solitary in axils of bracts, more rarely in few-flowered axillary clusters, polygamous or monoecious, rarely dioecious, upper flowers usually male and lower female, with intermediate flowers hermaphrodite. Perianth biseriate, tetramerous, or rarely calyx obsolete. Sepals small, minute in female flowers. Petals of male flowers concave, caducous, those of female flowers minute or absent. Stamens in one or two series of 4; anemophilous. Ovary 4-locular; styles 4, short; stigmas capitate. Fruit a schizocarp of 1-seeded nutlets.

Myriophyllum (from Gr. μύριο, *myrio*, very many, φύλλος, *phyllos*, leaf).
A cosmopolitan genus of about 45 species: two species in Iraq.

1. Upper bracts ovate, oblong or subrotund, entire or erose-denticulate, shorter than the flowers; filaments as long as the anthers . . . . 1. *M. spicatum*
   All bracts, even the uppermost, pectinate, longer than the flowers; anthers twice as long as the filaments. . . . . 2. *M. verticillatum*

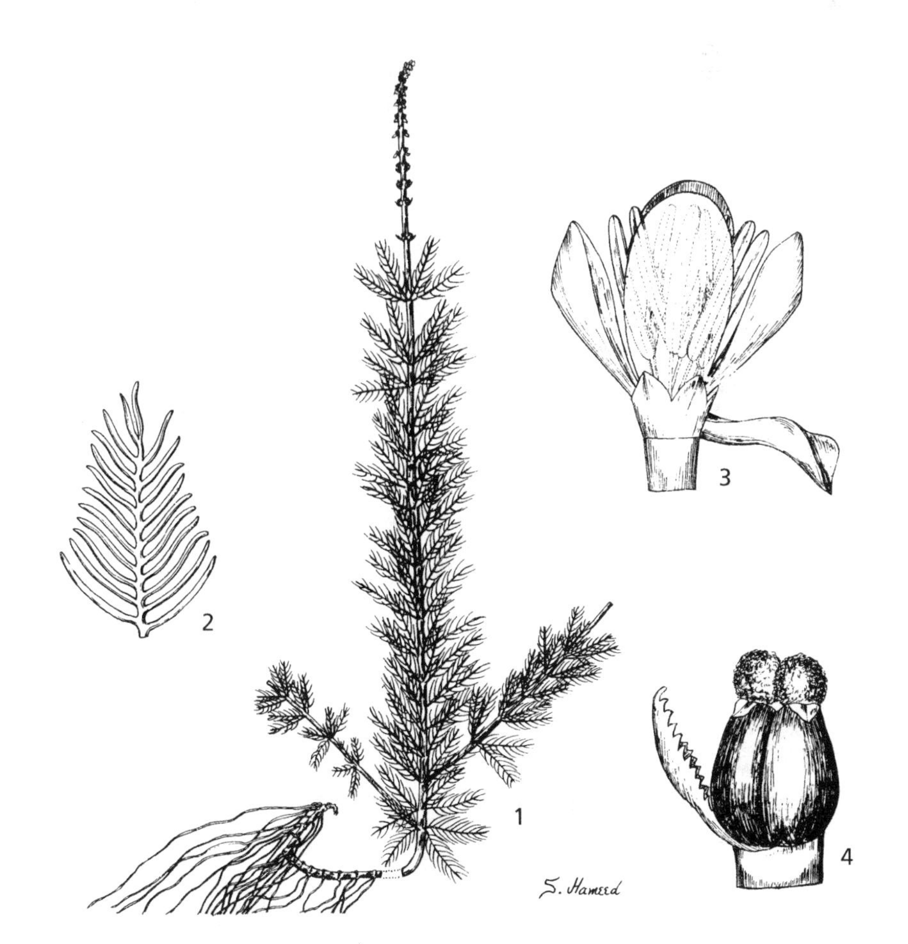

Fig. 8. **Myriophyllum spicatum**. 1, habit × 1/2; 2, leaf × 2; 3, male flower × 15; 4, female flower × 15. Reproduced from Fl. Pak. 113: 2 (1977) with permission.

1. **Myriophyllum spicatum** *L.*, Sp. Pl. ed. 1: 992 (1753); Boiss., Fl. Orient. 2: 755 (1872); Rech. f., Fl. Lowland Iraq: 446 (1964) & Fl. Iranica 18: 1 (1966); Cook in Fl. Europ. 2: 311 (1968); Chamberlain in Fl. Turk. 4: 198 (1972); Ghazanfar in Fl. Pak. 113: 2 (1977); Collenette, Wildflow. Saudi Ariabia: 428 (1999).

Rhizomatous, perennial aquatic herb, stems fragile and upper part of plant often found free-floating, more rarely subterrestrial on wet mud; lower part of the stem bare from the decay of the leaves. Submerged leaves 1.5–3 cm, in whorls of 3 or 4(–6), simple pinnate with (13–)19–37(–47) narrow-linear or filiform segments, ± collapsing when removed from water; subterrestrial forms with more rigid segments, as are aerial leaves (if present). Flowers in an interrupted spike, mostly in whorls of 4 in axils of bracts, upper (male) flowers often closer together than lower (female) flowers, 1 or more whorls of hermaphrodite flowers frequently present; bracts ovate, oblong or subrotund, 1.5–2 × 1–1.5 mm, entire or erose-denticulate, with a narrow reddish subcartilaginous margin, only the lowest pectinate and foliaceous; bracteoles 2, similar but paler, smaller (less than 1 mm) and more pronouncedly toothed. Male flowers with 4 calyx teeth distinct, broadly triangular, 0.5 mm or less, ± suffused reddish brown; petals 4, caducous, 1.5–2.5 × 0.4–0.6 mm, deeply concave, purplish-

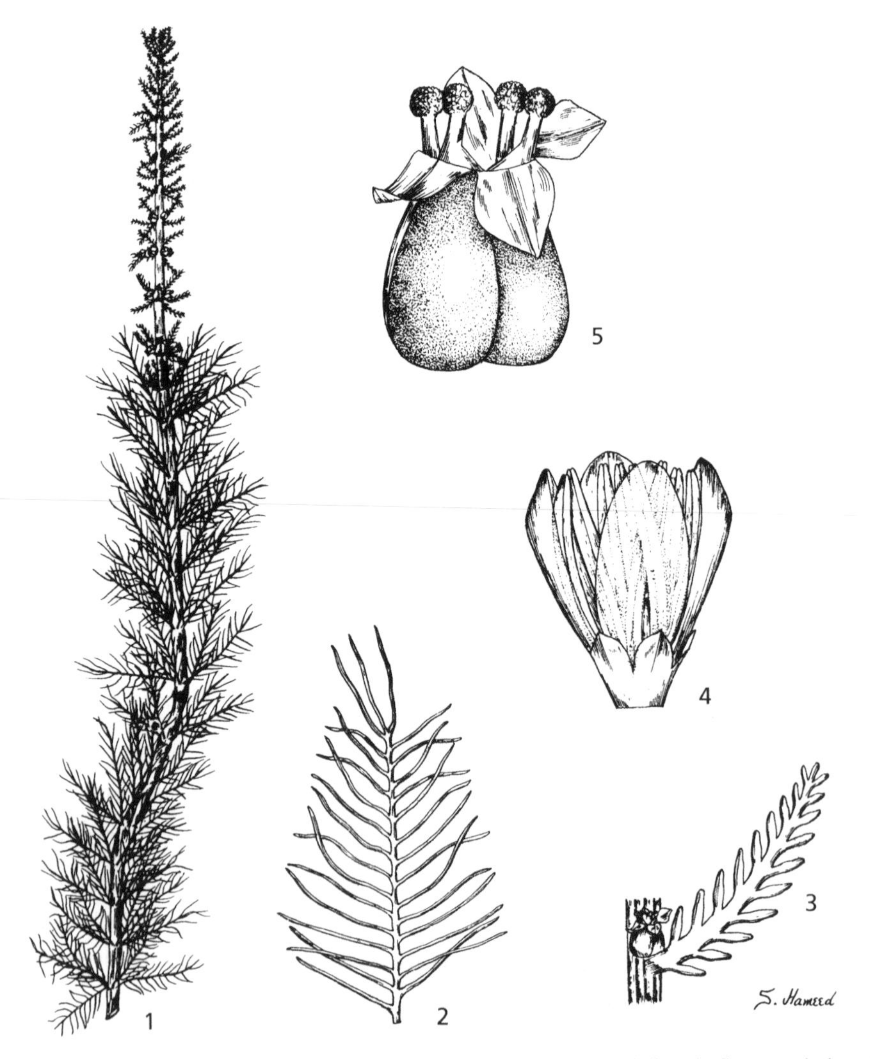

Fig. 9. **Myriophyllum verticillatum**. 1, habit × 1/2; 2, leaf × 2; 3, bract with female flower × 4; 4, male flower × 15; 5, female flower × 15. Reproduced from Fl. Pak. 113: 2 (1977) with permission.

red; stamens 8, ± 4 mm, filaments subequalling anthers. Female flowers with ± 0.5 mm broadly triangular sepals and minute, very early-caducous petals; ovary 4-angled, 4-sulcate, ± 1 mm; styles very short, stigmas strongly recurved, densely papillose. Fruit subglobose, ± 3 mm in diameter, separating into 4 oblong-ovoid nutlets; nutlets margined or narrowly winged at angles of rupture, smooth or sparsely verruculose on the dorsal surface, ventral surfaces flat. Fig. 8: 1–4.

HAB. Shallow water, marshes and permanently wet places; alt.: up to 800 m; fl. & fr.: Oct.–May.
DISTRIB. **MSU**: Jisr Ali, Penjwin Road, *Al Raddi* 5306!; **DLJ/LCA**: Saqlawiya, *s.coll.* 15690!; **LSM**: Shatt Al-Arab, *Al-Khayat* & *Th. Redha* 52331!

Throughout Europe and SW Asia, east to Afghanistan, Pakistan and N India; throughout N & C Asia, Sumatra, China and Japan, Egypt, NW Africa, S & Tropical Africa, Macaronesia, Australia, N America.

2. **Myriophyllum verticillatum** *L.*, Sp. Pl. ed. 1: 992 (1753); Boiss., Fl. Orient. 2: 755 (1872); Rech. f., Fl. Iranica 18: 1 (1966); Cook in Fl. Europ. 2: 312 (1968); Chamberlain in Fl. Turk. 4:198 (1972); Ghazanfar in Fl. Pak. 113: 2 (1977).

Vegetatively similar to *M. spicatum* and scarcely separable from it when sterile. All bracts pinnate or pectinate, longer than flowers; upper bracts pectinate and commonly 2–7× longer than flowers, lower passing gradually to a foliaceous form, larger. Male, female and hermaphrodite flowers disposed as in *M. spicatum*, but hermaphrodite usually more numerous than in that species; bracteoles 2, minute, ovate, acute, coarsely dentate. Male flowers with 4 calyx teeth broadly triangular, ± 0.5 mm; petals 4, caducous, 1.5–2.5 × 0.4–0.6 mm, deeply concave, greenish yellow, very rarely reddish; stamens 8, ± 3 mm, the filaments half as long as the anthers. Female flowers with 4 broadly triangular, ± 0.5 mm, persistent calyx teeth and minute, early-caducous petals; ovary 4-angled, 4-sulcate, ± 1 mm; styles very short, stigmas strongly recurved, densely papillose. Fruit subglobular, 4-furrowed, ± 3 mm in diameter, separating in 4 oblong-ovoid nutlets; nutlets margined or narrowly winged at angles of rupture, smooth or faintly rugose on the dorsal surface, ventral surfaces flat. Fig. 9: 1–5.

HAB. Lakes, streams, wet and moist soils; alt.: up to 920 m; fl. & fr.: Apr.–Aug.
DISTRIB. **FUJ**: N of Mosul, *F.A. Rogers* 0422!; **LSM**: Hor Al Hammar, Chabaish, *S. Omar* & *F. Karim* 36772!; ibid, *Thamar* 50100!

Throughout Europe; Turkey, Jordan, Iran, Caucasus, C Asia, E & W Siberia, Kamchatka, Mongolia, China, Japan, N India, NW Africa (Algeria, Morocco), N America.

102. GENTIANACEAE

Pflanzenfam. 4, 2: 50 (1895)

P. Taylor[5]

Annual or perennial herbs, rarely shrubs, usually glabrous. Leaves opposite, entire, exstipulate. Inflorescence usually cymose, cymes terminal or axillary. Bracts scale-like or absent. Flowers actinomorphic or slightly zygomorphic, bisexual. Calyx tube campanulate or absent, lobes 4–5, rarely up to 10. Corolla gamopetalous, campanulate or rotate, limb 4–5, rarely up to 12-lobed, lobes contorted in bud. Stamens equal in number to corolla lobes and alternating with them, inserted in the tube or in the throat. Glandular disc present at base of gynoecium. Ovary superior, 1- or incompletely 2-locular. Ovules numerous, placentas parietal. Fruit a dry capsule, rarely fleshy, 2-valvate or indehiscent. Seeds usually very small, angular.

A cosmopolitan family of over 1500 species, mainly temperate but well represented at higher altitudes in the tropics. Easily recognized among gamopetalous families by the combination of regular flowers, opposite leaves, and 1-locular ovary with parietal placentation.

[5] † Formerly Royal Botanic Gardens, Kew. Updated by Shahina A. Ghazanfar.

Fig. 10. **Swertia longifolia**. 1, habit × 2/3; 2, inflorescence × 2/3; 3, flower × 2; 4, capsule × 2. All from *Davis* 23985. Drawn by D. Erasmus.

Economically this family is of little importance, though it includes some rather striking flowering plants which are sometimes cultivated in gardens and the medicinal (tonic) properties of products derived from other species is well-known.

1. Corolla lobes with conspicuous basal fringed nectary or nectaries . . . . 1. **Swertia**
   Corolla lobes without nectary. 2
2. Leaves connate at base; corolla lobes 6–8, yellow . . . . .2. **Blackstonia**
   Leaves free at base; corolla lobes 5, white, pink or blue or rarely yellow. 3
3. Corolla pink; anthers at length twisted . . . . 3. **Centaurium**
   Corolla blue; anthers not twisted . . . . .4. **Gentiana**

1. SWERTIA L.

Sp. Pl., ed. 1, 226 (1753); Gen. Pl., ed. 5, 107 (1754)

Erect or rarely prostrate, annual or perennial herbs. Leaves opposite, usually sessile, or radical and long petiolate, entire. Flowers blue, white or rarely yellow, densely to loosely cymose, paniculate or corymbose. Calyx deeply 4 or 5-lobed. Corolla tube very short, lobes spreading, each provided near the base with 1 or 2 fringed nectaries. Stamens 4 or 5 inserted at base of corolla. Ovary 1-locular. Style short, stigma 2-lobed. Capsule dehiscing by 2 valves.

Swertia (named in honour of Emmanuel Swert of Haarlem, a well-known tulip-grower, c. 1620). Felwort.

According to Willis (1931), *S. perennis* is one of more commonly cultivated species, but Chittenden (1951) remarks that few species are of garden value. Uphof (1959) lists *S. chirata* of which the dried herb is used medicinally in the Himalaya region. It contains the bitter glucoside chiratin and is sold as a tonic in the bazaars of India as Chirata. Bederin (1936) gives the name RIKHA.

1. **Swertia longifolia** *Boiss.*, Diagn. Pl. Nov. ser. 1(5): 90 (1844); Edmondson in Fl. Turk. 6: 195 (1978); Omer in Fl. Pak. 197: 155 (1995).

S. aucheri Boiss., Diagn. ser. 1, 5: 90 (1844) & Fl. Orient. 4: 78 (1875); Nábělek in Publ. Fac. Sci. Univ. Masaryk 70: 13 (1926); Schiman-Czeika in Fl. Iranica 41: 27 (1967).
S. persica Griseb. in DC., Prodr. 9: 132 (1845).

Erect glabrous perennial up to 60 cm high. Stem simple, thick. Basal leaves ovate, oblong or oblanceolate, obtuse, long-petiolate, petiole about equal in length to the lamina, up to 20 cm total length, up to 5 cm wide, about 5-veined from the base. Cauline leaves few, smaller, lanceolate, acute, subsessile. Inflorescence a strict, usually many-flowered panicle, sometimes reduced and ± racemose. Flowers long pedicellate, tetramerous, white. Calyx lobes lanceolate, acute, ± 7 mm. Corolla lobes oblong or narrowly ovate, subacute, 10–12 × 4–5 mm; basal nectary solitary, orbicular. Capsule narrowly ovoid, acute, ± 15 mm, many seeded. Seeds angular, ± 1 mm. Fig. 10: 1–4.

HAB. Near stream; alt. ± 2250 m ; fl. Sep.
DISTRIB. Rare, only once found in Iraq, near the Iranian border in the central sector of the thorn-cushion zone; **MRO**: Kani Khanjar Khan, *Rawi & Serhang* 20211!

E Turkey, NW Iran, Afghanistan, Central Asia, Pakistan.

2. BLACKSTONIA Huds.

Fl. Angl. 146 (1762)
Chlora Adans. Fam. 2: 503 (1763)

Glabrous erect annual herbs. Leaves opposite, often connate at the base. Flowers few, in terminal cymes. Calyx deeply 6–8-lobed, lobes linear to narrowly lanceolate. Corolla yellow, tube short, limb spreading, 6–8-lobed, lobes ovate-oblong. Stamens 6–8, inserted in the throat. Anthers oblong or linear, sometimes slightly twisted. Style filiform with 2 deeply 2-lobed stigmas. Capsule ovoid.

Blackstonia (after John Blackstone, an 18th century English botanist).

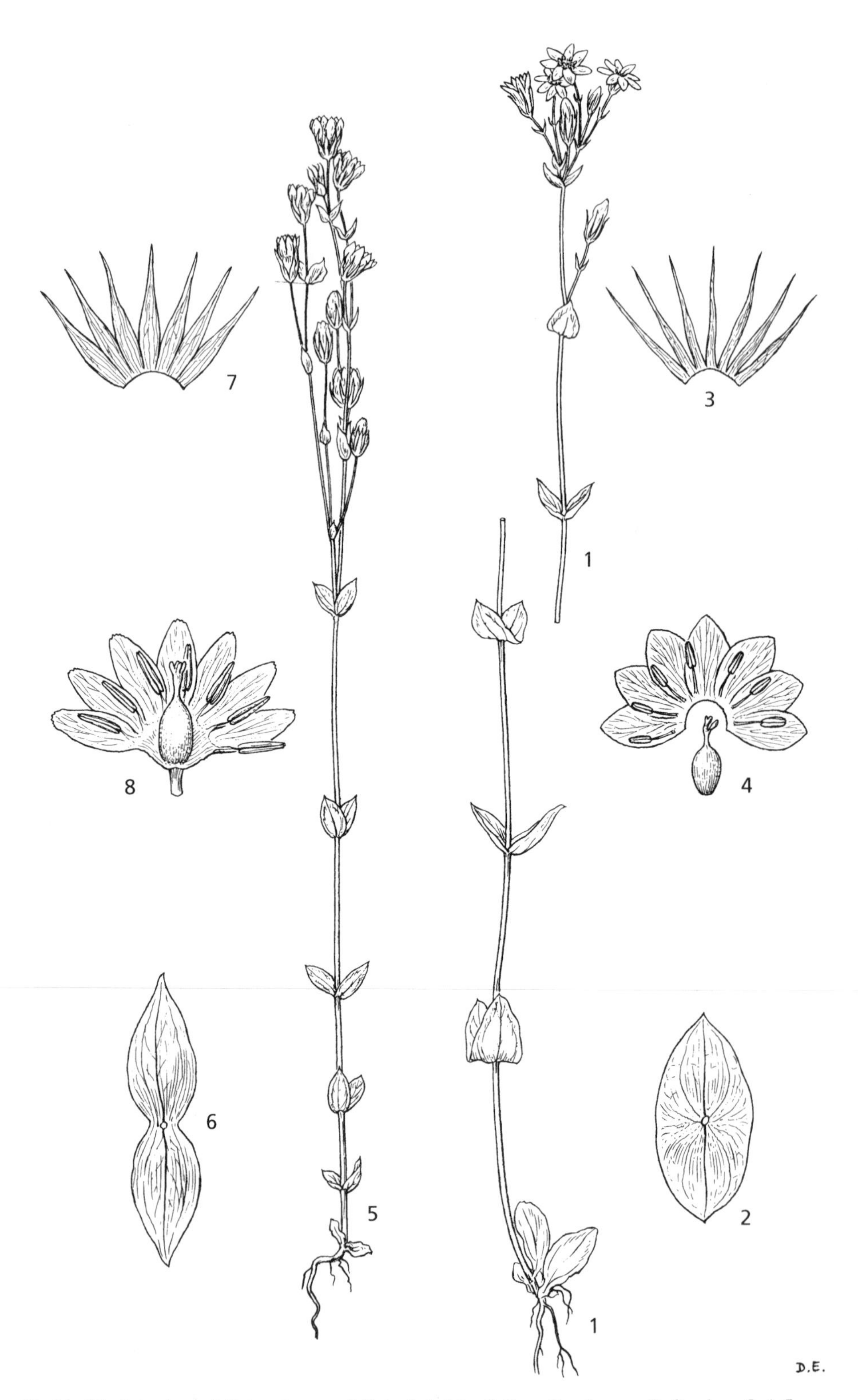

Fig. 11. **Blackstonia perfoliata** subsp. **perfoliata**. 1, habit × 2; 2, cauline leaves × 2 ; 3, calyx × 2; 4, flower, dissected × 2 . **Blackstonia perfoliata** subsp. **acuminata**. 5, habit × 2; 6, cauline leaves × 2 ; 7, calyx × 2; 8, flower, dissected × 2. Drawn from *Merton* 2241 & *McNeill* 224. Drawn by D. Erasmus.

1. **Blackstonia perfoliata** *(L.) Huds.*, Fl. Angl. 146 (1762); Nábělek in Publ. Fac. Sci. Univ. Masaryk 70: 12 (1926); Schiman-Czeika in Fl. Iranica 41: 27 (1967); Edmondson in Fl. Turk. 6: 177 (1978).

Erect, glabrous annual up to 45 cm high with a sparse basal rosette of leaves. Stem simple, terete. Basal leaves forming a rosette, obovate, obtuse, free at base. Cauline leaves opposite, larger, ovate to ovate-deltoid, acute to acuminate. Flowers in a loose cyme (in depauperate plants reduced to a single flower). Calyx lobed almost to base, lobes narrow lanceolate to linear-subulate, 7–9 × 0.5–1.5 mm, acute. Corolla 10–15 mm in diameter, lobes oblong, obtuse.

Cauline leaves broadly ovate-deltoid, broadest at or near the base, acute, up to 3 × 4 cm, each pair broadly connate at the base; calyx lobes ± 7 × 0.5 mm, linear-subulate .... a. subsp. *perfoliata*
Cauline leaves ovate to ovate-lanceolate, broadest about the middle, acute, 1–2 × 1 cm, each pair narrowly connate at the base; calyx lobes ± 9 × 1.5 mm, narrowly lanceolate .... b. subsp. *acuminata*

a. subsp. **perfoliata**
Fig. 11: 1–4.

Gentiana perfoliata L., Sp. Pl. ed. 1, 232 (1753); *Chlora perfoliata* (L.) L., Syst. ed. 12, 2: 267 (1767); Boiss., Fl. Orient. 4: 66 (1875); Zohary, Fl. Pal. ed. 2, 2: 197 (1933); Zohary in Dep. Agr. Iraq Bull. 31: 116 (1950).

HAB. Damp places and dry slopes in lower mountains and foothills.
DISTRIB. Rare in Iraq; recorded only once or twice (no specimens at BAG or K). **MRO**: Harir, *Nábělek* 2741; **MSU**: Sulaimaiya district, *Zohary*, (1950); **FNI**: Tal Kaif, *Nábělek* 2740.

Cyprus, Aegean Islands, Lebanon, Syria, Turkey, NW Africa and W C & S Europe.

subsp. a**cuminata** (*W.Koch & Ziz*) *Dostál* (1949).
Fig. 11: 5–8.

Chlora acuminata W.Koch & Ziz, Cat. Fl. Palat. 20 (1814).
Chlora serotina W.Koch ex Reichb., Pl. Crit. 3, 1: 6, t. 351 (1825); Boiss., Fl. Orient. 4: 66 (1875); Zohary, Fl. Pal. ed. 2, 2: 197 (1933).
Blackstonia serotina (W.Koch ex Reichb.) G. Beck, Fl. Nieder-Österr. 2:934 (1892); Nábělek in Publ. Fac. Sci. Univ. Masaryk 70: 12 (1926).
Blackstonia perfoliata subsp. *serotina* (W.Koch ex Reichb.) Vollmann, Fl. Bayern 594 (1914); Edmondson in Fl. Turk. 6: 177 (1978).
Blackstonia acuminata (W.Koch & Ziz) Domin in Bull. Internat. Acad. Sci. Prague 34: 25 (1933).

HAB. Damp ground, bank of stream; alt. 700–1200 m; fl. May–Aug.
DISTRIB. Occasional in the lower forest zone of Iraq. **MAM**: nr. Dinarti, *E. Chapman* 26113!; **MRO**: Qurnaqo, *Rawi & Serhang* 26614A!; **MSU**: Jarmo, *Helbaek* 1783!; *Haines* W. 290!

Aegean Isles, Cyprus, Lebanon, Syria, Turkey, C & S Europe.

3. **CENTAURIUM** Hill

Brit. Herb. 62 (1756)
Erythraea L.C. Rich. in Pers., Syn. 1: 283 (1805)

Erect or decumbent annual or perennial herbs. Leaves opposite, sessile, sometimes forming a basal rosette. Inflorescence cymose, laxly to densely corymbose or spiciform. Flowers pink or yellow. Calyx tubular, 5- or rarely 4-lobed, lobes keeled. Corolla tube short or elongated, limb 5- or rarely 4-lobed, lobes spreading. Stamens 5, rarely 4, inserted in the corolla tube. Anthers ± exserted, linear-oblong, becoming twisted. Ovary 1-locular. Style filiform. Stigma 2-lobed. Capsule narrowly oblong, 2-valved. Seeds numerous, reticulate.

Centaurium, so called, as *Cantaurea*, in Hippocrates after the Centaur, Chiron, reputed to have had great knowledge of herbs and, according to Pliny, to have discovered its medicinal properties. Centaury, Feverwort. No common name recorded in Iraq but the genus and others of the family have been referred to in Arabic literature as QANTARĪYA (*Sharif*, 1928), QANTARŪN (*Bedevian*,1936) and KANTARIYAN (*Hooper & Field*, 1937).

1. Flowers subsessile, in long lax simple or bifid unilateral spikes . . . . 1. *C. spicatum*
   Flowers distinctly pedicellate or subsessile, in fastigiate or corymbose cymes. 2
2. Cymes congested, corymbose; calyx glandular-ciliate . . . . 2. *C. turcicum*
   Cymes loose or dense, branches stiff or spreading; calyx glabrous 3
3. Branches of cymes spreading at a wide angle; internodes below the lowermost inflorescence branch 2–4 . . . . 3. *C. pulchellum*
   Branches of cymes stiff; internodes below the lowermost inflorescence branch 5–9 . . . . 4. *C. tenuiflorum*

1. **Centaurium spicatum** (*L.*) *Fritsch* in Mitt. Naturw. Ver. Wien 5: 97 (1907); Nábělek in Publ. Fac. Sci. Univ. Masaryk 70: 12 (1926); Zohary in Dep. Agr. Iraq Bull. 31: 116 (1950); V. Täckh., Stud. Fl. Egypt 186 (1956); Rech.f., Fl. Lowland Iraq: 477 (1964); Omer in Fl. Pak. 197: 8 (1995); Boulos, Fl. Egypt 2: 206 (2000).

Gentiana spicata L., Sp. Pl. 230 (1753).
Erythraea spicata (L.) Pers., Synops. 2: 283 (1805); Boiss., Fl. Orient. 4: 69 (1875); Zohary, Fl. Palest. ed. 2, 2: 199 (1933).
Erythraea babylonica Griseb. in DC., Prodr. 9: 60 (1845).

Erect glabrous annual (or biennial) up to 40 cm high. Stem 4-angled, narrowly winged, simple or branched above. Leaves basal and cauline, narrowly ovate-oblong, entire, obtuse or subacute, 1.5–3 cm long, up to 1 cm wide. Flowers pink, subsessile in long, lax, unilateral spikes up to 20 cm long. Calyx lobes 5, linear-subulate, acute, about as long as the corolla tube. Corolla tube ± 9 mm long, limb 5-lobed, lobes narrowly ovate-oblong, about 4–5 mm long. Stamens inserted in the throat. Stigma subcapitate, obscurely 2-lobed. Capsule about as long as the calyx. Fig. 12: 1–7.

HAB. Drains of damp salty depression (*Agnew & Haines* W. 2114); up to 20 m alt. or more; fl. Aug.–Oct.
DISTRIB. Occasional in the desert zone of Iraq (and probably in the steppe). **FUJ**: "Mesopot., Kurdistan & Mosul" *Kotschy* 401!); **DWD**: Shithatha, *Agnew & Haines* W.2114; **LCA**: Baghdad, *Aucher-Eloy* 2432!; **LSM**: E of Amara, *Rawi* 16637!

Mediterranean region, N Africa, Aegean Islands, Cyprus, Palestine, Syria, Turkey, Iran, Turkmenistan, Afghanistan to India, China and Polynesia.

2. **Centaurium turcicum** (*Vel.*) *Bornmüller* in Verh. Zoo. Bot. Ges. Wien 60: 150 (1910); Nábělek in Publ. Fac. Sci. Univ. Masaryk 70: 12 (1926); Zohary in Dep. Agr. Iraq Bull. 31: 116 (1950); Blakelock in Kew Bull. 4: 521 (1950).

Erythraea turcica Vel., Fl. Bulg. 384 (1891).

Erect annual up to 60 cm high. Stem simple, 4-angled, scarcely winged, glabrous. Leaves narrowly lanceolate to oblong-ovate, obtuse, prominently 3-veined, sometimes forming an obscure rosette at the base, 1.5–6 × 1 cm, basal and cauline leaves glabrous, those subtending the flowers glandular-ciliate. Flowers pink, subsessile, in congested corymbose cymes. Calyx lobes 5, narrowly lanceolate, acute, ± 4 mm, densely glandular-ciliate. Corolla tube ± 8 mm, limb 5-lobed, lobes narrowly ovate-oblong, ± 4.5 mm. Stamens inserted in throat. Stigma distinctly 2-branched. Capsule ± 6 mm.

HAB. Wet places near streams, weed by mountain roadside, in shady *Quercus* forest on metamorphic rock; alt. 700–1500 m; fl. May–Aug.
DISTRIB. Quite common in the forest zone of Iraq. **MAM**: Garin, 28 km NW of Zakho, Rawi 2368!; Sarsang, *Haines* W.529!; Zawita, *Hunting Aerosurveys* 128 (R.C. 842)!; *E. Chapman* 26286!; Atrush-Babakki, *G.W. Chapman* 9344!; **MRO**: Helgord, *Guest* 2822!; Gunda Shor-Darband, *Gillett* 12394!; N of Haji Umran, *Rawi* 24961!; Pushtashan, *Rawi & Serhang* 24205!; **MSU**: Jarmo, *Helbaek* 1811!; Sulaimani district, *Zohary* s.n. (1950).

Syria, Turkey, Iran, Afghanistan and SE Europe (Romania).

3. **Centaurium pulchellum** (*Sw.*) *Druce*, Fl. Berks. 342 (1897) as "*Centaurion*"; Nábělek in Publ. Fac. Sci. Univ. Masaryk 70: 12 (1926); Zohary in Dep. Agr. Iraq Bull. 31: 116 (1950); V. Täckh., Stud. Fl. Egypt 186 (1956); Rech. f., Fl. Lowland Iraq: 477 (1964); Jakobsen in Fl. Turk. 6: 180 (1978); Omer in Fl. Pak. 197: 9 (1995); Boulos, Fl. Egypt 2: 206 (2000); Thulin, Fl. Somalia 3: 115 (2006).

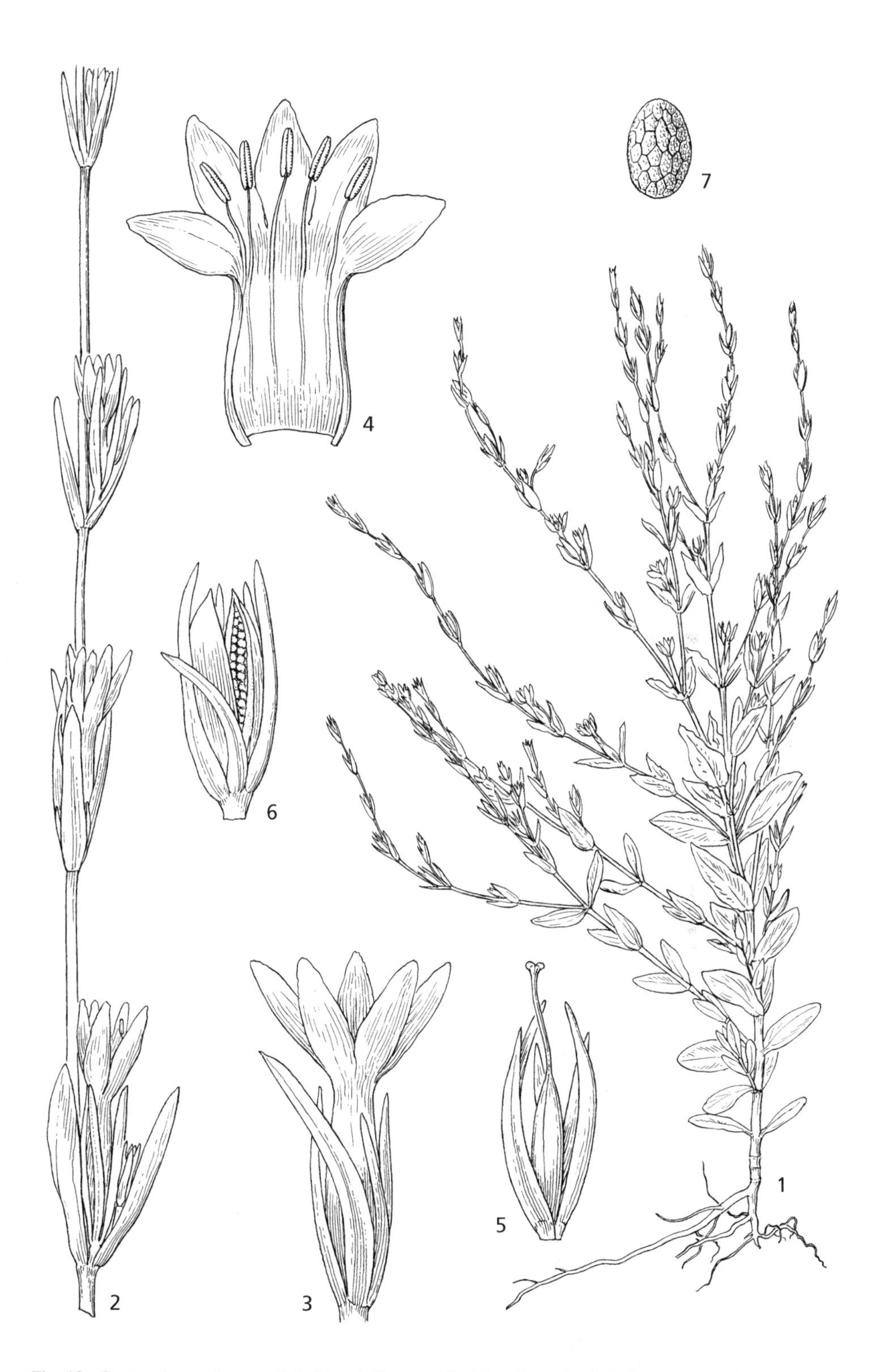

Fig. 12. **Centaurium spicatum**. 1, habit × 1; 2, part of habit enlarged × 3; 3, flower × 4; 4, flower opened × 4; 5, flower with petals removed × 4; 6, capsule × 4; 7, seed × 13. All from *Merton* 2431. Drawn by D. Erasmus.

Gentiana pulchella Sw., Kongl. Vetensk. Acad. Nya Handl. 84. t. 3. ff. 8, 9 (1783).
Centaurium pulchellum (Sw.) E.H.L. Krause in Sturm, Fl. Deutschland ed. 2, 14 (1903) nom. inval.
Centaurium pulchellum Hayek ex Hand.-Mazz., Stadlm., Janch. & Faltis, Österr. Bot. Z. 56: 70. (1906).

Erect glabrous annual up to 30 cm high. Stem simple, 4-angled, narrowly winged. Leaves in 2–4 (rarely up to 6) pairs below the lowermost inflorescence-branch, oblong to narrowly ovate-oblong, obtuse, 0.5–2 cm long, up to 0.6 cm wide. Flowers pink, pedicellate in a loose terminal cyme, cyme-branches spreading at a wide angle. Calyx lobes 5 (in depauperate plants sometimes 4), linear-subulate, ± 5 mm. Corolla-tube ± 10 mm long, limb 5- (rarely 4) -lobed, lobes narrowly ovate-oblong, ± 4.5 mm. Stamens inserted in throat. Stigma distinctly 2-branched. Capsule about as long as calyx.

HAB. Barren plain, sometimes salty, gypsaceous dry hills, fields, roadsides, banks of irrigation ditch, dry shady mountain slopes, sometimes by spring or near water; up to 1500 m alt.; fl. & fr. Apr.–Jul.
DISTRIB. Common in steppe region of Iraq, extending up into the forest zone and down into the desert region. **MAM**: Khaira, *Rawi* 23255!; **MRO**: Harir, *Nábělek* 2745; Koi Sanjaq, *Rawi, Nuri & Kass* 28140!; Saran, nr. Kani Kawan, *Nuri & Kass* 27266!; **MSU**: Jarmo, *Helbaek* 866!; Amoret, nr. Qaradagh, *Haines* W.1099!; **FUJ**: Sinjar, *Haussknecht* s.n.; Tal ash-Shur, *Anon.* 5717!; Qaiyara, *Bayliss* 86!; Kifri, *Gillett & Rawi* 7375!; **LEA**: Sada, *Alizzi & Nur* 19632!; **LCA**: Tarmiya, *Haines* W.124!; Baghdad, *Lazar* 3923!; **LSM**: Fakka Musaida, *Rawi* 14922!
NOTE. Dwarf Centaury. Bové records the name SEMOUM (SIMŪM) from Sinai (*Bové* 90).

Europe, N and NE Africa (Algeria, Morocco, Tunisia, Libya, Egypt, Sinai, Ethiopia, Somalia), Arabia, Socotra, Cyprus, Turkey, Caucasus, Iran, Baluchistan, Afghanistan, NW India to China and Mongolia; introduced in North America.

4. **Centaurium tenuiflorum** (*Hoffm. & Link*) *Fritsch* in Mitt. Naturw. Ver. Wien 5: 97 (1907); Hand.-Mazz. in Ann. Naturh. Mus. Wien 27: 424 (1913); Guest in Dep. Afr. Iraq Bull. 27: 21 (1933); Blakelock in Kew Bull. 4: 521 (1950); Zohary in Dep. Agr. Iraq Bull. 31: 116 (1950); Rech. f., Fl. Lowland Iraq: 477 (1964); Jakobsen in Fl. Turk. 6: 181 (1978); Omer in Fl. Pak. 197: 12 (1995).

Erythraea tenuiflora Hoffm. & Link, Fl. Port. 1: 354 t. 67 (1813–20).
E. latifolia Boiss., Fl. Orient. 4: 67 (1875) non Sm. (1824).

Erect, glabrous annual up to 35 cm tall. Stem simple, 4-angled, narrowly winged. Leaves in 5–9 pairs below the lowermost inflorescence branch, ovate to lanceolate, obtuse or subacute, 1–2.5 × 1 cm. Flowers pink, pedicellate in rather dense terminal cymes; cyme branches stiff. Calyx 5-lobed, lobes linear-subulate, acute, slightly shorter than corolla-tube. Corolla-tube ± 10 mm, limb 5-lobed, lobes narrowly ovate-oblong, ± 4.5 mm. Stamens inserted in throat. Stigma distinctly 2-branched.

HAB. Fields, wasteland, ditches; up to 1500 m alt.; fl. & fr. Apr.–May.
DISTRIB. Occasional in steppe and desert region of Iraq, extending up into lower forest zone. **MSU**: Jarmo, *Helbaek* 1714!; **FUJ**: Kalat Schargat (Shargat), *Handel-Mazzetti* 1087 (W); **FKI**: Kirkuk, *Rogers* O252!; **LEA**: nr. Baquba, *Rogers* 0252!; **LCA**: Baghdad, *Lazar* 1186! *Haines* W.124!; Daltawa (Khalis), *Guest* 2453!; **LBA**: Baquba to Basra, *Haussknecht* (in Rech. f., 1964).
NOTE. Jakobsen recognizes subsp. *acutiflorum* (Schott) Zeltner based on the basal rosette of leaves being absent, shorter internodes and a generally smaller corolla tube. Specimens from Iraq are quite variable and difficult to place in the subspecies recognized by Jakobsen.

Aegean Islands, Cyprus, Palestine, Jordan, Lebanon, Syria, Turkey, Armenia, Iran, Baluchistan, Turkmenistan, Egypt, N Africa (Morocco, Algeria, Libya), Cape Verde Islands, W & S Europe.

4. **GENTIANA** L.

Sp. Pl. ed. 1, 227 (1753); Gen. Pl. ed. 5, 285 (1754)

Annual, biennial or perennial herbs, decumbent or erect. Leaves opposite, sessile, entire. Flowers usually blue, axillary, cymose or solitary and terminal, sessile or shortly pedicellate. Calyx tubular, usually shortly, (4–)5(–7)-lobed. Corolla ± tubular-campanulate, lobes spreading, (4–)5(–7)-lobed, with a small tooth-like lobe in each sinus. Stamens shorter than the corolla. Ovary 1-locular. Style short or absent. Stigma 2-lobed. Capsule 2-valved. Seeds numerous, globose or lenticular, rarely winged.

Gentiana (name of plant in Pliny after Gentius, King of Illyria (W Balkans) who was reputed to have discovered its medicinal properties). Gentian. Ibn Baitar mentions the name JANTIANA, an obvious transliteration of the Latin name, BELCHELCHEKA (BALSHALSHIKA) (in the Arabic of Andalusia, an apparent corruption of the Greek 'basilika'), KUSHAD (Pers.) and SÎNANDIĀN (Greek): also KAFF ADH-DHÎB and DAWĀ'L-HAIYA (from its reputation as an antidote for snake bite). The root of *G. lutea*, Yellow Gentian, has long been known in Europe as a popular and innocuous bitter tonic, used to improve the appetite and stimulate digestion, this being the Gentian Root of commerce. Uphof (1959) says it is used in the manufacture of liqueurs and made into Gentian Bitter. He mentions that a decoction of the roots of another specices (*G. adsurgens*) is used as a stomachic and stimulant by the Maya Indians of S Mexico and that the flowers of a European species (*G. pneumonanthe*) provide the source of a blue dye.

Dwarf caespitose plant with solitary terminal flowers......................1. *G. verna*
Erect plant with flowers in a terminal corymbose cyme....................2. *G. olivieri*

1. **Gentiana verna** *L.*, Sp. Pl. 228 (1753); Boiss., Fl. Orient. 4: 73 (1875); Zohary in Dep. Agr. Iraq Bull. 31: 116 (1950).

G. verna L. var. *obtusa* Boiss., Fl. Orient. 4: 73 (1875); Guest in Dep. Agr. Iraq Bull. 27: 37 (1933); Blakelock in Kew Bull. 4: 521 (1950).

G. pontica Solt. in Österr. Bot. Zeitschr. 51: 168 (1901); Bornmüller in Beih. Bot. Centr. 61, Abt. B: 82 (1942); Rech. f. in Proc. Iraqi Sci. Soc. 1: 51 (1957); Schiman-Czeika in Fl. Iranica 41: 17 (1967); Pritchard in Fl. Turk. 6: 188 (1978).

Glabrous caespitose perennial herb with underground stems from a short rootstock, each ending in a rosette of leaves. Leaves ovate or ovate-oblong, obtuse to subacute, 5–15 × 5–10 mm. Flowers solitary, terminal. Calyx tube cylindrical, strongly 5-angled, about 1 cm, lobes 5, deltoid-lanceolate, much shorter than tube. Corolla brilliant deep blue, tube narrowly cylindrical, ± 2 cm, limb spreading, 5-lobed, 1.5–2 cm across, lobes ovate, obtuse. Capsule oblong.

HAB. On rocks, in damp alpine pastures near mountain summits; alt. 2600–3500 m; fl. Aug.–Sep.
DISTRIB. Rare in central sector of the alpine zone of Iraq. **MRO**: Helgord Dagh, *Bornmüller* 1545!; *Ludlow-Hewitt* 1520!, *Rawi* 24855!; Ser Kurawa, *Gillett* 9732!
NOTE. Spring Gentian. Cultivated as alpines in the rock gardens of Europe and other temperate regions; the large solitary flowers are usually of an intense blue, but Chittenden (1951) says there are many colour forms. A beautiful little plant.

Turkey, Caucasus, Iran, Turkmenistan, N Africa (Morocco), N & C Asia, Europe (Britain to Russia, Poland to Balkans).

2. **Gentiana olivieri** *Griseb.*, Gen. et Sp. Gent. 278 (1839); Boiss., Fl. Orient. 4: 77 (1875); Hand.-Mazz. in Ann. Naturh. Mus. Wien 27: 424 (1913); Nábělek in Publ. Fac. Sci. Univ. Masaryk 70: 13 (1926); Guest in Dep. Agr. Iraq Bull. 27: 37 (1933); Zohary, Fl. Palest. ed. 2, 2: 199 (1933); Bornmüller in Beih. Bot. Centralbl. 61, Abt. B: 82 ((1942); Blakelock in Kew Bull. 4: 521 (1950); Zohary in Dep. Agr. Iraq Bull. 31: 116 (1950); Schiman-Czeika in Fl. Iranica 41: 12 (1967); Pritchard in Fl. Turk. 6: 185 (1978).

Glabrous perennial herb with fleshy roots, a short erect rootstock clothed with fibrous leaf-bases and 1–3 simple, ascending angular stems up to 40 cm high. Basal leaves oblanceolate, obtuse, up to 12 cm long, 1–2.5 cm wide. Cauline leaves smaller, lanceolate, acute. Flowers sessile or shortly pedicellate in congested terminal corymbiform cymes. Bracts lanceolate-ovate, acute. Calyx tube campanulate, about 5 mm long, shorter than the lobes, lobes 5, narrowly lanceolate, acute. Corolla deep blue, tube campanulate, 2–2.5 cm long, lobes 5, erect, deltoid, acute or sub-acute. Fig. 13: 1–3.

HAB. Dry gypsaceous hills, sandy clays, stony slopes, rocky ledges; alt. 250–1400 m; fl. & fr. Apr.–Jul.
DISTRIB. Common in the steppe region and mountain zones of Iraq. **MAM**: Bokhair(?), *Rawi* 8549W!; Amadiya, *Majid Mustafa* 3605!; Suwara Tuka, *E. Chapman* 26366!; Zawita, *Guest* 2194!; 4775!; Bakirma, *Rawi* 8793!; Zakho pass, *Guest* 2268 **MRO**: Shaqlawa, *Bornmüller* 1546!, *Haines* W.783!, *Alkas, Nuri & Serhang* 27195!; Harir, *Nábělek* 2756; Balikian, *Cuckney* 3831!; Jabal Baradost, *Field & Lazar* 925!; Haibat Sultan Dagh, *Alkas, Nuri & Serhang* 28161!; **MSU**: Sulaimaniya province, *Mooney* 4343!; Jarmo, *Helbaek* 621!, *Haines* W.331!; Penjwin, *Rawi* 12252!, 22898!; Tawila, *Rawi* 21957!; **MJS**: Jabal Sinjar, *Handel-Mazzetti* 1484 (W); **FUJ**: Talash-Shur, *Field & Lazar* 571!; Sharqat, *Handel-Mazzetti* leg. *Maresch*, 1148 (W), *Thomas*, comm. *Graham* s.n.!; Jabal Makhul, *E. & C. Guest* 13194!, *Rawi & Gillett* 7225!; Hadhr, *Handel-Mazzetti s.n.* (W); **FUJ/FNI**: Mosul, *Radhi* 5550!; **FNI**: *Zohary* (1950); **FAR/MAM**: Mar Yaqub, *Nábělek* 2758; **FAR**, *Zohary* (HUJ); **FKI**: Baba Gurgur, *Guest* 1357!; nr. Darozna, *Poore* 377!

Fig. 13. **Gentiana olivieri**. 1, habit, × 2/3; 2, calyx opened × 2; 3, flower opened × 2. Drawn by D. Erasmus.

NOTE. Olivier Gentian (Am.). Majid Mustafa recorded the local name MĀMIRĀN (Kurd.) at Amadiya, but there is no record of its use in Iraq.

Syria, Turkey, Iran, Afghanistan, Baluchistan, Turkmenistan.

103. MENYANTHACEAE

(Dumortier) Dumortier
(Gentianaceae p.p. Pflanzenfam. 4, 2: 62 (1895)

P. Taylor[6]

Aquatic or marsh herbs. Leaves alternate or ± opposite, entire or 3-foliolate, sometimes peltate, sheathing at base. Inflorescence racemose, paniculate, fasciculate or solitary. Flowers actinomorphic, bisexual. Calyx 5-lobed. Corolla gamopetalous, 5-lobed, lobes valvate in bud, margins or inner surface often fimbriate or crested. Stamens 5, inserted near the base of corolla, alternating with lobes. Ovary superior, 1-locular with 2 parietal placentas. Fruit a 2-valved dry capsule, rarely fleshy and indehiscent. Seeds few or numerous, sometimes winged or carinate.

A small cosmopolitan family of aquatic plants formerly included in the Gentianaceae but differing in the aquatic habit, alternate leaves and valvate aestivation.

1. **NYMPHOIDES** Seguier

Pl. Veron. 3: 121 (1754)
Limnanthemum S.G.Gmel. in Nov. Act. Petrop. 14: 527 (1769)

Glabrous aquatic herbs. Leaves simple, orbicular, deeply cordate, petiolate. Flowers fasciculate in leaf axils, long pedicellate. Calyx lobes more or less lanceolate, slightly accrescent. Corolla yellow or white, lobes ± fimbriate-ciliate. Fruit opening irregularly.

Nymphoides (from *Nymphaea* and *-oides*, resembling); Floating Heart (Am.); KU'AIBA (Ir.). As the Latin name of the genus implies it bears a superficial resemblance to *Nymphaea* and, though not one of the true water lilies (Nymphaeaceae, q.v.) may sometimes be mis-called a "water lily". (Such a misconception may be the origin of the report mentioned by Guest (1933) of the supposed occurrence of the Yellow Water-Lily (*Nuphar luteum*) in the Muntafiq marshes, an edible plant said to be known locally as GUNGUL, though the occurrence of this species in Iraq has never been established. Handel-Mazzetti's specimen, which Guest mentions, was from the Khabur river beyond and northwards of the frontier of Iraq.)

Though not very palatable, these water plants probably provide some fodder for animals in the marshes. Uphof (1959) mentions a Sino-Malaysian species (*N. cristata*) used as food in China and an Australian species (*N. crenata*) of which the small round tubers are roasted and consumed as food by the aborigines of Queensland. Chittenden (1951) recommends a white-flowered East North American species (*N. aquatica*), sometimes known as Fairy Water Lily, for cultivation.

Calyx lobes 10–15 mm; corolla lobes 20–25 mm, yellow; capsule longer than the calyx, acute .... 1. *N. peltata*
Calyx lobes about 3–6 mm; corolla lobes 8–15 mm long, white; capsule shorter than the calyx, obtuse........ 2. *N. indica*

1. **Nymphoides peltata** (*S.G.Gmel.*) *O.Kuntze*, Rev. Gen. 2: 429 (1891); Schiman-Czeika in Fl. lranica 41: 2 (1967); Ting-nung Ho & Ornduff in Fl. China 16: 141 (1995); Qaiser in Fl. Pak. 111: 3 (1977).

Menyanthes nymphoides L., Sp. Pl. 145 (1753).
Limnanthemum peltatum S.G. Gmel., Nov. Comm. Acad. Sci. Imp. Petrop. 14(1): 527 (1770).
Menyanthes peltata Thunb., Nov. Act. Soc. Sc. Upsal. 7: 142 (1815).
Limnanthemum nymphoides (L.) Hoffg. & Link, Fl. Port. 1: 344 (1809); Boiss., Fl. Orient. 4: 65 (1875).

Leaves broadly ovate to orbicular, bright green often spotted and tinged with purple, 2.5–10 cm long, margin sinuate, petiole of inflorescence leaves 2–10 cm, basal sheath conspicuous, spotted with purple. Fascicles 2–5-flowered. Pedicels up to 10 cm. Calyx lobes

[6] † Formerly Royal Botanic Gardens, Kew. Updated by Shahina A. Ghazanfar.

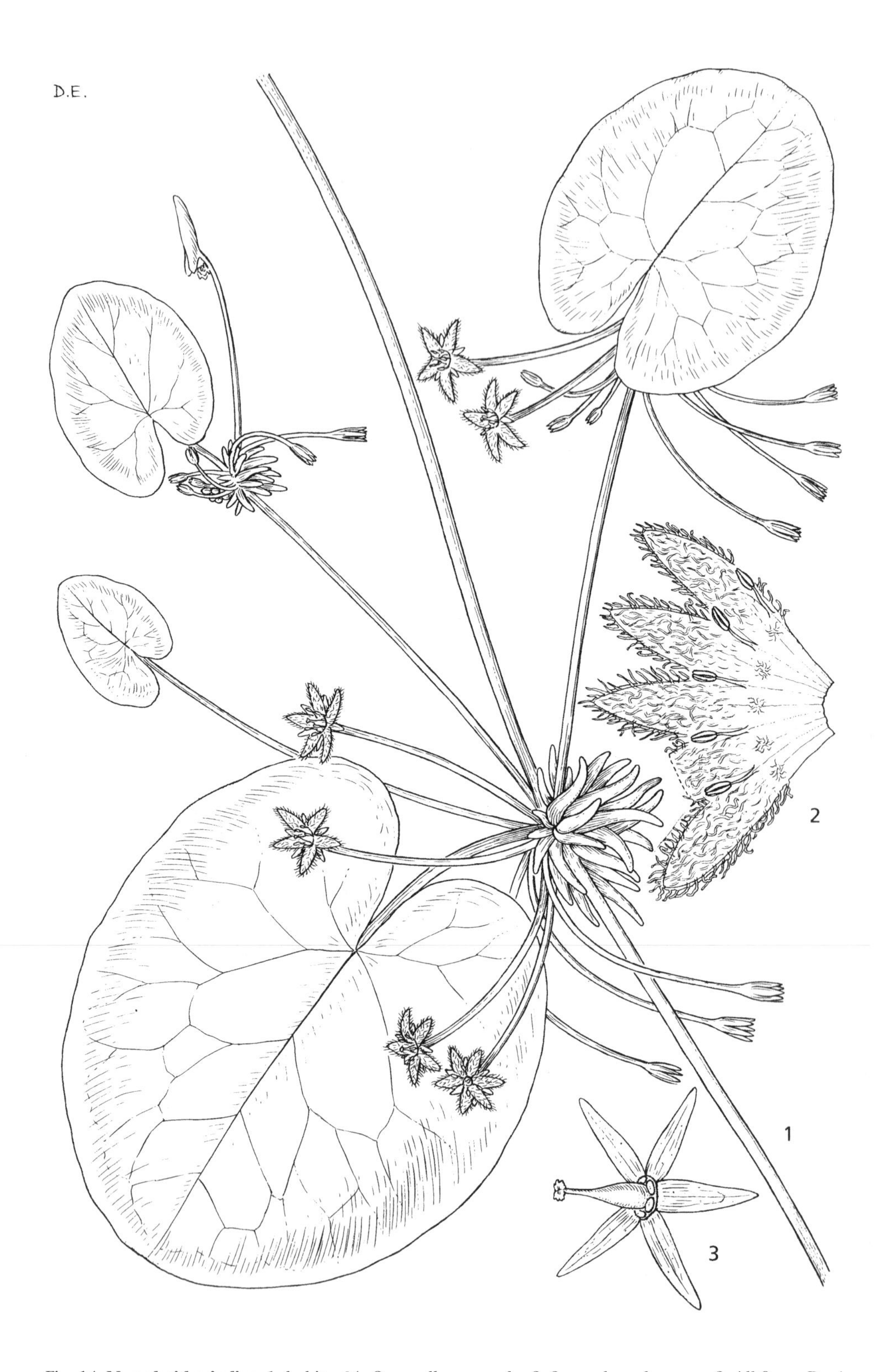

Fig. 14. **Nymphoides indica**. 1, habit × 2/3; 2, corolla opened × 3; 3, sepals and ovary × 3. All from *Rawi & Alizzi* 32400. Drawn by D. Erasmus.

oblong-lanceolate, acute, 10–15 × 3–5 mm. Corolla lobes bright yellow, obovate, 20–25 mm, margin finely denticulate-ciliate and with a basal fimbriate scale. Capsule narrowly ovoid, up to 25 mm, acute or acuminate. Seeds ovate, flat, margin ciliate, ± 4.5 × 2.5 mm.

HAB. In still or slow-flowing water; 0–10 m alt.; fl. Aug.
DISTRIB. Rare in Southern Marsh District of the desert region of Iraq. **LSM**: Garmat Bani Sa'ad, *Rawi* 16607!; Musaida, *Rawi* 16502!
NOTE. According to Chittenden (1951) this species, which is sometimes cultivated in water gardens, is known as Water Fringe.

Russia, Iran to Himalaya and Tibet, Central Europe, N Asia to China and N Mongolia; Korea, Japan.

2. **Nymphoides indica** (*L.*) *O. Kuntze*, Rev. Gen. 2: 429 (1891); Blakelock in Kew Bull. 4: 521 (1950); Ting-nung Ho & Ornduff in Fl. China 16: 141 (1995).

Menyanthes indica L., Sp. Pl. 145 (1753).
Limnanthemum indicum (L.) Griseb., Gen. et Sp. Gent. 343 (1838); Boiss., Fl. Orient. 4: 65 (1875); Nábělek in Publ. Fac. Sci. Univ. Masaryk 70: 12 (1926); Guest in Dep. Agr. Iraq Bull. 27: 57 (1933); Zohary in Dep. Agr. Iraq Bull. 31: 115 (1950); Rech. f., Fl. Lowland Iraq: 478 (1964).

Leaves broadly ovate to orbicular, bright green often tinged with reddish purple, 3–25 cm, margin entire or slightly sinuate, petiole of inflorescence leaves 1–2 cm, basal sheath inconspicuous. Fascicles 10–30-flowered. Pedicels up to 8 cm. Calyx lobes narrowly lanceolate, acute, 3–6 × 1–2 mm. Corolla lobes white with a yellow base, narrowly obovate, 8–15 mm, margin and inner surface densely bearded with long flexuous hairs. Capsule broadly ovoid-globose, up to 10 mm. Seeds lenticular, smooth or ± tuberculate, about 1.5 mm. Fig. 14: 1–3.

HAB. In still or slow-flowing water; 0–20 m alt.; fl. Apr.–Aug.
DISTRIB. Common in the southern marsh and other marshy districts of the desert region of Iraq. **LCA**: 5 km W of Shamiya, *Alizzi & Mohammad* 32141!; Amara marsh, *Guest* 1622!; Musaida, *Rawi* 16504!; Halfaya, *Field & Lazar* 30!; *Husham (Alizzi) & Mohammad* 29602!; **LSM**: Abu Ka'ida and Majarr marshes, *ibid.* 29611!; 20 km E of Alkahla, *Rawi & Alizzi* 32424!; Hor ach-Chiba, *ibid.* 32400!; near Nasiriya, *Gamal Abdin* 17!; 5 km S of Suq ash-Shiyukh, *Rawi* 26029!; Garmat Bani Sa'id, *Rawi* 16564! 26039!; **LBA**: Qarmat Ali, *Nábělek* 2739, *Guest, Rawi & Rechinger* 17379!; N Bank of Shatt al-Arab, opposite Ashar, *Gillett* 10031!

SW Iran, India, Nepal, Sri Lanka, Myanmar, Malaysia, Thailand, Vietnam, China, Taiwan, Australia, Pacific Islands, W & S Tropical Africa, Madagascar, Mauritius, S Africa.

104. PRIMULACEAE

F. Pax in Pflanzenfam. 4, 1: 98–116 (1890); F. Pax & R. Kunth in Pflanzenreich 4, 237: 1–386 (1905)

By P. Taylor[7]

Annual or usually perennial herbs, rarely undershrubs. Leaves alternate or opposite, simple or rarely dissected, exstipulate. Flowers actinomorphic, bisexual. Calyx gamosepalous, free or adnate to ovary, lobes (3–)5(–9). Corolla gamopetalous with a short to long tube and (3–)5(–9) lobes. Stamens equal in number to and inserted opposite to the corolla-lobes. Ovary superior or half inferior, ovules usually many on a free-central placenta. Fruit a capsule, circumscissile or valvate, rarely indehiscent. Seeds usually numerous, often trigonous.

The family is best represented in the temperate regions of the northern hemisphere but extends into the tropics (mainly at high altitudes). It is easily recognized among gamopetalous genera by the combination of free-central placentation and petal-opposed stamens.

According to Källersjö et al. (*Amer. J. Bot.* 87: 1325–1341, 2000), the following genera (found in Iraq), traditionally included in Primulaceae, should belong to the family Myrsinaceae: *Anagallis* L. (scarlet pimpernel), *Asterolinon* Hoffmans. & Link., *Samolus* L. (brookweed, water pimpernel) and *Lysimachia* L. (loosestrife, yellow loosestrife, yellow pimpernel). In the APG III (2009) classification, Myrsinaceae is not recognized, but sunk into the expanded Primulaceae.

[7] † Formerly Royal Botanic Gardens, Kew. Updated by Shahina A. Ghazanfar.

1. Ovary half inferior . . . . 7. **Samolus**
   Ovary superior 2
2. Plant with basal rosette of leaves and leafless scape (often with a whorl of bracts just below the flowers) or plants densely caespitose 3
   Plant with elongate leafy stems 5
3. Corolla annulate[8], tube shorter than calyx; annuals . . . . 1. **Androsace**
   Corolla annulate or exannulate, tube as long as or usually longer than calyx; perennials 4
4. Densely tufted subshrubs with stems covered with marcescent leaves; corolla tube usually several times longer than calyx; ovules usually few . . 2. **Dionysia**
   Rosulate herbs; corolla tube as long as or longer than the calyx; ovules usually numerous . . . . 3. **Primula**
5. Capsule circumscissile . . . . 6. **Anagallis**
   Capsule 5-valved 6
6 Corolla longer than calyx . . . . 4. **Lysimachia**
   Corolla much shorter than calyx . . . . 5. **Asterolinon**

1. ANDROSACE L.

Sp. Pl. ed. 1: 141 (1753) ; Gen. Pl. ed. 5: 69 (1754); R. Kunth in Pflanzenr. 4, 237: 172 (1905)

Annual or perennial herbs, often shortly stoloniferous and densely caespitose. Leaves usually basal, forming a rosette. Flowers solitary or umbellate. Bracts setaceous or foliose, rarely absent. Calyx campanulate or subglobose, 5-lobed. Stamens and style shorter than corolla tube. Capsule globose, dehiscing by 5 valves.

Distributed from Eurasia south to N Africa, Arabia, Himalaya and to New Guinea; North America excluding SE USA into Mexico; S Argentina.

Androsace (from Gr. ανδροσακες, *androsakes,* man's shield).

1. **Androsace maxima** *L.*, Sp. Pl.: 141 (1753); Boiss., Fl. Orient. 4: 18 (1875); Post, Fl. Syria, Pal. & Sinai ed. 2, 2: 180 (1933); Blakelock in Kew Bull. 4: 519 (1950); Zohary in Dep. Agr. Iraq Bull. 31: 114 (1950); Wendelbo in Fl. Iranica 9: 30 (1965).

Annual herb 2–15 cm high. Stems stiffly erect, solitary or up to 8, pilose. Leaves rosulate, broadly to narrowly ovate, 5–35 × 2–20 mm, ± pilose, sessile or stoutly petiolate, base cuneate, apex cute or obtuse, margins denticulate. Bracts leafy, about as long as pedicels or longer, obovate, denticulate. Flowers in umbels of up to 10. Pedicels 5–10 mm, pilose. Calyx urceolate-campanulate, 5–7 mm, accrescent, lobed to about middle; lobes ovate-deltoid to lanceolate, 4–5 mm, acute, denticulate, pilose. Corolla white or pink, annulate; tube ± 4 mm, lobes ± 1 mm. Capsule ± 4 mm in diameter. Seeds numerous, angular, ± 1.5 mm long. Fig. 15: 1–11.

HAB. Open hillside, stony and gravelly slopes, open *Poa bulbosa* steppe, edges of cultivated fields, on loamy soils; alt. 100–1450 m; Mar.– Apr.

DISTRIB. Common and widespread in Iraq. **MRO**: Haji Umran, *Mooney* 4277!; 42 km SW of Haji Umran, *Shahwani* 25261!; **MSU**: Sulaimaniya, *Graham* 773!; **FUJ**: Dohuk –Mosul, *Guest* 1325!; Tal Afar, *Guest* 13434!; 50 km SW of Hadar, *Chakrawarty et al.* 33073!; **FAR**: Arbil, *Shahwani* 25124!; *Guest* 629!; 4 km W of Arbil, *Polunin et al.* 78!; **FKI**: 25 km S of Kirkuk, *Rawi & Gillett* 10346!; 12 km S of Tuz Khurmatu, *Rawi & Gillett* 10271!; Kirkuk, *Rogers* 0710!; **FNI**: Table mountain (Mansur), *Graham* 715!; Jabal Hamrin, *Hemming* W307; **DLJ**: 4 km W of Haditha, *Rawi & Gillett* 6926!; **DGA**: Al-Dour, *Omar & Hamad* 17818!; Mussaige, Al Ghizal in Dour, *Kalaf et al.* 49922!; 4 km E Samarra, *Rawi* 20350!; **DWD**: Jabal Ana, *Al-Khayat & Hamad* 51724!; Ras Swaibat, *Al Khayat & Hamad* 51547!; 24 km N of Rawa to Shaabani, *Al Khayat & Hamad* 51799!; **DSD**: 4 km E Samawa, *Rawi* 20350!; **LCA**: 15 km WWN of Falluja, *Rawi* 30144!; **LEA**: 60 km N of Amara, Guest, *Rawi & Rechinger* 17482!; Bastura Chai, *Rawi & Gillett* 10493!

Caucasus, Turkey, Cyprus, Lebanon, Palestine, Jordan, Saudi Arabia, Iran.

[8] Annulate: with a ring-like structure at the mouth of corolla tube: see Fig. 15.

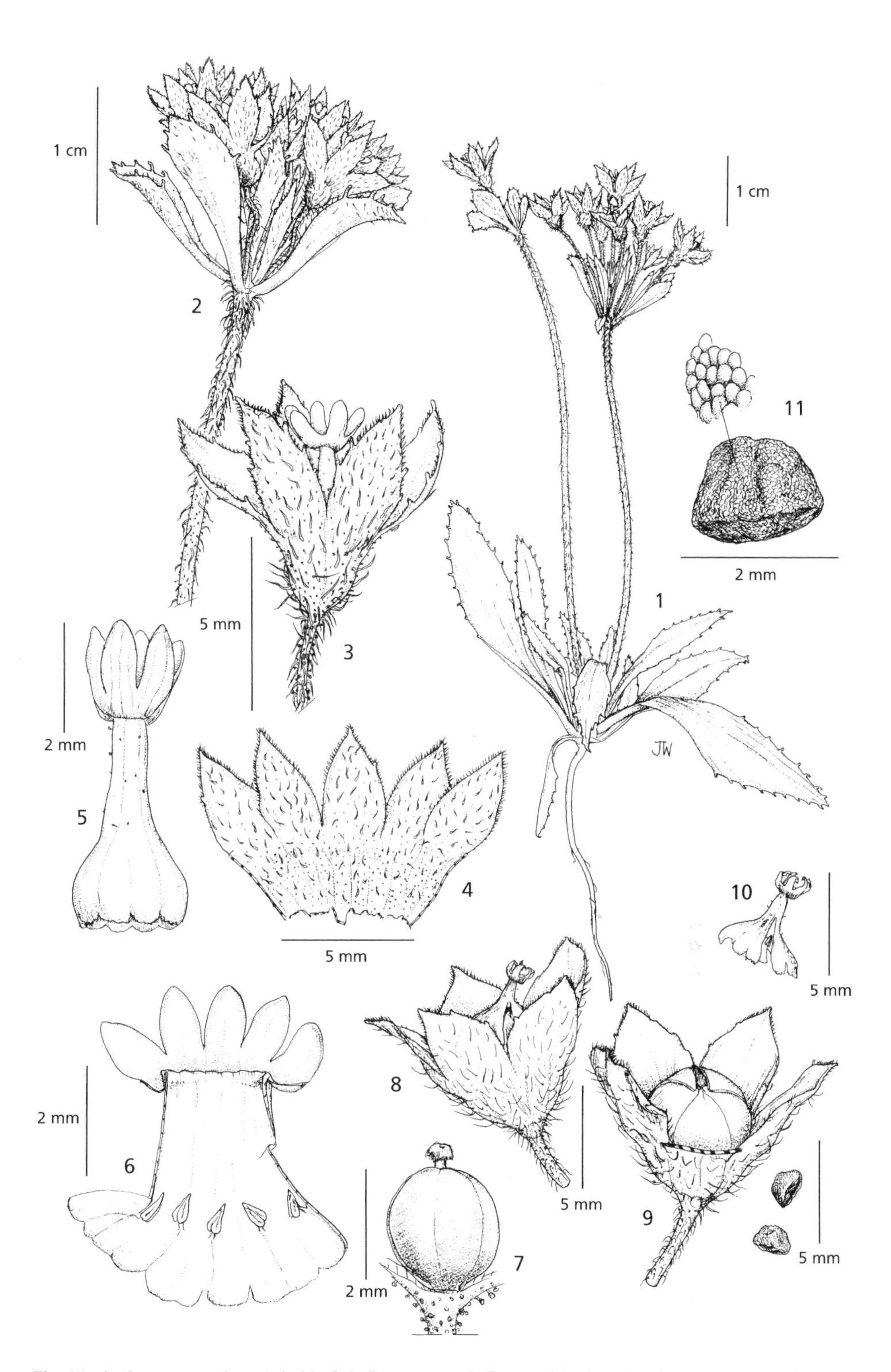

Fig. 15. **Androsace maxima**. 1, habit; 2, inflorescence; 3, flower, side view; 4, calyx opened, adaxial view; 5, corolla tube, side view; 6, corolla opened; 7, ovary and style; 8, capsue and calyx, coroll atube attached; 9, capsule and calyx with front lobe removed; 10, corolla tube detached; 11, seed with surface detail. 1, 2, 3 from *Barkley & Brahim* 4612; 4, 5, 6, 7 from *Barkley & Barkley* 4549; 8, 9, 10, 11 from *Khatab & Alizzi* 31618. © Drawn by J.E. Beentje (Feb. 2011).

2. **DIONYSIA** Fenzl

in Flora 26: 389 (1843)
Wendelbo in Acta Univ. Berg. 3: (1961); Lidén in Willdenowia 37: 37–61 (2007)
Gregoria Duby in DC., Prodr. 8: 45 (1844) p.p.

Dwarf shrubs forming ± dense tufts; old branches covered with marcescent leaves. Leaves variable from membranous and relatively large to coriaceous and minute. Inflorescence an umbel or of numerous superimposed verticils on a well-developed scape or reduced to a single escapose sessile flower. Bracts leafy or setaceous. Calyx tubular, 5-lobed. Corolla tube 3–4 times as long as calyx, limb 5-lobed. Capsule ± globose, dehiscing by 5 valves. Seeds usually few, angular.

Distributed in S Turkmenistan (Kopet Dagh), Turkey and Iran to Afghanistan and Pakistan.
Dionysia, from the God Dionysos, son of Semele, the Phoenician goddess Pen Samlath or Shemiramot. Dionysian festivals were a feature of the ancient culture of Iraq (specifically around Sinjar, the Biblical Land of Shinar).

Plant with scape 3–10 cm; leaves 1–6 cm .... 1. *D. bornmuelleri*
Plant with scape 0.5 mm; leaves 3–4.5 mm .... 2. *D. odora*

1. **Dionysia bornmuelleri** (*Pax*) *Clay*, The Present Day Rock Garden: 194 (1937); Wendelbo in Acta Univ. Berg. 3: 38 (1961); Wendelbo in Fl. Iranica 9: 15 (1965).

Primula bornmuelleri Pax, Schles. Ges. vaterl. Cult. 2, Abt. Naturw. Zool.-Bot. Sekt. 87: 19 (1909); Blakelock in Kew Bull. 4: 519 (1950).
[*Primula verticillata* (non Forssk.) Guest in Bull. Dep. Agr. Iraq 27: 77 (1933); Zohary in Dep. Agr. Iraq Bull. 31: 114 (1950)]

Aromatic low shrub forming large loose tufts; branches up to 4 mm in diameter. Leaves in dense rosettes, membranous, obovate to oblong-spatulate, 1–6 × 0.4–2 cm, attenuate at base, obtuse, margin dentate, ± pilose especially beneath. Scape erect, 3–10 cm, pilose. Flowers in 2–4 superimposed verticillasters of 3–5. Bracts foliaceous, ovate-lanceolate to lanceolate, 1–2.5 cm, obtuse or acute, dentate. Pedicels up to 10 mm. Calyx 6–8 mm, tubular-campanulate, lobes linear-lanceolate, acute. Corolla yellow, tube 18–30 mm, limb 6–8 mm across; lobes ovate, entire. Seeds numerous, angular, dark brown.

HAB. On cliff face and between rocks, in a narrow gorge by a stream; alt. 500–1000 m. Apr.– Aug.
DISTRIB. Endemic to western Iran (from where the type is collected) and northern Iraq. **MRO**: Rowanduz gorge, *Guest* 2139!, *Gillett* 8313!, *O. Polunin* 5069! Gali Ali Beg, *Shahwani* 25241!; nr. Sharanish, *Rechinger* 11485!

2. **Dionysia odora** *Fenzl* in Flora 26: 390 (1843); Nábělek in Publ. Fac. Sci. Univ. Masaryk 70: 9 (1926).

Gregoria aucheri Duby in DC., Prodr. 8: 46 (1844); Wendelbo in Fl. Iranica 9: 20 (1965).
Dionysia aucheri (Duby) Boiss., Fl. Orient. 4: 19 (1875); Zohary in Dep. Agr. Iraq Bull. 31: 114 (1950).
Primula odora (Fenzl) O. Kuntze, Revis. Gen. Pl. 2: 400 (1891).
Dionysia straussii Bornmüller & Hausskn. in Bull. Herb. Boiss. ser. 2, 3: 591 (1903).
D. sintenisii Bornmüller in Bull. Herb. Boiss. ser. 2, 3: 592 (1903).
D. odora subsp. *straussii* (Bornmüller & Hausskn.) Bornmüller in Beih. Bot. Centralbl. Abt. 2, 28: 462 (1911).

Aromatic glandular-pubescent low shrub forming large dense cushions or tufts up to 15 cm high. Branches densely covered with imbricate marcescent leaves. Leaves sessile, broadly obovate to narrowly oblong, 3–4.5 × 1–2 mm, obtuse, 3–7-dentate to almost entire, pubescent and glandular. Flowers solitary; scape 0.5 mm. Bract solitary, oblong, entire, pubescent and glandular. Calyx campanulate-tubular, 4–4.5 mm, lobes linear-oblong, glandular-pubescent. Corolla yellow, tube 12–25 mm, glandular, limb 7–10 mm across, lobes obovate, 2.5–3.5 mm, entire, rounded above. Seeds few (about 3). Fig. 16: 1–12.

HAB. Locally common on vertical and overhanging rocks and cliffs, in crevices in limestone; alt. 1400–2300 m; Apr.
DISTRIB. Found in north-east Iraq. **MSU**: Pira Omar Gudrun (Pira Magrun), *Gillett* 7819!; *Rawi* 12079!; *O. Polunin* 5155!

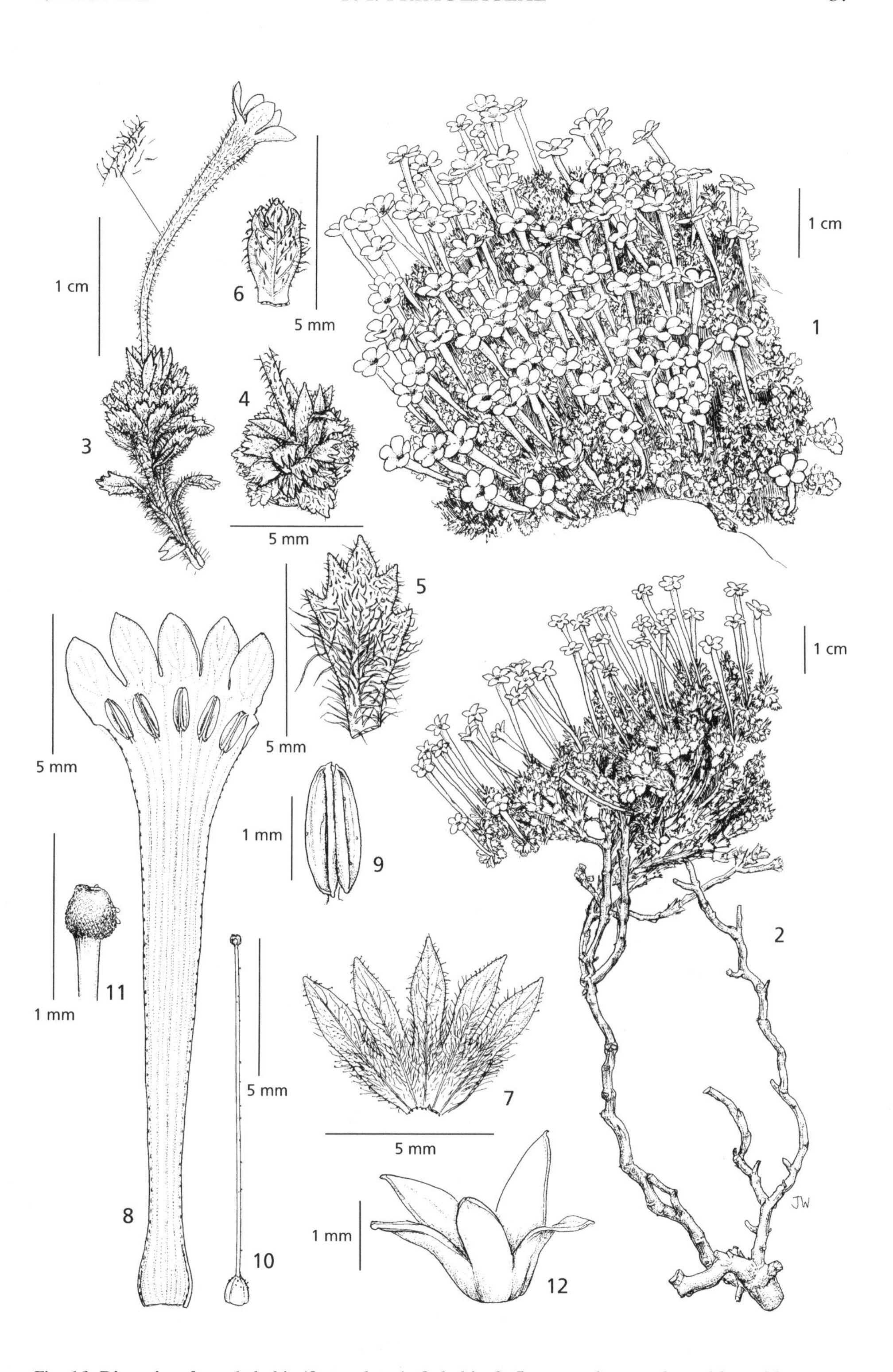

Fig. 16. **Dionysia odora**. 1, habit (from photo); 2, habit; 3, flower and young branchlets with rosette of leaves; 4, l eaf rosette; 5, leaf, detail top surface; 6, leaf, detail lower surface; 7, calyx opened, abaxial surface; 8, corolla tube opened; 9, stamen; 10, gynoecium; 11, stigma, side view; 12, capsule, dehisced. All from *O. Polunin* 5155. © Drawn by J.E. Beentje (Feb. 2011).

3. **PRIMULA** L.

Sp. Pl. ed. 1, 142 (1753); Gen. Pl. ed. 5: 70 (1754); Pax, Pflanzenr. 4, 237: 17 (1905)

Perennial rosulate scapose herbs. Leaves all basal, entire or lobed. Flowers heterostylous. Inflorescence an umbel or of numerous superimposed verticils or rarely racemose or spicate, very rarely solitary. Bracts foliaceous or setaceous. Calyx tubular or campanulate, sometimes inflated, persistent, lobes 5. Corolla tube as long as or longer than calyx, limb usually spreading, 5-lobed. Capsule globose to cylindrical, usually as long as calyx, dehiscing by 5–10 valves. Seeds numerous, angular.

Widespread from Eurasia south to India, Indo-China and to New Guinea, N Africa and Ethiopia; also found in Alaska, Canada and Greenland, N and W USA to New Mexico and Mexico; S Chile and Argentina and Falkland Is.

Primula (from Lat. *primulus*, very first, the diminutive of *primus*, first or foremost).

Corolla tube about twice as long as the calyx . . . . 1. *P. auriculata*
Corolla tube about as long as the calyx . . . . .2. *P. algida*

1. **Primula auriculata** *Lam.*, Ill. Gen. 1: 429, no. 1931 (1791); Boiss., Fl. Orient. 4: 28 (1875); Nábělek in Publ. Fac. Sci. Univ. Masaryk 70: 10 (1926); Blakelock in Kew Bull. 4: 4519 (1950); Zohary in Dep. Agr. Iraq Bull. 31: 115 (1950); Wendelbo in Fl. Iranica 9: 8 (1965).

Glabrous perennial herb. Leaves elliptic to oblong-spatulate or oblanceolate, 3–16 × ± 4 cm, base cuneate into a short petiole, apex obtuse or subacute, membranous, margins denticulate or almost entire. Scape 4–40 cm, longer than leaves. Flowers numerous, umbellate, in a dense head. Bracts linear-lanceolate, acuminate, base auriculate, usually

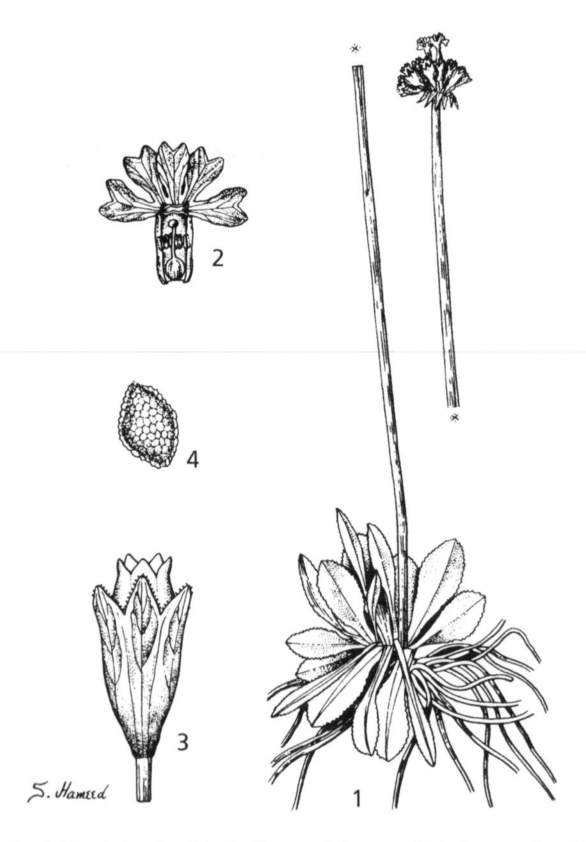

Fig. 17. **Primula algida**. 1, habit × 1/2; 2, dissected flower × 1 1/2; 3, capsule with calyx × 5; 4, seed × 30. Reproduced from Fl. Pakistan 157: 22 (1984), with permission.

shorter than pedicels. Pedicels shorter than to as long as calyx. Calyx campanulate, 5–6 mm long, lobes lanceolate, subacute, purplish. Corolla pink, violet or lilac, tube ± twice as long as the calyx; limb 15–20 mm across, lobes obcordate. Capsule ovoid, 5-valved, slightly longer than calyx.

HAB. In wet places on marshy turf, wet grassy places by a lake; alt. 1800–3000 m; Apr.
DISTRIB. **MRO**: Halgurd Dagh lake, *Ludlow-Hewitt* 1503; *Guest & Ludlow-Hewitt* 2875!; Chiya-i-Mandau near Walza, *Guest* 2687.[9]

Turkey, Caucasus, Iran, Afghanistan, Tien Shan.

2. **Primula algida** *Adam* in Weber & Mohr, Beitr. Naturkunde 1: 46 (1805); Lehman, Monogr. Prim. 68, t. 7 (1817); Boiss., Fl. Orient. 4: 29 (1879); Blakelock, Kew Bull. 4: 519 (1950); Wendelbo in Fl. Iranica 9: 11 (1965); Y. Nasir in Fl. Pak. 157: 22 (1984).

P. algida Adam var. *sibirica* (Ledeb.) Pax f. *colorata* Regel ex Pax, Pflanzenr. IV. 237, Primulaceae p. 73 (1905); Blakelock in Kew Bull. 3: 519 (1948).

Glabrous or occasionally farinose perennial herb. Leaves oblong-spatulate or oblong, 2–6 × 1–2 cm, base cuneate into a short petiole, apex obtuse, membranous, margin denticulate to almost entire, midrib prominent on the underside. Scape longer than leaves, 3–20 cm, elongating in fruit. Flowers numerous, umbellate in a dense head. Bracts linear-lanceolate, 6–8 mm, acuminate, about as long as the pedicels. Pedicels shorter than the calyx, elongating in fruit. Calyx 5–6 mm, campanulate, lobes lanceolate, subacute, blackish purple. Corolla violet, tube about as long as calyx, limb 8–10 mm across, lobes ± 5 mm, obcordate. Capsule oblong, about as long as calyx. Seeds angled, papillose. Fig. 17: 1–4.

HAB. Grassy ledges on cliff; alt. ± 3600 m; Apr.?
DISTRIB. **MRO**: Halgurd Dagh, *Guest* 2876.

Caucasus, N Iran, Altai, N Mongolia, Pamir Alai, N Pakistan, Afghanistan.

4. LYSIMACHIA L.

Sp. Pl. ed. 1, 146 (1753); Gen. Pl. ed. 5: 72 (1754); Kunth, Pflanzenr. 4, 237: 256 (1905)

Procumbent or erect, mostly perennial herbs or rarely subshrubs. Leaves alternate, opposite or verticillate, entire, often glandular. Flowers axillary or in spikes, racemes, corymbs or panicles. Calyx ± deeply 5–6-lobed. Corolla ± campanulate, 5–6-lobed. Ovary globose, style filiform. Stamens 5–6, ± adnate to corolla, sometimes alternating with staminodes. Capsule ovoid or globose, 5-valved, few- or many-seeded. Seeds oblong, orbicular or angular, sometimes winged.

Distributed from Europe east to the Far East and Eastern Asia, south through tropical Asia to Australia; Africa including Azores, Madagascar and Mascarenes; Hawai'ian Is., North America.

Lysimachia (from Gr. λύσις, *lysis,* release and μάχη, *machi,* (from) strife, or to commemorate King Lysimachus (c. 360–281 BCE).

1. **Lysimachia dubia** *Willd.* in Ait., Hort. Kew. 1: 199 (1789); Boiss., Fl. Orient. 4: 8 (1879); Nábělek in Publ. Fac. Sci. Univ. Masaryk 70: 9 (1926); Post, Fl. Syria, Pal. & Sinai ed. 2, 2: 178 (1933); Zohary in Dep. Agr. Iraq Bull. 31: 114 (1950); Wendelbo in Fl. Iranica 9: 32 (1965); Y. Nasir in Fl. Pak. 157: 90 (1984).

Erect, glabrous, simple or sparsely branched perennial herb up to 60 cm high; stem ribbed, young shoots glandulose. Leaves petiolate, alternate or subopposite, ± lanceolate to linear-lanceolate, 30–70 × 10–30 cm, acute, margins entire, undersurface with numerous small irregular dark brown immersed glands. Flowers racemose, congested or at length remote, bracteate; pedicel ± 2 mm, elongating to 5 mm in fruit. Bracts linear-subulate, acute, as long as flowering pedicels or longer. Calyx lobes 5, linear-lanceolate, ± 3 mm, acute, glandular, persistent. Corolla pink, ± 5 mm, tube broad and short ± 2 mm, lobes obovate or spatulate,

[9] Record taken from Blakelock (1948).

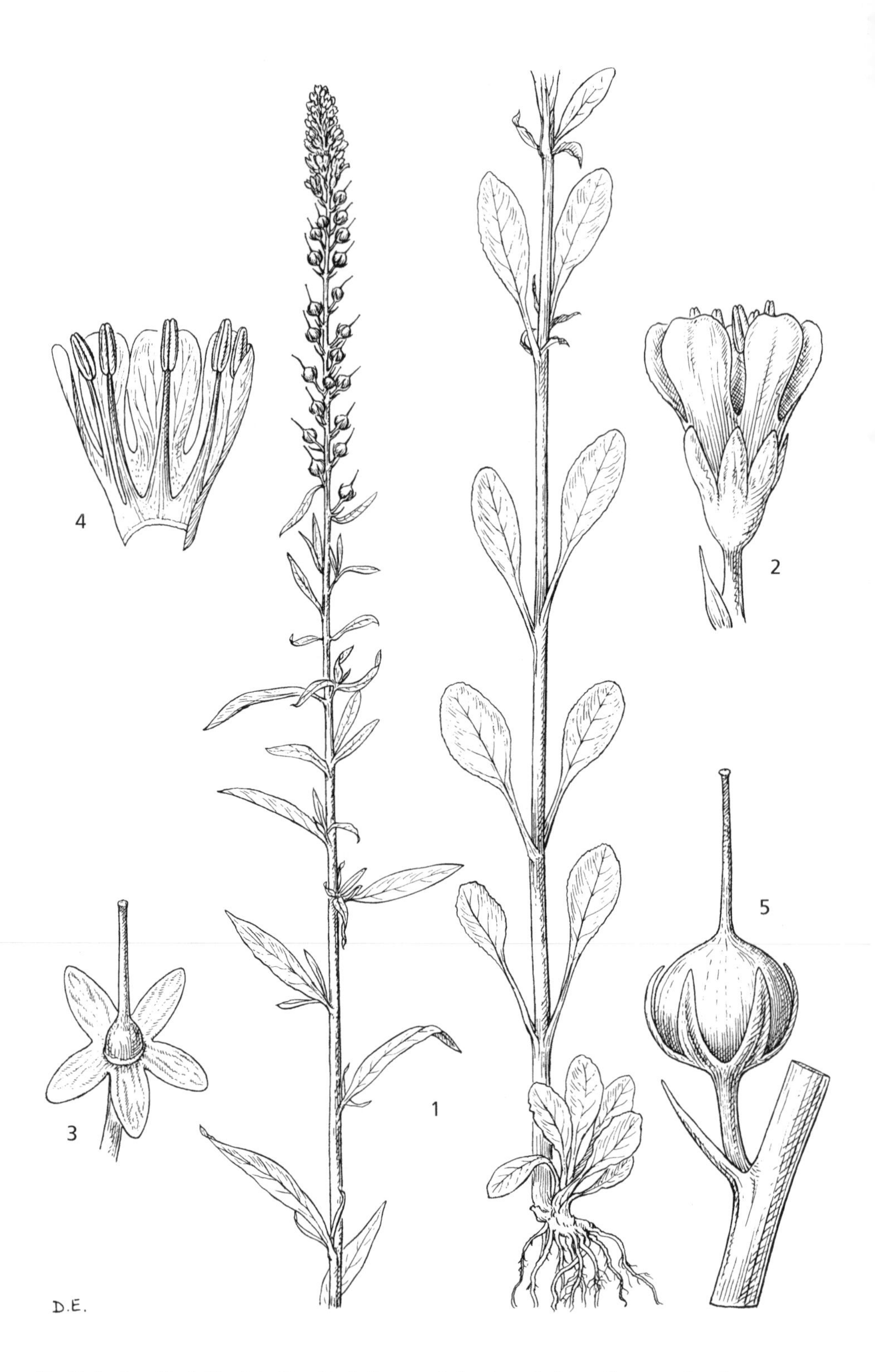

Fig 18. **Lysimachia dubia**. 1, habit × ²/₃; 2, flower × 6; 3, calyx × 6; 4, corolla opened to show stamens × 6; 5, fruit × 6. All from *Balansa* 1855. Drawn by D. Erasmus.

obtuse, glandulose. Stamens shorter than corolla. Style ± 4 mm, persistent. Capsule ± 3 mm in diameter. Seeds few, 3-angled, reticulate-vesiculose. Fig. 18: 1–5.

HAB. Roadsides, moist grassy meadows, near streams, wet places in wadis; alt.: 650–1100 m; fl. Jun.–Aug.
DISTRIB. **MAM**: Sarsang, *Haines* W1343!; Dinarta, *Chapman* 26119!; **MSU**: Old road to Dukan (Dokhan), *Robertson*, RC53!; 27 km NW of Sulaimaniyah, *Rawi* 22787!; Arbet, *Karim* 39302!

Balkan Peninsula, Caucasus, W Syria, Turkey, Iran, C Asia, Afghanistan, Pakistan.

5. **ASTEROLINON** Hoffm. & Link

Fl. Port. 1: 332 (1820); Pax & Kunth, Pflanzenr. 4, 237: 317 (1905)

Glabrous annual herbs. Stems erect, simple or branched. Leaves sessile, opposite, lanceolate or ovate, entire. Flowers axillary, solitary, pedicellate, erect at anthesis, strongly recurved in fruit. Calyx deeply 5-lobed. Corolla 5-lobed, lobes much shorter than calyx, white or yellow. Capsule globose, dehiscing by 5 valves. Seeds 2-many, tuberculate.

A genus of two species; Mediterranean region to Krim, Caucasus and Iran; in Africa, south to Tanzania. *Asterolinon* (from Gr. αστέρι, *asteri*, star, λινον, *linon*, flax).

1. **Asterolinon linum-stellatum** (*L.*) *Duby* in DC., Prodr. 8: 68 (1844); Boiss., Fl. Orient. 4: 10 (1879); Nábělek in Publ. Fac. Sci. Univ. Masaryk 70: 9 (1926); Post, Fl. Syria, Pal. & Sinai ed. 2, 2: 178 (1933); Zohary in Dep. Agr. Iraq Bull. 31: 114 (1950); Boulos, Fl. Egypt 2: 192 (2000).
Lysimachia linum-stellatum L., sp. Pl. 148 (1753); Wendelbo in Fl. Iranica 9: 32 (1965).

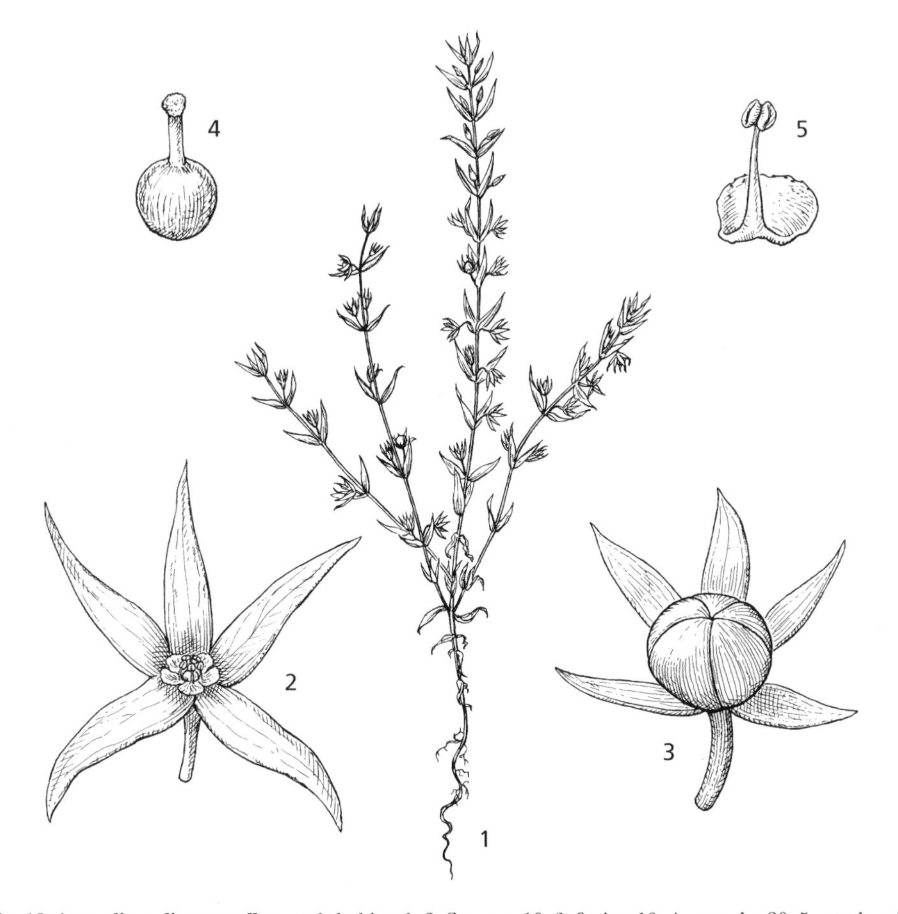

Fig. 19. **Asterolinon linum-stellatum**. 1, habit × 1; 2, flower × 10; 3, fruit × 10; 4, carpel × 20; 5, petal and stamen × 20. All from *Helbaek* 843. Drawn by D. Erasmus.

Small glabrous annual up to 10 cm high. Stems 4-angled. Leaves sessile, oblong-lanceolate or lanceolate, 0.3–7 × ±1 mm, base slightly clasping, acute, entire. Pedicel filiform, about as long as calyx. Calyx lobes linear-lanceolate, ± 3 mm, acuminate. Corolla white, ± 0.5 mm, scarcely 1 mm in diameter, lobes orbicular. Stamens and style about as long as corolla. Capsule globose, ± 2 mm in diameter. Seeds 2 or 3, depressed-ovoid, strongly tuberculate, dark brown. Fig. 19: 1–5.

HAB. Low limestone hills; alt. 50–100 m.
DISTRIB. **FNI**: Diyala, Jabal Hamrin, *Haines* W. 862.

Atlantic Isles; South and West Europe; Syria, Lebanon, Palestine, Iran, ?Afghanistan, Pakistan.

6. **ANAGALLIS** L.

Sp. Pl. ed. 1, 148 (1753); Gen. Pl. ed. 5, 73 (1754)

Glabrous annual or perennial herbs. Stems erect or prostrate and rooting at nodes. Leaves sessile or shortly petiolate, alternate or opposite, capillary to orbicular, entire. Flowers solitary and axillary or in terminal racemes. Calyx deeply (4–)5(–6)-lobed. Corolla ± deeply (4–)5(–6)-lobed. Filaments ± connate at base, often bearded. Ovary often with granular glands. Capsule globose, indehiscent and 1-seeded or usually circumscissile and many-seeded. Seeds usually 3-angled.

A largely Mediterranean and tropical African genus with the following species also throughout the cooler parts of the world as a weed of cultivation.
Anagallis (from Gr. αναγαλλις, *anagallis*, Pimpernel).

1. **Anagallis arvensis** *L.*, Sp. Pl. 148 (1753); Boiss., Fl. Orient. 4: 6 (1879); Post, Fl. Syria, Pal. & Sinai ed. 2, 2: 177 (1933); Wendelbo in Fl. Iranica 9: 33 (1965); Leblebici in Fl. Turk. 6: 141 (1978); Y. Nasir in Fl. Pak. 157: 94 (1984); Boulos, Fl. Egypt 2: 190 (2000).

Anagallis phoenicea Scop., Fl. Carn. ed. 2, 1: 139 (1772); *A. arvensis* subsp. *phoenicea* (Scop.) Vollman in Ber. Bayer. Bot. Ges. 9: 44 (1904); *A. arvensis* L. var. *phoenicea* Gouan, Fl. Mons. 29, 1765; Boiss. Fl. Orient. 4: 6 (1879).

Annual, simple or often much branched and diffuse herb. Stems 4-winged, 5–50 cm long. Leaves opposite, sessile, ovate, 9–25 × 7–10 mm, subcordate at the base, acute or obtuse, entire, gland-dotted. Pedicels filiform, longer than leaves, erect at anthesis, recurved in fruit. Calyx lobes 5, 3–4 mm, narrowly lanceolate, acuminate, with minute hairs on margins. Corolla orange-red to blue, up to 15 mm in diameter, lobes 4–5 mm, ± orbicular. Filaments bearded. Capsule ± 5 mm in diameter, circumscissile, many-seeded. Fig. 20: 1–5.

HAB. Common in cultivated fields, irrigated places and gardens, on silty, loamy and sandy/gravelly soils; alt.: 50–180 m. March-Apr.
DISTRIB. Found throughout Iraq; common. **MAM**: Gara Dagh, *Al-Dabbagh & Jasin* 46974!; **MRO**: Rowanduz, *Omar, Sahira, Karim & Hamid* 38347!; Koi Sanjaq, *Rawi, Nuni & Al-Kass* 28099!; **MSU**: Jarmo, camp above wadi, *Helbaek* 539!; Kharmal, *Rawi* 8423!; Lower Diyala, *P.E.N.* 137!; **FNI**: 37 km W of Khaikh [Khanaqin], *Cowan & Darlington* 266!; Pilkana, *Rawi* 12800!; **FUJ**: Mosul, *Bayliss* 32!; **FKI**: Fakka–Masaieda, *Rawi* 15000!; 6 km E of Kirkuk, Botany staff 42979!; **FNI**: Jabal al-Muwaila, *Guest, Rawi & Rechinger* 17562!; 2 km NE of Mandali, *Al Kaisi, Thamer & Salah* 51496!; Laqlaq, *Al Kaisi, Thamer & Salah* 51297!; **DLJ**: 5 km above Rawa, *Rawi & Guest* 7021!; Manaiyf, *Omar & Hamid* 36575!; **DGA**: 4 km E of Samarra, *Rawi* 20320!; Al-Dour, *Omar & Hamad* 47772!; **DWD**: S of Shitatha, *Rawi & Gillett* 6443!; 17 km N of Heditha, *Al-Kaisi* 44232!; Wadi al-Qaser, *Omar & Kaisi* 44359!; Wadi Fuhaimi, *Omar, Kaisi, Hamad & Hamid* 44675!; **DSD**: 5 km from Salman to Samawa, *Kaisi, Hamad & Hamid* 48227!; Karbala, *Rawi & Gillett* 6343!; E of As Salman, *Rawi & Gillett* 6231!; **LEA**: 55 km N of Amara, *Guest, Rawi & Rechinger* 17446!; Aziziyah, *Rawi* 15022!; 20 km N of Al-Lut, *Thamer* 47488!; **LCA**: Rustam, *Guest* 373!; Baghdad, *Guest* 1108!; Ghurfa plain nr. Daltawa, *Guest, Eig & Zohary* 5076!; **LSM**: 30 km E of Asmara, *Rawi, Bharucha & Tikrity* 29293!; **LBA**: 40 km S of Basra on road to Fao, *Alizzi & Sabah* 35043!
NOTE. A widespread and variable species which has been variously treated recognizing subspecies and varieties. In Iraq both the blue flowered (var. *arvensis*) and orange-red flowered (var. *coerulea* (L.) Gouan) colour forms occur showing no other consistent differences and no differences in range distribution.

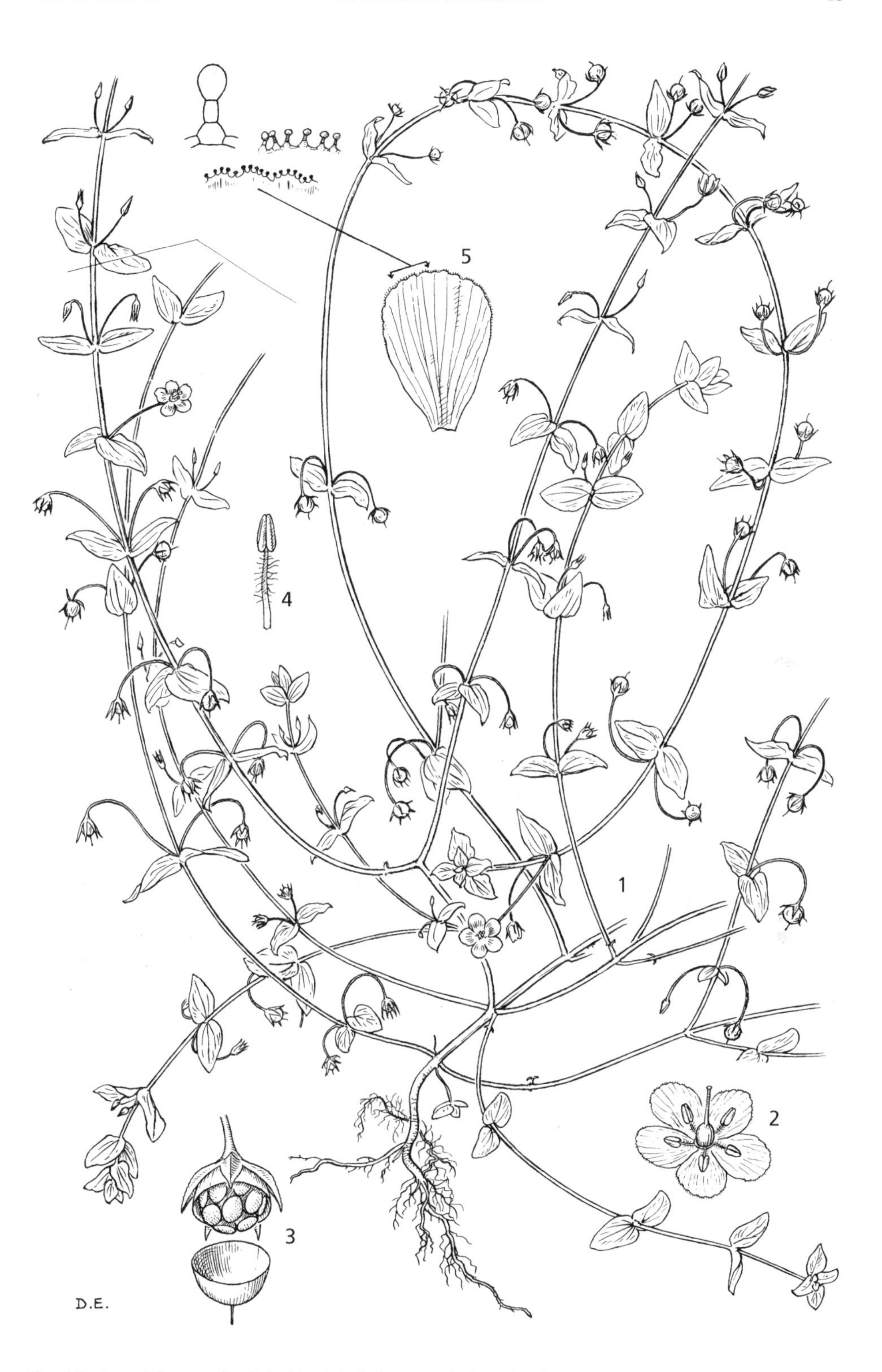

Fig. 20. **Anagallis arvensis**. 1, habit × ²/₃; 2, flower × 3; 3, fruit × 3; 4, stamen × 6; 5, petal × 6. Drawing from a live specimen. Drawn by D. Erasmus.

Fig. 21. **Samolus valerandi**. 1, habit × ²/₃; 2, habit with mature inflorescence × 2; 3, flower × 8; 4, corolla opened to show stamens × 8; 5, capsule with calyx × 8; 6, mature capsule with calyx × 8; 7, capsule with the calyx removed at front × 8. All from *Akhley* 1642. Drawn by D. Erasmus.

7. **SAMOLUS** L.

Sp. Pl. ed. 1, 171 (1753); Gen. Pl. ed. 5, 78 (1754); Pax & Kunth, Pflanzenr. 4, 37: 336 (1905)

Annual or perennial herbs. Stems erect or prostrate. Leaves alternate, cauline or forming a basal rosette, linear to spatulate, entire. Flowers in terminal racemes or corymbs, bracteate. Calyx tube adnate to ovary in its lower half, limb 5-lobed, persistent. Corolla perigynous, campanulate, limb 5-lobed. Staminodes often present, alternating with stamens. Capsule globose, many-seeded, dehiscing by 5 valves. Seeds numerous, angular.

Almost a cosmopolitan weed; introduced and naturalized in many warm countries.

1. **Samolus valerandi** *L.*, Sp. Pl. 171 (1753); Boiss., Fl. Orient. 4: 4 (1879); Nábělek in Publ. Fac. Sci. Univ. Masaryk 70: 9 (1926); Post, Fl. Syria, Pal. & Sinai ed. 2, 2: 177 (1933); Wendelbo in Fl. Iranica 9: 35 (1965); Y. Nasir in Fl. Pak. 157: 97 (1984); Boulos, Fl. Egypt 2: 190 (2000).

Glabrous erect annual up to 50 cm high. Leaves in a basal rosette and usually also numerous throughout the length of stem, the lowermost spatulate, those above obovate and smaller, 4–12 × 1–2 cm, base tapering, obtuse, entire. Flowers in terminal racemes up to 20 cm long. Pedicels up to 25 mm. Bract 2–3 mm, inserted at or above the middle of pedicel. Calyx ± 2 mm, lobes ovate-deltoid, acute. Corolla ± 2 mm, lobes obovate. Stamens inserted at base of corolla tube, about 1 mm long. Staminodes minute. Capsule valves strongly deflexed after dehiscence. Seeds ovoid, ± 0.6 mm long. Fig. 21: 1–7.

HAB. Moist locations, near streams and in water, in ditches; alt. 10–100 m; Feb.-Aug.
DISTRIB. Occasional in wetlands of lowland Iraq. **LSM**: Hor al Hawiza, *Thamen* 46674!; Um al Ward, 17 km from Al Kahla, *Noori* 41307!; **LBA**: Al Ashmah, *Al Khayat & Redha* 52298!; **DWD**: Shithatha, *Chakravarty & Rawi* 29790!

Mediterranean region, Balkans, Turkey, Syria, Lebanon, Palestine, Arabian Peninsula, Iran, Pakistan to China.

105. **PLUMBAGINACEAE**

F. Pax in Pflanzenfam. 4, 1: 116–125 (1890)

S. Mobayen[10]

Erect herbs, undershrubs or climbers, with alternate or basal rosette and sheathing leaves. Flowers regular, hermaphrodite, pentamerous. Calyx funnel-shaped or tubular with scarious or rarely herbaceous limb. Corolla 5, slightly united at base or forming a short tube. Stamens 5, fertile and epipetalous. Ovary 1-celled, superior, with one anatropous ovule; styles 5, free, united at base or up to middle or completely united, glabrous, hairy or warty; stigmas 5, filiform or capitate-peltate, papillate. Fruit included in the persistent calyx, coriaceous or membranous, indehiscent or dehiscent by a lid or irregular valves. Seeds with membranous testa.

A distinct family in the Order Caryophyllales of 29 genera and 730 species, cosmopolitan in distribution especially in maritime and saline locations.

1. Calyx herbaceous; styles connate to tip; fruit dehiscent by valves from the base . . . . 1. **Plumbago**
   Calyx scarious; styles free or fused only at base; fruit dehiscent by valves or opening by a lid or indehiscent 2
2. Erinaceous, subalpine or alpine shrubs; leaves subulate or needle-like; stigmas capitate . . . . 4. **Acantholimon**
   Shrubs or herbs, not erinaceous; stigmas filiform 3

[10] Formerly University of Tehran, Iran and University of Montpellier, France. Revised by J R Edmondson, Hon. Research Associate, Royal Botanic Gardens, Kew

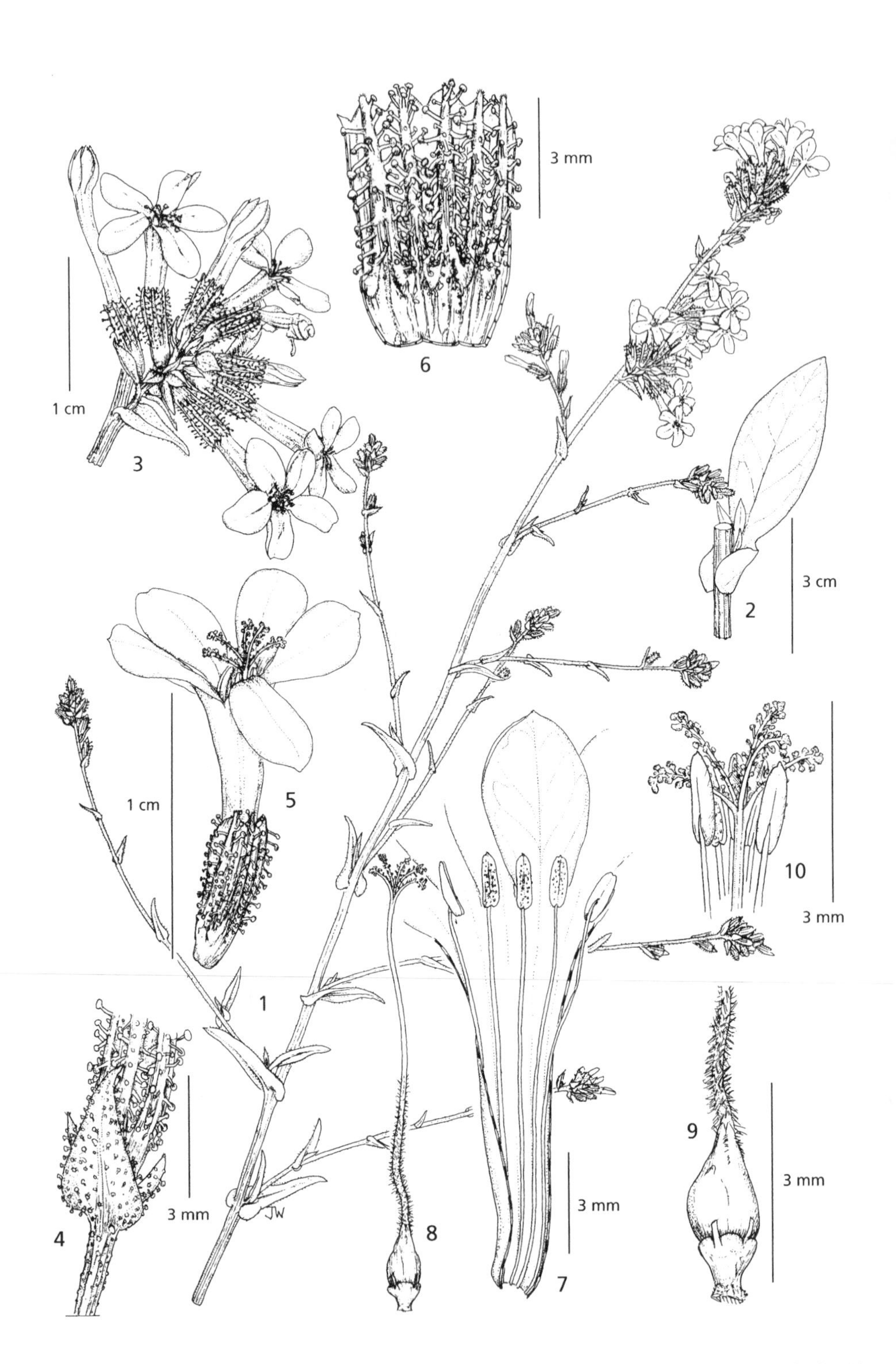

Fig. 22. **Plumbago europaea**. 1, habit; 2, leaf, lower stem; 3, inflorescence; 4, bracts at base of calyx; 5, flower, side view; 6, calyx opened, abaxial view; 7, corolla tube opened; 8, ovary and style; 9, ovary and style detail; 10, stigma and anthers detail. 1, 3–10 from *Furse* 4096; 2 from *Couter* 76. © Drawn by J.E. Beentje (Apr. 2012).

3. Annuals; corolla united; calyx tube with glandular hairs . . . . 2. **Psylliostachys**
Perennials; corolla free or connate slightly at base; calyx tube without glandular hairs . . . . 3. **Limonium**

1. PLUMBAGO

L., Sp. Pl. ed. 1, 151 (1753)

Perennial, shrubby or climbing plants. Flowers almost sessile, 3-bracteate, lilac or pink. Calyx united, tubular, stipitate-glandular. Corolla salver-shaped, limb 5-parted. Filaments free, dilated at the base into concave scales. Styles filiform, coherent; stigmas 5, filiform, glandular within. Utricle membranous, rupturing around base and splitting by longitudinal valves from base to middle.

Plumbago (from Latin *plumbum*, lead) is sometimes known as Leadwort on account of the glaucous-greyish hue of the leaves. A genus of about 20 species found in the warm temperate and tropical regions of the world.

1. **Plumbago europaea** *L.*, Sp. Pl. 3, 1, 151 (1753); Ledeb., Fl. Ross. 3: 471 (1850); Boiss. in DC., Prodr. 12: 691 (1848) & Fl. Orient. 4: 875 (1879); Kuznetsov, Fl. Cauc. Crit. 4, 1: 174 (1901); Guest in Dep. Agr. Iraq Bull. 29: 75 (1933); Zohary in Dep. Agr. Iraq Bull. 31: 115 (1952); Grossh., Fl. Kavk. ed. 2, 7: 190 (1967); Rech. f. & Schiman-Czeika in Fl. Iranica 108: 3 (1974); Zohary, Fl. Palaest. 3: 8 (1978); Bokhari & Edmondson in Fl. Turk. 7: 464 (1982).

P. lapathifolia Willd., Sp. Pl. ed. 4, 1(2): 837 (1798); Bieb., Fl. Taur.-Cauc. 1: 144 (1808); *P. angustifolia* Spaché,Veg. Phan. 10: 337 (1841).

Perennial with erect herbaceous stem 30–100 cm tall, branched from base or in its upper part; branches ± erect and narrow. Leaves glaucous or greenish, glabrous, with rare and very small calcareous dots; lower leaves elliptic or obovate to lanceolate, 5–8 × 3–4.5 cm, auriculate at base, slightly amplexicaul, with short petiole, margins dentate-glandular to entire; upper leaves sessile, narrowly lanceolate or linear, passing gradually into bracts. Inflorescence an elongated spike or capituliform; flowers with 2–3 leaf-like, greenish, lanceolate or ovoid-elongate glabrous bracts; margins glabrous or glandular, apex acute, shorter (3–4×) than calyx. Calyx tubular, 6–7 mm, with stipitate glands arranged along ribs. Corolla violet-pink, tube 2–3 × calyx, lobes obovate, truncate, obtuse or rounded, veins scarcely prominent. Fig. 22: 1–10.

HAB. Dry hillside, mountain sides, between Oak trees, beside road and eroded places, on stony ground and between limestones; alt. 650–2200 m; fl. May–Aug.
DISTRIB. Widespread in the thorn-cushion zone of the mountains of NE Iraq. **MAM**: nr. Amadiya, *Kotschy* 402!; Dohuk, *Guest* 1585 & 2702!; Shatch-i-Haus, *Guest* 13011!; Sharanish village, 25 km NE of Zakho, *Rawi, Tikriti & Nuri* 29036!; **MRO**: *Rust, Guest & Husham* 15847!; Rowanduz, *Guest* 455, 13005, 2982!; Bisht Ashan NE of Rania, *Serhang & Rawi* 26846 & 26492!; **MSU**: Darband-i-Bazian, *Rechinger* 10600!; Darband-i-Khan, *Noor & Ani* 9279; **FNI**: Wadi Aloka, *Barkley & Ani* 9015!; Bosakh pass, *Haley* 236!

W Mediterranean and S Europe to Turkey, Palestine, Upper Jordan valley, N and NW Iran, Azerbaijan, Turkmenistan.

2. PSYLLIOSTACHYS

Nevski, Tr. Inst. Bot. Ac. Sc. URSS, ser. 1, 4: 314 (1937); Fl. SSSR 18: 467 (1952)

Annual herbs, leaves simple or pinnate; scape simple or only once branched; flowers white or rose-coloured, spikelets 2–4-flowered, united in a dense cylindrical spike. Calyx tubular or funnel-shaped, slightly scarious, 5-lobed, 5-veined, herbaceous, glabrous inside. Calyx tube straight, glandular, hairy in the lower half. Corolla funnel-shaped, 5-lobed, with a long tube that remains erect after anthesis. Stamens with ± free filaments, adhering to corolla tube up to the length of ovary, glabrous. Styles free, glabrous; stigmas filiform-cylindrical; ovary elongate-obovate, or ± linear with clear ribs, narrowing slightly towards to apex, passing gradually into the styles. Fruit obovate or linear, opening with valves.

Seven or eight species known from Central Asia, Caucasus, Iran, Afghanistan, Syria, Jordan, Palestine. *Psylliostachys* (from Gr. Ψηλλιυμ, *Psyllium*, a name applied to one of the subgenera of *Plantago*, and σταχης, *stachys*, ear or spike (e.g. of wheat), referring to a resemblance to the inflorescence of *Plantago psyllium*).

Leaves generally pubescent, pinnatifid, margins runcinate; petiole up to 2 cm .... 1. **P. spicata**
Leaves generally glabrous, entire or pinnatisect, with triangular-rounded lobes; petiole up to 0.5 cm .... 2. **P. suworowii**

1. **Psylliostachys spicata** (*Willd.*) *Nevski*, Act. Inst. Bot. Acad. Sci. U.R.S.S. ser. 1, 4: 314 (1937); Rozhkova in Fl. SSSR 18: 471 (1952); Rech.f., Fl. Lowland Iraq: 475 (1964); Grossh., Fl. Kavk. ed. 2, 7: 195 (1967); Bokhari in Fl. Pak. 28: 12 (1972); Rech. f. & Schiman-Czeika in Fl. Iranica 108: 17 (1974); Zohary, Fl. Palaest. 3: 12 (1978).

Statice spicata Willd., Sp. Pl. ed. 4, 1: 1533 (1797); M. Bieb., Fl. Taur.-Cauc. 1: 253 (1808); Boiss. in DC., Prodr. 12: 669 (1848) & Fl. Orient. 4: 871 (1879); Handel-Mazzetti in Ann. Naturh. Mus. Wien 27: 2 (1913); Guest in Dep. Agr. Iraq Bull. 27: 96 (1933).
S. lyrata M. Bieb., Tab. Prov. Côte Occid. Mer Casp. 114 (1798); Beschreib. Länd. Terek Kasp. 166 (1800).
S. sisymbrifolia Jaub. & Spach, Ill. Pl. Or. 1: 158, t. 87 (1844).
S. plantaginiflora Jaub. & Spach, Ill. Pl. Or. 1: t. 88 (1844).
Limonium spicatum (Willd.) O. Kuntze, Rev. Gen. 2: 396 (1891); Post, Fl. Syria, Pal. & Sinai ed. 2, 2: 414 (1933).

Annual, 10–40 cm, glabrescent. Leaves in rosettes, papillose-pubescent, rarely glabrous, oblong-lanceolate, 5–15 × 1–3 cm, obtuse or mucronate, base attenute into a long petiole, margins entire or pinnatifid, runcinate, lower surface hairy, scape slightly projecting from leaves. Spikes 1-many, sessile, cylindrical, upper part dense and continuous, lower part interrupted; spikelets 2–4-flowered; bracts herbaceous, brownish, papillose, outer bract oblong-spatulate, curved inside, mucronate; inner bracts broader, shorter than the outer one, apex truncate. Calyx ± 3 mm, tube glandular-hairy in its lower $^2/_3$, with triangular aristate lobes. Corolla cream or bright rose, ± 4 mm, with ovate membranous lobes. Ovary oblong; styles free; stigmas filiform and glandular. Fig. 23: 1–10.

HAB. stony plains; alt. 50–450 m: fl. & fr.: Mar., Apr.
DISTRIB. Occasional in the Mesopotamian lowlands of Iraq. **FKI:** nr. Tuz, *Lazar* 1149; **FKI/FNI:** Jabal Hamrin, *Bornmüller* 1571 (fide FI); **FNI:** 25 km S.E. of Mandali, *Rawi* 20720!; **DLJ:** Jazira Desert, *Guest* 3793!, *Edmonds* 3795; **LEA:** Kut road, 3 km SE of Baghdad, *O. Polunin* 5002!; **LEA/FNI:** Fakka 70 km from Amara, *Rawi* 25208!; **LCA:** Rustam (Rustamiya), *Lazar* 3914; **LBA:** 35 km W of Samarra, *Rawi* 14875!; 20 km N by Basra NE of Shatt al-Arab, *Guest, Rawi & Rechinger* 16763!; Nabrwan (V R) between Basra and Fao, *Anon.* 6617!
NOTE. An entire-leaved form has also been recorded from the Jazira desert. ZIBAD AL BARRIYAH (Ir.-Baghdad, *Lazar* 3914).

Syria, Palestine, Saudi Arabia, Iran, Azerbaijan, Turkmenistan.

2. **Psylliostachys suworowii** (*Regel*) *Roshkova* in Fl. SSSR 18: 469 (1952); Bokhari in Fl. Pak. 28: 13 (1972); Rech. f. & Schiman-Czeika in Fl. Iranica 108: 16 (1974).

Statice suworowii Regel, Acta Horti Petrop. 7: 550 (1881).

Annual 10–50(–80) cm tall. Leaves in rosettes, glaucous, glabrous, oblanceolate or obovate-elongate, 5–15(–20) × 1.3 cm, entire or pinnatisect, with triangular-rounded obtuse or ± acute lobes, shortly petiolate. Scapes 1–5(–8), erect or slightly sinuate, 2–3(–4) × as long as leaves, cylindrical at base, glabrous, angled at points of ramification, densely hairy above, becoming less hairy below, simple or branched into a panicle in $^1/_2$–$^2/_3$ of the upper part; terminal spike very long, slightly sinuate, 10–20(–30) cm, ± 1.5 cm in diameter, with shorter lateral spikes 5–10(–20) cm, drooping with curved apex, united at the base of spike or slightly distant; spikelets 2–4-flowered, sessile, with 2 bracts, hairy in upper half, outer bract 3–5 mm long, up to 1.5–2 × as long as inner one, linear, pointed, or slightly broadening in its upper part before attenuating and with narrow membranous margins; inner bract broader, obovoid, concave, apex truncate, with 3 membranous, comparatively broad teeth, margins membranous. Calyx 3.5–4 mm, funnel-shaped, tube 10-ribbed in its lower part, herbaceous, densely covered with long glandular hairs; limb as long as tube, 5-lobed; lobes

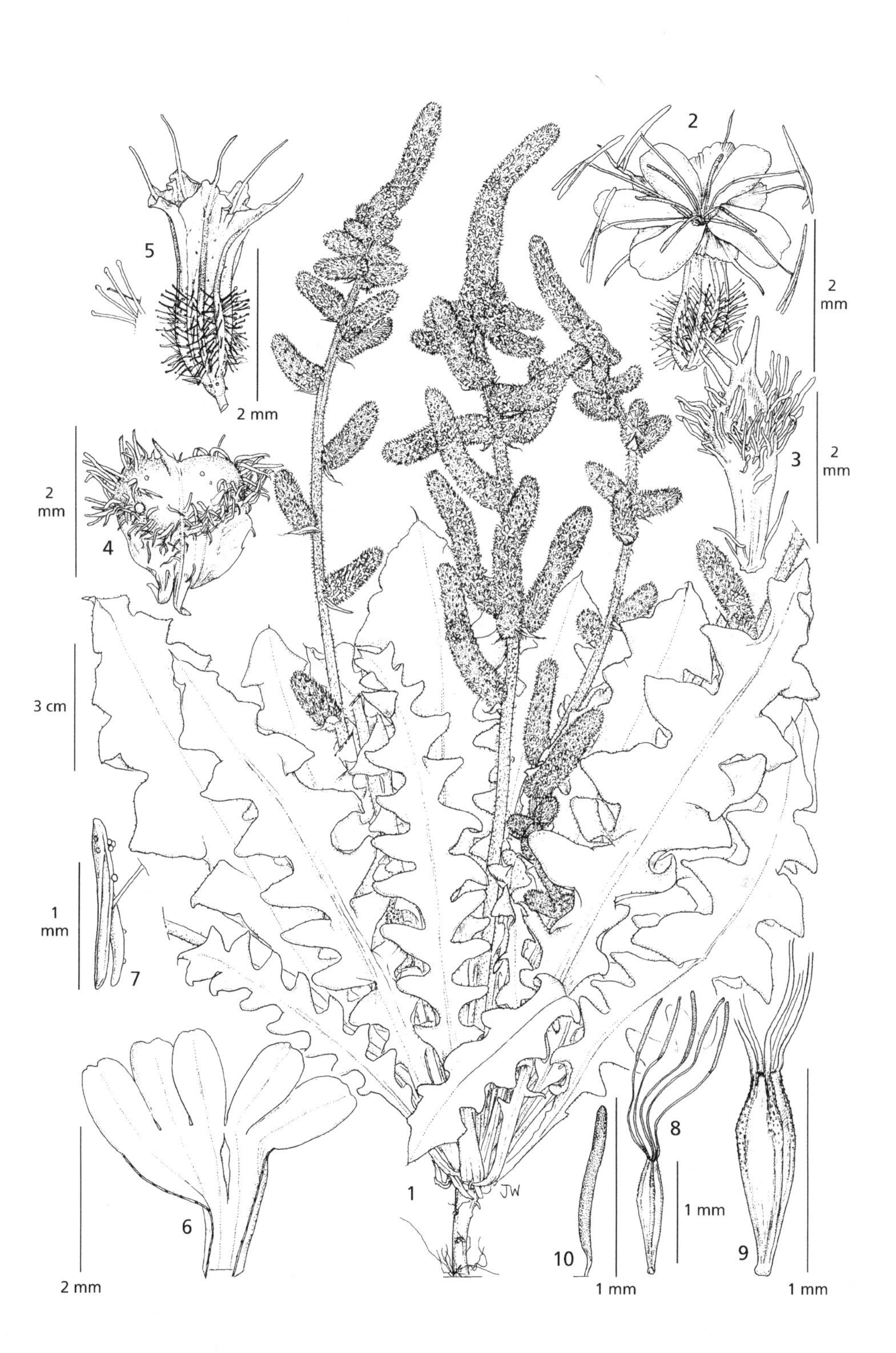

Fig. 23. **Psylliostachys spicata**. 1, habit; 2, flower; 3, outer bract; 4, inner bract; 5, calyx, side view; 6, corolla opened; 7, anther detail; 8, ovary and style; 9, ovary detail; 10, style. All from *Abbas Ali* 9865. © Drawn by J.E. Beentje (Apr. 2012).

triangular, with acute apex, and veins projecting 1 mm from lobes. Corolla rose, 5–6 mm in diameter, lobes ovate, obtuse, recurved.

HAB. Clayey soil, by roadside, in gardens; alt.: ± 30 m; fl. Apr.–May.
DISTRIB. Rare, possibly introduced; recorded from the central lowlands of Iraq. **LCA**: Rustam Farm, *Lazar* 1171A, 1172!; Baghdad, *Rogers* 6255! *Lazar* 3900!
NOTE. DHAIL AL BAZZUN (Ir.-Baghdad, *Lazar* 3900).

Afghanistan, C Asia, Turkmenistan, Iran, Pakistan (cult.).

3. **LIMONIUM** Mill.

nom. cons.

Gard. Dict. Ed. 4 (1754)

Statice L., Sp. Pl. 274 (1753), partim emend. Willd., Enum. Hort. Berol. 33 (1809)

Perennial, rarely annual, herbs or shrubs. Stems branched. Bracts 3. Spikelets 1–4-flowered. Calyx obconical-tubular or funnel-shaped, with scarious 5-veined 5–10-lobed limb. Corolla of 5 free petals, or united at base. Stamens adnate at base or up to middle of petals. Styles glabrous, free or short-connate at base; stigmas filiform, glandular over the whole surface. Utricle membranous below, undulated above, opening by a lid or irregular ruptures near base.

Linnaeus' views of the generic limits of *Statice* s.l. led to much confusion as they embraced both *Limonium* s.str., a name used by Tournefort, and other genera such as *Armeria*. The 6th International Botanical Congress, held in 1935, therefore conserved the name *Limonium* for the true sea-lavenders and suppressed the use of *Statice*. Despite this ruling, the name persists in horticultural and floristry circles for the everlasting inflorescences of certain species of *Limonium* and *Goniolimon*.
Limonium (from Gr. λειμών, *leimon*, meadow).

Annual; scape and rachis triquetrous, winged . . . . 1. *L. lobatum*
Perennial; scape and rachis cylindrical, not winged . . . . 2. *L. iranicum*

1. **Limonium lobatum** (*L.f.*) *Chaz.*, Suppl. Dict. Jard. 2: 36 (1786); Collenette, Wildflow. Saudi Ariabia: 606 (1999); Boulos, Fl. Egypt Checkl. 161 (2009).

Statice lobata L.f., Suppl. Pl. 187 (1782) [1781 publ. 1782].
S. thouinii Viv., Cat. Hort. Negr. 34 (1802) — Type: Planta culta ex Ana (Iraq) oriunda; Boiss., Fl. Orient. 4: 858 (1879); Handel-Mazzetti in Ann. Naturh. Mus. Wien 27: 2 (1913); Guest in Dep. Agr. Iraq Bull. 27: 76 (1933); Zohary in Dep. Agr. Iraq Bull. 31 (1956).
S.aegyptiaca Pers., Syn. Pl. 1: 334 (1805).
S.alata Willd., Enum. Pl. Suppl. 15 (1814).
S.tripteris H. Paris ex Poir. in Lam., Encycl. Suppl. 5: 237 (1817).
Limonium thouinii (*Viv.*) *O.Kuntze*, Rev. Gen. 2: 395 (1891); Rechinger, Fl. Lowland Iraq: 475 (1964); Rech. f. & Schiman-Czeika in Fl. Iranica 108: 5 (1974); Zohary, Fl. Palaest. 3: 9 (1978); Daud, Fl. Kuwait 1: 156 (1985).

Annual glaucescent herb, 10–25 cm tall. Leaves in rosette, calcareous-punctate, pinnatifid, lyrate with rounded-sinuate lobes, attenuate and reddish towards base, apex shortly mucronate, margins ciliate. Scape erect, glabrous, 2–3-bracteate, and ± triangular towards base, gradually becoming 3-winged and dichotomously branched into a corymbiform inflorescence; wings along the scape projecting at their ends giving rise to a triangular appendage much longer than axillary bracts. Rachis slightly triquetrous, but upper joints bearing spikes broadly flattened into a coriaceous 3-winged cladode prolonging into 3 lobes (one longer than others) covering the spikelets. Spike sessile, fasciculiform, with 2–3 spikelets attached at base, each of which is 2–3-flowered. Outer bracts membranous, cuspidate; inner one much longer than outer ones, dorsal side greenish, coriaceous, bicornate, margins scarious, apex with 2 triangular membranous lobes supported on their dorsal side with 2 unequal recurved hard horn-like appendages. Calyx funnel-shaped with glabrous ± compressed tube, constricted at its summit; limb blue, whitish or white, membranous, as long as or slightly longer than the tube, with 5 triangular lobes alternating 5 setose brownish veins separated from the limb. Corolla yellow. Fruit circumscissile. Fig. 24: 1–8.

HAB. Dry steppe, sandy and rocky soil, gravelly hillsides; alt.: 165–680 m; fl. Mar.–Jun., fr.: Jun.–Aug.
DISTRIB. Occasional in the moist steppe zone of Iraq. **MSU**: Sulaimaniya, *Graham* s.n.; **FKI**: Tuz, *Rogers* 0225!; **FKI/FNI**: J. Hamrin, *Rawi & Gillett* 7361!; **FNI**: Sa'diya, *Stutz* 1361; Makatu nr Mandali, *Guest* 877; Koma Sang police station nr. Mandali on Iranian border, *Rawi* 20661!; 10 km E. of Mandali,

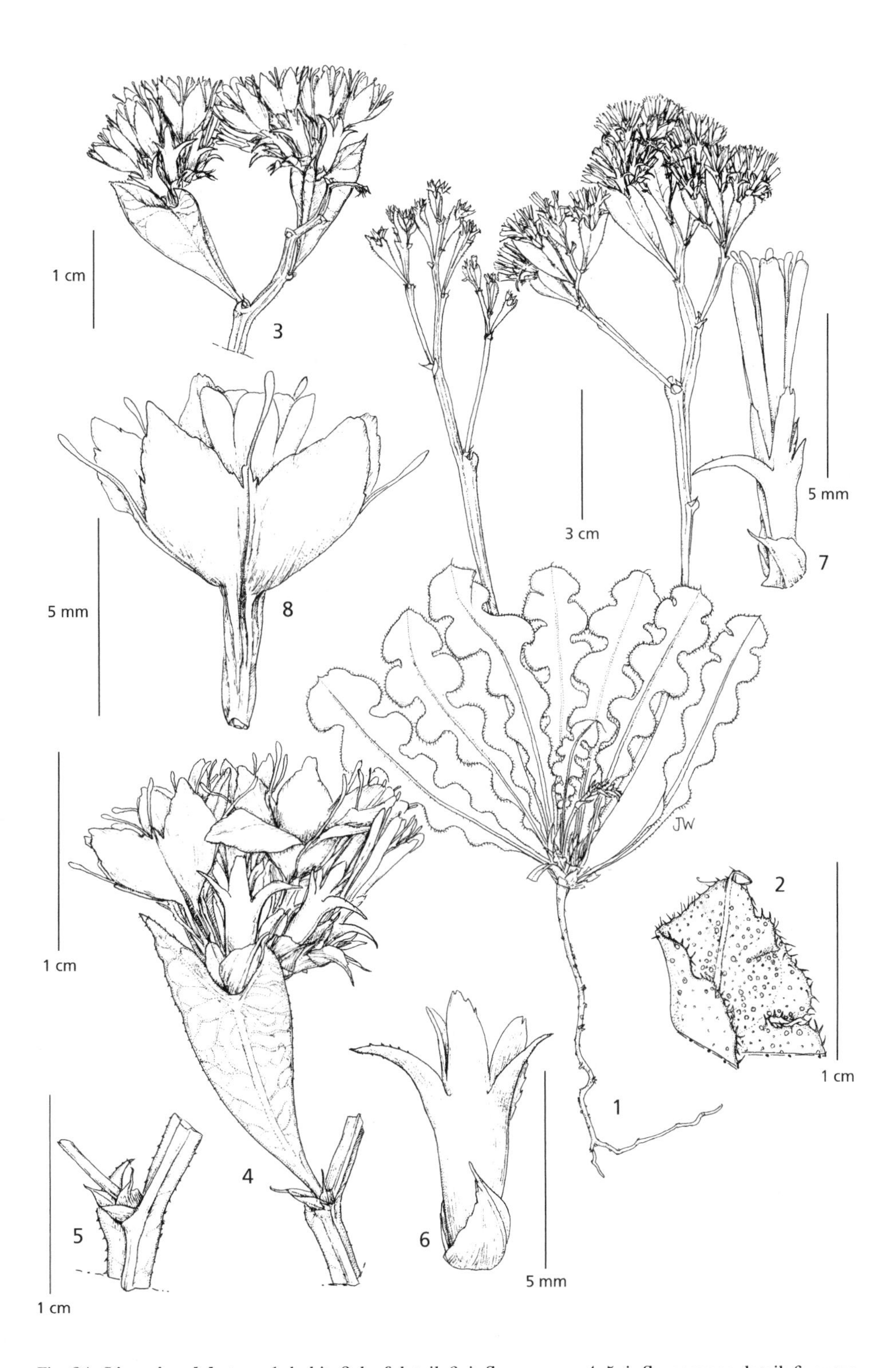

Fig. 24. **Limonium lobatum**. 1, habit; 2, leaf detail; 3, inflorescence; 4, 5, inflorescence detail; 6, outer and inner bracts; 7 flower bud, side view; 8, calyx and corolla, side view. 1 from *Chakrawarty et al.* 31710; 2–8 from *Al Kaisi et al.* 48315. © Drawn by J.E. Beentje (Apr. 2012).

Rechinger 9615; 30 km NE of Mandali, *Qaisi & Khayat* 50790; Qaraghan (Jalaula), *Rogers* 0224! 20 km N of Mandali, *Qaisi & Yahya* 45215; 15 km from Saadiya to Khanaqin, *Qaisi* 48757; 15 km N of Jalaula, *F. & E. Barkley* 4386, 33209; **DWD**: Tal an-Nisr, 48 km W of Rutba, *Qaisi & K. Hamad* 46589; 12 km N of T.1 pumping station, *Chakr., Rawi, Khatib & Alizzi* 31759; 65 km N of Rutba, *Rawi & Khatib* 32254; between Rutba and Syrian border, *Bharucha* s.n.; Qa'ra, *Alizzi & Husain* 34082; 70 km W of Nukhaib, *Rawi* 31008; Nukhaib, *Rawi* 14806; Wadi al-Ajrumiya, *Rawi* 31260; 200 km W of Ramadi, *Barkley & Palmatier* 2294; Nukhaib, *Rawi* 14806!; **DSD**: Al-Taamim between Jumaima and Ansab, *Qaisi, Hamad & H. Hamid* 48315. NOTE. JODHA (Jabal Hamrin, *Rawi & Gillett* 7361).

S Spain, N Africa, Greece, Syria, Lebanon, Palestine, Jordan, Iran, Egypt, Saudi Arabia, Kuwait.

2. **Limonium iranicum** (*Bornmüller*) *Lincz.* in Fl. SSSR 18: 461 (1952); Rech. f. & Schiman-Czeika in Fl. Iranica 108: 12 (1974).

Statice leptophylla Schrenk var. *iranica* Bornmüller, Beih. Bot. Centralbl. 22: 140 (1907).

Perennial, 10–(20–40) cm tall; lower part of stem thick, passing gradually upwards into a woody and branched raceme; branches erect, elongated. Leaves thick and fleshy, glaucous, linear-spatulate, 0.5–1 (–3 or 4) × 0.2–0.6 cm, base cuneate attenuated into a thick petiole, passing into the ochrea broadly covering half of the stem, disposed on annual branches and in a nearly dense rosette at base of scape, apex obtuse, rounded or subacute, margins membranous. Scapes 10–15, erect, cylindrical, or slightly sinuous, nearly glabrous, branched in upper half or along ¾ of its length into a panicle. Flowers on 5–10 spikelets on an almost narrow spike forming a lax panicle on tapering branches of the scape. Spikelets 2–3-flowered; outer bract 1.5–2.5 mm, covering partially flowers of the spikelet, half the length of first inner bract, ovoid-elongated, apex obtuse or subacute, distinctly membranous, margins glabrous; second inner bract smaller, membranous, 1-veined. Calyx (2.5–)3–5 mm, narrowly funnel-shaped; calyx tube 1–3 mm, glabrous or slightly hairy at base, limb 2 mm wide, 10-lobed, 5 bigger ones alternating with 5 smaller ones, distinctly ovate-rounded or ovate-acute, with glabrous veins. Corolla pale pink to pale violet.

HAB. Wet saline soil, farm land, semidesert grazed by camels; alt. 0–130 m; fl. Apr.–Jun. and Oct.-Nov.
DISTRIB. Occasional in the dry steppe and desert zones of Iraq. **FUJ/DLJ**: Jazira nr. Wadi Thirthar, *Guest* 3554!; **LEA**: Fakka, 70 km E of Amara, *Rawi* 25808! & 25782!; **FPF/LEA**: Shaikh Sa'id (nr. Iranian border), *Rawi* 12539!; **FPF**: Mandali, *Rawi* 43562!; Haj Yusuf village, *Rawi* 20686!; 25 km SE of Mandali, *Rawi* 20721!. **DWD**: Shithatha, 3 km from Bahr al-Mardh, *Rawi* 7184! & 16174; *Haines* W.2116!; N. of Shithatha, *Rechinger* 8335!.
NOTE. *Limonium iranicum* (Bornmüller) Lincz., was recorded for Iraq by Rechinger (Fl. Iranica 108: 12 (1974)) but the specimens he cited were immature and therefore of uncertain identification. However on the basis of the calyx measurements taken from material collected by Rawi (calyx tube 1–3 mm long) and comments by H. Akhani (in litt.) it would seem likely that all Iraq material previously identified as *L. carnosum* falls within the range of *L. iranicum* as recognised in Flora Iranica, as *L. carnosum* has the mature calyx 4–5 mm long according to Rechinger, op. cit. p. 13.
?SH(U)WAYWAH (Ir.-Jazira, *Guest* 3554, Bahr al-Milh, *Rawi* 26917). The plant is reported to be grazed by animals.

Saudi Arabia, Kuwait, Iran, Turkmenistan, Pakistan.

SPECIES EXPECTED TO OCCUR IN IRAQ

Limonium meyeri (Boiss.) O. Kuntze occurs in Iran close to its western frontier and can be expected to occur within Iraq, but no material has been seen. It has a basal rosette of broadly ovate leaves and a much-branched inflorescence.

4. ACANTHOLIMON Boiss.

Diagn. Pl. Or. Ser. 1, 7, 69 (1846); Bunge, Mem. Ac. Sc. Pétersb. ser. 7, 18 (2): 1–72 (1872); Doğan & Akaydın, Bot. J. Linn. Soc. 154: 397–419 (2007)
Statice subgenus *Armeriastrum* Jaub. & Spach, Ann. Sc. Nat. ser. 2, Botanique, 20: 248 (1843)

Ericaceous, cushion-forming high mountain shrublets with needle-like or subulate leaves. Inflorescences in axils of spring leaves of spike-like racemes (in Iraq); spikelets 3–9-bracteate, 1–5-flowered, showy. Calyx scarious, funnel-shaped, rarely tubular, 5-veined, 5–10-lobed or with truncate limb; limb purple, pink or white. Petals with short claws, free or adhering at base, incurved after anthesis. Stamens 5, with glabrous filiform or ribbon-like filament, adhering to base of claws of petals. Ovary linear-cylindrical, sometimes ovoid, ± compressed, 5-ribbed or pentagonal, tapering into 5, glabrous or rarely verrucose styles coherent at base. Utricle crustaceous-membranous, oblong, indehiscent or splitting at angles.

165 species from the Balkan Peninsula and Kriti through Western and Middle Asia to W China and W Himalaya.

Acantholimon (from Gr. άκανθα, *akantha*, spine and λειμών, *leimon*, meadow).

Species found in Iraq belong to two Sections:

Plants heterophyllous; spring leaves persisting; spikelets (1–)2–3(–5)-flowered, in capitula or congested spikes; inner bracts 2–6, suborbicular to lanceolate with a broad to narrow hyaline margin .... Sect. **Acantholimon**
Plants with spring and summer leaves ± similar, persistent; spikelets 1-flowered; inner bracts 2, ± equal .... Sect. **Staticopsis**

1. Most spikelets (1–)2–3-flowered; inner bracts 3–6 2
   All spikelets 1-flowered; inner bracts 2 5
2. Inflorescence hemispherical or capituliform; bracts as broad as long or broader, usually aristate .... 1. *A. bracteatum*
   Inflorescence oval to oblong, with or without spikelets interrupted; bracts longer than broad, usually acuminate 3
3. Inflorescence with widely separated spikelets; calyx tube hairy ...... 3. *A. bromifolium*
   Inflorescence with ± contracted spikelets; calyx tube glabrous 4
4. Scape robust, emerging above the leaves, glabrous; calyx glabrous, limb purple-brown, venation of the same colour as the limb .... 2. *A. latifolium*
   Scape thin, shorter than the leaves, densely hairy; calyx sparsely hairy, limb white, striated along the 5 veins with purplish large bands (giving the flower aspect of a hybrid *Petunia*) .... 4. *A. petuniiflorum*
5. Scape scabrid to pubescent 6
   Scape glabrous 13
6. All leaves with under surface keeled or leaves on lower part of stem flat, the upper subulate, triquetrous 7
   All leaves flat 8
7. Outer bract as long as the inner ones, inner bracts glabrous; calyx tube 10-veined, veins brownish pink not projecting from the limb margins .... 18. *A. astragalinum*
   Outer bracts shorter than the inner ones; inner bracts pubescent, rarely glabrous; calyx tube 5-veined, veins dark purple, projecting from the limb margins .... 20. *A. echinus*
8. Scape branched, glabrous, emerging above the leaves .... 7. *A. festucaceum*
   Scape unbranched, or if branched (sometimes bifurcate in *A. peronini*), not glabrous, emergent or not 9
9. Calyx limb white 10
   Calyx limb pale to bright pink 13
10. Scape shorter than leaves; leaves densely pubescent-hispid .... 21. *A. peronini*
    Scape as long as or longer than leaves; leaves glabrous or scabrid 11
11. Calyx limb 10-lobed, with purple veins normally reaching the margin .... 9. *A. haussknechtii*
    Calyx limb 5-lobed, margins folded, irregularly crenulate, veins slightly projecting from the limb margin .... 6. *A. brachystachyum*
12. Scape emerging from the leaves; spikelets 5–10 (or more) distichous or sub-secund .... 11. *A. venustum*
    Scapes long as or scarcely longer than leaves; spikelets 3–6, not distichous or sub-secund .... 13. *A. senganense*
13. Calyx limb white, veins purple 14
    Calyx pink to brown-pink or brownish purple 18
14. Scape with 2–4 lax 1-flowered separated spikelets .... 10. *A. blakelockii*
    Scape with many spikelets, not separated 15

15. Leaves with upper surface flat, lower surface keeled or leaves triquetrous; ribs of calyx projecting beyond the margins of limb. 16
    Leaves with both surfaces flat; ribs of calyx not projecting beyond the margins of limb 17
16. Scape shorter than the leaves, rarely longer; petals united along the length of claw . . . . 5. *A. genistioides*
    Scape longer than the leaves; petals connate only at base . . . . 14. *A acerosum*
17. Scape emerging from leaves; all bracts long-acuminate, spiny-mucronate at tip. . . . . 8. *A. armenum*
    Scape as long as leaves, not emerging; outer bract lanceolate toward apex, inner bracts acuminate . . . . 12. *A. caryophyllaceum*
18. Outer bracts longer than inner bracts; calyx limb 10-lobed. 19
    Outer bracts as long as or shorter than inner bracts; calyx limb 5-lobed 20
19. Lower leaves shorter and wider than the upper ones, usually recurved; upper leaves vertical-divergent; calyx tube finely and densely pilose, limb pale brownish pink, veins greenish, pilose at the base, turning the same colour as the limb upwards; petals united only at base.. . . . . 15. *A. olivierii*
    Lower and upper leaves similar in size, slightly folded, recurved; calyx tube with short spaced hairs, limb purple-brown, veins dark purple, glabrous; petals united along the length of the claw . . . 19. *A. laxiflorum*
20. Scape distinctly emerging above leaves 21
    Scape as long as or slightly longer than leaves but not distinctly emerging above leaves. 22
21. Spikelets widely separated, 1–3-flowered; outer bracts tomentose outside . . . . 3. *A. bromifolium*
    Spikelets not widely separated, 1-flowered; outer bracts glabrous outside . . . . 12. *A. caryophyllaceum*
22. Spikelets 5–7; calyx limb 10-lobed, 5 longer lobes (between the shorter ones) with ± glabrous veins; ovary linear, pentagonal, contracted at the base, elongated into a long neck. . . . . 17. *A. atropatanum*
    Spikelets 3–5; rachis ± flexuose; calyx limb truncate or slightly 5-lobed with 5 thick veins hairy towards the base, glabrous above; ovary oblong-ovate, with 5 acute ribs forming a characteristic swollen structure in its one-third upper portion, elongated into a 5-ribbed (star shaped in cross section) neck.. . . . . . . 16. *A. petraeum*

Sect. ACANTHOLIMON

1. **Acantholimon bracteatum** (*Girard*) *Boiss.*, Diagn. ser. 1, 7: 70 (1846); DC., Prodr. 12: 622 (1848); Bunge in Mem. Ac. Sc. Pétersb. ser. 7, 18: 18 (1872); Boiss., Fl. Orient. 4: 827 (1879); Parsa, Fl. de l'Iran 4: 898 (1949-50); Linczevskii, Fl. SSSR 18: 312 (1952); Grossh., Fl. Kavk. ed. 2, 7: 184 (1967); Rech. f. & Schiman-Czeika in Fl. Iranica 108: 45 (1974); Bokhari & Edmondson in Fl. Turk. 7: 482 (1982); Assadi, Fl. Iran 51: 70 (2005); Gabrielian & Fragman-Sapir, Fls. Transcauc. 300 (2008).

Statice bracteata Girard, Ann. Sci. Nat. sér. 3, 2: 330 (1844).
Acantholimon splendidum Bunge, Acanth. 17 (1846).
A. bracteatum var. *splendidum* (Bunge) Boiss., Fl. Orient. 4: 827 (1879).

Caespitose shrublet, glaucescent, almost woody; branches short, densely leafy. Leaves flat, the lower shorter than the upper ones, recurved, thick, slightly succulent, faintly wrinkled, margins scarious, apex provided with a spine; upper leaves 2–4 × 0.1–0.2 cm, faintly mottled, margins scabrous, curved towards the apex, venation parallel; basal sheath with narrow scarious margins. Scape usually short or sometimes projecting from the leaves, glabrous, with 1–4 scarious lanceolate scales; scales acuminate, brownish on the outer side; inflorescence terminal, hemispherical or capituliform, spikelets with 2–3 flowers; outer bract deltoid, pale brownish on the outer side, membranous, margins scarious, apex obtuse, provided with a short recurved awn; inner bracts scarious, ± membranous, midrib

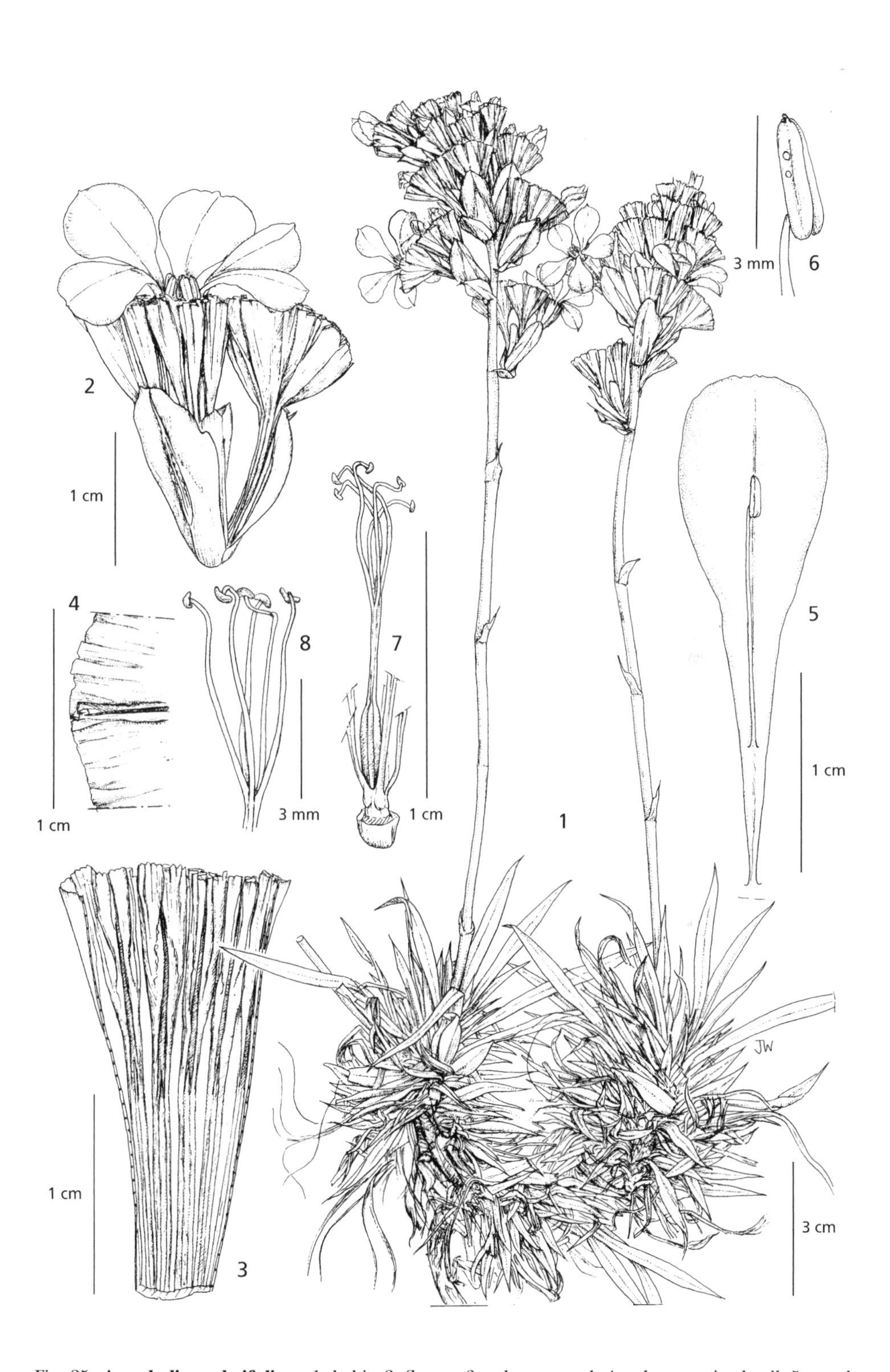

Fig. 25. **Acantholimon latifolium**. 1, habit; 2, flower; 3, calyx opened; 4, calyx margin detail; 5, petal with stamen; 6, anther; 7, ovary and style; 8, upper part of styles and stigmas. All from Rawi *et al.* 29354. © Drawn by J.E. Beentje (Apr. 2012).

projecting below the apex into a short purple awn. Calyx wide; calyx tube with 5 hairy ribs; limb funnel-shaped, wide, purplish pink, glabrous, margin faintly 5-lobed, finely crenulate, with 5 veins of the same colour as limb. Corolla bright pink, projecting from calyx. Ovary ellipsoid, elongated into a long neck terminating with 5 smooth styles ending with 5 capitate glandular stigmas.

HAB. Not recorded.
DISTRIB. One doubtful record from "Amadiya, Kurdistan" [Amediye], *Layard*. s.n. (K!).
NOTE. Layard's label is almost illegible, and as this very conspicuous and distinctive species has not been found since it must be considered as a doubtful record.

Turkey (SE Anatolia), NW Iran.

2. **Acantholimon latifolium** *Boiss. & Noë* in Boiss., Diagn. ser. 2, 4: 61 (1859). — Type: Iraq, nr. Mandali, *Noë* 185 (K!, holo.); Bunge in Mém. Ac. Sci. Petersb. 7, ser, 18: 24 (1872); Boiss;, Fl. Orient. 4: 828 (1879); Nábělek in Bull. Sci. Univ. Masaryk 105: 4 (1929); Zohary in Dep. Agr. Iraq Bull. 3: 115 (1950); Rech. f. & Schiman-Czeika in Fl. Iranica 108: 42 (1974); Bokhari & Edmondson in Fl. Turk. 7: 501 (1982); Assadi, Fl. Iran 51: 83 (2005).

Densely caespitose hemispherical prickly shrublet, branches short and entangled. Leaves in dense rosettes, ± flat, broadly lanceolate, gradually attenuated towards the base passing through a sheath, yellowish white, finely serrulate; leaves pale olive green, densely mottled, terminal and lower ones shorter than median ones which reach up to 6 cm long and up to 4 mm diameter. Scape robust, glabrous, projecting from the leaves, with 5–7 triangular lanceolate scales, ending in a dense (rarely opened at the base) oval-oblong distichous inflorescence with 5–12 spikelets, each with 2–3 flowers, and with 4 broad membranous bracts. Outer bracts ovate-triangular, glabrous, pale brown, membranous, apex mucronate; inner bracts longer than outer ones, obovate, broadly scarious-margined, apex obtuse. Calyx glabrous, funnel-shaped, 2 × 1.2 cm, limb shorter than the tube, purple-brown, faintly 5-lobed, margins slightly crenulate with venation of the same colour as the limb. Corolla pale pink; ovary ovoid, ± compressed at 2 ends, neck pentagonal; styles smooth, stigmas relatively small, subglobular-capitate. Fig. 25: 1–8.

HAB. Open patches in *Quercus* forest, on limestone, rocky places with clay soil; alt. (130–)1320–1900 m; fl. Jun.–Jul.
DISTRIB. Occasional in upper forest zone of Iraq. **MAM**: Dori 5 km E of Kani Mazi, *Omar & Qaisi* 45421!; **MRO**: between Darband and Haji Umran, *Hadač* 2352 (fide FI); Merga Droiga, *Rawi* 9159!; Sefin Dagh, *Gillett* 8113!; **MSU**: Mt. Avroman [Hawraman] (Hawara Birza), *Rawi, Alizzi & Nuri* 29354!; **FNI**: nr. Mandali, *Noë* 185 (type).

Turkey (SE Anatolia), W Iran.

3. **Acantholimon bromifolium** *Boiss.* ex *Bunge*, Mém. Ac. Sci. Petersb. sér. 7, 18: 23 (1872); Boiss., Fl. Orient. 4: 829 (1879); Rech. f. & Schiman-Czeika in Fl. Iranica 108: 43 (1974); Assadi, Fl. Iran 51: 97 (2005).

Hemispherical small shrub with short erect branches. Leaves flat, linear, grass-like, 2–5.5 × 0.2–0.4 cm, venation parallel, pale green, mottled, margins scabrous, broader towards the base forming sheaths that are narrowly and scariously margined, and finely ciliate, sometimes recurved, apex acuminate, prickly. Scape thick, hispid, distinctly projecting from the leaves, scales many, lanceolate, supporting the widely separated, 1–3-flowered spikelets, each with 4–7 bracts; spike lax, rachis erect, joints hispid, or with the inner side puberulent; outer bract shorter than inner ones, slightly thick, margins hyaline, attenuate towards an acuminate apex, outer side tomentose, but glabrous towards base. Inner bracts of a variable number, unequal, carinate, margins widely membranous, innermost dorsal side slightly hairy; apex truncate or mucronate, longer than calyx tube. Calyx long, tube longer than limb, with 10 longitudinal ribs at base, finely ciliate on either side of each rib, limb dirty purple, 3-lobed and 5-veined, veins darker in colour than limb and hispid on outer side. Corolla projecting from calyx, bright pink. Ovary ovoid-ellipsoid, pentagonal, finely scarious, styles smooth; stigmas capitate.

HAB. Stony hillsides, on limestone, and in rocky places above timberline among *Astragalus* plants on cliffs, alt. 1550–2010m; fl. Aug.

DISTRIB. Very rare; found only in a few isolated localities in the upper forest zone of Iraq. **MRO:** Mt. Safarin Dagh near Salah ad-Din, *Barkley* 3387; **MSU**: Penjwin, *Guest* 12959!; Mt. Hawraman *Saman* s.n.
NOTE. The Iraqi plants belong to var. *bromifolium*.

W Iran.

4. **Acantholimon petuniiflorum** *Mobayen*, Revis. Taxon. Acanthol. 305, fig. 49 (1964). — Type: Iraq, Mergan to near Bardanas, *Rawi & Serhang* 24365a (K!, holo.); Rech. f. & Schiman-Czeika in Fl. Iranica 108: 48 (1974); Bokhari & Edmondson in Fl. Turk. 7: 484 (1982).

Caespitose, densely hairy shrublet, basal branches stoloniferous, covered with remains of dead leaves, giving rise to 2 types of branches: the first more vigorous, higher, with 1-few scapes; the second short, usually vegetative and rarely with flowering scapes. Leaves prickly, 2–5 × 1.5 mm, regularly spreading, erect, covered with thin white hairs, ± canescent, apex mucronate, margins narrow, hairy, broadening into a hairy sheath. Lower leaves shorter, glabrous, recurved, margins scabrous. Scapes thin, shorter than leaves, densely hairy; spike terminal, dense, with 6–9 (sometimes more) comparatively long spikelets. Bracts glabrous, outer one ovate, half the length of inner ones, coriaceous, greenish, margins membranous, apex obtuse, the 2 inner bracts unequal, membranous except on the dorsal side, veins broadly herbaceous; apex retuse, slightly shorter than the calyx tube; tube 5-ribbed, sparsely hairy, ± viscous; limb funnel-shaped, white, striated along the 5 veins with purplish large bands, giving the flower the aspect of a hybrid *Petunia*, pentagonal, with 5 crenulate lobes. Corolla bright pink, petals united along their claws. Stamens shorter than styles; filaments broadening and adhering to petals towards base. Ovary short, pentagonal, elongated into a long neck; styles 5, smooth, stigmas capitate.

HAB. Mountain sides; alt. 2500–2600 m; fl. Aug.
DISTRIB. Very rare; only one record from the thorn-cushion zone of the mountains of NE Iraq. **MRO**: Mergan to near Bardanas, *Rawi & Serhang* 24365a (type, K!).
NOTE. Assadi in Fl. Iran 51: 82 (2005) discusses this species under *A. wendelboi*.

Turkey (SE Anatolia), W Iran.

Sect. STATICOPSIS Boiss.

5. **Acantholimon genistioides** (*Jaub. & Spach*) *Boiss.*, Diagn. ser. 1, 7: 76 (1846); DC., Prodr. 12: 28 (1848); Bunge in Mem. Ac. Sc. Pétersb. 7, ser. 18: 53 (1872) '*genistoides*'; Boiss., Fl. Orient. 4: 846 (1879); Rech. f. & Schiman-Czeika in Fl. Iranica 108: 87 (1974).

Statice genistioides Jaub. & Spach, Ann. Sci. Nat. sér. 2, 10: 252 (1843).

Hemispherical, prickly, glaucescent shrublet, with calcareous dots, branches elongated, covered with dead dry leaves. Leaves hispid, triangular in transverse section, upper surface flat, lower one keeled, densely covered with calcareous granules, base sheathed, broadly scariously margined, finely ciliate, apex pointed, prickly. Basal leaves on young branches short and broad, ± 10 × 2 mm; upper ones ± 25 × 1 mm, glabrous, keeled. Scape shorter than leaves, rarely longer, glabrous, with one scarious bract; spike provided with 1–2 short scales at the base; spikelets sessile, distichous, with one small flower; outer bract ovate-triangular, glabrous, margins membranous, purplish, apex mucronate, shorter than inner bracts, inner ones oblong, broadly scariously margined, apex obtuse. Calyx tube as long as inner bracts, with 5 purplish, glabrous ribs, which broaden towards limb and project from it; limb as long as the calyx tube, scarious, margins finely crenulate. Corolla pink, petals united along length of claw. Filaments ribbon-like, adhering to base of petals, longer than styles. Ovary ovate-oblong, pentagonal, with a short neck; styles filiform, slightly scabrid in upper half; stigmas compressed, excentric.

HAB. Mountain slopes; alt. 2210–3000 m; fl. Aug.
DISTRIB. Very rare; only two records from the thorn-cushion zone of the mountains of NE Iraq. **MRO**: Kermasur lake, Qandil range, *Rawi & Serhang* 24136!; Helgurd range, *Rawi & Serhang* 24154!
NOTE. Assadi in Fl. Iran 51: 141 (2005) discusses this species under *A. hohenackeri* (Jaub. & Spach.) Boiss.

W Iran.

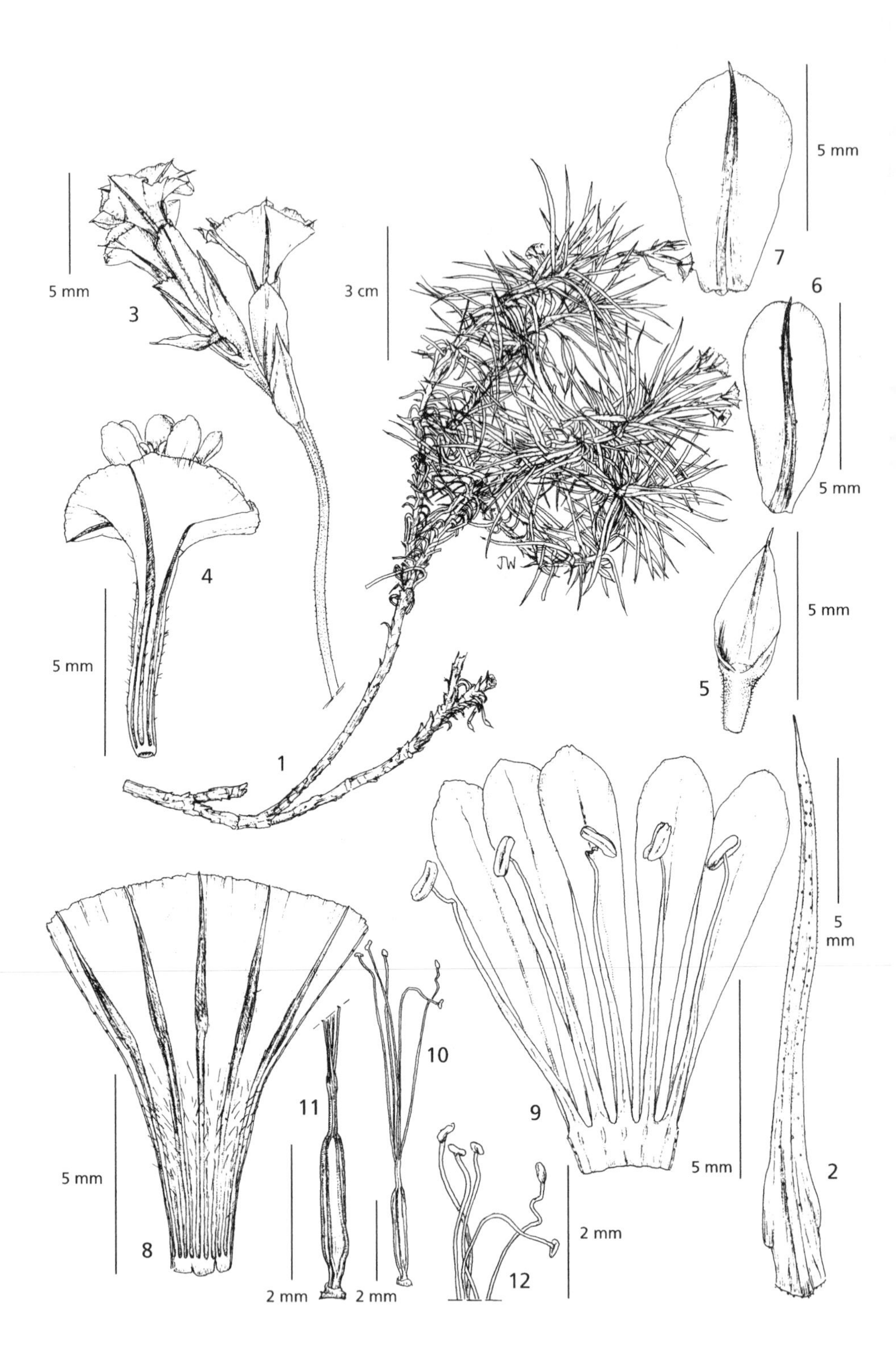

Fig. 26. **Acantholimon brachystachyum**. 1, habit; 2, leaf; 3, inflorescence; 4, flower; 5, outer bract; 6, middle bract; 7, inner bract; 8, calyx opened; 9, corolla opened; 10, ovary, styles and stigmas; 11, ovary; 12, styles and stigmas. 1, 3 from *Serhang & Rawi* 26731; 2, 4–12 from *Rawi & Serhang* 24365. © Drawn by J.E. Beentje (Apr. 2012).

6. **Acantholimon brachystachyum** *Boiss.* ex *Bunge* in Mém. Acad. Sci. Pétersb. ser. 7, 18: 51 (1872); Boiss., Fl. Orient. 4: 844 (1879); Zohary in Dep. Agr. Iraq Bull. 3: 115 (1950), Mobayen, Revis. Taxon. Acanthol. 179 (1964); Rech. f. & Schiman-Czeika in Fl. Iranica 108: 86 (1974); Assadi, Fl. Iran 51: 145 (2005).

Acantholimon kurdicum Bunge, l.c. 52 (1872).
A. acerosum (Willd.) Boiss. var. *brachystachyum* Boiss., Fl. Orient. 4: 845 (1879); Bokhari & Edmondson in Fl. Turk. 7: 487 (1982).

Greenish, caespitose shrublet, with scattered calcareous dots, branches 4–10 cm. Leaves flat, linear-lanceolate; lower ones recurved, some very short, flat, ± coriaceous, with ciliate margins, shortly mucronate, and some, as well as upper leaves, narrower, longer, ± triquetrous, broadening at base into a long sheath, widely scarious, margins scabrid, whitish, ± entire, apex mucronate. Scape as long as leaves, thin, pubescent, rarely glabrous; spike simple or bifurcate, few-flowered; rachis hispidulous. Outer bract shorter than inner ones, ovate-triangular, glabrous, scarious in its upper half, acute, shortly mucronate; inner bracts shorter than calyx tube, membranous, except at midrib, widely separated, apex obtuse or shortly mucronate. Calyx 8–10 mm; tube 10-ribbed, hispid between ribs, veins broadening and becoming darker in colour and glabrous towards the limb, slightly projecting from the limb-lobes; limb white, very short, 5-lobed, margins folded, irregularly crenulate. Corolla bright pink, petals united along length of claws. Stamens as long as or shorter than styles; filaments ribbon-like, adherent to claws and lower parts of petals. Ovary oblong, pentagonal, with fusiform apex and a short neck; styles filiform; stigmas elliptic, flattened, and excentric. Fig. 26: 1–12.

HAB. Rocky mountain slopes, south-facing cliff; alt. 2200–2600 m; fl. Jun.-Aug.
DISTRIB. Rare; occurs in isolated areas in the thorn-cushion zone. **MRO**: Mergan near Bardanas, *Rawi & Serhang* 24365!; Qandil Range, *Rawi & Serhang* 26731!; Warshanka on Magar range, *Rawi & Serhang* 24342!. **MSU**: (or Iran?), "Mt. Avroman & Schahu" [Hawraman], *Haussknecht* (type of *A. kurdicum*); Mt. Hawraman, *Rawi, Chakr., Nuri & Alizzi* 19777!

Turkey (C Anatolia), W Iran.

7. **Acantholimon festucaceum** (*Jaub. & Spach*) *Boiss.*, Diagn. ser. 1, 7: 76 (1846). — Type: 'Mesopotamia', *Aucher* 2511 (P); DC., Prodr. 12: 30 (1848); Bunge in Mem. Ac. Sc. Pétersb. 7, ser. 18: 53 (1872); Boiss., Fl. Orient. 4: 843 (1879); Zohary in Dep. Agr. Iraq Bull. 31: 115 (1953); Rech. f. & Schiman-Czeika in Fl. Iranica 108: 94 (1974); Assadi, Fl. Iran 51: 117 (2005).

Statice festucacea Jaub. & Spach in Ann. Sc. Nat. ser. 2, 20: 254 (1843).
Acantholimon flexuosum Boiss. ex Bunge var. *laxiflorum* Mobayen, Rev. Taxon. Acanth.: 312 (1964).

Caespitose, hemispherical, glaucous shrublet; branches densely leafy. Leaves equal or slightly caespitose unequal, flat, lower leaves spreading or recurved, upper ones forming spiny rosettes, erect, 15–25 × 1–2 mm. Scape branched, projecting from the leaves, sinuate, cylindrical, slightly swollen at nodes, usually with 3 triangular lanceolate scarious whitish bracts, apex mucronate, joint 2 cm long; spike sessile, with ± imbricate spikelets, rachis densely hispid. Outer bract thick, glabrous, triangular, 2.3 × length of inner ones, margins membranous; inner bracts brownish green on their dorsal surface, broadly membranous, apex shortly mucronate, glabrous or finely ciliate along dorsal midrib. Calyx 7–8 mm, obconical, slightly curved, marginal lines of ribs with long white hairs, ribs broadening towards limb, hispid; limb half the length of calyx tube, slightly lobed, with narrow glabrous slightly projecting veins, margins denticulate. Corolla whitish. Stamens with ribbon-like filaments. Ovary oblong-ovoid, developing into a capsule, covered with spiny prickles facilitating seed dissemination. In ripe fruit, the neck is easily detached from seed in the form of a pyramidal-like structure bearing the styles.

HAB. Mountain slopes.
DISTRIB. 'Mesopotamia'[11], *Aucher* 2511 (type).
NOTE. According to Rechinger (l.c.) Mobayen appears to have swapped the names 'festucaceum' and 'flexuosum' with their respective citations. There is an Iranian specimen at Kew labelled *A. festucaceum*

[11] The distribution of this species in most likely to be in the Zagros mountain range [Irano-Turanian region and Irano-Anatolian Sub-region)], not in Mesopotamia (comm. Saman Abdulrehram, 2012).

Boiss. var. *flexuosum* (B. & H.) Mobayen (on a Mobayen det. slip); this variety does not appear to have been validly published. The occurrence of the species in Iraq must be considered very doubtful.

N & W Iran.

8. **Acantholimon armenum** *Boiss. & Huet* in Boiss., Diagn. ser. 2, 4: 64 (1859); Bunge in Mem. Ac. Sc. Pétersb. 7, ser. 18: 36 (1872); Boiss., Fl. Orient. 4: 839 (1879); Zohary, Fl. Palest. ed. 2, 2: 411 (1933); Rech. f. & Schiman-Czeika in Fl. Iranica 108: 99 (1974); Bokhari & Edmondson in Fl. Turk. 7: 489 (1982); Gabrielian & Fragman-Sapir, Fls. Transcauc. 298 (2008).

Densely caespitose shrublet, with short woody branches, lower branches densely leafy, clad with recurved leaf remains. Leaves (10–)15–25(–40) × 1.5–2 mm, flat, white-dotted, glaucous, margin only scabrid, patent-erect becoming recurved, scarcely hyaline at basal margins. Scape unbranched, equalling the leaves, mostly with 3 leaves; cauline leaves similar to bracts but longer. Spikelets numerous 12–18; spikes ± fragile, erect or curved; bracteoles all lanceolate, long-acuminate, spiny-mucronate at tip, with broad hyaline margins, glabrous. Calyx tube ± 9 mm, hairy on veins; limb 4 or 5(–6) mm wide, obscurely 5-lobed, with ± purplish veins, not reaching margin, whitish. Corolla pink, scarcely exserted from calyx. Styles and filaments ± equalling length of calyx tube. [Description from Fl. Iranica].

HAB. Rocky limestone places; alt.: ± 3200 m; fl. Aug.
DISTRIB. Very rare; only gathered once, from the alpine zone of NE Iraq. **MRO**: Mt. Qandil above Gom-i Kirmosoran lake, *Rechinger* 11777!

W Syria, Turkey (S & E Anatolia), Armenia, Nachitchevan.

9. **Acantholimon haussknechtii** *Bunge*, Mém. Acad. Sci. Petersb. Ser. 7, 18 (2): 37 (1872); Rech. f. & Schiman-Czeika in Fl. Iranica 108: 101 (1974).

Densely caespitose shrublet, with short woody branches covered with persistent leaf remains. Spring and summer leaves similar, the latter slightly shorter; spring leaves 15–20 × ± 1.5 mm broad, linear-lanceolate, white-dotted, glaucous, glabrous, margins scabrid, patent-erect, later recurved, with narrow hyaline margins towards the base. Scape unbranched, glabrous, somewhat longer than leaves, with 2–3 leaves; cauline leaves similar to bracts but narrower and scarcely longer, always shorter than internodes. Spikelets 3–7, forming a short spike; internodes ± hairy, shorter than bracts, rarely the lowest equalling the bract. Outer bract ± 6 mm long, slightly shorter than inner bracts, ovate, acuminate-cuspidate, with hyaline margins. Inner bracts subequal, ± equalling the calyx tube, oblong, with broad hyaline margins, the mucro dark purple; bract and bracteoles glabrous. Calyx tube 7–8 mm long, hairy on veins; limb 4 or 5 mm broad, 10-lobed, white, with purple veins normally reaching the margin. Corolla pink, scarcely exserted from calyx. Filaments and styles equal, shorter than calyx.

HAB. Rocky mountain; alt. 900–2040 m. fl. ?Aug.
DISTRIB. Very rare; only gathered once, from the sub-alpine zone of NE Iraq. **MRO**: Jabal Baradost, near Diana in the Rowanduz gorge, *Lazar* 888.
NOTE. The specimen (*Lazar* 888) cited under *A. haussknechtii* Bunge was also cited under *A. acerosum* (Willd.) Boiss. in an earlier draft of this account. It is doubtful whether the two taxa are specifically distinct, as indeed is suggested by Mobayen (op. cit. 179, 1964).

Turkey, Armenia.

10. **Acantholimon blakelockii** *Mobayen*, Revis. Taxon. Acanthol. 309, f. 60 (1964) ["*blakelackii*"]. — Type: Iraq, Sulimaniyah District, Qaradagh, *Haines* W.1082 (K!, holo.); Rech. f. & Schiman-Czeika in Fl. Iranica 108: 97 (1974); Assadi in Fl. Iran 51: 158 (2005).

Densely caespitose shrublet, glaucescent, sparsely covered with calcareous dots; branches short, totally leafy. Lower leaves of young branches short, 1–15 × 2 mm, flat, oblong-lanceolate, recurved, with parallel-veined surface and scabrid finely ciliate margins, apex acute-mucronate; upper ones much longer, 20–30 × 1 mm, subulate, upper surface flat, lower surface keeled, broadening gradually at the base into a sheath with membranous and scabrid margins, becoming smooth from middle up to the apex, acuminate. Scape glabrous, with 1–2 herbaceous triangular lanceolate-acuminate scales and membranous margins, with 2–4 lax 1-flowered separated spikelets; rachis glabrous, lax, sinuate. Bracts glabrous, outer one oblong-lanceolate, margins membranous, longer than inner bracts and calyx tube; inner bracts oblong, widely membranous, with a dorsal greenish vein and mucronate apex, enclosing the calyx tube. Calyx funnel-shaped, 1.5 cm, with cylindrical tube, hairy along edges of veins; limb 8–10 mm wide, shorter than tube, 10-lobed with 5 puberulent veins, becoming glabrous towards apex of lobe. Corolla bright pink. Stamens with rubber-like filaments, adherent to base of petals, longer than styles. Ovary fusiform, short, with compressed neck, 5-ribbed; styles filiform, white, smooth; stigmas capitate.

HAB. Rocky mountain slopes, between eroded rocks beside road; alt. 1350–2000 m; fl. Jul.
DISTRIB. Occasional in the upper forest zone of Iraq. **MAM**: between Dohuk and Amadiya (Amadiye), above Sawara Tuka, *Rechinger* 11554; Jabal Khantur, *Rawi* 23357; *Rechinger* 10744; Zawita, *Rechinger* 10944; Jabal Khantur NE of Zakho, *Rawi* 23357!; **MRO**: Sakri Sakran, *Bornmüller* 1756, *Hadač* 5649; **MSU**: Qaradagh, *Haines* W.1082 (type)!; *Hadač et al.* 5193; Pira Magrun, *Haussknecht* 832; Azmir Dagh near Sulaimaniya, *Hadač* 1887; Malakawa pass near Penjwin, *Rechinger* 10434; 10 km E of Mandali, *Rechinger* 12766.

W Iran.

11. **Acantholimon venustum** *Boiss.*, Diagn. Ser. 1, 7: 80 (1846); Rech. f. & Schiman-Czeika in Fl. Iranica 108: 104 (1974).Bokhari & Edmondson in Fl. Turk. 7: 485 (1982).

Acantholimon assyriacum Boiss., l.c.: 81 (1846).

Laxly caespitose shrublet, forming spiny hemispherical tussocks, short-branched, brownish, the lower part of stem bare or laxly clad with dead leaves. Leaves 20–30 mm long, stiff, distinctly spiny, ± densely white-dotted, narrowly many-veined, flat, margins minutely serrulate, triquetrous towards tip, slightly broadened towards base with narrow membranous margins. Inflorescence 50–90 mm, above the leaves. Cauline leaves 3, scale-like, long mucronate, scarious-margined. Spike simple; rachis straight or slightly flexuous, ± hairy; spikelets 5–10 or sometimes more, distichous or sub-secund. Outer bract 1/3 length of inner bracts; inner bracts oblong, broadly membranous-margined, mucronulate, glabrous, slightly longer than calyx tube. Calyx ± 15 mm long, broadly funnel-shaped; tube cylindrical, laxly pilose on veins; veins reaching the margin of limb, which is brownish-purple, truncate, obscurely 5-angled, pilose along lower two-thirds of their length. Corolla deep pink, of almost free petals, joined only at base. Filaments swollen, membranous, connate to spur of petals, longer than styles. Styles filiform; stigma minutely capitate.

HAB. Mountain side and on dry hillside, on limestone and sandy soil; alt. 600–1500 m; ?June.
DISTRIB. Very rare; only recorded once from Iraq. **MRO**: near Baba Chichak (Shaqlawa) towards Kani Watman, *Nábělek* 2719.

Turkey (S, C & E Anatolia), W Syria.

12. **Acantholimon caryophyllaceum** *Boiss.*, Diagn. ser. 1, 7: 78 (1846). — Type: Mt. Gara (Gara Dagh), *Kotschy* 368 (K!, holo.); DC., Prodr. 12: 630 (1848); Bunge in Mem. Ac. Sc. Petersb. 7, ser. 18: 36 (1872); Boiss., Fl. Orient. 4: 838 (1879); Blakelock in Kew Bull. 3: 518 (1948); Zohary in Dep. Agr. Iraq Bull. 31: 113 (1950); Rech. f. & Schiman-Czeika in Fl. Iranica 108: 103 (1974); Bokhari & Edmondson in Fl. Turk. 7: 488 (1982); Assadi in Fl. Iran 51: 148 (2005).

Statice caryophyllacea Boiss. in Kotschy, Alepp. Kurd. Mos. no. 368 (in sched.).

Ceaspitose shrublet, glaucous, frequently with calcareous dots, branches leafless at base. Leaves linear, 2–5 cm, almost scabrid, sometimes pubescent; spring leaves shorter and more coriaceous than summer ones. Scape emerging above leaves, unbranched, finely puberulent; rachis straight, hispid on its inner side; spike elongated with numerous 1-flowered spikelets. Outer bract shorter than inner ones, comparatively wide, coriaceous, glabrous, lanceolate towards the apex, margins membranous; inner bracts as long as calyx tube, scarious, carinate, apex acuminate, dorsal side pubescent. Calyx funnel-shaped, wide, calyx tube sparsely hispid, as long as limb which is obscurely 5-lobed, with thin purple veins and blunt margins. Corolla bright pink, distinctly exserted from calyx. Stamens with ribbon-like filaments, adherent to claws of petals. Ovary oblong, rather short, with a long easily distinguishable neck (more so than in *A. acerosum* and *A. kotschyi*). Fig. 27: 1–10.

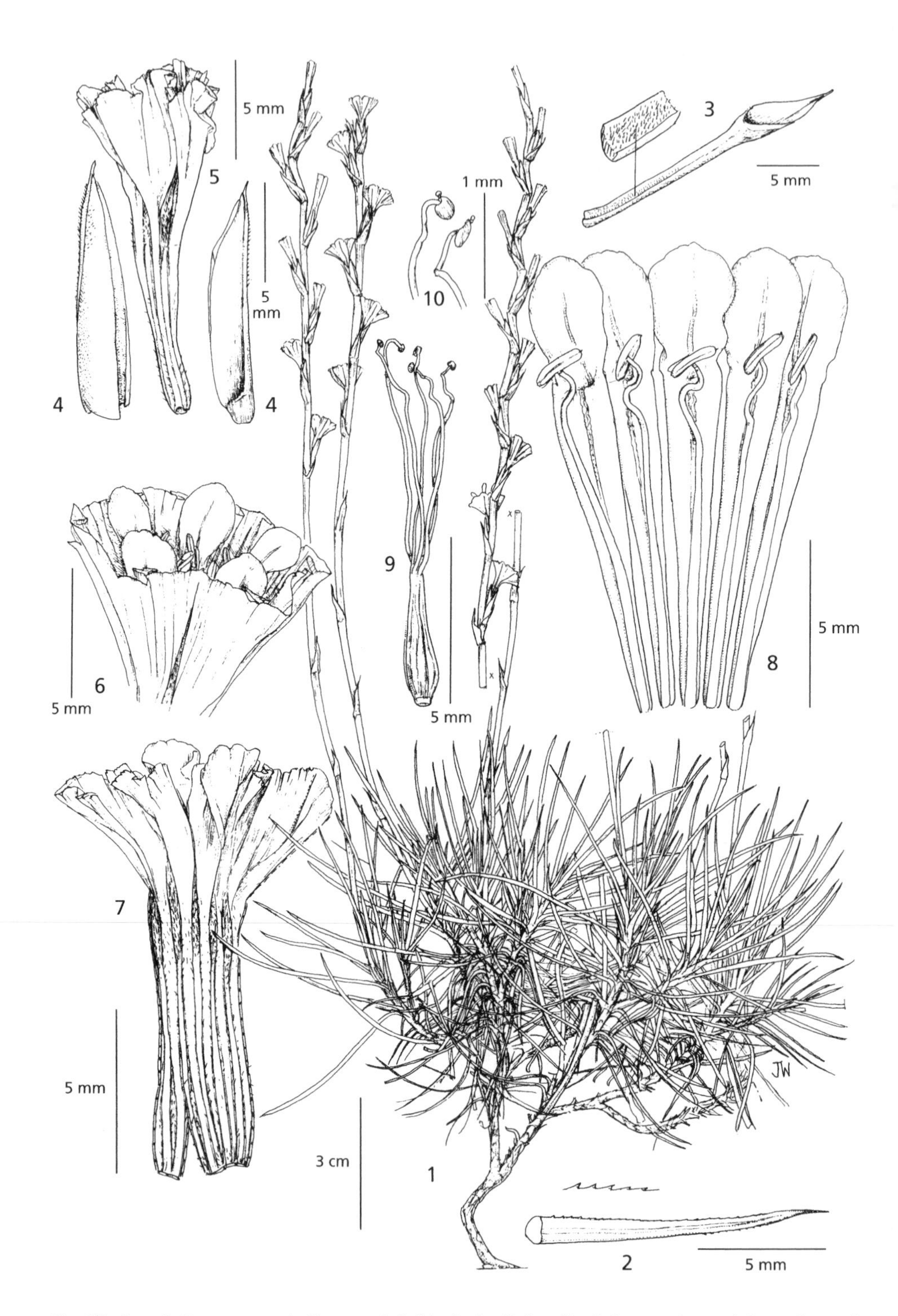

Fig. 27. **Acantholimon caryophyllaceum**. 1, habit; 2, detail tip of leaf; 3, outer bract; 4, inner bract; 5, calyx, side view; 6, flower, upper portion; 7, calyx; 8, corolla opened; 9, ovary and styles; 10, style and stigma detail. All from *Rawi* 23418. © Drawn by J.E. Beentje (Apr. 2012).

HAB. *Quercus* zone, open patches in *Quercus* forest, stony and gravelly hillsides, on limestone; alt. 1480–2600 m; fl. Jun.–Aug.
DISTRIB. Occasional in the upper forest and thorn-cushion zones of Iraq. **MAM**: Mt. Gara (Gara Dagh), *Kotschy* 368 (type)!; Jabal Khantur nr Sharanish, *Rechinger* 10836, *Rawi* 23418; Dori (Dudi) 5 km E of Kani Masi, *Omar & Qaisi* 45435; Sarsang, *Haines* W.491!; Gali Zawita, NE of Zakho, *Rawi* 23645; **MRO**: Chiya-i Mandau, nr. Walza, *Guest & Ludlow-Hewitt* 2664; 35 km NE of Mandali, *Qaisi & Khayat* 50807!; Mt. Helgord (Algurd Dagh), above Nawanda, *Rechinger* 11349!; Sefin Dagh, *Gillett* 8114!; **MSU**: W of Qala Diza, *Thesiger* 1139!; Mt. Hawraman, *Rawi, Chakr., Nuri & Alizzi* 19730!

Transcaucasus, N & W Iran, Turkey (E & SE Anatolia).

13. **Acantholimon senganense** *Bunge*, Mém. Acad. Sci. Pétersb. Ser. 7, 18 (2): 29 (1872); Rech. f. & Schiman-Czeika in Fl. Iranica 108: 102 (1974); Assadi in Fl. Iran 51: 156 (2005).

Densely caespitose shrublet, with short dense leafy branches, clad with persistent leaf remains towards base. Summer leaves somewhat shorter than spring leaves but twice as broad, spring leaves 10–20(–25) × 1 mm, flat or triquetrous, glaucous, glabrous, margins scabrid, lacking hyaline margin, patent or eventually recurved. Scape simple, mostly scarcely longer than leaves, including inflorescence 40–100 mm high, ± fragile, scabrid with dense rigidly patent hairs. Cauline leaves 2–3, similar to bracts. Spikelets mostly 3–6, forming a slender spike. Bracts glabrous; outer bract ± 5 mm long, triangular-acute, with hyaline margins; inner bracts equal in length, 7–8 mm long, ± ovate, with broad hyaline margins, subauriculate, veins extended into a mucro. Calyx tube 7–8 mm long, hispid on veins, limb ± 35 mm in diameter, truncate, pink, veins reaching the margin. Corolla pale pink, ± 5 mm, exserted from calyx. Styles and filaments exserted from calyx.

HAB. Not recorded.
DISTRIB. Very rare; only found once in Iraq. **MSU**: near Penjwin, *Rechinger* 10500.

Syria, W Iran, ?Turkey (E Anatolia).

14. **Acantholimon acerosum** (*Willd.*) *Boiss.*, Diagn. ser. 1, 7: 80 (1846); Bunge in Mém. Ac. Sc. Petérsb. 7, ser. 18: 33 (1872); Boiss., Fl. Orient. 4: 837 (1879); Zohary, Fl. Palest. ed. 2, 2: 410 (1933); Rech. f. & Schiman-Czeika in Fl. Iranica 108: 100 (1974); Bokhari & Edmondson in Fl. Turk. 7: 487 (1982).

Statice acerosa Willd. in Ges. Nat. Freund. 3: 420 (1801).

Densely caespitose shrublet, glaucescent, roughly punctuated with calcareous dots. Leaves flat or triquetrous, acuminate, spiny, parallel-veined, 15–30 × 1.5–2.5 mm, broadening at base into a short brown sheath. Scape unbranched, emerging above leaves, glabrous, with a few triangular-lanceolate scales. Spike lax, elongate, with 7–16 big spikelets; rachis flexuous, scabrid. Outer bract triangular, acuminate, shorter than inner ones, dorsal surface herbaceous; inner bracts narrowly membranous, as long as or sometimes longer than calyx tube, apex mucronate. Calyx comparatively long, with purplish pubescent-hispid ribs slightly projecting from margins of limbs, limb white, blunt or obscurely 5-lobed. Corolla bright pink. Stamens with filaments adherent to claws of petals. Ovary oblong linear, 5-ribbed, short-necked; styles filiform, smooth, stigmas capitate-peltate.

HAB. Stony and rocky hillsides, sometimes on metamorphic rocks; alt.: ?700 ft –2600 m; fl. June–Aug.
DISTRIB. Occasional in the upper forest and thorn-cushion zones of Iraq. **MAM**: Gali Zawita, *Guest* 4531!; Amadiya, *Guest* 4975!; **MRO**: Rayat, *Guest* 13120!; N of Haji Umran towards Iranian border, *Rawi & Serhang* 24970! *Guest & Alizzi* s.n!; Helgord Dagh, *Gillett* 9492!; Nawanda near Rust, *Guest & Alizzi* 15802!; nr. Ser Kurawa, *Gillett* 9796!; **FNI**: Koma Sang nr Mandali, *Guest* 794!.
NOTE. According to Guest & Alizzi, its vernacular (Kurdish) name is GIRGŪL.

W Syria, Turkey, Armenia, NW Iran.

15. **Acantholimon olivierii** (*Jaub. & Spach*) *Boiss.*, Diagn. ser. 1, 7: 80 (1846); Bunge in Mém. Ac. Sc. Pétersb. 7, ser. 18: 27 (1872); Rech. f. & Schiman-Czeika in Fl. Iranica 108: 98 (1974); Assadi in Fl. Iran 51: 152 (2005).

Statice oliveri Jaub. & Spach, Ill. Pl. Or. 163, t. 93 (1844).
Acantholimon venustum var. *olivieri* Boiss., Fl. Orient. 4: 832 (1879).

Loosely caespitose glaucescent shrublet, finely mottled, branches long, leafless in their lower parts. Leaves linear, narrow, slightly triangular in transverse section, up to 35 × 1 mm, slightly prickly; lower leaves shorter and wider than upper ones, usually recurved; upper leaves vertical-divergent, margins finely serrulate, broader towards base forming a sheath with scarious margins. Scape short, thin, glabrous or slightly puberulent, with scattered calcareous glands; spike lax, simple; rachis straight, few-flowered, slightly curved between one flower and the other; spikelets 1-flowered, with 3 bracts. Outer bract glabrous, green at base, dorsal surface pale yellow, narrowly whitish-margined, attenuate into a mucro, projecting from inner bracts and exceeding calyx tube; inner bracts oblong, veins green, apex short mucronate, glabrous or puberulent. Calyx funnel-shaped, tube greenish, finely and densely pilose, limb three quarters of length of tube, pale brownish pink, veins greenish, pilose at base, turning the same colour as limb upwards, calyx limb slightly folded, almost truncate. Corolla bright pink, with a short claw. Filaments compressed, adhering to lower part of petals including the claw. Ovary bottle-shaped, prolonged into a long pentagonal neck; styles filiform, smooth, stigmas capitate.

HAB. Among *Quercus* on dry slopes; alt.: ± 1250 m; fl. Jun.
DISTRIB. Upper forest zone; very rare. **MAM**: Ispindari Saddle, Sawara Tuka, *E. Chapman* 26323!

W Iran.

16. **Acantholimon petraeum** *Boiss. ex Bunge*, Mém. Ac. Sc. Pétersb. 7, ser. 18: 28 (1872). — Type: Iraq, Pir Omar Gudrun (Pira Magrun), *Haussknecht* 832 (syn.); Boiss., Fl. Orient. 4: 832 (1879); Zohary in Dep. Agr. Iraq Bull. 31: 115 (1950) Rech. f. & Schiman-Czeika in Fl. Iranica 108: 103 (1974); Bokhari & Edmondson in Fl. Turk. 7: 500 (1982).

Densely caespitose glaucous shrublet, with calcareous dots, branches short. Leaves short, up to 13(–20) mm, linear-lanceolate, flat in lower part, becoming keeled toward their pointed prickly apex. Lower leaves recurved spreading, glabrous with scabrous margins, obscurely ciliate, broadening towards base into a sheath, parallel-veined on its lower surface. Scape thin, slightly longer than leaves, fragile, glabrous; spike unbranched, flowers lax; spikelets 3–5, widely spaced, rachis ± flexuose. Bracts glabrous; outer bract yellowish, oblong-lanceolate, from three quarters the length of the inner ones to ± equalling them, margins narrowly whitish, apex acute or slightly acuminate; inner bracts obtuse with a short projecting pointed end (continuation of midrib). Calyx funnel-shaped; calyx tube as long as, or shorter than inner bracts, longer than limb, with 10 furrows, hairy; limb pinkish brown, paler towards the base, blunt or slightly 5-lobed, with 5 thick darker veins which are hairy towards the base and glabrous above, the veins spreading up to the margin of limb. Petals bright pink, with short claws. Stamens with ribbon-like filaments, adhering to the whole length of claws. Ovary oblong-ovate, with 5 acute ribs forming in its one-third upper portion a characteristic swollen structure, with a reticulate surface, elongated into a 5-ribbed (star shaped in cross section) neck. Styles filiform, smooth, long; stigmas disk-like, glandular.

HAB. Rocky mountain slopes, limestone and serpentinite; alt.: (700–)1200–1800 m; fl. Apr.–May.
DISTRIB. Frequent but local in the thorn cushion zone of the mountains of Iraq. **MAM**: Between Sandur and Zawita, *Alkas* 18633!; **MRO**: Karoukh mountain between Saran and Kilkil village, *Kass, Nuri & Serhang* 27348!; 2 km N of Chinaruk, N of Haibat-Sultan Dagh, *Rawi, Nuri & Kass* 28333!; Baski, mt. Hawaran, *Rawi & Serhang* 23977!; **MSU**: Darband Bazian, *Rawi & Gillett* 7632!; Pir Omar Gudrun (Pira Magrun), *Haussknecht* 832 (syn.); Penjwin, *Rawi* 17187!; Gigikka, in Gwaya Dagh, [Goyzha Dagh], above Sulaimaniya, *Rawi* 9419!; Kajan mountain near Penjwin, *Rawi* 22678!

Turkey (SE Anatolia).

17. **Acantholimon atropatanum** *Bunge*, Mém. Ac. Sc. Pétersb. 7, ser. 18: 29 (1872); Boiss., Fl. Orient. 4: 833 (1879); Rech. f. & Schiman-Czeika in Fl. Iranica 108: 99 (1974); Assadi in Fl. Iran 51: 155 (2005).

Densely caespitose, hemispherical, prickly, glaucous shrublet, slightly dotted, branches short. Leaves short, prickly, 12–18 × 1–1.5 mm, flat-triquetrous, broadening towards the base into a brownish sheath, with regularly serrulate margins, apex abruptly mucronate, prickly. Scape as long as leaves, glabrous, or obscurely puberulent; spike lax, distichous, rachis thin, ± sinuate, with 5–7 interrupted spikelets. Outer bract glumaceous, as long as, or slightly shorter than inner ones, but longer than the joints, pale brown, margins transparent, whitish, apex shortly mucronate; inner bracts obtuse or slightly mucronate. Calyx funnel-shaped, tube greenish, hairy, as long as or slightly longer than the wider limb, pale purplish, 10-lobed, 5 longer lobes (in between the shorter ones) with ± glabrous veins prolonging to tips of lobes. Corolla bright pink. Stamens with ribbon-like filaments adherent to the base of petals. Ovary linear, pentagonal, contracted at base, elongated into a rather long neck; styles smooth; stigmas capitate.

HAB. Sandy clay, between limestone rocks, silty soil, on serpentine area; alt. 1250–1500 m.; fl. ?May.
DISTRIB. Very rare, only recorded twice from Iraq. **MSU**: Mela Kowa (Malakawa), (on Sulaimaniya-Penjwin highway), *Rawi* 22434!; Gweija Dagh [Goyzha Dagh], above Sulaimaniya, *Rawi & Gillett* 11706!
NOTE. According to Assadi in Fl. Iran 51: 155 (2005), some material of this species was cited in Fl. Iranica under *A. blakelockii.*

NW Iran.

18. **Acantholimon astragalinum** *Mobayen*, Revis. Taxon. Acanthol. 301 (1964). — Type: Iraq, Diyala, Mandali, Koma Sang, *Guest* 794 (K!, holo.). Rech. f. & Schiman-Czeika in Fl. Iranica 108: 98 (1974)

Acantholimon forskalii Boiss. in sched.

Undershrub with thick base, lower branches densely noded, hollow, with the habit of perennial *Astragalus* plants, branches leafless at base, few leaved at the top. Leaves unequal, 20–40 × 1–1.5 mm, glaucous, densely roughly dotted, upper surface flat, lower surface keeled, triangular in transverse section, rigid, margins scarious. Scape shorter than leaves, slightly angled, scabrous, terminated with a few flowered spike; spikelets dense, with a single big flower. Outer bract as long as inner ones, coriaceous, glabrous with a few calcareous dots, margins scarious, inner bracts glabrous, scarious except along the midrib, apex shortly acuminate. Calyx tube slightly shorter than bracts, longitudinally 10-ribbed, 5 of which are hispidulous, broad near edge of limb. Limb wide, pale brownish rose, with veins of the same colour, bluntly margined or faintly 10-lobed, or sinuate. Corolla brownish pink. Ovary oblong, bladdery, 5-ribbed, passing into the pentagonal, longitudinally wrinkled neck; styles glabrous; stigmas capitate, slightly compressed.

HAB. Foothills; alt. 100–350 m; fl. Mar.
DISTRIB. Rare; only two records from Iraq. **FPF**: Mandali, Kauma to Saur, *Guest* 794! *Hadač et al.* 4614; between Makatu and Naft Khina, *Hadač et al.* 4634.

Endemic to Iraq.

19. **Acantholimon laxiflorum** *Boiss. ex Bunge*, Mém. Ac. Sc. Pétersb. Ser. 7, 18: 28 (1872); Boiss., Fl. Orient. 4: 832 (1879); Zohary, Fl. Palest. ed. 2, 2: 410 (1933).

Acantholimon venustum var. *laxiflorum* (Boiss ex Bunge) Bokhari in Notes R.B.G. Edinb. 32: 70 (1972); Bokhari & Edmondson in Fl. Turk. 7: 486 (1982).

Subshrub of a greenish purple colour, covered with dense calcareous dots. Leaves 15–40 × 1–2 mm, coriaceous, slightly folded, fragile when dry leaving the recurved lower parts attached to the branches, margins scarious ciliate, apex abruptly pointed, broadening towards base into a sheath with whitish scarious denticulate margins. Scape and rachis glabrous; spike lax, simple with an almost straight rachis; spikelets few, distichous. Bracts glabrous, outer ones ovate-oblong, narrowly scariously margined, with a mucronate purple apex, ± double the length of inner one, as long as calyx tube, with broader margins. Calyx funnel-shaped, tube cylindrical, longer than limb, slightly compressed, blackish, 5-veined, with short spaced hairs; limb purple brown, with 5 short lobes, provided with 5 dark purple glabrous veins, and alternating with 5 other smaller lobes without veins. Corolla bright pink, petals united along length of claw. Filaments membranous. Ovary oblong, 5-ribbed, apex elongated into a pentagonal neck; style filiform, stigmas capitate.

HAB. Mountain sides; alt. ± 1400 m; fl. Jun.
DISTRIB. Rare; only recorded twice from the mountains of N Iraq. **MAM**: Ispindari Saddle, Sawara Tuka, *E. Chapman* 26316!; Sulaf, *Omar* 37712!

Turkey (SE Anatolia), Syria.

20. **Acantholimon echinus** (*L.*) *Boiss. ex Bunge*, Mém. Ac. Sc. Pétersb. 7, ser. 18: 46 (1872); Boiss., Fl. Orient. 4: 840 (1879); Mobayen, Revis. Taxon. Acanthol. 213 (1964).

Statice echinus L., Sp. Pl. 276 (1753).
S. androsacea Jaub. & Spach, Ill. Pl. Or. 161 & 164, t. 89 (1844).
S. pauciflora Jaub. & Spach, Ill. Pl. Or. t. 162 (1844).
Acantholimon androsaceum (Jaub. & Spach) Boiss., Diagn. ser. 1, 7: 73 (1846); DC., Prodr. 12: 626 (1848); Rech. f. & Schiman-Czeika in Fl. Iranica 108: 78 (1974).

Densely caespitose greenish shrublet, up to 50 cm in diameter, with a few calcareous dots; branches short, leafy and prickly. Leaves dense, short; lower leaves withered, flat, recurved; upper ones subulate, triquetrous, usually glabrous or obscurely puberulent. Scape very short, pubescent, without scales, unbranched in the axils of upper leaves; spike short, with 3–7 spikelets in a fascicle; rachis pubescent. Bracts herbaceous, at least in basal part, finely hairy, or sometimes glabrous; outer bracts shorter than inner ones, acuminate-cuspidate, with narrowly membranous margins; inner bracts broadly scarious, as long as or slightly longer than calyx tube, pubescent, rarely glabrous. Calyx funnel-shaped, whitish; tube cylindrical, sparsely hairy, whitish-scarious, 5-veined; veins dark purple, glabrous, sometimes hairy at base, projecting from limb margins; limb pentagonal-pyramidal. Corolla pinkish white. Stamens slightly shorter than styles; filaments broadened towards base, adhering to short claws of the almost free petals. Ovary short, ovate-pentagonal, neck of the same length or slightly short than ovary; styles filiform, smooth; stigmas relatively small, capitate and glandular.

HAB. Mountain slopes. alt.: ± 3000 m; fl. ?Jul.
DISTRIB. Rare, in the thorn cushion zone of NE Iraq. **MRO**: Helgord Dagh, *Guest & Husham (Alizzi)* 15839!
NOTE. Also a possible Iraqi specimen between Berdick (sic) and Salmas, [Turco-Persian Frontier Delimitation Commission], *Major Cowie*, s.n.

Turkey, Armenia, Iran, W Syria.

21. **Acantholimon peronini** *Boiss.*, Fl. Orient. 4: 842 (1879).

Acantholimon puberulum var. *puberulum* Boiss. & Bal., Diagn. Ser. 2(4): 62 (1859); Bokhari & Edmondson in Fl. Turk. 7: 496 (1982).

Caespitose, glaucous shrublet, prickly; branches elongate, reddish at base. Leaves narrowly subulate, densely pubescent-hispid, splendidly recurved. Scape hairy, as long as or emerging above leaves, simple or bifurcate. Spike with 3–7 spikelets, contracted and distichous. Bracts and rachis densely hairy; outer bract oblong-lanceolate, margins membranous, puberulent, acuminate, longer than and covering the calyx tube; inner bracts longer than outer ones, finely and densely pubescent along dorsal vein, purplish or greenish. Calyx about 10 mm long; tube as long as limb, glabrescent at base, becoming hairy towards the limb; limb pentagonal with 5 hispidly veined lobes, alternating with 5 smaller ones. Corolla bright pink, petals shortly coherent at base. Stamens with broad and ribbon-like filaments, adhering to claws and basal part of petal, slightly longer than styles. Ovary oblong-ovoid, pentagonal, with pentagonal, comparatively long neck; styles smooth; stigmas capitate.

HAB. Mountain slopes; alt.: ± 1800 m; fl. Aug.
DISTRIB. Only known from one gathering in NE Iraq. **MRO**: Qandil, *Rawi* 9388!

Turkey (SE Anatolia).

106. PLANTAGINACEAE

Pilger in Engler, Pflanzenreich 102 (IV. 269) (1937)

Z. Chalabi Ka'bi[12]

Annual or perennial, mostly herbs, rarely subshrubs, usually acaulous, rarely branched. Leaves usually radical, alternate or opposite, often parallel-veined, exstipulate; inflorescence capitate or spicate, often scapose; flowers usually hermaphrodite, actinomorphic, bracteate, regularly 4-merous. Calyx of 3–4 imbricate sepals. Corolla 4-lobed, scarious. Stamens inserted on tube of corolla, alternating with lobes; anthers usually exserted, 2-celled and

[12] † Formerly Baghdad University, Iraq. Revised by Shahina A. Ghazanfar.

versatile. Ovary superior, carpels 2, locules (1–)2(–4), with (usually) 1, or more ovule in each locule; placentation usually axile; style 1, filiform, bifid; stigma hairy. Fruit a capsule or nutlet. Seed with small straight embryo, enveloped by fleshy endosperm.

Traditionally a family of three genera and about 270 species (as treated in this Flora), *Plantago*, a cosmopolitan genus, *Litorella* in Europe and the Antarctic, and *Bougueria* in the Andean region of South America, Colombia and Peru. Molecular evidence suggests that the family should be expanded to include several genera placed in the Scrophulariaceae and Globulariaceae. In the APG III (2009) classification, Scrophulariaceae s.l. is split with several genera moved to the Plantaginaceae. Under APG III, the Iraq Scrophulariaceae s.l., *Albraunia, Bacopa, Kickxia, Linaria, Penstemon* and *Veronica* would be included in the expanded Plantaginaceae, but are treated in this Flora in the Scrophulariaceae (Vol. 7).

Plantago is the only genus to occur in Iraq and is widely distributed. The mucilaginous seed coat of some species (*P. afra* and *P. ovata*) give an economic value to the family used as efficient laxatives and in the treatment of dysentery. A number of species are noxious lawn weeds. Several species make good fodder for livestock.

1. **PLANTAGO** L.

Gen. Pl. ed. 5: 52 (1754)

Herbs, annual or perennial with ± woody base and herbaceous shoots, usually with rosulate foliage. Inflorescence a spike on a scape or peduncle. Flowers all bisexual, sessile and bracteate. Sepals 4, imbricate, in some species the two anterior sepals connate. Corolla 4-lobed. Stamens usually 4. Ovary (in Iraq plants) usually 2-locular; ovules 1-many in each locule. Fruit (in Iraq plants) a capsule dehiscing by a transverse circular slit, or in some species a nutlet.

About 200 species, cosmopolitan in distribution. Several species are polymorphic showing great variations in their vegetative features which has led to the recognition of many infraspecific taxa. I have generally followed the treatment in Flora of Turkey for the circumspection of species and placed in synonymy many infraspecific taxa that were recognized in the original manuscript.

Plantago comes from the Latin "planta", a footprint.

1. Plants with a distinct stem and branches; leaves opposite .... 16. *P. afra*
   Plants without a stem; peduncles arising from a basal rosette, rarely shortly stemmed, then with alternate leaves 2
2. Capsule with 3-many seeds (in Iraq), but not more than 30 4
   Capsule usually with 2 seeds (rarely with 3–4 seeds as in *P. coronopus*) 3
3. Leaves broad, elliptic-ovate to orbicular .... 1. *P. major*
   Leaves narrowly linear................ 2. *P. tenuiflora*
4. Leaves ± amplexicaul, becoming dark green when dried; peduncle often elongate .... 4. *P. amplexicaulis*
   Leaves not amplexicaul; peduncle usually not elongate................ 5
5. Corolla tube hirsute; seeds usually 3(–4)................ 5. *P. coronopus*
   Corolla tube glabrous; seeds 2 6
6. Anterior sepals connate................ 7
   Anterior sepals free 8
7. Annual; bracts villous at back; corolla lobe with evident brown midrib ....7. *P. lagopus*
   Usually perennial; bract glabrous to glabrescent, in some forms villous at back; corolla lobe not as above, pale .... 6. *P. lanceolata*
8. Corolla lobes hairy 9
   Corolla lobes glabrous (rarely with a few short hairs below in *P. boissieri*)................ 10
9. Leaves obovate to lanceolate-spatulate, acute; corolla lobes narrowly ovate-lanceolate, acute................ 12. *P. ciliata*
   Leaves linear-lanceolate, acute; corolla lobes broad-lanceolate, acuminate................ 13. *P. psammophila*
10. Bracts obtuse or rounded or truncate 11
    Bracts acute or acuminate................ 15
11. Bracts usually hairy (sometimes glabrous in *P. atrata* and *P. boisseri*) 12
    Bracts glabrous................ 14
12. Corolla lobes ovate, obtuse .... 9. *P. notata*
    Corolla lobes ovate-lanceolate or elliptic, acute 13

13. Annual, growing at low altitudes; bracts ovate to broadly ovate, obtuse . . . 8. *P. albicans*
    Perennial, alpine; bracts ovate to orbicular, apex emarginate or crenate . . . . 3. *P. atrata*
14. Corolla lobes ovate to broadly ovate or orbicular-cordate, 2–3 mm long . . . 10. *P. ovata*
    Corolla lobes narrowly-lanceolate or elliptic, 1.5 mm long . . . . 11. *P. loeflingii*
15. Peduncles as long as or shorter than leaves; corolla lobes lanceolate-ovate . . . . 14. *P. bellardii*
    Peduncles much shorter than leaves; corolla lobes orbicular-ovate. . . . . . . .15. *P. cretica*

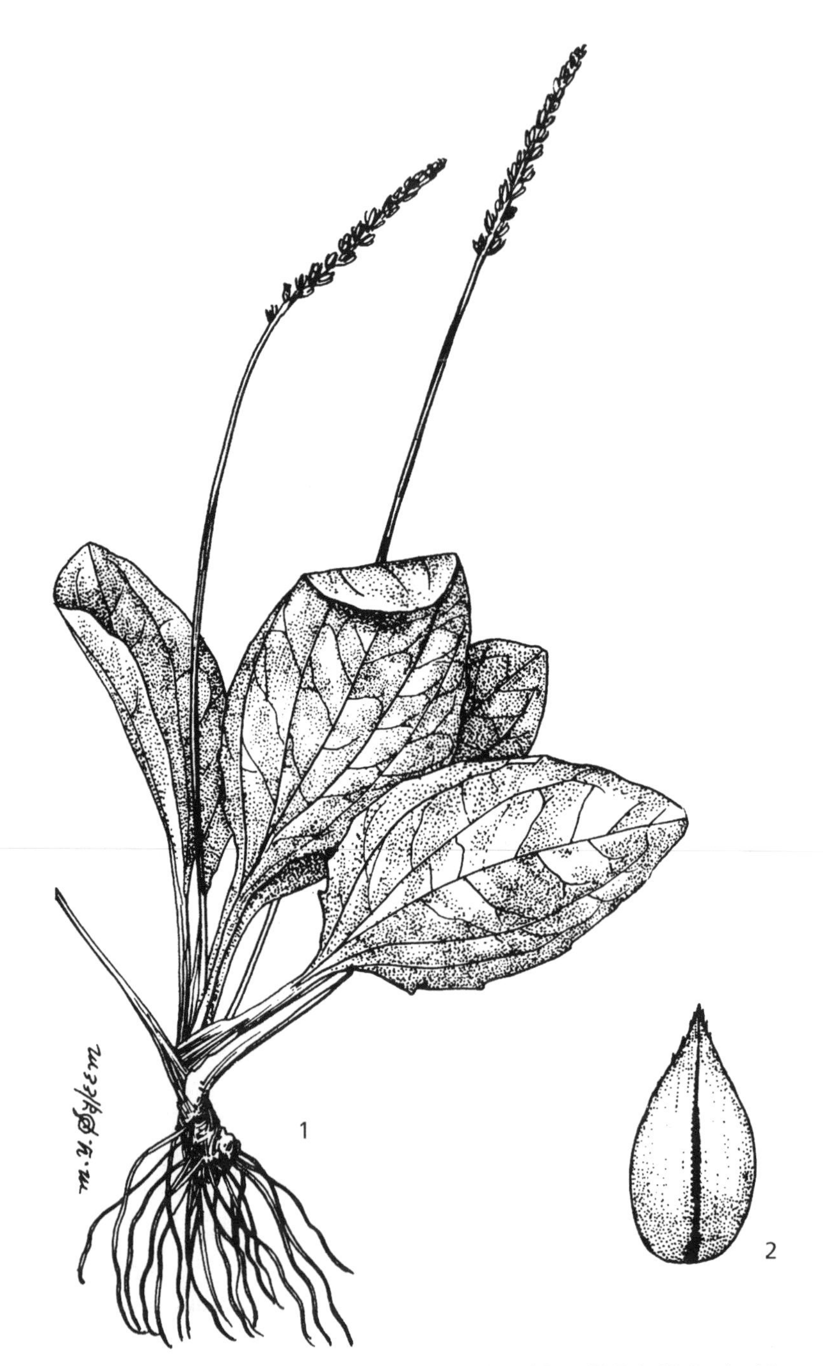

Fig. 28. **Plantago major**. 1, habit × 3/4; 2, bract × 18. Reproduced from Fl. Pak. 62: fig. 1, with permission.

1. **Plantago major** *L.*, Sp. Pl. ed. 1, 112 (1753); Boiss., Fl. Orient. 4: 8778 (1879); Pilger in Engl., Pflanzenr. 102 (IV. 269): 41 (1937); Zohary in Dep. Agr. Iraq Bull. 31: 136 (1950); Blakelock in Kew Bull. 3: 442 (1950); Chalabi Ka'bi in Bull. Iraq Nat. Hist. Mus. 1(2): 4 (1961); Rech.f., Fl. Lowland Iraq: 555 (1964); Patzak & Rech.f. in Fl. Iranica 15: 3 (1965); Verdc. in FTEA Plantag.: 2 (1971); Tutel in Fl. Turk. 7: 507 (1982); Collenette, Wildflow. Saudi Ariabia: 603 (1999); Boulos, Fl. Egypt 3: 117 (2002); Li, Wei & Hoggard in Fl. China 17: 497 (2011).

Plantago paludosa Turcz. ex Ledeb., Fl. Ross. 3: 478 (1849).

P. major L. var. *minor* Boiss., Fl. Orient. 4: 878 (1879); *P. major* L. var. *sinuata* (Lam.) Decne in DC., Prodr. 13 (1): 694 (1852); *P. major* L. var. *paludosa* Beguinot in Fiori & Paol. Fl. Anal. d'Italia 3: 94 (1903) emend. Fiori, Nuov. Fl. Anal. d'Italia 2: 477 (1926); *P. major* L. subvar. *brachyphylla* Pilger, Notizbl. Bot. Gart. Mus. Berlin-Dahlem 8: 274 (1922); *P. major* L. var. *pilgeri* Domin, Monogr. Studie Československ. in Vestn. Kral. České Společn. Nauk Tr. 2: Sep. 15 (1933); *P. major* L. forma *brevipedicellata* Pilger, Engl. Pfl. IV . 269: 48 (1937); *P. major* L. subsp. *pleiosperma* Pilger, Engl. Pfl. IV. 269: 48 (1937).

Perennial herb, acaulous, (10–)20–40(–70) cm tall with a short rootstock and many fibrous roots. Leaves large, elliptic-ovate to orbicular-ovate, (5–)10–25(–40) × (1–) 4–10 (–20) cm, base tapering into petiole about as long as lamina, apex rounded, margins entire or irregularly dentate or sinuate-dentate, strongly 5–9-veined, glabrous or rough with sparse minute hairs. Scape erect, ascending or arcuate, equal to leaves or shorter, pilose with sparse minute soft hairs. Spike dense or lax, linear or linear-cylindrical, (8–)5–30(–50) cm; bract ovate, as long as or usually shorter than calyx, 1–2 mm, along midrib herbaceous and slightly keeled, scarious at margins, entire, obtuse, glabrous. Sepals ovate-elliptic to orbicular, 1.5–2.5 mm, with a distinct midrib, margins scarious, obtuse, glabrous. Corolla greenish or yellowish, lobes ovate, 0.5–1 mm long, obtuse or acute. Capsule longer than calyx, globose or conical, tapering above, 2-celled with 4–30 seeds. Seeds angular, smooth. Fig. 28: 1–2.

HAB. In wet shaded places in farms and orchards, muddy places amongst rocks, near banks of shallow ponds and streams, in rich humus soil and on rich soil with limestone in upland forests; alt.: 150–900 m; fl. & fr. Mar.–Aug.

DISTRIB. Throughout Iraq except in the desert and plateau region; found in the upper plains and foothills and the alluvial plains. **MAM**: Amadiya, Robat Zier to Kubat Amien, *Mustafa* 15772!; Zawita, *Guest* 3753!; Pushtashan, NE. of Rania, *Rawi* 23196!; **MRO/MAM**: Ruwandiz Gorge, *Guest* 13011!; Gali Ali Beg, *Rawi* 24258; Rust Mountain, *Haley* 125!; near Shaqlawa, *Barkley* 3429B!; **FUJ**: Tel-Afar, *Allawi* 16366!; **LCA**: Abu Ghraib, *Al-Jibouri* 14586!; Chubayish, in Nasiriya Liwa, *Tolba* 1037!; Daltawa, *Guest* 2455!; Sa'ida, S of Baghdad city, *Chalabi* 2230, 2214!

NOTE. A species of some economic value whose seeds and leaves are used medicinally as a laxative and a coolant; also used for application on boils.

Several subspecific ranks were recognized in the original manuscript based mainly on leaf characters. I found it difficult to assign every collection to the subspecific rank recognized as many intermediates exist. The species is best treated as a single polymorphic species. The var., subvar. and f. are listed in synonymy.

BIZIR DINBIL (recorded by Blakelock, loc.cit.); TARAQUZ (recorded on label by Guest).

Europe to Siberia, from the Mediterranean region, Sinai, Arabian Peninsula to Iran and India; naturalized in N America and many other parts of the world.

2. **Plantago tenuiflora** *Waldst. & Kit.*, Descr. et Icon. Pl. Rar. Hungar. 1: 37 (1802); Boiss., Fl. Orient. 4: 879 (1879); Pilger in Engl., Pflanzenr. 102 (IV. 269): 68 (1937); Chalabi Ka'bi in Bull. Iraq Nat. Hist. Mus. 1(2): 5 (1961); Patzak & Rech.f. in Fl. Iranica 15: 4 (1965).

Annual, acaulous with thin short roots. Leaves rosulate, decumbent or ascending or erect, narrowly linear, 3–14 × 1–3 mm, obtuse, entire or minutely denticulate, glabrous or short-pilose at base. Peduncles shorter than leaves or rarely equal, erect or arcuate, 1–15 cm, glabrescent or sparsely short soft pilose. Spike elongate, narrow, linear or narrowly cylindrical, 2–7 cm long; bracts ovate-lanceolate, 2–2.5 mm, obtuse, glabrous, equal to calyx, but in the lower flowers of spike longer, concave, with a thick herbaceous midrib. Calyx with the anterior sepal elliptic-ovate, 1.5–2 mm, obtuse, midrib narrower than lateral lamina, glabrous. Posterior sepals obovate, 1.7 mm, acute, glabrous, with narrow midrib. Corolla lobes small, ovate, 0.7 mm, obtuse. Capsule conical, elongate, ± 3 mm, longer than calyx, attenuate and narrow at apex. Seeds 7–10, fusiform, reticulate-rugulose. Fig. 29: 9–13.

HAB. In saline areas flooded by rainwater; alt. ± 50 m; fl. & fr.: Mar.–Apr.

DISTRIB. Found in northeastern Iraq. **FNI**: Qizil Ribat, *Sutherland* 317!; *Graham s.n.*!

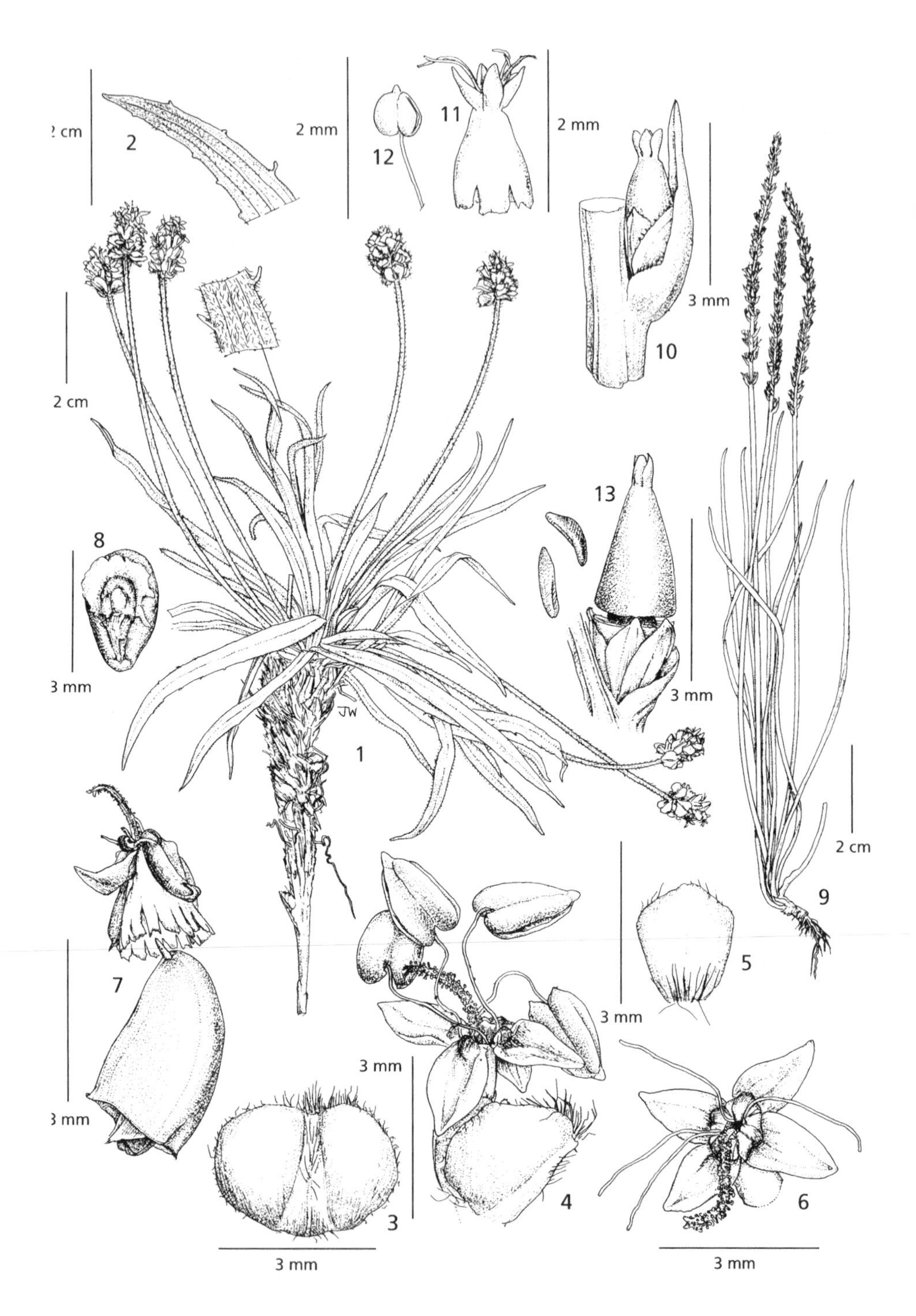

Fig. 29. **Plantago atrata**. 1, habit; 2, leaf tip, undersurface; 3, bract; 4, flower with bract; 5, sepal; 6, corolla; 7, capsule with old corolla; 8, seed. **Plantago tenuiflora.** 9, habit; 10, flower with bract; 11, corolla; 12, stamen; 13, capsule with 2 of 7–10 seeds. 1, 2, from *Rawi & Serhang* 24399; 3–6 from *Rustam* 2856; 7, 8 from *Rustam & Gillett* 9775; 9, 10 from *Lipsky* (25 May 1893); 11–13 from *J. Gay* s.n. Drawn by J.E. Beentje (Apr. 2012).

Hungary, southern Russia, Lithuania, Siberia and Iran.

3. **Plantago atrata** *Hoppe*, Botan. Taschenb. 85 (1799); Pilger in Engl., Pflanzenr. 102 (IV. 269): 313 (1937); Parsa, Fl. de l'Iran 4: 949 (1950); Patzak & Rech.f. in Fl. Iranica 15: 8 (1965); Kazmi in Fl. Pak. 62: 8 (1974); Tutel in Fl. Turk. 7: 511 (1982).

Plantago montana sensu Lam., Ill. Gen. 341 (1791) non Hudson (1762).

P. saxatilis Bieb., Fl. Taur.-Cauc. 1:109 (1808); Boiss., Fl. Orient. 4: 880 (1879); Nábělek in Publ. Fac. Sci. Univ. Masaryk 105: 5 (1929); Chalabi Ka'bi in Bull. Iraq Nat. Hist. Mus. 1 (2): 6 (1961); *P. saxatilis* M. Bieb. var. *angustifolia* Boiss., Fl. Orient. 4: 881 (1879).

P. fuscescens Jordan, Obs. Pl. Nouv. 3:231, t. 10a, f. 1–12 (1846).

P. spadicea Wallr. ex Ledeb., Fl. Ross. 3:484 (1846-51) nom. nud.

P. montana Hudson subsp. *spadicea* (Wallr. ex Ledeb.) Pilger in Feddes Rep. 23: 263 (1926).

P. saxatilis Bieb. var. *adnivalis* Náb. in Publ. Fac. Sci. Univ. Masaryk Brno 105:5 (1929); Chalabi Ka'bi in Bull. Iraq Nat. Hist. Mus. 1 (2): 6 (1961).

P. atrata Hoppe subsp. *saxatilis* (Bieb.) Pilger in Engl., Pflanzenr. 102 (IV. 269): 300 (1937); *P. atrata* subsp. *spadicea* (Wallr. ex Ledeb.) Pilger in Engl., Pflanzenr. 102 (IV. 269): 309 (1937); Patzak & Rech. in Rech. f., Fl. Iranica 15: 8 (1965); Kazmi in Fl. Pak. 62: 8 (1974); *P. atrata* Hoppe subsp. *spadicea* (Wallr. ex Ledeb.) Pilger var. *angustifolia* (Boiss.) Pilger in Engl., Pflanzenr. 102 (IV. 269): 303 (1937); *P. atrata* var. *spadicea* Pilger in Engl., Pflanzenr. 102 (IV. 269): 301 (1937).

Perennial herb, acaulous with thin or (in Iraq) thick rhizome, erect, 4–12(–20) cm tall. Leaves linear to linear-lanceolate to lanceolate, sessile, acute or acuminate, (1–)3–8(–14) × 0.1–0.4(–0.6) cm, glabrous or pilose, rarely villous at base, entire or shallowly dentate. Peduncles as long as leaves or often longer, 2–10(–17) cm long; spike globular-ovoid or short-cylindrical, 0.5–2(–3) cm, few-flowered; bracts ovate-orbicular, usually broader than calyx, glabrous or pilose, usually ciliate at apex, brown, scarious except for the herbaceous midrib, which sometimes protrudes at apex in a short mucro, or the apex usually emarginate, usually crenate; midrib ± pilose. Sepals broadly ovate, 2.5–3 mm, wholly scarious, keeled in the lower part, acute, glabrous, usually ciliate at apex. Corolla lobes ovate-lanceolate. Capsule narrow, conical, 2-celled, each 1-seeded. Seeds ovoid, rugose. Fig. 29: 1–8.

HAB. Found in between rocks and in snow-melt soil, rocky slopes of mountains and on the summit; alt.: 2800–3300 m; fl. & fr.: June–July, fruits still unripe at beginning of Aug.
DISTRIB. In the Mountain Region of NE Iraq. **MRO**: Al-Gird (Halgurd) Dagh, *Gillett* 9606 & *Guest* 2835!; 2846!; Halgurd Range, *Rawi & Serhang* 24923!; NE of Qandil range, *Rawi & Serhang* 24399!; 24511!; Qandil range, NE of Rania, *Rawi & Serhang* 18279!; 24462!; 26750A!; N of Helgard (Halgurd) Range, E of Bermasand lake, *Rawi & Serhang* 24766!; Ser Kurawa, *Gillett* 9775!; Halgurd Peak, *Rechinger* 11455!; Halgurd summit, *Agnew* 4961!; **FAR**: Koda, Erbil Liwa, *Rawi* 9210!

Balkans, Aegean Islands, Armenia, Caucasus, Turkey, Iran.

4. **Plantago amplexicaulis** *Cav.*, Icon. et Descr. Pl. Hisp. 2: 22 (1793); Roem. & Schult., Syst. Veg. 3: 142 (1818); Boiss., Fl. Orient. 4: 883 (1879); Hook f., Fl. Brit. Ind. 4: 706 (1885); Nábělek in Publ. Fac. Sci. Univ. Masaryk 105: 5 (1929); Pilger in Engl., Pflanzenr. 102 (IV. 269): 310 (1937); Chalabi Ka'bi in Bull. Iraq Nat. Hist. Mus. 1 (2): 13 (1961); Rech.f., Fl. Lowland Iraq: 556 (1964); Patzak & Rech.f. in Fl. Iranica 15: 11 (1965); Kazmi in Fl. Pak. 62: 10 (1974); Tutel in Fl. Turk. 7: 511 (1982); Boulos, Fl. Egypt 3: 117 (2002).

Annuals or perennials (annuals in Iraq), 4–15(–30) cm tall, acaulous, or mostly with a short and conspicuous stem, erect or oblique and arcuate, glabrescent or pilose; root system usually dense. Plant when dry becoming dark green especially leaves and bracts. Leaves alternate or spirally inserted, 2–12 × 0.3–1.2 cm, narrowly lanceolate, finely acuminate, entire or very sparsely toothed, tapering and sub-petiolate into a wider sheathing base, glabrescent or pilose, margins usually ciliate. Peduncle erect, straight or a little arcuate, equal to or usually longer than leaves, 5–15(–25) cm long, pilose. Spike ovate-globular or oblong, 1–2.5(–3.5) cm; bracts usually concave, orbicular-ovate, 4.5–5.5 mm, glabrous. Lateral lamina broad-scarious, midrib herbaceous, narrow. Sepals broadly ovate to ovate, glabrous or hirsute-villous on midrib (in ours), anterior sepals larger and with a broader more herbaceous midrib than posterior sepal, midrib of posterior sepal two-thirds the length of lobes. Corolla glabrous, lobes ovate-elliptic, acute, 3.2–3.7 mm, cymbiform. Capsule 1–2-seeded. Seed ovoid, smooth.

Canary Islands, Morocco, Algeria, Greece, Egypt, Syria to Iran and Afghanistan.

var. **bauphula** (Edgew.) Pilger in Notizbl. Bot. Gart. Mus. Berlin 9, no. 90: 1102 (1927) & in Engl., Pflanzenr. 102 (IV. 269): 312 (1937); Zohary in Pal. Journ. Bot. J. Ser. 1: 226 (1937); Collenette, Wildflow. Saudi Ariabia: 602 (1999). Fig. 30: 1–2.

Plantago bauphula Edgew. in Hook. J. Bot. 2: 285 (1840).

P. amplexicaulis Cav. subsp. *bauphula* (Edgew.) Rech.f., Fl. Iranica 15: 11 (1965); Zohary, Fl. Palaest. 3: 224(1978) & pl. 375 (1977).

HAB. Wadi beds, sandy depressions, sandy wadis in barren gravel desert, edges of rain cultivation in gypsum desert, hard silty flats, gravel plains, gravelly sandy slopes, sandy pockets in limestone cliffs, rocky slopes with sandy plains; alt.: 15–500 m; fl. & fr. Feb.-Apr.

DISTRIB. Mainly in the desert regions of Iraq. **FNI**: Kana Sank, near Mandali, *Rawi* 20586!; Chilat police station, near Iranian border, *Rawi & Haddad* 25700!; **DLJ**: Lake Thirthar, *Barkley & Agnew* 6036!; **DWD**: 80 km W of Shabicha, *Chakrovarty, Rawi, Khatib & Tikriti* 30060!; 77 km N of Rutba, Wadi Al-Ajraniya, *Rawi & Nuri* 27169!; 110 km E of Nukhaib, *Agnew, Haines & Al-Hashimi* 5465!; **DSD**: Zubair, *Rawi & Gillett* 6065!; 6931!; Jabal Sanam, *Guest, Rawi et al.* 14398!; Rowag, N of Ansab, southern desert, *Rechinger* 9342!; Salman cliffs, *Agnew & Haines* 4817!; 40 km NW of Shabicha, *Agnew, Haines, Hadač & Al-Hashimi* 5538!; **LCA**, Iskandariya, Baghdad Liwa, *Agnew & Shawqi* 181!

Algeria, Egypt, Palestine, Iran and Afghanistan to western India.

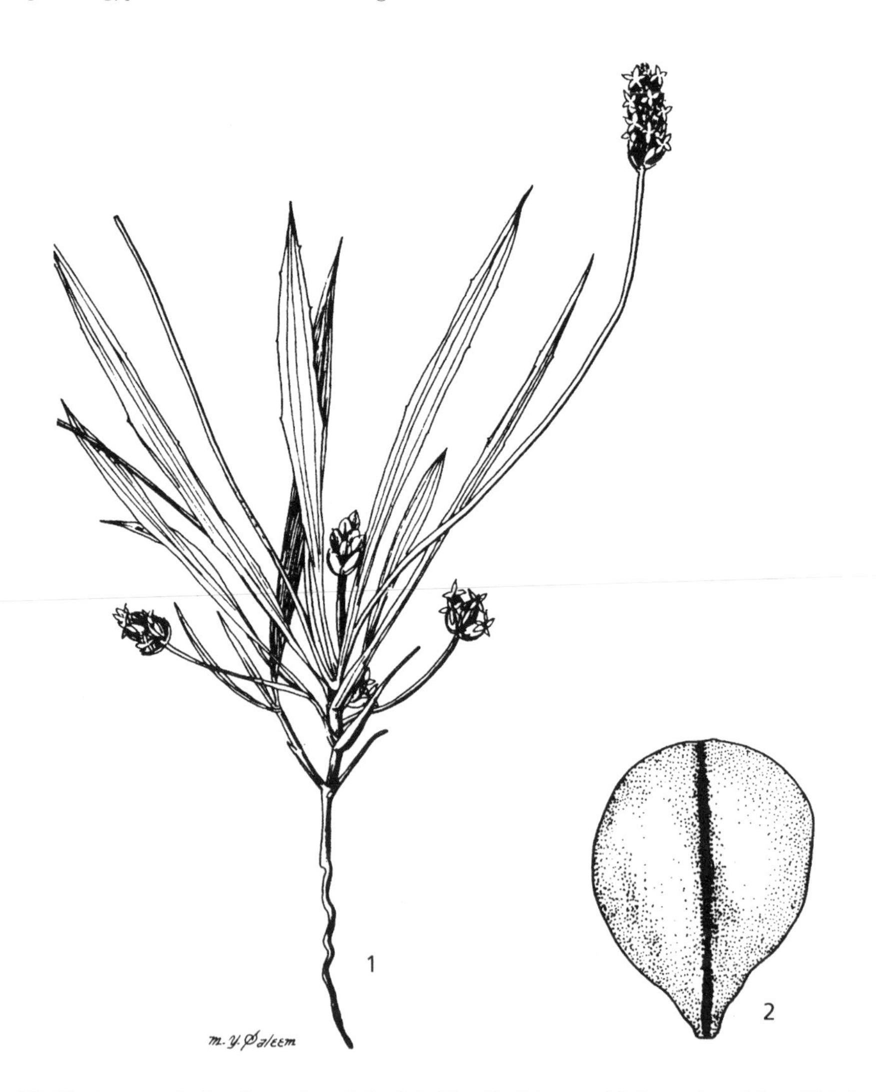

Fig. 30. **Plantago amplexicaulis** var. **bauphula**. 1, habit × 3/4; 2, bract × 10. Reproduced from Fl. Pak. 62: fig 1, with permission.

5. **Plantago coronopus** *L.*, Sp. Pl. 115 (1753); Boiss., Fl. Orient. 4: 888 (1879); Pilger in Engl., Pflanzenr. 102 (IV. 269): 126 (1937); Zohary in Dep. Agr. Iraq Bull. 31: 137 (1950); Blakelock in Kew Bull. 3: 441 (1950); Chalabi Ka'bi in Bull. Iraq Nat. Hist. Mus. 1 (2): 25 (1961); Rech.f., Fl. Lowland Iraq: 556 (1964); Patzak & Rech.f. in Fl. Iranica 15: 5 (1965); Kazmi in Fl. Pak. 62: 7 (1974); Tutel in Fl. Turk. 7: 508 (1982); Collenette, Wildflow. Saudi Ariabia: 602 (1999).

Annual or perennial herb (annual in Iraq), acaulous, 3–12(–17) cm tall, glabrous to pilose or unusually hirsute. Leaves oblong to linear in outline, 2–14 cm, entire, remotely lobed or mostly pinnatisect, the lobes 2–15 mm, linear or oblong, acute. Spike usually shorter than leaves, oblong-cylindrical, 1.5–4(–8) cm; bracts equalling or shorter and smaller than calyx, usually glabrous, herbaceous with narrow scarious lateral lamina, margin ciliate. Calyx 3–3.5 mm, the two anterior sepals similar to bracts, the two posterior sepals glabrous, scarious except for herbaceous keeled midrib. Corolla lobes glabrous, lanceolate-ovate, acuminate, tube hirsute. Seeds ovoid, with membranous, mucilaginous margin.

The main distinguishing characters for this species are the bracts that are equalling or smaller than calyx, posterior sepals with well keeled midribs, and a hairy corolla tube. The species shows much variation in the indumentum, leaf margin (entire, dentate or well lobed), length, thickness and number of spikes, the ratio of length of bract to calyx, apex of bract and the width of midrib of the anterior sepals.

Bract almost equalling calyx (or a little longer); lateral lamina of anterior sepals usually narrow. . . . . subsp. *coronopus*

Bract shorter than the calyx; lateral lamina of anterior sepals usually broad . . . . subsp. *commutata*

subsp. **coronopus**

Plantago coronopus L. subsp. *eucoronopus* Pilger in Feddes Repert. 28: 265 (1930).

No specimens at K or BAG, but Agnew noted the name on the label of a specimen from Mosul (*Birch* 579) at BM.

subsp. **commutata** *Pilger* in Feddes Repert. 28: 288 (1930) & in Engl., Pflanzenr. 102 (IV. 269): 126 (1937); Zohary, Fl. Palaest. 3: 223 (1978) & pl. 372 (1977).

Plantago coronopus L. var. *simplex* Boiss., Fl. Orient. 4: 888 (1879); Chalabi-Ka'bi, Bull, Iraq Nat. Hist. Mus. 1 (2): 27 (1961); *P. coronopus* L. var. *crassipes* Coss. & Dayneau in Bull. Soc. Bot. Fr. 36: 106 (1889); Pilger in Fedde Repert. 28: 296 (1930) & in Engl. Pfl. IV. 269: 149 (1937); Zohary in Pal. Journ. Bot. J. ser. 1: 228 (1938) & in Dep. Agr. Iraq Bull. 31: 137 (1950); Chalabi Ka'bi, Bull. Iraq Nat. Hist. Mus. 1 (2): 26 (1961); Zohary, Fl. Palaest. 3: 223 (1978) & pl. 373 (1977); *P. coronopus* L. subsp. *commutata* Pilger var. *stricta* Pilger in Fedde Repert. 28: 278 (1930); *P. coronopus* L. subsp. *commutata* Pilger var. *rigida* Pilger, Fedde Repert. 28: 292 (1930) & in Engl., Pflanzenr. (IV. 269): 146 (1937); *P. coronopus* L. subsp. *commutata* Pilger var. *rigida* Pilger subvar. *erecta* (K. Koch) Pilger forma *weldenii* (Reichb.) Pilger, Fedde Repert. 28: 293 (1930).

HAB. Sandy and loamy desert, alluvial land, gravel desert, by roadsides and deciduous orchards on loamy soil, slightly saline alluvial flats near edges of lakes, limestone desert plateau with sandy patches between rocks, river banks and islands; alt.: 150–550 m; fl. & fr.: Mar. –Apr.

DISTRIB. Frequent in the Lower Mesopotamian Region, less frequent in upper part of the Desert Plateau Region; rarely found in the Upper Plains and Foothills Region. **FUJ**: Gaisha Hill 78 km S of Sinjar, *Chakravarty, Rawi, Khatib & Al-Izzi* 32069!; **FKI**: Jabal Hamrin, on Tuz-Khalis road, *Rawi & Gillett* 10257!; Tauq Bridge, Kirkuk Liwa, *Polunin* 2014!; 19 km E of Kirkuk, *Rawi & Gillett* 10580!; **FNI**: Pilkana, 10 km E. of Khanaqin, *Rawi* 12754A!; near Mandali, *Guest* 911!; Badra, *Gillett* 6586!; Koma Sank Police Station, near Mandali, *Rawi* 20646!; & near Mandali, *Rawi* 12583!; **DLJ**: 65 km NW of Falluja, *Rawi* 30230!; **DGA**: 30 km N of Baghdad, *Barkley, Brahim & Abbas* 1591!; Daltawa (Khalis), *Guest* 2443!; between Ramadi and Hit, *Rawi & Gillett* 6793!; Wadi Horan, *Rawi & Gillett* 6892!; Fat-ha Gorge, *Guest* 427!; **DWD**: Al Kalah Island in the Euphrates River at Ana, *Barkley, Brahim & Abbas* 636!; Rutba wells vicinity, *Springfield & Al-Jaf* 21315!; 27 km W of Falluja, *Barkley et al.* 306!; 294!; 317!; **DSD**: near Nukhaib, *Springfield & Al-Jaf* 21361!; Al Batin 65 km SW of Basra, *Guest, Rawi & Rechinger* 17062!; near Jabal Sanam, *Guest, Rawi & Rechinger* 17032!; **LEA**: 20 km E of Amara, *Rawi* 15042!; between Fakka and Musaida, *Rawi* 15009!; Tarmiya, N of Baghdad, *Barkley, Dabbagh & Brahim* 1620!; 15 km N of Amara, *Guest, Rawi & Rechinger* 17424!; **LCA**: Baghdad, *Cowan & Darlington* 84!; Ali Gharbi, *Khudairi* 1157!; **LBA**: Maqil, near Basra, *Dept. of Bot. & Rechinger* 8431!; *Guest, Rawi & Rechinger* 16740!; near Aziar, *Rawi & Rechinger* 17387!; S of Zubair, *Rawi & Gillett* 6035!; Basra, *Rechinger* 8835!; 8518!; 8663!

NOTE. A fodder plant, grazed by sheep, camels and water-buffaloes. ANAIK (recorded by Blakelock, loc.cit.).

Europe and Mediterranean region, Turkey, Syria, Jordan, Palestine, Jordan, Saudi Arabia, Kuwait, Iran, Afghanistan, Pakistan.

6. **Plantago lanceolata** *L.*, Sp. Pl. 113 (1753); Boiss., Fl. Orient. 4: 881 (1879); Pilger in Engl., Pflanzenr. 102 (IV. 269): 313 (1937); Zohary in Dep. Agr. Iraq Bull. 42: 136 (1950); Blakelock in Kew Bull. 3: 441 (1950); Chalabi Ka'bi in Bull. Iraq Nat. Hist. Mus. 1 (2): 6 (1961); Rech.f., Fl. Lowland Iraq: 556 (1964); Patzak & Rech.f. in Fl. Iranica 12: 11 (1965); Verdc. in Fl. Trop. East Africa, Plantag. 6 (1971); Kazmi in Fl. Pak. 62: 11 (1974); Zohary, Fl. Palaest. 3: 222 (1978) & pl. 376 (1977); Tutel in Fl. Turk. 7: 513 (1982); Collenette, Wildflow. Saudi Ariabia: 603 (1999); Boulos, Fl. Egypt 3: 116 (2002).

Plantago altissima L., Sp. Pl. 164 (1753).
P. eriophora Hffmeg. & Link, Fl. Portug. 1: 423 (1809).
P. lanceolata L. var. *communis* Schl., Fl. Berol. 1: 109 (1823); *P. lanceolata* L. var. *dubia* (L.) Wahlenberg, Fl. Suecica 96 (1824) sensu lato; *P. lanceolata* L. var. *eriophora* Rap. in Ann. Soc. Linn. Paris 6: 458 (1827); P. *lanceolata* L. var. *altissima* (L.) Dcne. in DC., Prodr. 13: 714 (1852); *P. lanceolata* L. var. *eriophylla* Dcne. in DC., Prodr. 13: 715 (1852).
P. mediterranea Kerner in Österr. Bot. Zeitschr. 25: 59 (1875).
P. lanceolata L. var. *genuina* Boiss., Fl. Orient. 4: 881 (1879); *P. lanceolata* var. *altissima* (L.) Boiss., Fl. Orient. 4: 881 (1879); *P. lanceolata* L. var. *capitata* Presl in Boiss., Fl. Orient. 4: 881 (1879).
P. orientalis Stapf, Denkschr. Akad. Wiss. Wien. Math.Nat. Kl. 50: 33 (1885).
P. lanceolata L. var. *capitata* (Ten.) Presl in Fl. Sic. 69 (1886); *P. lanceolata* L. var. *communis* Schl. subvar. *genuina* (Dietrich) Pilger in Engl., Pflanzenr. 102 (IV. 269): 315 (1937); *P. lanceolata* L. var. *mediterranea* (Kerner) Pilger in Engl., Pflanzenr. 102 (IV, 269): 320 (1937); *P. lanceolata* L. var. *dubia* (L.) Wahlenb. subvar. *eriophora* (Hffmeg. & Link) Pilger in Engl., Pflanzenr. 102 (IV. 269): 323 (1937).

Perennial herb, acaulous, to 60 cm, glabrous to pilose or hirsute-lanate. Leaves alternate in a basal rosette, lanceolate to lanceolate-oblong, 5–35 × 0.5–5 cm, usually petiolate, acute, margins entire to minutely dentate, 3–5-veined, glabrous. Peduncles (15–)25–50(–100) cm, usually erect, sometimes arcuate-ascending until becoming erect, longer than leaves, tough, grooved, angled. Spike 1–8 cm, globose to cylindrical, dense, not woolly; bract as long as or somewhat longer than calyx, broadly ovate to ovate-lanceolate, usually glabrous to glabrescent, sometimes hirsute or villous in the lower flowers of the spike, scarious except midrib, acute to long-acuminate or caudate. Calyx 3–4 mm, scarious except midrib, obtuse, glabrous except apex which is shortly ciliate, or villous at midrib, the two anterior sepals united to form a two-midribbed lobe, the two posterior sepals normal. Corolla glabrous, lobes ovate-lanceolate, acute or shortly acuminate. Capsule 2-seeded. Seed ovoid, smooth. Fig. 31: 1–2.

HAB. In silty and clay soils, wet margins of shallow seepages from irrigation channels and irrigated cultivations, shady orchards, marshy areas, alluvial plains of rivers, along roadsides and mountain slopes, and shady limestone slopes in forest (in northern Iraq); alt.: 40–1700 m; fl. & fr.: Mar.–Aug.
DISTRIB. Frequent in the Mountain Regions and Upper Plains and Foothill Regions, less frequent in the Central Alluvial Plain Districts and in the Marsh District and in Basra; no specimens from the Southern Desert. **MRO**: Sei Rust, *Haley* 89!; 10–20 km from Darband, on Haji Umran road, *Chalabi* 2238!; 3391!; Sei Wake village, near S foot of Karoukh, *Kass, Nuri & Serhang* 27535!; Road between Kholan and Haji Umran, *Chalabi* 3405!; Haji Umran, *Chalabi* 3406!; Rowanduz Gorge, *Guest* 3017!; **MAM**: Khaira Mt. NE. of Zakho, *Rawi* 23250!; & *Rawi* 23510!; Aqra, *Rawi* 11371!; Ispindari Saddle, Sawara Tuka, *Chapman* 2634!; Pushtashan, 15 km NE of Rania, lower slopes of Qandil Range, *Rawi & Serhang* 23877!; Khalan, *Rawi* 13719!; Khalan, near Haji Umran, *Rawi* 2237 & 3389 & 13764!; Khalan, *Rawi* 13759!; N of Pushtashan, Qurnaqo Valley, *Rawi & Serhang* 26611!; 15 km on road to Raizan, *Chalabi* 3390!; near Rayat Police Station, *Chalabi* 3410!; **MSU**, Halabja, *Rawi* 8822!; 15 km NW of Sulaimaniya, *Rawi* 21786!; Bayara (Biyara), Rawi, *Husham & Nuri* 29485!; Penjwin Valley, *Rawi* 12143!; & 22582!; hills around Penjwin, *Rechinger* 10538!; Qaradagh, Ja'feran, in Sulaimaniya Liwa, *Gillett* 7868!; **FNI**: Mandali, *Guest* 1807!; **DGA**: Daltawa in Diyala Liwa, *Guest* 2463!; **LCA**: 15 km N of Baghdad, *Barkley, Dabbagh & Abbas* 1563!; Abu Ghraib, *Rawi* 10748A!; Khan Bani Sa'd, *A.F. & F.A.F.* 28!; Abu Ghraib, Baghdad Liwa, *Rawi* 10761!;**LSM**: Hubayish in Nasiriya Liwa, *Tolba* 1038!; road between Baquba and Khalis (Diltawa), *Dept. of Bot.* 180!. **LBA**: Maqil near Basra, on island of Shatt Al-Arab, *Rechinger* 8453!
NOTE. A very variable species with a wide distribution range. Several infraspecific taxa were recognized in the original manuscript, but I have found it difficult to maintain these as the differences are small and there are many specimens which show characters that are intermediate between taxa.

A spring and summer fodder plant eaten by all animals; used as poultice for humans. RIKESHA, ZIBAD, IDHAN AS SAKHAILAH (recorded by Blakelock, loc.cit.).

Europe to W Siberia and C Asia, Cyprus, Egypt, Sudan, Syria to Kuwait, Iran, India, naturalized in N America.

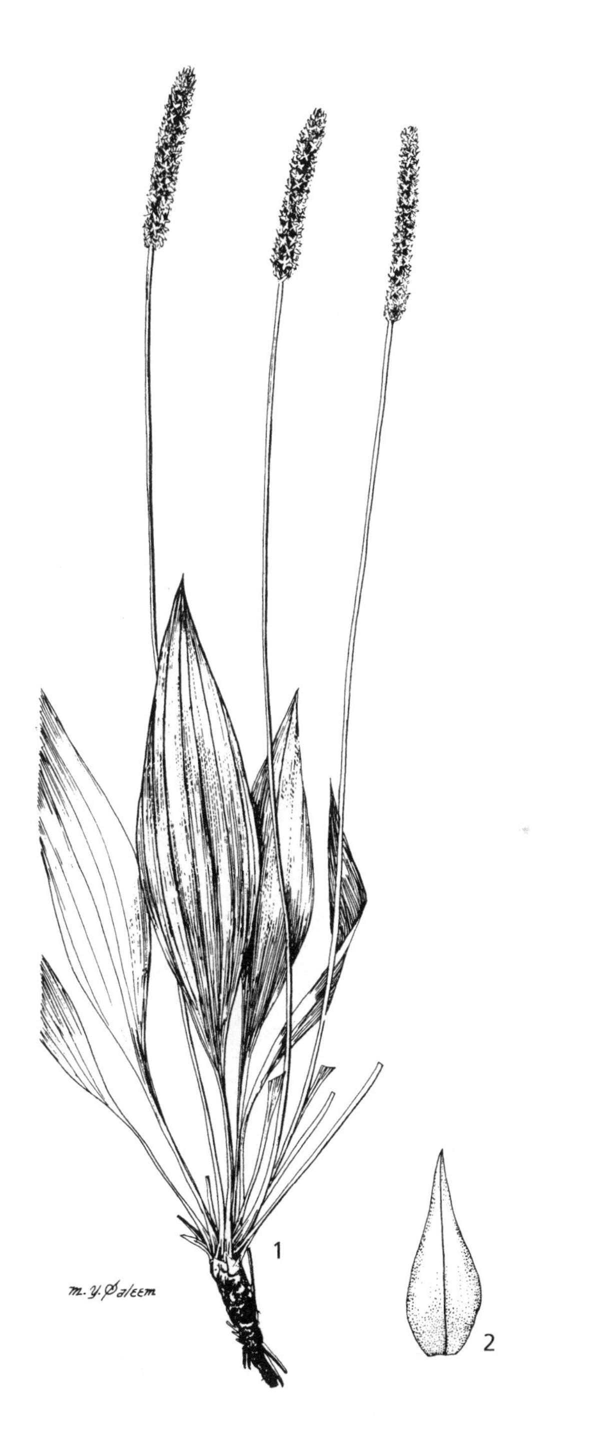

Fig. 31. **Plantago lanceolata**. 1, habit × 1/2; 2 bract × 7. Reproduced from Fl. Pak. 62: fig. 2, with permission.

7. **Plantago lagopus** *L.*, Sp. Pl. 114 (1753); Boiss., Fl. Orient. 4: 886 (1879); Pilger in Engl., Pflanzenr. 102 (IV. 269): t. 33, 332 (1937); Zohary in Dept. Agr. Iraq Bull. 31: 137 (1950); Blakelock in Kew Bull. 3: 441 (1950); Chalabi Ka'bi in Bull. Iraq Nat. Hist. Mus. 1 (2): 22 (1961); Patzak & Rech.f. in Fl. Iranica 12: 13 (1965); Kazmi in Fl. Pak. 62: 13 (1974); Zohary, Fl. Palaest. 3: 222 (1978) & pl. 377 (1977); Tutel in Fl. Turk. 7: 514 (1982); Boulos, Fl. Egypt 3: 116 (2002).

Plantago lagopus L. var. *genuina* Battandier in Batt. & Trabut, Fl. de l'Algérie 740 (1890); Pilger in Engl., Pflanzenr. 102 (IV. 269): 334 (1937); *P. lusitanica* L., Sp. Pl. ed. 2: 1667 (1763); *P. lagopus* L. β *major* Boiss., Fl. Orient. 4: 886 (1879); Chalabi-Kai'bi, Bull. Iraq Nat. Hist. Mus. 1, no. 2: 23 (1961); *P. lagopus* L. var. *gracilis* Webb & Berth., Phyt. Canar. sect. 3: 185 (1836–50); *P. lagopus* L. var. *cylindrica* Boiss., Voy. Bot. Midi de l'Espagne 2: 536 (1839–45); Pilger in Engl., Pflanzenr. 102 (IV. 269): 336 (1937); *P. lagopus* L. var. *lagopus* forma *minor* (Tenore) Pilger in Engl., Pflanzenr. 102 (IV. 269): 335 (1937).

Annual, rarely perennial herb, acaulous, (3–)10–30(–40) cm, tufted at base. Leaves erect, decumbent or ascending, lanceolate or lanceolate-elliptic, 5–12(–17) × (0.2–)0.6–1.5(–2) cm, base tapering to a petiole, acute, margins entire or mostly remotely denticulate, usually 3-veined, usually glabrous-glabrescent, rarely pilose or pubescent-villous. Peduncle sulcate, much longer than leaves, 10–20(–40) cm, hirsute. Spike globular-ovoid or ovoid-cylindrical,

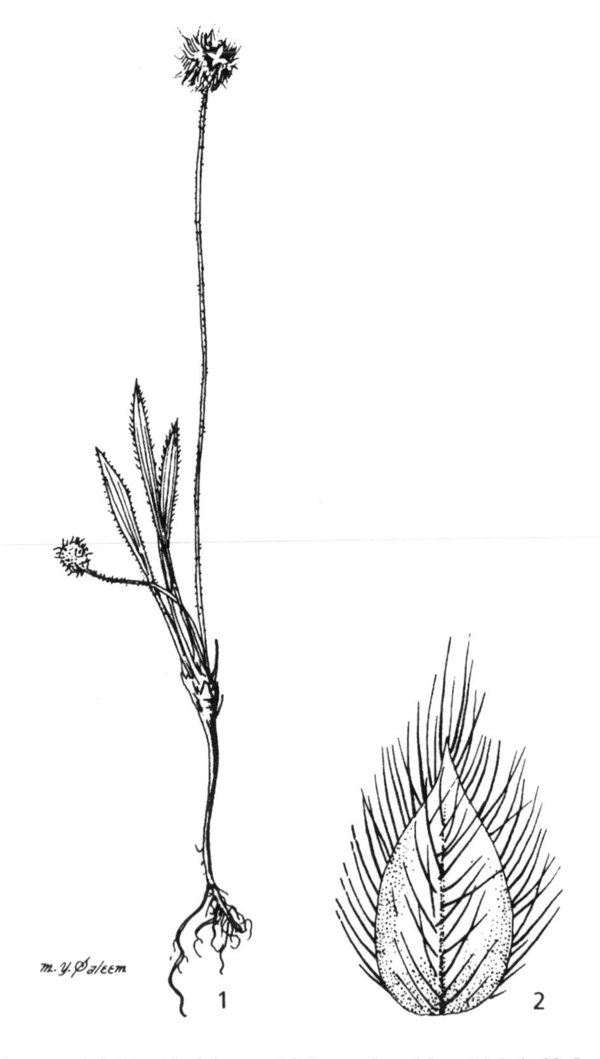

Fig. 32. **Plantago lagopus**. 1, habit × 3/4; 2, bract × 10. Reproduced from Fl. Pak. 62: fig. 2, with permission.

0.5–1.5(–6) cm; bracts ovate-lanceolate, acuminate, 2–4 mm long, villous outside, scarious except the herbaceous midrib. Calyx as long as bract but narrower, membranous except for the narrow herbaceous midrib, villous at apex and margins, the two anterior sepals united, posterior ovate. Corolla lobes with brown midrib, usually glabrous, rarely sparsely pilose, 2.0–2.5 mm, ovate-lanceolate, acute to acuminate, corolla tube glabrous. Capsule 2-seeded. Seed ovoid. Fig. 32: 1–2.

HAB. In depressions in desert by irrigation canals, damp sandy soils, by roadside in damp soil, cultivated fields, stony river banks, rocky and pebbly slopes of hills and in shade of *Quercus* forest near streams; alt. 50–2000 m; fl. & fr. Feb.–May.

DISTRIB. Frequent in the Upper Plains and Foothills Region. Less frequent in the lower Mesopotamian Region and in the Mountain Region. **MAM**: Zawita, *Emberger et mult. al.* 15389!; near Dohuk, *Emberger et al.* 15336!; Zinta Gorge, east end, *Chapman* 26116!; **MRO**: Pushtashan, NE of Rania, lower slopes of Qandil Range, *Rawi & Serhang* 26561!; **FNI**: Jabal Maqlub, Na'ib 2108!; near Eski Kellek, *Na'ib & Shakir* 2105!; **FAR**: 30 km W. of Erbi to Mosul, near Eski Kellek, *Shahwani* 25167!; **FKI**: a few km N of Tuz Khurmatu, *Polunin & Shahwani* 2106!; Injana, *Polunin & family* 2017!; Altun Kupri, *Hadač, Luckman & Waleed* 1336!; **FNI**: Pilkana, 10 km E of Khanaqin, *Rawi* 12752!; Qizil Robat, (Sadiya) *Guest* 1767!; Khanaqin, *Rechinger* 9045!; 20 km to Khanaqin, *Chalabi* 3471!; **DLJ**: Sumaicha, near Samarra, *Guest* 387!; **DGA**: Ba'quba, Diyala Liwa, *Regel* 620!; 397!; N of Diltawa, *Polunin & family* 2012!; **LCA**: near Falluja, *Guest* 963!; near Abu Ghraib, *Bharucha* 554!; **LEA**: Kut (Debouni), *Guest* 200!; Khan Bani Sa'ad, *Challabi* 1368 & Shahraban, *Chalabi* 3480!; road between Shaikh Sa'ad and Ali Gharbi, Amara Liwa, *Khudairi* 1185!; **LSM**: 30 km E of Amara, *Rawi, Tikriti & Bharucha* 29289!

NOTE. Grazed by livestock. IDHAN AS SAKHAILAH (recorded by Blakelock, loc.cit.).

S Europe, N Africa, Mediterranean, Palestine, Syria to Kuwait, Caucasus, Iran, Afghanistan to Pakistan.

8. **Plantago albicans** *L.*, Sp. Pl. 114 (1753); Ic: Bonnier, Fl. Compl. Fr., Suisse et Belg. 9: t. 500 f. 2295 (1927); Pilger in Engl. Pflanzenr. 102 (IV.269): t. 34 f. 1–7 (1937); Zohary in Pal. J. Bot. J. Ser. 1: 226 (1938) & Dep. Agr. Iraq Bull. 31: 136 (1950); Blakelock in Kew Bull. 3: 441 (1950); Zohary, Fl. Palaest. 3: 222 (1978) & pl. 378 (1977); Boulos, Fl. Egypt 3: 119 (2002).

Plantago boissieri Hausskn. & Bornmüller in Mitt. Thur. Bot. Ver., N Folg., 6: 60 (1894); Pilger in Engl. Pflanzenr. 102 (IV. 269): 345 (1937); Chalabi Ka'bi, Bull. Iraq Nat. Hist. Museum 1, no. 2: 9 (1961); Collenette, Wildflow. Saudi Ariabia: 602 (1999); *P. boissieri* Hausskn. & Bornmüller var. *tenera* Pilger in Feddes Repert. 24: 318 (1928) & in Engl. Pflanzenr. 102 (IV. 269): 345 (1937).

Annual, acaulous or short-stemmed, (5–)10–20(–30) cm tall, silky villous or more often lanate. Leaves narrowly lanceolate to linear-spatulate, 2–8(–12) × (0.1–)0.3–0.7(–1) cm, base cuneate, acute, margins entire, villous. Peduncle (3–)5–10(–16) cm, usually longer than leaves, or equal to them. Spike oblong-cylindrical or cylindrical, (0.5–)4–8(–14) cm long; bracts broadly ovate to ovate, 1–3 × 1–2 mm, glabrous or pilose or sparsely long-ciliate on the herbaceous midrib and base, long-ciliate on margins. Anterior sepals oblong-ovate, 2–3 mm, with lateral lamina at one side wide, usually glabrous, short-ciliate, the other side narrow, usually long-pilose and long-ciliate; posterior sepals elliptic with narrow midrib, the two lateral laminae equal, glabrous or pilose on midrib. Corolla glabrous, rarely the lobes pilose (with few hairs), lobes ovate-elliptic, acute, 1–2 × 0.8–1 mm. Capsule 2-locular, each locule 1-seeded.

HAB. Sandy soil in desert, plains, on dunes and low hills and on compact soil in open wadis, on sandy gravel hillsides and on barren conglomerate hills; alt. 50–500 m; fl. & fr. Feb.-May.

DISTRIB. Frequent in Desert Plateau Region, less so in the Lower Mesopotamian Region. **FNI**: 3 km E of Shahraban (Muqdadia), *Agnew* 1514!; 25 km E of Badra, *Rawi* 20773!; 16 km and 22 km SE of Badra, *Rechinger* 9219!; 92323!; **DLJ**: 70 km NW of Falluja, *Rawi* 20273!; **DGA**: 50 km E of Samarra, Asila Village, *Rawi* 20428!; Beiji, *Agnew & Na'ib* 6516!; N of Beiji, *Barkley & Brahim* 1302!; N of Lake Habbaniya, *Guest & Haider* 15996!; along E side of Wadi Thirthar, *Rechinger* 95341!; 50 km E of Samarra, *Rawi* 20441!; Shahraban, *Rechinger* 9720!; **DWD**: Wadi al-Khirr, 90 km NW by W Shabicha, *Guest, Rawi & Rechinger* 19422!; 26 km W of Ramadi, *Rechinger* 9843!; 5 km W of Karbala, *Rawi & Gillett* 10716!; Nukhaib and Shabichah, *Springfield & Al-Jaf* 21350!; near Al-Aidaha (c. 110 km SW of Al-Salman), *Guest, Rawi & Rechinger* 19136!; 15 km E by N of Busaiya, *Guest & Rawi* 14221!; **DSD**: 35 km E by N Busaiya, *Guest & Rawi* 14188!; near Najaf, *Guest, Rawi & Lang* 14003! **LCA**: Iskandariya, S of Baghdad, *Rechinger & Haines* 8281! 25 km N of Taji Camp, N of Baghdad, *Agnew, Hadač & Haines* 5763!; **LEA**: between Kut and Amara, *Bharucha & Abbas* 1957!; between Kut and Amara, *Bharucha & Abbas* 2273!; **LBA**: Jabal Sanam, in Basra Liwa, *Rechinger* 8719!; between Shaib al Batin and the railway W of Basra, *Rechinger* 8792!

NOTE. In the original manuscript *P. boissieri* Hausskn. & Bornmüller var. *tenera* Pilger was separated on its bracts being pilose in the upper part and in var. *boissieri* being glabrous to short pilose. The differences in indumentum are slight and it is difficult to distinguish the two varieties on that basis. *P. albicans* is recorded to be abundant and frequent in the Desert Plateau Region especially in the Southern Desert District; rarer in the Lower Mesopotamian Region.

Saudi Arabia, Palestine, Iraq and Iran.

9. **Plantago notata** *Lagasca*, Gen. et Spec. Plant. 7 (1816). Boiss., Fl. Orient. 4: 885 (1879); Pilger in Engl., Pflanzenr. 102 (IV. 269): 354 (1937); Zohary in Pal. J. Bot. ser. 1: 227 (1937); Zohary in Dep. Agr. Iraq Bull. 31: 137 (1950); Blakelock in Kew Bull. 3: 442 (1950); Chalabi Ka'bi, Bull. Iraq Nat. Hist. Mus. 1, no. 2: 21 (1961); Zohary, Fl. Palaest. 3: 227 (1978) & pl. 381 (1977); Boulos, Fl. Egypt 3: 117 (2002).

Plantago polystachya Viv., Pl. Aegypt. Dec. 22 (1831); Boiss., Fl. Orient. 4: 893 (1879).
P. haussknechtii Vatke in Verh. Bot. Ver. Prov. Brandenburg 16 (1874): 53 (1875). — Type: Baghdad, 1868, *Haussknecht* s.n. (type, JE); Boiss., Fl. Orient. 4: 884 (1879); Nábělek in Publ. Fac. Sci. Univ. Masaryk 105: 7 (1929); Zohary in Pal. Journ. Bot. J. Ser. 1: 227 (1938).
P. notata Lagasca var. *haussknechtii* (Vatke) Pilger, Feddes Repert. 24: 321 (1928) & in Engl. Pfl. IV. 269: 355 (1937); *P. notata* Lagasca var. *alba* Pilger, Fedde Repert. 18: 461 (1932) in Engl. Pfl. IV. 269: 355 (1937).
P. phaeopus Payne in Proc. Palest. Expl. Soc. App. 3, 120 (1875); Boiss., Fl. Orient. 4: 884 (1879); Zohary in Pal. Journ. Bot. J. Ser. 1: 227 (1938).

Annual, acaulous, 2–10(–20) cm tall, tufted, hirsute at base. Leaves linear or narrowly linear in outline, 2–8(–17) × 0.1–0.3(–1) cm, glabrescent to sparsely pilose then with tuft of hairs at base of lobes, but usually pilose-villous, entire or mostly furnished with linear acute lobes, lobes 1–8(–15) mm, glabrescent or usually sparsely pilose, rarely pilose-villous. Peduncle thin, ascending or erect, usually equal to leaves or longer, rarely shorter, long-pilose. Spike dense, globose-oblong or short-cylindrical, 0.5–1.5(–3) cm; bracts ovate-orbicular, usually broader than long, ± 3 × 3–4 mm, obtuse, scarious except for herbaceous midrib, long-pilose or lanate, margin usually long-ciliate. Sepals rotund-elliptic, obtuse, scarious with short midrib, pilose or fleecy at base, otherwise glabrescent, ciliate at apex. Corolla lobes rotund to broadly ovate, 1.5–2.2 mm, obtuse, usually brown-veined on the throat (the base of the lobe), tube a little longer than lobes. Capsule 2-seeded. Seeds elliptic.

HAB. Wadi beds, depressions in deserts and at margins of desert, amongst ruins, sand and limestone hills in desert, north slope of mounds, banks of irrigation channels and on damp loamy and clay soils on plains; alt. 10–230 m; fl. Feb.-May.
DISTRIB. Frequent in the Desert Plateau Region, the Lower Mesopotamian Region and the Upper Plains. **FKI**: Tuz-Kifri, *Guest* 408!; **DLJ**: SE of Ana, *Rawi & Gillett* 6940!; 3 km W of Thirthar, *Rawi & Gillett* 7140!; Hadhr, Jazira, *Agnew & Hadač* 5198!; Wadi Bogaa, Jazira, *Barkley & Brahim* 4206!; ruins of Hadhr, *Barkley & Brahim* 4134!; **DGA**: N of Deltawa, *Guest* 509!; **DWD**: 80 km W of Karbala, *Chakravarty & Rawi* 29821!; Nukhaib, *Springfield & Al-Jaf* 21366A!; Sharqat, *Guest* 399!; As Sudur, *Rawi* 20018A!; 20 km S of Ana, *Agnew & Na'ib* 6606!; between Nukhaib and the Saudi Arabian border, *Rawi & Barucha* 1909!; 1949!; Rutba, *Springfield & Al-Jaf* 21312!; Falluja, desert, *Agnew* 4996!; **DSD**: Shabicha, *Rawi & Gillett* 6297!; 87 km from Busaiya on road to Salman, *Chakrovarty, Rawi, Khatib & Tikriti* 29993!; Samawa-al-Salman, *Rawi & Gillett* 6115!; **LCA**: near Baghdad, *Lawand & Hashimi* 2840!; N of Diyala Bridge, S of Baghdad, *Hadač, Agnew & Haines* 180!; Birs Nimrud, Hilla Liwa, *Agnew* 5190!; Aqqur Quf NW of Baghdad, *Hadač* 493!; Abu Ghraib Farm, Baghdad Liwa, *Agnew* 1388!; Salman Pak, *Su'ud al Dhahir* 1947!; N of Khalis, *Polunin* 2010!; **LEA**: 60 km N of Amara, *Guest, Rawi & Rechinger* 17488!; 15 km from Baghdad, on Kut Road, *Haines* 2636!; Muqdadiya, *Guest* 13165!
NOTE. Grazed by livestock. IDHAN AS SAKHAILAH (recorded by Blakelock, loc.cit.).

Spain and Algeria through Mediterranean region to Cyprus, Caucasus and Iran.

10. **Plantago ovata** *Forssk.*, Fl. Aegypt.-Arab. 31 (1775); Pilger in Engl., Pflanzenr. 102 (IV. 269): 347 (1937); Hook. f., Fl. Brit. Ind. 4: 707 (1885); Zohary in Dep. Agr. Iraq Bull. 31: 442 (1950); Blakelock in Kew Bull. 3: 442 (1950); Chalabi Ka'bi in Bull. Iraq Nat. Hist. Mus. 1 (2): 18 (1961); Patzak & Rech. f. in Fl. Iranica 15:15 (1965); Kazmi in Fl. Pak. 62: 14 (1974); Zohary, Fl. Palaest. 3: 222 (1978) & pl. 380 (1977); Collenette, Wildflow. Saudi Ariabia: 604 (1999); Boulos, Fl. Egypt, 3: 210 (2002); Li, Wei & Hoggard in Fl. China 17: 496 (2011).

Plantago decumbens Forssk., Fl. Aegypt.-Arab. 30 (1775).
P. ispaghula Roxburgh ex Fleming, Asiat. Res. 11: 174 (1810).

P. trichophylla Nábělek in Publ. Fac. Sci. Univ. Masaryk) 105: 6 (1929); Chalabi Ka'bi in Bull. Iraq Nat. Hist. Mus. 1, no. 2: 20 (1961).
P. ovata Forssk. var. *decumbens* (Forssk.) Zohary, Palest. J. Bot. Jerusalem 1: 227 (1938).

Annual, acaulous or short-stemmed, 3–10(–16) cm tall, branched at base, white villous-lanate. Leaves linear to narrowly lanceolate, 3–5 × 0.1–0.6 cm, 3-veined, base attenuate tapering into the petiole, margins entire or remotely denticulate, apex acute, densely to sparsely white villous. Peduncle as long as leaves or longer, rarely shorter than leaves, arcuate-ascending or decumbent, rarely erect, 1–9(–12) cm, villous. Spike globose to ovoid to narrowly ovoid, 0.5–2.5(–3) cm, densely flowered; bracts glabrous, broadly ovate, 2–3 × 3–4 mm, obtuse, scarious except for the herbaceous midrib, lateral lamina much wider than midrib. Calyx 2.5–3 mm, glabrous, scarious except midrib; anterior sepals obovate-elliptic, posterior sepals elliptic. Corolla white, lobes broadly ovate or orbicular-cordate, 2–3 × 2 mm, acute, glabrous. Capsule 2-seeded. Seeds ovoid. Fig. 33: 1–2.

HAB. In desert and wadi beds, on sandy gypsophilous and calcareous soils; on plateau in sandy patches, limestone and conglomerate hillsides and hard silty plains; alt. 10–150 m; fl. & fr.: Feb.–Jun.
DISTRIB. Common in the Desert Plateau Region and lower Mesopotamian Region; frequent in Upper Plains and Foothills Region. **FAR**: Arbil Road, about 7 km NW of Altun Kupri, *Polunin* 2025!; **FKI**: Jabal Hamrin, *Guest* 1831!; Jabal Maqlub, near Beiji, *Guest et al.* 13223!; Kirkuk, *Rawi & Gillett* 1481!; Injana, *Polunin et al.* 2023!; **FNI**: Badra, *Gillett* 6661!; Pilkana, 10 km E. of Khanaqin, *Rawi* 12755!; near Mandali District, *Rawi* 12535!; 16 km S of Badra, *Rechinger* 9218!. **DLJ**: 4 km W of Thirthar, *Rawi & Gillett* 7134!; Jazira, 25 km NNW of Falluja, *Guest & Rawi* 13664!; Hadhr, *Agnew & Na'ib* 6536!; **DGA**: 4 km E of Samarra, *Rawi* 20334!; 25 km N of Taji Camp in Baghdad Liwa, *Agnew, Hadač & Haines* 5767!; **DWD**: 260 km NW of Ramadi, *Rawi* 20992!; 65 km N of Rutba, *Rawi & Khatib* 32271!;12 km N of road to Husbah, *Chakr., Rawi, Khatib & Izzi* 31736!; Falluja, *Guest* 958!; 5 km W of Rutba, *Rawi & Nuri* 27072!; Nukhaib, *Springfield, Rawi & Al-Jaf* 21366!; Wadi al-Khurr (90 km NW by W of Shabicha), *Guest, Rawi & Rechinger* 19424!; 20 km S of Ana, *Agnew & Na'ib* 6603!; Imhafer to Rutba, *Rawi* 13792!; between Najaf and Karbala, *Rawi* 20153!; Abu Ghraib, *Rawi* 16499!; **DSD**: Shabicha, *Rawi & Gillett* 6178!; 65 km W.

Fig. 33. **Plantago ovata.** 1, habit × 3/4; 2, bract × 6. Reproduced from Fl. Pak. 62: fig. 2, with permission.

by N Busaiya, *Guest, Rawi et al.* 14136!; 62 km WNW of Ansab (c. 130 km SSW of Salman), *Guest, Rawi & Rechinger* 19080!; 100 km WSW of Basra, near Safai Al-Maghif, *Guest, Rawi & Rechinger* 17282!; N of Salman, *Rawi & Al-Jaf* 21431!; 6 km S of Salman, *Rawi, Khatib & Tikriti* 29215!; 10 km S by W Busaiya, *Guest et al.* 15262!; **LCA**: near Hilla, *Roess* 439!; Iskandariya, *Rawi & Walter* 18370!; Imam Ibrahim, near Suwaira, *Hadač* 1689!; Babylon, *Agnew* 1391!; **LEA**: 70 km E of Amara, *Rawi & Hadad* 25781!; 55 km N of Amara, *Guest, Rawi & Rechinger* 17458!; Diltawa road, near police station, *Chalabi* 2204!; c. 30 km S E of Badra, *Rechinger* 9149!; **LBA**: the new Athel of Zubair, *Rechinger* 8519!; between Umm Qasr and Safwan, on Kuwait border, *Rechinger* 8706!; Zubair, *Mustafa* 3162!; WNW of Umm Qasr, *Guest, Rawi & Rechinger* 16926!

NOTE. Chalabi recognized *P. trichophylla* Nábělek & subsp. *decumbens* Chalabi as separate taxa in the original manuscript, but said that many specimens of both taxa examined in Iraq integrated with *P. ovata* and were difficult to tell apart. Blakelock (op. cit., 444) considered *P. trichophylla* a depauperate form of *P. ovata*. I have placed both (*P. trichophylla* and *P. decumbens*) under *P. ovata*.

The seeds are used medicinally, and the plant is recorded as spring fodder in the Iraq-Nejd border areas. RIBLAH (recorded by Blakelock, loc.cit.).

Mediterranean region, southwest Asia to Pakistan, N America, Atlantic Islands.

11. **Plantago loeflingii** L., Sp. Pl. 115 (1753); Boiss., Fl. Orient. 4: 883 (1879); Pilger in Engl., Pflanzenr. 102 (IV. 269): 352 (1937); Zohary in Dep. Agr. Iraq Bull. 31: 136 (1950); Blakelock in Kew Bull. 3: 442 (1950); Chalabi Ka'bi in Bull. Iraq Nat. Hist. Mus. 1(2): 15 (1961); Patzak & Rechinger in Fl. Iranica 15: 16 (1965); Kazmi in Fl. Pak. 62: 16 (1974); Tutel in Fl. Turk. 7: 516 (1982). Incl. *P. loeflingii* L. var. *caspia* Fisch. & Mey. in Ind. Sem. Horti Petrop. 111: 46 (1836); Pilger in Engl., Pflanzenr. 102 (IV. 269): 353 (1937).

Annual, acaulous, 2–8(–20) cm tall. Leaves linear or narrowly lanceolate, 2–8(–12) cm × 0.1–0.6(–1) mm, entire or remotely long-denticulate, pilose to velutinous. Peduncle usually equal to but also longer or shorter than leaves, pilose or villous. Spike 0.4–2 cm long, globose to or ovoid-oblong; bracts equal to sepals, broadly ovate to broadly elliptic, 2 × 3 mm, concave and somewhat keeled, obtuse, midrib narrow, herbaceous protruding at tip, glabrous. Sepals ovate-orbicular, 2 × 1–2 mm, glabrous, thin membranous except for the short basal somewhat herbaceous midrib. Corolla glabrous, lobes narrowly lanceolate or elliptic, 1.5 × 0.5 mm, acute. Capsule 2-seeded. Seeds ovoid.

HAB. Fallow fields and grasslands, ditches, depression in desert area, muddy river banks, cultivated areas, and in wooded areas at foot of hills, on clay, alluvial and light sandy soils; alt. 30–350; fl. & fr. Feb.–May.

DISTRIB. Frequent in Upper Plains and Foothills Region and in the Eastern Alluvial Plain District; less frequent in other Districts of the Lower Mesopotamian Region. **FNI**: Mosul road, 1.5 km before Eski Kellek, *Shahwani* 25153!; Shaqlawa road bridge, Bastura Chai, *Polunin* 2029!; **FUJ**: 48 km NW of Beiji, *E. & F. Barkley* 5411!; **FAR**: W face of Mountain below Salah ad-Din, *Barkley & Haddad* 5738!; **FKI**: 10 km N of Altun-Kupri, *Guest* 527A!; Jabal Hamrin, *Gillett & Rawi* 10245!; Beiji, *Agnew* 5185!; Gau Gossik on r. Zab, *Rawi & Gillett* 10514!; Arbil road ± 7 km NW of Altun Kupri, *Polunin* 2019!; S of Kirkuk, *Barkley* 4655!; **DGA**: As-Sudor, *Rawi* 20018A!; N of Diltawa (Khalis), *Polunin* 2011!; near Diltawa, *Guest* 527!; Diltawa (Khalis) road, *Chalabi-Ka'bi* 2203!; **LCA**: Latifiya, S of Baghdad, *Guest* 1681!; **LEA**: Hafariya 60 km SE of Baghdad, *Rechinger* 9089!; road between Shaikh Sa'ad and Ali Gharbi, *Khidairi* 1183!

Portugal, Mediterranean region, through Turkey and Iran to Pakistan.

12. **Plantago ciliata** *Desf.*, Fl. Atlant. 1: 137, t. 39 f. 3 (1758). Boiss., Fl. Orient. 4: 887 (1879); Pilger in Engl., Pflanzenr. 102 (IV. 269): 357 (1937); Zohary in Dep. Agr. Iraq Bull. 31: 137 (1950); Blakelock in Kew Bull. 3: 441 (1950); Chalabi Ka'bi in Bull. Iraq Nat. Hist. Mus. 1(2): 24 (1961); Patzak & Rechinger in Fl. Iranica 15: 18 (1965); Kazmi in Fl. Pak. 62: 17 (1974); Zohary, Fl. Palaest. 3: 227 (1978) & pl. 382 (1977); Collenette, Wildflow. Saudi Ariabia: 602 (1999); Boulos, Fl. Egypt 3: 119 (2002).

Plantago ciliata Desf. var. *lanata* Boiss., Fl. Orient. 4: 887 (1879); Pilger in Engl., Pflanzenr. 102 (IV. 269): 358 (1937).

P. ciliata Desf. var. *angustifolia* Pilger in Feddes, Repert. 18: 467 (1922) & in Engl. Pfl. IV. 269: 338 (1937).

P. ciliata Desf. subsp. *lanata* (Boiss.) Rech.f., Fl. Iranica 15: 18 (1965).

Annual, acaulous or short-stemmed, branched from base, 2–10(–18) cm tall, densely silvery-silky hairy. Leaves rosulate, obovate to lanceolate-spatulate, 3–6 × 0.7–1.5 cm, base tapering abruptly into a long petiole, acute, margins usually entire. Peduncles stout, about as long as leaves. Spike ovoid to oblong-cylindrical, 0.5–3.5(–4) cm long; bracts concave, ovate, ± acute, midrib herbaceous, broader than the membranous lateral lamina, long-villose, as long as sepals. Sepals ovate-elliptic, 3–3.5 mm, villous-ciliate, anterior sepals with midrib broader than lateral lamina, posterior sepals with midrib narrower than lateral lamina. Corolla lobes narrowly ovate-lanceolate, 2–3 × 0.5–1 mm, acute, ciliate, pilose on the outside. Capsule 2-seeded. Seeds glossy. Fig. 34: 1–6.

HAB. In loamy and silty soils, sand dunes, in gypsum quarry in desert, gravel or sandy soil in depression or desert wadis; alt. 10–140 m; fl. & fr. Feb.–June.

DISTRIB. Frequent in the Desert Plateau Region; also found in the Lower Mesopotamian Region. **FKI**: Jabal Hamrin, Sudur, *Agnew & Abdul-Wahab* 6503!; **DLJ**: 5 km SW. of Lake Thirthar, *Barkley, Abbas & Inman* 1717!; Jazira, Wadi Thirthar canal, near Samarra, *Guest, Rawi & Haider* 18836!; 70 km NNW of Falluja, *Rawi* 20266!; near Lake Thirthar, *Barucha & Abbas* 1903!; **DWD**: 90 km N of Shabicha, *Rawi* 14829!; 35 km W of Mukhaib, *Rawi & Chalabi* 1387!; 36 km SW of Shithatha, *Barkley & Abbas* 1947!; 33 km NE of Nukhaib, *Barkley & Abbas* 1994!; Ukhaider (c. 50 km SWS of Karbala), *Guest, Rawi & Lang* 14012!; 100 km W of Ramadi, *Hadač* 3006!; on Falluja road, W of Abu Ghraib, *Polunin, Haines, Rawi, Kittani et al.* 90!; 110 km E. of Nukhaib, *Agnew, Haines & Hashimi* 5468!; 80 km W of Shabicha, *Chakrovarty et al.* 30063!; **DSD**: c. 15 km E by N Busaiya, *Guest & Rawi* 14208!; Al-Batin, SW of Basra, *Rawi, Guest & Rechinger* 17076!; 99 km NE of Shabicha, *Rawi, Khatib & Tikriti* 29266!; ± 6 km W of Safwan, *Guest, Rawi & Rechinger* 16948!; 33 km WNW of Ansab (c. 135 km S by W. of Salman), *Guest, Rawi & Rechinger* 19036!; between Safwan and Shaib al-Batin, *Rechinger* 8742!; Al-Aidaha (± 110 km SW of Salman), *Guest, Rawi & Rechinger* 191441!; 10 km SE by S of Jiraibiyat (± 125 km SW by S of Basra), *Guest, Rawi & Rechinger* 17160!; between Shaib al-Batin and the railway W of Basra, *Rechinger* 8816!; **LCA**: 25 km N of Taji Camp, Baghdad Liwa, *Agnew, Hadač & Haines* 5766!; **LEA**: between Kut and Amara, *Bharucha & Abbas* 1956!; between Kut and Amara, *Bharucha & Abbas* s.n.!; **LBA**: Umm Qasr (± 40 km SE by S of Zubair), *Guest, Rawi & Rechinger* 16909!

NOTE. Chalabi recognized var. *lanata* Boiss. and var. *angustifolia* Pilger in the original manuscript. Var. *angustifolia* was separated on its narrowly lanceolate leaves and more hairy corolla lobes and var. *lanata* on its pilose bracts and marginally broader anterior sepals. Rechinger (loc. cit.) raised the rank of var. *lanata* to subspecies with a note to say that the type subsp. *ciliata* is found from North Africa west to Sinai, and subsp. *lanata* from Syria and Saudi Arabia to Pakistan. Both taxa occur in Iraq, and according to Chalabi var. *lanata* is as common as the type variety, and morphologically not easy to distinguish from it.

North Africa east to Saudi Arabia, Syria and Iran to Pakistan.

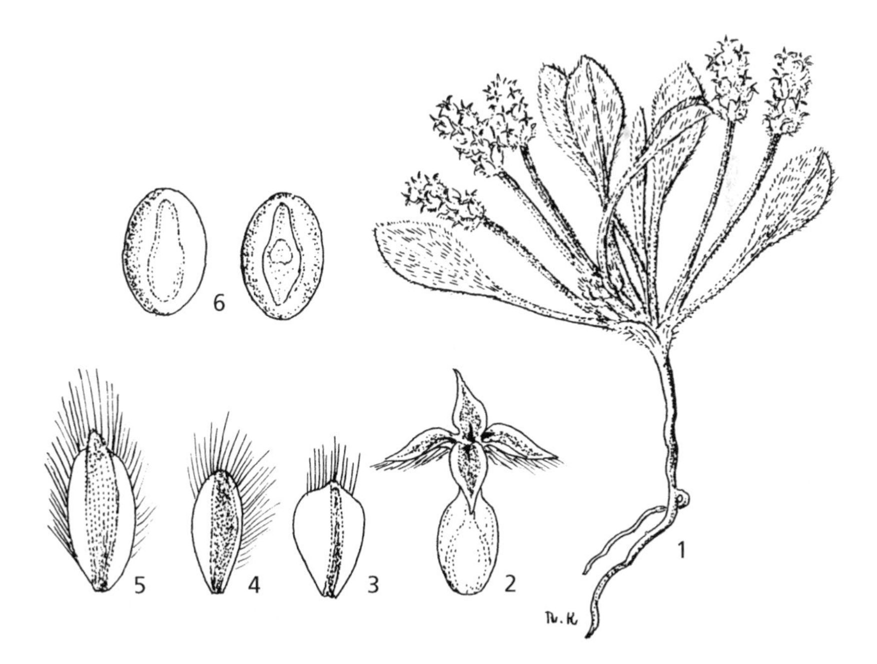

Fig. 34. **Plantago ciliata**. 1, habit × 1; 2, flower × 5; 3–5 sepals × 5; 6, seed (both sides) × 8. From Naomi Feinbrun-Dothan, *Flora Palaestina*, Part Three: *Ericaceae* to *Compositae*, Plates, Jerusalem: The Israel Academy of Sciences & Humanities, 1977, pl. 382. © The Israel Academy of Sciences and Humanities. Reproduced with permission.

13. **Plantago psammophila** *Agnew & Chalabi Ka'bi* in Notes R.B.G. Edinb. 25 (1): 56 (1963). — Type: Iraq, 25 km N of Taji Camp, Jazira, *Agnew* 5048 (BUH, holo.; BM, E, K!, iso.); Patzak & Rech.f. in Fl. Iranica 15: 19 (1965); Collenette, Wildflow. Saudi Ariabia: 604 (1999).

Annual, acaulous or mostly short-stemmed, 5–20 cm tall, densely hairy. Leaves alternate, linear-lanceolate, 3–6 × 0.5–1.5 cm, acute, base gradually attenuate to petiole. Spike 1–10 cm, erect, oblong-cylindrical; bracts ovate-oblong, 4–5 × 2–5 mm, obtuse, dorsally tomentose, ciliate, midrib herbaceous, broader than the membranous lateral lamina. Sepals as long as bract or shorter, ovate-oblong, obtuse, dorsally tomentose, margin and apex long-ciliate, midrib herbaceous, as broad as membranous lateral lamina or narrower. Corolla lobes broad-lanceolate, 3 × 1.5 mm, acuminate, dorsally densely hairy, ± glabrous on margin. Capsule ovoid, 2-seeded. Seeds cymbiform.

HAB. Sandy clay soil, sandy desert, sandy depression, sandy wadi, associated with *Haloxylon salicornicum* in sandy desert, in gravel lands and in soft sandy ridge with large pebbles; alt. 15–130 m; fl. & fr.: Mar.–May.

DISTRIB. Frequent in the Desert Plateau Region, also found in the Lower Mesopotamian Region and on the edge of the Desert Plateau. **DLJ**: 25 km N of Taji Camp, Jazira, *Agnew* 5048! (type); 60 km N of Taji Camp, *Agnew & Hadač* 5209!; **DSD**: Um Qasr, *Guest, Rawi & Rechinger* 16909!; 60 km SW by W of Zubair, *Guest & Rawi* 14310!; Zubair, *Haines* 1231!; 10 km SE of Busaiya [Bussiya], *Rawi* 26005!; 33 km S of Samawa, *Rawi & Gillett* 6522!; Al Baniyah, 65 km WSW of Basra, *Guest, Rawi & Rechinger* 17327A!; **DWD**: Nukhaib, *Rawi* 15672!; 135 km NW of Ramadi, *Rawi* 21278!; **LCA**: 5 km W of Tel Ibrahim, *Barkley & Bullard* 4108!; between Musaiyib and Zuwaira, *Hadač, Agnew & Haines* 1645!; Falluja desert, *Haines* W204.

A regional endemic found in the desert areas of Iraq and Saudi Arabia.

14. **Plantago bellardii** *All.*, Fl. Pedem. 1: 82, t. 85, f. 3 (1785); Boiss., Fl. Orient. 4: 884 (1879); Pilger in Engl., Pflanzenr. 102 (IV. 269): 411 (1937); Zohary in Dep. Agr. Iraq Bull. 31: 137 (1950); Blakelock in Kew Bull. 3: 441 (1950); Chalabi Ka'bi in Bull. Iraq Nat. Hist. Mus. 1(2): 16 (1961); Patzak & Rech.f. in Fl. Iranica 15: 19 (1965); Tutel in Fl. Turk. 7: 517 (1982); Zohary, Fl. Palaest. 3: 228 (1978) & pl. 384 (1977).

(Incl. var. **deflexa** *Pilger* in Feddes Repert. 18: 474 (1922) & Engl. Pflanzenr. 2 (IV. 269): 415 (1937); Patzak & Rech.f. in Fl. Iranica 15: 19 (1965) as subsp. *deflexa* (Pilger) Rech.f.)

Annual, acaulous, (2–)5–10 cm tall; root fibrous, filiform. Leaves erect, linear or linear-lanceolate, or linear-spatulate to oblanceolate, 5–8(–12) × 0.3–0.4(–0.6) cm, long thin pilose or yellow silky villous, usually entire or at upper part few-denticulate. Peduncle usually erect or arcuate-ascending, equal to or slighty shorter than leaves, (0.5–)2–5(–7) cm, pilose. Spike ovoid-globose or oblong, dense, 1–2(–4) cm long; bracts herbaceous, in lower flowers equal to calyx or usually longer, 4–7 mm, those at base ovate, then becoming long narrow or linear-acuminate, midrib well formed, long-pilose and with long-ciliate margins, lateral lamina narrower than midrib; bract of upper flowers of spike shorter, ± 4 mm, sparsely short-pilose. Anterior sepals lanceolate-ovate, 3.5–4.5 mm (longer than posterior sepals), acute, with midrib broader than lateral lamina and many-veined, long-pilose; posterior sepals elliptic, ± 3 mm, midrib as broad as lateral lamina or narrower. Corolla lobes lanceolate-ovate, 1.5–2 mm, acute or acuminate, reflexed, glabrous. Capsule globose, 2 mm long, 2-locular, each 1-seeded. Seeds oblong-ellipsoid, reticulate, deeply sulcate at middle. Fig. 35: 1–4.

HAB. Upland and hillsides, limestone hill, open grasslands, along roads, cultivated wheat fields, river banks and sides of streams, dry stony pastures, in sandy, gravelly and loamy soils; alt. 20–250 m; fl. & fr. Feb.–Mar.

DISTRIB. Common in the Upper Plains and Foothills Region; scarce in the Lower Mesopotamian Region. **FNI**: Mosul to Atrush road, *Broadbume* 93!; Jabal Maqlub, Mosul Liwa, *Agnew* 1462!; 1.5 km before Eski-Kellek, *Agnew* 1908!; *Agnew & Polunin* 2026!; Jabal Hamrin, *Rawi & Gillett* 10258!; Kirkuk, *Guest* 1366!; 25 km S of Kirkuk, *Rawi & Gillett* 10347!; Tauq Bridge, *Polunin* 2027!; Jabal Hamrin, middle range, *Haynes* 721!; Jabal Muwaila, near Kumait, ± 70 km N of Amara, *Guest, Rawi & Rechinger* 17550C!; **FAR**: 10–15 km from Arbil, on road to Darband, *Shahwani* 25280!; **FKI**: 7 km NE of Kirkuk, *Rawi, Al-Kuss & Nuri* 27928!; Kirkuk, *Rogers* 215B!; Adhaim river, Jabal Hamrin, *Agnew* 5110!; Tuz Khurmatu, *Agnew, Hadač & Al-Hashimi* 5440!; **FNI**: Qara Los, near Mandali, *Rawi* 12636!; Jabal Darawishka, near Khanaqin, *Guest* 13970!; Pilkana, 10 km E of Khanaqin, *Rawi* 12753!; 3 km SE of Badra, *Rechinger* 9204!; Jabal Darawishka, near Khanaqin, *Guest* 1789!; Badra, *Gillett* 6617!; Zorbatiya, *Gillett* 6697!; **LCA**: Sechir, E Falluja, *Hadač &*

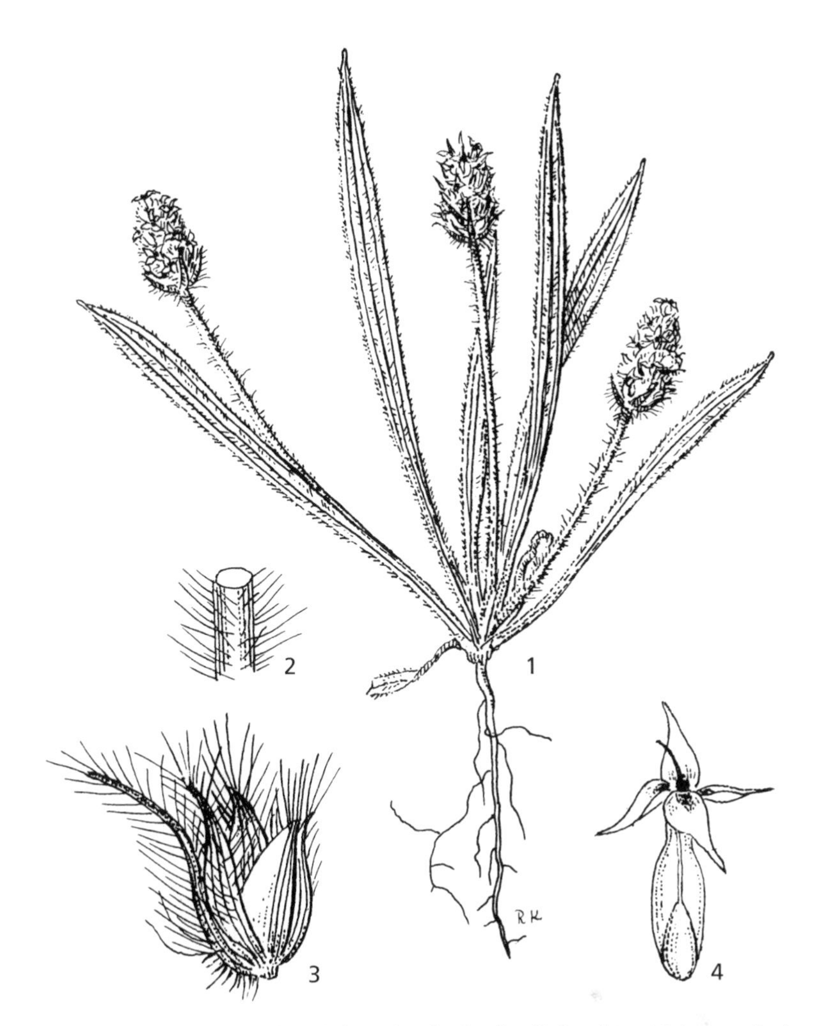

Fig. 35. **Plantago bellardii**. 1, habit × 1; 2, peduncle detail × 5; 3, calyx with bract × 5; 4, corolla × 5. From Naomi Feinbrun-Dothan, *Flora Palaestina*, Part Three: *Ericaceae* to *Compositae*, Plates, Jerusalem: The Israel Academy of Sciences & Humanities, 1977, pl. 384. © The Israel Academy of Sciences and Humanities. Reproduced with permission.

Agnew 513!; LEA: low foothills near Wadi Tib police station, Amara Liwa, *Rechinger* 8897!
NOTE. A common and widespread species. Pilger recognized var. *deflexa* Pilger (Feddes Repert. 18: 474. 1922) based on smaller plants with smaller leaves, shorter peduncles and spikes usually less than 1 cm long. Both taxa are found in Iraq in the Upper Plains and Foothills Region, with var. *deflexa* more abundant in the Lower Mesopotamian Region in the Eastern and Central alluvial plains Districts.

Mediterranean region, and from Turkey south to Jordan and eastwards to SW Iran.

15. **Plantago cretica** *L.*, Sp. Pl. 114 (1753) and ed. 2: 165 (1762); Boiss., Fl. Orient. 4: 884 (1879); Pilger in Engl., Pflanzenr. 102 (IV. 269): 410, t. 41 f. 9 (1937); Blakelock in Kew Bull. 3: 441 (1950); Chalabi Ka'bi in Bull. Iraq Nat. Hist. Mus. 1 (2): 17 (1961); Zohary, Fl. Palaest. 3: 228 (1978) & pl. 383 (1977); Tutel in Fl. Turk. 7: 516 (1982). (Incl. f. *albido-pilosa* Pilger in Engl., Pflanzenr. 102 (IV. 269): 411 (1937).

Annual, acaulous, 5–15 cm tall, villous, especially peduncles and flowers; roots filiform. Leaves linear-lanceolate or lanceolate to spatulate-lanceolate, (2–)4–12 × (0.2–)0.5–0.7 cm, apex obtuse, base gradually narrowed, margins usually entire or with a few short teeth, hirsute or villous, at base lanate. Peduncle much shorter than leaves, 1–2.5 cm, arcuate-reflexed, lanate-villous. Spike globose to ovoid-globose, 8–12 mm long; bract elliptic-lanceolate, acute or shortly acuminate, a little concave, gradually narrowed to apex, ± 6 mm long in the lower flowers of spike, shorter in upper flowers, 4–4.5 mm, and not narrowed to apex, long yellow-pilose, hirsute, or short rigid-pilose, lateral lamina narrower than the distinct midrib. Sepals 3.5–4 mm long, anterior elliptic, obtuse, unequal, with broad herbaceous midrib, and narrow thin scarious lateral lamina, long pilose-villous, posterior sepals broadly ovate-elliptic, membranous, midrib narrower than lateral lamina, upper margin and midrib sparsely rigid-pilose. Corolla lobes broad, orbicular-ovate, 2–2.5 × 1.2–1.5 mm, without veins, reflexed, acute or shortly acuminate, corolla tube 3 mm long, glabrous. Apex of anthers broad, membranous. Capsule globose, 2-locular, each locule 1-seeded. Seeds ovoid. Fig. 36: 1–8.

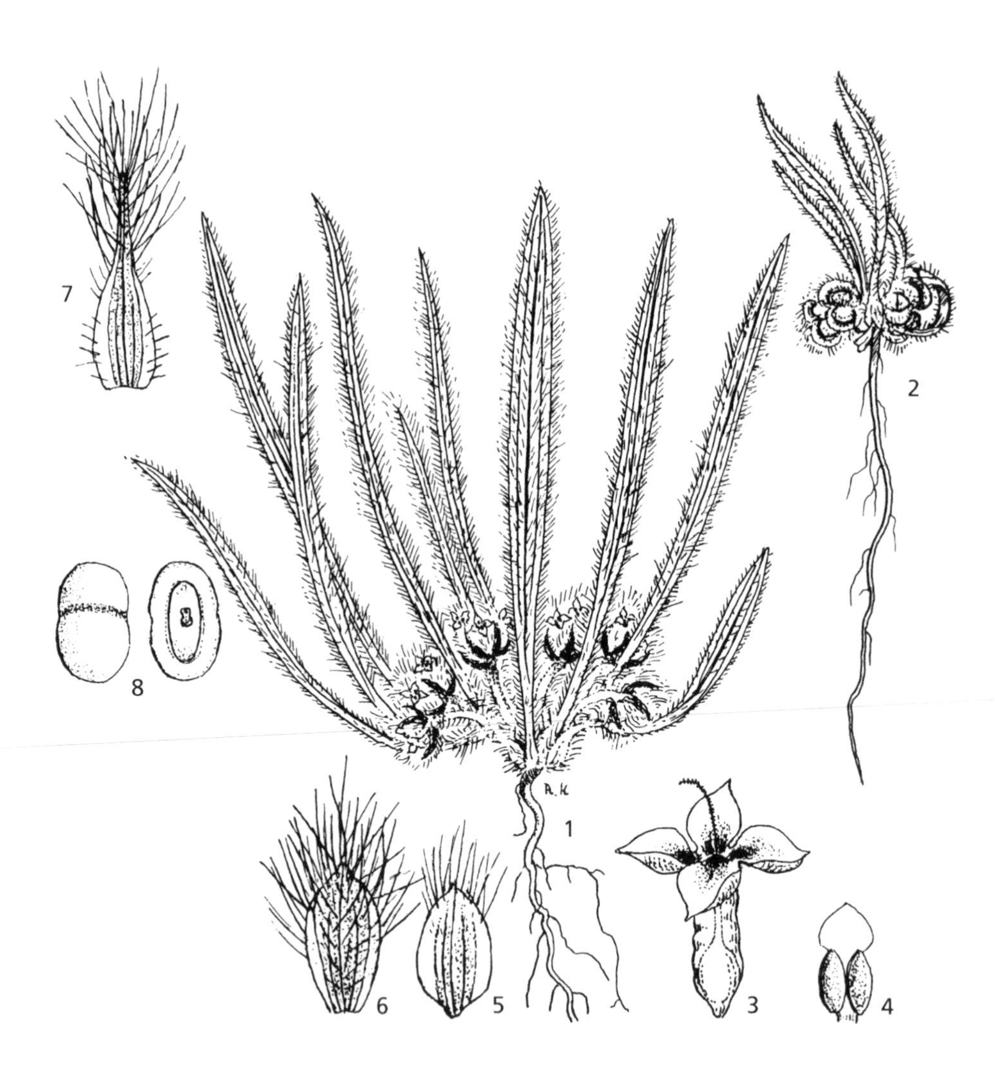

Fig. 36. **Plantago cretica.** 1, 2, habit (in flower and fruit) × 1; 3, flower (showing corolla) × 5; 4, anther × 8; 5, 6, calyx lobes × 5; 7, bract × 5; 8 seed (dorsal and ventral view) × 8. From Naomi Feinbrun-Dothan, *Flora Palaestina*, Part Three: *Ericaceae* to *Compositae*, Plates, Jerusalem: The Israel Academy of Sciences & Humanities, 1977, pl. 383. © The Israel Academy of Sciences and Humanities. Reproduced with permission.

HAB. Fallow land, clay soil between rocks on mountain slopes; alt. ± 450 m; fl. ?Apr.–Jun.
DISTRIB. Upper Plains and Foothills Region and in the Mountain Region. **MSU**: Chewa Rash, NE of Rania, *Rawi, Nuri & Kass* 28445!; **FNI**: Ain Sifni, *Effendi* 2561!
NOTE. Recorded to be good fodder for goats (Blakelock, loc. cit.).

Greece, Cyprus and Turkey east to Syria, Lebanon, Palestine and eastern Iraq.

16. **Plantago afra** *L.*, Sp. Pl. ed. 2: 168 (1762); Pilger in Engl., Pflanzenr. 102 (IV. 269): 422 (1937); Boiss., Fl. Orient. 4: 691 (1879); Verdc., Kew Bull. 23: 509 (1969) & in FTEA Plantag.: 6 (1971); Kazmi in Fl. Pak. 62: 19 (1974); Zohary, Fl. Palaest. 3: 231 (1978) & pl. 389 (1977); Tutel in Fl. Turk. 7: 519 (1982); Collenette, Wildflow. Saudi Ariabia: 601 (1999); Boulos, Fl. Egypt 3: 113 (2002).

Plantago psyllium L., Sp. Pl. ed. 2: 167 (1762), non L. (1753); Bonnier, Fl. Compl. Fr., Suisse et Belg. 9: t. 498 f. 2288 (1927), as *P. psyllium*; Boiss., Fl. Orient. 4: 891 (1879); Pilger in Engl., Pflanzenr. 102 (IV. 269): t. 43A f. 1–9 (1937), as *P. psyllium*; Zohary in Dep. Agr. Iraq Bull. 31: 137 (1950); Chalabi Ka'bi in Bull. Iraq Nat. Hist. Mus. 1 (2): 27 (1961); Patzak & Rech.f. in Fl. Iranica 15: 21 (1965).

P. stricta Schousb., Jagtt. Vextr. Marokko 69 (1800); Boiss., Fl. Orient. 4: 891 (1879); *P stricta* Schousb var. *divaricata* (Zucc.) Barneoud, Monogr. Plantag. 48, n. 105 (1845); *P. stricta* var. *stricta* (Schousb.) Maire in Jah. & Maire, Cat. Pl. Maroc 3: 706 (1934); Pilger in Engl., Pflanzenr. 102 (IV. 269): 424 (1937).

Annual, 2–15(–25) cm tall, stem distinct, erect or ascending, elongated and branched or branched at base, glandular, hirsute and pubescent or short-pilose; internodes (2–)3–6(–7) cm long. Leaves opposite, with fascicles in axils hence appearing whorled, narrowly linear to linear-lanceolate, 10–40(–50) × (0.5–)1–2 mm, entire or bidentate, flat or with somewhat recurved edges, short-pilose, hirsute or pubescent, ± glandular. Spike round-ellipsoid or shortly ovoid-subglobose, (3–)6–20 mm long; bract of lower flowers ovate-lanceolate, long- or short-acuminate, 3–5 mm, or with long point up to 8 mm long, midrib herbaceous, gradually narrowed to a point, lateral lamina distinct in base of bract, but bracts of the upper flowers of spike lanceolate, short-acuminate or acute, 3–4 mm long, short-pilose or hispid, pubescent or not, margins ciliate. Sepals lanceolate to narrowly ovate, 3–3.5(–4) mm, midrib distinct, herbaceous, lateral lamina scarious, sparsely short-pilose, pubescent or not, margins ciliate. Corolla lobes narrowly ovate-lanceolate, 2–2.5 mm, acute, glabrous; corolla tube glabrous, 3.5–5 mm, rugulose. Anthers slightly exserted. Capsule oblong-subglobose, 2-locular, each locule 1-seeded. Seeds cymbiform, glabrous. Fig. 37: 1–2.

HAB. Cultivated and waste ground, fallow fields on silty or sandy soils, stony and gravelly hillsides, in between rocks on mountains and on gravelly banks near river and in wooded hillside and gullies; alt. 10–550 m; fl. Feb.–Jun.
DISTRIB. Frequent in Upper Plains and Foothills Region, less frequent in the Lower Mesopotamian Region and Desert Plateau Region; also found in the Mountain Region. **MSU**: ± 10 km W of Tawila, *Rawi* 2213!; Chewa Rash, NE Rania, *Rawi & Nuri* 28453!; near the Silwan (Sirwan) river, 7 km S of Darband-i-Khan, Sulaimaniya Liwa, *F. & E. Barkley* 5150!; **FUJ**: 90 km SW of Sinjar, *Chakrovarty, Rawi, Khatib & Alizzi* 32008!; 40 km S of Sinjar, *Chakrovarty, Rawi, Khatib & Al-Azzi* 32118!; N of Sinjar, *Barkley* 6670!; **FNI**: Jabal Maqlub, Mosul Liwa, *Agnew* 1398!; Dohuk, *Sidqi & Shahabi* 3268!; Eski Kellek, *Rawi & Gillett* 10400!; **FAR**: Haibat Sultan Dagh, N of Koi Sanjaq, *Rawi, Nuri & Al-Kuss* 28163!; Salah Al-Din Mt., *Raeder* R, 353!; 10–15 km E of Arbil to Darband, *Shahwani* 25285!; **FKI**: Tuz, *Guest* 641!; Kirkuk, Baba-Gurgur, *Guest* 1362!; 36 km W of Kirkuk, *Barkley* 4548!; **FNI**: Pilkana, near Khanaqin, *Rawi* 12801!; Badra, Kut Liwa, *Rechinger* 9172!; E of Zurbatiya, Kut Liwa, *Gillett* 6710!; Badra, *Gillett* 6660!; Qara Las, near Mandali, *Rawi* 12635!;. Pilkana 10 km E of Khanaqin, *Rawi* 12754A!; **DLJ**: Rawa, *Rawi & Gillett* 7006!; 7028!; Jazira, between Beiji and Talail Al-Nil, *Guest* 13422!; Jazira at Samarra, *Agnew* 5258!; Al Oaka Wadi in Anah, *Barkley* 675!; Jabal Makhal, near Ain Dibs, *Rawi & Gillett* 7218!; **DSD**: Busaiya, *Rawi & Haddad* 25596!; **LEA**: Jabal al Muwaila near Kumait, 70 km N of Amara, *Guest, Rawi & Rechinger* 17550!; near Wadi al-Tib police station, ± 70 km N of Amara, *Guest, Rawi & Rechinger* 17511!; **LBA**: Zubair, *Rawi & Gillett* 6027!

Morocco, Algeria, Egypt, Ethiopia, Palestine, Syria, Turkey, Mediterranean Islands to Iran, E of India and Middle Asia; introduced into C Europe and N America.

Fig. 37. **Plantago afra**. 1, habit × 1/2; 2, bract × 4. Reproduced from Fl. Pak. fig. 62, with permission.

107. CRASSULACEAE

DC. in Lam. & DC., Fl. Fr. ed. 3, 4 (1): 382 (1805)

Marie Bywater[13]

Annual, biennial or perennial, succulent, usually xerophytic, herbs. Leaves opposite, alternate or rosulate, exstipulate. Inflorescence usually cymose, sometimes paniculate, racemose or spicate, occasionally present as solitary flowers. Flowers actinomorphic, bisexual, (2–)3–32-merous. Sepals free or united basaly, persistent. Petals free or united basally. Stamens equal in number and alternating with the petals, or twice as many. Scale-like nectaries present near or at the base of carpels. Follicles free or basally connate, 1-many-seeded. Seeds small; endosperm scanty or not developed.

A cosmopolitan family of some 1500 species, with centres of diversity in South Africa, Mexico and Central Asia. Of the 35 genera in the family six are native to Iraq; *Bryophyllum* from Madagascar is cultivated in gardens and sometimes found as an escape.

Crassulaceae systematics have undergone considerable change after the last monographic treatment of this family. I (SAG) have followed Eggli, Hart & Nyffeler (1995), except for *Telmissa microcarpa* which remains an unresolved name, and which I have maintained in the monotypic genus *Telmissa* (not included in the Crassulaceae database).

Ref.: Toward a consensus classification of the Crassulaceae. In: Hart, H., Eggli, U. (eds). Evolution and systematics of the Crassulaceae. pp. 173–192. Backhuys: Leiden.

A database of the Crassulaceae is maintained at www.crassulaceae.com

1. Basal leaves orbicular, peltate to cuneate .... 4. **Umbilicus**
   Basal leaves linear to elliptic to pinnate 2
2. Leaves basally rosulate 3
   Leaves not rosulate 4
3. Caudex present .... 5. **Rosularia**
   Caudex absent .... 6. **Prometheum**
4. Leaves opposite 5
   Leaves alternate or verticillate 6
5 Flowers 4-merous; petals free .... 1. **Tillaea**
   Flowers 5-merous; petals united into a tube (in Iraq cultivated) ...... 7. **Bryophyllum**
6. Petals united into a tube (in Iraq cultivated) .... 7. **Bryophyllum**
   Petals free 7
7. Stamens 3–5; carpels 1-seeded .... 2. **Telmissa**
   Stamens 10–14; carpels 5–many-seeded .... 3. **Sedum**

1. TILLAEA L.

Sp. Pl. ed. 1: 128 (1753); Eggli et al. 190 (1995)
Crassula L., Sp. Pl. ed. 1: 282 (1753)

Minute annual herbs to succulent perennial subshrubs with fibrous roots. Stems erect or ascending. Leaves opposite, sessile, connate at base to form a sheath. Inflorescense axillary or terminal, cymose. Flowers (2–)4(–5)-merous. Sepals free. Petals free. Stamens equal in number to petals. Carpels free; style short; stigma minute. Follicles 1–many-seeded.

A genus of about 20 species, worldwide, except Antartica. One species in Iraq.

1. **Tillaea alata** *Viv.*, Pl. Aegypt. Dec. 4: 16 (1831); Fu Kunjun & Gilbert in Fl. China 8: 203 (2001); Sarwar in Fl. Pak. 209: 3 (2002).

subsp. **alata**

Crassula alata (Viv.) Berger in Engl. & Prantl, Pflanzenfam. ed. 2, 18a: 389 (1930); Jansson & Rechinger in Fl. Iranica 72: 1 (1970); Gilbert in Fl. Ethiop. & Eritr. 3: 8 (1989); Collenette, Wildflow. Saudi Ariabia: 239 (1999); Boulos, Fl. Egypt 1: 239 (1999); Akhani in Fl. Iran 32: 4 (2000).
Tillaea trichopoda Fenzl ex Boiss., Fl. Orient. 2: 767 (1872).
T. muscosa L. var. *trichopoda* (Fenzl) Post, Fl. Syria: 3141 (1896).

[13] Royal Botanic Gardens, Kew. Revised by Shahina A. Ghazanfar.

Small, erect or ascending, glabrous annual, (2–)3–5(–8) cm tall; roots fibrous. Stems terete. Leaves subsucculent, triangular to lanceolate, 1–3 × ± 1 mm, entire, acuminate, glabrous. Pedicels ± 1 mm. Flowers 3 or 4-merous, axillary, 1 or 2 per node. Sepals ovate, ± 1 mm, acuminate. Petals ovate, acute. Stamens 3 or 4, opposite the sepals. Nectary scales filamentous, slightly spatulate, < 1 mm. Carpels obtuse, 1.0 × 0.5 mm, 2–seeded. Seeds minute, ellipsoid, ± smooth. Fig. 38: 1–8.

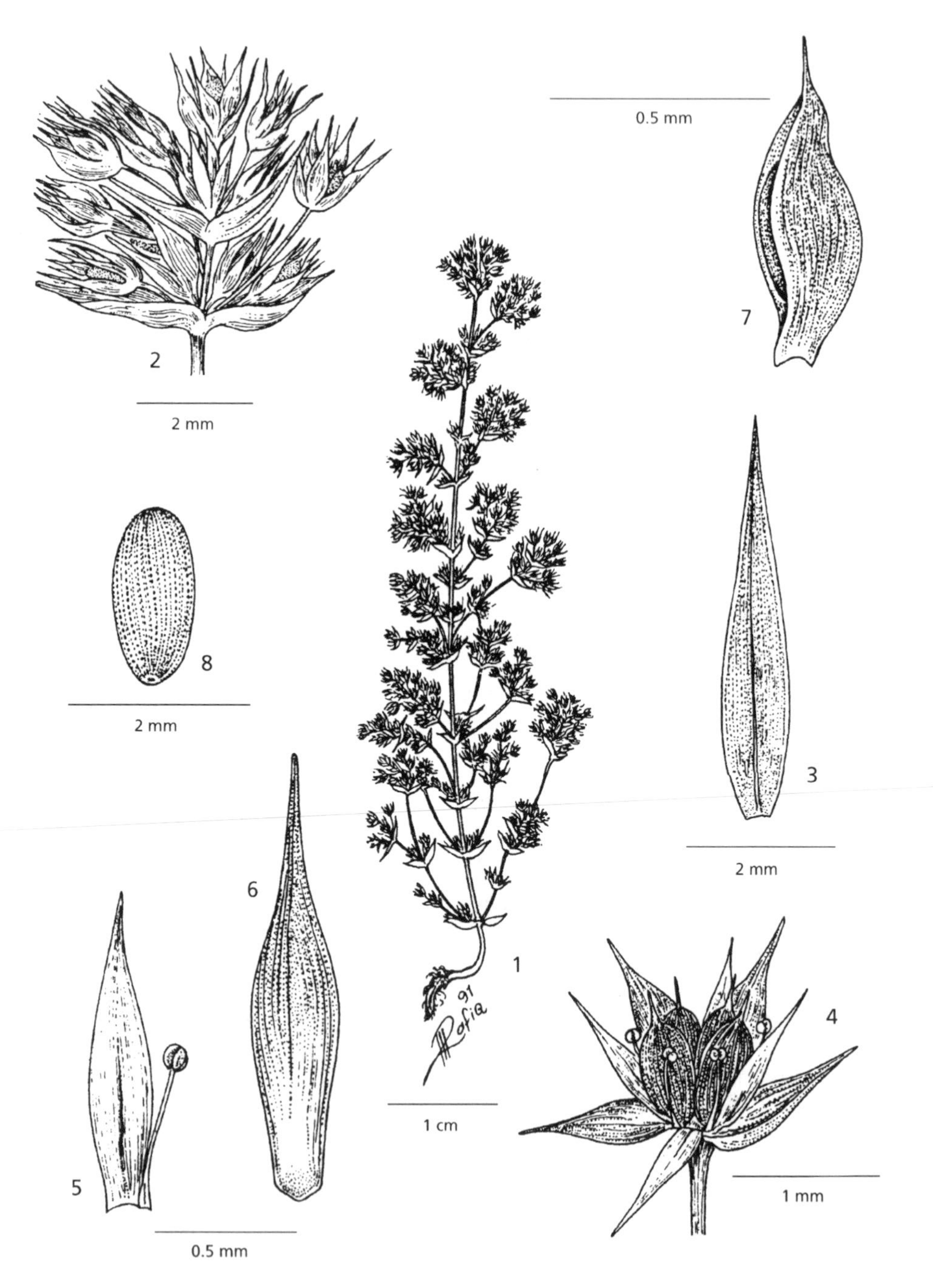

Fig. 38. **Tillaea alata**. 1, habit; 2, part of flowering branch; 3, leaf; 4, flower; 5, sepals; 6, petal with stamen; 7, fruit; 8, seed. Reproduced from Fl. Pak. 209: f. 2, with permission.

Hab. Seasonally inundated water courses and flats, damp rocky slopes; alt.: ± 300 m; fl. Mar./Apr.
Distrib. Southern Desert, very rare. **DSD**: Shabicha, *Gillett & Rawi* 6282!
Note. Subsp. *pharnaceoides* (Fisch. & Mey.) Wickens & Bywater, not recorded for Iraq, has 5–merous flowers.

Greece, Crete, Cyprus, Egypt, Syria, Palestine, Ethiopia, Saudi Arabia, W & S Iran.

2. TELMISSA Fenzl

Pugill. Pl. Nov. Syr. 14 (1842)

Small, annual, subaquatic herbs. Stems simple, erect. Leaves alternate, sessile, fleshy, ± terete. Inflorescence branched, cymose. Flowers sessile, 3–5-merous. Sepals free. Petals free. Stamens 3–5. Nectary scales small, as many as carpels. Carpels free, erect; style short. Follicles 1-seeded.

Monotypic. Distinguished from *Sedum* by its single-seeded carpels and number of stamens, and from *Tillaea* by its alternate leaves.

1. **Telmissa microcarpa** (*Sm.*) *Boiss.*, Fl. Orient. 2: 795 (1872); Chamberlain in Fl. Turk. 4: 244 (1972).

Crassula microcarpa Sm. in Sibth. & Sm., Prodr. Fl. Graec. 1: 217 (1806).
Telmissa sedoides Fenzl, Pugill. Nov. Syr. 15 (1842).

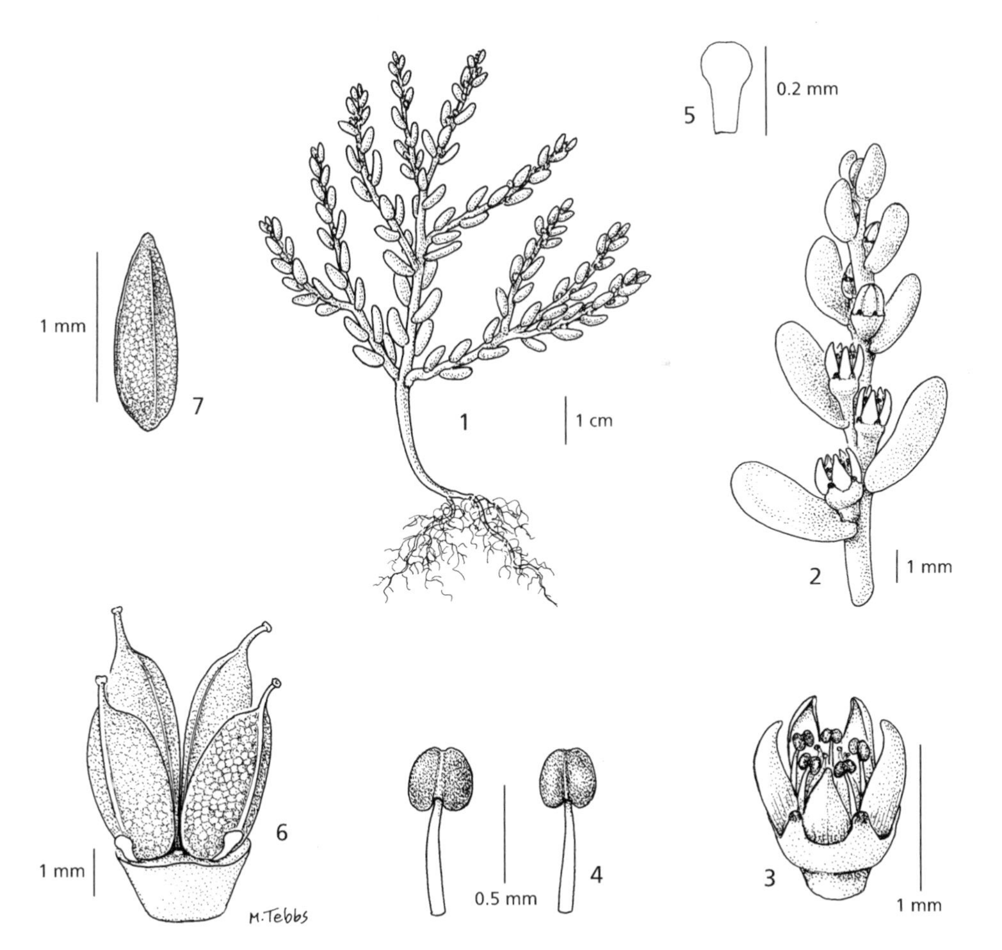

Fig. 39. **Telmissa microcarpa**. 1, habit; 2, part of inflorescense; 3, flower; 4, stamens; 5, nectary, 6, carpels; 7, seed. 1–4 from *Meikle & Casey* 982; 5–7 from *Meikle* 2989. © Drawn by M. Tebbs (Oct. 2012).

Small, glabrous, erect annual, up to 11 cm tall; roots fibrous. Stems succulent, erect; branches simple. Leaves succulent, ± terete, oblanceolate, 3–10 × ± 1 mm, obtuse, glabrous. Inflorescence 2 or 3 (or 4)-branched; flowering branches 4–10(–20)-flowered; bracts succulent, sessile, linear to oblanceolate, 1–4 × ± 1 mm, obtuse. Flowers alternate, sessile. Sepals triangular, < 1 mm, obtuse. Petals white tinged pink, ovate, ± 1 mm, hooded, ± acute. Stamens 3–5, ± 1 mm; nectary scales minute, triangular-oblong, slightly emarginate. Carpels free, oblong, 2–3 × 0.5–1 mm, distinctly angled. Follicles erect, reticulate, 1-seeded. Seeds narrowly ovoid, reticulate, acute. Fig. 39: 1–7.

HAB. In wadis and on rocky, often limestone, hillsides; alt. up to 600 m; fl. Mar.–Apr.
DISTRIB. Very rare in the extreme NW of Iraq. **MSU**: Jabal Sinjar, *Qaisi & Wedad* 52078! **FUJ**: N of Sinjar, *Barkley* 6684!

Cyprus, Turkey, Syria, Jordan.

3. **SEDUM** L.

Sp. Pl. ed. 1: 430 (1753); Eggli et al. 190 (1995)

Annual or perennial herbs, rarely subshrubs, mostly succulent. Stems erect or ascending or creeping. Leaves subsucculent or succulent, alternate; rosulate leaves absent. Cauline leaves sessile or petiolate, alternate, opposite or sometimes whorled, entire or toothed or lobed. Inflorescence cymose, occasionally spicate or paniculate or flowers solitary. Flowers sessile or pedicellate, (4–)5(–9)-merous. Sepals free or basally united, fleshy. Petals usually free. Stamens usually twice the number of petals. Nectary scales small, as many as carpels. Carpels as many as petals, free or basally united; style short. Follicles erect or spreading. Seeds few to many per folicle.

A cosmopolitan genus of nearly 600 species, six of which occur in Iraq. The name is taken from the Latin name for "house leek", which is derived from the verb *sedo* (I sit) because it colonizes flat surfaces such as roofs.

1. Leaves and sepals ciliate . . . . 1. *S. aetnense*
   Leaves and sepals glabrous 2
2. Flowers yellow . . . . 2. *S. assyriacum*
   Flowers white or pink 3
3. Plant seldom exceeding 3 cm; flowers 4 or 5-merous; stamens 4 or 5 . . 5. *S. caespitosum*
   Plant 3–15 cm; flowers 5–7(–9)-merous; stamens 10–14. 4
4. Carpels stellate-spreading . . . . 4. *S. hispanicum*
   Carpels erect to spreading 5
5. Caespitose perennial; leaves linear to oblong 3–6(–8) × 2 mm; inflorescence dense, 1–3-flowered, glandular . . . . 6. *S. tenellum*
   Erect, annual or perennial; leaves lanceolate (4–)9–12(–20) × 1–2 (–3) mm; inflorescence with 4–8(–11)-flowered branches, glabrous or glandular . . . . 3. *S. pallidum*

1. **Sedum aetnense** *Tineo* in Guss., Fl. Sic. Syn. 2: 826 (1845); Chamberlain in Fl. Turk. 4: 239 (1972).

Sedum tetramerum Trautv. in Acta Horti Petrop. 7: 2 (1881); Jansson & Rechinger in Fl. Iranica 72: 16 (1970); Chamberlain in Fl. Turk. 4: 240 (1972); Akhani in Fl. Iran 32: 39 (2000).
S. skorpilii Velen. in Sitz. Böhm. Ges. Wiss., Math.–Nat. Kl. 40: 29 (1899).
S. albanicum Beck in Ann. Nat. Hofmus. Wien 19: 74 (1904).
S. aetnense Tineo var. *tetramerum* (Trautv.) Hamet in Bull. Jard. Bot. 1: 143 (1914); *S. aetnense* Tineo var. *genuinum* Hamet in Annali di Bot. Roma 16, fasc. 2: 213 (1924).

Annual (2–)3–4(–6) cm. Stems erect, terete, subsucculent, glabrous. Leaves succulent, sessile, ovate to elliptic, (2–)3–4(–5) × 1–2 mm, denticulate, ciliate. Inflorescence spicate, (1–)4–11-flowered; bracts foliaceous, ovate to lanceolate, 2–3(–5) × 1–2 mm, spurred, ciliate; pedicels < 1 mm. Flowers 4(–5)-merous. Sepals lanceolate, 2–4 × 0.5–1 mm, spurred, ciliate. Petals white, lanceolate to ovate, ± 2 × 1 mm, acuminate. Stamens 4 (or 5); scales linear to spatulate, < 1 mm. Carpels spreading, 2 × 1 mm, papillate. Seeds 5–7 per carpel, ellipsoid, smooth to papillate.

HAB. On sandy clay soil and gravel; alt. ± 200 m.
DISTRIB. Very rare, recorded only once from near the ancient ruins of the Parthian city of Hatra, now Hadhr. **FUJ**: Abu Tina, E of Hadhr, *Omar & Hamid* 36342!

Spain, Sicily, SE Europe, Crimea, Turkey, Caucasus, Syria, Iran, Afghanistan, Tien Shan.

2. **Sedum assyriacum** *Boiss.*, Diagn. ser. 1 (6): 55 (1845). — Type: Iraq, Mosul, Pesh Khabur, *Kotschy* 172! (syn.); Jabal Taktak, *Haussknecht* s.n. (JE, syn.).

Erect glabrous annual 10–15 cm. Stems subsucculent, terete, glabrous, up to 3 mm across. Leaves succulent, sessile, lanceolate, 9–12 × 1–2(–3) mm, acute. Inflorescence cymose with 4–11-flowered branches; bracts foliaceous, linear to lanceolate, (3–)6–7(–20) × 1–2(–3) mm, acute; pedicels ± 1 mm long. Flowers 5–6-merous. Sepals narrowly triangular to lanceolate, 1–2 mm, acute. Petals yellow, free, lanceolate, 3–4 mm, acute to acuminate. Stamens 10 or 12; anthers yellow; nectary scales obtriangular, < 1 mm, emarginate. Carpels erect, 2–3 × ± 1 mm, glabrous; ovules numerous. Seeds not seen.

HAB. Marshland; alt. 1600–2000; fl. ?Jun.
DISTRIB. Very rare, possibly now extinct in Iraq as the records are ancient. **FNI**: Mosul, Pesh Khabur, *Kotschy* 172! (syn.); Jabal Taktak, *Haussknecht* s.n. (syn.).
NOTE. This species has been maintained purely on the strength of the colour references made by Boissier in his original description, "petalis intense luteis" and "antheris flavis". Close examination of the isotypes at Kew, however, shows a slight darkening down the mid-vein in some of the petals, implying a white petal with a pink vein. The anthers, though generally appearing yellow, show definite purple pigmentation in the younger flowers. Further collections within the type locality are needed to verify the existence of a yellow-flowered *Sedum* in the area. Without such evidence, *S. assyriacum* may be better placed in synonymy with *S. pallidum.*

Lebanon, Syria according to Eggli (no material at K from Lebanon or Syria).

3. **Sedum pallidum** *Bieb.*, Fl. Taur.-Cauc. 1: 358 (1808); Chamberlain in Fl. Turk. 4: 242 (1972); Akhani in Fl. Iran 32: 32 (2000).

Erect glabrous annual or perennial 5–15 cm tall. Stems ± succulent, terete, up to 3 mm across. Leaves succulent, sessile, lanceolate, (4–)9–12(–20) × 1–2(–3) mm, acute. Inflorescence cymose with 4–8(–11)-flowered branches, glabrous or glandular; bracts foliaceous, linear to lanceolate, (3–)6–7(–20) × 1–2(–3) mm, acute; pedicels ± 1 mm. Flowers 5-merous. Sepals narrowly triangular to lanceolate, 1–2 × 0.5–1 mm, acute. Petals free, lanceolate, 4–5 × 1 mm, white or pale pink with a pink mid-vein, acute to acuminate. Stamens 10, anthers purple; scales obtriangular, < 1 mm, emarginate. Carpels erect or spreading, 3–5 × 1 mm, glabrous or glandular. Seeds numerous, elliptic, striate.

var. **pallidum**

S. sanguineum Boiss. & Hausskn. in Boiss, Fl. Orient. 2: 790 (1872) non Ortega 1800. — Type. Iraq, Jabal Taktak, *Haussknecht* s.n.
S. hispanicum L., Jansson & Rech. f. in Fl. Iranica 72: 14 (1970).

Plants annual.

HAB. In damp ground on rocky limestone slopes; alt. ± 950 m; fl. May.
DISTRIB. Very rare; only two recent records from northern Iraq. **MRO**: E side of Jabal Karokh, between Sei–Waka (Sewok) and Dargala villages, *Kass & Nuri* 27622!; **MSU**: Qopi Qaradagh, *Haines* W. 1150!; **FAR**: Jabal Taktak, *Haussknecht* s.n. (type of *S. sanguineum*)
NOTE. Var. *bithynicum,* with perennial vegetative shoots is found in the Aegean Islands, Turkey and Armenia.

Turkey, Crimea, Caucasus, Cyprus, Iran.

4. **Sedum hispanicum** *L.* in Jusl. Cent. 1: 12 (1755) & Amoen. Acad. 4: 273 (1759); Jansson & Rech. f. in Fl. Iranica 72: 14 (1970); Chamberlain in Fl. Turk. 4: 241 (1972); Miller & Cope, Fl. Arab. Penin. & Soc. 1: 467 (1996); Akhani in Fl. Iran 32: 34 (2000); Sarwar in Fl. Pak. 209: 12 (2002); Ghazanfar, Fl. Oman 1: 1 (2003).

Sedum glaucum Waldst. & Kit., Pl. Rar. Hung. 2: 198 (1805).
S. sexfidum M. Bieb., Fl. Taur.-Cauc. 1: 354 (1808).

S. orientale Boiss., Diagn. ser. 1 (10): 17 (1849).
S. armenum Boiss., Diagn. ser. 2 (2): 61 (1856).

Erect or ascending annual, sometimes biennial or perennial with small sterile shoots at base, (2–)5–15 cm tall. Stems subsucculent to ± woody, terete. Leaves succulent, sessile, lanceolate to oblong, 3–20 × 1–3 mm, acute. Inflorescence cymose, (1–)2(–9)-branched, glaucous to glandular; bracts foliaceous, lanceolate to oblong, 4–16 × 1–3 mm, ± acute; pedicels ± 1 mm. Flowers 5–6(–9)-merous. Sepals triangular, 1–3 × 1 mm, often glandular, acute. Petals free, white, sometimes with a red mid-vein, narrowly ovate to triangular, 2–4(–6) × ± 1 mm, mucronate. Stamens 10 or 12 (–18); nectary scales obtriangular, < 1 mm, emarginate. Carpels stellate-spreading, 2–5 × 1–2 mm, connate, glaucous or glandular. Seeds numerous, ellipsoid, striate.

HAB. In cracks between rocks on dry limestone slopes; alt. 300–1100 m; fl. Apr.–May.
DISTRIB. **MJS**: Jabal Sinjar, N slope opposite Andalus village, *Widad & Al-Khayat* 53481!; Jabal Sinjar, *Qaisi & Wedad* 52030! Jabal Sinjar, Kursi, *Gillett* 10887!; **MAM**: Aqra, *Rawi* 11316!; Jabal Makhul (Mosul Liwa), *Rawi & Gillett* 7205!; **MRO**: Rowanduz, *Gillett* 8331!; Warta Village, 30 km NW by N of Rania, *Rawi, Nuni & Kass* 28767!; 70 km NW of Shaqlawa, towards Haji Umran, *Shahwani* 25223!; Abu Ghraib, near Harir, 25 km N of Shaqlawa, *Barkley* 8330!; Abu Gharaib, near Salah ad Din, *Barkley* 5716!; Jabal Makloub

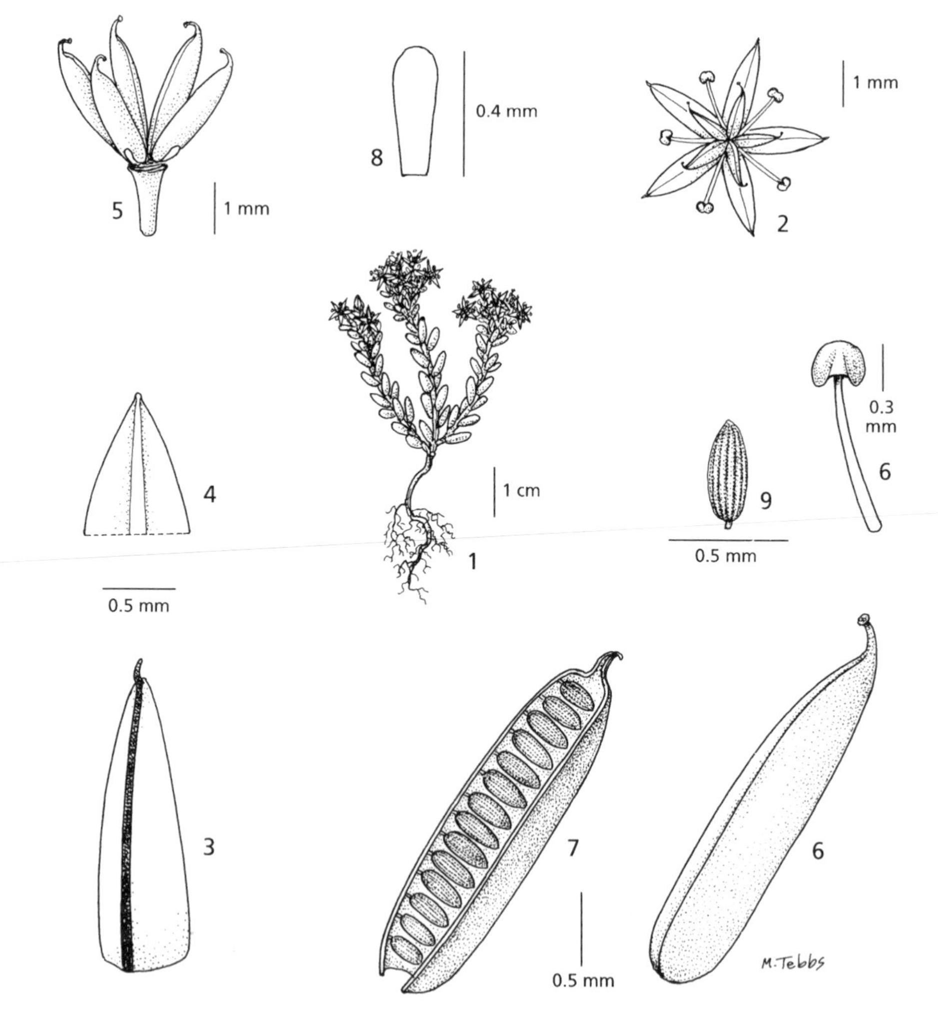

Fig. 40. **Sedum caespitosum**. 1, habit; 2, flower; 3, petal; 4, sepal apical portion; 5, carpels; 6, single carpel; 7, carpel opened; 8, nectary; 9, seed. 1 from *Kaisi & Hamad*; 2–9 from *Ravi & Gillett* 10407. © Drawn by M. Tebbs (Oct. 2012).

(?Makhul), *Haines* W.1407!; Kuh Sefin, *Bornmüller* 1219B; **MSU**: Abu Ghraib, near the Silwan river 7 km S of Darband-i-Khan, *Barkley & Shalli* 5156! Tawila, *Rawi* 21930!; **FNI**: Eski Kellek, *Rawi & Gillett* 10406!; **FNI**: Jabal Muwaila, near Kuwait, E of Jabal Hamrin, ± 70 km N of Amara, *Guest, Rawi & Rechinger* 17639!; 16 km S of Badra, *Rawi* 18172!; Zurbatiya, *Gillett* 671!; **FUJ**: Qaiyara, *Bayliss* 113!

C & SE Europe, S Russia, Turkey to Pakistan, Arabia, Eritrea.

5. **Sedum caespitosum** (*Cav.*) *DC.*, Prodr. 3: 405 (1828); Jansson & Rech. f., Fl. Iranica 72: 12 (1970); Chamberlain in Fl. Turk. 4: 240 (1972).

Tillaea rubra L., Sp. Pl. 129 (1753).
Crassula caespitosa Cav., Ic. et Descr. Pl. 1: 50, t. 69, f. 2 (1791).
Sedum rubrum (L.) Thell. in Feddes Rep. 10: 290 (1912) non Royle ex Edgew. (1839); Akhani in Fl. Iran 32: 39 (2000).

Small, erect, glabrous annual (1–)2–3(–6) cm tall. Stems succulent, terete. Leaves succulent, sessile, oblong to obovate, imbricate, (2–)3–5(–6) × 1–2(–3) mm, obtuse. Inflorescence cymose, 1–9-branched, frequently reduced to a single branch; bracts foliaceous, oblong to obovate, 2–5 × 1–2 mm, obtuse; pedicels ± 1 mm. Flowers 4–5-merous. Sepals ovate to triangular, 1 × 0.5–1 mm, ± acute. Petals free, white with a red mid-vein, narrowly ovate to triangular, slightly keeled, 1–2 mm, mucronate. Stamens 4 or 5; nectary scales spatulate, < 1 mm. Carpels spreading, 2–3 × 1 mm, usually glabrous but a few glandular hairs may be present towards base of suture. Seeds numerous, ellipsoid, striate. Fig. 40: 1–9.

HAB. In silt and clay on rocky hillsides and dry river beds; alt. 500–600 m; fl. Mar.–Apr.
DISTRIB. Scattered and occasional in northern parts of Iraq. **MRO**: 80 km SW Haji Umran, *Shahwani* 25270!; **MSU**: Jarmo, *Helbaek* 710!; **MJS**: Jabal Sinjar, *Qaisi & Wedad* 52126 & *Qaisi & Hamad* 49216!; **FUJ**: Abu Tina, E of Hadhr, *Omar & Hamis* 36343!; **FNI**: Mosul Liwa, Eski Kellek, *Rawi & Gillett* 10407!; **FAR**: Bastura chai, *Rawi & Gillett* 10494!; **FKI**: Kirkuk, *Bornmüller* 1219; **FNI**: Badra, *Gillett* 6611!; **DWD**: Wadi Fuhaimi, *Omar, Qaisi, Hamad & Hamid* 44669!

C & SE Europe, N Africa, Crimea, Caucasus, Turkey, Syria.

6. **Sedum tenellum** *M. Bieb.*, Fl. Taur.–Cauc. 3: 315 (1820); Jansson & Rech. f., Fl. Iranica 72: 12 (1970); Chamberlain in Fl. Turk. 4: 235 (1972); Akhani in Fl. Iran 32: 26 (2000).

Caespitose perennial, ascending to erect, (2–)3–7(–10) cm tall. Stems subsucculent to ± woody, terete. Leaves succulent, sessile, subterete, linear to oblong, 3–6(–8) × 2 mm, obtuse, imbricate. Inflorescence cymose, dense, 1–3-flowered, glandular; bracts foliaceous, linear to lanceolate, 2–5 × 1 mm; pedicels ± 2 mm long. Flowers 5–merous. Sepals triangular, 2–3 × 1 mm, obtuse. Petals free, white to deep pink, ovate, slightly keeled, 3–4(–6) × 1 mm, acute. Stamens 10; nectary scales spatulate, < 1 mm. Carpels erect, 2–3 × 1 mm, obtuse. Seeds numerous, minute, oblong, striate, minutely winged, apex acute.

HAB. Alpine slopes; alt. 3000–3400(–3800) m.; fl. Aug., Sep.
DISTRIB. **MRO**: Halgurd Range, *Rawi & Serhang* 24824!; *Rechinger* 6502; *Guest & Hewitt* 3047!; *Rechinger* 11882; Ser Kurawa, *Gillett* 9725!

Turkey, Caucasus, N Iran.

4. **UMBILICUS** DC.

Bull. Soc. Philom. Paris 49, 3: 1 (1801)

Perennial fleshy succulent glabrous herbs with tuberous roots or rhizomatous. Basal leaves alternate, fleshy, orbicular, peltate, long petiole. Inflorescence terminal, racemose or paniculate, simple or branched, many-flowered. Flowers sessile or pedicellate, 5-merous, urceolate to campanulate. Sepals basally united, fleshy. Petals united at base. Stamens (5–)10, in 2 whorls, epipetalous. Nectary scales oblong. Carpels free. Follicles erect, several-seeded. Seeds minute, striate.

About 12 species, from the mediterranean region in W Europe, west to Macaronesia, east to Iran and south to Cameroon and Tanzania.
Umbilicus (from the Lat. for navel, refers to the central depression in the peltate leaves.)

Inflorescence a simple raceme; pedicels shorter than flowers. . . . . 1. *U. horizontalis*
Inflorescence paniculate; pedicels equalling or longer than flowers . . 2. *U. tropaeolifolius*

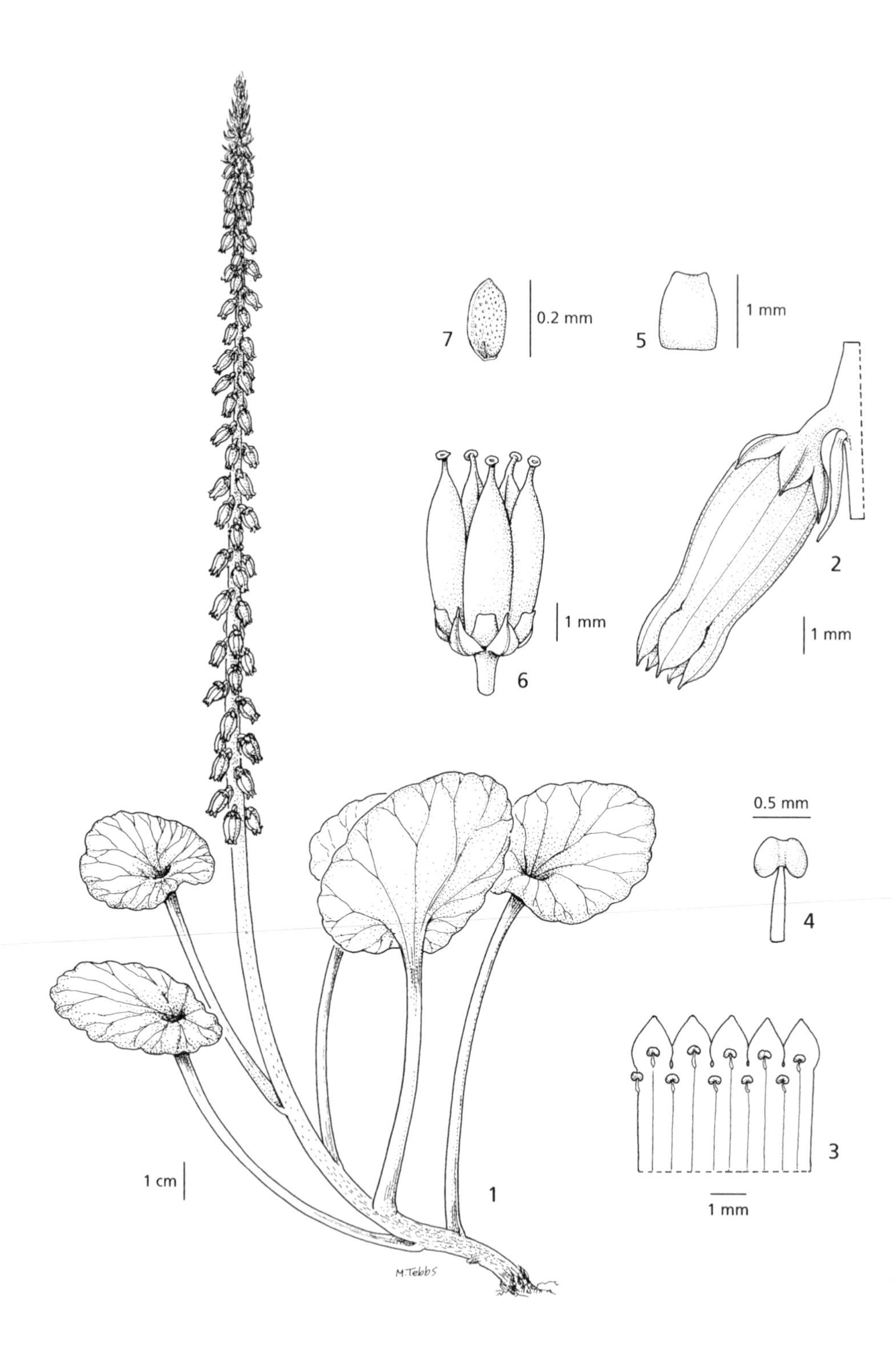

Fig. 41. **Umbilicus horizontalis** var. **intermedius**. 1, habit; 2, flower; 3, corolla opened; 4, stamen; 5, nectary, 6, carpels; 7, seed. 1 from *Rawi* 31365 (flowering stem) & *Widad & al Khayat* 53457 (leaves); 2–7 from *Rawi* 31365. © Drawn by M. Tebbs (Oct. 2012).

1. **Umbilicus horizontalis** DC., Prodr. 3: 400 (1828).

Glabrous, erect perennial up to 57 cm tall; roots tuberous, ± woody. Stems succulent, terete. Leaves orbicular and peltate at base, (1–)3–5(–9) cm across, undulate to crenate, with petioles of (3–)6–10 cm long. Inflorescence racemose, (8–)13–43(–57) × (1–)3–5(–9) cm; bracts (2–)10–25(–65) × 2–6 mm, undulate to crenate, becoming reduced and graduating up the stem from peltate, orbicular to cuneate, suborbicular to spatulate to narrowly lanceolate. Pedicels 1–2 mm. Flowers ± pendulous or horizontally spreading. Sepals triangular, ± 1 mm. Corolla greenish-white, urceolate or cylindrical; tube 4–6 × 2–4 mm; lobes ovate or lanceolate, 1–2 × 1 mm, emarginate. Follicle obtuse, ± 4 × 1 mm. Seeds numerous, minute, ellipsoid, papillate.

var. **intermedius** (Boiss.) Chamberlain in Fl. Turk. 4: 213 (1972).

Cotyledon intermedia (Boiss.) Bornmüller in Bot. Jahrb. 33: 427 (1904).
Umbilicus intermedius Boiss., Fl. Orient. 2: 769 (1872); Jansson & Rech. f., Fl. Iranica 72: 3 (1970); Akhani in Fl. Iran 32: 8 (2000).
U. pendulinus DC. var. *intermedius* (Boiss.) Post, Fl. Syria ed. 1: 314 (1896).

Flowers ± pendulous. Corolla tube urceolate; lobes ovate. Fig. 41: 1–7.

HAB. Crevices in limestone rocks; alt. 800–1500 m; fl. Apr.–May.
DISTRIB. Widespread, mainly in northern Iraq. **MAM**: Agra, *Rawi* 11484!; Bakirma (Bakurman), *Rawi* 8794!; 2 km W of Zakho, *Barkley* 9199; Dohuk gorge, 18 km NE of Chemchamal, *Barkley* 5842!, *Emberger, Guest, Lond, Schwan & Yusuf* 15432!, *Chapman* 26270!; Sarsang, *Shrief & Hamad* 50310!; Makhlat, *Guest* (*Rustam* 4275)!; **MRO**: Zawita mountain between Rania and Shaqlawa, *Karim, Hamid & Tasim* 40804!; Kuh-i-Sefin, by Shaqlawa village, *Bornmüller* 1215!; Haji Umran, *Chapman* 11927!; Chinaruk (Haibat Sultan), *Omar, Qaisi & Wedad* 49467!; Jabal Baradost, *Thesiger* 808; **MSU**: Khormal, *Rawi* 8914!; Pira Magrun, *Rawi* 12106!; Tainal (Sulaimaniya to Kirkuk), *Omar & Karim* 37893!; Serkabkhan, 7 km NW of Rania, *Rawi, Nuri & Kass* 28548!; Kani Henjeer (Khanjar) village, *Omar* 42911!; Jarmo, *Haines* W257!; Qarachitan (western foot of Pira Magrun), *Gillett* 7707!; Chemchemal, *Rechinger* 12197; Ahmad Awa, *Omar, Qaisi & Wedad* 49399!; c. 7 km W of Tawila to Halabja, Baicha village, *Rawi* 221855!; Sulaimaniya, Qara Dagh, *Bowles* 619!; **MJS**: Jabal Sinjar, N slope opposite Andalus village, *Wedad & Alkhayat* 53457!; Jabal Sinjar, SE slope, curve no. 126, *Wedad & Alkhayat* 53359!; Kursi, Jabal Sinjar, *Gillett* 10888!; Tawila, *Rawi* 21907!, *Rechinger* 10261; **DWD**: Wadi Ajrumiya, *Rawi* 31239!; 160–180 km NW of Ramadi to Rutba, *Rawi* 31365!
NOTE. Var. *horizontalis* with horizontally spreading flowers, cylindrical corolla tubes and lanceolate corolla lobes, is found in S Europe, Egypt, W Syria, and W Iran; not found in Iraq.

Turkey, Syria, Lebanon, Jordan, Egypt, Arabia, Iran.

2. **Umbilicus tropaeolifolius** *Boiss.*, Diagn. ser. 1, 3: 14 (1843). — Type: N Iraq, Ninevam (Mosul) *Aucher* 2685 (K!, P, syn.); Jansson & Rech. f. in Fl. Iranica 72: 2 (1970); Chamberlain in Fl. Turk. 4: 212 (1972); Akhani in Fl. Iran 32: 6 (2000).

Umbilicus oxypetalus Boiss., Diagn. ser. 1, 3: 15 (1843).
Cotyledon tropaeolifolius (Boiss.) Bornmüller, Beih. Bot. Centralbl. 32 (2): 389 (1914).

Glabrous, erect perennial 7–25 cm tall; roots tuberous, ± woody. Stems succulent, terete. Leaves orbicular, peltate, 2–5(–7) cm across, undulate to crenate; petioles 6–12(–22) cm. Inflorescence racemose, lax, 7–15(–26) cm; bracts suborbicular, cuneate to peltate, 6–24 × 1–4 cm, undulate to crenate; pedicels 2–8 mm long; bracteoles minute, lanceolate, ± 1 mm long. Flowers spreading. Sepals triangular, ± 1 mm. Corolla cream, obconical; tube ± 1 mm; lobes triangular, 3–5 × 1–2 mm, emarginate. Stamens inserted on corolla at mouth of tube. Nectary scales oblong, < 1 mm. Carpels tapering, 5 × 1 mm. Seeds numerous, minute, ellipsoid, minutely papillate.

HAB. Deep shaded limestone crevices; alt. 1000–1200 m.; fl. May–June.
DISTRIB. Widespread in the mountains and foothills of northern Iraq. **MAM**: Aloka valley, 8 km SW of Dohok, *Barkley* 8296!; Chewa Rash, NE of Rania, *Rawi, Nuri & Kass* 28479!. **MRO**: Shaqlawa, *Haines* W619!; Jabal Baradost nr. Shanidar, *Erdtman & Goedemans* (in *Rechinger* 15632); Haji Umran, *Chapman* 11932!; Kuh Sefin, by Shaqlawa, *Bornmüller* 1216!; Sefin Dagh, *Gillett* 8100!; Shakh-i-Harir (near Harir, Rowanduz area), *Emberger, Guest, Long, Schwan & Yusuf* 15438!; Rowanduz gorge, *Guest* 2096!; **MSU**: Qarachitan (W foot of Pira Magrun), *Gillett* 7709!; E of Darband-i-Khan, *Hussein & Barkley* 7726!; *Barkley & Haddad* 7757!; Pira Magrun, Kurdul, *Haussknecht* s.n.; Tainal, *Omar & Karim* 37894!; Palegawra, W of Sagirma Dagh, 25 km SW of Pira Magrun, *Helbaek* 651!; Agra, *Rawi* 11382!; **MJS**: Kursi, Jabal Sinjar, *Gillett* 10941!; **FUJ**: Mosul, *Bowles* 670!

Turkey, Lebanon and Syria, W & S Iran.

5. **ROSULARIA** (DC.) Stapf

Bot. Mag. 149, sub tab. 8985 (1923); Eggli, Bradleya 6: 1–118 (1988)
Umbilicus sect. *Rosularia* DC., Prodr. 3: 399 (1828)

Perennial succulent herbs with fibrous or tuberous roots. Stems erect. Radical leaves forming basal rosettes, sessile, fleshy, oblong to spatulate to obovate, linear or elliptic, attenuate at base; cauline leaves sessile. Flowering stems arising from axils of rosulate leaves, simple, erect. Inflorescence terminal or lateral, racemose or paniculate. Flowers 5(–6)-merous, tubular, urceolate to campanulate. Sepals basally united, fleshy. Petals united at base, lobes membranous. Stamens 10–18, epipetalous in two whorls. Nectary scales oblong. Carpels united basally. Follicles erect, free, many-seeded.

About 40 species in the arid and semi-arid regions of Asia, from Turkey to the Himalayas, and one in Ethiopia. Several species and subspecies and varieties have been recognized from Iraq. Following Eggli, only one species with three subspecies are recognized.

The name is taken from the rosulate arrangement of the leaves. *Rosularia* (from *rosa*, a rose and the diminutive *-ula*, because the leaves are arranged in a rosette).

Ref.: Rosularia systematics: http://sempervivum.aforumfree.com/t2636-rosularia-systematics, and http://iphylo.org/~rpage/theplantlist/A/Crassulaceae/Rosularia.

1. **Rosularia sempervivum** (*Bieb.*) *Berger* in Engler & Prantl., Nat. Pfl. 18a: 466 (1930).

Cotyledon sempervivum Bieb., Beschr. Land. Flüss Terek Kur Kasp. Meere 176 (1800); Chamberlain & Muirhead in Fl. Turk. 4: 216 (1972); Akhani in Fl. Iran 32: 47 (2000).
Umbilicus sempervivum (Bieb.) DC., Prodr. 3: 400 (1929).
U. libanoticus DC., Prodr. 3: 399 (1828); *U. libanoticus* DC. var *steudelii* Boiss., Fl. Orient. 2: 772 (1872).
Rosularia radiciflora Boriss. Fl. U.S.S.R. 9: 120 (1939) subsp. *radiciflora* ; Chamberlain & Muirhead in Fl. Turk. 4: 220 (1972).

Erect perennial, (2–)12–20(–28) cm tall; roots tuberous; caudex present; rosettes (2–)4–6(–10) cm across, open, rounded, with no apparent resting phase; offsets on runners, 1–10 cm long. Leaves succulent, oblong to obovate to spatulate, ciliate, 10–50(–88) × 4–18 mm. Inflorescence paniculate, 14–20(–23)-flowered, glandular-pubescent; bracts obovate to spatulate, 8–20 × 1–3 mm, obtuse or emarginate, ciliate, spurred; pedicels 1–3 mm. Flowers 5-merous, 7–10 × 6–8 mm. Sepals triangular, 4 × 1 mm, acute or obtuse. Corolla pink or white, veined with pink; tube urceolate to campanulate, 3–5 × 3–5 mm; lobes ovate to triangular, 3–5 × 1–3 mm, acuminate. Stamens 10, adnate to the lower half of the corolla tube; anthers purple; nectary scales elliptic. Carpels 7–8 × 1 mm. Seeds numerous, ovoid–ellipsoid, acute, striate.

1. Inflorescence ± glabrous . . . . c. subsp. *persica*
   Inflorescence glandular-pubescent 2
2. Leaves oblong, 10–25(–40) × 2–6(–8) mm; flowers predominantly pink or white veined with pink . . . . a. subsp. *sempervivum*
   Leaves obovate, (13–)16–50(–88) × (2.5–)7–11(–18) mm; flowers white or white with pink or purple veins . . . . b. subsp. *kurdica*

a. subsp. **sempervivum**

Leaves oblong, 10–25(–40) × 2–6(–8) mm; scape glandular, 14–20-flowered; flowers predominantly pink or white veined with pink.

HAB. In damp rocky places; alt. 900–2400 m; fl. Apr.–Jul.
DISTRIB. Widespread in the mountain region of Iraq. **MAM**: Zawita, *Rechinger* 12011; Mt. Gara, *Kotschy* 332; Dohuk gorge, 2 km from the town, *Chapman* 26273!; Gali Zawita, NE of Zakho, near Turkish border, *Rawi* 23596!; Matina, *Rawi* 8579B!. **MRO**: Bisht Ashan, 15 km NE of Rania, lower slope of Qandil range, *Rawi & Serhang* 24217!; Gali Ali Beg, Kasim, *Hamid & Jasim* 40912!; Gali Warta, 30 km NW by N of Rania, *Rawi, Nuri & Kass* 28839!; Gali Ali Beg, *Omar, Sahira, Karim & Hamid* 38322!; Rowanduz gorge, *Guest & Long* 13624!; Haji Umran, *Chapman* 11930!; Mt. Bisht Ashan, *Rechinger* 11803; Masis village, 25 km NE of Zakho, *Rawi* 23449!; Halgurd Dagh, Gali Ali Beg, near last bridge on the way to Rowanduz, *Kass, Nuri & Serhang*!. **MSU**: upper parts of Qara Dagh, *Kotschy* 332!; Pira Magrun, *Haussknecht* s.n., *Rawi* 12103!; Tawila, *Rawi* 21962! 21938!; Hawraman, above Biyara, *Gillett* 11773!; Chalki Adi, *Guest* 3681; Pir Omar Gudrun (Pira Magrun), *Haussknecht* 397 (syn. of *U. libanoticus* var. *steudelii*).

Turkey, Armenia, Caucasus, NW, W & S Iran.

b. subsp. **kurdica** Eggli, Bradleya 6 Suppl.: 91 (1988).— Type: Iraq, Kuh-i-Serfin beim Dorfe Schaklawa [Shaqlawa], 1060 m, 20/5/1893, *Bornmüller* 1217 (HT, JE, IT, B, K!, LE, W, syn.).

Cotyledon libanotica Labill. var. *kurdica* (Bornmüller) Chamberlain & Muirhead in Fl. Turk. 4: 221 (1972); *Rosularia radiciflora* subsp. *kurdica* Chamberlain & Muirhead in Fl. Turk. 4: 221 (1972).

Scape glandular, (6–)11–43-flowered; flowers campanulate, generally larger than those of the other two varieties, predominantly white with spreading lobes. Leaves obovate, (13–)16–50(–88) × (2.5–)7–11(–18) mm, larger than those of other two varieties. Fig. 42: 1–8.

Hab. In crevices of rocks and cliffs, on slopes of limestone mountains; alt. 700–2000 m; fl. May–Jul. Distrib. Widespread in the mountain region of Iraq. mam: Dohuk gorge, *Emberger, Guest, Long, Schwan & Serkahia* 15433!; Bakurman, *Rawi* 8575!; Khantur mt., NE of Zakho, *Rawi* 23498!; Sharanish, *Rawi* 8590!;

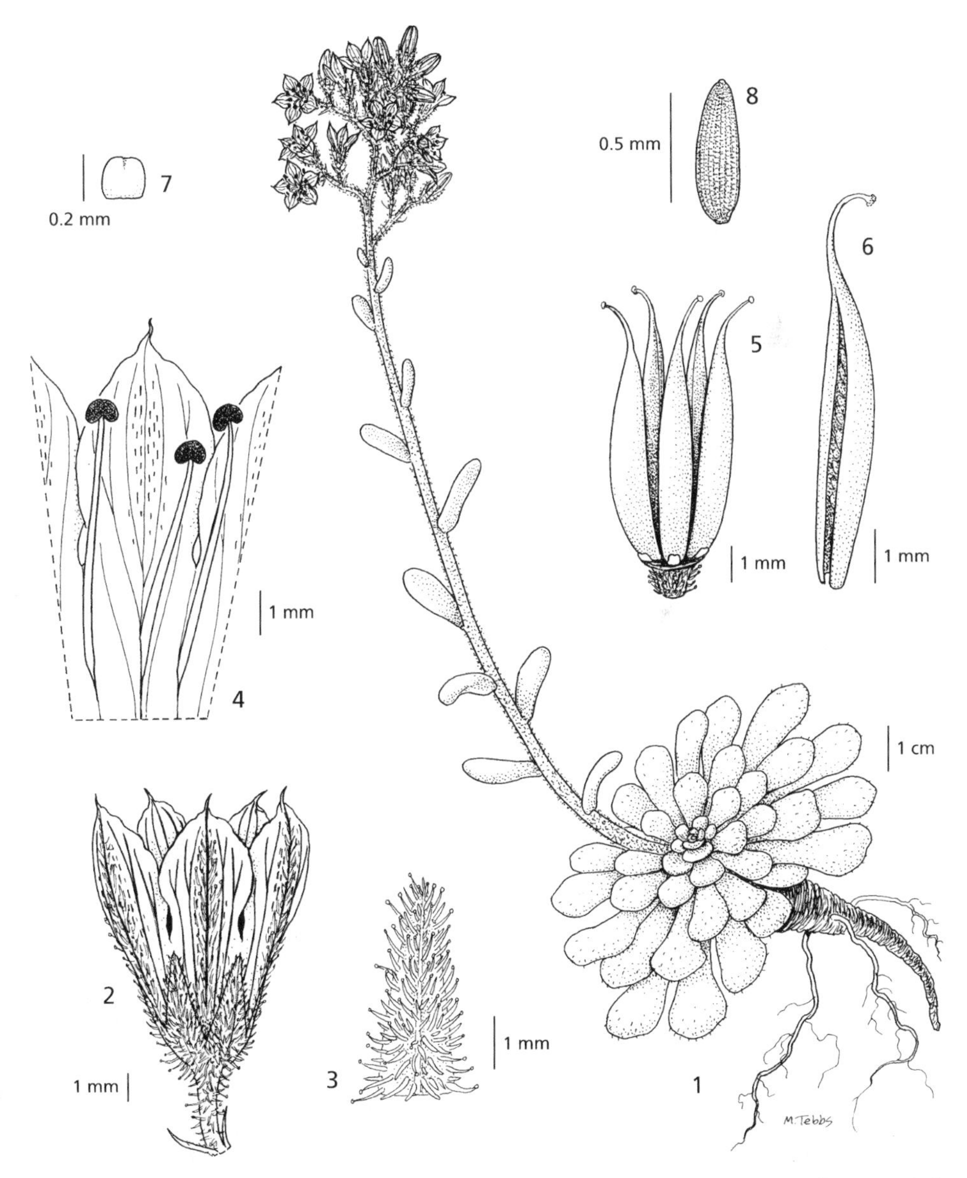

Fig. 42. **Rosularia sempervivum** subsp. **kurdica**. 1, habit; 2, flower; 3, sepal; 4, corolla opened; 5, carpels; 6, single carpel; 7, nectary; 8, seed. 1, 6–8 from *Emberger et al.* 15433; 2–5 from *Kass et al.* 27682. © Drawn by M. Tebbs (Oct. 2012).

Dohuk, *Rawi* 8706!; Gali Mazurka, 2 km SW of Amadiya, *Al-Dabbagh & Jasim* 46912!; Amadiya, *Guest* 13276!; Gali Mazurka, 2 km NW of Amadiya, *Al-Dabbagh & Jasim* 46918!. **MRO**: Kuh-i-Sefin, lower slopes by Shaqlawa village, *Bornmüller* 1217!; Sefin Dagh, above Shaqlawa, *Gillett* 8081!; Kawriesh, E side of Karokh mt., *Kass, Nuri & Serhang* 27682!; Jabal Karoukh, between Sewok village and Kilkil, *Kass, Nuri & Serhang* 27352!; Shaqlawa, *Haines* W.618!; Chew ar Rash, NW of Rania, *Rawi, Nuri & Kass* 28476!; mt. Hawara Blinda NE of Haj Umran, *Rawi, Nuri & Kass* 27825!; S slope of Karoukh mountain, *Kass, Nuri & Serhang* 27591!; Sarsang, slopes of Gara Dagh, *Chapman* 26412!, 26428!; Gara Dagh, *Rawi* 9284!; Serkupkan, 7 km NW of Rania, *Rawi, Nuri & Kass* 28538!; Hindian near Rowanduz, *Guest, Eig & Zohary* Rustam 5510!. **MSU**: Kajan near Penjwin, *Rawi* 22663!; Sulaimaniya, *Rawi* 12189!; Pira Magrun, *Rawi* 11523!; Kamarspa, on the road between Halabja and Tawila, *Rawi* 22198!. **FNI**: Sheikh Adi, near Ain Sifni, *Field & Lazar* 704!
NOTE. Akhani, in Fl. Iran 32: 47 (2000) places *S. radiciflora* in synonymy with *R. sempervivum*, but it is not clear which variety of the former species grows in Iran. ?GIA PAPINA (Sulaimaniya, *Rawi* 12189).

Turkey, ?Iran.

c. subsp. **persica** (Boiss.) Eggli, Bradleya 6 Suppl.: 95 (1988).

Umbilicus persicus Boiss., Diagn. ser. 1, 3: 3 (1843); *U. libanoticus* DC. var. *glaber* Boiss., Fl. Orient. 2: 272 (1872).
Rosularia persica (Boiss.) Berger in Engler & Prantl, Nat. Pfl. 18a: 466 (1930).
R. radiciflora Boriss. subsp. *glabra* (Boiss.) Chamberlain & Muirhead in Fl. Turkey 4: 221 (1972).
R. sempervivum var. *glabra* (Hamet) Assadi (2000) (nom. inval.).

Leaves oblong, (11–)25–40(–65) × 2–8(–12) mm. Inflorescence glabrous or very sparsely pubescent, 16–34-flowered. Scape ± glabrous; flowers urceolate to campanulate, pink.

HAB. Not recorded.
DISTRIB. **MAM**: Aqra, *Rawi* 11491!; **MRO**: Jindian, near Rowanduz, *Guest* 739!

Turkey, Lebanon, Syria.

6. **PROMETHEUM** (Berger) H.Obha

in J. Fac. Sci. Univ.Tokyo Ser. III, 12, 168 (1978)

Perennial succulent herbs, ± glabrous, erect, with fibrous roots, stoloniferous; caudex absent. Leaves succulent, rosulate. Cauline leaves alternate. Inflorescence terminal, cymose, 1–7-flowered. Flowers sessile or pedicellate, 5(–6)-merous, tubular, urceolate to campanulate. Sepals free. Petals united at base to form a tube. Stamens 10 (or 12), adnate to the corolla tube. Nectary scales oblong. Carpels free. Follicles erect, many-seeded. Seeds striate.

A small genus of about 6 species. The main features that distinguishes this genus from *Rosularia* is the absence of caudex, large chromosomes and a base number x=7. A typical Anatolian element, distributed from Central Greece, Turkey, eastward to the Caucasus, northern Iran, and northern Iraq. A single species in Iraq treated by many authors under *Rosularia*.

1. **Prometheum rechingeri** (C.-A. Jansson) 't Hart in H. 't Hart & U. Eggli (eds), Evol. & Syst. Crassulac. 170 (1995).

Rosularia rechingeri Jansson in Acta Horti Gothob. 28: 182 (1966); Jansson & Rech. f. in Fl. Iranica 72: 24 (1970); Chamberlain & Muirhead in Fl. Turk. 4: 223 (1972); Eggli, Bradleya 6: 1–118 (1988).

Erect perennial, (1–)3–7 cm tall; roots fibrous; caudex absent. Leaf rosettes 18–27 mm across. Lateral offsets present on short leafy stolons. Leaves succulent, oblong-spatulate, 6–11 × 1–2(–3) mm, sparsely ciliate. Inflorescence cymose, dense, 1–7-flowered, 20–60 mm high, glandular. Bracts oblong, (2–)4–6 × 1–2 mm, spurred, glandular. Pedicels 2–4) mm, glandular. Flowers 5-merous. Sepals ovate to triangular, 2–3 × 1 mm, acute, glandular. Corolla tube, pale pink, campanulate, 2–3 × 3–5 mm; lobes pink with red streaks, oblong, 6–7 × 2–3 mm, acute, glandular. Stamens 10, adnate to corolla tube; anthers yellow; nectary scales oblong, < 1 mm; Follicles 4–6 × 1 mm, glandular. Seeds numerous, linear, striate. Fig. 43: 1–7.

HAB. In rock crevices, on serpentine and other metamorphic rocks; alt.: 2400–3800 m.; fl. Aug.
DISTRIB. Very rare in the thorn-cushion zone of the mountain region of Iraq. **MRO**: Chia-i Mandau, *Guest* 2802!; Halgurd Range, *Rawi & Serhang* 24831! & *Rechinger* 11883.

Turkey, Iran.

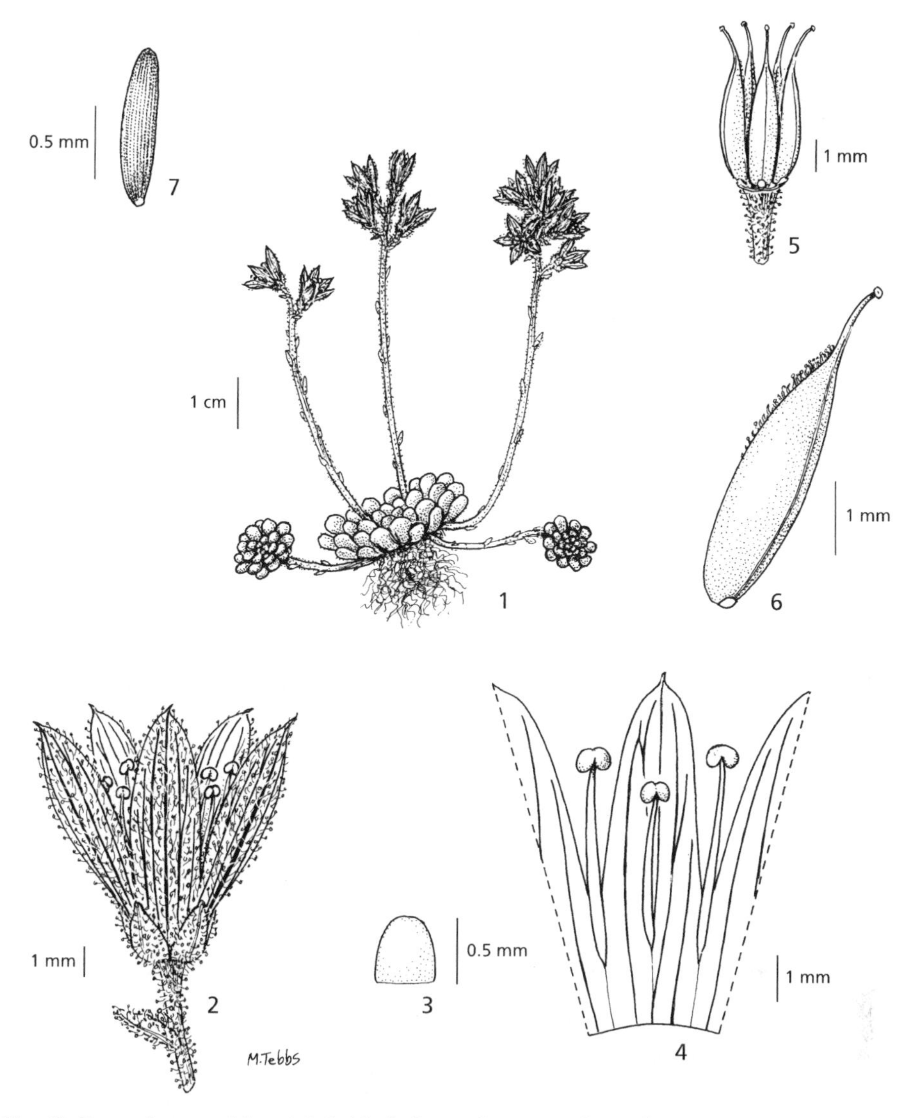

Fig. 43. **Prometheum rechingeri**. 1, habit; 2, flower; 3, nectary; 4, corolla opened; 5, carpels; 6, single carpel; 7, seed. 1 from *Ravi & Serhang* 24831; 2–7 from *Fadden* 60. © Drawn by M. Tebbs (Oct. 2012).

7. **BRYOPHYLLUM** Salisb.

Parad. Lond. t. 3 (1805)

Succulent plants, ± erect. Leaves fleshy, opposite or verticillate, simple, lobed or pinnate, usually petiolate, producing adventitious plantlets usually along the margins of the leaves. Inflorescence terminal, corymbose. Flowers 4-merous, pendulous, pedicellate. Sepals usually united into a calyx tube, not exceeding corolla. Corolla tubular, lobes spreading or reflexed. Stamens 8, usually inserted in lower half of base of corolla tube. Scales rounded or rectangular. Style longer than ovary. Seeds numerous, ± oblong, angular, rugulose.

Originating in Madagascar, there are more than 30 species of *Bryophyllum*, many of which are cultivated. *Bryophyllum* (from the Gr. βρηο *bryo* to sprout and φηλλον *phyllon* a leaf, the name refers to the adventitious plantlets produced on the leaves. The production of these numerous propagules accounts for many escapes into the wild.)

To date, two species have been recorded as being in cultivation or as escapes in Iraq.

Leaves linear, never pinnate, apex dentate .... 1. *B. tubiflorum*
Leaves elliptic, simple or pinnate, crenate .... 2. *B. pinnatum*

1. **Bryophyllum tubiflorum** *Harv.* in Harv. & Sond., Fl. Capensis 2: 380 (1862).

Kalanchoe delagoensis Eckl. & Zeyh., Enum. Pl. Afr. Austr. 3: 305 (1837) nom. nud. (see note).
K. verticillata Scott Elliott in Ann. Bot. 5: 354 (1891).
Bryophyllum delagoense (Eckl. & Zeyh.) Schinz in Mem. Herb. Boiss. 10: 38 (1900).
Kalanchoe tubiflora (Harv.) R.-Hamet. in Beih. Bot. Centr. Beih. 29, 2: 41 (1912).

Glabrous, ± glaucous, erect perennial (0.2–)0.4–1(–1.2) m tall. Roots slightly woody. Stem ± succulent, erect, unbranched, ± terete. Leaves succulent, sessile, alternate or verticillate, rarely opposite, linear, subterete, canaliculate, marked with large, irregular purplish blotches, (13–)40–60(–94) × (1–)2–5(–7) mm; apex with 4–8 teeth from which arise plantlets with entire, obovate leaves. Inflorescence corymbose, 4–16 cm across. Bracts foliaceous, 1–8 × 0.5–1 mm. Pedicels slender, (6–)8–16(–23) mm. Flowers ± pendulous, 21–39 × 7–15 mm. Calyx campanulate, green, 9–11 × 5–12 mm; lobes triangular, 6–7 × 3–4 mm, acute. Corolla tube 21–28 × 6–10 mm, lobes obovate to spatulate, 7–11 × 5–9 mm, apiculate. Stamens inserted in lower half of corolla tube, not exceeding mouth. Scales oblong, slightly emarginate. Carpels 25 × 2 mm; style 3 × ovary. Seeds numerous, angular, minutely ridged.

HAB. On rocks and in exposed, sunny places, occasionally in shade.
DISTRIB. Cultivated or occasionally found as an escape. **LCA**: Baghdad, *Sahira* C. 473!; Abu Ghraib, College of Agriculture, *Janan, Sahira & Raja* C. 893!
NOTE. The original description of *Kalanchoe delagoensis* carries the three word diagnosis, "Flores saturate rosei", and a good argument may be made in favour of this name. For this work, however, I have followed the treatment used in *Flora Zambesiaca* 7, 1: 67 (1983) which considers this to be insufficient and maintains the name *Bryophyllum tubiflorum* which is more widely used in African literature.

Madagascar, E Africa. Naturalised in parts of Africa, Australia and the Samoan Islands. Also introduced to C America.

2. **Bryophyllum pinnatum** (*Lam.*) *Oken* in Allg. Naturgesch. 3, 3: 1966 (1841).

Cotyledon pinnata Lam., Encycl. Méth., Bot. 2: 141 (1786).
C. pinnata Lam. var. β Lam., loc. cit. (1786).
Kalanchoe pinnata (Lam.) Pers., Synops. Sp. Pl. 1: 446 (1805).
K. pinnata (Lam.) Pers. var. *floripendula* Pers., loc. cit. (1805).
Bryophyllum calycinum Salisb., Parad. Lond. t. 3 (1805).
Coltyledon rhizophylla Roxb., Hort. Beng. 34 (1814).

Glabrous, erect, perennial (0.2–)1–1.5(–3) m tall. Roots woody. Stem subsucculent, erect, little branched, ± terete. Leaves succulent, simple or imparipinnate with 3–5 leaflets, elliptic, crenate, (40–)45–90(–130) × (16–)35–65(–90) mm, obtuse; plantlets arising from sides of crenations; petioles (15–)20–50(–82) mm long. Inflorescence thyrsoid, 20–30 × 6–27 cm. Bracts foliaceous, simple or 3-foliolate, 20–35(–43) × 3–9(–14) mm. Pedicels slender, 5–15 mm long. Flowers pendulous. Calyx cylindrical, green ± tinged red, 27–37 × 12–17 mm; lobes triangular, reddish, 6–11 × 6–9 mm, acute. Corolla 35–53 × 8–15 mm, narrowly constricted above the ovary, remainder narrowly urceolate, 1.3 × calyx; tube green, tinged red, 21–40 × 8–15 mm; lobes narrowly triangular to ovate, red or purple, 7–13 × 4–6 mm, acuminate. Filaments inserted in lower half of corolla tube, reaching the mouth; anthers exserted. Scales oblong. Carpels 41 × 4 mm; style 1.8 × ovary. Seeds numerous, angular, minutely ridged.

HAB. Coastal rocks to mountain forest, from sea level to 1800 m.
DISTRIB. Cultivated. **LCA**: Baghdad, *Sahira* C. 586.
NOTE. Pantropical; probably distributed through its cultivation as an ornamental and for the saponins in its leaves.

108. SAXIFRAGACEAE

A. Engler in Pflanzenfam. ed. 1, 18a: 74–166 (1930)
(Tribe Saxifrageae only, excluding subtribe Vahliinae)

C. C. Townsend[14]

Annual or perennial herbs. Leaves alternate (rarely opposite), exstipulate. Flowers pedicellate or sessile, actinomorphic or somewhat zygomorphic, hermaphrodite; inflorescence various, racemose or cymose. Sepals (4–)5, imbricate or valvate. Petals (4–)5, imbricate or valvate, rarely absent. Receptacle flat, or concave and adnate to the ovary so that the petals and sepals may be hypogynous, perigynous or epigynous. Stamens isomerous and alternate with petals, or often twice as many and obdiplostemonous[15]; filaments free, filiform. Carpels commonly 2, more rarely 3–5, ± fused below or rarely free, tapering into styles; stigmas capitate. Fruit a capsule. Seeds numerous; embryo small and straight, with abundant endosperm.

About 30 genera and 600 species, cosmopolitan but chiefly in the N Temperate region; one genus in Iraq.

1. SAXIFRAGA L.

Sp. Pl. ed. 1: 398 (1753); Gen. Pl. ed. 5: 189 (1754); Engler & E. Irmscher in Pflanzenr. 4. 117: 1–709 (1919)

Perennial, frequently caespitose herbs, more rarely annual, biennial or suffruticose, chiefly alpine. Leaves alternate (rarely opposite), simple and entire, toothed or lobed, or tripartite with lobed segments, often thick in texture. Sepals (4–)5, imbricate. Petals 5 (rarely 4 or absent), cuneate but scarcely clawed below, equal or more rarely unequal, imbricate; receptacle flat or concave, petals and stamens ± hypogynous, perigynous or epigynous. Stamens 10, rarely 8. Carpels 2 (rarely 3–5), fused below. Styles free, at first connivent, later ± patent or reflexed. Capsule opening along the upper inner suture of the carpels. Seeds numerous, small, usually ± verruculose.

About 375 species, chiefly in the N Temperate regions and the Arctic; also on the Andes and one species in the Ethiopian mountains.
Saxifraga (from Lat. *saxis*, stones, *fractus*, broken, alluding to the plant's tendency to cling to rock crevices in alpine areas).

1. Perennial; base of stem and rootstock with large globose or ovoid bulbils; petals 7–16 mm . . . . 3. *S. sibirica*
   Annual; base of stem and rootstock without bulbils; petals up to 4 mm 2
2. Weak, scrambling plant; lamina of lower leaves orbicular in outline with (3–)5–7 broadly rounded-deltoid lobes and brown tannin-filled cells; ovary and fruit two thirds free from the receptacle. . . . . 1. *S. hederacea*
   Erect plant; lamina of lower leaves spatulate or cuneate-obovate in outline, 3–5-lobed, with no brown tannin-filled epidermal cells; ovary and fruit almost entirely enclosed by the adnate, campanulate receptacle. . . . .2. *S. tridactylites*

1. **Saxifraga hederacea** *L.*, Sp. Pl. ed. 1: 405 (1753); Boiss., Fl. Orient. 2: 812 (1872); Schönbeck-Temesy in Fl. Iranica 42: 6 (1967); Matthews in Fl. Turk. 4: 258 (1972).

Annual herb, glabrous or glandular-pilose, with weak, fragile, scarmbling stems up to 30 cm but often much less, much-branched from base upwards when well-developed, smaller plants simple. Leaves simple, peltate, orbicular in outline, lower (3–)5–7-dentate with large, subacute or obtuse, broadly rounded-deltoid lobes, upper commonly 3-dentate to entire; lower leaves truncate to cordate at base, 4–30 × 3–30 mm, long-petiolate with petioles up to 3 cm long and 3 times as long as the lamina, uppermost sometimes cuneate

[14] Royal Botanic Gardens, Kew. Updated by Shahina A. Ghazanfar.
[15] With the stamens in two whorls, twice as many as the petals, and the outer series of stamens opposite to the petals.

at base, shortly petiolate or sessile, much smaller; all leaves, and also sepals, with dark, elongate tannin-filled epidermal cells scattered over the surface. Cymes lax and rather few-flowered, flowers on long slender spreading pedicels up to 4 cm, much exceeding the subtending bract, frequently glandular-pilose. Sepals deltoid-ovate, 2–2.5 mm, subacute; receptacle shallowly hemispherical. Petals yellow to white, 2.5–4 mm, oblong to cuneate-obovate. Stamens about two-thirds as long as the petals. Carpels one-third inferior, the apical two-thirds free, abruptly narrowed to the short style; stigmas large, strongly papillose. Capsule globose to ovoid, 3.5–4 mm, yellowish-brown, with a raised vein joining the styles, parallel to the suture, the persistent sepals patent-erect. Seeds ovoid, dark reddish brown, strongly papillose.

Flowers white . . . . var. **hederacea** (not found in Iraq)
Flowers yellow . . . . var. **scotophila**

var. **scotophila** (*Boiss.*) *Engl. & Irmsch.*, l.c.: 201 (1919).

S. scotophila Boiss., Diagn. ser. 1, 3: 23 (1842) & Fl. Orient. 2: 842 (1872).
S. scotophila Boiss. var. *libanotica* Bornmüller in Beih. Bot. Centralbl. 31(2): 216 (1914).

HAB. On cliffs, on rocks by a waterfall, shady limestone rocks; alt. 500–1050 m; fl. Apr.–May.
DISTRIB. Very rare in the mountain district of NE Iraq. **MRO**: Rowanduz gorge, *Guest* 2052!; 'Kurdistan', *Lowe* 202 p.p.
NOTE. The taxonomic value of this variety is somewhat doubtful; it is certainly not possible to correlate petal colour and petal shape and size as done by Engler & Irmscher, and one specimen cited by them as var. *hederacea* (*Kotschy* 364, from Cyprus) was annotated by the collector as "floribus aureis". The variation in flower colour is unusual in wild *Saxifraga* as opposed, for example, to such a genus as *Anemone*; but even this needs testing in the field, as the writer's limited experience of *S. hederacea* has not hitherto included material with pure white flowers, and the yellow-flowered forms pale on drying.

Sicily, Balkans, Crete, Cyprus, Aegean Is., Turkey, Syria, Lebanon, Palestine, Iran (not listed in Jamzad, 1995), Libya.

2. **Saxifraga tridactylites** *L.*, Sp. Pl. ed. 1: 404 (1753); Boiss., Fl. Orient. 2: 808 (1872); Schönbeck-Temesy in Fl. Iranica 42: 7 (1967); Jamzad, Fl. Iran 12: 18 (1995).

Slender annual herb, ± glandular-hairy throughout, erect, simple or branched, (2)5–10(–15) cm. Basal leaves forming a rosette (but often ± withered by flowering time), spatulate or cuneate-obovate, lamina up to 1 cm, palmately 3–5-lobed or -fid or sometimes entire, tapering and pseudopetiolate below; upper leaves few, soon becoming entire and passing into bracts; tannin-filled epidermal cells absent. Cymes lax, few-flowered, flowers axillary on long, slender, ascending pedicels up to 3 cm long, much exceeding the small bracts, glandular-pilose. Sepals oblong-ovate, ± 1 mm, but deltoid and acute in fruit; receptacle campanulate, accrescent. Petals 2–3 mm, white, cuneate-obovate. Stamens about half as long as petals. Carpels fused almost to the apex, abruptly narrowed to the short styles, ovary almost completely inferior; stigmas decurrent on inner surface of styles, papillose. Capsule subglobose, 3–4 mm, almost completely enclosed by the adnate receptacle, with a raised vein close to suture joining the very divaricate styles. Seeds ovoid, yellowish with reddish papillae.

HAB. Between shady rocks, forming a turf; alt. ± 1220 m; fl. May.
DISTRIB. Very rare; only one confirmed and two doubtful localities perhaps from NE Iraq.'Kurdistan', *Low* 202 p.p.; **MRO**: Shaqlawa, *Haines* W.617!; village of Gorluk, fissures of a gypsum rock, "Lindley 1837" (Col. Chesney's expedition to the Euphrates no. 52)!
NOTE. The last record may be from Turkey, as the name is possibly of Turkish etymology and no locality with that name can be traced in Iraq.

Throughout Europe except the extreme north and C & E Europe, Aegean Is., Turkey, Syria, Lebanon, Palestine, Caucasia, Iran, Morocco, Algeria, Tunisia, Libya.

3. **Saxifraga sibirica** *L.*, Syst. Nat. ed. 10, 2: 1027 (1759) & Sp. Pl. ed. 2: 577 (1762); Boiss., Fl. Orient. 2: 807 (1872); Matthews in Fl. Turk. 4: 255 (1972); Ghazanfar in Fl. Pak. 108: 10 (1977); Jamzad, Fl. Iran 12: 14 (1995); Jintang et al. in Fl. China 8: 337 (2001).

S. mollis Sm. in Rees, Cyclop. 31 (1819) and in Sternb., Rev. Sax. Suppl.: 37 (1832); Loz.-Lozinsk. in Fl. URSS 9: 172 (1939); Schönbeck-Temesy in Fl. Iranica 42: 8 (1967).

Perennial herb, (5–)8–15(–25) cm, stems solitary to caespitose, vegetative parts glabrous to ± densely glandular-pilose. Base of stem and upper part of rootstock with numerous roundish to ovoid bulbils which are clad in sparsely long-hairy to densely villous scales, globular and frequently naked bulbils also usually present on the fibrous roots. Basal leaves reniform, cordate at the base, 5–7(–9)-lobed with obtuse or bluntly mucronate, ovate or subrotund segments which are usually ± constricted below, lamina 10–30 × 8–40 mm, petioles long and slender, up to 6 cm; stem leaves becoming less and more narrowly divided, passing into the entire bracts, uppermost only shortly petiolate; tannin-filled epidermal cells absent. Flowers solitary or in few-flowered (rarely more than 7) cymes; peduncles long, ascending or flexuous, up to 2.5 cm, ± glabrous to densely glandular-pilose. Sepals linear-oblong, ± glabrous or ± glandular-pilose, 3–4 mm, rounded; receptacle almost flat. Petals white, oblong-obovate, 7–16 × 4–6 mm, cuneate below. Stamens about one-third as long as petals or a little more. Carpels free in upper one-third, abruptly narrowed to the short styles, ovary practically entirely superior; stigmas pale, shortly papillose. Capsule oblong, 6–9 × 3–4 mm, clasped by the erect, persistent sepals, with a raised vein joining the divaricate styles, with or without more slender transverse veins below. Seeds linear-oblong, blackish brown, verruculose-papillose. Fig. 44: 1–6.

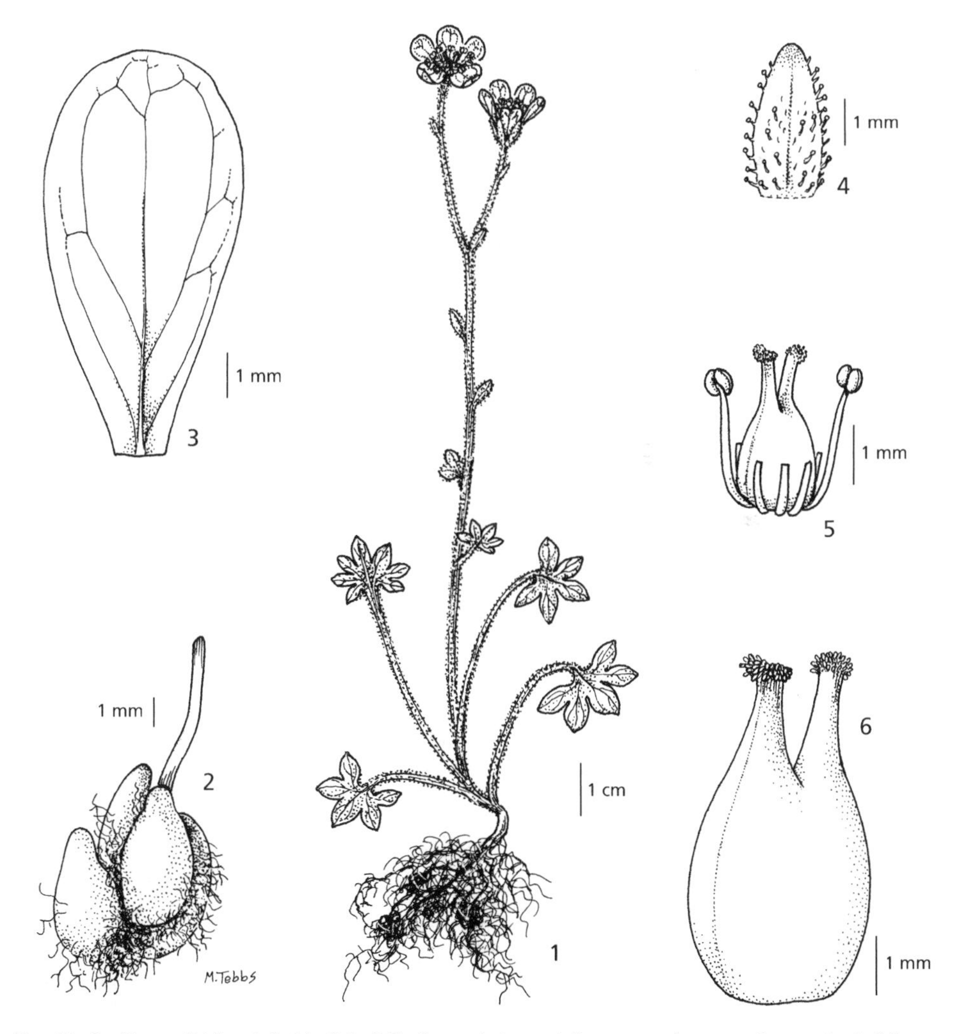

Fig. 44. **Saxifraga sibirica**. 1, habit; 2, bulbils; 3, petal; 4, sepal; 5, ovary and stames; 6, capsule. 1–5 from *Rawi* 24806; 6 from *Rawi* 2964. © Drawn by M. Tebbs (Aug. 2012).

HAB. Among rocks by spring, damp scree below snow, alpine meadow, under rocks in wet places; alt. 2800–3350 m; fl. Jul.-Sep.
DISTRIB. Rare and very local in the alpine zone of NE Iraq. **MRO**: Halgurd Dagh, *Rechinger* 11870, *Gillett* 12346!, *Guest* 2964!, *Haley* 132 , *Ludlow-Hewitt* 1519!, *Rawi* 24806!
NOTE. I (Townsend) have been unable to separate *S. mollis* from *S. sibirica* on the characters given in Fl. URSS. The petals of *S. sibirica* are always basically 3-veined from the base, but the lateral veins may or may not divide; I have seen plants with petals 3-veined above from arctic Russia and the Himalayas, and plants with 5-veined petals from Samos and Turkey. Similarly, petal size is very variable in the species, though I have not seen petals as large as 20 mm, the upper limit claimed for *S. mollis* by Lozina-Lozinskaya.

Turkey, Russia, W Iran, E Afghanistan, Pakistan, C Asia, China.

109. VAHLIACEAE

A. Engler in Pflanzenfam. ed. 2, 18a: 166 (1930), as subtribe Vahliinae of Saxifragaceae.

C.C. Townsend[16]

Erect annual herbs. Leaves opposite, entire, linear to ovate, exstipulate. Flowers in axillary pairs, pedicellate or subsessile, actinomorphic, hermaphrodite. Sepals 5, joined below, valvate. Corolla with 5 free petals, equalling or shorter than calyx, imbricate or valvate. Stamens 5, alternating with petals, inserted on margin of an epigynous disk; filaments free, filiform; anthers longitudinally dehiscent. Receptacle concave, adnate to ovary. Ovary inferior, unilocular, 2(or 3)-carpellate; styles 2 or 3, long or short, stigmas capitate. Ovules numerous. Fruit a subglobose or ovoid 2-valved capsule, dehiscing from the apex. Seeds numerous, minute, endospermic.

A single genus with eight species distributed from SW Asia to Pakistan and India, and in tropical and S Africa. This family had previously been placed in the Order Saxifragales, in family Saxifragaceae, but is currently left unplaced as to order in the asterids clade.

1. **VAHLIA** Thunb.

Nov. Gen. Pl. 2: 36 (1782) *nom. cons.*
Bistella Adans., Fam. Pl. 2: 226 (1763)

Characters as for the family.

1. **Vahlia digyna** (Retz.) O.Kuntze, Rev. Gen. 227 (1891); Fischer in Kew Bull. 1932: 56 (1932) pro parte; Keay in Fl. W. Trop. Africa ed. 2, 1: 120 (1954) pro parte; Schönbeck-Temesy in Fl. Iranica 23: 4 (1966); Bridson in Kew Bull. 30, 1: 177, fig. 1 A–H (1975).

Oldenlandia digyna Retz., Obs. Bot. 4:23 (1786).
Vahlia weddenii Rchb., Iconogr. Bot. Exot. 1: 62, t. 91 (1827); Boiss., Fl. Orient. 2: 799 (1782).
V. viscosa Roxb., Fl. Ind. 2: 89 (1832).
Bistella digyna (Retz.) Bullock in Act. Bot. Neerl. 15: 85 (1966); Siddiqi in Fl. Pak. 10: 1 (1971).

Annual herb, 15–40 cm, erect or with basal branches decumbent or ascending, much-branched with lower branches ± equal and opposite, upper branches unequally opposite or opposite to a pair of flowers; vegetative parts and calyx densely furnished with an indumentum of multicellular glandular cells. Leaves of main stem oblong-ovate to lanceolate, sessile, 10–40 × 2–4 mm; branch and inflorescence leaves gradually becoming shorter and proportionally broader, oblong- to orbicular-ovate. Flowers in axillary pairs, upper sessile, some lower frequently on glandular pedicels up to 4 mm long, with or without a short common peduncle. Calyx ± 3 mm, campanulate, tube 1.5–2 mm, densely furnished with long, whitish hairs, strongly 5-ribbed, ribs green or reddish; lobes variable in length, ovate or lanceolate-ovate, acute, green, more shortly and sparingly glandular-

[16] Royal Botanic Gardens, Kew. Updated by Shahina A. Ghazanfar.

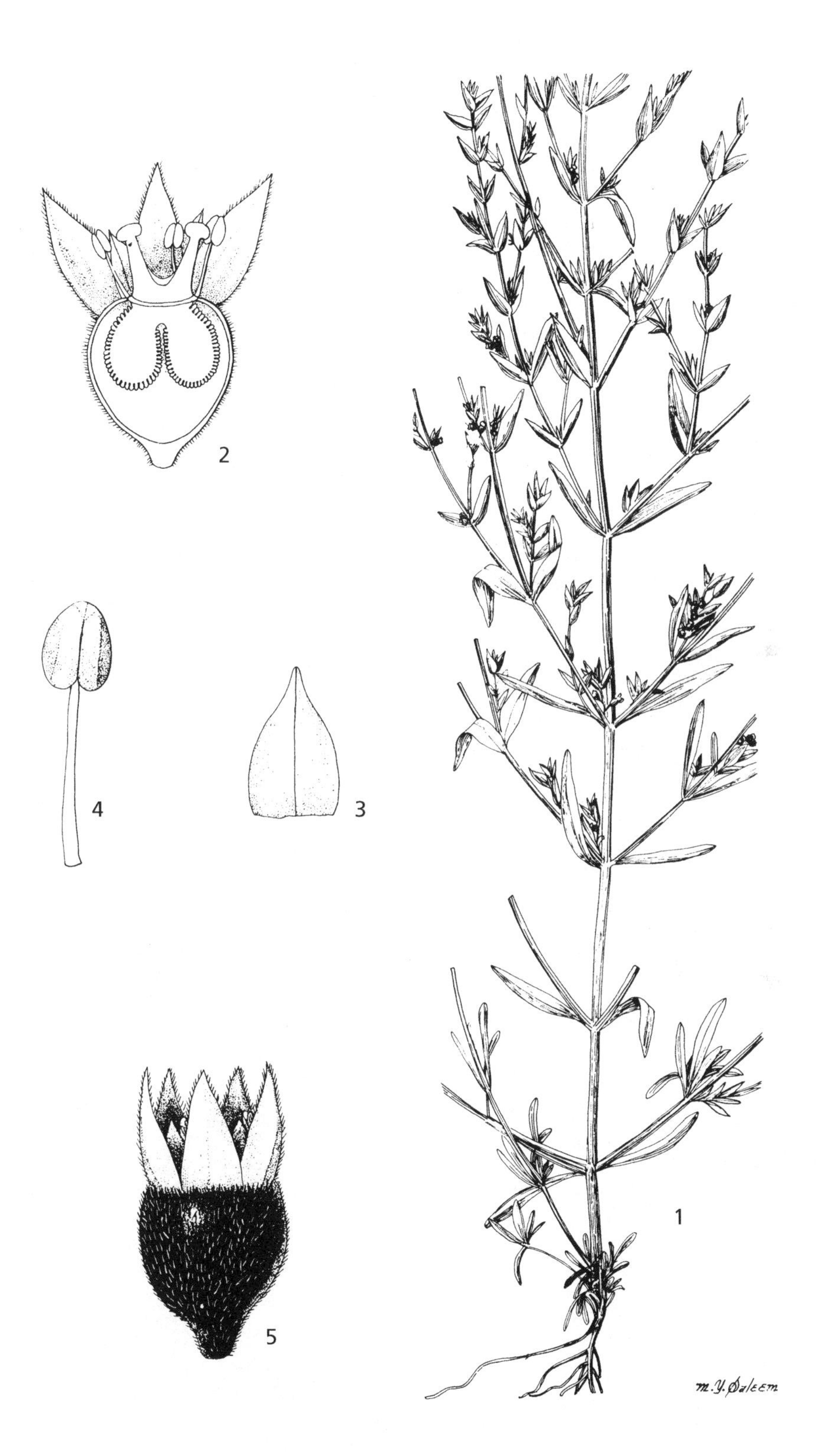

Fig. 45. **Vahlia digyna**. 1, habit × 1; 2, flower, vertical section (enlarged); 3, petal (enlarged); 4, stamen (enlarged); 5, flower × 12. Reproduced from Fl. Pak. 10: f. 1 (1971), with permission.

hairy. Petals yellow (fading white), shorter than to about equalling sepals, ovate to obovate. Stamens subequalling petals; filaments glabrous or patent-hairy, with a minute hairy scale at the inner base. Styles about as long as stamens, slightly divergent, pilose. Capsule subglobose, 2–3 mm, opening by 2 or 3 short teeth; seeds very small, reddish, minutely reticulate. Fig. 45: 1–5.

Hab. Sandy river beds, swampy ground; alt. ± 50 m; fl. ?Apr.
Distrib. **fuj**: Diyala, *Sutherland* 286.
Note. The Iraqi plant resembles the common African form of this variable species rather than the Indian, having larger petals and longer, pilose filaments than the latter.

West Africa eastwards to India.

110. PARNASSIACEAE

S.F. Gray, Nat. Arr. Br. Pl. 2 : 670 (1821)

J.R. Edmondson[17]

Rosette-forming rhizomatous perennials with simple long-petiolate entire leaves. Flowering stem normally an almost leafless scape with a single terminal flower. Flowers hermaphrodite, pentamerous. Sepals 5, ± free. Petals 5, white or greenish white, entire, ± marcescent. Stamens 5, alternating with 5 much-divided ± glandular staminodes. Ovary ± superior, unilocular, with parietal placentation and a single very short style; stigmas 3-4; ovules numerous. Fruit a 3–4-valved capsule. Seeds many.

A family of two genera and about 70 species with a close affinity to Saxifragaceae, in which it was often included. In APG III (2009) classification, it is subsumed in the family Celastraceae. It is distributed across the whole N Temperate region, with a concentration of species in China.

1. PARNASSIA L.

Sp. Pl. 273 (1753)

Characters as for the family.

Parnassia, named for Mt. Parnassus in Greece, the haunt of Apollo and other mythical deities.

1. **Parnassia palustris** *L.*, Sp. Pl. 273 (1753); Boiss., Fl. Or. Suppl. 249 (1888); Losinsk. in Fl. URSS. 9: 216 (1939); Webb in Fl. Europaea 1: 381 (1964); Kitamura, Pl. W. Pak. and Afghan. 3: 79 (1964); Matthews in Fl. Turk. 4: 260 (1972); Schönbeck-Temesy in Fl. Iranica 20: 1 (1966); Siddiqi in Fl. Pak. 31: 1 (1973); Gu Cuizhi & Hultgård in Fl. China 8: 379 (2001).

Parnassus vanensis Azn. in Magyar Bot. Lapok 1917, 16: 17 (1918).

A glabrous perennial herb with 2–4 slender erect scapes to 25(–35) cm. Basal leaves ovate to suborbicular or triangular-ovate, 1.5–3 × 1–2.5 cm, base cordate or truncate, apex rounded, long-petiolate, the blade somewhat decurrent. Cauline leaves absent. Flower terminal, solitary, 2.5–3.0 cm in diameter. Sepals 4–6 × 1.5–3 mm, apex obtuse. Petals broadly ovate to oblong, ± 1.5 × 0.8 cm, white with dark green veins. Anthers ellipsoid, ± 3 mm; filaments 3–7 mm; staminodes spatulate, divided into 8–12(–20) linear gland-tipped segments. Capsule loculicidal, ovoid to subglobose. Seeds brown, oblong, glossy. Fig. 46: 1–3.

Hab. Side of mountain spring; alt. ± 2440 m; fl. Sep.
Distrib. Very rare, having been recorded only once from the alpine zone of NE Iraq. **mro:** Baradost Dagh, Lohlan, *Thesiger* 334 (BM!).

Europe, N Africa, Turkey (SE Anatolia), Russia, W Iran, Afghanistan, Pakistan, Kazakstan, Himalaya eastwards to China and Japan; N America.

[17] Honorary Research Associate, Royal Botanic Gardens, Kew.

Fig. 46. **Parnassia palustris.** 1, habit × 1; 2, flower spread open × 3 ; 3, staminode × 9. Reproduced from Fl. Pak. 31: 2, f. 1 (1973), with permission.

111. APIACEAE

(Umbelliferae)

C.C. Townsend[18]

Annual, biennial or perennial herbs, occasionally suffrutescent but rarely shrubs. Stem often furrowed or striate, with a large pith which shrinks at maturity and leaves the internodes hollow. Leaves alternate (very rarely opposite), simply pinnate to pinnately decomposite, rarely simple or entire, petioles usually sheathing. Inflorescence a simple, or more commonly compound umbel, sometimes reduced to a capitulum or reduced and cymose. Bracts and bracteoles (involucre and involucel) present beneath the primary and secondary ("partial") umbels or not, simple or divided. Flowers usually bisexual, more rarely unisexual, actinomorphic, the perianth biseriate, the calyx adnate to the ovary, the 5 teeth free and obvious or much reduced or obsolete. Petals 5, usually with the tips inflexed, the inflexed lobule short to very long and subequalling the lamina; petals of outer flowers frequently larger than the remainder, the flowers thus radiant; colour usually white or pinkish, frequently yellow, rarely deep red or blue. Stamens 5, alternate with the petals, inflexed in bud. Ovary inferior, 2–(rarely 1)-locular; ovules pendulous, solitary in each loculus. Styles 2, often with an enlarged, swollen base (stylopodium) and set on a small or dilated or crenulate-margined disk. Fruit a schizocarp of 2 indehiscent carpels which may be dorsally or laterally compressed and are separated by a broad or narrow commissure; carpels adnate to or suspended from a slender simple or bifid wiry stalk or carpophore in those species in which they separate at maturity (the great majority). Each mericarp typically with 5 evident primary ribs, occasionally with 4 intermediate secondary ribs; some or all of the ribs may be dilated into a wing, more commonly they are filiform, or prominent and corky. Oil tubes (vittae) present or absent in the valleculae (intercostal spaces) or under the ribs, and on the commissural face. Embryo minute, endosperm abundant, firm, its shape furnishing a valuable character in the major subdivisions of the family.

A family of over 200 genera (though opinions on generic delimitation vary), and 3000 species, occurring mainly in the northern hemisphere. Of considerable economic importance. Many species contain lethal alkaloid poisons (e.g. *Conium, Cicuta, Oenanthe* spp.); others are valued food plants (e.g. *Daucus* (carrot), *Pastinaca* (parsnip), *Apium* (celery), or used for culinary or medical purposes (*Carum, Foeniculum, Pimpinella, Anethum* spp.). Ornamental species, or plants grown as curiosities, include *Eryngium, Trachymene,* and some of the huge *Heracleum* spp.

Sexuality in the Umbelliferae is very difficult of description. Many genera have "polygamous" inflorescences, usually with the terminal umbel hermaphrodite or mixed male and hermaphrodite and the lateral umbels entirely male or mixed. However, it is often difficult to distinguish between functionally male flowers which have clear styles and stigmas but do not produce fruit and hermaphrodite flowers which abort. Individuals of normally hermaphrodite species which are at any rate apparently polygamous occur, and species normally polygamous will produce entirely hermaphrodite plants. Genuine male flowers with styles completely absent, such as occur in *Ferula,* present no difficulty; but the character is by no means as straightforward as might be judged from the accounts of earlier authors, and virtually impossible to make use of in a key.

1. Leaves simple, undivided, margins entire . . . . 13. **Bupleurum**
   Leaves (at least the lower) either spiny, dentate or variously dissected – often pinnatisect (twiggy plants of deserts and mountains in which all the evident leaves are reduced to sheaths may be caught here) 2
2. Bracts and bracteoles large, rigid, with very broad, white margins; umbels very small, the pedicels filiform; central flower alone fertile; fruit deeply immersed in the receptacle-like tip of the peduncle . . . . 4. **Ergocarpon**
   Plant not as above 3

[18] Royal Botanic Gardens, Kew. Updated by J.R. Edmondson.

3. Umbel with a central female flower giving rise to a partially immersed bicarpellate fruit, some of the outer flowers producing unicarpellate fruits, their pedicels hard and incrassate at maturity; bracts and sepals spinous or accrescent-foliaceous; partial umbels falling entire . . . . . 3. **Anisosciadium**
   Plant not as above 4
4. Bracts spinous; umbels forming dense capitula 5
   Bracts, if present, not spinous. 6
5. Leaves spatulate, broad, cuneate and entire below, long-spinous around the upper margin . . . . . 5. **Pycnocycla**
   Leaves not spatulate, palmatifid or pinnatisect. . . . . 1. **Eryngium**
6. Umbel simple, globose; sepals pinnatifid into filiform segments . . . . . 2. **Lagoecia**
   Plant with compound umbel, or sepals not pinnatifid 7
7. Wiry desert plant with terete, striate stems; leaf sheaths broadly bordered with white; umbels shortly 3-4-radiate; fruit with dense lanate hairs . . . . . 15. **Pituranthos**
   Plant without the above combination of characters 8
8. Involucre of the simple umbel leafy, finely aristate-dentate at the margin; fruit tuberculate; annuals . . . . . 10. **Actinolema**
   Plant not as above 9
9. Flowers strongly radiant, the umbel with a penicillate tuft of purple hairs at the centre; fruit with spatulate scales around the marginal ribs . . . . . 56. **Artedia**
   Plant not as above 10
10. Involucre of pinnate, trifid or dentate bracts or bracteoles (some species, e.g. *Pimpinella olivierioides*, occasionally have one or two bracts trifid at the apex; they are not included here) 11
    Involucre of simple bracts or absent 19
11. Flowers yellow (*papillaris*) or grey-green . . . . . 6. **Cachrys**
    Flowers white 12
12. Leaves simply pinnate with coarsely serrate, lanceolate segments; plant of wet places . . . . . 22. **Berula**
    Leaves not simply pinnate. 13
13. Fruit with a beak at least as long as the carpels . . . . . 31. **Scandix**
    Fruit not beaked . . . . . 14.
14. Petals densely hairy, umbels with ± 3 very short rays, compact, not more than 2 cm diameter, villous. . . . . 17. **Oliveria**
    Petals not hairy, umbels of considerable size, > 2 cm in diameter 15
15. Ovary and fruit with secondary ridges bearing glochidiate spines, the filiform primary ridges with rows of divergent bristles. . . . . 57. **Daucus**
    Ovary and fruit glabrous or with simple hairs. 16
16. Fruit covered with long, shining hairs much exceeding its own diameter . . . . . 55. **Cuminum** [*setifolium*]
    Fruit and ovary glabrous 17
17. Leaf segments capillary. . . . . 28. **Grammosciadium**
    Leaf segments linear to lanceolate, flat 18
18. Dwarf perennial, alpine, stem leaves very few; radical leaves pinnate, the pinnae further dissected . . . . . 18. **Carum**
    Tall plants, stem leaves many; radical leaves mostly withered, multisect . . . . . 16. **Ammi**[19]
19. Flowers yellow 20
    Flowers white, pink, red or greenish 41
20. Leaves simple, only obscurely lobed, cordate at the base, with large acuminate teeth; densely cinereous small rock plant . . . . . 29. **Pimpinella**
    Leaves variously dissected, not simple 21

[19] One exceptional plant of *Scaligeria assyriaca* has been seen with broadly pinnatisect bracts, and would key out here.
It differs from *Ammi* in its tuberous root and fruit not distinctly longer than broad.

21. Leaf segments terete, rush-like, elongate and resembling the petiole 22
    Leaf segments not cylindrical and rush-like 23
22. Segments of lower leaves flat, the upper reduced to sheaths; involucel present ....51. **Dichoropetalum**
    Leaf-segments all terete, cylindrical, rush-like; involucre and involucel absent .... 40. **Leutea**
23. Leaf segments large and broad, broadly lanceolate to ovate 24
    Leaf segments narrow, oblanceolate to capillary............................. 29
24. Fruit dorsally planocompressed, suborbicular, the lateral ribs thickened, contiguous to form a tumid[20] border to the fruit................. 25
    Fruit not dorsally planocompressed, without a tumid border 26
25. Involucre present... 44. **Opopanax**
    Involucre absent .... 45. **Malabaila**
26. Fruit much larger than broad, the lateral ribs contiguous, dilated to form a flat, thin wing; intermediate ribs winged above only ......35. **Heptaptera**
    Fruit with none of the ribs winged.. 27
27. Involucre and involucel absent; fruit didymous, black when ripe; upper leaves cordate-amplexicaul...............................9. **Smyrnium**
    Involucre and involucel present; fruit not didymous; upper leaves not cordate-amplexicaul 28
28. Plant relatively slender in all its parts; the terminal leaflets of the lower leaves about 4–5 cm long; main umbels up to 7–8 cm diameter; mature fruit about as long as broad, ± 5 × 5 mm, ribs not keeled.. 11. **Smyrniopsis**
    Plant very stout in all its parts; the terminal leaflet of the lower leaves up to 15 cm long or more; main umbels about 20–25 cm diameter; mature fruit about twice as long as broad, ± 10 × 6 mm, the ribs very prominent and keeled................ 12. **Petroedmondia**
29. Ovary and fruit papillose 30
    Ovary and fruit glabrous, or hairy (not papillose) 31
30. Glabrous mountain plant; rootstock crowned with a mass of the remains of the previous years' leaf-sheaths; leaf segments not rigid .... 8. **Pseudotrachydium**
    Robust plant, hairy throughout (including petals); leaf segments rigid, divaricate.. 6. **Cachrys**
31. Ribs of fruit *all* winged, the wings often very undulate; leaf segments linear, rigid or capillary.............................. 36. **Prangos**
    Ribs of fruit not winged, or if with a narrow erose crisped with the leaf segments narrowly lanceolate, flat (lateral ribs only may be narrowly winged) 32
32. Stem bearing conspicuous stellate, papillate hairs .... 6. **Cachrys** (*C. papillaris*)
    Stems without stellate hairs.. 33
33. Involucre and involucel absent.. 34
    Involucre and/or involucel present.. 36
34. Wiry perennial, usually with only a few basal leaves visible; most of the upper leaves reduced to sheaths, the lower with the main axis appearing jointed at the junction of the lower pinnae.. 39. **Peucedanum**
    Aromatic annual, biennial or perennial; rarely one or two uppermost leaves reduced to sheaths, the lower not jointed along the main axes 35
35. Marginal ribs of fruit expanded to form a narrow, flat wing, fruit dorsally compressed ....41. **Anethum**
    Fruit not narrowly winged, laterally compressed.................. 33. **Foeniculum**
36. Involucre absent; fruit with a pale tumid border, ± 4 mm .... 50. **Johrenia**
    Involucre or involucel present; fruit without a pale, tumid border, often larger 37

---

[20] Inflated, swollen.

37. Fruit small, ± 4 mm; ultimate leaf-segments broadly oblanceolate .. 30. **Hellenocarum**
Fruit larger, at least 8 mm when mature; ultimate leaf-segments narrowly lanceolate, linear or capillary 38
38. Fruit laterally compressed, the ribs carinate 39
Fruit dorsally compressed, the ribs thickened and prominent but not carinate 40
39. Stylopodia deeply immersed in the apical emargination of the mericarp, the two halves almost vertical in ripe fruit ..... 6. **Cachrys** (*C. papillaris*)
Stylopodia of mature fruit with the two halves flat or slightly inclined, set on top of fruit .... 36. **Prangos**
40. Involucre absent; umbel peduncles never very short and crowded with irregular bracts, giving a racemose inflorescence .... 37. **Ferula**
Involucre present, bracts usually reflexed, or peduncles of umbels very short and crowded with bracts, inflorescence thus raceme-like .... 38. **Ferulago**
41. Fruit strongly dorsally compressed, margin pale and tumid, interrupted to form large transversely oblong, flattened tuberculiform processes .... 53. **Ormosciadium**
Fruit not as above; if margin pale and tumid then smooth to crenulate-rugose 42
42. Fruit furnished with long, white, soft hairs longer than its own diameter........ 43
Fruit with stout glochidiate or uncinate prickles, or hairs shorter than its diameter, or glabrous 45
43. Hairs of fruit glochidiate at tip .... 59. **Psammogeton**
Hairs of fruit simple or uncinate 44
44. Calyx teeth distinct, persistent on the ± 4 mm ripe fruit; hairs of fruit often purplish when young .... 66. **Chaetosciadium**
Calyx teeth very small, fruit 3 mm or less; young fruit with hairs never purplish .... 29. **Pimpinella**
45. Fruit with stout glochidiate prickles 46
Fruit with simple hairs, or tuberculate, or glabrous 52
46. Flowers extremely conspicuously radiate; stem with glochidiate bristles .... 61. **Lisaea**
Flowers not very conspicuously radiate, or if so the stem with simple bristles 47
47. Leaves simply pinnate with coarsely dentate pinnae .... 64. **Turgenia**
Leaves bi- or tri-pinnate 48
48. Bracts and bracteoles broadly scarious-margined .... 58. **Orlaya**
Bracts and bracteoles, when present, without broad, scarious margins 49
49. Leaves bi- or tri-pinnatisect, the segments lanceolate to linear-lanceolate; mature fruit not exceeding 5 mm, the secondary ribs narrow .... 65. **Torilis**
Leaves pinnately decomposite; fruit ± 8 mm at maturity 50
50. Rays 6 or more .... 63. **Astrodaucus**
Rays 2–5 51
51. Leaf segments long- and narrowly-linear; secondary ribs very broad, flat, with several series of spines .... 60. **Turgeniopsis**
Leaf segments short; secondary ribs narrow, with a single row of spines .... 62. **Caucalis**
52. Fruit with simple hairs or tuberculate 53
Fruit glabrous........ 64
53. Fruit furnished with a beak considerably exceeding the length of the carpels .... 31. **Scandix**
Fruit without or with only a short beak 54
54. Fruit at least twice as long as broad 55
Fruit not twice as long as broad, obcordate to suborbicular 57

55. Short plant with long, narrowly linear leaf segments; flowers often red, the fruit crowned with the persistent, conspicuous calyx teeth . . . . . 55. **Cuminum**
Tall plants with ovate or lanceolate leaf segments 56
56. Stem often ± swollen just below the nodes; stigma sessile, fruit not beaked . . . . . 26. **Myrrhoides**
Stem not swollen below the nodes; styles clearly developed; fruit with a short beak . . . . . 32. **Anthriscus**
57. Fruit terete or laterally compressed, the marginal ribs not thickened and contiguous to form a turgid border to the fruit 58
Fruit strongly dorsally planocompressed, the marginal ribs thickened and contiguous, forming a tumid border to the fruit 61
58. Calyx persistent; fruit furnished with white papillae; annual plant with glabrous leaves, inflorescence not corymbose . . . . . 23. **Trachyspermum**
Calyx absent; fruit pubescent, or if papillose the plant a stout biennial with a corymbose inflorescence and hairy leaves . . . . . 29. **Pimpinella**
59. Surface of at least one mericarp of the fruit with minute white pruinose papillae 60
Surface of fruit not pruinose-papillose 61
60. Margin of fruit crenulate-rugose; leaves subtripinnatisect . . . . . 48. **Tordylium**
Margin of fruit smooth; leaves simple or simply pinnate . . . . . 49. **Ainsworthia**
61. Leaves glabrous, ternate or pinnatisect; petals pubescent dorsally . . . . . 46. **Ducrosia**
Leaves hairy; petals glabrous, or if hairy (*Heracleum lasiopetalum*) the leaves merely lobed, roundish in outline 62
62. Leaves tripinnatisect with the pinnules deeply toothed; flowers not radiant . . . . . 47. **Zosima**
Leaves merely shallowly lobed or singly pinnate with lanceolate to ovate, ± deeply toothed segments; flowers radiant 63
63. Calyx absent; root tuberous . . . . . 42. **Trigonosciadium**
Calyx present; root not tuberous . . . . . 43. **Heracleum**
64. Rigid mountain plant with terete and subcylindrical or angular segments to the lower leaves; stem leaves mostly reduced to sheaths . . . . . 34. **Sclerochorton**
Plant not as above 65
65. Fruit didymous, broader than long 66
Fruit not didymous, not broader than long 67
66. Flowers radiant, umbels with only 2–3 rays . . . . . 54. **Bifora**
Flowers not radiant, umbels with 8–12(–15) rays . . . . . 24. **Korshinskia**
67. Flowers conspicuously radiant, the fruit shortly ovoid, not narrowed to the commissure, hard, crowned with the conspicuous, unequal, recurved calyx teeth . . . . . 52. **Coriandrum**
Flowers only rarely radiant; fruit not ovoid, the commissure distinct 68
68. Fruit at least twice as long as broad 69
Fruit scarcely longer than broad 73
69. Fruit with a short, broad, sulcate beak $^1/_3$ – $^1/_2$ the length of the mericarps . . . . . 32. **Anthriscus**
Fruit not beaked 70
70. Leaves hairy, stem with dense downwardly directed whitish hairs; fruit linear, 7–10 × as long as broad or more; involucre absent . . 27. **Chaerophyllum**
Leaves glabrous save at most some scabridity on the principal veins beneath; stem glabrous; fruit 2–4(–8) × as long as broad; involucre present on at least some umbels 71
71. Leaves ternate-pinnate, the pinnae elongate, linear-lanceolate, with regular cartilaginous margined teeth, decurrent along the axes to form a broad serrate wing . . . . . 20. **Falcaria**
Leaves not as above 72
72. Root tuberous-globose; leaf segments long, linear, flat, entire . . . . . 19. **Bunium**
Root not tuberous; leaves deltate, pinnate to sub-bipinnate with oblong-lanceolate segments resembling those of an *Anthriscus* . . . . . 29. **Pimpinella**

73. Leaves simply pinnate with lanceolate finely or coarsely serrate leaflets 74
Leaves not simply pinnate with serrate leaflets 76
74. Lower leaves with rarely fewer than 8 pairs of coarsely and often irregularly serrate leaflets . . . . . 22. **Berula**
Lower leaves with rarely more than 5 pairs of finely and regularly serrate leaflets 75
75. Procumbent or ascending plant with sessile or shortly pedunculate, usually ebracteate umbels . . . . . 14. **Apium**
Erect plant with long-pedunculate umbels; involucre present . . . . . 21. **Sium**
76. Leaf-segments broad; involucre and involucel absent; plant with the characteristic smell of celery . . . . . 14. **Apium**
Leaf-segments narrow; involucre and involucel present; plant not smelling of celery 77
77. Slender, much-branched plant, the upper branches and umbel rays almost capillary; lower leaves often withered at flowering, the upper reduced to small sheaths . . . . . 25. **Elaeosticta**
Stouter plants with well-developed stem leaves 78
78. Tall plant (to 2 m) with purple-spotted stems; leaves graceful, large, bi- to tri-pinnatisect with short, lanceolate segments . . . . . 7. **Conium**
Shorter plants to 50 cm, stems not purple-spotted; leaves smaller, either less divided or with the ultimate segments long 79
79. Short mountain plant rarely attaining 20 cm in height; root not tuberous; leaves pinnate with short, deeply-divided pinnae . . . . . 18. **Carum**
Tall plant 30–50 cm, with globose-tuberous roots; leaves with elongate-linear, entire segments . . . . . 19. **Bunium**

1. ERYNGIUM L.

Sp. Pl. ed. 1: 232 (1753); Gen. Pl. ed. 5: 108 (1754); Wolff in Pflanzenr. IV. 228, 61: 106–271 (1913)

Annual or usually perennial herbs (always perennial in Iraq), very rarely shrubs. Leaves entire or variously divided, usually ciliate or spinose and glabrous. Flowers bracteolate, sessile in dense heads surrounded by leafy, usually rigid and spinous bracts. Calyx teeth present, acute or spiny. Petals white or coloured, narrow, deeply notched, with a tightly incurved lobule often almost as long as the petal lamina. Fruit ovoid or obovoid, usually ± flattened but occasionally almost globose, 5-ribbed, often furnished with prominent chaffy scales. Commissure broad. Mericarps obscurely ribbed. Vittae slender, occasionally subreticulate or absent. Commissural face of endosperm concave.

About 230 tropical and temperate species (excluding tropical and S Africa); 7–8 species in Iraq.

Eryngium (from ηρυγγιον, *ēruggion*, a Gr. name in Theophrastus for a plant with spinous leaves which is thought to have been *E. campestre*; Eryngo.

Bailey (1939) lists some 30 species as horticultural subjects, remarking that they are excellent in borders and rock gardens, and Chittenden (1951) mentions over 25. They are particularly prized for their coloured stems and involucres which vary in hue from shades of greenish blue to purplish or even brownish. As mentioned by Guest (1933) certain European species have long been known as medicinal plants. In this connection Wren (1956) mentions Field Eryngo (*E. arvense*, q.v.). Sea Holly (*E. maritimum*), which is found in coastal habitats round the Mediterranean and in Britain but does not occur in our territory, was formerly highly regarded for its supposed aphrodisiac effect and as mentioned by A.W Smith (1963), its roots were collected and candied, recipes for doing this being found in old cookery books up till well into the 18th century. Ibn al-Baitar (c. 1240 AD) records that Eryngo roots, called SHAQĀQIL in Syria, were imported into Iraq from Bokhara, Samarkand and even from as far away as China. Watt (1890) states that the root of *E. planum* L. (sub *E. coeruleum*) – strangely enough known as SHAQĀQIL MISRI ("Egyptian eryngo", Ar.) – was also considered as an aphrodisiac and vein tonic and that its seed were said to be officinal at Kandahar in Afghanistan. Finally a useful fibre, known as Garaguata fibre, is obtained from the leaves of the subtropical South American species, *E. pandanifolius*.

1. Lower stem leaves entire or trisect, the segments serrate but undivided, very large, up to 10 cm broad and the terminal to 20 cm long; plant very robust . . . . 2. *E. pyramidale*
   Plant not as above 2
2. Inflorescence thyrsoid 3
   Inflorescence corymbose 5
3. Leaves of the inflorescence region broad-based, at least the lower with reticulate venation . . . . 5. *E. thyrsoideum*
   Leaves of the inflorescence with narrow bases, venation parallel 4
4. Venation of bases of median stem-leaves, which are all narrow, parallel throughout; lower stem-leaves 1–2-pinnatifid . . . . 3. *E. glomeratum*
   Median stem-leaves with broader bases, with reticulate venation at least towards the margins; lower stem-leaves trisect or only the very lowest pinnatisect, with few pinnae . . . . 4. *E. hainesii*
5. Sepals broadly ovate, truncate or even emarginate at the tip, the vein percurrent to form a very short mucro; inflorescence repeatedly forked with long branches often ± as wide as long; bracteoles all tricuspidate . . . . 1. *E. creticum*
   Sepals lanceolate, tapering to the long mucro formed by the excurrent vein; inflorescence usually less expansive and the heads usually larger (12–15 mm); bracteoles simple, or only the outermost tricuspidate 6
6. Amethyst blue suffusion present on plant, usually widespread but sometimes confined to the developing capitula; inflexed lobes of petals dilated at tip . . . . 6. *E. billardieri*
   Plant yellowish or glaucous-green; inflexed lobes of petals parallel-sided 7
7. Plant smaller, usually to 60 cm; ultimate branches of the inflorescence shorter, rendering it more compact at the extremities; secondary divisions of leaves with deeper, more narrowly based leaves; stem and branches rarely pure white, usually greenish-tinged . . . . 7. *E. campestre*
   Plant very robust, attaining 1.5 m or more in height; final branches of inflorescence long, making this very broad; leaves large, with strong, broad-based teeth; stem and branches pure white . . . . 8. *E. noeanum*

1. **Eryngium creticum** *Lam.*, Encycl. Meth. Bot. 4: 754 (1797); DC., Prodr. 4: 89 (1830); Boiss., Fl. Orient. 2: 827 (1872); Hand.-Mazz. in Ann. Naturh. Mus. Wien 27: 87 (1913); Wolff, l.c. 131 (1913); Nábělek in Publ. Fac. Sci. Univ. Masaryk 35: 119 (1923); Hayek, Prodr. Fl. Balc. 1: 968 (1927); Guest in Dep. Agr. Iraq Bull. 27: 33 (1933); Anth. in Notes Roy. Bot. Gard. Edinb. 18: 289 (1935); Bornmüller in Beih. Bot. Centralbl. 58B: 274 (1938); Blakelock in Kew Bull. 3: 433 (1948); Zohary in Dep. Agr. Iraq Bull. 31: 33 (1950); H. Riedl in Fl. Lowland Iraq: 450 (1964); Rawi in Dep. Agr. Iraq Tech. Bull. 14: 91 (1964); Rawi & Chakr., ibid. 15: 41 (1964); Mouterde, Nouv. Fl. Lib. et Syr. 2: 580 (1970); Davis, Fl. Turk. 4: 298 (1972); Zohary, Fl. Palaest. 2: 385 (1972); Meikle, Fl. Cyprus 1: 690 (1977); Pimenov & Tamamschian in Fl. Iranica 162: 59 (1987); Boulos, Fl. Egypt 2: 155 (2000).

E. dichotomum DC., Prodr. 4: 90 (1830) nec Desf. (1798).

Plant attaining 1 m, but commonly about half this height. Oldest leaves undivided and crenate, but these are usually withered by flowering time and not present in most herbarium material. Lower cauline leaves very deeply palmatipartite into narrow segments, closely and finely spinous. Upper stem-leaves and the 4–6 involucral bracts very narrow, very little lamina visible between the thickened margins and the vein, entire for the greater part of their length but with 1-several pairs of lateral spines especially at the base. Bracteoles broad and rigid, all tricuspidate above, exceeding the flower. Inflorescence broad and repeatedly dichotomous, often bluish, the ultimate branches long. Capitula small, up to 10 mm. Sepals broad and obtuse, sometimes truncate or even emarginate, with a very short mucro. Petals very narrow, oblong. Fruit ± 3 mm, the broad flat face smooth, the convex dorsal face with 3 prominent ribs. Scales very small and much less conspicuous than in species 5–7.

HAB. Dry stony hillsides, sometimes in *Quercus* forest, on borders of fields and wasteland; alt. (150)300–900(–1150) m; fl. & fr. Jun.–Aug. or later.
DISTRIB. Common, locally abundant, in the lower forest zone of Iraq, occasional in the steppe region. ?**MAM**: (or more probably in Turkey), between Zakho and Jazira Ibn Omar (Cizre), *Nábělek* 578; Atrush, *Guest* 3645!; Sarsang, *Haines* W448!; Dinarta, *E Chapman* 26140!; 26153!; **MRO**: 1 km from Salah ad-Din, along the road to Shaqlawa, *M. Nuri & K. Hamid* 41145!; mountain nr Shaqlawa, *Bornmüller* 1225!; Sefin Dagh, above Shaqlawa, *Rawi* 9091!; 'Ain Shaikh Ismail, Shaqlawa, *Sahira* 37570?; between Shaqlawa and Harir, *Omar* 37624?; NW of Haruna, *Gillett* 9805!; Haibat Sultan Dagh, *Omar, Sahira, Karim & Hamid* 38231?; **MRO/MSU**: Talan district, nr Doka, *F. Karim* 39288?; **MSU**: Darband-i Khan, *id.* 39246?; **MJS/FUJ**: between Sinjar and Ain Ghazal, *Handel-Mazzetti* 1352; **FUJ/MAM**: nr Filda police post, 51 km NW of Mosul, *Rawi* 22940!; **FNI**: below Mar Ya'qub, *Nábělek* in obs.; 1 km NE of Mindan bridge, *E Chapman* 26207!; **FAR**: Ankawa, nr Arbil, *Bornmüller* 1224!; **FKI**: Kor Mor, nr Tuz, *Rogers* 0314!; **FNI**: 3 km E of Badra, *Haddad* 9967?; **LEA**: Ba'quba, *Omar, Janan al-Mukhtar & Sahira Abdully* 37013; Balad Ruz, in *Anth.*, l.c.; **LCA**: Daltawa (Khalis), *Guest* 980!
NOTE. As Meikle (1977) suggests, this species is probably much common than the records suggest; as it flowers late it tends to be missed by collectors.
Cretan Eryngo; ?ZABĪB AR-RŪMI ('Frank's currants', Ir. – Khalis, *Guest* 2466) or, more probably, KASSŪB, a general name for thistly plants. According to Rawi & Chakravarty (1964) the leaves and roots of this plant are used medicinally as an alleviant in anaemia, dropsy, colic and other complaints.
In the *E. campestre* group radical leaves are usually present at flowering time and are deeply ternate-pinnatifid or pinnatifid. There is frequently (but not constantly) amethyst-blue suffusion in *E. creticum.*

SE Europe (Balkans, Greece), Crete, Aegean Isles, Cyprus, Syria, Lebanon, Palestine, Jordan, Egypt, Turkey, Iran.

2. **Eryngium pyramidale** *Boiss. et Hausskn.* ex *Boiss.*, Fl. Orient. 2: 829 (1872); Hand.-Mazz. in Ann. Naturh. Mus. Wien 27: 87 (1913); Wolff, l.c. 136 (1913); Nábělek in Publ. Fac. Sci. Univ. Masaryk 35: 119 (1923); Bornmüller in Beih. Bot. Centralbl. 58B: 275 (1938); Rawi in Dep. Agr. Iraq Tech. Bull. 14: 91 (1964); Mouterde, Nouv. Fl. Lib. et Syr. 2: 581 (1970); Davis, Fl. Turk. 4: 299 (1972); Pimenov & Tamamschian in Fl. Iranica 162: 48 (1987).

Plant very robust, glaucescent, 1.5 m, with a pyramidal habit, stem very stout. Leaves coriaceous, reticulate-veined, the radical to 50 cm long, entire or trisect, the segments broad and strongly regularly serrate-spinulose but otherwise undivided. Lower cauline leaves similar, trisect; upper cauline leaves rapidly reducing, very shortly petiolate or sessile-amplexicaul, undivided but undulate and closely spinose. Capitula medium, 2–2.5 cm; the 5–9 involucral bracts rather short, narrow, in length at most twice the length of the capitulum and often scarcely exceeding it, their margins remotely spinulose below or entire. Outer bracteoles trisect to less than halfway, the inner linear, scarcely exceeding the flowers. Sepals oblong-lanceolate, the midrib excurrent in a long mucro almost equalling the lamina. Petals 2–2.5 mm, oblong. Fruit 8 mm, somewhat gibbous, with acute, lanceolate scales.

HAB. Rocky mountain slopes, on limestone, sometimes in open *Quercus* forest; alt. 700–1000 m; fl. & fr. Jun.–Jul.
DISTRIB. Occasional in the lower forest zone of Iraq. **MAM**: Jabal Bekher, nr Zakho, *Rawi* 23044; 8 km S of Zakho, *Rechinger* 10704!; **MRO**: Kuh-i Sefin, *Bornmüller* 1227!; Chinaruk, *Rawi & Serhang* 25350!; Rush (Kew-ar Rash), *id.* 23780!; **MSU**: Dukan, *Agnew & Haines* W2131!

E Syria, SE Turkey, NW Iran.

3. **Eryngium glomeratum** *Lam.*, Encycl. méth. Bot. 4: 755 (1796); DC., Prodr. 4: 89 (1830); Boiss., Fl. Orient. 2: 823 (1872); Hand.-Mazz. in Ann. Naturh. Mus. Wien 27: 87 (1913); Wolff, l.c. 146 (1913); Hayek, Prodr. Fl. Balc. 1: 970 (1927); Guest in Dep. Agr. Iraq Bull. 27: 33 (1933); Blakelock in Kew Bull. 3: 433 (1948); Zohary in Dep. Agr. Iraq Bull. 31: 108 (1950); Rawi in Dep. Agr. Iraq Tech. Bull. 14: 91 (1964); Mouterde, Nouv. Fl. Lib. et Syr. 2: 578 (1970); Davis, Fl. Turk. 4: 301 (1972); Zohary, Fl. Palaest. 2: 385 (1972); Meikle, Fl. Cyprus 1: 691 (1977); Pimenov & Tamamschian in Fl. Iranica 162: 50 (1987); Boulos, Fl. Egypt 2: 155 (2000).

[*E amethystinum* (non L.) V. Täckholm, Stud. Fl. Egypt ed. 2: 385 (1974).]

Plant attaining 75 cm, commonly about 50 cm, glaucescent. Stem rather slender, densely foliose, basal leaves 1–2-pinnatifid, withered at flowering, small; stem leaves similar, large below but decreasing upwards in size and number of pinnae, all pungently spinose, the segments narrow and rarely exceeding 4 mm wide, the sheathing basal part of the lamina with prominent parallel veins which are very rarely connected transversally. Inflorescence thyrsoid, rather compact, capitula small (8–10 mm diameter) and ± globose. Involucral bracts, 5–6, narrow, stout and rigidly pointed, much exceeding the capitula (2–4 cm long), entire or very remotely spinulose, with single small spines at the base. Bracteoles lanceolate, tapering to a short stout mucro formed by the excurrent midrib. Sepals lanceolate-ovate, shortly mucronate, 1.0–1.5 mm long. Petals oblong-ovate, ± 1.5 mm. Fruit 4–5 mm, covered with narrow, acute scales. Fig. 47: 1–3.

HAB. Rocky mountain slopes and rough dry hillside pastures; alt. 400–1050 m; fl. & fr. Aug.–Oct.
DISTRIB. Occasional in the NW sector of the lower forest zone of Iraq, only found once in the moist-steppe zone. **MAM**: nr Zawita, *Guest* 1657A!; Sarsang, *Haines* W1237!, *Anders* 233; Mar Ya'qub, nr Simel, *Handel-Mazzetti* 3094; Sawara Tuka, *Barclay* 9097; **MJS**: Jabal Sinjar, *Alizzi & Omar* 35318?; **FKI**: nr Kirkuk, *Guest, Eig & Zohary* 5096!

Crete, Aegean Isles, Cyprus, Syria, Lebanon, Palestine, Jordan, NE Mediterranean, Egypt, Turkey. (Probably more common than would appear from the records; as suggested for *E creticum*, q.v., this late-flowering species tends to be overlooked by collectors).

4. **Eryngium hainesii** *C.C. Townsend* in Kew Bull. 19 (1): 69 (1964). — Type: Iraq, Sarsang, on dry rocky slopes in oak scrub, 1050 m, 29 Aug.1957, *Haines* W1238 (K!, holo.); Pimenov & Tamamschian in Fl. Iranica 162: 60 (1987).

Plant glaucous, stiff, with stems solitary or few together, about 65 cm in height. Stem finely striate, white, laxly foliose. Leaves rigid, coriaceous, the basal to 20 cm, long-petiolate, pinnatisect with 2 pairs of pinnae which are about 10–15 mm in width excluding the spines and strongly spinous. Cauline leaves rather distant, the median trisect with broad bases, reticulate-veined at least towards the margins, the upper with narrower segments; leaves of the inflorescence region very narrowly divided, the narrow base parallel-veined. Inflorescence ± 30 cm long, thyrsoid, cylindric. Capitula 10 mm in diameter. Involucral bracts subulate, very rigid, sharply pointed, entire or with a few lateral spines, 2–3 × exceeding the capitula; bracteoles rigid, 6.5–7(–12) mm long, entire, subulate, mucronate. Sepals lanceolate, acuminate, 3 mm long. Petals oblong, narrowed into a linear lobule subequalling the lamina in length. Fruit covered throughout with lanceolate scales which are largest towards the apex, 2.5 mm long.

HAB. Dry rocky mountain slopes among *Quercus* scrub; alt. 1150–1800 m; fl. Aug.
DISTRIB. Very rare, only found once in the NW sector of the upper forest zone of Iraq. **MAM**: at and above Sarsang, *Haines* W1238 (type).
NOTE. There is much to be said for Davis' view (Fl. Turk. 4: 302 (1972)) that this species comes very close to *E. billardieri* in spite of its thyrsoid inflorescence, and it needs further study. For the moment it can only be kept distinct. The basal leaves are less divided than in Iraqi material of *E. billardieri* seen.

Endemic.

5. **Eryngium thyrsoideum** *Boiss.*, Ann. Sc. Nat. sér. 3, 1: 121 (1844); Boiss., Fl. Orient. 2: 822 (1872); Wolff, l.c.: 146 (1913); Nábělek in Publ. Fac. Sci. Univ. Masaryk 35: 118 (1923); Zohary in Dep. Agr. Iraq Bull. 31: 108 (1950); Fl. Turk. 4: 301 (1972); Pimenov & Tamamschian in Fl. Iranica 162: 49 (1987).

Plant glaucous or yellowish-green, stem stout, 30–40 cm. Basal leaves trisect-pinnartipartite, up to 35 × 35 cm with stout broad teeth all round and along the decurrent posterior margin, petiolate. Lower stem leaves ternate-pinnatisect, upper trisect, uppermost narrower and simple, all leaves strongly dentate, sessile-amplexicaul, the sheathing base broad, venation strong and conspicuously reticulate except in the centre. Inflorescence thyrsoid, the capitula moderate, 10–14 cm in diameter. Involucral bracts 7–9, stout, with one or two pairs of lateral spines below the middle, up to twice as long as the capitula or a little more. Bracteoles subulate, mucronate with the excurrent midrib, rigid, exceeding flowers. Sepals lanceolate, ± 3 mm, vein percurrent into a stout arista up to 1/4 as long as lamina. Petals 2 mm, white or pale blue. Fruit compressed-obovoid, with lanceolate-acute or somewhat obtuse scales.

HAB. Rocky mountain slopes, in *Quercus* forest and scrub on limestone, under *Juglans* trees by a stream, treeless steppic hillsides and barren denuded earthy hills; alt. 400–1700 m; fl. & fr. Jul.–Sept.

Fig. 47. **Eryngium glomeratum**. 1, habit × ²/₃; 2, flower × 6; 3, fruit cross section × 8. All from *Wheeler Haines* W1237. Drawn by D. Erasmus (1964).

DISTRIB. Occasional, locally frequent in the forest zone of Iraq. **MAM**: "Kurdish hills", *Mrs A. Low* 385!; **MRO**: Salah ad-Din, *Gillett* 12425!; Musaiyif Nawanda, nr Rost, *Guest & Husham (Alizzi)* 15803!; in valle inter Rayat et Haji Umran, *Rechinger* 11838!; Qandil Mts. between Pushtashan and Shahidan, *Rechinger* 11729!; nr Rayat, *Zohary* s.n.!; Pushtashan, 15 km NE of Rania, *Rawi* 23884!; **MSU**: near Sulaimaniya, first hill range on road to Choarta (Chuwarta), *Haines* W2092!; Jarmo, *Helbaek* 1993; Tawila, *Rechinger* 10224; Penjwin, *Rechinger* 10527; Kani Takht Bardu Serish, Qara Dagh, *Eig, Feinbr. & Zohary* s.n.!; Pira Magrun, *Haines* s.n.!; between Kirkuk and Sulaimaniya, *id.* s.n.!; Mt. Hawraman above Khurmal, *Hadač* 5035; **FKI**: nr Kirkuk, *Guest, Eig & Zohary* 5096!
NOTE. ?TAISŪ (Kurd.-Rayat, *Guest & Alizzi* 15803, "said to be a good fodder").

Guest & Husham 15803 is atypical, being grazed or diseased, while *Haines* W2092 is immature; the identity of neither of these specimens may be regarded as fully established, but they both probably belong to this species.

SE Turkey, NW & W Iran.

6. **Eryngium billardieri** *Del.*, Eryng. Hist. 25, t. 2 (1808); DC., Prodr. 4: 88 (1830); Boiss., Fl. Orient. 2: 825 (1872); Wolff, l.c. 149 (1913); Zohary, Fl. Palest. ed. 2, 1: 506 (1932); Bornmüller in Beih. Bot. Centralbl. 58B: 274 (1938); Blakelock in Kew Bull. 3: 433 (1948); Mouterde, Nouv. Fl. Lib. et Syr. 2: 579 (1970); Fl. Turk. 4: 301 (1972); E. Nasir in Fl. Pak. 20: 18 (1972); Pimenov & Tamamschian in Fl. Iranica 162: 50 (1987).

E. nigromontanum Boiss. & Buhse, Aufz. Pfl. Reise Transk. Pers.: 95 (1860); Schischkin in Fl. SSSR 16: 78 (1950); *E. billardieri* var. *meiocephalum* Boiss., Fl. Orient. 2: 825 (1872); Hand.-Mazz. in Ann. Naturh. Mus. Wien 27: 87 (1913); Rawi in Dep. Agr. Iraq Tech. Bull. 14: 91 (1964); *E. billardieri* subsp. *nigromontanum* (Boiss. et Buhse) Wolff, l.c. 150 (1913); Blakelock in Kew Bull. 3: 433 (1948); Rawi in Dep. Agr. Iraq Tech. Bull. 14: 91 (1964).

Plant 30–60(–70) cm. Radical leaves persistent at flowering, ternate-bipinnatifid, margins decurrent from one pinna to next, densely set with long, stout spinous teeth and long-petiolate. Cauline leaves sessile, trisect or palmatipartite, decreasing upwards and similarly stoutly spinous. Whole plant often suffused with a beautiful amethyst-blue coloration, which is occasionally only to be seen in young capitula. Inflorescence a narrow or more expansive corymb, capitula 12–15 mm in diameter. Bracts 5–7, considerably (2–3 ×) exceeding the capitula, stoutly spinose with two or three pairs of slender marginal spines in the lower half. Bracteoles entire or bluntly shouldered near the apex, to ± 0.8 mm long. Sepals lanceolate-ovate, with a distinct mucro into which the lamina tapers. Petals violet blue, the inflexed lobes dilated at apex, ± 1.5 mm. Fruit compressed-turbinate, furnished particularly dorsally with acute lanceolate scales.

HAB. Rocky mountainsides; alt. 1300–2600 m; fl. & fr. Jul.-Aug.
DISTRIB. Occasional in the lower thorn-cushion and upper forest zones of Iraq, particularly common in the Halgurd Dagh district. **MAM**: Khantur, *Rawi* 23499!, *Rawi, Tikriti & Nuri* 28982!; Ser Amadiya, *Guest* 4977!; nr Dohuk, *Haines* 1569; **MRO**: Handren, *Bornmüller* 1226; Halgurd Dagh, *Gillett* 9498!, *Rechinger* 11861; between Halgurd and Lalan (Lolan), *Gillett* 12466!; Rayat, *Guest* 13084!; N of Haji Umran, nr Iranian frontier, *Rawi* 24971!; between Bardanes and Qandil, *Rawi & Serhang* 24625!; Qandil Dagh above Pushtashan, *Rechinger* 11084; ?**MSU**: (or only in Iran?), "Avroman and Schahu", *Haussknecht* s.n.

Lebanon, Syria, Turkey, Caucasus, Iran, Afghanistan, Pakistan, Kashmir.

7. **Eryngium campestre** *L.*, Sp. Pl. ed. 1: 233 (1753); DC., Prodr. 4: 88 (1830); Boiss., Fl. Orient. 2: 824 (1872); Hand.-Mazz. in Ann. Naturh. Mus. Wien 27: 87 (1913); Wolff, l.c. 50 (1913); Nábělek in Publ. Fac. Sci. Univ. Masaryk 35: 118 (1923); Zohary, Fl. Palest. ed. 2, 1: 506 (1932); Schischkin in Fl. SSSR 16: 78 (1950); Mouterde, Nouv. Fl. Lib. et Syr. 2: 579 (1970); Davis, Fl. Turk. 4: 303 (1972); Meikle, Fl. Cyprus 1: 691 (1977); Boulos, Fl. Egypt: 2: 155 (2000).

E. trifidum L., Amoen. Acad. 3: 405 (1756); *E. vulgare* Lam., Fl. Franç. 3: 401 (1778); *E. noeanum* (non Boiss.) Blakelock in Kew Bull. 3: 434 (1948).

Plant yellowish or glaucescent-green, commonly up to 60 cm high. Radical leaves pinnatifid or ternate-pinnatifid, each pinna decurrent along the axis to the next, pinnae and wings with strong, long spines commonly longer than their width at base. Cauline leaves sessile-amplexicaul from a cordate base, decreasing upwards and similarly spinose. Inflorescence corymbose, final branches shorter. Capitula small, 8–12 mm in diameter, roundish. Involucral bracts 2 or 3 × exceeding capitula, stoutly spiny, entire or with one or two weak spines at the base. Bracteoles entire and subulate, sharply spiny, much exceeding flowers. Sepals lanceolate, the excurrent vein forming a ± long mucro into which the lamina tapers. Petals oblong, narrow. Fruit compressed-ovoid, with acute lanceolate scales.

HAB. Lower mountains, steppic hills and plains, a weed in fields; alt. 350–900 m; fl. & fr. (Apr.–)May–Jul.
DISTRIB. Occasional in the NW sectors of the lower forest and moist steppe zones of Iraq. **MAM**: 12 km N E of Zakho, *Rawi* 23162!; Sandur, *Haines* W1569!; **MJS**: Jabal Golat (Qilat) between Balad Sinjar and Ain Tellawi, *Field & Lazar* 499!; ? **FUJ/MJS**: (and in Syria), "everywhere on stony steppe" between R. Khabur and Jabal Sinjar, *Handel-Mazzetti* 3094; **FUJ**: 'Ain Ghazal, *Guest* 4084!; between Sinjar and Tal Afar, *Guest* 13631!
NOTE. Field Eryngo. The root of this species has long been collected in Britain and other parts of Europe for its medicinal properties which are summed up by Wren (1956) as diaphoretic, diuretic and expectorant, it having been used in folk medicine as an alleviant of uterine irritation and bladder afflictions.

The Iraqi plant is the "var. *virens* Link"; but the variety hardly seems sufficiently well-defined to perpetuate here. Pimenov & Tamamschian in Fl. Iranica 162: 53 (1987) do not recognise this species as occurring in the Flora Iranica area, preferring to include it and the following species within *E. billardieri*, q.v.

W, C & S Europe (from Britain to W Russia, Greece and Crimea), Syria, Palestine, Egypt, Turkey, Caucasus, Iran, Siberia, N Africa (Morocco to Libya).

8. **Eryngium noeanum** *Boiss.*, Diagn. ser. 2, 2: 72 (1856); Boiss., Fl. Orient. 2: 825 (1872); Wolff, l.c.: 152 (1913); Blakelock in Kew Bull. 43: 434 (1948); Schischkin in Fl. SSSR 16: 76 (1950); Field in Papers of Peabody Mus. of Archaeology & Ethn. Harvard Univ. 45 (2): 179 (1960); Rawi in Dep. Agr. Iraq Tech. Bull. 14: 91 (1964); Fl. Turk. 4: 303 (1972).

Plant yellowish-green to glaucescent, very robust, attaining 1.5 m or more, with a stout stem and long branches, both stem and branches white. Basal leaves large, with a long petiole subequalling the lamina, ternate-pinnatisect or -bipinnatisect with strong, broad teeth all round and on the decurrent posterior margin. Stem leaves sessile, clasping the stem, also with strong, broad spines, diminishing upwards. Inflorescence very expansive, corymbose, ultimate branches long and thus giving a more open appearance than that of *E. campestre*. Capitula moderate, 12–15 mm in diameter. Involucral bracts 5 or 6, strong and spinous, simple or with one or two weak teeth below. Bracteoles subulate, 6–7 mm, exceeding the flowers. Sepals lanceolate, the lamina tapering to the excurrent midrib, which forms a long mucro. Petals oblong, ± 2 mm. Fruit compressed-obovoid with numerous acute, lanceolate scales.

HAB. & DISTRIB. The occurrence of this species in Iraq is doubtful. It has been recorded by Field (l.c.), but the specimen on which the record is based (*Field & Lazar* 499) appears to be young *E. campestre*, as is probably *Guest* 4084, mentioned by Blakelock (l.c.) as possibly referable here. Nevertheless, it occurs over the border in Iran (Luristan etc.) and may perhaps yet be found in Iraq.

2. **LAGOECIA** L.

Sp. Pl. ed. 1: 203 (1753); Gen. Pl. ed. 5: 95 (1754); Wolff in Pflanznr. IV. 228, 61: 271 (1913)

Glabrous annual herbs, leaves pinnatisect. Umbels simple, subglobose. Involucre present, bracts pectinate-pinnatifid. Flowers bracteolate, white. Calyx lobes aristate-dentate, persistent in fruit. Petals minute, shortly oblong, with a short incurved lobule, the main lamina emarginate at the apex, the margins each bearing an arista about 2–3 × as long as the main lamina itself. Ovary unilocular, with 1 ovule. Fruit ovoid, somewhat compressed, the ribs inconspicuous, setose. Pericarp thin. Stylopodium single, conic, surmounted by the short persistent style. Vittae very faint, apical only. Carpophore absent. Endosperm scarcely compressed, face somewhat flattened.

Monotypic.
Lagoecia (Linnaeus in *Hortus Cliffortianus* p. 73 states that this is derived from the Gr. λαγώς, *lagos*, a hare, and οικος, *oikos*, a house, after a locality on the island of Lemnos).

1. **Lagoecia cuminoides** *L.*, Sp. Pl. ed. 1: 203 (1753); Boiss., Fl. Orient. 2: 833 (1872); Hand.-Mazz. in Ann. Naturh. Mus. Wien 27: 88 (1913); Wolff, l.c. 272 (1913); Nábělek in Publ. Fac. Sci. Univ. Masaryk 35: 120(1923); Hayek, Prodr. Fl. Balc. 1: 970 (1927); Guest in Dep. Agr. Iraq Bull. 27: 54 (1933); Bornmüller in Bot. Centralbl. 58B: 2756 (1938); Blakelock in Kew Bull. 3: 434 (1948); Zohary in Dep. Agr. Iraq Bull. 31: 109 (1950); H Riedl in Fl. Lowland Iraq: 451 (1964); Rawi in Dep. Agr. Iraq Tech. Bull. 14: 93 (1964); Mouterde, Nouv. Fl. Lib. et. Syr. 2: 577 (1970); Zohary, Fl. Palaest. 2: 386 (1972); Hedge & Lamond in Fl. Turk. 4: 304 (1972); Meikle, Fl. Cyprus 1: 692 (1977); Rechinger, Fl. Iranica 162: 63 (1987).

Glabrous annual 20–25(–40) cm. Stem ivory, terete, lightly sulcate, alternately branched from about the middle. Stem leaves all similar, reducing in size upwards, in outline linear-oblong, long-petiolate, pinnate with as many as 25 pairs of pinnae in the lower leaves; pinnae

Fig. 48. **Lagoecia cuminoides**. 1, habit × 2/3; 2, leaf × 30; 3, fruit × 6; 4, fruit cross section × 2. All from *Gillett* 10880. Drawn by D. Erasmus (1964).

obscurely trilobed, the lobes deeply dentate, teeth minutely serrulate and terminating in a fine seta which is largest in the upper leaves. Umbels nodding at first, later erect, globular, 15–18 mm in diameter when fully developed. Involucral bracts 10–12, 5–7 mm long, pinnate, with the pinnae denticulate all round and bi- or tri-furcate at the apex, each segment long-aristate. Bracteoles pectinate, segments mostly bifurcate at the apex. Sepals simply pectinate, exceeding bracteoles. Overall length of the petals (including aristae) ± 2 mm. Fruit ± 2 mm long. Fig. 48: 1–4.

Hab. Mountainsides, in denuded *Quercus* forest, on a steep limestone slope, dry steppic gypsiferous hills, fallow fields in foothills, bed of wadi; alt. 250–1500 m; fl. & fr. Apr.–May (–Jun.).
Distrib. Common in the middle and lower forest zone of Iraq, occasional in the steppe region. **mam**: Khantur, *Rawi* 23316!; Dohuk, *Makki* (*Sidqi ash-Sharbati*) 3286A!; Aqra, *Qaimaqam of Aqra* 3083!, *Rawi* 11481!; **mro**: Der Harir, *Nábělek* 404; Kuh-i Sefin, at Shaqlawa, *Bornmüller* 1223; between Karoukh and Dargala, *Kas & Nuri* 278710!; Koi Sanjaq, *Rawi, Nuri & Kass* 28111!; **msu**: Jarmo, *Helbaek* 1014!; Darband-i Bazian, *Rawi* 22798!; Dokan, *Omar & Karim* 38026!; Khormal, *Rawi* 8850!; Pira Magrun, *Rawi* 12122!; 17 km SW of Sulaimaniya, *Rawi* 21703!; Timara, Qara Dagh, *Omar & Karim* 40644?; Darband-i Khan, *id.* 38136!; 31 km N of Kirkuk, on road to Koi Sanjaq, *Rawi, Nuri & Kass* 27914!; **mjs**: Kursi, *Gillett* 10880!; Asi, *Rawi* 8484!; above Sinjar, *Handel-Mazzetti* 1360; **fni**: Maltai, between Mar Ya'qub and Tal Kaif, *Nábělek* 350; Ain Sifni, *Salim Effendi & Guest* 2562!; **far**: 2 km E of Eski Kellek, *Fred A. & Elizabeth Barkley* 5545?; **fki**: Kifri, *Gillett & Rawi* 7407!
Note. GIYĀ-I GHAZALAH (Kurd.-Ain Sifni, *Salim* 2562); GIYĀ GUMINK (Kurd.-Dohuk, *Makki* 3286).

Mediterranean Europe (Spain, Italy, Balkans, Greece), Aegean Isles, Cyprus, Turkey, Syria, Lebanon, Palestine, Jordan, Iran, N Africa.

3. **ANISOSCIADIUM** DC.

Mém. Omb. t. 15 (1829)

Erect to decumbent annual herbs with bi- to subtri-pinnatisect leaves. Involucre and involucel present. Flowers polygamous, the central immersed in the receptacle to the point of greatest width at ripe fruit, and here fused to the rim in which the outer flowers are set by the thin pericarp. The remainder of the flowers pedicellate and mostly male, but some fertile, free. Petals white, radiant or not. Calyx present in outer flowers, present or absent in central. Central fruit with two tapering sulcate carpels, the remaining fruits with one only. Umbel rays disarticulating at their meeting point above the involucre and falling with the whole partial umbel at maturity, pedicels indurate. Pericarp of the central fruit membranaceous, of the remainder crustaceous. Vittae solitary or 3 in the valleculae, commissure bivittate. Endosperm with the face deeply excavate, the margins involute.

3 species in SW Asia, all found in Iraq.
Anisosciadium (from the Gr. ανισος, *anisos*, unequal or uneven, σκιαδειον, *skiadeion*, an umbrella or parasol, referring to the dissimilarities within the umbels).

1. Two exterior calyx segments of the outer row of flowers, and the bracts of the involucel, not becoming accrescent and foliose in fruit, gradually tapering to a rigid yellowish mucro .... 3. *A. lanatum*
   Two exterior calyx segments of the outer row of flowers, and the bracts of the involucel, becoming accrescent in fruit and foliose 2
2. Outer row of flowers conspicuously radiate; accrescent calyx segments very unequal in size, the larger oblong-lanceolate; surface of leaf segments almost glabrous .... 1. *A. orientale*
   Outer row of flowers not conspicuously radiate; acccrescent calyx segments ± equal in size, roundish to oval; leaves with abundant short, stiff, white hairs .... 2. *A. isosciadium*

Subgenus 1. Anisosciadium. Bracts of the involucel and outer calyx segments accrescent in fruit. Some of the *outer* row of outer flowers fertile.

1. **Anisosciadium orientale** *DC.*, Mém. Omb. 5: 63 (1829); DC., Prodr. 4: 234 (1830); Boiss., Fl. Orient. 2: 950 (1872); Hand.-Mazz. in Ann. Naturh. Mus. Wien 27: 88(1913); Nábělek in Publ. Fac. Sci. Univ. Masaryk 35: 126 (1923); Guest in Dep. Agr. Iraq Bull. 27:

7 (1933); Anth. in Notes Roy. Bot. Gard. Edinb. 18: 289 (1935); Bornmüller in Beih. Bot. Centralbl. 58B: 281 (1938); Blakelock in Kew Bull. 3: 432 (1948); Zohary in Dep. Agr. Iraq Bull. 31: 112 (1950); Townsend in Kew Bull. 3: 432 (1948); H. Riedl in Fl. Lowland Iraq: 451 (1964); Rawi in Dep. Agr. Iraq Tech. Bull. 14: 88 (1964); Mouterde, Nouv. Fl. Lib. et Syr. 2: 582 (1970); Rech.f., Fl. Iranica 162: 66 (1987).

Dicyclophora morphologica Vel., Mém. Soc. Sci. Bohème, Cl. Sci. 1921–22, Art. 6: 5 (1923).

Diffuse herb, to 15(–25) cm high. Stem rigid, sulcate, glabrous or downy, branching dichotomously from about halfway. Leaves tripinnatisect, oblong in outline, decreasing upwards, with lanceolate segments, glabrous or with a thin sprinkling of short bristles. Umbels opposite the leaves, few (3–5) -radiate, covered throughout with a very short crisped down even in plants which are otherwise glabrous. Umbel branches becoming broad and much flattened when in ripe fruit, turbinate below the partial umbel. Involucral bracts 3–6, lanceolate, those of the involucel accrescent in fruit, becoming elongate oblong and foliose. Outer calyx lobes of the outer row of flowers accrescent in fruit, very unequal, the larger often having 4 × the surface area of the smaller or more, oblong-lanceolate; the remaining calyx lobes rigid and uncinate. Outer flowers subtended by a bract of the involucel, usually gibbous within at ripe fruit with the developed carpel.

HAB. Lower mountain slopes and eroded steppic foothills, on sandstone, conglomerate, gravel and clay, sometimes gypsaceous, also on sandy or gravelly desert soils; alt. 50–450(–800) m; fl. & fr. Mar./Apr.–May/Jun.

DISTRIB. Very common in the steppe region of Iraq, rarely penetrating up into the SE sector of the lower forest zone and down onto the desert region. **MRO**: Koi Sanjaq, *Rawi, Nuri & Kass* 28141!; Salah ad-Din, *Haines* s.n.; **MSU**: between Kirkuk and Jarmo, *Haines* W279!; Jarmo, *Helbaek* 548 1!;1!; Qaranjir, *Rawi* 21607!; **FUJ/FNI**: Mosul, *Aucher* 3598!; **FUJ**: Qaiyara, *Baylis* 110!; Sharqat, *Handel-Mazzetti* 116; Ain Diba, *Rawi & Gillett* 7147!; **FNI**: nr Dohuk, *Emberger, Guest, Long, Schwan & Yusuf* (*Serkahia*) 15341!; Ain Sifni, *Salim Effendi* 2566!; **FKI**: 13 km N of Kirkuk, *Gillett & Rawi* 10575!; **FKI/MSU**: 7 km NE of Kirkuk, *Rawi, Nuri & Kass* 27919!; 8 km E of Kirkuk, *Gillett & Rawi* 11591!; **FKI**: Kirkuk, *Bornmüller* 1268!; Tuz, *Rogers* 0363!; Darozna, N of Kifri, *Poore* 378!; Kifri, *Paranjpye* in Graham 635!; Jabal Hamrin, nr Injana, *Bornmüller* 1267! 1268! *Guest* 681!, 1428!, *Omar, Karim, Hamza & Hamid* 37045?; Adhaim, *Agnew* 1436; **FNI**: between Tanura and Gor-i Shala, *Poore* 494!; Qaraghan (Jalaula), *Rogers* 0223!; Jabal Hamrin, by Table Mt., between Khanaqin and Baghdad, *Nábělek* 387, *Paranjpye* in Graham 690! *Gillett & Rawi* 7322! *Rechinger* 14242!; Naft Khana, *Guest* 1858!; Koma Sang, *Rawi* 20657!; 10 km E of Mandali, *Rechinger* 12793; 20 km N of Badra, *Rechinger* 9687!; 6 km S of Badra, *Rawi* 18211!; 30 km SE of Badra, *Rechinger* 9147, 14012; Bagsaiya, *Rawi & Haddad* 25598!; Jabal Muwaila, 70 km N of Amara, *Guest, Rawi & Rechinger* 17563!, *Rechinger* 8887; **DLJ**: 90 km N of Rawa, on road to Shbaichan, *Chakr., Rawi, Khatib & Alizzi* 31970!; Ajuba, in Anthony, l.c.; by R. Tigris at Tikrit, *Handel-Mazzetti* 1005; Jazira desert, *C.J. Edmonds* 3821!; 18 km W of Khamrana, *Alizzi & Hussain* 33790?; 85 km NNW of Falluja, *Rawi* 20295!; between Wadi Thirthar and Falluja, *Rechinger* 13509!; **DWD/DLJ**: by R. Euphrates, below Hit, *Handel-Mazzetti* 820; **DSD**: Jabal Sanam, in Anthony, l.c.; **LEA**: Shahraban (Muqdadiya), *Haines* W917!; 60 km N of Amara, nr Wadi Tib, *Guest, Rawi & Rechinger* 17491!, *Rechinger* 14279; Sudur Mansuriya, *Barkley, Abdul Wahab & Danail Oraha* 6837? *Yalda Matti & Jack Isaac* 325?

NOTE. ? GIYĀ BUHUN NA KHUSH (Kurd.-Ain Sifni, *Salim Effendi* 2566).

Syria, ?Turkey, Kuwait, W & C Iran, Afghanistan.

2. **Anisosciadium isosciadium** *Bornmüller*, Fedde Rep. 10: 468 (1912); Zohary, Fl. Palest. ed. 2, 1: 541 (1932); Townsend in Kew Bull. 17: 427 (1964); Rawi in Dep. Agr. Iraq Tech. Bull. 14: 88 (1964); Zohary, Mouterde, Nouv. Fl. Lib. et Syr. 2: 582 (1970); Zohary, Fl. Palaest. 2: 388 (1972).

Dicyclophora caucalioides Vel., Mém. Soc. Sci. Bohème, Cl. Sci. 1921–22, Art. 6: 5 (1923).

Diffuse herb, to 10(–30) cm high, stem striate, hairy, branching from about halfway. Leaves tripinnate, similar to those of *A. orientale* but triangular-ovate in outline, ± copiously clothed with short, patent, white hairs. Umbels opposite the leaves, few (3–8) -radiate, covered throughout with a very short crisped down. Involucral bracts 3–6, lanceolate. Umbel-branches thickened and turbinate below the partial umbels in ripe fruit, sulcate but not broadly flattened. Bracts of the involucel accrescent in fruit, not greatly elongating and remaining as a rule ± rotund-acuminate or oval. Outer calyx lobes of outer flowers accrescent but equal in size or almost so, roundish to oval, the remaining calyx lobes rigid and uncinate. Outer flowers subtended by a bract of involucel, gibbous within at ripe fruit with the developed carpel. Fig. 49: 1–6.

HAB. Sandstone and gravelly hills, often gypsaceous, sandy and gravelly places in stony deserts, sometimes in silty depressions; alt. (50–)200–700 m; fl. & fr. Apr.–May or early Jun.
DISTRIB. Quite common in the NW sector of the desert region of Iraq, occasionally in the SE sector. **DWD**: 5 km S of H.3, *Rawi* 21205!; Wadi Hauran, 12 km SE of H.2, *Chakr., Rawi, Khatib & Alizzi* 31694!; Qa'ra, N of Rutba, *Rawi* 31282?; 100 km NE of Rutba, *id.* 31523!; 20 km WNW of Rutba, *Rawi* 21117!; 15 km N of Rutba, *Rawi & Nuri* 27077!; nr Rutba, *Alizzi & Hussain* 34111?; 45 km S of Rutba, *Rawi* 21255!; 45 km E of Rutba, *Rechinger* 170! 9960!; on Rutba – H.2 road, *Omar & Hamid* 36718?; 100 km E of Rutba, *Omar* 34316? *Botany Staff* 41816?; 260 km W of Ramadi, *Rawi* 20944! *Rawi & Nuri* 27060!; 230 km W of Ramadi, *Rawi* 20926!; 210 km W of Ramadi, *Rechinger* 9812!; 120 km N of Nukhaib, *Rawi* 31108!; 70 km N of Nukhaib, *Rawi* 31389?; 35 km W of Nukhaib, *Rawi* 15741!; 20 km W of Nukhaib, *Rawi* 3097!; 12 km W of Ukhaidhir, *Rawi* 30852!; 114 km W of Karbala, *Rawi* 30948?; **DWD/DSD**: Wadi al-Khirr, *Guest, Rawi & Rechinger 19425*!; **DSD**: c. 115 km S.E of Ashuriya, *id.* 19365!; ± 40 km WNW of Shabicha, *id.* 19324! *Rechinger* 9457!; 10–15 km SW of Shabicha, *Alizzi & Omar* 35448?; 50 km N of Salman, *Rawi* 14870!; between Aidaha and Salman, *Alizzi & Omar* 35613?; between Aidaha and Ansab, id. 35647?; **LCA**: between Baghdad and Karbala, *Rawi & Bharucha* s.n.?

Syria (SE of Palmyra), Palestine, Jordan.

Fig. 49. **Anisosciadium isosciadium.** 1, habit × 2/3; 2, leaf × 4; 3, petal × 8; 4, cross section of umbel × 2; 5, longitudnal section of umbel × 1; 6, central fruit cross section × 6. 1, 4–6 from *Rawi* 21117; 2–3 from *Rawi* 20944. Drawn by D. Erasmus (1964).

Subgenus 2. ECHINUM C.C. Townsend. Bracteoles and calyx teeth spinous, scarcely accrescent in fruit. Inner row of outer flowers partially fertile.

3. **Anisosciadium lanatum** *Boiss.*, Fl. Orient. Suppl.: 261 (1888); Blakelock in Kew Bull. 3: 432 (1948); V. Dickson, Wild Flowers of Kuwait & Bahrain: 19 (1955); Townsend in Kew Bull. 17: 427 (1964); H. Riedl in Fl. Lowland Iraq: 452 (1964); Rawi in Dep. Agr. Iraq Tech. Bull. 14: 88 (1964).

Echinosciadium arabicum Zohary in Pal. Journ. Bot. J. ser. 4: 175 (1948); Riedl in Fl. Lowland Iraq: 453 (1964).

Diffuse herb, to 20 cm but usually shorter. Stem rigid, sulcate, downy, branching from near the base. Leaves bi- or tri-pinnatisect, decreasing upwards, triangular-ovate in outline, the segments lanceolate and ± thickly clothed with short whitish hairs. Umbels opposite leaves, few- (usually 7) radiate, rays and bracts of involucel lanate, especially when young, with long, white hairs. Umbel branches flattened, becoming somewhat broader and turbinate below the partial umbel when in ripe fruit. Involucral bracts 5–7, lanceolate, those of the involucel enlarging only a little in ripe fruit, reflexed and tapering to a rigid yellowish mucro. Calyx teeth all similar though somewhat diverse in size, rigid and recurved-uncinate. Inner row of outer flowers usually fertile and gibbous with the developed carpel at ripe fruit.

HAB. Sandy desert, low dunes and sandy ridges, sandy gravel desert, sandy pockets in rocky limestone desert; alt. up to 400(–600) m; fl. & fr. Mar.–May.
DISTRIB. Common in the SE sector of the desert region of Iraq, occasional in the NW sector. **DWD**: 20–25 km W of Rutba, *Rechinger* 12546; 3 km SE of Rutba, *id.* 12616; 45 km E of Rutba, *id.* 9960; 260, 230 and 210 km W of Ramadi respectively, *id.* 9840, 9829, and 9812; **DWD/DSD:** Wadi al-Khirr, 90 km NNW of Shabicha, *id.* 13578; **DSD**: 40 km WNW of Shabicha, *id.* 9547; 13 km NE of Shabicha, *id.* 13673; Jumaima, *id.* 13766; 27–46 km WNW of Ansab, *Guest, Rawi & Rechinger* 19008! 19026!; *Rechinger* 9369; *Ichrishi, Guest & Rawi* 14184!; nr Chilawa, *id.* 17225! *Rechinger* 8808; NE of Ghazlani, *Rechinger* 14353!; 81 km ESE of Busaiya, *Guest & Rawi* 14278!; between Jaliba and Zubair, *Eig, Guest & Zohary* 5058!; nr Zubair, *Guest & Rawi* 14329!; Jabal Sanam, *Watson* s.n. comm. Sharples, *Guest, Rawi & Schwan* 14377! *Guest, Rawi & Rechinger* 16965! *Rechinger* 8724!; SW of Safwan, *Rechinger* 14485; 12 km NE of Safwan, *Rawi* 25980!
NOTE. CHÎRSH (Ar.-Ichrishi, *Guest & Rawi* 14184, "sheep and camels eat it". According to Dickson (1955) the young green leaves of this plant are eaten by Badawin children in Kuwait and it is known there as BISBĀS or UM DRUS (for UM DURŪS or "mother of teeth"), on account of its hard spiky seeds, and it seems likely that one or other of these two names may also be known in the Southern Desert of Iraq. A note on the label of one of the specimens gathered by Mrs Dickson in Kuwait bears the information that the plant is grazed by sheep, goats and other animals, which suggests that the plant is a useful component of the desert spring herbage, confirming the statement of Guest & Rawi's informant at Ichrishi as above.
Carter (1917), who published details of Sir Percy Cox's collection of plants from the Zor hills in Kuwait, however attaches the names BISBĀS and GARSH (c.f. our CHÎRSH above) to *Erodium cicutarium*, so it may be that they are general names used by the Badawin for several palatable desert forage plants.

Kuwait (incl. Falaicha Island), N & E Arabia.

4. ERGOCARPON C. C. Townsend

Kew Bull. 17: 437 (1964)

Annual or biennial herbs with pinnately dissected early-marcescent lower leaves, the upper chartaceous and bract-like. Umbels compound. Involucre and involucel with large bracts and bracteoles which are broadly white-margined, the bracteoles much exceeding the very small partial umbels. Partial umbels with a central, sessile, fertile, female flower, surrounded by a ring of sterile, male flowers which are themselves surrounded by a ring of reduced claviform flowers. Calyx teeth obsolete. Petals equal, ovate, with a long, inflexed acumen, emarginate-incised at the midrib above. Central fruit deeply immersed, bicarpellate (or rarely one carpel abortive), carpels subterete, primary ribs filiform and inconspicuous. Disk narrowly explanate, stylopodia conical. Styles short, not hard and horn-like. Carpophore absent. Vittae absent (?). Seed deeply excavate on the commissural face, margins strongly involute.

A monotypic genus in SW Asia.

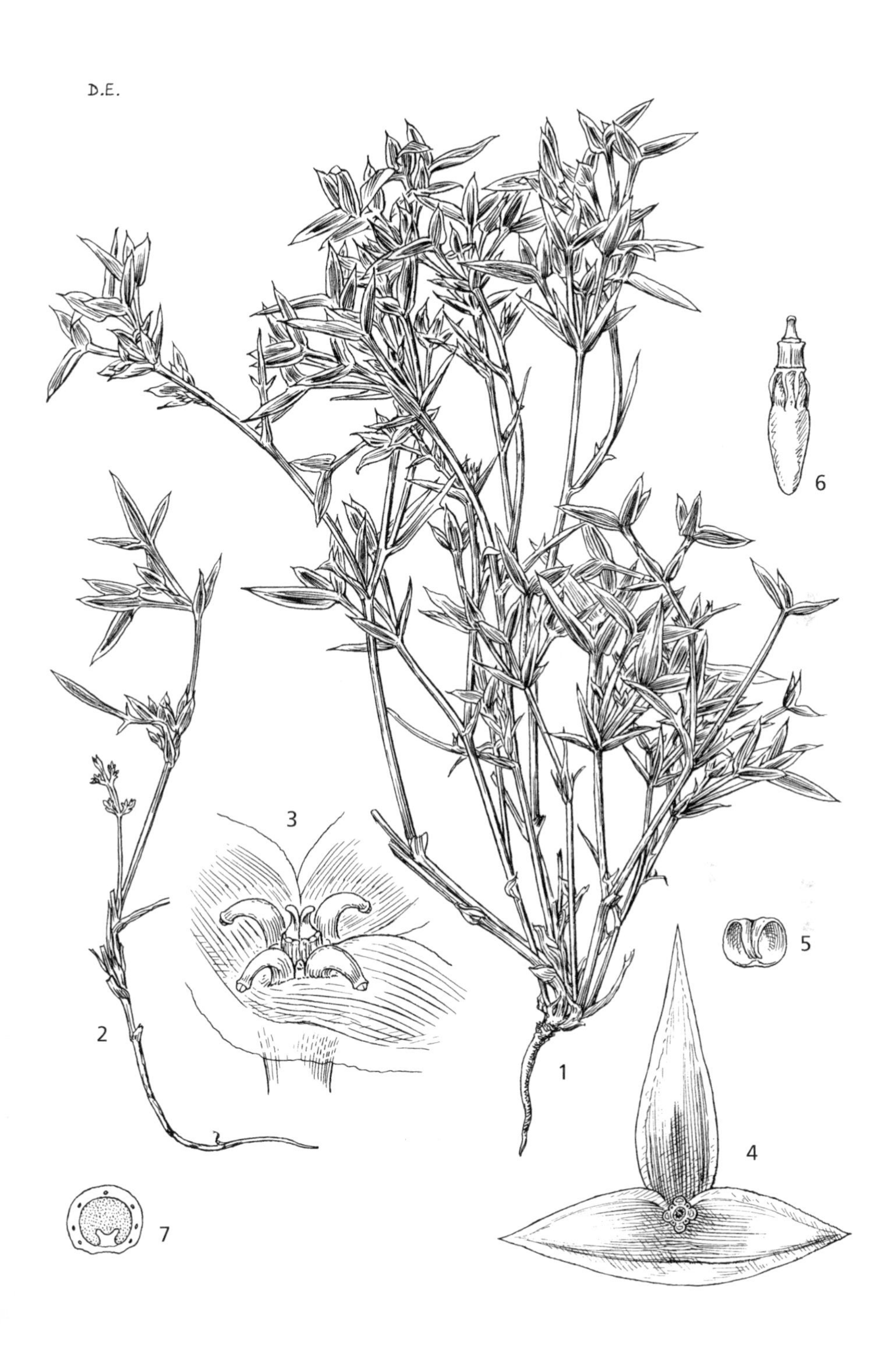

Fig. 50. **Ergocarpon cryptanthum.** 1, habit × ²/₃; 2, habit with lower portion of stem × ²/₃; 3, bracts opened to show styles × 4; 4, bracts flattened seen from above × 2; 5, petal × 8; 6, fruit × 4; 7, fruit cross section × 10. 1, 2, 5 from *Rechinger* 9648; 3–4, 6–7 from *Rawi* 20631. Drawn by D. Erasmus (1964).

Ergocarpon (from the Gr. εργο, *ergo*, I imprison, καρπος, *karpos*, a fruit – descriptive of the immersed central fruit).

1. **Ergocarpon cryptanthum** (*Rech.f.*) *C.C. Townsend* in Kew Bull. 17: 438 (1964).

Exoacantha cryptantha Rech.f., Umbell. nov. Iranica 2, Anz. math.-nat. Klasse Österr. Akad. Wiss. 12: 7 (1952); Riedl in Fl. Lowland Iraq: 470 (1964); Rech.f., Fl. Iranica 162: 80 (1987).

Plant short, 8–20 cm, light bluish-green, quite glabrous, branched from the base with many ascending stems. Stems and branches striate. Radical and lower cauline leaves oblong in outline, remotely bipinnate, pinnae very short, sessile, in 2 or 3 pairs, the pinnules incised. Cauline leaves becoming chartaceous and bract-like above, broadly white-margined, trisect with two shorter lateral and a long terminal segment, and finally simple. Sheaths shortly expanded, white-margined. Involucre and involucel of 3 tapering-lanceolate bracts and bracteoles which are broadly white-margined and have prominent midrib excurrent in a sharp mucro. Umbel rays ± 6, very unequal, 5–20 mm. Partial umbels very small, ± 4 mm in diameter, almost concealed within the broad bases of bracteoles. Petals very small, equal. Fruit deeply immersed, only a small portion below the short (0.7 mm) styles appearing above the "receptacle". Disk flat, crenulate-margined. Pedicels of sterile flowers very slender, unchanged in fruit, withering. Fig. 50: 1–7.

HAB. Steppic foothills, on a conglomerate hillside; alt. 100–200 m; fl. & fr. Apr.–May.
DISTRIB. Rare in Iraq; only found in one district near the Iranian frontier in the eastern sector of the steppe region. **FNI**: On Iranian frontier, 10 km E of Mandali, *Rechinger* 9648!; 54 km SE of Mandali, *Rechinger* 12742; Mandali, *Hadač et al.* s.n.; Mandali to Naft-i Shah, *Hadač et al.* 4627; Koma Sang nr Mandali, *Rawi* 20631.

W Iran.

5. **PYCNOCYCLA** Lindl.

in Royle, Illustr. Bot. Himal.: 232, t. 51 (1834)

Perennial, rigid herbs, glabrous or shortly hairy, with divaricate subaphyllous branches. Leaves pinnately dissected, palmate, or entire and palmately veined, when dissected the segments few, filiform or rigidly spinescent. Umbels compound, usually very compact. Involucre and involucel present, spinescent or not. Central flower hermaphrodite and fertile, the remainder pedicellate and male. Petals white, hairy, emarginate and ± 2–lobed above, the tips inflexed. Calyx present, rather small, sometimes conspicuous at the tips of the male flowers in the fruiting stage. Fruit ovate-oblong, attenuate above, not partially immersed in the receptacle, which does not become accrescent; ribs not conspicuous, filiform. Vittae 2–3 in the valleculae. Stylopodia conical, entire. Styles long, filiform, erect with spreading tips. Carpels solitary or 2, the endosperm face deeply excavate, the margins involute. Carpophore absent.

Twelve species distributed between tropical W Africa and NW India; a single species in Iraq.
Pycnocycla (from Gr. πυκνός, *pyknos*, dense, κυκλος, *kyklos*, a ring or circle, referring to the congested inflorescence).
The only species in Iraq (*P. flabellifolia*, q.v.) is too rare and of too limited a distribution to be of economic interest, but another species *P. aucheriana*, which has not been found in Iraq, is reported as providing grazing for camels in Baluchistan (note on label of *Crookshank* 148 in Herb. Kew) while the roots are used as fuel in Afghanistan (*Aitchison* 817) and exude a foetid odour when burned. The vernacular name of another species (*P. tomentosa*) has been noted as SESEBAN (that is no doubt for SAISBĀN) in Sinai (*Schimper* 148), indicating that it is probably of local use there.

1. **Pycnocycla flabellifolia** *Boiss.*, Diagn. ser. 2, 2: 105 (1856) & Fl. Orient. 2: 953 (1872); Riedl in Fl. Lowland Iraq: 453 (1964).

Perennial, up to ± 25 cm, stiff and rigid. Stem terete, striate, minutely downy. Branches few, arising from upper leaf axils. Basal leaves rigid and coriaceous, lamina ± 2 cm long and ± orbicular-spatulate in outline, furnished in upper ⅔ of its length with 12–20 long, very unequal spinous teeth each separated by a ± obtuse and usually wide and curving sinus. Lower third of lamina entire, attenuately curving into the long petiole, which is furnished with a short, rigid, expanded sheath. Stem leaves reducing in size, becoming sessile, toothed all round, teeth longer in proportion to the breadth of the undivided portion of lamina than is the case with basal leaves. All leaves shortly pubescent when young, becoming glabrous with age. Umbels long-pedunculate, axillary, small and compact, ± 10 mm in diameter, rays of main and partial umbels very short, much exceeded by reddish spine-tipped bracts of the involucre and involucel, the latter surpassing the main. Rays and pedicels tomentose, with dense white hairs. Petals and filaments hairy. Fruit shortly oblong, hairy. Fig. 51: 1–4.

Fig. 51. **Pycnocycla flabellifolia.** 1, habit × ²/₃; 2, bract × ²/₃; 3, petal × 4; 4, fruit × 4. All from *Rawi* 20783. Drawn by D. Erasmus (1964).

HAB. Steppic foothills, in a wadi, on gravelly soil; alt. 110–150 m; fl. May–Jun., fr. Aug.?
DISTRIB. Rare in Iraq, though locally common; found in one district near the Iranian frontier in the eastern sector of the steppe region. **FNI**: Makatu, *Guest* s.n.; Khir-i Charmig, a few km NE of Mandali, *Qaisi & K. Hamad* 43565; 55 km SE of Mandali, *Rawi* 20733! *Rechinger* 9672.

W Iran.

6. **CACHRYS** L.

Sp. Pl. ed. 1: 246 (1753); Gen. Pl. ed. 5: 117 (1754)
Hippomarathrum Hoffmans. & Link, Fl. Port. 2: 411 (1820) non Gaertn., Mey. & Schreb., Fl. Wetterau 1: 249 (1799)

Much-branched, robust perennial herbs with deeply dissected leaves. Involucre and involucel present on the compound umbel. Flowers hermaphrodite. Calyx with 5 short but distinct teeth. Petals obovate and entire with the apex inflexed, usually ± yellow. Fruit ovoid to subglobose, somewhat compressed laterally, crowned with a broad disk. Stylopodia small. Mericarps with 5 thick ridges which may be cristate or verrucose, commissure broad. Pericarp crustaceous. Commissural face of the endosperm deeply excavate, margins involute. Vittae dense, copious, slender.

Some 7–8 species distributed in the Mediterranean, W & C Asia; two species in Iraq. *Cachrys microcarpum* Bieb. (syn. *C. crispum*) has been recorded from the extreme SE corner of Turkey close to the Iraqi border; it can therefore be expected to occur in N Iraq, but no material has so far been seen.

Cachrys (a name said to be derived from a Greek word indicative of the hot or carminative properties of the fruit, though Liddell & Scott give κάχρυ, *kahru*, for the fruit of *Libanotis*). According to Lindley & Moore (1870) the seeds of one species of this genus (*C. odontalgica*), which is found in Caucasus and elsewhere in Russia, used to be chewed by the Cossacks to alleviate toothache, this effect being due to the salivation they induce.

Plant scabrid and finely pubescent; fruit densely verruculose, verrucae
usually scabridulous. . . . . 1. *C. scabra*
Plant stellate-hairy; fruit glabrous . . . . 2. *C. papillaris*

1. **Cachrys scabra** (*Fenzl*) *Meikle*, Fl. Cyprus 1: 727 (1977).

Ferula scabra Fenzl, Flora 26: 461 (1843); *Hippomarathrum scabrum* (Fenzl) Boiss., Fl. Orient. 2: 933 (1872); Hand.-Mazz. in Ann. Naturh. Mus. Wien 27: 89 (1913); Zohary in Dep. Agr. Iraq Bull. 31: 112 (1950); Rawi, ibid. Tech. Bull. 14: 93 1964); Mouterde, Nouv. Fl. Lib. et Syr. 2: 611 (1970); Chamberlain in Fl. Turk. 4: 392 (1972); *Bilacunaria scabra* (Fenzl) Pimenov & V.N. Tikhom. in Feddes Repert. 94: 152 (1982).

Stem aromatic, very rigid, angulate-sulcate, with fine, close pubescence (rarely glabrous) and sparingly scabrid, especially below and in the region of the nodes, much-branched, robust, to ± 1.5 m. Lower leaves with broad sheaths, long-petiolate, triangular-ovate in outline, repeatedly 3–5-sect in several places with the final divisions usually trisect, up to 40 cm long and 25 cm wide; all leaf-segments solid, ribbed, canaliculate-filiform, shortly mucronate; long flattened sheaths and primary divisions of leaves very scabrid with densely hirtellous verrucae and with the fine, close hirtellous pubescence which covers all the vegetative parts, the scabridity much less pronounced on the ultimate segments. Upper leaves reducing, less divided, becoming sterile on sheaths. Branches mostly opposite or in whorls of 3, subtended by prominent bracts. Bracts of the involucre and involucel mostly 5, linear to subulate. Primary umbel 5–9-radiate, rays 1–4(–7) cm; partial umbel 3–15-flowered, pedicels 2–10 mm. Peduncles 2.5–10 cm. Petals appearing grey-green from the close downy indumentum which covers them, ± 1 mm; filaments conspicuous between the petals from being incurved over the anthers. Fruit subrotund, ± 5 mm, densely verruculose, the verrucae usually scabridulous, the contrasting chocolate-coloured, crenate-edged disk smooth; styles 1.5–2 mm, slender, erect or flexuose, finally reflexed. Fig. 52: 1–8.

HAB. Steppic hills and plains, stream-bed; alt. 500–600 m; fl. & fr. May–Jun.
DISTRIB. Occasional in steppic parts of the lowest forest zone of Iraq, found only once in the western corner of the desert region. **MSU**: 6 km E of Qaranjir, *Gillett* 11907!; **MJS**: above Balad Sinjar, *Handel-*

Fig. 52. **Cachrys scabra**. 1, habit base × ²/₃; 2, habit leaf × ²/₃; 3, habit × ²/₃; 4, part of leaf with apex and stem detail × ²/₃; 5 petal × 6; 6, 7, fruit × 4; 8, fruit cross section × 4. 1, 6–8 from *Gillett* 11907; 2–5 from *Staph* 907. Drawn by D. Erasmus (1964).

Mazzetti 1408; **FUJ**: Jabal Khatchra, nr Balad Sinjar, *Field & Lazar* 667!; **DWD**: Al-Masad depression, S of Rutba, *Alizzi & Omar* 46185!; illegible locality, 'Mesopotamia' (or in Syria or Turkey), *Haussknecht* 574!

Cyprus, Syria, Turkey.

2. **Cachrys papillaris** *Boiss.*, Ann. Sc. Nat. sér. 3, 2: 75 (1844); Boiss., Fl. Orient. 2: 936 (1872); Herrnstadt & Heyn in Fl. Turk. 4: 387 (1972) & in Boissiera 26: 84 (1977).

Plant stout, erect, height not known. Stem fistular, sulcate, corymbosely branched above, the branches opposite. Stem, branches, leaf sheaths and primary leaf divisions ± densely clothed with stellate and, above, simple papillate hairs. Lower cauline leaves large, sessile, broadly ovate to oblong in outline, to 35 cm or more in length, pinnately decompound into narrowly linear, 3–15 × 0.5 mm segments which are crisped-hairy, hairs up to as long as width of segments. Upper leaves similar in form, reducing in size. All leaves with a short, broad sheath. Peduncles 3.5–11 cm. Umbels several, compound, terminal on the main stem and branches, the former the largest, polygamous, the laterals male; fructiferous rays 15–25, glabrous or very sparsely hairy, 5–11 cm, male rays shorter. Involucre of 4–6 simple or pinnate bracts to 3 cm long. Partial umbels 8–24-flowered, up to 10 flowers of the central umbel with fruit. Involucel of 6 subulate bracteoles to ± 12 mm long. Flowers yellow. Fruit globose, didymous, glabrous and smooth, ± 4 mm, on 2–3 cm pedicels. Disk broad and thin, somewhat immersed between the commissural faces of the mericarps and thus V-shaped in lateral view, undulate-margined, rugose over the surface. Stylopodia minute. Styles about equalling the diameter of the fruit, reflexed, flattened.

HAB. & DISTRIB. Only one collection of this plant is known to have been made positively within Iraq, viz., **FUJ**: 27 km S of Khanaq, *Eig & Zohary* s.n. on 24 Apr. 1933, alt. 500 m. Of the other three gatherings on which its occurrence has been claimed ("Mesopotamia", *Aucher* 3582!!; "in deserto Kotchassar, Assyriae", *Haussknecht* s.n.! and "in deserto proper Messhkok", *Haussknecht* s.n.!) both the second and third are in Turkey or Syria and, as to the first, "Mesopotamia", literally "the territory between the two rivers" (R. Tigris and R. Euphrates) covers parts of Syria and Turkey as well as Iraq.
NOTE. This plant is certainly not a *Cachrys* as the genus is at present defined, and is only placed here in the Flora for ease of reference and indexing. Its affinity appears to be with *Prangos* (including *Cryptodiscus* – see Pimenov & Tikhomirov in Feddes Rep. 94: 145–164, 1983, though these authors do not deal with the present species), but it has been excluded from that genus by Herrnstadt & Heyn in their revision [*Boissiera* 26: 1–91 (1977)]. It is known only with immature fruit, and until ripe fruit is collected its generic position will remain in doubt.
Pimenov's list also includes *Bilacunaria microcarpa* (M.Bieb.) Pimenov & V.N. Tikhomirov, which belongs in *Cachrys* as here defined, but no material has been seen from Iraq.

Turkey.

7. CONIUM L.

Sp. Pl. ed. 1: 243 (1753) & Gen. Pl. ed. 5: 114 (1754)

Tall, branched biennial herbs, glabrous, with pinnately compound leaves. Umbels compound, involucre and involucel present, small. Calyx teeth obsolete. Petals obovate or cuneate, sometimes acuminate with the tip inflexed, sometimes obtuse. Fruit broadly ovoid to suborbicular, laterally compressed; mericarps strongly 5-ribbed, ± pentagonal in section, commissure rather narrow. Vittae obsolete. Stylopodia flattened, ± entire, persistent styles short, recurved. Inner face of endosperm with narrow grooves.

Five North temperate Eurasian and one S African species; only one species in Iraq.
Conium from the Gr. κοειον, *koeion,* the name of the Common Hemlock plant; hemlock.

1. **Conium maculatum** *L.*, Sp. Pl. ed. 1: 243 (1753); DC., Prodr. 4: 242 (1830); Boiss., Fl. Orient. 2: 922 (1872); Hayek, Prodr. Fl. Balc. 1: 1069 (1927); Zohary, Fl. Palest. ed. 2, 1: 534 (1932); Schischkin in Fl. SSSR 16: 225 (1950); Rawi in Dep. Agr. Iraq Tech. Bull. 14: 91 (1964); Rawi & Chakr., ibid. 15: 29 (1964); Mouterde, Nouv. Fl. Lib. et Syr.Syr. 2: 610 (1970); Zohary, Fl. Palaest. 2: 405 (1972); Stevens in Fl. Turk. 4: 380 (1972); E. Nasir in Fl. Pak. 20: 27 (1972); Meikle, Fl. Cyprus 1: 691 (1977); Leute in Fl. Iranica 162: 171 (1987); Pan Zehui & Watson in Fl. China 14: 58 (2005).

Fig. 53. **Conium maculatum**. 1, 2, habit with young and mature umbels × ²/₃; 3, leaf × ²/₃; 4, petal × 6; 5, fruits × 3; 6, fruit cross section × 3. 1, 4 from *Davis & Hedge* D28990; 2–3, 5–6 from *Wheeler Haines* W1166. Drawn by D. Erasmus (1964).

Tall but graceful biennial with a foetid odour, to 2 m. Stem very hollow, especially below, striate to lightly sulcate, purple-spotted, branching above. Lower leaves long-petiolate, up to 60 cm broad and long, the lamina ovate or triangular-ovate in outline, delicate in appearance, tripinnatisect with narrowly lanceolate or oblong-lanceolate segments; cauline leaves diminishing in size and less dissected but also petiolate except for the uppermost. Branches opposite, umbels axillary and terminal. Involucre and involucel delicate, with broad membranous margins, often withered and not apparent at ripe fruit, of 6–7 lanceolate bracts, those of the involucel disposed unilaterally on the partial umbels. Peduncles 1.5–6 cm. Main umbel of 10–18 rays, 0.5–3 cm long; partial umbels of (6–)10–25(–20) flowers of 2–3 mm in diameter, pedicels 2–7 mm. Petals obovate, the tip slightly inflexed. Fruit broadly ovoid, 3–3.5 mm in diameter; mericarps with 5 strong pale ribs which contrast with the dark body of the fruit and are strongly crenulate in the typical plant, particularly near the apex of the fruit; styles ± 0.75 mm. Fig. 53: 1–6.

HAB. In the mountains, near water, by a stream, on the edge of a damp thicket, in *Quercus* forest; alt. 900–1750 m; fl. & fr. Jun.–Jul.
DISTRIB. Occasional but locally abundant, in the forest zone of Iraq. **MRO**: Haji Umran, Omar, *Karim, Hamid & Sahira* 38493!; Rowanduz above Zeita, *Hadač* 5948; **MSU**: Sarchinar, nr Sulaimaniya, *Gillett* 11724! *Omar & Karim* 37946! *Karim, Hamid & M. Jasim* 40703? Qopi Qaradagh, *Hadač* 5184.
NOTE. Represented in Iraq by the var. *leiocarpum* Boiss., which has the ribs of fruit quite smooth and devoid of crenulations.

Common Hemlock (Eng.), Poison Hemlock (Am.); SHAUKARĀN (Ar., Watt, Bedevian et al.). This was the State Poison of ancient Athens for the execution of criminals, so tragically used to eliminate the great philosopher Socrates, as Plato described.

As Schischkin mentions, all parts of the plant, and especially the fruit, contain toxic alkaloids and in former times preparations containing these substances have been widely used to treat various diseases; but in view of side effects and the occurrence of poisoning, such preparations are no longer prescribed. The plant has a disagreeable odour in warm weather, and when dried is reminiscent of mice; its taste is sharp and bitter.

Atlantic, C & S Europe, Cyprus, Syria, Lebanon, Palestine, Jordan, Turkey, Caucasus, Iran, Afghanistan, C Asia, Siberia, N Africa (Algeria, Tunisia, Libya), Macaronesia (Canary Is., Azores). Introduced into China, Australia, S Africa, C & S America and in some places subspontaneous.

8. **PSEUDOTRACHYDIUM** (Kljuykov et al.) Pimenov & Kljuykov, Feddes Rep. 111: 526 (2000)

Trachydium Lindl. in Royle, Illustr. Bot. Himal.: 232 (1835), p.p.
Aulacospermum Ledeb. sect. *Pseudotrachydium* Kljuykov et al., Bull.Soc. Nat. Mosc. 81 (5): 65 (1976)

Caulescent perennial herbs. Leaves 2 or 3-pinnatisect with sessile primary segments; sheaths of cauline leaves not inflated. Umbel 3–9-rayed, involucre and involucel present. Flowers yellow. Calyx teeth small but broad, subulate. Petals yellow, ovate or obovate, the tip ± acuminate and inflexed. Stylopodia shortly conical. Fruit ovoid, dorsally not compressed, commissure narrow, mericarps with 5 blunt but prominent ± equal ribs which are covered with smooth, obtuse, ± vesicular scales; valleculae papillose. Vittae in the valleculae 1 or 2 (rarely 3 or 4), commissural vittae 2–8. Endosperm pentagonal, deeply grooved on the commissural face, the margins slightly involute.

Five species from SW & C Asia to W China; only one species in Iraq.
Pseudotrachydium (from the Gr. πσευδο, *pseudo*, false, τραχύς, *trachys*, rough and the noun base *-idium* – from the rough covering of the fruits.

1. **Pseudotrachydium depressum** (*Boiss.*) *Pimenov & Kljuykov* in Feddes Rep. 111: 527 (2000).

Trachydium depressum Boiss., Fl. Orient. 2: 929 (1872); Blakelock in Kew Bull. 3: 437 (1948); Rawi in Dep. Agr. Iraq Tech. Bull. 14: 96 (1964); Hedge & Lamond in Fl. Turk. 4: 380 (1972); Rech.f., Fl. Iranica 162: 186 (1987).

Dwarf, glabrous, perennial herb with a very stout rootstock, 8–20(–30) cm. Stem slender, almost terete, at the base surrounded by a dense fibrous mass of previous years' persistent stem-bases and leaf-sheaths, little-branched (one or two alternate branches) below. Petioles broad, widening to a long sheathing base. Radical leaves oblong-lanceolate in outline, 5–19 × 0.8–5 cm, bipinnatisect, the segments lanceolate, acuminate. Cauline leaves mostly 1 or 2, much reduced. Umbels shortly and very unequally 3–5-rayed, rays 0–25 mm; involucre and involucel present, bracts and bracteoles 3–5, rather delicate, 2–3 mm, lanceolate with pale margins. Partial umbels 3–5-flowered, pedicels ± 0.7 mm; petals with the tips strongly inflexed and acuminate. Peduncles (1.5–)3–11(–19) cm. Fruits bluntly but distinctly 5-ribbed, the ribs thickly covered with obtuse ± vesicular scales; valleculae 1-vittate, commissure 2-vittate. Stylopodium disciform, the persistent styles rather long (± 1.5.mm). Fig. 54: 1–5.

HAB. High mountain slope and summit, on rocks; alt. 2600–3800 m; fl. & fr. Jun.-Aug.
DISTRIB. Occasional in the alpine region of Iraq. **MRO**: Halgurd Dagh, *Guest* 3057A!, *Rechinger* 11444; Kodo (Kudu), *Rawi* 9209!; Qandil range, *Rawi & Serhang* 18269!; Qandil range, NE of Qandil, *id.* 24423! *Rechinger* 11154.

Turkey (E Anatolia), Iran, Afghanistan.

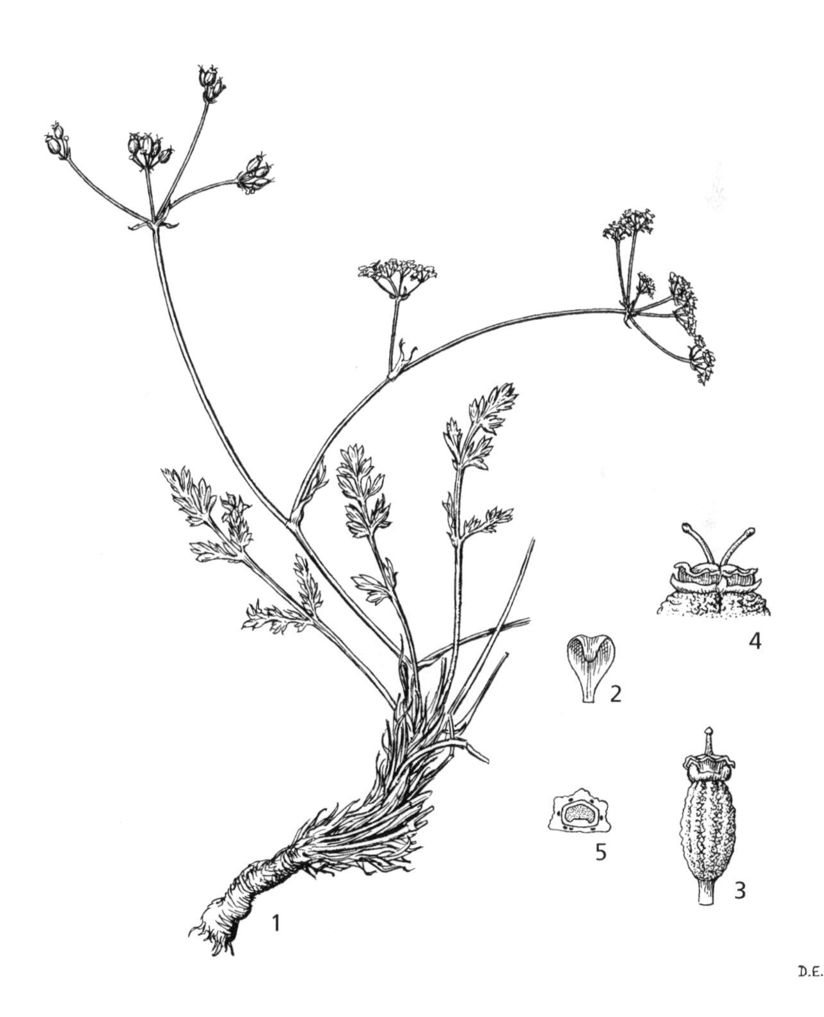

Fig. 54. **Pseudotrachydium depressum**. 1, habit × 2/3; 2, petal × 6; 3, fruit × 4; 4, top of fruit showing stylopodium and styles enlarged; 5, fruit cross section × 4. All from *Davis & O. Polunin* D24193. Drawn by D. Erasmus (1964).

9. **SMYRNIUM** L.

Sp. Pl. ed. 1: 262 (1753) & Gen. Pl. ed. 5: 127 (1754)

Erect glabrous biennial or perennial herbs. Radical and lower leaves ternate-pinnately dissected, the upper undivided, or at least much less so. Involucre and involucel small or absent on the compound umbels. Flowers yellow, hermaphrodite or polygamous. Calyx teeth absent. Petals obovate or oblong, with a short inflexed tip. Fruit ovoid or broader than long, laterally compressed, ± didymous, commissure narrow. Mericarps subterete or angular with 3 (rarely 5) distinct ridges. Vittae many, in valleculae and commissure. Commissural face of endosperm excavate, with the margins involute. Stylopodia conical or finally flattened, the margins entire or almost so.

Eight European and Mediterranean species; only one species in Iraq.

Smyrnium (from the Gr. name for myrrh, σμύρνα, *smyrna*, alluding to the aroma of the species *S. olusatrum*). Ibn al-Baitar (c. 1240 AD) mentions the name SMIRNIŪN and calls the plant "wild celery". Alisander or Alexanders, as it has been called (*S. olusatrum*) was indeed formerly used as a potherb, its taste being not unlike that of celery, though much stronger and less agreeable. Bedevian (1936) gives the name KARAFS BARRI (Ar. "wild celery") to the above species and also to another species with similar properties (*S. perfoliatum*).

1. **Smyrnium cordifolium** *Boiss.*, Diagn. ser. 1, 6: 64 (1845) & Fl. Orient. 2: 926 (1872); Bornmüller in Beih. Bot. Centralbl. 58B: 280 (1938); Schischkin in Fl. SSSR 16: 221 (1950); Zohary in Dep. Agr. Iraq Bull. 31: 112 (1950); Rawi, ibid. 14: 96 (1964); Mouterde, Nouv. Fl. Lib. et Syr. 2: 614 (1970); Stevens in Fl. Turk. 4: 338 (1972).

Very stout, glabrous biennial, 0.5–1.5 m. Stem hollow, terete, finely striate, to 1.8 cm wide at base, the upper branches opposite and forming a terminal corymb. Radical leaves large, with a broad sheathing base, triternate, segments 2.5–6 × 2–5.5 cm, petiolate, subrotund with broad, shallow crenulations; lower stem leaves similar but smaller, becoming pinnate, with the segments ovate or deltoid-ovate with an amplexicaul base, entire, the lowest to ± 12 cm long. Peduncles 3.5–6 cm. Primary umbel 14–18(–20)-radiate, rays 1.5–3 cm; secondary umbel with about the same number of flowers, pedicels ± 5–8 mm. Involucre and involucel absent. Petals obovate-spatulate, tip shortly inflexed. Fruit broader than long (± 5 × 2.5–3 mm when ripe), broad and long, deep black. Mericarps subglobose with the primary ridges shallow but obvious, and finely rugose with many intermediate ridges and wrinkles; styles long and flexuose to recurved, stylopodia flat. Fig. 55: 1–6.

HAB. In the mountains in denuded *Quercus* forest on limestone, on a limestone scree, in a ravine, in a soil-filled gully on a N-facing cliff; alt. (500–)950–2200 m; fl. & fr. (Apr.–)May-Jun.

DISTRIB. Common in the central and upper forest zones of Iraq, occasional in the lower thorn-cushion zone. ?**MAM**: Alho, *Rawi* 8567!; SW of Aqra, *Anders* 1283; **MRO**: Halgurd (Algurd Dagh), *Bornmüller* 1261! *Rawi* 13737!; Haji Umran, *Omar, Sahira, Karim & Hamid* 38494?; between Dargala and Karokh Mt., *Rawi, Nuri & Kass* 27727!; Serkabkhan, 7 km NW of Rania, *id.* 28547!; **MSU**: Darband-i Bazian, *Rawi & Gillett* 7651!; Gweija Dagh, behind Sulaimaniya, *Gillett & Rawi* 11683!; Qopi Qaradagh, *Haines* W1556!; Penjwin, *Rawi* 8918A!; Biyara, *Gillett* 11745!; **MJS**: above Balad Sinjar, *Gillett* 11261.

NOTE. QALANDŪR (Kurd.-Halgurd, *Rawi* 13737), ?GUNOR (Kurd.-Darband-i Bazian, *Rawi & Gillett* 7651), NINUR (Kurd.-Hawraman (Diyara), *Gillett* 11745); though none of these local common names have been confirmed, that it should have been recognised and named in three separate mountain districts suggests that it has some use in these places. In Afghanistan, Aitchison noted that it occurred in damp situations and was eaten, both raw and cooked, by the local people (*Aitchison* 399 in Herb. Kew).

NE Syria (Jabal Abdul Aziz), Turkey (C & E Anatolia), Iran, Afghanistan, C Asia (Turkmenistan).

10. **ACTINOLEMA** Fenzl

Pugill. pl. nov. Syr.: 16 (1842)

Astrantia Baill. Hist. Pl. 7: 241 (1880) p.p.; Wolff in Pflanzenr. IV. 228, 61: 92 (1913)

Annuals, repeatedly di- or tri-chotomously branched above. Leaves undivided. Umbels simple, few-flowered, involucral bracts large and exceeding the umbel, rather thin and papery in texture, persistent. Flowers polygamous, ± sessile central flower hermaphrodite, the outer pedicellate and male. Calyx teeth leafy, 3–5-spinulose. Petals with a long, inflexed

Fig. 55. **Smyrnium cordifolium**. 1, habit × ²/₃; 2, leaf × ²/₃; 3, petal × 6; 4, fruit × 4; 5, top of fruit × 4; 6, fruit cross section × 4. 1, 3–6 from *Kamil* 314; 2 from *Bornmüller* 1261. Drawn by D. Erasmus (1964).

terminal lobule. Fruit ovoid-oblong, sessile or very shortly pedicellate, with 5 equally sharp ribs which are dorsally scabrid with tooth-like papillae; interstices between ribs densely tuberculate-nodulose. Vittae solitary in the valleculae. Carpophore fused to mericarps.

Two species in SW Asia, both found in Iraq.
Actinolema (from the Gr. ακτις, *aktis,* a ray and ειλημα, *eilema,* a wrapping, from the large regular involucre).

Involucral bracts up to 14 mm long, elliptic-lanceolate; ripe fruit 3.5–4 mm long .... 1. *A. eryngioides*
Involucral bracts up to 20 mm long, ovate-lanceolate to broadly ovate; ripe fruit 6–7 mm long .... 2. *A. macrolema*

1. **Actinolema eryngioides** *Fenzl,* Pugil. pl. nov. Syr. 16, t. 12. (1842); Boiss., Fl. Orient. 2: 831 (1872); Wolff, l.c.: 93 (1913); Nábělek in Publ. Fac. Sci. Univ. Masaryk 35: 119 (1923); Zohary, Fl. Palest. ed. 2, 1: 508 (1932); Bornmüller in Beih. Bot. Centralbl. 58B: 275 (1938); Blakelock in Kew Bull. 3: 431 (1948); Zohary in Dep. Agr. Iraq Bull. 31: 109 (1950); Schischkin in Fl. SSSR 16: 72 (1950); H. Riedl in Fl. Lowland Iraq: 450 (1964); Rawi in Dep. Agr. Iraq Tech. Bull. 14: 88 (1964); Mouterde, Nouv. Fl. Lib. et Syr. 2: 576 (1970); Hedge & Lamond in Fl. Turk. 4: 292 (1972); Rech.f., Fl. Iranica 162: 44 (1987).

Glabrous annual, attaining 30 cm but often much smaller. Stem subterete-sulcate below, more angular above, corymbosely branched from near the base, 2- or 3-furcate above. Stem leaves 15–55 × 7–20 mm, lower broadly ovate and long-petiolate, upper narrower and much more shortly petiolate, often withered by fruiting period, undivided but with regular subacute teeth; uppermost reduced leaves ± trifid. Involucral bracts 5 or 6, violet-tinged, elliptic-lanceolate, regularly spinulose-dentate, thin, with delicate but obvious reticulate venation and with a scarious border which narrows towards the apex; those of axillary umbels 10–14 mm, of lateral umbels 7–10 mm, all exceeding umbels. Flowers greenish, calyx segments obcuneate, apex truncate and with a short arista formed by the excurrent midrib; angles acute and usually spinulose. Petals deltoid with a subequal similarly shaped inflexed lobule; perianth segments of central hermaphrodite flower slightly larger than those of outer male flowers. Fruit 3.5–4 mm, the dorsal teeth of ribs rather short and blunt, upwardly directed, alternate teeth often (but not constantly) staggered, alternately on each side of rib so as to give a biseriate appearance.

HAB. In the mountains and occasionally on sub-montane plains, on stony hillsides, in fields and on abandoned cultivation; alt. (300–)600–1500 m; fl. & fr. Apr.–Jun.
DISTRIB. Common in the middle and lower forest zones of Iraq, occasional in the moist-steppe zone. **MAM**: Ain Nuni, *Nábělek* 407; between Amadiya and Sarsang, *Karim & Hamid* 40983; **MAM/FNI**: between Aqra and Mosul, *Rawi* 11359!; **MAM/FUJ/FKI**: Mosul district, *Noë* s.n.; **MRO**: Kuh-Sefin, above Shaqlawa, *Bornmüller* 1228!; Saran, nr Kani Kawan spring, in Karoukh district, *Nuri & Kass* 27296A!; 5 km S W of Rania, *Rawi, Nuri & Kass* 28423!; **MRO/FAR**: Arbil province, *Radhi* 3851!; **MSU**: 4 km E of Qaranjir, *Rawi* 21622!; Qaranjir, *Rechinger* 10018; Khurda Luk, 51 km NW of Sulaimaniya, *Rawi* 21781!; Pira Magrun, *Haussknecht* s.n.; **MSU/FKI**: Kirkuk district, *Haussknecht* s.n.; **FUJ**: Tal Afar, *Ahmad & Jabar* 50154; 10 km N of Mosul, *Hossain* s.n.; **FAR**: Ankawa, *Bornmüller* 1229.
NOTE. ?HASSACH (Kurd.-Arbil, *Radhi* 3851, "grazed by animals").

Syria (Aleppo), Turkey (E Anatolia), Caucasus, W Iran.

2. **Actinolema macrolema** *Boiss.,* Fl. Orient. 2: 832 (1872); Wolff, l.c.: 93 (1913); Nábělek in Publ. Fac. Sci. Univ. Masaryk 35: 119 (1923); Schischkin in Fl. SSSR 16: 72 (1950); Rawi in Dep. Agr. Iraq Tech. Bull. 14: 88 (1964); Mouterde, Nouv. Fl. Lib. et Syr. 2: 576 (1970); Hedge & Lamond in Fl. Turk. 4: 291 (1972); Rech.f., Fl. Iranica 162: 43 (1987).

Glabrous annual, resembling the previous species in general appearance but larger in its parts though smaller in stature, usually 10–15(–30) cm. Stem leaves similar to those of *A. eryngioides,* but involucral bracts 15–20 mm, ovate-lanceolate to broadly ovate, broader and longer than those of the terminal umbels, secondary venation more pronounced, at least beneath, and beautifully reticulate, scarious border less pronounced and narrower even at base. Flowers greenish, calyx segments similar in shape to those of *A. eryngioides* but longer in proportion to their breadth and with antical angles with a distinct arista at least half as long as that formed by the excurrent midrib. Petals yellow, obdeltoid, with a subequal triangular inflexed lobule, rather narrower than in *A. eryngioides*. Perianth segments of the central hermaphrodite flower slightly larger. Fruits 6–7 mm, dorsal teeth longer and rather falcate, upwardly directed, clearly uniseriate on the rather more delicate rib. Fig. 56: 1–6.

HAB. Rocky mountain slopes, in coppiced *Quercus* on limestone, in dry fields and on abandoned cultivation; alt. 700–1700 m; fl. May-Jun., fr. Jun.–Jul.
DISTRIB. Rare but locally frequent in Iraq; scattered in the forest zone. **MJS**: Jabal Sinjar, N slope, *Qaisi, Khayat & Karim* 50925; **MAM**: Bekher, *Rawi* 8450!; **MRO**: Handren Dagh, above Jindian, *Nábělek* 425, 513; *Gillett* 8281!

Turkey, Armenia, NW Iran.

Fig. 56. **Actinolema macrolema**. 1, habit × 2/3; 2, petal × 6; 3, flowers ×1; 4, fruit × 3; 5, styles × 6; 6, fruit cross section × 6. 1 from *Rawi* 8450; 2–6 from *Balansa* 233. Drawn by D. Erasmus (1964).

Fig. 57. **Smyrniopsis aucheri**. 1, habit × ²/₃; 2, habit leaves × ²/₃; 3, habit mature umbel with fruits × ²/₃; 4, petal × 6; 5, fruit × 3; 6, styles × 3; 7 fruit cross section × 3. **Petroedmondia syriaca**. 8, habit mature umbel with fruits × ²/₃; 9, styles × 3; 10, fruit cross section × 3. 1 from *Gillett* 11110; 2 from *Polak* 1882; 3–7 from *Haussknecht* 1868; 8–10 from *Haussknecht* 1867. Drawn by D. Erasmus (1964).

11. SMYRNIOPSIS Boiss.

Ann. Sc. Nat. sér. 3, 2: 72 (1844)

Erect biennial or perennial herbs with the general appearance of *Smyrnium* and pinnate or pinnate-ternate leaves. Umbels compound. Involucre and involucel present or absent (present in Iraqi species). Flowers yellow, hermaphrodite. Calyx teeth obsolete. Petals oblong, acute, with inflexed tips. Fruit variously shaped, from oblong to suborbicular-ovate, laterally compressed, with broad, flat stylopodia and rather long deflexed styles; mericarps sharply pentagonal in section with 5 very acute ridges. Valleculae 1or 2 or multi-vittate, commissural vittae 4-many. Endosperm with involute margins.

Four species from the E Mediterranean to Iran. Originally two species in Iraq, but *S. cachroides* is now placed in a separate genus *Petroedmondia*.

Smyrniopsis (from *Smyrnium*, q.v. and the Gr. suffix -οπσις, *-opsis*, having the appearance of, from the resemblance of the plants to *Smyrnium* species.

Section 1. SMYRNIOPSIS. Ribs of fruit slender; vittae in the valleculae superficial, 1or 2 (rarely 3), thick and prominent; commissure 4-vittate.

1. **Smyrniopsis aucheri** *Boiss.*, Ann. Sc. Nat. sér. 3, 2: 72 (1844) & Fl. Orient. 2: 928 (1872); Nábělek in Publ. Fac. Sci. Univ. Masaryk 35: 119 (1923); Bornmüller in Beih. Bot. Centralbl. 58B: 280 (1938); Zohary in Dep. Agr. Iraq Bull. 31: 112 (1948); Rawi, ibid. Tech. Bull. 14: 96 (1964); Stevens in Fl. Turk. 4: 340 (1972); Tamamschian in Fl. Iranica 162: 164 (1987).

Smyrniopsis munzurdaghensis Yıldırımlı in Ot Sistematik Botanik Dergisi 17 (2): 13 (2010).

Yellowish biennial herb, 1–2 m, glabrous in all its parts save for sparse cylindrical- or conical-papillose hairs on petioles, lower surface of midribs of lower leaves and leaflets. Stem rather slender, striate, large (to about 50 × 40 cm), lamina deltoid-ovate in outline, broadly sheathing, bipinnate with pinnules of lower pinnae often ternate, all segments regularly obtusely serrate, oblong-ovate, 2.4–7 × 1.2–4(–5) cm. Upper leaves reducing first to three obovate leaflets, then simple, subentire. Umbels 7–14-radiate, rays 1.8–4.5 cm; partial umbels with 9–12 flowers of ± 3 mm diameter, pedicels 1–9 mm. Peduncles 2–7 cm. Involucre and involucel soon withering, of 5 or 6 short, linear-lanceolate bracts. Fruit about as broad as long, ± 5 × 5 mm, at least half as long as the pedicel at maturity, with acute but not carinate ridges. Stylopodia broad, discoid or shallowly conical, crenate-edged; style long, deflexed, exceeding the width of the fruit. Fig. 57: 1–7.

HAB. In the mountains, among *Quercus* forest and scrub, on limestone; alt. (800–)1000–1800 m; fl. & fr. May–Jun.

DISTRIB. Occasional, locally common, in the forest zone of Iraq. ? **MAM** "Mesopot., Kurdistan & Mossul", *Kotschy* 247; **MRO**: Serderian, between Arbil and Rowanduz, *Nábělek* 432; Kuh-i Sefin, above Shaqlawa, *Bornmüller* 1260!; Shaqlawa, *Haines* W647!; between Dargala and Karouk Mt., *Nuri & Kass* 27745!; between Shaikhan and Sakri-Sakran, *Hadač* 5408; **MSU**: Pira Magrun, *Haussknecht* 511!; Zewiya, *Rawi* 12074!; Dukan, *Agnew & Haines* W2094!; N of Biyara, *Gillett* 11783!; Mt. Kalkou (Kal Kuh?), *Aucher* 3689; **MJS**: above Sinjar, *Gillett* 11110!

NOTE. The record for 'monte Kalkou Assyriae', *Aucher* 4591 (syntype) is possibly not Iraqi; although 'Assyria' is largely confined to modern Iraq, this locality has not been traced and is not mentioned in our Gazetteer (Fl. Iraq vol. 1). Stevens, in Fl. Turk. 4: 340 (1972), places this locality in Iraq. The locality is also spelled 'Valkou' under *Chaerophyllum macropodum*, q.v. (but there is no 'v' in modern Arabic).

?GENDOR (Kurd.-Sulaimaniya, *Rawi* 12074), ?TALINDOS (Kurd.-Hawraman, *Gillett* 11783 – "leaves collected for winter fodder").

Turkey (SE Anatolia), Iran.

12. PETROEDMONDIA Tamamschian

Fl. Iranica 162: 167 (1987)

Erect perennial herbs with pinnatisect leaves. Stems glabrous, branched to form a corymbose inflorescence. Umbels compound, with numerous rays; involucre and involucel somewhat leathery, lanceolate. Calyx teeth inconspicuous. Flowers yellow, hermaphrodite; petals with inflexed tips; stylopodium flattened into a disc, styles short. Mericarps prismatic, keeled or ridged, triangular in section. Valleculae with minute vittae.

Monotypic.

1. **Petroedmondia syriaca** (*Boiss.*) *S.G. Tamamschian* in Fl. Iranica 162: 167 (1987).

Smyrniopsis cachrioides Boiss., Fl. Orient. 2: 928 (1872); Zohary, Fl. Palest. ed. 2, 1: 536 (1932); Blakelock in Kew Bull. 3: 4317 (1948); Zohary in Dep. Agr. Iraq Bull. 31: 112 (1950); Rawi, ibid. Tech. Bull. 14: 96 (1964); Mouterde, Nouv. Fl. Lib. et Syr. 2: 616 (1970); Stevens in Fl. Turk. 4: 342 (1972)
Colladonia syriaca Boiss., Ann. Sc. Nat., Bot. sér. 3, 2: 86 (1844).
[*Prangos* (*Colladonia*) *crenata* (non (Boiss.) Fenzl) Guest in Dep. Agr. Iraq Bull. 27: 77 (1933.]

Very stout perennial (or biennial?) herb, 0.6–1.05 m, glabrous throughout save for the lower leaves, which have the petioles and lower margins and veins of leaflets shortly scabrid. Lower leaves with lamina deltoid-ovate in outline, large, to ± 35 × 35 cm, broadly sheathing, simple pinnate or with one or two pinnae bipartite, upper pair at least decurrent along the rhachis, all obtusely crenate with the crenations frequently overlapping at base; segments very large, thick and rigid, terminal leaflet attaining ± 24 × 8 cm; uppermost leaves of three leaflets or simple; branches few, verticillately corymbose. Umbels very large, of 10–20(–35) rays (the terminal largest), rays 4–24 cm; partial umbels with a similar number of flowers of 5–8 mm diameter and fully as large as the compound umbels of *Smyrniopsis aucheri*; pedicels 12–50 mm. Involucre and involucel of 3–8 lanceolate bracts. Fruit ± 10 × 6 mm, at most 1/4 the length of the pedicel, jugae very pronouncedly carinate and strong. Stylopodium broad, disciform, margins crenate. Styles weakly deflexed, about spanning the width of the fruit. Fig. 57: 8–10.

HAB. On lower mountainsides, among destroyed *Quercus* on limestone, in fields and abandoned cultivation, on steppic hills and plains; alt. 350–850(–1100) m; fl. & fr. Apr.–Jun.
DISTRIB. Occasional, locally frequent, in the lower forest and moist-steppe zones of Iraq. **MAM**: nr Dohuk, *Guest* 2233!; **MRO**: 12 km N of Haibat Sultan Dagh, *Nuri & Kass* 28396; **MSU**: 6 km E of Qaranjir, *Gillett & Rawi* 7557!; 20–25 km E of Qaranjir, *Rawi* 21655!; 5 km from Chemchamal on the road from Kirkuk, *Omar & Karim* 37878!; Tainal, *Gillett & Rawi* 11624!; Zewiya, *Rawi* 12074!; Surdash, *Omar, Karim, Hamza & Hamid* 37371!; 20 km NW of Sulaimaniya, on road to Dukan, *Rawi* 21731!; **FUJ**: Tal Afar, *Guest* 12439!; **FNI**: between Mosul and Zakho, *Christabel & Evan Guest* 13241!; **FNI/MAM**: Ain Sifni, *Salim Effendi* 2557!; **FAR**: nr Arbil, *Bornmüller* 1269, *Guest* 2170!; **FKI**: Kirkuk, *Haussknecht* s.n.; 37 km E of Kirkuk, *Rechinger* 12506.
NOTE. GENDOR (Kurd.-Zewiya, *Rawi* 12074), HAFJĀR KALÎSHAT (Kurd.-Ain Sifni, *Salim* 2557).

Syria, Palestine, Jordan, Turkey, Iran.

13. **BUPLEURUM** L.

Sp. Pl. ed. 1: 236 (1753); Gen. Pl. ed. 5: 110 (1754); Wolff in Engler & Prantl, Pflanzenr. 4, 228: 36 (1910); Neves & Watson in Ann. Bot. 93: 379 (2004)

Annual or perennial glabrous herbs, or more rarely shrubs. Leaves simple and entire, sometimes perfoliate. Umbels compound, of few to many often very unequal rays. Involucre and involucel present or involucre absent. Calyx teeth obsolete (very rarely obvious, and not in Iraqi species). Flowers yellow or yellowish, petals with tips inflexed, cucullate, variable in shape, smooth or occasionally finely papillose. Mericarps with 5 prominent primary ridges, fruit laterally compressed, usually ovoid to oblong, rarely subglobose, commissure broad. Vittae in valleculae 1–5, variously developed. Commissural face of endosperm flat or shallowly excavate, rarely more deeply excavate. Stylopodium depressed, flat, margin usually entire. Styles short, reflexed.

Some 150 species in Europe, Asia, Africa and N America; nine species in Iraq.
Although traditional infrageneric headings have been inserted, recent work by Neves & Watson (2004) has shown that these do not conform to the phylogeny of this genus.
Bupleurum (Gr. name of plant from βους, *bous*, ox, πλευρα, *pleura*, rib, used for a plant of this affinity by Nicander; Hare's Ear, Thorow-wax or Throw-away etc. (Eng.). Bedevian gives the name UDHN AL-ARNAB ("hare's ear", Ar.) which appears to be merely an Arabic translation of the English common name.

1. Upper leaves perfoliate 2
   Upper leaves not perfoliate, linear or lanceolate 3
2. Upper leaves broadly ovate to orbicular; fruit smooth . . . . 1. *B. croceum*
   Upper leaves ovate-lanceolate; fruit verrucose-granulate . . . . . . 2. *B. lancifolium*

3. Plant pseudo-dichotomously branched from base with a subsessile, very unequally-rayed umbel at each division; fruiting involucel becoming hard and horny, contiguous to conceal ripe fruit ....4. *B. brevicaule*
   Characters not as above.. 4
4. Bracteoles of involucel large, broadly ovate (± 8 × 4 mm minimum), much exceeding the partial umbel, glumaceous................... 3. *B. aleppicum*
   Bracteoles of involucel lanceolate or linear to almost setaceous, not glumaceous 5
5. Fruit with distinct small white papillae (dwarf annual) .... 8. *B. semicompositum*
   Fruit either smooth on and between ribs, or papillae not white 6
6. Fruit papillose between ribs (stems often whitish); petals papillose . . .9. *B. leucocladum*
   Fruit smooth between ribs; petals smooth.. 7
7. Much-branched delicate annual, inflorescence branches capillary; fruit scarcely longer than broad, rounded at sides .... 7. *B. cappadocicum*
   Less-branched annuals or perennials, inflorescence branches not finely capillary; fruit obviously longer than broad, oblong and ± parallel-sided.. 8
8. Perennial; bracts of involucel much shorter than branches of the partial umbel .... 10. *B. falcatum*
   Annuals; bracts of involucel at least equalling branches of the partial umbel 9
9. Tall plant normally attaining at least 30 cm; bracts of involucel with very narrow but distinct hyaline, denticulate margins, usually about equalling the fruiting partial umbel ....5. *B. kurdicum*
   Shorter plant, rarely attaining 25 cm; bracts of involucel without a narrow, pellucid, serrulate margin, usually exceeding the fruiting partial umbel .... 6. *B. gerardii*

SECT. 1 PERFOLIATA Godr. Upper and middle leaves perfoliate; involucre absent.

1. **Bupleurum croceum** *Fenzl*, Pugillus Plant. Nov. Syr. 16 (1842); E. Nasir in Fl. Pak. 20: 43, f. 12, (1972); Rech.f. & Snogerup in Fl. Iranica 162: 273 (1987).

Tall, erect annual up to 65 cm, branched only towards the upper part. Lower leaves broadly ovate, amplexicaul and perfoliate, mucronate; upper leaves broadly ovate to orbicular, perfoliate. Umbels with 8–18 rays, outer subequal, inner shorter; involucre absent; involucel of 5 herbaceous segments, three larger, two much smaller, golden-yellow when in flower. Fruit pentagonal-elliptic, ± 4 × 2 mm, smooth, valleculae very narrow.

HAB. Limestone rocks, ± 1400 m.
DISTRIB. Very rare in Iraq, only one record from close to the Iranian frontier. MSU: Tawila, *Rechinger* 12372.

Turkey (NW & C Anatolia), Syria, W & S Iran.

2. **Bupleurum lancifolium** *Hornem.*, Cat. hort. Hafn. 1: 267 (1813); Hayek, Prodr. Fl. Balc. 1: 982 (1927); Zohary in Dep. Agr. Iraq Bull. 31: 109 (1950); I.A. Linchevsky in Fl. SSSR 16: 286 (1950); H. Riedl in Fl. Lowland Iraq: 459 (1964); Mouterde, Nouv. Fl. Lib. et Syr. 2: 634 (1970); S. Snogerup in Fl. Turk. 4: 399 (1972); Meikle, Fl. Cyprus 1: 729 (1977); Boulos, Fl. Egypt: 2: 160 (2000).

B. subovatum Link ap. Spreng., Spec. Umb. Min Cogn: 19 (1820); Wolff, l.c.: 46 (1910); *B. protractum* Hoffmgg. & Link, Fl. Portug. 2: 387 (1820); DC., Prodr. 4: 129 (1830); Anth. in Notes Roy. Bot. Gard. Edinb. 18: 289 (1935).
B. rotundifolium L. var. *intermedium* Lois. in Lam. & DC., Fl. Fr. 6, t. 5: 514 (1815).

Erect, glaucous or yellowish green annual, up to 30 cm, but more usually ± 20 cm. Stem subterete, alternately branched from near base. Lower leaves oblong-lanceolate, attenuate toward the base, the upper ovate-lanceolate, obtuse with a sharp apiculus, perfoliate. Umbels with 2–6 rays, the partial umbels with 10–15 flowers. Involucre absent, involucel of 5 or 6 yellowish green orbicular-ovate, sharply and shortly acuminate bracts, twice as long as the flowering pedicels. Fruit ovate-globose, 2–3 mm long, sharply 5-ridged, the valleculae densely verrucose-granulate.

var. **heterophyllum** (*Link*) *Boiss.*, Fl. Orient. 2: 836 (1876); Bornm in Beih. Bot. Centralbl. 58B: 275 (1938); Blakelock in Kew Bull. 3: 432 (1948); Rawi in Dep. Agr. Iraq Tech. Bull. 14: 89 (1964); Mouterde, Nouv. Fl. Lib. et Syr. 2: 634 (1970); V. Täckh., Stud. Fl. Egypt ed. 2: 387 (1974).

B. perfoliatum L. var. *longifolium* Desv., Journ. Bot. 2: 315 (1809)
B. protractum Hoffmgg. & Link subsp. *heterophyllum* (Link) Munby, Contr. Fl. N ouest Afr.: 84 (1897)

Differs from the typical variety in that its lower leaves are linear or narrowly linear-lanceolate, almost grass-like, long acuminate; upper leaves from a narrowly orbicular-ovate base gradually produced into a long, narrow, pointed acumen. Fig. 58: 1–7.

HAB. Fields and fallow in the lower mountains, steppic plains and on irrigated desert, usually on clay, silt, calcareous loam etc.; alt. 50–600(–1000) m; fl. & fr. Apr.–May.

Fig. 58. **Bupleurum lancifolium** var. **heterophyllum**. 1, habit × 2/3; 2, habit showing leaf and inflorescence × 2/3; 3, petal × 6; 4, fruit × 4; 5, top of fruit enlarged; 6, mericarp × 4; 7, fruit cross section × 8. 1–3 from *Rawi & Gillett* 7573; 4–7 from *Gillett* 7995. Drawn by D. Erasmus (1964).

Distrib. Occasional in the lower forest zone and steppe region of Iraq and on the alluvial plain in the desert region. **MRO**: 20 km S of Rania, Rawi, *Nuri & Kass* 28414!; **MSU**: 5 km SW of Chemchamal, *Gillett & Rawi* 7573!; 10 km "before" (i.e. SW of?) Dukan Dam, *Omar* 42791?; Halabja, *Rawi* 88271!; **MJS**: Jabal Sinjar, *Haussknecht* s.n.; **FUJ**: nr Balad Sinjar, *Guest* 4118!; **FNI**: Nimrud, *Helbaek* 934!; **FAR**: Arbil, *Gillett* 7995!; **FKI**: 9 km from Kirkuk on road to Altun Kupri, *Erdtman & Goedemans* in *Rechinger* 15531; Kirkuk, *Bornmüller* 1230; Jabal Hamrin, *Bornmüller* 1230b; **FNI**: nr Khanaqin, *Cowan & Darlington* 241!; Qizil Robat (Sa'diya), in Anthony, l.c.; Badra, *Rechinger* 13991; 1 km S of Badra, *Anon.* (?*Rawi*) 18142!; **DLJ**: 5 km above Rawa, *Rawi & Gillett* 7023!; **DGA**: 4 km E of Samarra, *Rechinger* 13484; **LEA**: Ba'quba, *Paranjpye* s.n.!; comm. *Graham, Hunting Aerosurveys* 4!; **LEA**: Shahraban (Muqdadiya), *Haines* W850! **LCA**: Baghdad, *Schafli* 30!; Daltawa (Khalis), *Guest* 2472!
Note. GULA BEWUZHNA (Kurd.-Arbil, *Gillett* 7995), is "said to be eaten by sheep and other animals" (Khalis, *Guest* 2472). A troublesome weed among field crops.

S Europe, Aegean Isles, Cyprus, Syria, Lebanon, Palestine, Jordan, Egypt, Kuwait, Turkey, Armenia, Iran, C Asia (Turkmenistan), N Africa (Morocco, Algeria, Libya), Macaronesia (Madeira).

Sect. 2 Bupleurum. Leaves not perfoliate; involucre present; annual, biennial or perennial herbs.

3. **Bupleurum aleppicum** *Boiss.* in Ann. Sc. Nat. 3 ser., Bot. 1: 48 (1844); Boiss., Fl. Orient. 2: 840 (1876); Wolff, l.c. 72 (1910); Nábělek in Publ. Fac. Sci Univ. Masaryk 35: 120 (1923); Zohary, Fl. Palest. ed. 2, 1: 511 (1932); Zohary in Dep. Agr. Iraq Bull. 31: 109 (1950) sub "*B. leppicum*", sphalm.; Rawi, ibid., Tech. Bull. 14: 89 (1964); Mouterde, Nouv. Fl. Lib. et Syr. 2: 635 (1970); S. Snogerup in Fl. Turk. 4: 404 (1972); S. Snogerup in Fl. Iranica 162: 275 (1987).

Erect annual herb, to ± 40 cm. Stem rather wiry, subterete below, sharply 4-angled above, alternately branched from near base. Lowest leaves attenuate into a petiole, narrowly linear-lanceolate, long acuminate at the apex, 5–9-veined. Upper leaves reducing in size and becoming narrowly linear, sessile. Umbels 2–4-radiate, on slender peduncles with an involucre of 2 or 3 lanceolate long-acuminate herbaceous bracts, ± 10 mm long and subequalling the rays. Partial umbels of mostly 8–10 flowers, on short peduncles at most half as long as the involucel of 5–6 pale yellow or straw-coloured conspicuous, glumaceous bracteoles, which are broadly ovate, may attain about 14 × 7 mm and are semipellucid and beautifully veined, entire or subserrulate at margins and abruptly and shortly acuminate at apex. Fruit shortly oblong-ovoid, ± 2 mm long, sharply 5-ridged, valleculae smooth.

Hab. In the mountains, among *Quercus* trees on limestone, in dry *Quercus* forest on a steep stony slope; alt. 450–1450 m; fl. & fr. May–Jul.
Distrib. Occasional in the middle and lower forest zones of Iraq. **MAM**: Jabal Bekhair (Bekher), nr Zakho, *Field & Lazar* 773! *Rawi* 23057!; Bekma gorge, *Gillett* 8226!; **MRO**: Der Harir, *Nábělek* 403; N of Shaqlawa, *Anders* 1449; **MSU**: Gweija Dagh, *Gillett & Rawi* 11694!; Amoret, nr Qaradagh, *Haines* W1137!

Syria, Turkey, W Iran.

4. **Bupleurum brevicaule** *Schlecht.* in Linnaea 17: 124 (1843); Boiss., Fl. Orient. 2: 840 (1872); Wolff, l.c. 78 (1910); Zohary, Fl. Palest. ed. 2, 1: 511 (1932); Blakelock in Kew Bull. 3: 432 (1948); Rawi in Dep. Agr. Iraq Tech. Bull. 14: 89 (1964); Mouterde, Nouv. Fl. Lib. et Syr. 2: 635 (1970); S. Snogerup in Fl. Turk. 4: 400 (1972); S. Snogerup in Fl. Iranica 162: 276 (1987).

Annual, 10–30 cm, stem wiry, with raised lines, repeatedly pseudo-dichotomously branched from at or near base, branches somewhat angular. A subsessile umbel arises near the base of one branch at each division, and thus appears to be axillary in the fork. Branches elongated between divisions. Leaves linear-lanceolate, amplexicaul, gradually tapering above to a fine acumen, margins very narrowly scarious and minutely serrulate, reducing in size upwards and subtending the branches. Umbels many, lateral as described and also terminal and long-pedunculate all with 4 or 5 very unequal rays. Involucre of 4–7 lanceolate and long-acuminate, 3–5-veined, minutely serrulate, herbaceous bracts. Partial umbels 8–10-flowered, pedicels very unequal and becoming incrassate in fruit. Bracts of involucel 4, equal, 5-veined with the lateral veins marginal, herbaceous, remotely denticulate above, spreading in flower but becoming closely connivent to conceal the ripe fruit, their texture also becoming firm and horny. Fruit ± 2 × 1.5 mm, smooth, the ribs narrow, pale, scarcely prominent.

HAB. In the lower mountains, on stony clay soil, steppic plains and hills; alt. 300–800 m; fl. & fr. May–Jun.
DISTRIB. Rather rare in the lower forest and moist-steppe zones of Iraq. **MRO**: between Rowanduz and Agoyan, *Kass & Nuri* 27239! 27256!; **MSU**: between Sulaimaniya and Arbat, *Omar & Karim* 38119!; **FNI**: Baban (Beiban), nr Alqosh, *Field & Lazar* 767!; Mahad, nr Shaikhan, *Salim Effendi* 2614!
NOTE. GIYĀ BIRNÎK (Kurd.-Mosul, *Salim* 2614, "a good fodder plant").

Syria, Lebanon, Palestine, Jordan, S Turkey.

5. **Bupleurum kurdicum** *Boiss.*, Ann. Sc. Nat. sér. 3, Bot. 1: 146 (1844); Boiss., Fl. Orient. 2: 844 (1872); Wolff, l.c.: 87 (1910); Nábělek in Publ. Fac. Sci Univ. Masaryk 35: 120 (1923); Zohary, Fl. Palest. ed. 2, 1: 512 (1932); Bornmüller in Beih. Bot. Centralbl. 58B: 275 (1938); Blakelock in Kew Bull. 3: 432 (1948); Zohary in Dep. Agr. Iraq Bull. 31: 109 (1950); Rawi, ibid. Tech. Bull. 14: 89 (1964); Mouterde, Nouv. Fl. Lib. et Syr. 2: 636 (1970); Snogerup in Fl. Turk. 4: 413 (1972).

Annual herb, 30(–60) cm high. Stem wiry, subterete or with longitudinal ridges, alternately branched from halfway or below, branches widely divaricate, strongly ridged or quadrangular. Leaves somewhat few, at least above, lower broadly linear and ± 8-veined, usually withered at flowering time, upper decreasing in size, uppermost bract-like, all reducing somewhat abruptly to a short mucro formed by the excurrent midrib, sessile and amplexicaul. Umbels ± 6–rayed, rays at least twice as long as involucre. Involucre of 3–4 linear-acuminate bracts. Partial umbels bearing 6–10 flowers. Bracts of involucel 4–6, equalling the partial umbels, 3-veined, with a very narrow but usually obvious denticulate hyaline margin. Fruit oblong, parallel-sided, smooth, ribs very prominent and subobtuse or rounded (often so broad as to conceal the valleculae), ± 2 mm.

HAB. On lower mountain slopes, in cleared *Quercus* scrub, on a rocky plateau sometimes by streams, often in fields and on waste land; alt. (250–)400–800(–1300) m; fl. & fr. Jun.–Jul.
DISTRIB. Common in the NW sector of the lower forest zone of Iraq, occasional in the moist steppe zone. **MAM**: Bekhair Mt. (Bekher) *Rawi* 22973!; 5 km S of Zakho, *Rawi* 23106!; *Rechinger* 10709!; 84–85 km NW of Mosul, *Rawi* 22955!; 23130!; Shaikh Mama, *Nábělek* 529; Zawita, *Guest* 4579!; Sarsang, *Haines* W433!; nr Dinarta, *E Chapman* 26121?; ? **MAM/FNI**: (or in Syria/Turkey), "Mesopot., Kurdistan & Mossul", *Kotschy* 342!; **MRO**: Salah ad-Din, *Nuri & Hamid* 41169!; between Arbil and Rowanduz, *Bornmüller* 1232!; Handren, nr Rowanduz, *Bornmüller* 1231!; Galala, *Omar, Sahira, Karim & Hamid* 38441!; Rayat, *Jamil Afghani* 5218!; Koorak, nr Shahidan, *Rawi & Serhang* 23804!; **FNI**: Maltai, *Nábělek* 504; Gerwona, nr Ain Sifni, *Field & Lazar* 733!; Eski Kellek, *Olaf Anderson* 41399 (=985); **FAR**: Arbil, *Haussknecht* s.n.; Ankawa, *Nábělek* 504.

Syria, Lebanon, Turkey, Iran.

6. **Bupleurum gerardii** *All.*, Auct. Syn. 81 (1774); Willd., Spec. Pl. 1: 2 (1798); DC. Prodr. 4: 128 (1830); Boiss., Fl. Orient. 2: 845 (1872); Wolff, l.c. 88 (1910); Hand.-Mazz. in Ann. Naturh. Mus. Wien 27: 89 (1913); Hayek, Prodr. Fl. Balc. 1: 975 (1927); Zohary, Fl. Palest. ed. 2, 1: 512 (1932); Zohary in Dep. Agr. Iraq Bull. 31: 109 (1950); Linczevski in Fl. SSSR 16: 334 (1950); Rawi in Dep. Agr. Iraq Tech. Bull. 14: 89 (1964); Mouterde, Nouv. Fl. Lib. et Syr. 2: 637 (1970); Snogerup in Fl. Turk. 4: 414 (1972); Meikle, Fl. Cyprus 1: 735 (1977); Snogerup in Fl. Iranica 162: 281 (1987).

B. aristatum Nábělek in Publ. Fac. Sci Univ. Masaryk 35: 120 (1923) = *B. rohlenae* Nab., ibid. 52: 57 (1925) non *B. aristatum* Bartt. (1824)

B. gerardii All. var. *trichopodioides* Wolff, l.c.: 90 (1910); Bornmüller in Beih. Bot. Centralbl. 58B: 276 (1938)

Short annual herb, ± 25 cm or less (in Iraq). Stem wiry, subterete or more often longitudinally ridged, branched alternately from near the base; branches divaricate, strongly ridged or quadrangular. Lower leaves linear-lanceolate and 5-veined, somewhat shortly contracted to a short mucro formed by the excurrent midrib. Middle and upper leaves linear, gradually tapering to a very fine acumen, the uppermost bract-like, all sessile and amplexicaul. Umbels of 3–9 very unequal rays, some partial umbels occasionally almost sessile while others are on rays 6 × as long as long or more. Involucre of 3–5 lanceolate-subulate bracts. Partial umbels of 8–10 flowers, exceeded (often considerably so) by the

4–6 linear-subulate bracts of involucel, margins entire. Fruit oblong, parallel-sided, ± 2 mm, smooth, vittae prominent, narrow, acute, undulate, much narrower than valleculae.

HAB. In the mountains among denuded *Quercus* trees on limestone, on dry open overgrazed stony ridges and slopes; alt. 800–1600 m; fl. & fr. May–Jun.
DISTRIB. Occasional in the forest zone of Iraq. **MAM**: above Ain Nuni, *Nábělek* 411, 508; Matina (Chiya-i Matin?), *Rawi* 8749!; **MRO**: Kuh-i Sefin, above Shaqlawa, *Bornmüller* 1233!; Shaqlawa, *Gillett* 8068! 11554! *Haines* W762!; **MSU**: Tawila, *Rawi* 21910!; 19 km W of Sulaimaniya, *Rechinger* 10073; **MJS**: Kursi, *Gillett* 10953; Tschil (Shil?), Miran, *Handel-Mazzetti* 1520.
NOTE. KHANJARŪK (Kurd.-Amadiya, *Rawi* 8749).

W Mediterranean (Spain, France) & S Central Europe (to Hungary, Bulgaria and Crimea), Aegean Isles, Cyprus, Syria, Lebanon, Palestine, Turkey, Caucasus, W Iran, Turkmenistan.

7. **Bupleurum cappadocicum** *Boiss.*, Ann. Sci. Nat. sér. 3, Bot. 1: 146 (1844); Boiss., Fl. Orient. 2: 847 (1872); Wolff, l.c.: 94 (1910); Nábělek in Publ. Fac. Sci Univ. Masaryk 35: 121 (1923); Zohary, Fl. Palest. ed. 2, 1: 513 (1932); Zohary in Dep. Agr. Iraq Bull. 31: 109 (1950) sub "*B. lappadocicum*" sphalm.; Rawi in Dep. Agr. Iraq Tech. Bull. 14: 89 (1964); Mouterde, Nouv. Fl. Lib. et Syr. 2: 639 (1970); S. Snogerup in Fl. Turk. 4: 411 (1972).

Very graceful annual herb, to 60 cm. Stem slender, wiry, subterete or somewhat longitudinally striate, repeatedly subdichotomously branched from near the base, the ultimate divisions and the rays of the abundant inflorescences capillary, giving the plant a *Gypsophila*-like habit. Leaves very narrowly linear with a long fine acumen, reducing in size above to subulate-bract-like. Umbels very numerous, 4–6-rayed, the capillary rays to 2 cm long and much exceeding the 3–6 subulate involucral bracts. Partial umbels small and compact, only 4 mm in diameter when in fruit, of 3–7 flowers on short pedicels, subequalled by the 4–6 subulate bracts of the involucel. Fruit shortly ovoid, scarcely longer than broad, ± 1 mm long, smooth, ribs not very prominent. A very distinct and beautiful plant.

HAB. On a steppic mountainside; alt. ± 900 m; fl. & fr. Jun.
DISTRIB. Rare in Iraq; only found in one district near the western margin of the lower forest zone. **MAM**: above Babushki, between Zakho and Amadiya, *Náběkek* 537; Nazarki, nr Mar Ya'qub, *Náběkek* 438, 499, 506; "Mesopot., Kurdistan & Mossul", *Kotschy* 97!

Syria, Turkey.

8. **Bupleurum semicompositum** *L.*, Dissert. Demonstr. Pl.: 7 (1753); Sp. Pl. ed. 2, 342 (1762); DC., Prodr. 4: 128 (1830); Boiss., Fl. Orient. 2: 842 (1872); Wolff, l.c. 106 (1910); Hayek, Prodr. Fl. Balc. 1: 977 (1927); Bornmüller in Beih. Bot. Centralbl. 58B: 275 (1938); Linchevsky in Fl. SSSR 16: 347 (1950); Zohary in Dep. Agr. Iraq Bull. 31: 109 (1950); Riedl in Fl. Lowland Iraq: 459 (1964); Rawi, ibid. Tech. Bull. 14: 89 (1964); Mouterde, Nouv. Fl. Lib. et Syr. 2: 635 (1970); Snogerup in Fl. Turk. 4: 411 (1972); Meikle, Fl. Cyprus 1: 733 (1977); Snogerup in Fl. Iranica 162: 277 (1987); Boulos, Fl. Egypt 2: 162 (2000).

B. glaucum Robill. et Cast. in Lam. & DC., Fl. Franç. ed. 3, 5: 515 (1815); Grossh., Fl. Kavk. ed. 2, 7: 68 (1967).

Deep glaucous green, dwarf, decumbent annual herb, up to 15 cm but often much less. Stem rigid, wiry, subterete, more sharply ridged or quadrangular above, branched from the base (or above only in smaller plants), branches decumbent, further branched below, and with axillary umbels above. Lowest leaves lanceolate, subacute, diminishing upwards to lanceolate-acuminate and linear, sessile. Umbels small, ± 5-rayed, rays very unequal, some normally sessile. Involucral bracts 3–5, lanceolate, margin denticulate. Partial umbels of 5–10 flowers, exceeded by involucel. Bracts of involucel 3–5, lanceolate, denticulate at margins and also often dorsally on midrib. Fruit small, less than 1 mm long, covered with minute white papillae, subdidymous and subglobose (being rounded into the minute stylopodium), ribs not prominent.

HAB. Barren desert land, often on sandy soils, sometimes on semi-saline soils, dry-farmed fields, gardens etc.; alt. 25–150 m; fl. & fr. Mar.–Apr.
DISTRIB. Quite common in the E & SE sectors of the desert region of Iraq, rare in the steppe region. **FNI**: 5 km W of Mandali, *Rechinger* 13397!; 30 km SE of Badra, *id.* 14035; **DLJ**: 14 km NW of Rawa, *Gillett & Rawi* 7097!; on pipeline, 2 km E of Wadi Thirthar, *Rawi & Gillett* 7146!; **DSD**: Jabal Sanam, *Rechinger*

15829; ± 15 km S by W of Zubair, *Guest, Rawi & Schwann* 14349!; S of Zubair, *Rawi* 25859!; 18 km SSE of Zubair, *Rechinger* 14552; 23 km S by E of Zubair, *Guest, Rawi & Rechinger* 16831!; Um Qasr, *Alizzi & Omar* 34987!; **LEA**: 60 km N of Amara, *Rechinger* 14286; 25 km E of Chabbab's 13 ridges, *Anon.* 17959!; Muqdadiya, *Haines* W845!; Fakka, 70 km E of Amara, *Rawi* 25818!; 23 km W of Shaikh Sa'ad, *Rawi & Haddad* 25508!

NOTE. The above description refers to the var. **glaucum** (Robill. et Cast.) Wolff, to which the Iraq material is referable. The var. **semicompositum** is a larger plant with longer branches, the lower leaves distinctly petiolate, ± spatulate and obtuse, and bracteoles with margins usually entire.

Mediterranean Europe (to Greece), Crete, Aegean Isles, Cyprus, Syria, Palestine, Egypt, Arabia, Kuwait, Bahrain, Turkey, Azerbaijan, N & W Iran, Turkmenistan.

9. **Bupleurum leucocladum** *Boiss.* in Ann. Sci. Nat. sér. 3, Bot. 1: 144 (1844); Boiss., Fl. Orient. 2: 843 (1872); Wolff, l.c.: 129 (1910); Hayek, Prodr. Fl. Balc. 1: 971 (1927); Bornmüller in Beih. Bot. Centralbl. 58B: 275 (1938); H. Riedl in Fl. Lowland Iraq: 459 (1964); Rawi in Dep. Agr. Iraq Tech. Bull. 14: 89 (1964); S. Snogerup in Fl. Turk. 4: 411 (1972).

Tall, much-branched annual herb, 30–50 cm. Stem slender, wiry, whitish, round and smooth, only very finely striated even on the upper branches, branched alternately from near the base. Branches long, rigid or somewhat flexuose, divaricate. Lower leaves narrowly linear, prominently 3–5-veined, acuminate at the tip, base amplexicaul. Upper leaves diminishing and becoming bract-like, much shorter than the long internodes. Umbels 1–3(usually 2)-rayed, often apparently simple through the considerable elongation of one branch. Bracts of the involucre 1 or 2, lanceolate, rigid, green, 5-veined, the tip with a short yellowish-brown mucro, shorter than the umbel branches or subequalling one of them. Partial umbel much condensed, ± 5 mm in diameter, with 8–12 flowers on very short pedicels, exceeded by bracts of the involucel. Petals distinctly granular-papillose dorsally, upper margins denticulate-papillose. Bracts of involucel broadly lanceolate, rigid, 5-veined, with a subcucullate mucronate tip and minutely serrulate margin. Fruit ± 1.5 mm, papillose throughout, shortly oblong, ribs inconspicuous. Fig. 59: 1–7.

DISTRIB. (of species): Occasional in the lower forest zone and steppe region of Iraq; for details, see under vars. below.

Syria, ?Turkey, Iran.

var. **leucocladum**

HAB. Dry steppic hills, gravelly slopes; alt. 150–200 m; fl. & fr. May–Jun.

DISTRIB. Occasional in the SE sector of the dry-steppe zone of Iraq. **FPF**: 10 km E of Mandali, *Rechinger* 12791; Inaiza, nr Tursak, *Haines* W995!; 20 km NW of Badra, *Rechinger* 9692!; 23 km E of Badra, *Rawi* 20771!

Syria, ?Turkey, Iran.

var. **haussknechtii** (*Boiss.*) *Wolff*, l.c. 109 (1910); Hand.-Mazz. in Ann. Naturh. Mus. Wien 27: 89 (1913); Bornmüller in Beih. Bot. Centralbl. 58B: 275 (1938); Zohary in Dep. Agr. Iraq Bull. 31: 109 (1950) – sub "*B. leucosciadium*" sphalm.

B. haussknechtii Boiss., Fl. Orient. 2: 843 (1872); H. Riedl in Fl. Lowland Iraq: 459 (1964); Snogerup in Fl. Iranica 162: 279 (1987).

Stem 20–30 cm, not whitish, strongly striate, the upper branches often sharply quadrangular and concave between angles. Bracts of the involucre scabrous dorsally near the tip. Papillae of fruit somewhat larger.

HAB. In the lower mountains, in grassy places, on overgrazed slopes by a town; alt. 300–1200 m; fl. & fr. May–Jul.

DISTRIB. Occasional in the SE sector of the lower forest zone of Iraq, rarer in the moist steppe zone. **MRO**: in the mountains, near Rowanduz, *Bornmüller* 1234; **MSU**: Qaranjir, *Rawi* 21613!; Pira Magrun, *Haussknecht* s.n. (type); Sulaimaniya, *Haines* W2091!; **FUJ**: between Ain Ghazal and Tal Afar, *Handel-Mazzetti* 1324; **FNI**: 20 km N of Badra towards Mandali, *Rechinger* 9692.

NOTE. A distinct and apparently constant variety. Mature fruit has been collected of neither variety nor typical plant. When this is available the former may well prove worthy of the specific rank given to it by Boissier.

S & W Iran.

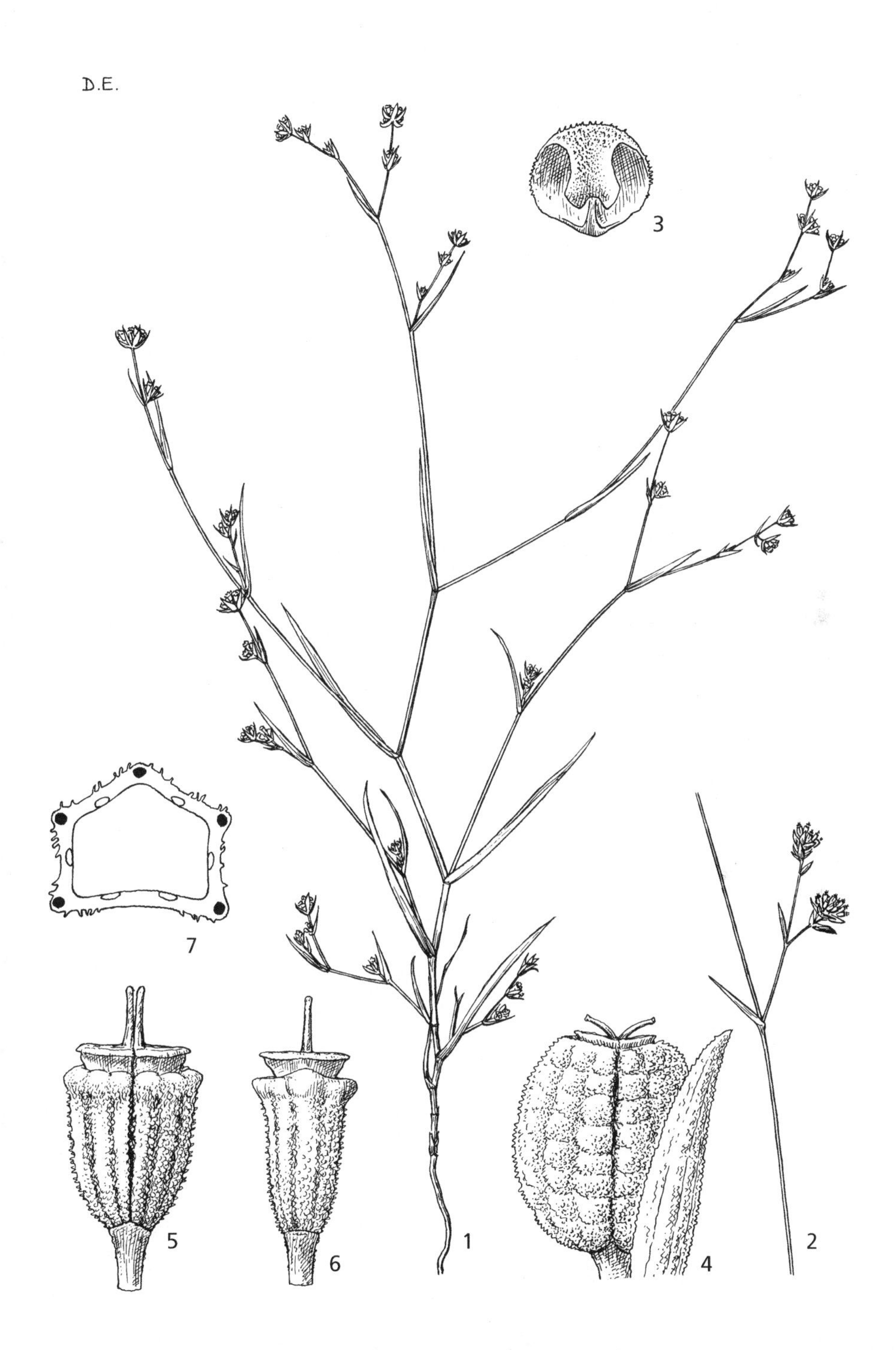

Fig. 59. **Bupleurum leucocladum** var. **leucocladum** 1, habit × ²/₃; 2, habit showing leaf and inflorescence × ²/₃; 3, petal × 12; 4, 5, 6, fruit × 12; 7, fruit cross section × 8. 1–3 from *Rawi* 21613; 4–7 from *Rechinger* 9692. Drawn by D. Erasmus (1964).

10. **Bupleurum falcatum** *L.*, Sp. Pl. ed. 1: 237 (1753); DC., Prodr. 2: 132 (1830); Boiss., Fl. Orient. 2: 850 (1872); Wolff, l.c. 106 (1910); Hayek, Prodr. Fl. Balc. 1: 971 (1927); Linchevsky in Fl. SSSR 16: 310 (1950); Snogerup in Fl. Turk. 4: 416 (1972).

var. **linearifolium** (*DC.*) *Wolff*, l.c.: 135 (1910); V. Täckh., Stud. Fl. Egypt ed. 2: 387 (1974).

B. linearifolium DC., Prodr. 4: 132 (1830); Boiss., Fl. Orient. 2: 849 (1872); Zohary, Fl. Palest. ed. 2, 1: 514 (1932); Zohary, in Dep. Agr. Iraq Bull. 31: 109 (1950); Rawi, ibid. Tech. Bull. 14: 89 (1964).

B. exaltatum M. Bieb., Tabl. Prov. Mer Casp. 113 (1798); Grossh., Fl. Kavk. ed. 2, 7: 62 (1967); Rech.f., Fl. Iranica 162: 287 (1987); She Menglan & Watson in Fl. China 14: 69 (2005); *B. exaltatum* M. Bieb. var. *linearifolium* (DC.) Boiss., Ann. Sc. Nat. sér. 3, Bot. 1: 150 (1844).

B. falcatum L. subsp. *exaltatum* (M. Bieb.) Briq. emend. Wolff var. *linearifolium* (DC.) Wolff., l.c.: 135 (1910); Blakelock in Kew Bull. 3: 432 (1948); Rawi in Dep. Agr. Iraq Tech. Bull. 14: 98 (1964).

Perennial, at the base often twiggy and suffruticose, 25–70 cm (usually ± 40 cm) high, much branched at the base with rigid, wiry, erect stems. Stem subterete, striate, or slightly angled above, with many alternate, divaricate branches. Lower leaves broadly linear, somewhat narrowed below, sessile and amplexicaul, prominently 5-veined and with a distinct white margin. Upper leaves gradually reducing in size, becoming linear-acuminate and ultimately bract-like. Umbels with 2–7 unequal rays, much exceeding the 2–5 shortly linear involucral bracts. Partial umbels with 5–9 flowers, the pedicels of which much exceed the involucel of 3–5 very shortly linear bracts. Fruit ± 3.5 mm, oblong, the sides slightly rounded or parallel, smooth throughout; ribs narrow, prominent or slightly winged.

HAB. Rocky mountain slopes; alt. 1050–2400(–2800) m; fl. & fr. Jul.–Aug.

DISTRIB. A highly polymorphic species, quite common in the central sectors of the thorn-cushion and upper forest zones of Iraq. **MRO**: Chiya-i Mandau, *Guest* 2706!; Halgurd Dagh, *Gillett* 9621!; Sula Khal, *Rawi & Serhang* 24693!; Baski Hawaran, *id.* 23965!; Qurnaqo, *id.* 26633!; Qandil range, above Pushtashan, *Rechinger* 11808!; Pushtashan, *Rawi & Serhang* 24212!; Serin, *id.* 24015!; Nurobar, *id.* 24156!; Sakri Sakran, *Hadač* 5425.

NOTE. This is the most frequent form of the species in most parts of the Middle East and C Asia. Typical *B. falcatum*, which is the usual European form, is not twiggy at the base; the lower leaves have a distinctly expanded lamina which may be broadly lanceolate to ovate, and are thus obviously petiolate; and the bract of the involucel exceed the partial umbel. A considerable number of infraspecific taxa of varying worth have been described.

The distribution of the species extends over C & S Europe (to Crimea) and from SW Asia (Syria, Lebanon, Palestine, Sinai, Turkey, Caucasia, Iran, Afghanistan, Pakistan) to C Asia, India, China and Japan.

According to Uphof (1968) and Usher (1974) the leaves of this species are used in E Asia, and especially in Chinese medicine, to stimulate perspiration.

Turkey, Iran, Afghanistan, Turkmenistan, Kazakhstan, Kyrgyzstan, Tajikistan, China (Xinjiang).

Bupleurum papillosum has been recorded from Iraq on the basis of some old specimens from 'Euphratem superior'; these are likely to have come from Turkey, and no well-localised material has been seen.

14. **APIUM** L.

Sp. Pl. ed. 1: 264 (1753); Gen. Pl. ed. 5: 128 (1754)

Helosciadium (L.) W.D.J. Koch in Nov. Ac. Acad. Caes. Leop. Carol. 12, 1: 125 (1824); Wolff in Pflanzenreich IV. 228, 90: 26 (1927)

Glabrous annual, biennial or perennial herbs. Leaves pinnate or pinnate-ternate. Umbels compound, often leaf-opposed. Involucre of few bracts or absent; involucel of numerous bracts or absent. Flowers white, hermaphrodite. Calyx teeth obsolete or very small. Petals entire, acute, occasionally with the tip shortly inflexed. Fruit broadly ovoid to elliptic-oblong or subrotund, laterally compressed, commissure narrow, glabrous or (not in Iraq) rarely setulose. Mericarps with 5 equal or rarely somewhat unequal ridges. Stylopodia flat or shortly conical, the margins entire. Valleculae univittate, commissure bivittate. Endosperm gibbous or teretely convex, the commissural face rather flat. Carpophore entire or shortly bifid at the apex.

A genus of about 30 species, of almost cosmopolitan distribution. Two species in Iraq.

Apium (name used by Latin authors for several umbelliferous plants, according to Gilbert-Carter; Stearn suggests that the name may have been derived from *apis*, a bee); Celery.

Characteristic smell of celery; umbels axillary and pseudo-terminal; involucel absent .... 1. *A. graveolens*
Not smelling of celery; umbels leaf-opposed only; involucel of 4–6 bracts present .... 2. *A. nodiflorum*

1. **Apium graveolens** *L.*, Sp. Pl. ed. 1: 264 (1753); DC., Prodr. 4: 101 (1830); Boiss., Fl. Orient. 2: 856 (1872); Hayek, Prodr. Fl. Balc. 1: 985 (1927); Zohary, Fl. Palest. ed. 2, 1: 515 (1932); Guest in Dep. Agr. Iraq Bull. 27: 10 (1933); Anth. in Notes Roy. Bot. Gard. Edinb. 18: 289 (1935); Zohary in Dep. Agr. Iraq Bull. 31: 109 (1950); Schischkin in Fl. SSSR 16: 371 (1950); H. Riedl in Fl. Lowland Iraq: 460 (1964); Rawi & Chakr. in Dep. Agr. Iraq Tech. Bull. 15: 14 (1964); Grossh., Fl. Kavk. ed. 2, 7: 72 (1967); Mouterde, Nouv. Fl. Lib. et Syr. 2: 621 (1970); Peşmen in Fl. Turk. 4: 422 (1972); Zohary, Fl. Palaest. 2: 416 (1972); E. Nasir in Nasir & Ali, Fl. Pak. 20: 32, f. 10, (1972); Täckh., Stud. Fl. Egypt ed. 2: 388 (1974); Husain & Kasim, Cult. Pl. Iraq 105 (1975); Meikle, Fl. Cyprus 1: 737 (1977); Rech.f., Fl. Iranica 162: 298 (1987); She Menglan & Watson in Fl. China 14: 76 (2005).

Erect, yellowish-green, glabrous biennial herb, 30–60 cm, with a strong and characteristic smell. Stem rather weak, strongly sulcate, much-branched above. Radical leaves simply pinnate with (3–)5(–7) cuneate-ovate, 5–30 × 4–35 mm incised leaflets, long-petiolate; stem leaves becoming pinnate with trisect segments which are obtusely or subacutely dentate, the teeth shortly mucronate with a percurrent vein; uppermost leaves sessile on the sheaths, small, simply trisect, the segments lanceolate and entire. Umbels axillary and pseudo-terminal, shortly pedunculate or sessile in the axils of a small ternate leaf or at the junction of two or more main branches. Main umbel with 5–12 unequal rays 0.5–2.5 cm, involucre absent. Partial umbels with 10–20 small (± 1 mm diameter) greenish-white flowers; pedicels 1–3 mm, involucel absent. Fruit broadly ovoid, ± 1.5 mm long, blackish when ripe, the ribs prominent, pale, narrow; stylopodia shortly conical, subequalled by the deflexed styles. Fig. 60: 1–6.

HAB. Damp places, along streams, ditches in irrigated gardens; often cultivated or subspontaneous as a garden escape; alt. up to ± 500 m; fl. & fr. Apr.–May/Jun.
DISTRIB. Quite common on the alluvial plain in the desert region of Iraq, occasional in the steppe and in other parts of the desert region. **FUJ**: 6 km E of Balad Sinjar, *Gillett* 11074!; **DSD**: Busaiya (cult.), *Fawzi, Hazim & Hamid* 38936; **LCA**: Baghdad (cult.), *Paranjpye* s.n.! comm. *Graham, Haines* W1440!; *Sahira* C.520!; **LSM**: Amara, in Anth., l.c.; *Musaida, Field & Lazar* 617!; **LBA**: 80 km SE of Basra, along road to Fao, *Rawi* 25904!; "Babylonia", *Haussknecht* s.n.
NOTE. Wild Celery, formerly also known as Smallage; KARAFS (Arabic, a name mentioned by Ibn al-Baitar (c. 1240) who quotes the opinions of Galen, Dioscorides and others on its properties), often colloquially called KRAFAS in Iraq. Medicinally, its properties are listed by Wren (1956) and Rawi & Chakravarty (1964) and may briefly be summed up as carminative, diuretic, tonic and aphrodisiac. According to Watt (1809), it has also been used by local doctors in India as an emmenagogue for the expulsion of stone.

Burkill (1935) points out that celery has certainly been in cultivation since fairly remote times, its oldest use probably being for flavouring. It has been found in Egypt in a tomb of the 20th Pharaonic dynasty woven into a garland; the Romans also used it for garlands and the ancient Greeks as a potherb, while the Chinese had it some 19 centuries ago. Wild forms of the plant contain a glucoside, apiumoside, also found in the leaves of parsley and allied herbs, which may, he suggests, be the cause of a certain measure of indigestibility, and a volatile oil is also present, most abundantly in the seeds.

The modern cultivated forms of the plant are innocuous, the edible blanched leaf-stalks being used as a vegetable in salads, for flavouring soups and so on.

Most of Europe, especially in coastal regions (from Britain & Scandinavia to W Russia, Greece & Crimea), Crete, Cyprus, Syria, Lebanon, Palestine, Jordan, Egypt, Arabia, Kuwait, Turkey, Caucasus, Iran, Afghanistan, Pakistan, NW India to China & Korea, C Asia (Turkmenistan, Tajikistan), N Africa (Morocco, Algeria, Libya), Macaronesia (Canary Is.).

2. **Apium nodiflorum** (*L.*) *Lag.*, Amen. Nat. Espan. 1(2): 101 (1821); Hayek, Prodr. Fl. Balc. 1: 986 (1927); Zohary, Fl. Palest. ed. 2, 1: 515 (1932); Rawi in Dep. Agr. Iraq Tech. Bull. 14: 88 (1964); Peşmen in Fl. Turk. 4: 422 (1972); Zohary, Fl. Palaest. 2: 416 (1972); Meikle, Fl. Cyprus 1: 737 (1977); Boulos, Fl. Egypt: 2: 163 (2000).

Sium nodiflorum L., Sp. Pl. 1: 251 (1753).

Fig. 60. **Apium graveolens**. 1, habit × ²/₃; 2, lower portion with roots × ²/₃; 3, petal × 12; 4, fruit × 8; 5, upper portion of fruit showing styles × 8; 6, fruit cross section × 8. 1, 3 from *L.F.H. Merton* 2315; 2, 4, 6 from *Gillett* 11074; 5, 7 from *Rawi* 21898. Drawn by D. Erasmus (1964).

Helosciadium nodiflorum (L.) W.D.J. Koch, Nov. Act. Acad. Caes. Leop. Carol. 12 (1): 126 (1824); DC., Prodr. 4: 104 (1830); Boiss., Fl. Orient. 4: 856 (1872); Schischkin in Fl. SSSR 16: 372 (1950); Zohary in Dep. Agr. Iraq Bull. 31: 109 (1950); H. Riedl in Fl. Lowland Iraq: 461 (1964); Rawi in Dep. Agr. Iraq Tech. Bull. 14: 92 (1964); Mouterde, Nouv. Fl. Lib. et Syr. 2: 621 (1970).

Procumbent or ascending perennial herb, bright shining green, (10–)30–80(–100) cm. Stem finely striate throughout, stout or slender, hollow and very brittle, rooting at nodes below. Leaves all simply pinnate, lamina oblong in outline, segments sessile, lanceolate to ovate, in (2–)3–5(–6) pairs, often shallowly lobed but never dissected, regularly serrate or crenate; both leaves and leaflets very variable in size. Umbels all leaf-opposed, sessile or shortly pedunculate; rays 6–12, unequal, 5–25 mm; involucre absent. Partial umbels with 12–20 small (± 1 mm diameter) white flowers, pedicels 1–3 mm; involucel of 4–6 narrowly ovate or lanceolate bracts which subequal or exceed flowers. Fruit 1.5–2 mm, broadly ovoid or oblong-ellipsoid, dark when ripe, ribs prominent, pale, narrow. Styles slender, flexuose, subequalling the shortly conical stylopodia.

HAB. In moist places, beside streams, along irrigation channels in gardens; alt. 250–920 m; fl. & fr. May–Jun.

DISTRIB. Locally abundant in the moist-steppe zone of Iraq. **MSU**: foot of Mt. Pira Magrun, *Hadač et al.* 4707; Saraw Subhm Aga, 35 km SE of Sulaimaniya, *Salah & K. Hamid* 52653!; **FUJ**: 6 km E of Balad Sinjar, *Gillett* 11072!; between Balad Sinjar and Tal Afar, *Field & Lazar* 518!; 15 km from Tal Afar to Sinjar, *Qaisi, Khayat & Karim* 50836!; **FNI**: by bridge at Tell before Dohuk, *E Chapman* 26256!; Mindan bridge, *id.* 26226!; **FKI/FUJ**: between Altun Kupri and Ain Dibs, *Hadač et al.* 5742; **FKI/MSU**: along Kirkuk-Chamchamal road, *Haines* W353!; **LCA**: Khan Dhari, 20 km W of Baghdad, Omar & Al-Khayat 53927!; 50 km N of Diwaniya, *Thamer* 47656!; "Mesopotamia", *Haussknecht* s.n.

Much of Europe (particularly in the west to W Russia, Greece), Aegean Isles, Cyprus, Syria, Lebanon, Palestine, Jordan, Egypt, Arabia, Turkey, Iran, Afghanistan, C Asia (Turkmenistan, Tajikistan), N Africa (Morocco, Algeria, Libya), Macaronesia (Madeira, Canary Is., Azores). Locally naturalized in parts of Africa (Ethiopia), N & S America.

15. **PITURANTHOS** Viv.

Fl. Lib. Spec.: 15, t. 7 (1824)

Deverra DC., Prodr. 4: 143 (1830); Wolff in Pflanzenr. IV. 228, 90: 97 (1927)

Rigid perennial herbs, usually glabrous in the vegetative parts, ± suffruticose below. Radical and lower stem-leaves ternate-pinnate, segments linear; upper leaves reduced to scales. Umbels compound, of few (to 10) rays. Involucre and involucel of few caducous bracts, or absent. Calyx teeth obsolete. Petals ovate, scarcely notched above, tip inflexed, white or greenish yellow, ± villous or tomentose, rarely subglabrous. Fruit ovoid or shortly ovoid-oblong, laterally compressed, hairy or subglabrous, very rarely (not in Iraq) vesicular-papillose. Ribs 5, narrow, prominent, usually ± concealed by hairs. Vittae solitary in the valleculae, commissure bivittate. Stylopodia conical, expanded below into an undulate-margined disk. Styles rather short, finally spreading or reflexed. Endosperm subterete. Carpophore bipartite.

Ten species distributed in N Africa and SW Asia, two in S Africa; one species in Iraq.

1. **Pituranthos triradiatus** (*Hochst.*) *Aschers. & Schweinf.*, Illustr. fl. de l'Egypte: 80 (1887); Wolff, l.c.: 29 (1927); Zohary, Fl. Palest. ed. 2, 1: 517 (1932); V. Dickson, Wild Fl. Kuwait: 75 (1955); H. Riedl in Fl. Lowland Iraq: 462 (1964); Rawi in Dep. Agr. Iraq Tech. Bull. 14: 95 (1964); V. Täckh., Stud. Fl. Egypt ed. 2: 390 (1974); Boules, Fl. Egypt 2: 167 (2000).

Deverra triradiata Hochst. in sched. Schimp. Pl. Arab. Petr. ed. 2: 454, nomen ex Boiss., Fl. Orient. 2: 861 (1872).

Stem rigid, terete, pronouncedly sulcate, 0.3–1(–1.8) m, much branched, suffruticose at the base. Branches rigid, sulcate, divaricate-ascending. Leaves few, lowest stem-leaves pinnate-ternate, sheaths amplexicaul, broadly white-margined below, segments long and narrowly linear, to ± 60 × 0.5–1 mm wide; lowest branch-leaves ternate or simple, uppermost short and simple or reduced to whitish-margined scales. Umbels 3–6-rayed, rays 0.5–3 cm, on rigid 1–5 cm peduncles; bracts of involucre 2–4, soon caducous. Partial umbels compact, of 3–12 flowers on 1–2 mm shortly hairy pedicels. Bracts of the involucel 2–4, broadly ovate,

Fig. 61. **Pituranthos triradiatus**. 1, habit, lower part × 2/3; 2, habit, flowering branch × 2/3; 3, habit, fruiting branch × 2/3; 4, petal × 9; 5, fruit × 4; 6, top of fruit enlarged; 7, fruit cross section × 4. 1 *Rawi* 14776; 2, 4 from *Schweinfurth* 76; 3, 5, 7 from *Wheeler Haines* W1645. Drawn by D. Erasmus (1964).

scarious-margined, subglabrous or ± densely hairy, caducous. Petals ± 1.5 mm, with a broad area of yellowish-brown hairs dorsally, the margins fimbriate. Fruit 3–4 mm long, ovate-oblong, mericarps often curved, hairy throughout, the ribs prominent and not concealed by the hairs at maturity. Fig. 61: 1–7.

HAB. Sandy desert wadis, gypsaceous substrate, sandy gravel desert; alt. 50–270 m or more; fl. & fr. Aug.–Oct.?
DISTRIB. Occasional in the desert region of Iraq. **DWD**: Wadi al-Ghadaf, between Rutba and Nukhaib, *Rawi* 14740; Wadi at-Tibul, 60 km N of Nukhaib, *Rawi* 31078?; Nukhaib, Rawi 14776!; 24 km from Nukhaib, along road to Shabicha, *Serkahia & Hussain* 16477!; between Shithatha and Karbala, *Agnew & Haines* W2120!; E of Wadi al-Ubaiyidh, *Gillett & Rawi* 6403!; **DSD**: near Najaf on pilgrim route to Shabicha, *Agnew, Haines & Rawi* W1645!; 30 km W of Ma'niya, *Hazim* 32525?
NOTE. Recorded by *Agnew et al.* as "tastes of Celery".

Egypt, Sinai, Arabia.

16. **AMMI** L.

Sp. Pl. ed. 1: 243 (1753); Gen. Pl. ed. 5: 113 (1754); Wolff in Pflanzenr. IV. 228, 90: 115–123 (1927)

Annual or biennial herbs with pinnate or pinnatisect leaves. Umbels compound. Bracts of the involucre numerous, large, pinnatisect. Involucel of many simple bracteoles. Flowers white, hermaphrodite, the petals obovate to obcordate, 2-lobed with the tips inflexed. Calyx obsolete. Fruit oblong-ovoid, somewhat laterally compressed, carpels with 5 pale, narrow, obtuse ridges. Vittae solitary in the valleculae, commissure bivittate. Stylopodia flat or shortly conical. Endosperm subterete, the commissural face flattened or somewhat concave. Carpophore entire or deeply bipartite.

Ten species distributed from Madeira and the Azores to the Mediterranean region and W Asia; two species in Iraq.
Ammi (name for an umbelliferous plant in both Greek and Latin authors, this plant being *Trachyspermum ammi* (syn. *T. copticum, Carum copticum*), according to Gilbert-Carter); Bishop's Weed.

Primary rays of the umbel confluent at the branch on a firm, prominent disk; segments of the upper leaves narrow linear-filiform, entire .... 1. *A. visnaga*
Primary rays of the umbel not set on a prominent disk; segments of the upper leaves linear-lanceolate, serrate-dentate .... 2. *A. majus*

1. **Ammi visnaga** (*L.*) *Lam.*, Fl. Franç. 3: 462 (1778); DC., Prodr. 4: 113 (1830); Boiss., Fl. Orient. 2: 892 (1872); Hand.-Mazz. in Ann. Naturh. Mus. Wien 27: 90 (1913); Nábělek in Publ. Fac. Sci. Univ. Masaryk 35: 124 (1923); Hayek, Prodr. Fl. Balc. 1: 987 (1927); Wolff, l.c. 116 (1927); Zohary, Fl. Palest. ed. 2, 1: 526 (1932); Guest in Dep. Agr. Iraq Bull. 27: 6 (1933); Anth. in Notes Roy. Bot. Gard. Edinb. 18: 289 (1935); Blakelock in Kew Bull. 3: 431 (1948); Schischkin in Fl. SSSR 16: 381 (1950); H. Riedl in Fl. Lowland Iraq: 463 (1964); Rawi in Dep. Agr. Iraq Tech. Bull. 14: 88 (1964); Grossh., Fl. Kavk. ed. 2, 7: 74 (1967); Mouterde, Nouv. Fl. Lib. et Syr. 2: 628(1970); Rawi & Chakr. in Dep. Agr. Iraq Tech. Bull. 15: 12 (1972); Peşmen in Fl. Turk. 4: 426 (1972); Zohary, Fl. Palaest. 2: 418 (1972); E. Nasir in Fl. Pak. 20: 30 (1972); V. Täckh., Stud. Fl. Egypt ed. 2: 390 (1974); Meikle, Fl. Cyprus 1: 740 (1977); Rech.f., Fl. Iranica 162: 302 (1987); Boulos, Fl. Egypt 2: 168 (2000); She Menglan & Watson in Fl. China 14: 80 (2005).

Visnaga daucoides Gaertn., Fruct. Sem. 1: 192 (1788); *Sium visnaga* Stokes, Bot. Math. Med. 2: 106 (1812).

Erect, glabrous annual (sometimes biennial?), 25–75(–100) cm. Stem very stout, terete, striate, erect, branched from near base. Branches long, ascending. Plant very leafy, radical leaves 3 or 4-pinnate, with a deltoid-ovate lamina (withered at flowering); cauline leaves all similar, 2 or 3-pinnatisect into narrow linear-filiform entire segments, mostly 5–20 × 0.5–0.75 mm, each with a mucronate tip. Umbels large, with very many (up to 100) stout rays; rays spreading 10–30 cm in flower, closing up together in fruit and elongating to as much as 8 cm, fusing at the base on to a prominent, firm disk which surmounts the parent branch; bracts of the involucre large, deflexed at base and then spreading, 2 or 3-pinnatisect, subequalling the length of the rays. Partial umbels many (± 50)-flowered, pedicels 1–7 mm, also set on a disk. Peduncles 5–12(–16) cm. Bracts of the involucel numerous (12–20), linear-filiform, entire, subequalling the partial umbels. Fruit glabrous, ovoid or ovoid-oblong, 2 mm long, ribs narrow, not very prominent. Styles long and reflexed, much exceeding the conical stylopodia. Fig. 62: 1–7.

HAB. Moist places, by streams, banks of ditches, a common weed in fields, among cereal stubble, on an abandoned hill, rice fields, etc.; alt. 50–600(–1500) m; fl. & fr. May–Jul.(–Aug.).
DISTRIB. Rare in the forest zone, but quite common in the steppe region of Iraq; also on the alluvial plain in the desert region. **MAM**: Sevara Gaurik, NE of Zakho, *Rawi* 23654!; **MSU**: Halabja, *Guest* 12924!; **FNI**: Maltai, *Nábělek* 354; Nineveh, *Anders* 1567; between Mosul and Ain Sifni, *Guest* 4033!; **FUJ/FNI**: on an island in R. Tigris, nr Mosul, *Kotschy* 446!; **FAR**: Ankawa, *Bornmüller* 1247!; **FKI/MSU**: 25 km N of Bibas, *Rawi, Hosham (Alizzi) & Nuri* 29432! **FKI**: nr river 75 km W of Kirkuk, *Khatib & Tikriti* 33153?; Hawija, *Rawi* 15174!; **FNI**: 4 km N of Sa'diya, *Qaisi* 42888; **DGA**: Adhaim, *Hunting Aerosurveys* 252, R.B. 18!; **LEA**: between Ba'quba and Muqdadiya, *Rechinger* 8039; Shahraban (Muqdadiya), *Haines* W1005! *Rechinger* 9732; Baladrus, *Sutherland* 289; **LCA**: Karrada cotton farm, Baghdad, *Paranjpye* s.n! comm. *Graham*; Daltawa (Khalis), *Rogers* 0272!, 0286! *Guest* 2433!; "Babylonia", *Aucher* s.n.
NOTE. Toothpick Bishop's-weed; also sometimes called Toothpick, Spanish Carrot, Toothpick Ammi (Am.), etc.; ? KASRI (Ir.-Mosul, *Nábělek* 354 – "dried rays of umbels used as toothpicks"); KHAIZARAN (*Guest* (1932), Ir.-Baghdad district, *Guest* 2433 – "dry heads used as toothpicks; seeds boiled and decoction given to children with bad breath"; Ir.-Kirkuk, *Rawi* 15174 – "weed used as fuel"); ?DAIRAM (Ir.-Mosul, *Guest* 4033); ?KURAILA (Kurd.-Halabja, *Guest* 1294). V. Täckholm (1974) who has added to the list of vernacular Egyptian Arabic plant-names drawn up by Drar and published in Stud. Fl. Egypt ed. 1 (1956), gives KHILLA, KHILLĀL (Eg.) and other variants, based on the Arabic roots KHALIA ("pierce") and KHALLALA ("to pick the teeth") as the common vernacular names for this plant in Egypt. Handel-Mazzetti (1913) noted the Arabic name HULL, from Syria (probably for KHALL, cf. KHALLA above) as recorded on the label of a specimen collected at Aleppo (*Hakim* 22). The name KHILLA was also recorded from the Levant by Post (1896), though whether from Egypt, Lebanon or Palestine is not mentioned. Zohary (1972) states that the seeds of this plant are the main source of the drug Khellin, their medicinal properties being as a diuretic and emmenagogue; he also says that they are used to make red dye in Iraq, though the source of this information is not quoted. Lindley & Moore (1870) noted that the rays or stalks of the main umbel of this species shrunk after flowering and became so hard that they form convenient tooth-picks in Spain and, after they had fulfilled that purpose, that they are chewed and held to be of service in cleaning the gums, while the leaves have a pleasant aromatic flavour in the mouth.

Mediterranean Europe (Portugal & Spain to Greece), Aegean Isles, Syria, Lebanon, Palestine, Egypt, Arabia, Turkey, Caucasus, Iran, N Africa (Morocco, Algeria), Macaronesia (Madeira, Canary Is.). Introduced elsewhere and subspontaneous as a weed (e.g. Britain, China, N, C & S America).

2. **Ammi majus** *L.*, Sp. Pl. 1: 243 (1753); DC., Prodr. 4: 112 (1830); Boiss., Fl. Orient. 2: 891 (1872); Hand.-Mazz. in Ann. Naturh. Mus. Wien 27: 90 (1913); Nábělek in Publ. Fac. Sci. Univ. Masaryk 35: 123 (1923); Wolff, l.c.: 117 (1927); Hayek, Prodr. Fl. Balc. 1: 987 (1927); Zohary, Fl. Palest. ed. 2, 1: 417 (1932); Guest in Dep. Agr. Iraq Bull. 27: 6 (1933); Anth. in Notes Roy. Bot. Gard. Edinb. 18: 289 (1935); Blakelock in Kew Bull. 3: 431 (1948); Zohary in Dep. Agr. Iraq Bull. 31: 111 (1950); H. Riedl in Fl. Lowland Iraq: 463 (1964); Rawi in Dep. Agr. Iraq Tech. Bull. 14: 88 (1964); Rawi & Chakr., ibid. 15: 11 (1964); Mouterde, Nouv. Fl. Lib. et Syr. 2: 628 (1970); Peşmen in Fl. Turk. 4: 427 (1972); Zohary, Fl. Palaest. 2: 417 (1972); E. Nasir in Fl. Pak. 20: 29 (1972); V. Täckh., Stud. Fl. Egypt ed. 2: 390 (1974); Meikle, Fl. Cyprus 1: 741 (1977); Rech., Fl. Iranica 162: 301 (1987); Boulos, Fl. Egypt 2: 168 (2000); She Menglan & Watson in Fl. China 14: 81 (2005).

Erect, glabrous annual (8–)25–100 cm, stems less stout than those of the preceding species, terete, striate, branched from base. Branches long, ascending. Lowest leaves pinnate or pinnate-ternate with few segments, often persisting until flowering time, segments very variable in size, regularly and sharply serrate, oblanceolate to broadly obovate. Upper leaves reducing in size, pinnate-ternate to variously pinnatisect, but the narrower segments remaining broadest at or above the middle, and almost always serrate-dentate. Umbels somewhat unequally (10–)15–50(–75)-rayed, the rays 2–6 cm, shorter in the centre of umbel, not set on a disk and not closing up together in fruit. Involucral bracts large, pinnatifid, shorter than the rays. Partial umbels ± 30-flowered; pedicels 2–7(–9) mm, the longer subequalled by the 8–12 linear-subulate, entire bracteoles. Peduncles ± 4–16 cm.

Fig. 62. **Ammi visnaga**. 1, 2, 3, habit × ⅔; 4, petal × 6; 5, 6, fruit × 3; 7 fruit cross section × 3. 1 from *Rawi* 23654; 2 from *Kotschy* 448; 3, 4–7 from *Guest* 12924. Drawn by D. Erasmus (1964).

Fruit glabrous, ovoid-oblong, 2 mm long, ribs narrow and not prominent. Styles reflexed, slightly shorter than those of *A. visnaga*, but still exceeding the conical stylopodia.

HAB. Gravelly hillsides, riverine thickets, along ditches in fields, damp places, weed in fields and gardens, on sandy gravel soils; alt. 50–400 m; fl. & fr. Apr.–Jun.
DISTRIB. Common in the steppe region of Iraq, and also on irrigated alluvial plain in the desert region. **FUJ**: Wadi Charab, between Sinjar and Tal Afar, *Handel-Mazzetti* in obs.; 44 km W of Balad Sinjar, *Gillett* 11164!; Hadhr, *Handel-Mazzetti* 1110. **FUJ/FNI**: bank of R. Tigris at Mosul, *Nábělek* 489; by Telegraph pole M.90, between Mosul and Baiji, *Field & Lazar* s.n!. **FNI/MAM**: (or in Syria/Turkey), "Mesopot., Kurdistan & Mossul", *Kotschy* 432!; **FNI**: Nineveh, *E Chapman* 25365!; Mindan, *id.* 26199! 26222! 26234! **FAR**: between Arbil and Altun Kupri, *Nábělek* 393; Makhmur, *Robertson* 250, RB60. **FKI**: on banks of R. Zab, nr Kirkuk, *Mrs Pamela Grigg* 75!; Hawija, *Rawi* 15168!; **FNI**: Jabal Hamrin, between Khanaqin and Baghdad, *Nábělek* 388, 400; Koma Sang, *Rawi* 26676!; Lajama, *Rawi* 20716!; Shihabi, *Rawi & Haddad* 25165!; **DLJ**: Sumaicha, *Handel-Mazzetti* 957. **LEA**: 5 km NW of Ba'quba, *Alkas* 18560!; Plots W9 & 10, Ba'quba district, *Hunting Technical Services* 2!; Weed plot no. 9, Ba'quba district, *P.E.N.* Kew no. 59; Salman Pak, *Gillett* 8406!; Aziziya, *Guest* 0262!; 30 km from Aziziya, on road to Kut, *Alizzi & Omar* 34637?; between Kut and Badra, *Naji Sofair* 5531!; Sha'di, nr Kut, *Guest* 3577! **LCA**: Bad'a , *Guest* 2508!; Dujaila, *B. Anderson* 1985?; Daltawa (Khalis), *Guest & Darwish al Haidari* 2230!; Baghdad, *Aucher* 3677! 3742! *Colvill* s.n.! *Schlafli* 116! *Rogers* 056!; Kadhimiya, *Graham* 605; Adhimiya, Baghdad, *Rechinger* 11704!; Abu Ghraib, *Rawi* 10743! *Samir Munim* 16492! (cult.) *Omar, Janan & Sahira* C.182; Drug Gardens, near Aqurquf (cult.), *Barkley & Hikmat Abbas* (*al-'Ani*) 2521; Karrada Mariam, *H. Ahmad* 9820! *Haines* W177; Rustamiya, *Rogers* 0262! *Guest* 759A! *Lazar* 1164!; 15 km SW of Kut, *Dabbagh & Qaisi* 41933? 41936? **LSM**: 40 km from Amara, along road to Basra, *Khatib & Alizzi* 32619!
NOTE. Common Bishop's weed; also Great Ammi (Am.), GARAIR (for GHURAIR), (Ir.-Baghdad district, *Guest* 2508, *Guest & Darwish al-Haidari* 2230); ZAND AL-ARŪS (Ir.-Khalis, *Lazar* 1164); KHAIZARĀN Baghdad, H. Ahmad 9820). Ibn al-Baitar (c. 1240) mentioned AKHALLA as the name of an umbelliferous plant, the seeds of which were used to make a soothing mouthwash and which LeClerck (1877–83) refers to this species, though the passage seems more likely to refer to *A. visnaga*, q.v. As regards the present species, a note on a specimen of it communicated to Herb. Kew by E Gouldring in 1929 noted that the plant had been reported as obnoxious and was said to cause an affection to the eyes of animals browsing on it, which led to an opacity of the cornea in their eyes. Guest (1938) noted that the plant was said to be poisonous and to cause blindness in horses which grazed it.

S & Mediterranean Europe (France, Portugal & Spain to Balkans & Greece), Aegean Isles, Cyprus, Syria, Lebanon, Palestine, Egypt, Arabia, Kuwait, Turkey, Iran, Pakistan, N Africa (Morocco to Libya), Macaronesia (Madeira, Tenerife), Ethiopia; cultivated and adventive in China.

17. **OLIVERIA** Vent.

Hort. Cels. 21, t. 21 (1801); Wolff in Pflanzenr. IV. 228, 90: 97 (1927)

Erect, branched annual herb, leaves variously pinnatisect. Umbels compound. Involucre and involucel present, ± dissected or dentate. Flowers white or pink, hermaphrodite. Petals broadly obcordate, concave, with crenate margins, hirsute dorsally, deeply incised and 2-lobed above, tip longly inflexed almost to the base of petal, adnate. Calyx teeth prominent, setose. Fruit ovoid-oblong, distinctly laterally compressed, densely covered with white hairs. Ribs 5, ± conspicuous, but mostly hidden by hairy covering. Commissure narrow, bivittate. Vittae solitary in the valleculae. Stylopodium elongate-conic, margin entire. Styles spreading or loosely reflexed. Carpophore undivided. Endosperm subterete dorsally, face somewhat concave.

One species only, which also occurs in Iraq.
Oliveria (so named in honour of Guillaume Antoine Olivier, 1756–1814), French naturalist and plant collector who travelled the Orient from 1794–98.

1. **Oliveria decumbens** *Vent.*, l.c.; Jaub. & Spach, Ill. Pl. Orient. 5 (44): t. 431 (1854); Hand.-Mazz. in Ann. Naturh. Mus. Wien 27: 90 (1913); Nábělek in Publ. Fac. Sci. Univ. Masaryk 35: 124 (1923); Blakelock in Kew Bull. 3: 434 (1948); Zohary in Dep. Agr. Iraq Bull. 31: 51 (1950); H. Riedl in Fl. Lowland Iraq: 461 (1964); Rawi in Dep. Agr. Iraq Tech. Bull. 14: 93 (1964); Mouterde, Nouv. Fl. Lib. et Syr. 2: 629 (1970); Matthews in Fl. Turk. 4: 427 (1972); Rech.f., Fl. Iranica 162: 209 (1987).

O. orientalis DC., Prodr. 4: 234 (1830); Boiss., Fl. Orient. 2: 894 (1872); Guest in Dep. Agr. Iraq Bull. 27: 68 (1933); Anth. in Notes Roy. Bot. Gard. Edinb. 18: 289 (1935); Bornmüller in Beih. Bot. Centralbl. 58B: 279 (1938).
Callistroma erubescens Fenzl in Endl., Gen. Pl. Suppl. 5: 34 (1843).

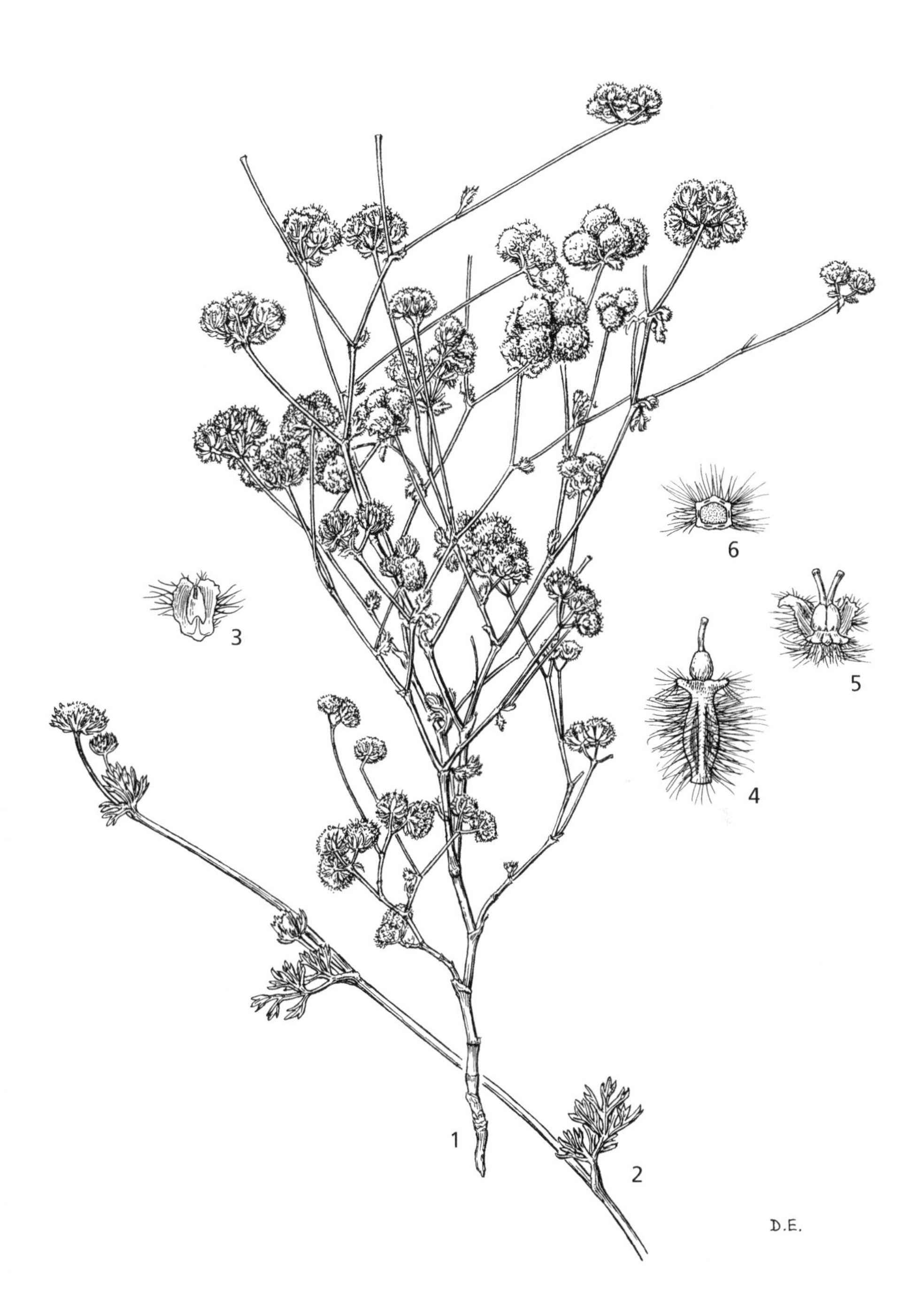

Fig. 63. **Oliveria decumbens**. 1, habit × ²/₃; 2, habit showing leaves and inflorescence × ²/₃; 3, petal × 6; 4, fruit × 6; 5, top of fruit enlarged × 6; 6, fruit cross section × 6. 1, 3–6 from *Guest* 4001; 2 from *Field* 527. Drawn by D. Erasmus (1964).

Erect annual herb, 10–50 cm, branched from base upwards. Stem and branches shining, whitish, subterete, very finely striate. Radical leaves small, tripinnatisect, withered at flowering time; cauline leaves shortly petiolate, lamina oblong-lanceolate in outline, ± 2 cm wide and up to 10 cm long, segments obovate-lanceolate, upper leaves rapidly reducing in size, sessile and becoming deltoid-ovate in outline, much shorter than internodes, variously pinnatisect, segments lanceolate. All leaves furnished with long and spreading hairs, increasingly so upwards, radical often subglabrous. Umbels on ± elongate, 2–5 cm long, peduncles, dense, largest usually ± 2.5 cm across; bracts of involucre trisect or pinnatisect, segments narrowly oblong-lanceolate. Rays 3–5, at most 1 cm long, densely patent-hairy. Partial umbels very dense, many-flowered, ± globose; bracts of involucel equalling short rays, variously trifid. Fruit narrow, ± 2 mm, all but the marginal ribs concealed in the indumentum of long, spreading white hairs. Fig. 63: 1–5.

HAB. Steppic hills and plains, arid barren hills, dry gravel slopes, on the margins of depressions, sometimes in fields and in fallow; alt. 150–400 m; fl. & fr. (Apr.–) May–Jun (Jul.).
DISTRIB. Common in the steppe region of Iraq; very rare in the desert. **FUJ**: ? "in desertis Mesopotamiae et Assyriae", *Aucher* 3597; between Ain Ghazal and Tal Afar, *Handel-Mazzetti* 1343; Tal as-Shur, *Field & Lazar* 569!; Ain Tellawi, *id.* 527!; Tal Afar, *H. Mastuf* 5176!; FUJ, Qaiyara, *Bayliss*, 124!; between Jabal Makhul and Jabal Khanuqa, *Handel-Mazzetti* 1054. **FUJ/FNI**: nr Mosul, *Noë* s.n.; **FNI**: banks of R. Tigris nr Pesh Khabur, *Kotschy* 418 (type of *C. erubescens*)!; **FNI/FAR/FKI**: between Mosul and Kirkuk, *Davis* 711!; **FAR**: Arbil province, *Radhi* 3842!; FAR, Arbil, *Haussknecht* s.n; Ankawa, *Bornmüller* 1222!; Makhmur, *Gillett* 11242! **FAR/FKI**: Altun Kupri, *Bornmüller* 1223. **FKI**: Kirkuk province, *Ali al-Haidari* 3937!; 5 km NW of Daquq, *Rawi* 22862!; Kor Mor, nr Tuz, *Rogers* 0315! **FNI**: Jabal Hamrin, between Khanaqin and Baghdad, *Nábělek* 384; *Sutherland* 291; Jabal Hamrin, between Jalaula and Shahraban (Muqdadiya), *Rechinger* 9962!; Naft Khina, *Rogers* 0315!; 20 km N of Badra, *Rechinger* 12718. **FNI/LEA**: S of Daima, nr Mandali, *Guest* 858!; **DGA**: Ghura plain N of Daltawa (Khalis), *Guest* 858!; **LCA**: Baghdad, *Olivier & Bruguière* s.n.
NOTE. ALAICH AL-GHAZĀL ("gazelle's nourishment", Ir.-Arbil, *Radhi* 3842, "grazed by livestock"), NUÎNŪA (Kurd.-Kirkuk, *Ali al-Haidari* 3937, "useful grazing plant"), GÎYĀ ASKILA (Kurd.-Khanaqin, *Nábělek* 384); though none of these names has been confirmed it appears likely that this species is regarded as a useful forage plant for livestock. Handel-Mazzetti (1913) noted the local name "NA-NA MAL GHAZAL" on the R. Khabur near Jabal Abd al-Aziz in Syria; although he called it an Arabic name it appears more likely to be a Kurdish colloquial name similar to ALAICH AL-GHAZĀL, NĀN being the Kurdish name for bread.

SE Turkey (nr Syrian frontier), Syria, Jordan, Iran.

18. **CARUM** L.

Sp. Pl. ed. 1: 263 (1753); Gen. Pl. ed. 5: 127 (1754)

Biennial or perennial herbs with fusiform or fibrous roots. Leaves variously 2–4-pinnatisect. Umbels compound. Involucre absent or of few bracts; involucel constantly present. Flowers usually white or pink, rarely (not in Iraq) dark red or cream-coloured, mostly hermaphrodite but apparent male flowers sometimes present. Petals obovate or obcordate, the tip emarginate, with an inflexed lobule. Fruit oblong-ovoid or ovoid-rotund, laterally compressed (sometimes considerably so), smooth and glabrous; ribs 5, narrow, pale, prominent and conspicuous; mericarps often falcately curved. Valleculae 1–3-vittate (usually with one broad vitta), commissure 2–4-vittate. Stylopodia shortly conical or flattened, styles somewhat longer, ± spreading or reflexed. Carpophore bifid or deeply bipartite. Endosperm subterete, commissural face flat or almost so.

About 30 species in the temperate regions of the Old World; only one native species in Iraq and perhaps another sometimes cultivated.

Carum (probably from the district of Caria in Asia Minor where caraway was much grown; Stearn points out that it is the Latin form of the Greek name for caraway, καρον, *karon*, the name used for caraway by Dioscorides).

Dwarf mountain perennial with only one stem leaf or another set very low on stem; fruit ovoid-oblong to broadly ellipsoid, 2–4 × 1.5–2 mm; involucre and involucel present . . . . 1. *C. caucasicum*
Tall biennial or perennial with many stem leaves; fruit ovoid to narrowly oblong, 3–5 × 2–3 mm; involucre and involucel usually absent . . . . 2. *C. carvi*

1. **Carum caucasicum** (*M. Bieb.*) *Boiss.*, Fl. Orient. 2: 880 (1872); Wolff, l.c.: 150 (1927); Blakelock in Kew Bull. 3: 432 (1948); Schischkin in Fl. SSSR 16: 393 (1950); Rawi in Dep. Agr. Iraq Tech. Bull. 14: 90 (1964); Grossh., Fl. Kavk. ed. 2, 7: 78 (1967); Hedge & Lamond in Fl. Turk. 4: 349 (1972); Rech.f., Fl. Iranica 162: 265 (1987).

Laserpitium caucasicum M. Bieb., Fl. Taur.-Cauc. 1: 222 (1808).
Cnidium carvifolium M. Bieb., Fl. Taur.-Cauc. 3: 212 (1819); DC., Prodr. 4: 153 (1830)

Dwarf glabrous perennial, 5–20 cm, with a stout tapering rootstock often many times as long as the aerial part of the plant. Leaves mainly radical, stem leaves one or with a second positioned very low on the stem, 2–13 × 0.5–3.5 cm, all pinnate with 3–8 pairs of ternate to pinnatisect pinnae, the ultimate segments oblong-lanceolate; lamina outline oblong-lanceolate, all leaves petiolate, with a broad membranous-bordered sheath. Umbels solitary or few, with 3–8(–11) very unequal rays (2–)5–25(–35) mm. Bracts of the involucre 1–4, simple, trifid or rarely somewhat pinnate. Partial umbels of 10–15 white or pink-tinged flowers, 1–4 mm, the pedicels finally somewhat longer than the 4–6 bracts of the involucel. Peduncles 2–8(–12) cm. Fruit ovoid-oblong to broadly ellipsoid, 2–4 × 1.5–2 mm; stylopodium flattened, with undulate, pale margins. Styles straightish or spreading, not reflexed, 0.6–1 mm. Fig. 64: 1–6.

HAB. In the high mountains, on wet grassy slopes, swampy ground in a mossy peat mat; alt. 2800–3300 m; fl. & fr. Jul.–Sep.
DISTRIB. Rare in Iraq; so far only found in three localities in the central sector of the alpine region. **MRO**: Halgurd Dagh, *Guest* 2857!; Hisar-i Rost, *Guest & Husham* (*Alizzi*) 15817!; Ser Kurawa, *Gillett* 9735!

NE & SE Turkey, Caucasus, Iran.

2. **Carum carvi** *L.*, Sp. Pl. ed. 1: 263 (1753); Boiss., Fl. Orient. 2: 879 (1872); Wolff, l.c. 145 (1927); Zohary, Fl. Palest. ed. 2, 2: 522 (1933); Guest in Dep. Agr. Iraq Bull. 27: 19 (1933); Schischkin in Fl. SSSR 16: 386 (1950); Rawi & Chakr. in Dep. Agr. Iraq Tech. Bull. 15: 23 (1964); Hedge & Lamond in Fl. Turk. 4: 347 (1972); Husain & Kasim, in Cult. Pl. Iraq 105 (1972); E. Nasir in Fl. Pak. 20: 87 (1972); Meikle, Fl. Cyprus 1: 743 (1977); Rech.f., Fl. Iranica 162: 263 (1987); She Menglan & Watson in Fl. China 14: 81 (2005).

Glabrous biennial or perennial herb with a short incrassate-fusiform rootstock, branched considerably from base and also above, (10–)25–80 cm; stem and branches ± striate with pale, raised lines. Leaves bipinnate with pinnatisect pinnae, ultimate segments linear-lanceolate, mucronate, 1–1.5 mm wide. Leaf outline deltoid-lanceolate to linear-oblong, all lower leaves petiolate with a long, narrowly oblong, membranous-margined sheath, uppermost leaves sessile and rapidly reducing with fewer and narrower segments. Umbels numerous, terminal and leaf-opposed, with 5(–15) very unequal (2–)5–40 mm rays; bracts of involucre none or up to ± 5, small and membranous. Partial umbels of 8–15 white or pink-tinged flowers, involucel absent or of 1–3 small bracteoles much shorter than the longest of the very unequal 1–10 mm pedicels. Peduncles (15–)3–9 cm. Fruit narrowly oblong to ovoid, dark brown with paler ribs, 3–5 × 2–3 mm, stylopodia conical, small, ovoid. Styles short, divaricate or incurved, slightly exceeding stylopodia.

HAB. Perhaps sometimes cultivated and occasionally naturalized; alt. ± 230 m; fl. & fr. Apr.
DISTRIB. Apparently very rare in Iraq; only found once in the moist-steppe zone. **FKI**: 5 km E of the stone bridge on R. Tawukshi, Kirkuk province, on red sandy clay and sandstone, *Barkley & Jum'a Brahim* 1395b! There is no such river as "R. Tawukshi" in Iraq but this name is clearly a phonetic corruption of Tauq Chai (Tauq stream or rivulet, Turk.) which flows by the village of Tauq where there is a stone bridge over which the main Kirkuk-Tuz-Kifri road passes. Since there is no indication on the collectors' label that the specimen was from a cultivated plant the presumption is that the plant was growing wild.
NOTE. Caraway, Common Caraway; KARĀWĪYĀ (as spelled by Ibn al-Baitar), or KARAUYĀ (as more usually spelled in Arabic). The dry aromatic fruits of this species, known as caraway seeds, have been widely used as a spice in many countries from time immemorial. Caraway oil is obtained by steam distillation of the crushed fruits; it is used in medicine as a carminative and stimulant and also as a perfume for soap.

Most of N & C Europe to N Spain, N Italy, W Russia, Cyprus (probably a naturalized escape from gardens), N Africa, Palestine (cult.), Egypt (cult.), Turkey (widely cultivated and often naturalized), Caucasus, N Iran, Afghanistan, C Asia (Turkmenistan to Tian Shan), Siberia, Himalaya, China; introduced into New Zealand, N America.

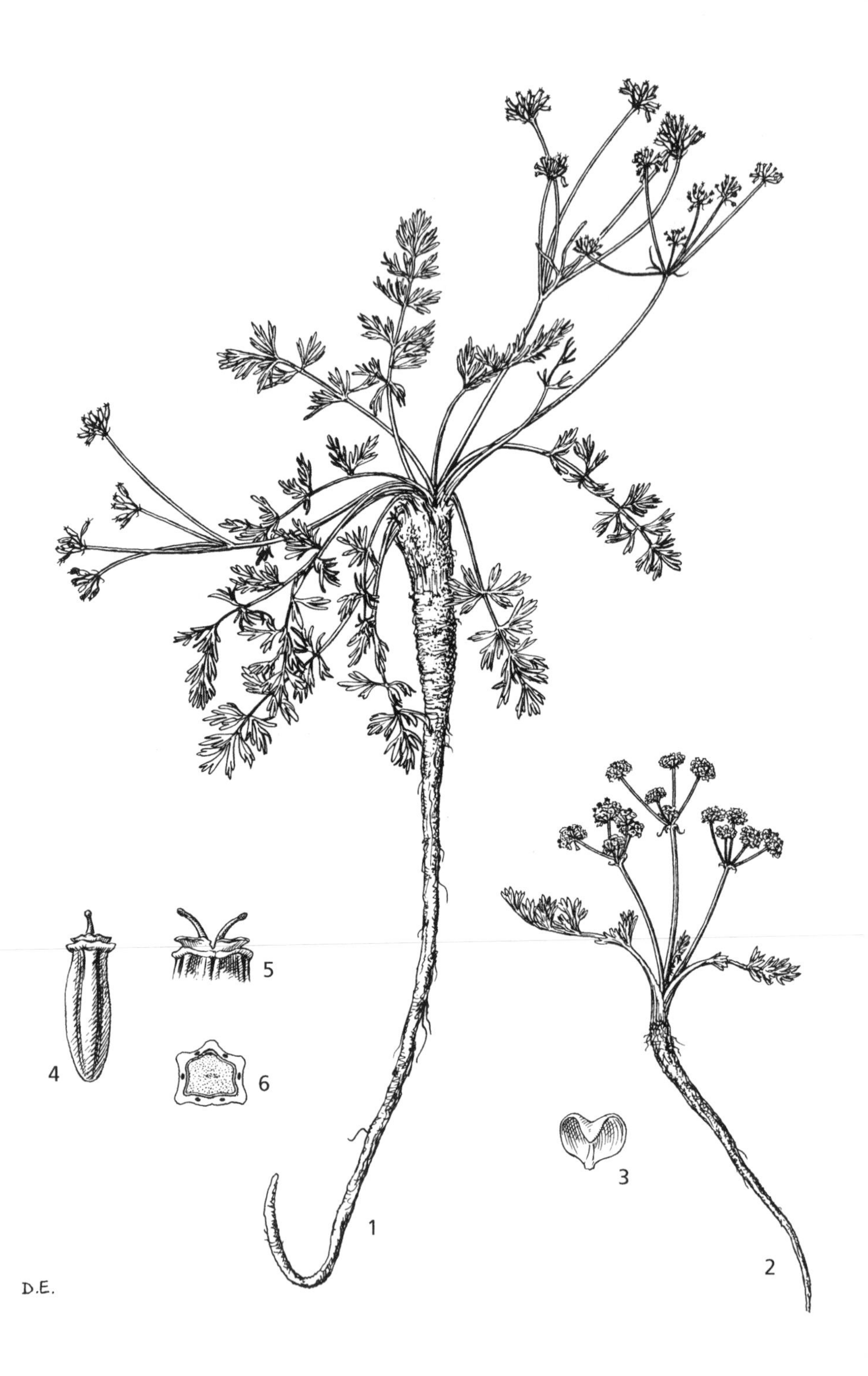

Fig. 64. **Carum caucasicum**. 1, habit × 2/3; 2, habit, flowering × 2/3; 3, petal × 6; 4, fruit × 6; 5, upper portion of fruit showing styles × 6; 6, fruit cross section × 6. 1, 4–6 from *Davis* 24164; 2–3 from *Guest* 2857. Drawn by D. Erasmus (1964).

19. **BUNIUM** L.

Sp. Pl. ed. 1: 243 (1753); Gen. Pl. ed. 5: 114 (1754)
Bulbocastanum Adans., Fam. Pl. 2: 97 (1765); *Carum* Sect. *Bulbocastanum* (Adans.) DC., Prodr. 4: 115 (1830), p.p.
Carum Sect. *Bunium* (L.) Benth. in Benth. & Hook., Genera Plantarum 1 (3): 891 (1867); Wolff in Pflanzenr. IV. 228, 90: 186–212 (1927)

Tall or dwarf, glabrous perennial herbs with globose-tuberous roots, underground portion of stem short or commonly slender and flexuose. Leaves variously 2–3-pinnatisect. Umbels compound. Involucre present or absent, involucel constantly present. Flowers white to deep pink, hermaphrodite or polygamous, fruiting pedicels often incrassate. Sepals obsolete. Petals obcordate, ± emarginate, with an inflexed tip. Fruit ovoid to oblong-ovoid or linear-prismatic, glabrous, smooth or very rarely minutely verruculose; primary ribs 5, narrow, pale, conspicuous, prominent or only feebly so. Valleculae 1-3-vittate (univittate in Iraq). Stylopodia various, flat, convex or conical; styles long or short, slender or stout, sometimes becoming indurate in fruit. Carpophore bifid. Endosperm subterete or sulcate, commissural face flat or almost so.

Some 40 species in the N Temperate zone of the Old World: eight species in Iraq.
Bunium (Gr. name in Dioscorides for *B. ferulaceum*, according to Gilbert-Carter, βουνιας being a kind of turnip; Stearn however suggests that it is derived from βούνιον, *bounion*, the Gr. name of the earth nut, *Conopodium denudatum*, another umbelliferous plant with an edible tuber).

1. Surface of fruit minutely verruculose .... 7. *B. verruculosum*
   Surface of fruit smooth 2
2. Ultimate leaf segments with the terminal ones elongate relative to the lateral ones .... 8. *B. rectangulum*
   Ultimate leaf segments equal 3
3. Fruit roundish to ovoid; styles not indurate and horn-like in ripe fruit, deflexed or flexuose 4
   Fruit linear-oblong; styles indurate and rigid, erect or diverging, horn-like .... 1. *B. paucifolium*
4. Styles > 1 mm, erect to erect-divergent, exceeding the flattened disk 5
   Styles very short (< 1 mm), deflexed, scarcely exceeding the flattened disk .... 2. *B. microcarpum*
5. Ribs of fruit very narrow, pale and prominent with narrow wings; mericarps separating at maturity, endosperm somewhat sulcate .... 3. *B. cornigerum*
   Ribs of fruit quite unwinged; mericarps separating at maturity or not, but if so then the endosperm not sulcate 6
6. Partial umbels 10–15-flowered .... 4. *B. cylindricum*
   Partial umbels 4–10-flowered 7
7. Endosperm terete, not sulcate; mericarps separating at maturity ..... 5. *B. avromanum*
   Endosperm deeply grooved; mericarps not separating at maturity .... 6. *B. caroides*

1. **Bunium paucifolium** *DC.*, Prodr. 4: 117 (1830); Grossh., Fl. Kavk. ed. 2, 7: 79 (1967); Hedge & Lamond in Fl. Turk. 4: 343 (1972); Rech.f., Fl. Iranica 162: 238 (1987).

B. elegans (Fenzl) Freyn, Öst. Bot. Zeit. 45: 83 (1892); Nábělek in Publ. Fac. Sci. Univ. Masaryk 35: 123 (1923); Wolff, l.c.: 188 (1927); Zohary, Fl. Palest. ed. 2, 2: 523 (1933); Guest in Dep. Agr. Iraq Bull. 27: 17 (1933); Bornmüller in Beih. Bot. Centralbl. 58B: 278 (1938); Korovin in Fl. SSSR: 16: 402 (1950); Zohary in Dep. Agr. Iraq Bull. 31: 110 (1950); H. Riedl in Fl. Lowland Iraq: 464 (1964); Grossh., Fl. Kavk. ed. 2, 7: 78 (1967); Mouterde, Nouv. Fl. Lib. et Syr. 2: 630 (1970); Zohary, Fl. Palaest. 2: 422 (1972).
Carum elegans Fenzl, Pugill. pl. nov. Syr.: 16 (1842); Boiss., Fl. Orient. 2: 883 (1872); Blakelock in Kew Bull. 3: 432 (1948); Rawi in Dep. Agr. Iraq Tech. Bull. 14: 90 (1964).
Carum noeanum Boiss., Diagn. ser. 2, 2: 77 (1856).
[*Bunium brachyactis* (non (Post) Wolff) Agnew, Hadač & Haines in Bull. Iraq Nat. Hist. Inst. 2, 2: 6 (1962)].

Very variable, erect, glabrous herb, (10–)20–60 cm, with a large globose tuber, much branched from a little above the base. Stem and lower branches subterete, weakly striate, uppermost branches more strongly sulcate or quadrangular. Lower leaves long-petiolate, triangular-ovate in outline, 2–3-pinnate with short (2–4 mm) or ± elongate, narrowly or broadly linear, entire segments which are mucronate at tip. Upper leaves ternate or ternate-pinnate, similar but mostly sessile on sheaths and segments rather more elongate; all leaves with an expanded, sheathing base of variable length. Umbels many, with (3–)6–13(–15) ± equal 0.6–3.5 cm rays which much exceed the mostly 5–9 small, linear-lanceolate, membranous-bordered bracts of the involucre. Partial umbels many (5–)9–16(–20)-flowered, pedicels 2–6 mm, not incrassate or flattened, about twice as long as the involucel of 4–6 linear bracteoles. Peduncles 1.3–11 cm. Flowers rather large, pure white to pinkish, to 4 or 5 mm in diameter, petals long-persistent. Fruit oblong, 2.5–3.5 × 2–3 mm, ribs narrow but conspicuous. Stylopodium depressed, styles filiform, very long, flexuose or finally reflexed, in this position extending to about ²/₃ the length of fruit. Endosperm ± sulcate dorsally, vittae large, bundles under primary ribs small. Fig. 65: 1–7.

Hab. On mountain sides, in coppiced *Quercus*, on silty eroded slopes, in hill pastures, in foothills and upper plains, sometimes in fields, a weed in irrigated crops; alt. 50–1800 m; fl. & fr. (Apr.–)May–Jun.
Distrib. Common in the forest zone and steppe of Iraq, occasional as a weed in irrigated fields on the alluvial plain in the desert region. **MAM/FNI**: 60 km from Mosul on the road to Dohuk, *F. Karim* 37496!; **MRO**: Salah ad-Din, *Gillett* 11289!; Kuh-i Sefin, *Bornmüller* 1253!; Helgord (Algurd Dagh), *Rawi* 13730!; between Helgord and Kholan, *Rawi* 13761!; Sakri Sakran, *Bornmüller* 1255; Serkabkhan, NW of Rania, *Rawi, Nuri & Kass* 40513!; **MSU**: 30 km from Kirkuk on the road to Sulaimaniya, *F. Karim, H. Hamid & M. Jasim* 40513!; 20–25 km E of Qaranjir, on Kirkuk-Sulaimaniya highway, *Rawi* 2146!; nr Chemchemal, *Rogers* 0196!; Jarmo, *Helbaek* 1546!; Pira Magrun, *Haussknecht* s.n.!; Zewiya, *Rawi* 12045!; Qopi Qaradagh, *Haines* W1176! W1557!; Tanjaro valley, *C.H. Gowan* 2397!; 4 km N of Penjwin, *Rawi* 22887?; Hawraman Mts, N of Halabja, *Rawi* 22081!; Tawila, *Rawi* 21898?; **MJS**: Jabal Sinjar, *Haussknecht* s.n.; **FUJ**: Tal ash-Shur, between Sinjar and Tal Afar, *Field & Lazar* 587!; between Balad Sinjar and Ba'aj, *Ani* 8790?; nr Hamman Ali, *id.* 9730!; **FNI**: Tal Kuchek, *Gillett* 10804!; **FNI**: Tal Kaif, *Mudir of Tal Kaif* 3213!; **FAR**: Arbil, *Bornmüller* 1256, *Gillett* 7996!; between Arbil and Altun Kupri, *Bornmüller* 1254; **FKI**: Arbil province, *Radhi* 3852!; Qara Lu, *Ali Burhan ad-Din* 5513!; 4 km S of Taktak bridge, between Kirkuk and Koi Sanjak, *Rawi, Nuri & Kass* 28092!; Tuz, *Rogers* 0228!; **FNI**: Qizil Robat (Sa'diya), *Guest* 1961!; nr Zurbatiya, *Ani & Bharucha* 7637!; **DGA**: 15 km E of Samarra, *Rawi* 20363!; **LEA**: 6 km N of Ba'quba, *Gillett & Rawi* 7281!; W6 district, nr Ba'quba, *Hunting Technical Services* 20!; Ba'quba, *Gillett & Rawi* 7281!; Shahraban (Muqdadiya), *Haines* W890!; Ra'aya, *Alizzi & Hazim* 33707!; **LCA**: Daltara (Khalis), *Guest* 1437!
Note. A number of colloquial names for this plant have been noted by collectors, though none of them has yet been confirmed except, to a limited extent, the name JŌZ BAWA (Ir.-Khalis, *Guest* 1437 & Ir.-Sa'diya, *Guest* 1961, "the tuber is dug up and eaten by the people"); nevertheless it is strange that none of the many other collectors who have visited these localities over the last 45 years or more have again noted that name. What has however been sufficiently confirmed is that this species is recognised by the local people over a considerable part of our territory as a useful forage plant for sheep and other grazing animals; also that the tuber of the plant is dug up and eaten by the people. The following other names have also been noted: BINHOZALA (Kurd.-Sulaimaniya, *Goran* 2397 – "on "daim" (dry-farmed) land throughout the province, spring plant eaten by all animals. It later forms a tuber like the ground artichoke which is eaten by the people"); UGHRARIRA (Mudir of Tal Kaif 3213 – "useful for sheep grazing"); ZEREA ZAWI (Kurd.-Arbil, *Gillett* 7996); LIZZAIJ (Arbil province, *Radhi* 3852 – "grazed by livestock") this being an unlikely Arabic name to be current in a mainly Kurdish-speaking province; GÎYĀ PEN (?R)OSA (Kurd.-Rowanduz, *Rawi* 13730); BISBAAS (Tal Kochak, *Gillett* 10804). However, even if some of these colloquial names are incorrect or only of limited local application it seems clear that the species is generally recognised as a useful component of the vegetation.

Turkey, Caucasus, Iran, Syria, Lebanon, Palestine, Jordan.

2. **Bunium microcarpum** (*Boiss.*) *Freyn et Sint.* ex *Freyn*, Öst. Bot. Zeit. 44: 99 (1894); Wolff, l.c. 195 (1927); Hedge & Lamond in Fl. Turk. 4: 345 (1972); Rech.f., Fl. Iranica 162: 247 (1987).

Carum microcarpum Boiss. in Ann. Sci. Nat. sér. 3, Bot. 1: 137 (1844); Boiss., Fl. Orient. 2: 885 (1872).

Erect or ascending herb, (5–)10–35 cm, divaricately branched almost from base. Stem and branches terete, finely striate. Radical leaves long-petiolate with narrow long-attenuate sheaths, triangular in outline, bipinnate or ternate-bipinnate, the ultimate segments to ± 10 × 1 mm, mucronate at tip. Upper stem leaves sessile, with fewer narrower segments, finally reducing to sheaths, the latter scarious-margined above. Umbels several, 4–8-radiate; rays somewhat unequal, 0.8–5.5 cm, outer curving inwards slightly in fruit. Involucre absent or of 1 or 2 short linear caducous bracts. Partial umbels 5–15(–20)-flowered, the pedicels unequal, short, 2–4 mm, the longest scarcely exceeding the fruit in length. Peduncles

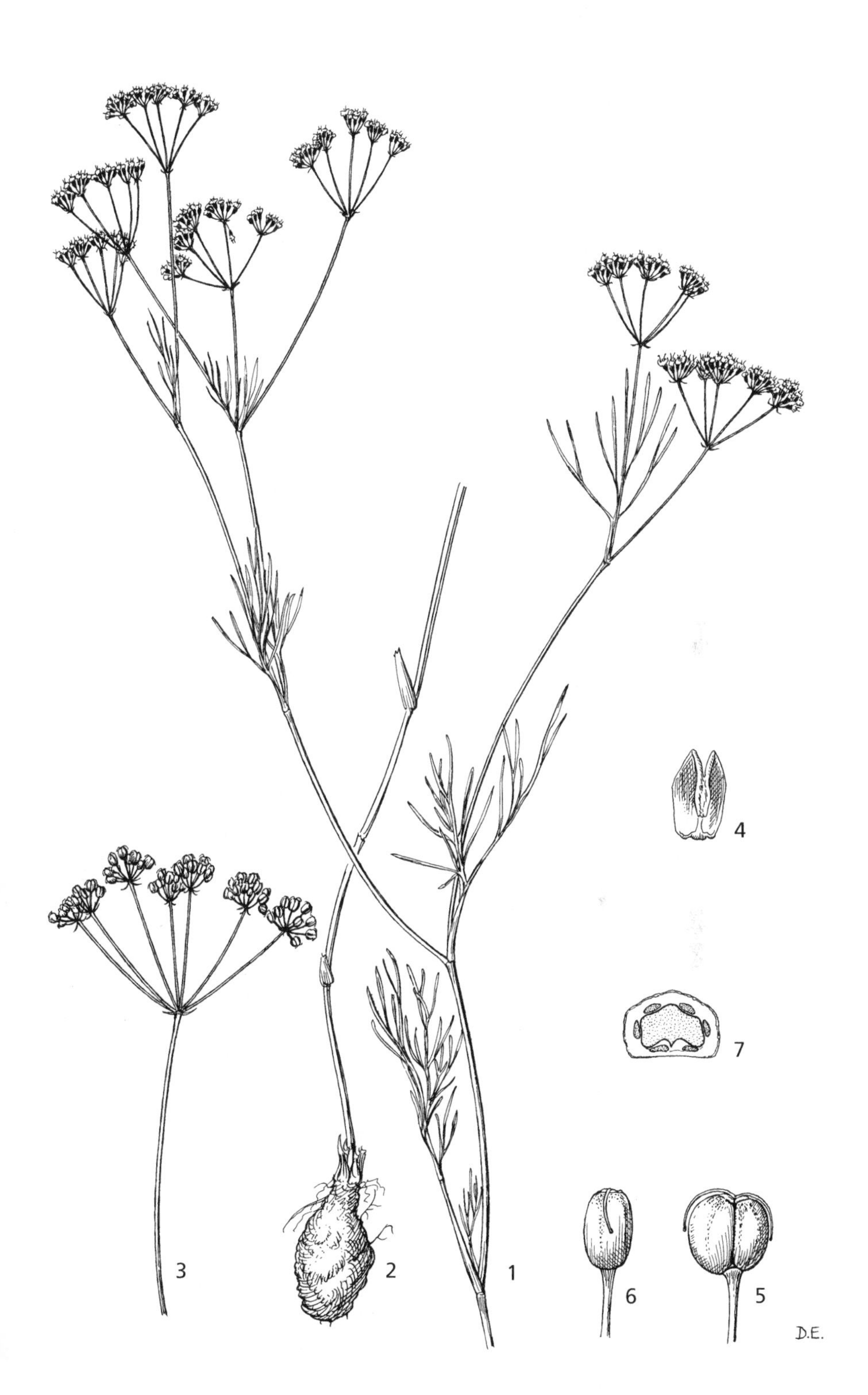

Fig. 65. **Bunium paucifolium**. 1, habit × ²/₃; 2, lower portion with root × ²/₃; 3, mature umbel × ²/₃; 4, petal × 8; 5, fruit × 4; 6, mericarp × 4; 7, fruit cross section × 8. 1–3 from *Helbaek* 1546; 4 from *Rawi* 22887; 5–7 from *Rawi* 21898. Drawn by D. Erasmus (1964).

2–5.5 cm. Involucel of 4–6 shortly lanceolate bracteoles, subequalling the shortest pedicels. Flowers small, 2.5 mm in diameter. Fruit shortly ovoid to elliptic-oblong, 2.5–4 × 1.5–2 mm, ribs pale and conspicuous. Stylopodium small, convex, disk flattened. Styles very short, ± 0.5 mm, slender, deflexed, scarcely exceeding to somewhat longer than stylopodia. Endosperm sulcate; vittae very large; bundles under primary ribs small.

HAB. On a mountain summit; alt. 2060 m; fl. & fr. Jul.
DISTRIB. Very rare in Iraq; only found once near the Turkish frontier, in the north-west sector of the thorn-cushion zone of Iraq. **MAM**: Mt. Zawita, NE of Zakho, *Rawi* 23595.

Turkey (mainly in W & S Anatolia), W Syria, NW Iran.

3. **Bunium cornigerum** (*Boiss. et Hausskn.*) *Drude* in Pflanzenfam. 3, 8: 194 (1898); Nábělek in Publ. Fac. Sci. Univ. Masaryk 35: 123 (1923); Wolff, l.c. 206 (1927); Bornmüller in Beih. Bot. Centralbl. 58B: 278 (1938); Zohary in Dep. Agr. Iraq Bull. 31: 110 (1950).

Carum cornigerum Boiss. et Hausskn. in Boiss., Fl. Orient. 2: 887 (1872); Blakelock in Kew Bull. 3: 432 (1948); Rawi in Dep. Agr. Iraq Tech. Bull. 14: 90 (1964).

Erect, glabrous herb, (20–)30–40 cm, with long branches from near base. Stem and lower branches ± strongly ridged, uppermost branches and rays of inflorescence may be sharply quadrangular. Lower leaves petiolate, ovate or triangular-ovate in outline, bi- to subtri-pinnatisect into oblong-linear, entire, mucronate segments up to 35 × 4 mm. Upper leaves ternately or variously pinnate, similar, but becoming sessile on the sheaths, segments more elongate; all leaves with a sheathing base. Umbels several, 5–9-rayed, rays 0.5–3.3 cm, much exceeding the 4–6 linear-lanceolate, membranous-margined involucral bracts. Partial umbels 9–15-flowered; pedicels 2–7 mm, thickening in fruit, the shortest compressed. Flowers ± 3 mm in diameter, white or pinkish. Peduncles 3.5–8 cm. Fruit oblong-prismatic, ± 4–5 × 1–1.5 mm wide, separating into mericarps at maturity; ribs very narrow, much narrower than the valleculae, but pale and prominent, narrowly winged. Stylopodia elongate-conic, gradually attenuate into the styles, which are erect, ± rigid, ± 2 mm long, and purplish in ripe fruit. Endosperm somewhat sulcate, vittae small to moderate, bundles under primary ribs small.

HAB. In the mountains, on rocky slopes, under *Quercus* forest, among *Quercus* scrub on a steep N-facing slope, among serpentinite rocks; alt. (600–)1000–2000(–2500) m; fl. & fr. (Apr.–)May–Jun(–Sep.).
DISTRIB. Quite common in the SE sectors of the lower thorn-cushion and upper forest zones of Iraq. **MRO**: nr summit of Kuh-i Sefin, *Bornmüller* 1292!; Shaqlawa, *Haines* W761!; Rowanduz gorge, *Guest* 2041!; Kurruk-Dar (Kurek Dagh), nr Rowanduz, *Nábělek* 515a; Handren, above Rowanduz, *Nábělek* 424; Karouk, *Kass & Nuri* 27400! 27270! 27525!; Sakri-Sakran, *Bornmüller* 1293b; Gali Warta, *Rawi & M. Jasim* 40678!; **MSU**: Kajan, near Penjwin, *Rawi* 22693!; Kamarspa, between Halabja and Tawila, *Rawi* 22260!

Endemic, but likely to occur also over the frontier in Iran.

4. **Bunium cylindricum** (*Boiss. & Hohen. ex Boiss.*) *Drude* in Engler & Prantl, Nat. Pflanzenfam. 3, 8: 194 (1898); Grossh., Fl. Kavk. ed. 2, 7: 79 (1967); E. Nasir in Fl. Pak. 20: 85, f. 26 (1972).

Carum cylindricum Boiss. & Hohen ex Boiss., Diagn ser. 1, 10: 23 (1849); Boiss., Fl. Orient. 2: 885 (1872).

Rootstock swollen, globose. Plant glaucous. Stems 15–35 cm tall, shallowly grooved, branched from base. Leaves bipinnatisect, the segments oblong-linear, 2–3-partite; basal leaves broadly triangular in outline. Cauline leaves shortly petiolate or sessile, segments narrowly linear. Umbels with 5–10 rays, 25–70 cm; bracteoles usually absent, rarely 1–4, linear-subulate. Partial umbels 10–15-flowered, with unequal pedicals 2–6 mm, scarcely enlarged in fruit. Inner flowers staminate, outer hermaphrodite. Bracteoles 2–5, ± lanceolate, equalling pedicels. Petals broadly rounded. Fruits 5–6 mm, cylindrical, with a single vitta. Stylopodium flattened.

HAB. On hills; alt. ± 600 m.
DISTRIB. Very rare in Iraq. **MRO**: mt. Potin (Botin), *Hadač* 6161; **FKI**: 37 km E of Kirkuk, *Rechinger* 10031.

Turkey (E Anatolia), Transcaucasia, Iran, Afghanistan, Pakistan, C Asia.

5. **Bunium avromanum** (*Boiss. et Hausskn.*) *Wolff*, l.c.: 207 (1927); Zohary in Dep. Agr. Iraq Bull. 31: 110 (1950); Rech.f., Fl. Iranica 162: 252 (1987). — Type: Iraq, (or in Iran), 'in rupestribus Mt. Avroman", 1867 *Haussknecht* s.n. (K!, syn.).

Carum avromanum Boiss. et Hausskn. in Boiss., Fl. Orient. 2: 888 (1872); Rawi in Dep. Agr. Iraq Tech. Bull. 14: 89 (1964).

Erect, glabrous herb, 20–30 cm, with a few long branches from near the base and above. Stem and lower branches terete, striate, rays of inflorescence terete to ± quadrangular. Leaves bipinnate, segments very long and narrowly linear, straight, mucronate. Umbels rather few, terminal on the main stem and branches, 6–10-rayed, rays 1.2–2.8 cm, exceeding the ± 8 linear-lanceolate, caducous involucral bracts. Partial umbels 6–10-flowered, pedicels unequal, 2–5 mm, not incrassate in fruit. Involucel of 5–8 bracts, which are similar in form to those of the involucre. Peduncles 3–7 cm. Fruit narrowly oblong, parallel-sided or somewhat curved above, 3–5 × 1 mm, separating into mericarps at maturity; ribs narrow, much narrower than the valleculae, but pale and conspicuous. Stylopodia elongate-conic, gradually attenuate into the ± 1.5 mm styles, which are erect-divergent or somewhat flexuous. Endosperm subterete dorsally, with two shallow grooves at the commissure only; vittae small, shallow; bundles under the scarcely prominent primary ribs small.

HAB. Rocky places on mountains, clearings in *Quercus* woods, on limestone; alt. 1070–2750 m; fl. & fr. Jun.
DISTRIB. Very rare in Iraq. **MAM**: between Dohuk and Amadiya, above Suwara Tuka, *Rechinger* 11964; **MRO**: Mergasur, *Hadač* 5840; foot of Mt. Potin (Botin), *Hadač* 5981; ?**MSU**: (or in Iran), 'in rupestribus Mt. Avroman", 1867 *Haussknecht* s.n. (syn.)!

Endemic (or possibly just over the frontier in Iran); not indicated for Iran in Flora Iranica, l.c.

6. **Bunium caroides** (*Boiss.*) *Bornmüller*, Beih. Bot. Centralbl. 19 (2): 259 (1906) & 58B: 279 (1938); Nábělek in Publ. Fac. Sci. Univ. Masaryk 35: 123 (1923); Zohary in Dep. Agr. Iraq Bull. 31: 110 (1950); Rech.f., Fl. Iranica 162: 253 (1987).

B. elvendia (Boiss.) Drude in Pflanzenfam. 3, 8: 194 (1898).
Elwendia caroides Boiss., Ann. Sci. Nat. sér. 3, Bot. 1: 140 (1844).
Carum elvendia Boiss., Fl. Orient. 2: 888 (1872); [*C. caroides* (non (Boiss.) Bornmüller) Blakelock in Kew Bull. 3: 432 (1948); Rawi in Dep. Agr. Iraq Tech. Bull. 14: 90 (1964)].

Erect, glabrous herb, rather short (10–25 cm), with long branches from near base. Stem and lower branches ± strongly ridged, upper branches and rays of inflorescence often sharply quadrangular. Lower leaves broadly triangular-ovate in outline, ternate–bipinnatisect, ultimate segments entire, linear, mucronate. Upper leaves ternate or variously pinnatisect, similar but sessile and less divided; all leaves sheathing at base. Umbels several, 5–7-rayed, rays 1–35 mm, much exceeding the involucre of 1–3(–6) linear-lanceolate, membranous-margined bracts, incrassate and widely divaricate to somewhat reflexed at fruiting stage. Partial umbels of 4–10 flowers, pedicels 1–3 mm, incrassate and compressed, subequalled by the 5–7 bracts of the involucel, which are similar to those of involucre. Flowers ± 2 mm in diameter, white to pinkish. Peduncles 1.5–11 cm. Fruit linear-prismatic, curved, 5–7 × 1 mm broad, not separating into mericarps at maturity; ribs broad, obtuse, equalling or exceeding the valleculae in width. Stylopodia elongate-conic, gradually attenuate into the ± 1 mm styles, which are erect or divaricate and ± rigid. Endosperm deeply grooved, vittae shallow, curving round the sinus of each vallecula, bundles under primary ribs large, semilunar.

HAB. In the mountains on stony slopes, among *Quercus* scrub, on gentle slopes and on screes above tree line; alt. 950–2500 m; fl. & fr. Apr.–Jun. or later.
DISTRIB. Occasional in lower thorn-cushion and upper forest zones of Iraq. **MAM**: Sarsang, *Haines* W999!; **MRO**: Kuh-i Sefin, *Bornmüller* 1294; Kurruk Dar (Kurek Dagh) nr Rowanduz, *Nábělek* 515; Mt. Handren, above Rowanduz, *Nábělek* 426, 512; Putin (Botin), nr Shirwan Mazna, *Agnew, Hadač & Haines* W2099!; Marmarut mts., *Guest* 2084.

NW, W & S Iran.

7. **Bunium verruculosum** *C.C.Townsend* in Kew Bull. 22: 429 (1968); Rech.f., Fl. Iranica 162: 240 (1987). — Type: Iraq, Penjwin, *Rawi* 22576 (K!, holo.).

Glabrous, erect herb, ± 20 cm, branched from base upwards. Stem and branches subterete and striate below, above more strongly angular with raised lines. Lower leaves shortly petiolate, deltoid-ovate in outline, tripinnate with narrowly linear segments up to 20 × 0.7 mm, mucronate at the tip. Upper leaves ternate-pinnate, similar but sessile; all leaves with an expanded, membranous-margined base. Umbels rather few, with 8–13 somewhat unequal rays 8–15(–20) mm long, about twice as long as the 8–10 linear-setaceous bracts of involucre reaching almost or quite to the apex of the young fruit. Flowers white, 3.5 mm in diameter, petals rather long-persistent. Peduncles 5.5–10 cm. Fruit linear-prismatic, 3–5 mm, pericarp minutely verruculose, ribs narrow but conspicuous and rather prominent, pedicel not incrassate. Stylopodium depressed, styles filiform, very long, finally flexuose or reflexed, about 2/3 of the length of the fruit. Endosperm bluntly angled, sulcate only on the commissural face, vittae small, bundles under primary ribs small.

HAB. On a cultivated hillside, on bare slatey substrate; alt. 1280–1500 m; fl. & fr. Jun.
DISTRIB. Very rare, only found twice near the Iranian frontier, in the SE sector of the middle forest zone. **MSU**: Penjwin, *Rawi* 22576 (type)!; near Penjwin, *Rechinger* 10505!

W Iran.

8. **Bunium rectangulum** *Boiss. & Hausskn.* in Boiss., Fl. Orient. 2: 884 (1872).

Rootstock forming a tuber 10–20 mm in diameter, globose. Stems 20–40 cm tall, weakly grooved, branched almost from base. Leaves oblong-triangular in outline, 2- or 3-pinnatisect, the segments ovate-oblong, crenate or oblong-lobate, the ultimate lobes linear, terminal segments longer relative to the lateral ones; upper leaves with narrowly linear segments. Umbel rays 10–12, unequal, slender, 20–30 cm, not thickened in fruit. Bracts subulate, 5–15 mm, membranous. Pedicels up to 20, elongating in fruit to 10–15 mm. Petals white, minute, ± 1 mm. Fruits oblong, 2–3 × 1–1.5 mm; stylopodium flattened; styles deflexed.

DISTRIB. Very rare, only a single record from Iraq. **MRO**: Mt. Baradost near Shanidar, *Erdtman & Goedemans* (*Rechinger* 15668)!
NOTE. A Kotschy gathering of *B. papillosum* DC. from the 1841 series "Pl. Mesopot., Kurdistan et Mossul" (*Kotschy* 10, K!) is doubtfully recorded from Iraq; it is more likely to have been collected on the upper Euphrates within Turkey. *B. croceum* Fenzl is widespread in Turkey but no material has been seen from Iraq.

W & SW Iran.

20. **FALCARIA** Fabr.

Enum. meth. Helmst. ed. 1, 34 (1759), *nom. conserv.* non Riv. ex Rupp. (1745);
Wolff in Pflanzenr. IV. 228, 90: 129–133 (1927)

Biennial or perennial herbs. Leaves undivided, ternatisect or pinnatisect, sharply and closely serrate. Umbels compound, involucre and involucel present or absent. Flowers white, with obovate petals which are irregularly 2-lobed above, with a long inflexed lobule; polygamous or hermaphrodite. Calyx conspicuous, acute, persistent. Fruit ± oblong to ovoid, somewhat laterally compressed, commissure rather narrow; ribs 5, obtuse, broader than the valleculae. Vittae solitary or 3 or 4 in the valleculae, 2–8 on the commissure. Stylopodia short, obtusely conical, the disk crenulate-edged. Endosperm subterete, commissural face ± flat. Carpophore bipartite.

Three or four species from C & Mediterranean Europe to W & C Asia and N Africa; only one species in Iraq.
Falcaria (from the Lat. *falx*, *falcis*, a sickle, *falcarius*, a sickle-maker – referring to the form of the leaf segments.)

1. **Falcaria vulgaris** *Bernh.*, Syst. Verzeichn. Pfl. Erfurt: 176 (1800); Hayek, Prodr. Fl. Balc.: 989 (1927); Zohary, Fl. Palest. ed. 2, 2: 526 (1933); Rawi in Dep. Agr. Iraq Tech. Bull. 14: 92 (1964); Grossh., Fl. Kavk. ed. 2, 7: 75 (1967); Mouterde, Nouv. Fl. Lib. et Syr. 2: 629 (1970); Zohary, Fl. Palaest. 2: 420 (1972); Rech.f., Fl. Iranica 162: 306 (1987).

Fig. 66. **Falcaria vulgaris**. 1, habit × 2/3; 2, leaf × 2/3; 3, petal × 8; 4, fruits (both faces) × 4; 5, fruit cross section × 8. 1–3 from *Chapman* 26235; 4–5 from *Wheeler Haines* W415. Drawn by D. Erasmus (1964).

Sium falcaria L., Sp. Pl. ed. 1: 252 (1753).
Bunium falcaria (L.) M. Bieb., Fl. Taur.-Cauc. 1: 211 (1808).
Falcaria rivini Host, Fl. Austr. 1: 381 (1827); DC. Prodr. 4: 110 (1830); Boiss., Fl. Orient. 2: 892 (1872); Guest in Dep. Agr. Iraq Bull. 27: 34 (1933); *F. sioides* (Wib.) Aschers., Fl. Prov. Brandenb.: 241 (1864); Wolff, l.c. 130 (1927); Schischkin in Fl. SSSR 16: 383 (1950); Blakelock in Kew Bull. 3: 434 (1953); Rawi in Dep. Agr. Iraq Tech. Bull. 14: 92 (1964).

Glaucous perennial, 30–100 cm, branched from somewhat above the middle into a spreading compound inflorescence. Stem and branches wiry and tough, subterete, striate. Radical leaves petiolate, biternate or ternate-pinnate, segments up to ± 30 × 2 cm, the segments long, linear-lanceolate or broadly linear, regularly sharply serrate with forward-directed, long-acuminate, cartilaginous-margined teeth; upper leaves decreasing in size, pinnate or biternate to ternate, upper reduced and sessile; all leaves ± sheathing, glabrous or minutely puberulous on lower surface. Umbels very numerous, (5–)8–10(–15)–radiate, rays 0.8–2.5(–3) cm; bracts of involucre 4–8, linear-subulate, much shorter than the rays. Partial umbels usually with 8–10 flowers, pedicels 2–12 mm, exceeding bracts of the involucel, which are similar in number and form to those of the involucre. Flowers ± 2.5 mm in diameter. Peduncles 2–4 cm. Fruit linear-oblong, 3–4 × 0.75–1 mm, glabrous or somewhat puberulent when young. Styles slender, ± 1 mm, spreading or reflexed. Vittae solitary in the valleculae, 2 on the commissure. Fig. 66: 1–5.

HAB. Mountain slopes and valleys among *Quercus* forest, in fields and on waste land, in a *Populus* plantation; alt. 250–1650 m; fl. & fr. Jun.–Sep.
DISTRIB. Occasional in the forest zone of Iraq, less common in the moist-steppe zone. **MAM**: (or in Turkey?): "Mesopot., Kurdistan & Mossul", *Kotschy* 354; "Mts. of Assyria & Mesopotamia", *Haussknecht* s.n.; Zawita, *Guest* 3749!; Sarsang, *Haines* W415!; **MRO/FAR**: Arbil province, *Radhi* 3871!; **MRO**: Haji Umran, *Rawi* 24284!, *Rechinger* 11848; **MSU**: Jarmo, *Helbaek* 1949!; Susa, *Karim* 39264!; 8 km N of Kani Spi, *Rawi* 22391!; Hawraman, *Haussknecht* s.n.; **FNI**: Nineveh, *E Chapman* 23595!; Mindan, *id.* 26235!
NOTE. ASYĀGH (Ir.-Arbil province, *Radhi* 3871 – "grazed by animals"); but Guest noted that it had been reported in Kut province as poisoning sheep, affecting their eyes and causing temporary blindness, being known there as ZAND AL-AL'ARŪS – a general name in Lower Iraq covering various white-flowered umbellifers. These two reports on the palatability of the plant are incompatible. It now seems probable that the plant on which Guest's note was based had not been correctly determined since no specimen of *F. vulgaris* can now be traced from Lower Iraq, nor would this species of the mountains and foothills be likely to occur so far south as on the alluvial plain in Kut province. Nevertheless Radhi's note requires confirmation before *F. vulgaris* may be assumed to be a wholesome forage plant.

C Europe (France to Greece and W Russia), Turkey, Caucasus, Iran, C Asia (Kazakhstan, Siberia), Syria, Lebanon, Palestine, N Africa.

21. **SIUM** L.

Sp. Pl. ed. 1: 251 (1753), p.p.; Gen. Pl. ed. 5: 120 (1754); Wolff in Pflanzenr. IV. 228, 90: 341 (1927)

Glabrous perennial herbs growing in colonies in wet or damp places, very rarely in dry situations, the lower leaves often submerged and variously pinnatifid into narrow segments; aerial leaves simply pinnate with broad, toothed segments. Umbels compound, lateral or terminal. Involucre and involucel present, of many bracts. Flowers hermaphrodite, white, the petals broadly obovate-rotund or obcordate, retuse or emarginate above with an inflexed lobule. Calyx present, of 5 minute to ± conspicuous sepals. Fruit globular-ovoid to oblong, slightly compressed laterally, commissure narrow. Primary ribs 5, obtuse or thickened, conspicuous, the lateral marginal; all narrower than the 1–3-vittate valleculae. Stylopodia stout, depressed-conic, margin of the disk entire or crenate. Carpophore bipartite, adnate or free, or entire. Endosperm convex, the commissural face flat.

10–15 species, almost cosmopolitan, not in Australia or S America; one species in Iraq.
Sium (a name thought by some to have been derived from the Celtic word *siu*, water, alluding to the aquatic habitat of the species; but considered by most authors to have derived from the Greek σιον, *sion*, the name for an umbellifer used by Dioscorides and others); Water Parsnip.

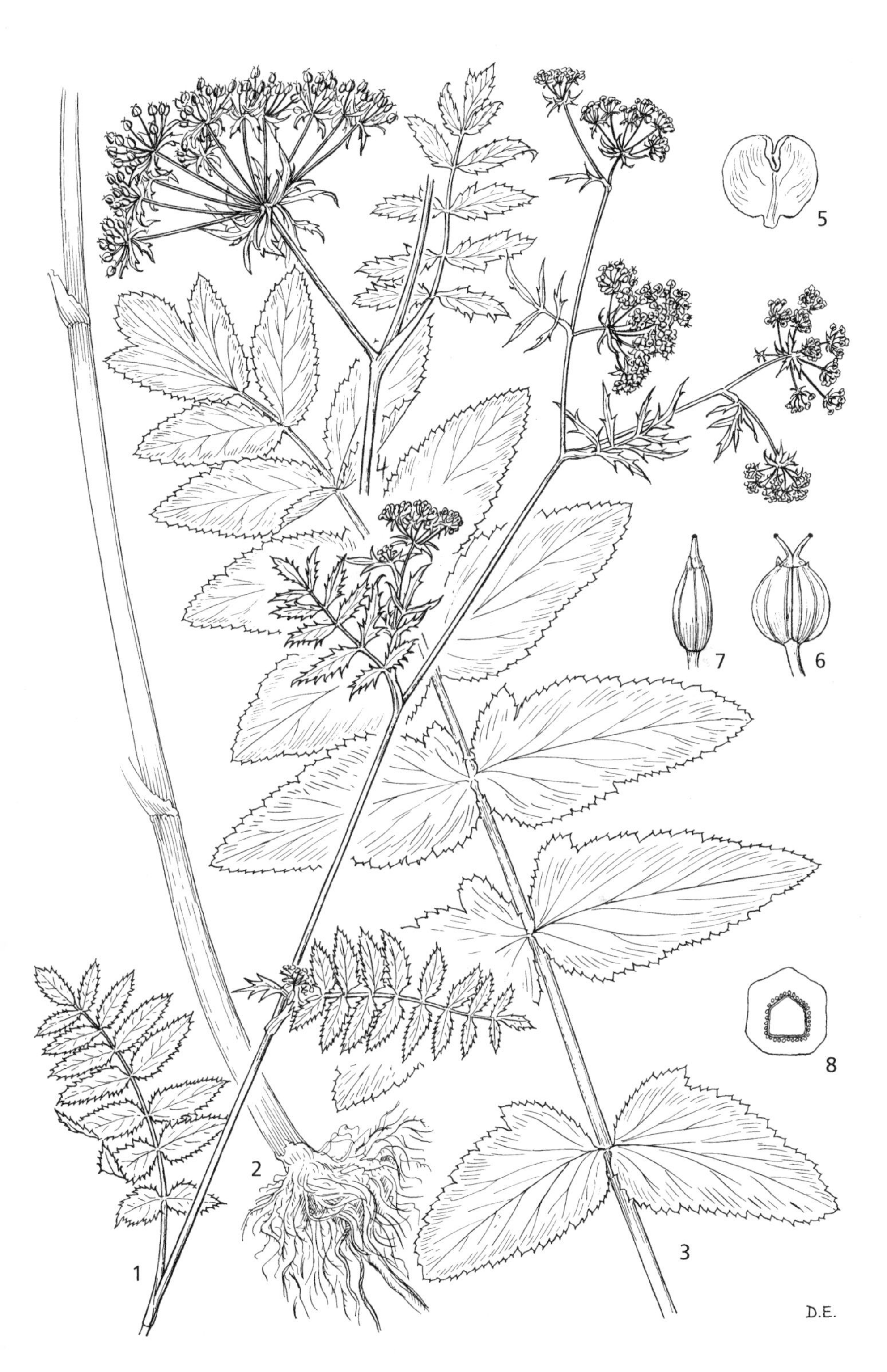

Fig. 67. **Sium sisarum** var. **lancifolium**. 1, habit × ²/₃; 2, habit, lower part with root × ²/₃; 3, leaf × ²/₃; 4, mature umbel × ²/₃; 5, petal × 4; 6, fruit × 4; 7, mericarp × 4; 8, fruit cross section × 8. 1–2 from *Lowne* 690; 3–8 from *Garland*. Drawn by D. Erasmus (1964).

1. **Sium sisarum** *L.*, Sp. Pl. ed. 1: 251 (1753).

The typical form of this species (var. *sisarum*), with a tuberous rootstock, is now only known in cultivation; this is the Skirret, originally brought to the west from China over 400 years ago, according to Lindley & Moore (1870). It is a vegetable of minor importance sometimes cultivated for its edible tubers. The wild form of the plant, as it occurs in Iraq, is the following:

var. **lancifolium** (*M. Bieb.*) *Thell.* in Hegi, Illustr. Fl. Mitteleurop. 5 (2): 1223 (1926); Wolff, l.c. 349 (1927); Peşmen in Fl. Turk. 4: 365 (1972).

S. lancifolium M. Bieb., Fl. Taur.-Cauc. 3: 230 (1918) non Schrank (1789); Boiss., Fl. Orient. 2: 888 (1872); Hayek, Prodr. Fl. Balc. 1: 1001 (1927); Zohary, Fl. Palest. ed. 2, 1: 525 (1932); Blakelock in Kew Bull. 3: 437 (1948); Rawi in Dep. Agr. Iraq Tech. Bull. 14: 96 (1964); *S. sisaroideum* DC., Prodr. 4: 124 (1830); Schischkin in Fl. SSSR 16: 430 (1950); Grossh., Fl. Kavk. ed. 2, 7: 87 (1967); Rech.f., Fl. Iranica 162: 303 (1987); Pu Fading & Watson in Fl. China 14: 116 (2005).

A stout, erect, glabrous perennial, 0.4–1.5 m high, with a shortly stoloniferous rootstock. Stem simple below, branched above, branches long and slender, stem and branches strongly sulcate-striate. All aerial leaves similar, sheathing, simply pinnate, lower ovate-lanceolate in outline with 3–5(–6) pairs of ovate-lanceolate to elliptic or ovate, sharply and regularly toothed 1.5–8.5(–11) × 0.8–3.5 cm leaflets; the upper reducing in size, breadth and number of leaflets, becoming sessile on the sheaths, the uppermost very small and with only one pair of lateral leaflets. Umbels 7–17-rayed, rays very unequal, 0.6–3.5 cm; involucre of 4–6 lanceolate bracts, which are narrowly but distinctly white-margined and frequently ± reflexed. Partial umbels of 8–20 flowers, pedicels 1–7 mm, exceeding the bracts of involucel, which resemble those of the involucre but are generally more numerous (± 8). Peduncles 2.5–10 cm. Fruit ovoid or ovoid-oblong, 1.3–3.5 mm long, ribs narrow, pale and prominent, much narrower than the trivittate valleculae. Styles short, scarcely or not exceeding the narrow disk, which has a crenulate-undulate margin, stylopodia small, conical. Carpophore small and slender, free, divided to the base. Fig. 67: 1–8.

HAB. In the mountains by streams and springs, under shade of *Juglans* trees, near a damp shady streamside; alt. 700–1500 m; fl. & fr. Aug.–Sep.(–Oct.).
DISTRIB. Occasional in the forest zone of Iraq. **MAM**: Sarsang, *Haines* W480!; **MRO**: Gali Ali Beg, *Hadač et al.* 6241; Rowanduz gorge, *Guest* 452!, *Gillett* 12417!, *Omar, Sahira, Karim & Hamid* 38303!; Haji Umran, *Rawi* 26831!; Pushtashan, *Rawi & Serhang* 26494!, *Rechinger* 11047; **MSU**: Hawraman Mts, *Rawi, Chakr., Nuri & Alizzi* 19826!

C & S Europe (Italy, Hungary, Greece, S Russia, Crimea), Turkey, Caucasus, Iran, Afghanistan, C Asia, China (Xinjiang).

22. **BERULA** W.D.J. Koch

Deutsch. Fl. ed. 3, 2: 25, 433 (1826); Wolff in Pflanzenr. IV. 228, 90: 336–341 (1927)

Erect, branched, glabrous, stoloniferous aquatic herbs. Submerged lower leaves 2–4-pinnatisect, aerial leaves simply pinnate with large, toothed, crenate or incised segments. Umbels compound. Involucre and involucel present, of many bracts. Flowers hermaphrodite, white, petals broadly obcordate with a distinct incurved lobule. Calyx present, evident. Fruit almost didymous, shortly ovoid to sub-orbicular, glabrous and smooth, ribs obtuse, not prominent, rather obscure, laterals not marginal. Vittae many, slender, ± contiguous or confluent, deeply embedded. Stylopodia shortly conical, bordered with a very narrow disk. Carpophore with the divisions adnate to mericarps. Endosperm obtusely pentagonal to subrotund, commissural faces flat, often cristate centrally.

Two species in Europe, temperate Asia and Africa; one species in Iraq.
Berula (according to Gilbert-Carter, the Latin name of a plant in Marcellus Empricus; however this may be, Koch's name seems to be merely a latinisation of the German vernacular name which he gives for *Berula angustifolia* (= *B. erecta* – Berle).

1. **Berula erecta** (Huds.) Colville in Contr. U.S. Nat. Herb. 4: 115 (1893); Hayek, Prodr. Fl. Balc. 1: 1001 (1927); Wolff, l.c. 337 (1927); Schischkin in Fl. SSSR 16: 466 (1950); Grossh.,

Fl. Kavk. ed. 2, 7: 87 (1967); Mouterde, Nouv. Fl. Lib. et Syr. 2: 623 (1970); Peşmen in Fl. Turk. 4: 366 (1970); Zohary, Fl. Palaest. 2: 429 (1972); E. Nasir in Fl. Pak. 20: 22 (1972);V. Täckh., Stud. Fl. Egypt ed. 2: 390 (1974); Pu Fading & Watson in Fl. China 14: 115 (2005).

Berula angustifolia (L.) Mertens & Koch, Deutschl. Fl. ed. 3, 2: 25 (1826); Grossh., Opred. Fl. Kavk. 229 (1949); Hedge & Lamond in Fl. Iranica 162: 305 (1987).

Sium erectum Huds., Fl. Angl. ed. 1: 103 (1762); Zohary, Fl. Palest. ed. 2, 1: 525 (1932); *S. latifolium* L. β, Sp. Pl. ed. 2: 361 (1762); *S. angustifolium* L., l.c. 672 (1763); DC. Prodr. 4: 125 (1830); Hand.-Mazz. in Ann. Naturh. Mus. Wien 27: 91 (1913); Nábělek in Publ. Fac. Sci. Univ. Masaryk 35: 123 (1929); Zohary in Dep. Agr. Iraq Bull. 31: 111 (1950) sub "Srem angustifolia", sphalm.

Erect or decumbent perennial herb with a creeping stoloniferous rootstock, 30–100 cm. Stem hollow, much-branched, sulcate, striate. Aerial leaves all simply pinnate, very variable in size, segments oblong-lanceolate to ovate, sessile or lowest pair very shortly stalked. Lowest leaves above surface of water long-petiolate, lanceolate in outline, with (5–)7–12(–15) pairs of leaflets; upper shorter, more shortly petiolate, and uppermost sessile with fewer leaflets; all sheathing at base. All leaf-segments sharply and irregularly incised-dentate, those of upper leaves more sharply so than those of the lower. Umbels leaf-opposed, the uppermost pseudo-terminal. Peduncles 1–4.5 cm. Bracts of involucre 6–10, leafy, simple or trifid, shorter than the 10–14 rays of the main umbel, rays slender, 0.4–3.5 mm. Partial umbels with 15–20 flowers, ± 2 mm in diameter, exceeding bracts of involucel, which are similar in form to those of involucre but smaller, fewer and more rarely trifid, pedicels 1–8 mm. Fruit subdidymous, ± 2 mm long, ribs concolorous with the body of fruit and not conspicuous. Styles spreading, slender, not or scarcely spanning the width of fruit. Vittae many, slender, ± contiguous but not confluent.

HAB. On lower mountains, by streams; alt. 650–800 m; fl. & fr. Jun.
DISTRIB. Very rare in Iraq, only found in two localities in the NW sector of the lower forest zone some 70–80 years ago, and not collected or recorded recently. **MAM**: Aradin, W of Amadiya, *Nábělek* 534; **MJS**: Bara on NW slope of Jabal Sinjar, *Handel-Mazzetti* 1568.

Atlantic & C Europe (Britain & Sweden to W Russia, Romania, Italy & the Balkans), Syria, Palestine, Jordan, Egypt, Turkey, Caucasus, Iran, Afghanistan, Siberia, China (Xinjiang), tropical & S Africa; introduced into Australia, N America and Mexico.

23. **TRACHYSPERMUM** Link

Enum. hort. bot. reg. Berol. 1: 267 (1821); Wolff in Pflanzenr. IV. 228, 90: 87–92 (1927)

Annual herbs, ± tall and branched. Leaves 2–4-pinnatisect, the ultimate segments narrow, more rarely ternate with broader segments. Umbels compound, involucre and involucel present. Flowers white, the petals oval, deeply emarginate above and ± 2-lobed, with a broad and obtuse inflexed lobule. Calyx teeth small. Fruit narrowly or broadly ovoid, narrowed to the commissure, somewhat compressed laterally, shortly and often densely hairy or papillose. Ribs filiform but prominent, lateral marginal, mericarps thus obviously pentagonal in section. Stylopodia conical, styles reflexed, varying in length. Valleculae univittate, commissure bivittate. Endosperm similar in shape to mericarp, commissural face flat. Carpophore bipartite to the base.

Some 20 tropical and subtropical species from NE Africa to C Asia, India and China; one species in Iraq.
Trachyspermum (from the Gr. τραχύς, *trachys*, rough and σπερμα, *sperma*, seed, from the hairy or verrucose fruits found in this genus.)

1. **Trachyspermum ammi** (*L.*) *Sprague* in Turrill, Kew Bull. 1929 (7): 228 (1929); Zohary, Fl. Palest. ed. 2, 1: 525 (1932); Schischkin in Fl. SSSR 16: 379 (1950); Zohary, Fl. Palaest. 2: 417 (1972); E. Nasir in Fl. Pak. 20: 72 (1972).

Sison ammi L., Sp. Pl. ed. 1: 252 (1753); *Ammi copticum* L., Mant. 1: 56 (1767); Boiss., Fl. Orient. 2: 891 (1872); Zohary, Fl. Palest. ed. 2, 1: 525 (1932); Rawi in Dep. Agr. Iraq Tech. Bull. 14: 88 (1964); *Trachyspermum copticum* (L.) Link, Enum. hort. bot. reg. berol. 1: 267 (1821); Wolff, l.c. 87 (1927); H. Riedl in Fl. Lowland Iraq: 462 (1964); Hedge & Lamond in Fl. Iranica 162: 337 (1987).

Fig. 68. **Trachyspermum ammi**. 1, habit × ²/₃; 2, habit, lower part with root × ²/₃; 3, petal × 10; 4, fruits (both faces) × 4; 5, fruit cross section × 6. 1–2 from *Haussknecht* 6/70; 3–5 from *Hay* 431. Drawn by D. Erasmus (1964).

Erect annual to 60 cm or more, usually ± 40 cm, branched from near base, glabrous except in the fruit and inflorescence, and occasionally margins of upper leaf-sheaths ciliate. Stem and branches terete, striate. Lower leaves long-petiolate, lamina oblong-lanceolate in outline, 2–3-pinnatisect, ultimate segments narrow-linear to almost capillary, up to ± 17 cm long, < 1 mm wide, mucronate; upper leaves similar, only uppermost simply pinnate, becoming sessile, all leaves with a short amplexicaul sheath. Peduncles 1.5–6 cm. Umbels many, 4–12–rayed, rays 5–15 mm. Involucre of 3–6 bracts, which are most frequently narrow and entire but occasionally dentate or trifid, shorter than rays. Involucel of 3–6 linear, simple bracteoles, subequalling rays of partial umbel. Partial umbel rather close, 10–20-flowered, flowers ± 2 mm in diameter, petals pilose dorsally; anthers purplish black; pedicels 1–3 mm. Fruit ovoid, 1.5 mm long, broadest at or below middle, ± densely muricate with very short broad-based papillae. Styles short, spreading or reflexed, shorter than to about equalling breadth of fruit. Fig. 68: 1–5.

HAB. On the upper plains in fields; alt. 250–350 m; fl. & fr. May.
DISTRIB. Very rare in Iraq, only found, well over 100 years ago in two places in the moist steppe zone; no recent collections, **FUJ/FNI**: Mosul, *Noë* s.n.; **FKI**: nr Kirkuk, *Haussknecht* s.n.!; **LCA**: (cult.), medicinal plant garden at Abu Ghraib, *Janan, Sahira & Omar* c. 1978.
NOTE. Ajawan Caraway (Am.). Though occasionally found in ruderal habitat, in out of way places (e.g. a shipping yard in the Swedish island of Götberg, on a refuse heap near granaries in Scotland at Glasgow) it is basically known only as a cultivated plant or as an adventive in secondary habitats such as fields, gardens, irrigation ditches etc. Burkill (1935) states that there was long a trade in the export of its seeds from India where it was widely cultivated for centuries. Its name in Sanskrit was YAMANI and, at the beginning of the Christian era, it had reached Greece, where it was called AMMI, via Egypt. In India the seeds were much used as a carminative, to alleviate pain and for several other uses in folk medicine, and in China (Xinjiang) they are reputed to have medicinal value.

Palestine (very rare), Arabia, Iran, Afghanistan, Pakistan, India, China.

24. **KORSHINSKIA** Lipsky

in Trud. Bot. Sada 18: 60 (1900)

Erect, branched perennial herbs, rootstock tuberous or rhizomatous or a taproot. Leaves variously divided to tripinnate with capillary segments. Umbels compound. Involucre and involucel present or absent. Flowers greenish-yellow, petals ± obcordate, emarginate-2-lobed above with an inflexed lobule. Calyx obsolete. Fruit smooth and glabrous, ovoid to globose, or broader than long, constricted at the commissure and ± didymous, mericarps rounded dorsally. Ribs 5, very slender and filiform, not prominent and often indistinct. Valleculae with 1–2 vittae (rarely coalescent into 1), which may be evident or very slender and interrupted. Stylopodia flattened to depressed, margin of the disk entire and narrow or crenate-undulate. Carpophore deeply bipartite. Endosperm convex, commissural face concave, excavate, bisulcate.

Five species distributed in SW and C Asia; a single species in Iraq.

1. **Korshinskia assyriaca** (*Freyn & Bornmüller*) *Pimenov & Kljuykov* in Edinb. J. Bot. 52 (3): 339 (1995).

Scaligera assyriaca Freyn et Bornmüller in Bull. Herb. Boiss. 5: 611 (1897); Rawi in Dep. Agr. Iraq Tech. Bull. 14: 95 (1964); Rech.f., Fl. Iranica 162: 223 (1987).

Glabrous perennial, 40–80 cm high, with a napiform rootstock; branched above middle of stem, branches divaricate, not or very little further divided. Stem and branches terete, striate. Radical and lower stem leaves ovate in outline, petiolate, sheathing, bi- or tri-pinnatisect, ultimate segments flat, blunt, oblanceolate, to ± 12 × 3.5 mm; upper stem leaves becoming sessile, reducing in size and number of divisions, not reduced to bract-like sheaths. Peduncles 2–6.5(–12) cm. Umbels few, large, with 8–12(–15) unequal sharply angled 1–6.5 cm rays. Bracts of involucre ± 8, oblong-lanceolate to broadly pinnatisect, ± 1 cm long, herbaceous or with a narrow membranous border, much shorter than rays. Partial umbels with 12–20(–30) flowers, the slender unequal 2–12 mm pedicels mostly

Fig. 69. **Korshinskia assyriaca.** 1, habit × 2/3; 2, mature umbel × 2/3; 3, petal × 6; 4, fruit × 4; 5, mericarp × 4; 6, fruit cross section × 6. 1, 3 from *Rawi et al.* 28251; 2, 4–6 from *Wheeler Haines* W 1053. Drawn by D. Erasmus (1964).

exceeding bracts of involucel, which are similar in form and number to those of involucre. Fruit much broader than long, when ripe 4 × 2 mm, didymous, each mericarp subglobose, ribs filiform and inconspicuous. Stylopodia broad, shortly conic, less conspicuous in ripe fruit as the mericarps become gibbous above; disk with a pale, strongly undulate margin. Styles reflexed, attaining ⅔ of the width of the ripe fruit; valleculae with a single large vitta. Fig. 69: 1–6.

HAB. On mountain slopes in deep soil among *Quercus* trees, on limestone; alt. 450–1300(–1450) m; fl. & fr. May–Jun.
DISTRIB. Occasional in the middle and lower forest zones of Iraq. **MAM**: Sarsang, *E Chapman* 26406!; Bekma gorge, *Gillett* 8240!; Kuh-i Sefin, *Bornmüller* 1276 (type)!; Shaqlawa, *Haines* W759!; Haibat Sultan Dagh, *Rawi, Nuri & Kass* 28251!; **MSU**: Amoret, nr Qaradagh, *Haines* W1053!; Nalparaiz, *Hadač et al.* 4808.

SW Iran.

25. ELAEOSTICTA Fenzl

in Flora 26: 458 (1843)

Erect, branched perennial herbs with a shallow tuber-like rootstock; leaves tripartite with the ultimate segments broadly lanceolate to filiform. Umbels compound. Involucre and involucel present. Flowers white, petals obovate to ± obcordate, emarginate-2-lobed above with an inflexed lobule. Calyx obsolete. Fruit smooth and glabrous, ovoid to ellipsoid, not strongly constricted to the commissure and didymous, primary ribs very slender or indistinct; cells of the exocarp large and thin-walled, radially elongated, many times larger than cells of mesocarp. Vittae small, 2–5 in valleculae and 4–8 on commissure, forming an almost complete ring round endosperm, eseptate. Stylopodia shortly conical or depressed, constricted above mericarps. Carpophore deeply bipartite. Endosperm dorsally compressed, not or scarcely furrowed dorsally, concave and bisulcate at commissure.

Elaeosticta (from the Gr. ελαιόν, *elaion*, oil, στικτός, *stiktos*, spotted or tattoed – a reference to the numerous small vittae).
Ref. E.V. Klyukov, M.G. Pimenov & V.N. Tikhomirov in Bull. Soc. Nat. Mosc. N.S. 81 (6): 83–94 (1976) and ibid., N.S. 83 (6): 100–107 (1978).

1. **Elaeosticta meifolia** *Fenzl* in Flora 26: 458 (1843); Kljuykov et al. 92 (1976).
Scaligeria meifolia (Fenzl.) Boiss., Fl. Orient. 2: 877 (1872); Zohary, Fl. Palest. ed. 2, 1: 521 (1932); Bornmüller in Beih. Bot. Centralbl. 58B: 278 (1938); Mouterde, Nouv. Fl. Lib. et Syr. 2: 605 (1970); Hedge & Lamond in Fl. Turk. 4: 335 (1972); Rech.f., Fl. Iranica 162: 222 (1987).

Tall, glabrous perennial, attaining 1 m or more with a sub-tuberous rootstock; much branched from near base, branches divaricate, long and slender, with many graceful secondary branches and peduncles. Stem and branches terete, finely striate. Radical and lower stem-leaves shortly sheathing, sessile, oblong-lanceolate, tripinnatisect into filiform segments up to 3 × 0.2 mm, withered at advanced flowering stage; upper leaves reduced to bract-like sheaths. Peduncles 0.8–3.5 cm. Umbels many, small, with 3–6 very unequal, up to ± 2.2 cm, rays, some of which are so short that partial umbels are sessile or almost so. Bracts of involucre 2–4, very shortly lanceolate with widely membranous margins, central green band not attaining the apex of each bract. Bracts of involucel similar in form and number, shorter than the filiform, 1–6 mm pedicels of the 6–11-flowered partial umbels. Fruit shortly ovoid to subglobose, ± 2 × 2 mm, ribs pale, filiform. Stylopodium depressed, disk crenulate-edged, somewhat exceeded by short, reflexed styles.

HAB. On the mountains, in light *Quercus* scrub on a dry rocky hillside; alt. ± 1450 m; fl. & fr. Jun.
DISTRIB. Rare in Iraq; only found in a few places in the forest zone. **MAM**: (or in Turkey?), "Mesopot. Kurdistan & Mossul", *Kotschy* 325!; **MRO**: Mt. Qandil between Shahidan and Pushtashan, *Rechinger* 11018; between Shaqlawa and Rowanduz, *Bornmüller* 1275; **MSU**: Qaradagh, *Haines* W1510!

Turkey, Syria, Lebanon, NW, C & S Iran, Afghanistan, C Asia.

26. **MYRRHOIDES** Heister ex Fabr.

Enum. Pl. Hort. Helmst. ed. 2, 66 (1763)
Chaerophyllum sect. *Physocaulis* DC., Coll. Mem. 5: 59 (1829)
Physocaulis (DC.) Tausch in Bot. Zeit. 1: 342 (1834)

Erect annual herb, ± roughly hairy, stems swollen below the nodes at maturity. Leaves 2–3-pinnatisect with broad incised-dentate segments. Umbels compound. Involucre absent; involucel of several bracts. Flowers white or pinkish, petals obovate, entire or very shallowly emarginate, with an inflexed lobule. Calyx absent. Fruit linear-oblong, somewhat laterally compressed, commissure narrow, hispid. Ribs 5, broad and flat, much wider than the univittate valleculae, commissure bivittate. Stylopodia small, conical, stigmas sessile or on short erect styles at their tips. Carpophore bifid at apex. Commissural face of endosperm deeply sulcate, dorsal surface rounded.

A monotypic genus.
Myrrhoides (from *Myrrhis* and *-oides*, resembling).

1. **Myrrhoides nodosa** (*L.*) *Cannon* in Fedde Rep. 79: 65 (1968) & Fl. Europaea 2: 324 (1968); Hedge & Lamond in Fl. Turk. 4: 310 (1972); Zohary, Fl. Palaest. 2: 389 (1972).

Scandix nodosus L., Sp. Pl. ed. 1: 257 (1753).
Physocaulis nodosus (L.) W.D.J.Koch, Syn. Fl. Germ. Helv. ed. 2, 348 (1843); Boiss., Fl. Orient. 2: 909 (1872); Hayek, Prodr. Fl. Pen. Balc. 1: 1058 (1927); Bornmüller in Beih. Bot. Centralbl. 58B: 279 (1938); Schischkin in Fl. SSSR 16: 93 (1950); Zohary in Dep. Agr. Iraq Bull. 31: 111 (1950); Rawi, ibid. Tech. Bull. 14: 94 (1964); Mouterde, Nouv. Fl. Lib. et Syr. 2: 583 (1972); Hedge & Lamond in Fl. Iranica 162: 101 (1987).
Chaerophyllum nodosum (L.) Crantz, Cl. Umbell. emend.: 76 (1767); DC., Prodr. 4: 225 (1830).

Erect annual herb, to 60 cm or sometimes more, branched (often rather sparingly) from near base, or above in less robust individuals. Stem and branches terete, finely striated, fistular with hollow fusiform swellings below nodes at maturity (not always evident in less well-grown plants); indumentum variable, from quite glabrous to an abundant clothing of patent or deflexed, stiff, swollen-based bristles. Radical and lower stem-leaves similar, triangular-ovate in outline, up to 25 × 15 cm, petiolate, bi- to tri-pinnatisect, hairy, with broad, ovate or ovate-lanceolate segments; upper leaves reducing in size and number of divisions, sessile on diminishing sheaths, ternate-pinnate, uppermost scarcely sheathing. Umbels few, leaf-opposed and terminal, shortly (1–3.5 cm) pedunculate, lower subequalling the subtending leaves, 2–3-rayed, rays 1–3.5 cm. Partial umbels 4–9-flowered, on short filiform pedicels which elongate to as much as 1 cm and become considerably incrassate in ripe fruit, though some fruits may be sessile. Bracts of involucel 5–7, lanceolate, shorter than pedicels. Fruit when ripe 8–12 × 2–3 mm, broadest about or below middle and gradually tapering with rounded margins to the apex, densely clothed with swollen-based bristles which are very diverse in size, the longer pronouncedly upward-falcate. Fig. 70: 1–7

HAB. On the lower mountain slopes, in *Quercus* shade on limestone, on stony clay soil, in a N-facing ravine, by streams in *Quercus* woods etc.; alt. 700–1400 m; fl. & fr. May–Jun.
DISTRIB. Occasional in the middle forest zone of Iraq. **MAM**: Shaikh Adi, *Field & Lazar* 700!; **MRO**: Sefin Dagh, above Shaqlawa, *Gillett* 8137!; Shaqlawa, *Bornmüller* 1252, *Haines* W648!; between Rowanduz and Agoyan, *Kass & Nuri* 27249!; Mt. Qandil nr Pushtashan, *Rechinger* 11744; **MSU**: Qarachitan, *Gillett* 77141!; 7 km W of Tawila, *Rawi* 21850!, 22356!; **MJS**: Singara (Jabal Sinjar), *Haussknecht* s.n!

W, C & E Europe (Portugal, Spain, Corsica, Italy, Hungary, Bulgaria, mountains of Greece to Crimea), Syria, Palestine, Turkey, Caucasus, Iran, C Asia (Turkmenistan, Kazakhstan, Tadjikistan), N Africa (Morocco, Algeria).

27. **CHAEROPHYLLUM** L.

Sp. Pl. ed. 1: 258 (1753); Gen. Pl. ed. 5: 125 (1754)

Annual, biennial or perennial herbs, sometimes with a fusiform or tuberous rootstock, usually ± hairy, rarely glabrous. Leaves usually 2–4-pinnate, more rarely ternately decomposite, segments broad or narrow. Umbels compound. Involucre of very few bracts or absent; involucel of many narrow to broad and herbaceous bracts. Flowers white (rarely

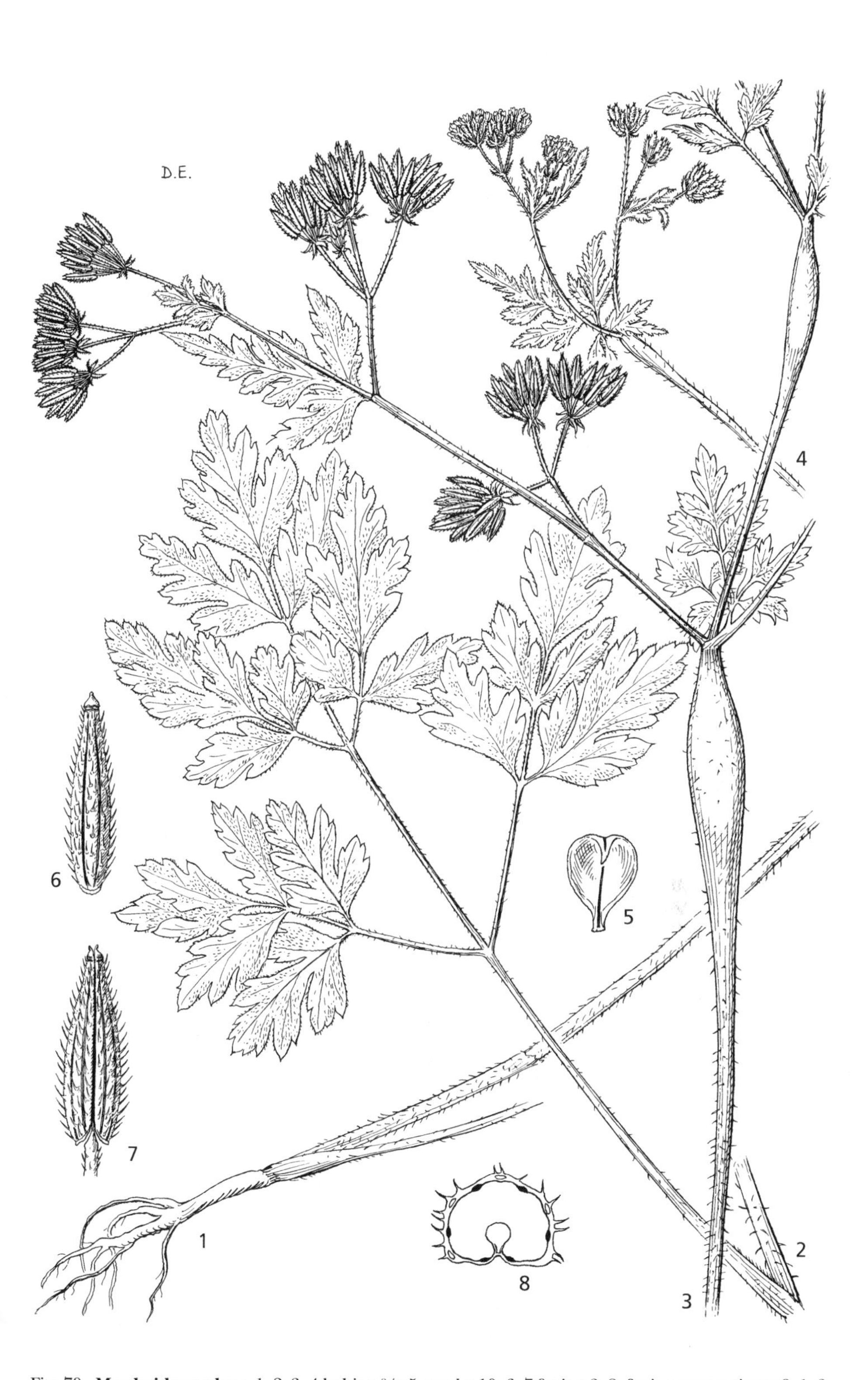

Fig. 70. **Myrrhoides nodosa**. 1, 2, 3, 4 habit × 2/3; 5 petal × 10; 6, 7 fruit × 3; 8, fruit cross section × 8. 1–3, 6–8 from *Davis* 19613; 4–5 from *Bengamin* 3/78. Drawn by D. Erasmus (1964).

yellowish), usually polygamous; petals oblong or cuneate-obovate, emarginate or subentire above, with an inflexed lobule, or outer obcordate and radiant. Calyx teeth obsolete, or subulate and persistent. Fruit oblong to linear, slightly compressed laterally, commissure narrow, ribs 5, very obtuse, mericarps most often subterete and only rarely pentagonal in section; valleculae univittate, commissure bivittate. Stylopodia small, conical, usually entire, rarely expanded into a crenate-margined disk. Carpophore simple or bifid. Endosperm subterete or somewhat dorsally compressed, with a deeply sulcate commissural face.

40 species in the N temperate region; two species in Iraq.

Chaerophyllum (for Greek χάριστος, *kharistos,* pleasant and φύλλον, *phyllon,* leaf, alluding to its agreeably scented foliage); Chervil.

One or two species of this genus are sometimes cultivated as herbs in gardens. The Bulbous-rooted Chervil (*C. bulbosum*) has turnip-like roots which are eaten as a vegetable like carrots, in southern Europe. The aromatic leaves of the Common Chervil (*Anthriscus cerifolium,* syn. *C. sativum*) are used as a salad in S Russia and the Caucasus.

Plant, 25–60 cm; stem ± 3 mm wide at base; radical leaves rarely 12 cm; flowers conspicuously radiate; ripe fruit 8–14 cm, much exceeding the short, 2–3 mm pedicel . . . . 1. *C. crinitum*

Plant ± 1 m; stem to ± 1.5 cm wide at base; radical leaves to 40 cm; flowers not or much less conspicuously radiate; ripe fruit 17–30 mm, subequalling the long, 18–30(–40) mm pedicel . . . . 2. *C. macropodum*

1. **Chaerophyllum crinitum** *Boiss.*, Ann. Sc. Nat. sér. 3, 2: 63 (1844); Boiss., Fl. Orient. 2: 904 (1872); Bornmüller in Beih. Bot. Centralbl. 58B: 279 (1938); Schischkin in Fl. SSSR 16: 112 (1950); Rawi in Dep. Agr. Iraq Tech. Bull. 14: 90 (1964); Grossh., Fl. Kavk. ed. 2, 7: 28 (1967); Hedge & Lamond in Fl. Turk. 4: 301 (1972); Hedge & Lamond in Fl. Iranica 162: 92 (1987).

C. gracile Freyn et Sint. in Freyn, Österr. Bot. Zeitschr. 42: 120 (1892); Nábělek in Publ. Fac. Sci. Univ. Masaryk 35: 124 (1923); Zohary in Dep. Agr. Iraq Bull. 31: 111 (1950); Rawi, ibid. Tech. Bull. 14: 90 (1964), non Bess. (1826).

Erect biennial herb with a ± globose tuberous rootstock, 25–60 cm high. Stem branched from near ± 3 mm wide base or about the middle, branches rather few, ascending, both stem and branches subterete, weakly striate; indumentum of spreading or deflexed long white hairs, copious or rather sparse. Radical and lower stem leaves finely tripinnate, oblong- or triangular-ovate in outline, densely and often white-hairy, long-petiolate, sheathing, rarely attaining 12 cm in length, ultimate segments 2–5 mm, narrow, shortly linear or lanceolate. Upper stem leaves similar, sessile, segments much longer and narrower. Peduncles 2.5–9(–13) cm. Umbels few or many, with 10–15(–18) long rays, rays 0.7–7 cm. Involucre absent, rarely 1–2 bracts. Partial umbels up to 20-flowered, most or commonly only 3–10 flowers producing ripe fruit, the outer strongly radiate, radiate petals 3–5 mm long; pedicel 2–3 mm. Bracteoles narrowly ovate, ciliate, 5–7 mm. Fruit 8–14 × 1.5–2 mm, linear, parallel-sided or somewhat curved above; ribs very broad and flat, much broader than valleculae. Stylopodia conical, with a crenate-margined flat disk overlapping edges of apex of fruit; styles erect, divergent, rigid, reddish-purple, ± 1.5 mm.

HAB. Mountain sides, in *Quercus* woodland and also above the tree line, on a dry limestone slope among *Quercus* scrub; alt. (900–)1100–1600(–2200) m; fl. & fr. May–Jun. or later.

DISTRIB. Occasional in the upper forest zone of Iraq, rare in the lower thorn-cushion zone. **MAM**: Sarsang, *Haines* W998!; **MRO**: Jabal Baradost, nr Diana, *Field & Lazar* 885! 902!; Kurruk Dar (Kurek Dagh), nr Rowanduz, *Nábělek* 509; on Handren, nr Rowanduz, *Nábělek* 423, 514; Sakri-Sakran, *Bornmüller* 1251, *Hadač et al.* 5553b; **MSU**: Zewiya, *Rawi* 11525!; Qopi Qaradagh, *Poore* 606!

SE Turkey, Caucasus, NW Iran.

2. **Chaerophyllum macropodum** *Boiss.*, Ann. Sc. Nat. sér. 3, 2: 64 (1844). — Type: ?Iraq, Mt. Valkou (?Kalkou), *Aucher* 3612 (K!, syn.); Boiss., Fl. Orient. 2: 904 (1872); Nábělek in Publ. Fac. Sci. Univ. Masaryk 35: 124 (1923); Bornmüller in Beih. Bot. Centralbl. 58B: 279 (1938); Schischkin in Fl. SSSR 16: 111 (1950); Rawi in Dep. Agr. Iraq Tech. Bull. 14: 90 (1964); Hedge & Lamond in Fl. Turk. 4: 315 (1972); Hedge & Lamond in Fl. Iranica 162: 89 (1987).

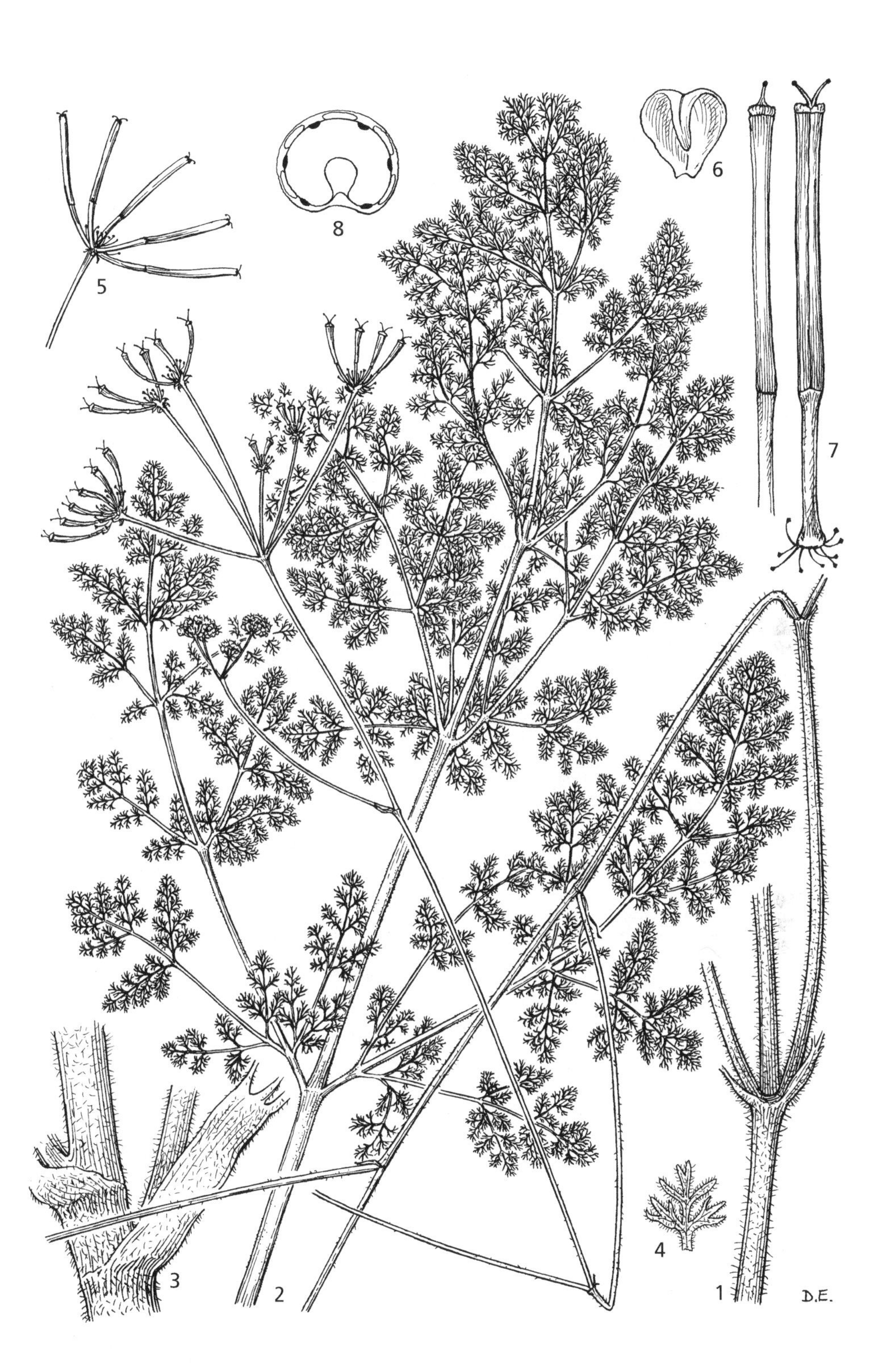

Fig. 71. **Chaerophyllum macropodum**. 1, habit × ²/₃; 2, portion with of branch with leaves × ²/₃; 3, leaf bases × ²/₃; 4, apical segments of leaf × ²/₃; 5, mature umbel × ²/₃; 6, petal ×14; 7, fruits × 2; 8, fruit cross section × 12. 1, 3 from *Furse & Synge* 429; 2, 4–8 from *Rawi* 1834. Drawn by D. Erasmus (1964).

Tall, erect (?perennial) herb attaining 1 m or more. Stem much branched from below, branches many, divaricate or ascending; upper part of stem and branches terete and finely striate, thinly clothed with spreading white hairs; lower part of stem fistular, more angular, to ± 1.5 cm wide, ± densely hairy. Radical and lower stem leaves broadly oblong- or triangular-ovate in outline, 3–4-pinnate, long-petiolate, broadly sheathing, often very large (to 40 cm long or more), generally whitish-lanate, ultimate segments very shortly and narrowly linear; upper stem-leaves reducing in size and number of divisions with a reduced quantity of indumentum. Peduncles 4–10 cm. Umbels many, with 6–10 rays 1.3–9 cm long. Involucre absent. Partial umbels 12–20-flowered, only 2–7 flowers producing ripe fruit, outer petals to ± 2.5 mm, not conspicuously radiate. Bracteoles 5–6, lanceolate-ovate to narrowly elliptic, ciliate, short. Fruit 17–30 mm long when mature, when only semi-developed distinctly broader above and clavate, later ± parallel-sided, its basal width equalled or exceeded by that of the apex of the incrassate, finally 18–30(–40) mm long pedicels. Ribs equalling the valleculae, not prominent, the fruit appearing regularly striate. Stylopodia depressed, with a flat disk not exceeding the width of the fruit. Styles not rigid, spreading or reflexed, 1–2 mm. Fig. 71: 1–8.

HAB. In the mountains and valleys, on stony slopes, in oakwoods, near a stream in valley; alt. 1000–1750 m; fl. & fr. (May–)Jun.–Jul.
DISTRIB. Quite common in the upper forest zone of Iraq. **MAM**: nr Sharanish, *Rawi* 23285!; 30 km NE of Zakho, *Rawi* 23515!; between Aradin and Ain Nuni, *Nábělek* 412; **MRO**: Zeita, nr Shirwan Mazin, *Hadač & Haines* W2413!; on Handren, nr Rowanduz, *Bornmüller* 1290!; Khalana (Kholan), *Rawi* 13834!; Haji Umran, nr Iranian frontier, *Alkas* 18591!, *Rechinger* 11328; **MSU**: 13 km N of Kani Spi, *Rawi* 22427!; Mela Kowa (Malakawa), *Rawi* 22504!, *Rawi, Husham & Nuri* 29471!; Penjwin, *Rawi* 22520!; Naragora, nr Balkha, *Rawi, Husham & Nuri* 29545!; Hawraman, above Darimar, *Gillett* 11845!; Tawila, *Rawi* 22288!; Hawraman, *Haussknecht* s.n.!; ?**MSU**: Mt. Valkou (?Kalkou), *Aucher* 3612 (type).
NOTE. MANDAKA (Kurd.-Rowanduz, *Rawi* 13834), BANAKILA (Kurd.-Hawraman).

Turkey (SE Anatolia), W & C Iran.

28. **GRAMMOSCIADIUM** DC.

Mém. Omb. (Coll. Mém. 5): 62 (1829)

Glabrous to scabrid perennial herbs with long, stout taproots. Leaves pinnately decomposite into linear segments. Umbels compound. Involucre and involucel present; bracts and bracteoles pinnately divided or the latter entire. Pedicels and sometimes rays incrassate in fruit, rays then often indurate at the point of junction with peduncle. Flowers white, polygamous, usually outer flowers of partial umbels or only those of the terminal umbel forming fruit. Petals obcordate with an inflexed lobule, outer shortly radiate. Calyx teeth present, small to conspicuous, persistent on mature fruit. Fruit linear- to oblong-cylindrical, sometimes curved; ribs 5, prominent, laterals marginal, obtuse or the laterals winged. Commissure broad, bivittate. Stylopodia shortly to long-conical; styles erect or divergent, finally ± rigid. Carpophore bipartite. Endosperm rounded or shallowly grooved dorsally, concave on the commissural face.

Seven to eight species in SW Asia, four of which occur in Iraq.
Grammosciadium (from the Gr. γραμμή, *gramma*, a line, σκιας, *skias*, an umbel, probably a reference to the grooved fruits with alternating dark vittae and pale ridges, or perhaps the long umbel rays).
Koso-Poliansky (1915), K revisy vostochnago roda *Grammosciadium* DC., Journ. Russe Bot. 1915 (1–2): 1–22; Aleshina & Vinogradova in Bot. Zhurn. 56: 963–969 (1971) gives an account of the pollen grain morphology of the species of *Grammosciadium*, with a key for pollen identification.

1. Marginal primary ribs of fruit winged .... 1. *G. platycarpum*
   Marginal primary ribs of fruit unwinged, similar to dorsal ribs 2
2. Mericarps in section with 9 vascular bundles, 5 in primary ribs and 4 above the vallecular vittae; calyx teeth in fruit long, recurved, uncinately hooked at top.................................. 5. *G. cornutum*
   Mericarps in section with 5 vascular bundles in primary ribs only; calyx teeth straight.. 3

3. Calyx teeth very strong and stout in ripe fruit, at least some subequalling styles; ribs of fruit broad and flat, mericarps in section with broad, flat, subcontiguous vascular bundles with very small white vittae in between . . . . 4. *G. macrodon*
   Calyx teeth shorter, rarely attaining half the length of styles and usually less; ribs of fruit narrow, mericarps in section with narrower vascular bundles with vittae about equally wide in between 4
4. Oil ducts of petals narrowly linear, extending from the apical notch to base of each petal; fruit mostly 12–15(–18) mm long . . . . 2. *G. scabridum*
   Oil ducts of petals elliptical or narrowly obovate, extending a short way down from the apical notch; fruit 8–10(–12) mm long . . . . 3. *G. daucoides*

Subgenus 1. CAROPODIUM (Stapf et Wettst.) Tamamshyan & Vinogradova. Marginal ribs of fruit winged. Calyx teeth small, non-accrescent.

1. **Grammosciadium platycarpum** *Boiss. et Hausskn.*, Fl. Orient. 2: 901 (1872); Tamamshyan et Vinogradova in Bot. Zhurn. 54: 1208 (1969) & in Taxon 18: 547 (1969); Hedge & Lamond in Fl. Turk. 4: 320 (1972).

Caropodium meoides Stapf & Wettst. in Denkschr. Akad. Wiss. Wien, math.-nat. Kl. 51: 317 (1886), repr. Bot. Ergeb. Polak'schen Exped. Pers. 1882, Th. 2: 49 (1886).
C. platycarpum (Boiss. et Hausskn.) Schischkin in Not. Syst. Herb. Hort. Petrop. 4: 30 (1923) & Fl. SSSR 16: 124 (1950); Rawi in Dep. Agr. Iraq Tech. Bull. 14: 89 (1964).

Plant erect, to ± 45 cm high but more often 20–30 cm, much-branched from near base. Stem and branches strongly striate or sulcate, often angled below, glabrous. Basal leaves shortly petiolate with a long, expanded sheath, lanceolate to narrowly oblong-ovate in outline, 6–12 × 0.8–2 cm, bi- or tri-pinnate with ultimate segments setaceous and ± congested, ± 2–5 mm long; upper leaves reducing, soon sessile on the shorter sheaths, broader and bushier with segments up to ± 15 mm long. Peduncles 3–13 cm. Umbels numerous, large, 9–19-rayed, rays bearing fruit incrassate, 1–7 cm; bracts of involucre 4 or 5(–7), pinnate or bipinnate, half the length of fruiting rays or less. Partial umbels 15–20-flowered, some or all producing 3–12 fruit or some with no fructiferous flower; pedicels 1–3 mm, those bearing fruit incrassate; bracts of the involucel 3–5, pinnate or setaceous. Fruit narrowly linear-oblong, 10–18(–22) mm long with a narrow (± 0.5 mm) wing, width across the major axis 2–3 mm; persistent sepals minute. Stylopodia conical; styles divergent or deflexed, ± 1 mm. Fig. 72: 1–4.

HAB. In mountains, in *Quercus* forest, on cultivated ground at the foot of hill, near stream; alt. 1300–1650 m; fl. & fr. Jun.
DISTRIB. Rare in Iraq, only found in two districts, both near the Iranian frontier, in the central and SW sectors respectively of the upper forest zone. **MRO**: Haji Umran, *Rawi, Nuri & Kass* 27771!; **MSU**: Mela Kowa (Malakawa Pass) nr Penjwin, *Rawi* 22487!; Penjwin, *Rawi* 12222! 22631! *Rechinger* 10522!

E Turkey, Caucasus, NW Iran. According to Schischkin (1950) this plant was found to have a high yield of essential oils, more even than coriander (*Coriandrum sativum* L., q.v.) and its widespread cultivation in Azerbaijan has been recommended.

Subgenus 2. GRAMMOSCIADIUM. Marginal ribs of fruit not winged; calyx teeth accrescent and conspicuous in fruit.

2. **Grammosciadium scabridum** *Boiss.* in Ann. Sc. Nat. sér. 3, 2: 66 (1844); Boiss., Fl. Orient. 2: 822 (1872); Tamamschian et Vinogradova in Bot. Zhurn. 54: 1204 (1969) & in Taxon 18: 547 (1969); Tamamschian in Fl. Iranica 162: 97 (1987).

G. longilobum Boiss. et Hausskn. in Boiss., Fl. Orient. 2: 900 (1872). — Type; Iraq, Pira Magrun, *Haussknecht* s.n.; Nábělek in Publ. Fac. Sci. Univ. Masaryk 35: 124 (1923); Bornmüller in Beih. Bot. Centralbl. 58B: 279 (1938); Parsa, Fl. Iran 2; 748 (1948); Zohary in Dep. Agr. Iraq Bull. 31: 111 (1950); Rawi, ibid. Tech. Bull. 14: 92 (1964).

Plant erect, 25–65 cm, sparingly to much-branched from near base or only above. Stem and branches striate but not angled, glabrous. Basal leaves shortly petiolate with a rather

Fig. 72. **Grammosciadium platycarpum**. 1, habit × ²/₃; 2, petal × 6; 3, fruits (both faces) × 2; 4, fruit cross section × 6. 1–3 from *Rawi* 27771; 4 from *Davis & Hedge* D29316. Drawn by D. Erasmus (1964).

narrow sheath, lanceolate to narrowly oblong-ovate in outline, 8–18 × 1.5–3 cm, bi- or tripinnate with ultimate segments setaceous and ± congested, 2–5 mm long; upper leaves reducing, soon sessile on the shorter sheaths, broader and bushier with segments up to 15 mm long. Peduncles 5–14 cm long. Umbels several, large, 6–16-rayed, rays bearing fruit not incrassate, 3–11 cm; bracts of involucre 2–5, pinnate or bipinnate, less than half the length of fruiting rays. Partial umbels 10–20-flowered, some producing 1–7 fruit, some non-fructiferous; pedicels 2–6 mm, those bearing fruit incrassate; bracts of involucel 3–6, pinnate or setaceous. Fruit linear, 12–15(18) × ± 1.5 mm; persistent sepals 0.5–1(–2) mm. Stylopodia conical; styles rigid, divergent, 2–3 mm.

HAB. Dry rocky mountain slopes, in *Quercus* forest on limestone, in coppiced *Quercus* shade and grassy places, on stony red soil above the tree line; alt. (800–) 1200–1900(–2200) m; fl. & fr. May–Jun.
DISTRIB. Quite common in the upper forest zone of Iraq, rare in the lower thorn-cushion zone. **MAM**: Matina, *Rawi* 8678!; Ispindari saddle, Sawara Tuka, *E Chapman* *26319!; Sarsang, *Haines* *W1000!; **MRO**: Kuh-i Sefin, *Bornmüller* 1249! *Gillett* 8106!; Kurruk Dar (Kurek Dagh), nr Rowanduz, *Nábělek* 510; Handren, above Rowanduz, *Nábělek* 428; S slope of Karoukh, *Kass & Nuri* 27503!; Gali Warta, ± 30 km NW by N of Rania, *Rawi, Nuri & Kass* 38773!; foot of Mt. Botin, by Kani Shirin spring nr Zeyta, *Agnew et al.* 5973; **MSU**: Pira Magrun, *Haussknecht* s.n. (type of *G. longilobum*), *Rawi* 12109!; between Ja'faran and Qaradagh, *Gillett* 7883!; Hawraman, above Darimar, *Gillett* 11850! Hawraman, nr Tawila, *Rechinger* 12437.
NOTE. The two specimens marked * probably belong here but lack fruit.

Turkey, Caucasus, NW Iran.

3. **Grammosciadium daucoides** *DC.*, Mém. Omb.: 62 (1829) & Prodr. 4: 233 (1830); Boiss., Fl. Orient. 2: 899 (1872); Koso-Poliansky in Journ. . Russ. Bot. 1915 (1–2): 12, f. 5 (1915); Schischkin in Fl. SSSR 16: 120 (1950); Rawi in Dep. Agr. Iraq Tech. Bull. 14: 92 (1964); Grossh., Fl. Kavk. ed. 2, 7: 32 (1967); Tamamshyan et Vinogradova in Bot. Zhurn. 54: 1203 (1969) & in Taxon 18: 546 (1969); Hedge & Lamond in Fl. Turk. 4: 319 (1972).

G. szowitsii Boiss. in Ann. Sc. Nat. sér. 3, 2: 67 (1844).
G. aucheri Boiss., ibid.: 67 (1844); *G. aucheri* subsp. *pauciradiatum* Freyn & Sint. ex Freyn in Öst. Bot Zeit. 42: 128 (1892).

Differs from *G. scabridum* in the characters given in the key and in the length of umbel-rays, which rarely attain as much as 7 cm; the overlap in this character is however considerable. The fructiferous partial umbels of *G. daucoides* produce more (mostly 6–18) fruits.

NOTE. Tamamshyan et Vinogradova in Taxon 18: 546 (1969) corrected my error in performing one too many reductions by reducing *G. scabridum* (including *G. longilobum*) to the synonymy of *G. daucoides* [Townsend in Kew Bull. 20: 83–85 (1966)]. However, Hedge & Lamond in the Flora of Turkey continue to cite *G. daucoides* as an Iraqi species. All the material of this affinity which I have seen from Iraq is referable to *G. scabridum*, but *G. daucoides* may occur and is included here in order that it may be sought by future collectors. At present *G. daucoides* is known only from Turkey, Syria and Transcaucasia (see also Fl. Iranica 162: 97, 1987).

4. **Grammosciadium macrodon** *Boiss.* in Ann. Sc. Nat. sér. 3, 2: 67 (1844); Boiss., Fl. Orient. 2: 900 (1872); Koso-Poliansky in Journ. Russ. Bot. 1915 (1–2): 12, f. 2(1915); Rawi in Dep. Agr. Iraq Tech. Bull. 14: 92 (1964); Tamamshyan et Vinogradova in Bot. Zhurn. 54: 1205 (1969) & in Taxon 19: 652 (1970); Hedge & Lamond in Fl. Turk. 4: 319 (1972); Tamamschian in Rech.f., Fl. Iranica 162: 98 (1987).

Similar to *G. scabridum* and *G. daucoides* in all but fruit anatomy. Fructiferous partial umbels with 2–6(–8) fruits 12–20 cm in length as in *G. scabridum*, fruits crowned with the stout, straight teeth of calyx, at least one or two of which equal or exceed styles. Mericarps with 5 low, broad and obtuse primary ribs, pericarp in section with 5 broad, flat vascular bundles separated only by a small space containing a single tiny vitta; a similar small vitta is recessed into the dorsal surface of each vascular bundle. Styles 0.3–4 mm. Fig. 73: 1–4.

HAB. not recorded.
DISTRIB. Very rare, if at all it occurs in Iraq; only one rather dubious record, well over a century ago. ?**MAM** (or in Turkey): "Pl. Mesopot., Kurdistan & Mossul", *Kotschy* 178!

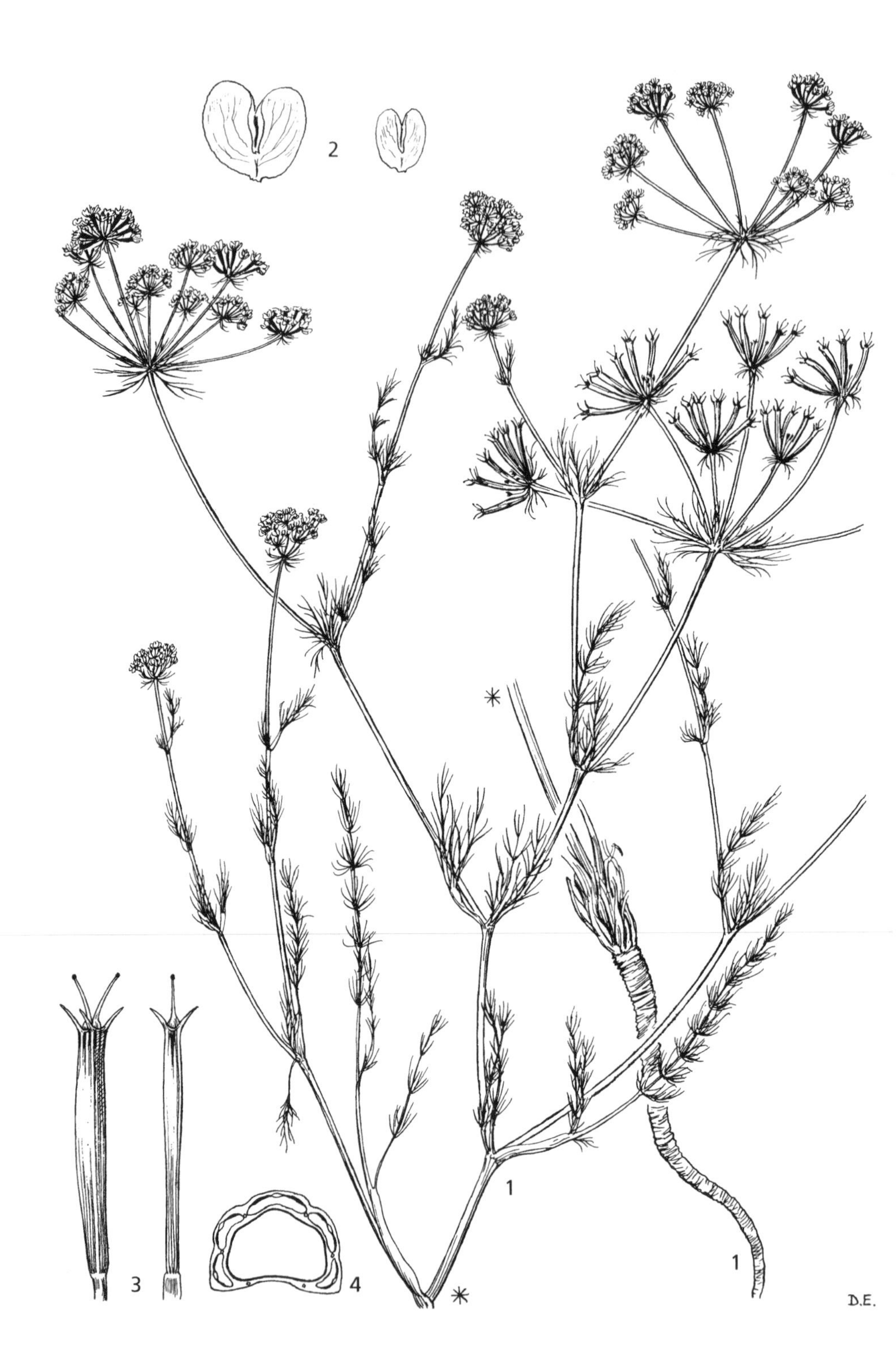

Fig. 73. **Grammosciadium macrodon**. 1, habit × 2/3; 2, petal × 6; 3, fruits (both faces) × 2; 4, fruit cross section × 6. 1from *Sintences* 1888; 2–4 from no. 1517. Drawn by D. Erasmus (1964).

5. **Grammosciadium cornutum** (*Nab.*) *C.C. Townsend* in Kew Bull. 20: 83 (1966); Tamamshyan & Vinogradova in Bot. Zhurn. 54: 1205 (1969) & in Taxon 18: 547 (1969); Hedge & Lamond in Fl. Turk. 4: 319 (1972); Tamamschian in Fl. Iranica 162: 98 (1987).

G. macrodon Boiss. var. *cornutum* Nábělek in Publ. Fac. Sci. Univ. Masaryk 35: 124 (1923); Blakelock in Kew Bull. 3: 434 (1948); Rawi in Dep. Agr. Iraq Tech. Bull. 14: 92 (1964).

Similar to *G. macrodon, G. scabridum* and *G. daucoides* in all but fruiting characters. Fructiferous partial umbels bearing 2–17 fruits, 11–17 mm long, crowned with the greatly elongated, recurved calyx teeth, which may attain half the length of fruit or more and are uncinately hooked at apex. Mericarps with 5 sharply defined primary ribs and 4 more slender lines in the furrows between, in section with 9 subequal vascular bundles, 5 ± rounded in the primary ribs, each with a small dorsal vitta, and 4 more angled in the valleculae, each with a ventral vitta.

HAB. Oakwoods in the mountains, alt. 1200–1500 m; fl. & fr. Jun.–Jul.
DISTRIB. Rare in Iraq; only found in the NW corner of the upper forest zone. **MAM**: Sevara Gaurik valley NE of Zakho, *Rawi* 23677!; nr Amadiya, *Majid Mustafa* 3606!; nr Sharanish in mts. N of Zakho towards Jabal Khantur, *Rechinger* 10815.
NOTE. KAMMUN (Kurd.-Amadiya, *Majid Mustafa* 3606).

Turkey (SE Anatolia).

29. PIMPINELLA L.

Sp. Pl. ed. 1: 263 (1753); Gen. Pl. ed. 5: 128 (1754); Wolff in Pflanzenr. IV. 228, 90: 219 (1927)

Perennial, more rarely annual or biennial herbs, diverse as to stature, leaves pinnately divided, the lower often entire or simply pinnate, the upper usually multisect. Umbels compound. Involucre absent, involucel absent or present. Flowers white, pink, yellow or more rarely deep red. Calyx teeth obsolete or small. Petals variable in form, above deeply emarginate to almost entire, with an inflexed lobule variously developed, glabrous or hairy dorsally. Fruit ovoid to ovoid-globose, ovoid-oblong or oblong, usually broadest below middle and ± attenuate toward apex, glabrous or furnished with diverse forms of indumentum, laterally compressed. Commissure broad. Mericarps easily separating, ± didymous, subterete, ribs filiform, not or scarcely prominent, usually pale, narrower than valleculae but often hidden beneath hairy or other indumentum. Vittae most frequently 3 in the valleculae, occasionally 2 or 4 (rarely more), slender or conspicuous. Stylopodia broad and rounded-convex to shortly- or elongate-conic, the styles variable in length, margin of disk entire. Carpophore bifid or bipartite. Endosperm convex, ± terete or bluntly pentagonal dorsally, the commissural face flat or slightly concave.

Some 150 species in N temperate Europe and Asia, N Africa, with one N American Pacific species and a few species in S America; 17 species in Iraq.

Ref.: Axenov E.S., Tikhomirov, V.N. (1972). Key for the identification of Pimpinella species on the basis of fruit characters. Byull. Glavn. Bot. Sada (Moskva) 85: 35–45.

1. Flowers yellow; all leaves with broad, acuminate teeth, ± reniform or orbicular, undivided or at most somewhat trilobed, the upper reduced and sheath-like .... 1. *P. nephrophylla*
   Flowers white or pinkish or reddish; at most the very lowest leaves undivided, the remainder variously pinnatisect........ 2
2. Receptacle and fruit glabrous........ 3
   Receptacle and fruit hairy or papillose 4
3. Basal leaves 2–3-pinnate; fruit ovoid-oblong .... 17. *P. anthriscoides*
   Basal leaves simply pinnate; fruit ovoid .... 14. *P. saxifraga*
4. Plant annual........ 5
   Plant biennial or perennial........ 9
5. All leaves, including the lowest, bipinnatisect or pinnate into long, narrowly linear segments........ 6. *P. barbata*
   Lower leaves pinnate or ternate with broad, toothed segments; only the upper stem leaves, if any, narrowly dissected........ 6

6. Fruit 3–5 mm long at maturity, flask-shaped and broadest below the middle, with short, appressed hairs . . . . .2. *P. anisum*
Fruit smaller, ± 1.5 mm or less 7
7. Stylopodia long-conic, ± attenuate into the styles (typically with the hairs of the fruit long and spreading, longer than the diameter of the fruit, deflexed and then upwardly arcuate . . . . 5. *P. eriocarpa*
Stylopodia convex, obtuse, not attenuate into the styles (hairs of the fruit always shorter than its diameter, ± appressed upwardly) 8
8. Plant cinereous with white pubescence; upper leaves bi- or tri-pinnatisect, the segments very widely divaricate (almost at 90°) with a short broad sheathing base not as long as the lamina; fruit ± densely covered with longer white hairs; rays of main umbel patent-hairy, at least below . . . . 4. *P. puberula*
Plant green; uppermost leaves trisect with narrowly linear non-divaricate segments and a long narrow sheath subequalling or sometimes exceeding the lamina; fruit rather thinly covered with short hairs; rays of main umbel almost invariably quite glabrous. . . . . .3. *P. cretica*
9. Diameter of fruiting umbel not exceeding 2.5 cm, on short secondary branches 10
Diameter of fruiting umbel exceeding 2.5 cm 11
10. Plant clothed with ash-white indumentum; fruit patent-hairy . . . . 11. *P. brachyclada*
Plant sparsely pubescent, glabrous above; fruit glabrous or slightly pilose . . . . .12. *P. hadacii*
11. Involucre and involucel absent. 12
Involucre and involucel distinct 15
12. Plant densely cinereous-pubescent, especially in the region of the inflorescence, the hairs of the umbel rays exceeding the diameter of the rays . . . . .13. *P. kurdica*
Plant green to slightly whitish, the pubescence short, hairs of the rays not exceeding their diameter in length 13
13. Fruit very small, ± 1 mm, as broad as long or almost so . . . . 7. *P. peregrina*
Fruit larger, longer than broad, at least 2–3 cm long 14
14. Umbel 6–10-rayed . . . . 15. *P. tragium*
Umbel 15–20-rayed . . . . 16. *P. zagrosica*
15. Involucel very conspicuous, of 10–12 broadly linear bracts, as long as the rays of the partial umbel, ± 7 mm long; petals chartaceous with age, densely tomentose over the entire dorsal surface . . . . .9. *P. olivierioides*
Involucel not so conspicuous, bracteoles fewer and shorter; petals when fully developed hairy only basally or along the centre line, not chartaceous with age. 16
16. Fruit hairy; all inflorescence rays hairy; styles hairy . . . . 8. *P. kotschyana*
Fruit papillose; inflorescence rays glabrous or minutely papillose; styles glabrous. . . . . 10. *P. olivieri*

Section 1. REUTERA (Boiss.) Benth.

Flowers yellow or yellowish. Petals either involute or inflexed at the tip, entire or emarginate above, or with a well-developed inflexed lobule and deeply excised.

1. **Pimpinella nephrophylla** *Rech.f. et Riedl*, Anz. Math.-Nat. Kl. Österr. Akad. Wiss. 98: 226 (1961). — Type: Iraq, Jabal Khantur near Turkish frontier, *Rechinger* 10737a (W!, holo.); Townsend in Kew Bull. 17: 428 (1964); Engstrand in Fl. Iranica 162: 313 (1987).

P. sintenisii Wolff var. *cinerea* Blakelock in Kew Bull. 3: 435 (1948). — Type: Iraq, Mazurka gorge, *Guest* 3783 (type of var. *cinerea*); Rawi in Dep. Agr. Iraq Tech. Bull. 14: 94 (1964).

Perennial, to 20 cm, with a suffrutescent base, leaf-bases of the previous year's growth persistent; ashy-white-pubescent throughout, only the fruit becoming subglabrous or glabrous as it matures. Stem and the few branches wiry, terete, striate. Basal leaves long-petiolate, lamina 7–25 × 8–25 mm, cordate at the base, ± reniform, orbicular or slightly longer than broad in outline, sometimes obscurely 3-lobed, but sinus between lobes scarcely more than a usually deep incision, coarsely and irregularly toothed, teeth long-acuminate, mucronate, up to 24 in number. Lower stem leaves 1 or 2(rarely 3), long-petiolate, lamina very small, roundish, with 3–5 large teeth. Upper stem leaves reduced to sheaths. Peduncles 7–15 mm. Umbels triradiate, small, rays equal, 4–7 mm, involucre absent. Partial umbels 7-flowered, pedicels 1–2 mm; involucel absent. Flowers yellow, small (1.5 mm in diameter). Fruit ovoid, ± 2 mm, ribs scarcely prominent, becoming glabrous. Stylopodia long-conical, tapering into ± rigid, erect or slightly divergent styles.

HAB. Rocky places in the mountains, fissures in limestone rock, cleft in cliff in a gorge; alt. 1200–1650 m; fl. & fr. Aug.
DISTRIB. Rare in Iraq; only found three times (two of which were in neighbouring localities within 1–2 km of Amadiya) in the NW sector of the upper forest zone. **MAM**: Mazurka gorge, *Guest* 3783 (type of var. *cinerea*)!; Sulaf, *Haines* W2065; **MAM**: Jabal Khantur nr Turkish frontier, *Rechinger* 10737a (type).

Endemic.

Section 2. TRAGIUM (Spreng.) DC.

Flowers usually white, rarely pink or reddish. Petals always deeply excised above with a well-developed inflexed lobule. Fruit furnished with hairs, granules, tubercles, papillae or other form of indumentum, rarely glabrescent at maturity (not in Iraqi species).

2. **Pimpinella anisum** *L.*, Sp. Pl. ed. 1: 264 (1753); DC., Prodr. 4: 122 (1830); Boiss., Fl. Orient. 2: 866 (1872); Wolff in Pflanzenr. IV. 228, 90: 234 (1927); Hayek, Prodr. Fl. Balc. 1: 999 (1927); Zohary, Fl. Palest. ed. 2, 1: 528 (1933); Guest in Dep. Agr. Iraq Bull. 27: 73 (1933); Anth. in Notes Roy. Bot. Gard. Edinb. 289 (1935); Rawi & Chakr. in Dep. Agr. Iraq Tech. Bull. 15: 74 (1964); Matthews in Fl. Turk. 4: 355 (1972); Zohary, Fl. Palaest. 2: 425 (1972); Husain & Kasim, Cult. Pl. Iraq 106 (1975); Meikle, Fl. Cyprus 1: 745 (1977); Pu Fading & Watson in Fl. China 14: 95 (2005).

Anisum vulgare Gaertn., De Fruct. 1: 102 (1788); Schischkin in Fl. SSSR 16: 445 (1950).

Robust annual, 18–75 cm, considerably branched above and with a few branches nearer the base. Stem and branches terete, striate or sulcate, minutely downy. Extreme basal leaves undivided or shortly trilobed, ovate-rotund to cuneate-obovate in outline, coarsely and irregularly toothed, cuneate at base; other basal leaves ternate with similar broad, deeply-toothed and cuneate-based segments; all long-petiolate. Lower stem leaves pinnate with 1 or 2 pairs of incised, cuneate-obovate leaflets; upper stem leaves sessile, pinnate or bipinnate, with narrow ultimate segments, sheathing, with a ± well-developed membranous web between the sheath and the lowest pair of pinnae. All leaves green, subglabrous or sparsely and minutely puberulent. Peduncles 2–8 cm, with 10–20 finally ± incurved, subequal, minutely downy 1–3.5 cm rays. Partial umbels 10–20-flowered, flowers ± 1.5 mm in diameter, petals glabrous, not radiate, pedicels 1.5–4 mm, downy but glabrescent. Involucre and involucel usually absent, but 1 subulate bract or bracteole usually to be found on a few inflorescences. Fruit flask-shaped, broadest below the middle and attenuate above, expanding again slightly at the tip, 3–4 mm long and brownish when ripe, covered with short, appressed grey pubescence; ribs pale and rather conspicuous. Vittae 4–8 in the valleculae, almost forming an annulus. Stylopodia shortly conical, the margins rugose below; styles slender, about half as long as fruit or more, flexuose-divergent or reflexed.

HAB. & DISTRIB. Apparently very rare in Iraq, the only record of its ever having been found presumably growing spontaneously in our territory being almost 90 years ago (**FKI/FNI**: Jabal Hamrin, *Anthony*, l.c.). The specimen has not been located and all records based on Anthony's list alone are doubtful. There is however evidence that the species has been cultivated in comparatively recent years on the Agricultural Experimental Station nr Baghdad (**LCA**: Abu Ghraib (cult.), *Janan, Sahira & Omar* C.188!).
NOTE. This species is widespread in Europe, the Mediterranean region, western Asia and China (Xinjiang), but at least for the most part as an escape – though it is believed to be of Asiatic origin; it also occurs as an adventive in many other parts of the world. Shishkin (1950) reported that this species was widely cultivated in C Asia, as well as in W & SW Russia and Caucasus, adding that it is known only in cultivation there. So far as our own territory is concerned Guest (1933) stated that it was "said to be cultivated sometimes in the north, at Mosul and elsewhere" while Hussain and Kasim (1975) were perhaps nearer the mark in noting that there is "no definite report for the region of its cultivation" in Iraq. It is significant that in spite of the many botanists who have collected specimens all over our

territory during the past fifty years or more this species has not been found, whether in cultivation or adventive. Thus, until and unless specimens are collected in different parts of the country, we doubt whether the plant occurs in Iraq at all.

Anise, ĀNĪSŪN (Ar., Pers. – sometimes also known as ĀNISŪN or YĀNÎSŪN), RAZYANA (Kurd., Wahby & Edmonds). The fruits of the plant, Aniseed, are used in flavouring, particularly to flavour the local alcoholic grape-spirit in the N; this is known as ZAHLĀWI (after the Lebanese town of Zahle, a famous centre for the production of this type of 'araq') in distinction with the date-spirit of southern Iraq called MUSTAKIA which is flavoured with mastic, obtained from *Pistacia lentiscus* (see Fl. Iraq, vol. 4). Guest (1933) mentioned that, apart from any anise which may be grown locally, aniseed was imported into Iraq and on sale in the local markets. There was however some confusion between aniseed and the seed of the dill (*Anisum graveolens* q.v.), a plant of similar smell and appearance, also used as a condiment, as a vegetable and, medicinally; though the two plants can readily be distinguished by the layman by the white flowers of anise and the yellow flowers of dill.

According to Wren (1956) aniseed was known to the ancients in many parts of the world and valued as a carminative. It is also used in cough medicines and lozenges as a pectoral and the powdered seeds are employed in conditioning condiments for horses. Rawi & Chakravarty (1964) add that it also possesses diuretic and diaphoretic properties and is used as a flavourant in some bakery products. Campbell Thompson (1949) considers that the name SAMRĀNU in the ancient Assyrian Medical Texts probably refers to anise.

3. **Pimpinella cretica** *Poir.* in Lam., Encycl. Méth. Bot. Suppl. 1: 684 (1810); DC., Prodr. 4: 122 (1830); Boiss., Fl. Orient. 2: 866 (1872); Wolff in Pflanzenr. IV. 228, 90: 234 (1927); Hayek, Prodr. Fl. Balc. 1: 999 (1927); Zohary, Fl. Palest. ed. 2, 1: 518 (1933); Anth. in Notes Roy. Bot. Gard. Edinb. 18: 289 (1935); Rawi in Dep. Agr. Iraq Tech. Bull. 14: 94 (1964); Mouterde, Nouv. Fl. Lib. et Syr. 2: 606 (1970); Matthews in Fl. Turk. 4: 356 (1972); Zohary, Fl. Palaest. 2: 426 (1972); V. Täckh., Stud. Fl. Egypt. ed. 2: 391 (1974); Meikle, Fl. Cyprus 1: 745 (1977).

Slender, wiry annual (6–)25–30(–50) cm, green, sparsely and minutely puberulous or subglabrous. Stem branched from base upwards, branches long, ascending. Stem and branches terete, striate. Basal and often lower stem leaves entire or obscurely 3–lobed, ovate-rotund in outline, base cuneate to subcordate, margins crenate-dentate; middle stem leaves ternate, segments dentate or occasionally 3-lobed; upper stem leaves ternate, segments narrowly linear, entire, sessile, base with a long narrow sheath subequalling or exceeding the length of lamina. Peduncles 0.6–3(–4) cm. Umbels (4–)6–10(–12)-rayed, the rays 0.6–3 cm, almost always quite glabrous. Partial umbels 7–15-flowered, pedicels glabrous, (0–)1–4 mm, the outer flowers shortly radiant, ± 2 mm in diameter; petals glabrous. Involucre and involucel absent. Fruit ± 1.5 mm, ovoid to subrotund, with appressed, very short whitish hairs; ribs obsolete, inconspicuous. Stylopodia depressed-conical, not tapering into the fine, flexuous styles which are caducous and rarely to be found on ripe fruit. Vittae 3–5 in the valleculae.

HAB. & DISTRIB. Very rare in Iraq, if in fact it occurs. There is only one record of its occurrence, in the NE sector of the steppe region (**FNI**: Kizil Robat (Sa'diya), *Anth.*, l.c.) but no specimen has been seen and as with the previous species this may be a misidentification.

Greece, Crete, Aegean Isles, Cyprus, Turkey, Syria, Lebanon, Palestine, Jordan, Sinai, Arabia.

4. **Pimpinella puberula** (*DC.*) *Boiss.*, Ann. Sci. Nat. sér. Bot. 3: 129 (1844); Boiss., Fl. Orient. 2: 873 (1872); Wolff in Pflanzenr. IV. 228, 90: 235 (1927); Zohary, Fl. Palest. ed. 2, 1: 519 (1933); Guest in Dep. Agr. Iraq Bull. 27: 73 (1933); Anth. in Notes Roy. Bot. Gard. Edinb. 18: 289 (1935); Bornmüller in Beih. Bot. Centralbl. 58B: 276 (1938); Blakelock in Kew Bull. 3: 435 (1948); Schischkin in Fl. SSSR 16: 442 (1950); Zohary in Dep. Agr. Iraq Bull. 31: 110 (1950); V. Dickson, Wild Flowers of Kuwait & Bahrain: 494 (1955); Riedl in Fl. Lowland Iraq: 465 (1964); Rawi in Dep. Agr. Iraq Tech. Bull. 14: 94 (1964); Mouterde, Nouv. Fl. Lib. et Syr. 2: 607 (1970); Matthews in Fl. Turk. 4: 356 (1972); E. Nasir in Fl. Pak. 20: 62 (1972); Engstrand in Fl. Iranica 162: 317 (1987); Pu Fading & Watson in Fl. China 14: 95 (2005).

Ptychotis puberula DC., Prodr. 4: 109 (1830).

Slender, wiry annual, (8–)15–60 cm, ± dusty whitish from the short tomentellous indumentum with which the plant is furnished throughout, branched from below the middle and frequently from near the base. Stem and branches terete, striate. Basal leaves entire or obscurely 3–lobed, ovate-rotund in outline, toothed all round, usually cordate at

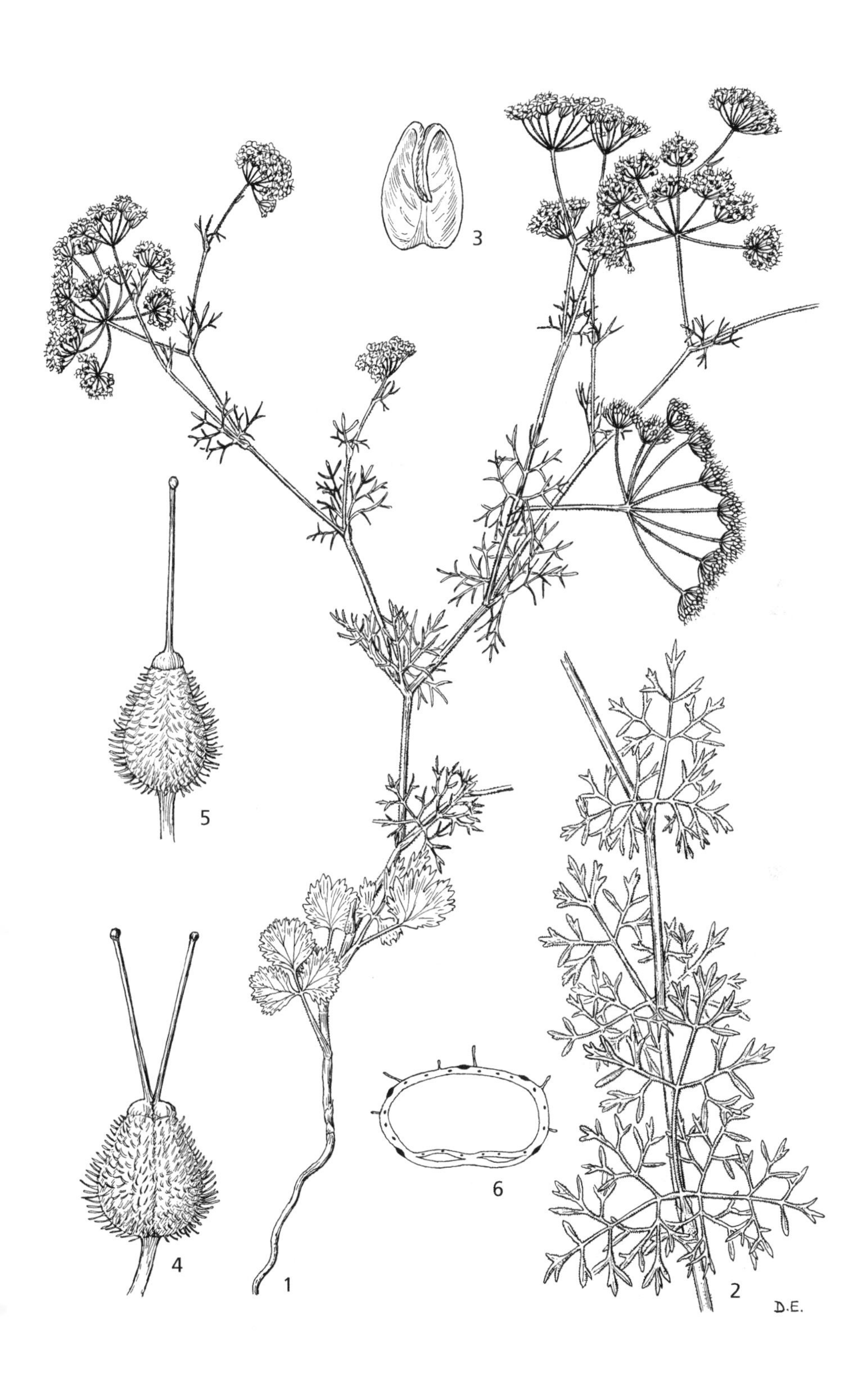

Fig. 74. **Pimpinella puberula.** 1, habit × ²/₃; 2, leaf × ²/₃; 3, petal × 12; 4, fruit × 12; 5, mericarp × 12; 6, fruit cross section × 18. All from *Rechinger* 9798. Drawn by D. Erasmus (1964).

base, long-petiolate. Stem leaves first long-petiolate, ternate, with toothed but undivided, cuneate-based segments; becoming shortly petiolate and finally sessile, bi- or subtri-pinnate, segments widely divaricate (almost at 90°) from their axes. All leaves sheathing, sheaths broad, oblong, even in the uppermost leaves much shorter than lamina. Peduncles 1.6–4.5 cm. Umbels (4–)10–15-rayed; rays 1–2(–2.5) cm, very shortly patent-hirtellous at least at the base and usually throughout, as are the pedicels of the 7–15-flowered partial umbels, outer flowers shortly radiate, 2 mm in diameter; petals glabrous or slightly hairy dorsally. Involucre and involucel absent. Fruit 1–1.25 mm, ovoid-subrotund slightly narrowed above, ± densely covered with rather short loosely adpressed or subpatent white hairs; ribs indistinct. Stylopodia mammillate; styles long, flexuose, considerably longer than fruit, persistent. Fig. 74: 1–6.

HAB. Lower mountains in light *Quercus* scrub, by spring, near waterfalls, in fields, on the upper plains, in steppic grassland, sandy gravel hillside, on sandy patches and dry wadi beds in the desert, in fields and along ditches on irrigated alluvial plains; alt. (up to 200–)250–1150(–1400) m; fl. & fr. (Apr.–) May–Jun.
DISTRIB. Occasional in the lower forest zone and steppe region of Iraq, and also in the NW and SE sectors of the desert region. **MAM**: "Pl. Mesopot., Kurdistan & Mossul", *Kotschy* 343!; Khaira, NE of Zakho, *Rawi* 23257!; nr Sharanish, *Rawi* 23267!; Sarsang, *Haines* W419!; 15 km W of Amadiya, *Rechinger* 11632; **MRO**: between Arbil and Rowanduz, *Bornmüller* 124!; Rowanduz gorge, *Guest* 487!, 2990!, *Rechinger* 11834; 4 km SE of Mergasur, *Gillett* 9626!; **MSU**: (or in Iran), "Montes Avroman et Schahu", *Haussknecht* s.n; ?"deserto Assyriae", *Aucher* 2682; **FUJ**: Tal Kochak, *Gillett* 10817!; **FNI**: by Mindan bridge, E *Chapman* 26228!; **FNI**: between Haji Lar and Kani Kirmaj, *Poore* 434!; Jabal Hamrin, *Sutherland* 292; Koma Sang, *Rawi* 20651!; 10 km E of Mandali, *Rechinger* 12779; 54–55 km SE of Mandali, *Rawi* 20730! *Rechinger* 9677; Chlat (Chilat), *Rawi & Haddad* 25699!; **DWD**: 2 km E of Rutba, *Rechinger* 12813; 260 km W of Ramadi along road to Rutba, *Rawi* 20990!; 160–190 km W of Ramadi, *Rawi* 20901! 31340! *Rechinger* 9788, 9798!; Wadi al-Khirr, *Guest, Rawi & Rechinger* 19415!; **DSD**: 15 km W of Lussuf, *Rechinger* 13573; 15 km SE of Ashuriya, *Guest, Rawi & Rechinger* 19355!, *Rechinger* 13573; 40 km NNW of Shabicha, *Rechinger* 13636; 13 km W by N of Shabicha, *Guest, Rawi & Rechinger* 19285!; c. 28 km W by N of Busaiya, *Guest & Rawi* 14171A!; 27–46 km WNW of Ansab, *Rechinger* 13800; 25 km N of Ansab, *id.* 9343; **LCA**: nr Baghdad, *Olivier & Bruguière* s.n.; **LBA**: Basra, *Memerian* 5158!; Dura village, between Basra and Fao, *Alizzi & Omar* 35866!; 17 km NW of Fao, *Alizzi & Khatib* 33407!
NOTE. Said to be "eaten by camels" in Arabia (*Dickson* 494 in K).

Turkey, Syria, Iran, Kuwait, Arabia, Afghanistan, C Asia (Turkmenistan, Kazakhstan, Siberia).

5. **Pimpinella eriocarpa** *Soland.* in Russell, Nat. Hist. Aleppo ed. 2, 2: 249 (1794); Boiss., Fl. Orient. 2: 867 (1872); Nábělek in Publ. Fac. Sci Univ. Masaryk 35: 122 (1923); Wolff in Pflanzenr. IV. 228, 90: 236 (1927); Zohary, Fl. Palest. ed. 2, 1: 519 (1933); Anth. in Notes Roy. Bot. Gard. Edinb. 18: 289 (1935); Bornmüller in Beih. Bot. Centralbl. 58B: 277 (1938); Zohary in Dep. Agr. Iraq Bull. 31: 110 (1950); H. Riedl in Fl. Lowland Iraq: 465 (1964); Rawi in Dep. Agr. Iraq Tech. Bull. 14: 94 (1964); Mouterde, Nouv. Fl. Lib. et Syr. 2: 607 (1970); Matthews in Fl. Turk. 4: 356 (1972); Zohary, Fl. Palaest. 2: 426 (1972); Engstrand in Fl. Iranica 162: 316 (1987).

Chesneya daucoides Bertol., Nov. Comm. Bonon. 5: 427 (1842) & Miscell. Bot. 1: 17 (1842).

Slender annual, (5–)10–30(–40) cm, shortly and usually sparsely puberulent, branched from the base in small individuals or around the middle in better developed plants. Stem and branches terete, striate or occasionally quadrangular in upper part of plant. Basal leaves entire or obscurely 3-lobed, orbicular-ovate in outline, toothed all round, cordate at base, long-petiolate; lower stem leaves tripartite to ternate or pinnatisect, becoming pinnate with few lanceolate to elliptic segments; upper stem leaves sessile, bi- to tri-pinnate with linear segments, basal sheath short and broad, shorter than lamina. Peduncles 1–2.8 cm. Umbels (3–)5–10-rayed, rays 4–15 mm, sparsely and shortly patent-hairy, finally glabrescent. Partial umbels 10–20-flowered, the 2–5 mm pedicels more copiously hairy than those of the main umbel. Outer flowers scarcely radiate, ± 2 mm in diameter; petals glabrous or sparsely hairy dorsally. Involucre and involucel absent. Fruit 1.5 mm, ellipsoid, densely covered with long hairs subequalling or exceeding width of fruit; hairs first deflexed, arcuate-ascending toward tips; fruit shortly hairy, rarely ribs indistinct. Stylopodia long-conical, tapering into persistent, divergent or somewhat flexuose styles which equal or slightly exceed the length of fruit.

var. **eriocarpa**

Hairs of fruit long, equalling or exceeding the width of the mericarps, arcuate-ascending.

HAB. On lower mountain slopes, in coppiced *Quercus* on limestone, in open *Quercus* forest on eroded slopes, on heavily eroded sandstone ridges, grassy places and fields on steppic hills and plains, silty desert plains, in depressions on a rocky plateau, etc.; alt. (50–)200–1100(–1800) m; fl. & fr. May–Jun. in the mountains, Apr.–May on the plains.
DISTRIB. Common in the lower forest zone, steppe and desert regions of Iraq. **MAM**: nr Aradin, W of Amadiya, *Nábělek* 347; Mar Ya'qub, above Simel, *Nábělek* 352; **MRO**: Kuh-i Sefin, above Shaqlawa, *Bornmüller* 1244!; Shaqlawa, *Gillett* 11296!; **MSU**: 4 km E of Qaranjir, on road to Sulaimaniya, *Rawi* 21635?; Jarmo, *Helbaek* 1131! 1770!; 12 km E of Chemchamal, *Gillett & Rawi* 11604!; Darband-i Bazian, *Rawi* 22834!; Pir Omar Gudrun (Pira Magrun), *Haussknecht* s.n?; Qopi Qaradagh, *Haines* W1566!; c. 10 km W of Tawila, *Rawi* 22923!; **FUJ**: Tal Kochak, *Gillett* 10817!; 20 km SW of Ilaimid (Tal Julaimid), *Alizzi & Hussain* 33956?; **FNI**: Nineveh, *Aucher* 3683 , *Haussknecht* s.n; **FAR**: Ankawa, nr Arbil, *Bornmüller* 1246; **FAR/FKI**: between Arbil and Kirkuk, *Haussknecht* s.n; **FKI**: between Tuz and Tuq, *Gillett & Rawi* 7452!; Jabal Hamrin, between Kirkuk and Baghdad, *Sutherland* 297; **FNI**: Sa'diya, *Sutherland* 295; **DLJ**: 10 km W of Tikrit, *Alizzi & Hussain* 33731!; Balad, *Haines* W1490!; **DWD**: 20–25 km W of Rutba, *Rechinger* 12544!; 50 km N of Rutba, *Chakr., Rawi, Khatib & Alizzi* 32902!; Wadi al-Ajrumiya, *Rawi* 31266! *Alizzi & Hussain* 34046?; *'Ud an-Nisr, Alizzi & Omar* 36196; 62 km W of H.1, *Barkley* 5375!; Masad, S of Rutba, *Alizzi* 35197?; 230–235 km W of Ramadi, on road from Rutba, *Rawi* 20929! *Rechinger* 9827; 160–190 km W of Ramadi, *Rawi* 31347!; **DWD/DSD**: Wadi al-Khirr, *Guest, Rawi & Rechinger* 19444!; **DSD**: 52 km E by S of Salman, *Guest, Long & Rawi* 14124!; 25 km N of Ansab, *Guest, Rawi & Rechinger* 18924!; 12 km WNW of Ansab, *id.* 18981! *Rechinger* 9356; 80 km W by N of Busaiya, *Guest, Long & Rawi* 14130!; 30 km W by N of Busaiya, *id.* 14171!

Syria, Palestine, Jordan, Arabia, Turkey, W & S Iran.

var. **brevihirta** *C.C. Townsend* in Kew Bull. 19: 72 (1964).

Hairs of fruit very short, as short as in *P. puberula*, loosely adpressed, not nearly as long as the width of the fruit.

HAB. On silty loam plains, waste ground near irrigation ditches and low hills; alt. 20–194 m; fl. & fr. Apr.–May.
DISTRIB. Occasional in the steppe region and subdesert. **FUJ**: Qaiyara, *Bayliss* 122!; **FKI**: Jabal Hamrin, between Kirkuk and Baghdad, *Bornmüller* 1245!; **LCA**: Tarmiya, Baghdad, *Haines* W109!; **LSM**: between Musaida and Fakka, *Rawi* 14984!

Saudi Arabia.

6. **Pimpinella barbata** (*DC.*) *Boiss.*, Ann. Sc. Nat. sér. 3, 1: 129 (1844). — Type: ?Iraq, "deserta Assyriae", *Aucher* 3682 (K!, syn.); Baghdad to Kermanshah, *G.A. Oliv.* s.n. (K!, syn.); Boiss., Fl. Orient. 2: 867 (1872); Hand.-Mazz. in Ann. Naturh. Mus. Wien 27: 91 (1913); Nábělek in Publ. Fac. Sci. Univ. Masaryk 35: 122 (1923); Wolff in Pflanzenr. IV. 228, 90: 238 (1927); Anth. in Notes Roy. Bot. Gard. Edinb. 18: 289 (1935); Bornmüller in Beih. Bot. Centralbl. 58B: 277 (1938); Blakelock in Kew Bull. 3: 435 (1948); Zohary in Dep. Agr. Iraq Bull. 31: 110 (1950); H. Riedl in Fl. Lowland Iraq: 466 (1964); Rawi in Dep. Agr. Iraq Tech. Bull. 14: 94 (1964); Engstrand in Fl. Iranica 162: 319 (1987).

Ptychotis barbata DC., Prodr. 4: 109 (1830); *Pimpinella glaucescens* Boiss., Ann. Sc. Nat. sér. 3, 1: 130 (1844).

Wiry annual, 10–40 cm high, quite glabrous save for the fruits and occasionally a few hairs on the rays of the partial umbels. Stem and branches (which are numerous and arise from near the base upwards) very slender, terete, finely or not striate, often pale and shining whitish or pale brownish green. All leaves divided into long, remote, setaceous, mostly 0.8–3 cm segments, the lowest leaves very shortly petiolate and bipinnate, the upper pinnate or ternate and sessile, uppermost often reduced and bract-like, all longly and rather narrowly sheathing. Peduncles 1.2–3 cm, sometimes geniculate close to attachment of the umbel. Umbels with 4–6 filiform, 1–3 cm rays. Partial umbels 10–15-flowered, pedicels 1–4 mm. Outer flowers conspicuously radiate, ± 2.5 mm in diameter; petals glabrous. Involucre and involucel absent. Fruit ± 2 mm, ovoid, densely furnished with rather long white hairs which are uncinate at tip. Ribs obscure, hidden by hairs. Stylopodia very shortly conic, styles long and flexuose, ± 2 mm, rarely persisting unbroken in the ripe fruit.

HAB. Steppic hills and plains, grassy places on eroded sandstone and on gravelly banks, on sandy soil in wadis; alt. 150–800 m; fl. & fr. Apr.–Jun.
DISTRIB. Quite common in the steppe region of Iraq, rare in the lower forest zone. **MRO**: Harir, between Arbil and Rowanduz, *Bornmüller* 1242!; **MSU**: Jarmo, *Helbaek* 1797b! *Haines* s.n.!; Pira Magrun, *Haussknecht* s.n.; **FUJ**: "deserta Assyriae", *Aucher* 3682 (syn.)! (or Iran), Baghdad to Kermanshah, *G. A. Oliv.* s.n. (syn.); Qaiyara, *Bayliss* 125!; Qala'a Sharqat (As-shur), *Handel-Mazzetti* 1118; **FUJ/FKI**: between Arbil and Kirkuk, *Haussknecht* s.n.; **FKI**: 7 km NW of Kirkuk, *Rawi, Nuri & Kass* 27937!; 8 km E of Kirkuk, *Gillett* 11592!; Shaikh Salih, 30 km S of Tuz, *Anthony*, l.c.; 5 km S of Daquq, *Rawi* 22856!; **FKI/FNI**: Jabal Hamrin, *Haines* W719!; between Tanura and Garashala, *Poore* 590!; Jabal Hamrin, between Khanaqin and Baghdad, *Nábělek* 386 , *Sutherland* 294; Mandali, *Guest* 793!; Koma Sang nr Mandali, *Rawi* 20602!, *Agnew & Hadač* 4589; 10 km E of Mandali, *Rechinger* 9651; 54 km SE of Mandali, *Rechinger* 9680; 10 km N of Badra, *id.* 9693; 16 km SE of Badra, *id.* 13960; 25 km SE of Badra, 20772!; Buksaya (Bugsaiya), *Rawi & Haddad* 25581!; **DLJ**: Tikrit, *Handel-Mazzetti* 1004; between Samarra and Balad, *id.* 992.

W & S Iran.

7. **Pimpinella peregrina** *L.*, Mant. 2: 357 (1771); DC., Prodr. 4: 121 (1830); Boiss., Fl. Orient. 2: 867 (1872); Wolff in Pflanzenr. IV. 228, 90: 239 (1927); Hayek in Fl. Penins. Balc. 1: 999 (1927); Zohary, Fl. Palest. ed. 2, 1: 519 (1932); Townsend in Kew Bull. 17: 428 (1964); Mouterde, Nouv. Fl. Lib. et Syr. 2: 607 (1972); Zohary, Fl. Palaest. 2: 424 (1972); Matthews in Fl. Turk. 4: 356 (1972); Meikle, Fl. Cyprus 1: 746 (1977).

P. affinis Ledeb., Fl. Ross. 2: 257 (1844); Boiss., Fl. Orient. 2: 868 (1872), incl. var. *multiradiata* Boiss.; Wolff. in Pflanzenr. IV. 228, 90: 242 (1927); Schischkin in Fl. SSSR 16: 441 (1950); Zohary in Dep. Agr. Iraq Bull. 31: 110 (1950); Rawi in Dep. Agr. Iraq Tech. Bull. 14: 94 (1964); Matthews in Fl Turk. 3: 358 (1972); *P. affinis* Ledeb. var. *multiradiata* Boiss., Fl. Orient. 2: 868 (1872); Wolff in Pflanzenr. IV. 228, 90: 243 (1927); Bornmüller in Beih. Bot. Centralbl. 58B: 277 (1938).

Erect, green biennial, (30–)40–75(–100) cm. Stem terete, wiry, smooth or finely striate, glabrous or moderately to densely deflexed-pubescent below. Branches few, from about the middle or below, long and ascending, naked or usually with a few small leaves. Lowest leaves ternate or entire, cordate-ovate and crenate, soon withering; remaining basal leaves long-petiolate, pinnate with 2–4 pairs of ovate to cordate-ovate subsessile segments which vary from obtusely crenate to ± deeply incised, terminal segment largest, segments subcordate to broadly cuneate at base. Upper leaves rapidly reducing, variously pinnate and pinnatisect into lanceolate or linear-lanceolate segments, sessile on long sheaths. Leaf surfaces varying from glabrous to ± densely pubescent after the manner of the stem. Peduncles (3–)5–18(–27) cm. Umbels nodding in bud, usually rather few, extremely variable as to number of rays, (6–)12–35(–50), rays unequal, 1.5–7 cm, shortly patent greenish-hairy to subglabrous, contracted in fruit or not. Involucre and involucel absent. Partial umbels 25–30-flowered, pedicels 2–10 mm, indumentum similar to that of rays. Flowers not noticeably radiate, 1.5–2 mm in diameter. Fruit densely to thinly patent-hairy with long or short hairs, small, ± 1 mm long, subrotund to pyriform. Stylopodia tumid, elongate-conic, styles rigidly divergent (in European material apparently almost always so) to variously flexuose or deflexed, ± 1 mm.

HAB. In the mountains along streams under *Quercus* and *Juglans* shade, on a rocky limestone slope; alt. 950–1150 m; fl. & fr. Jun.–Aug.
DISTRIB. Rather rare in Iraq; only found in three or four localities in the middle forest zone. **MAM**: Sarsang, *Haines* W1239!; **MRO**: Kuh-i Sefin, above Shaqlawa, *Bornmüller* 1240; Pushtashan, *Rawi & Serhang* 23825! 23857! 26575! 26575A!; **MSU**: (or in Iran), "montes Avroman et Schahu", *Haussknecht* s.n.
NOTE. The plants of this affinity need special investigation in the Middle East. All Iraq material has small fruits with thinly hairy mericarps and is "*P. affinis* Ledeb.". These differences in fruit are apparently not constant over the range of the species as a whole, and my reasons for combining these two species are given in Kew Bull. 17: 428–429 (1964). Iraqi material has variably flexuose to closely reflexed styles, but this is not particularly an *affinis* character, although Boissier employs it. A specimen from "Iberia" (Georgia), Herb. Ledebour No. 2597 (LE) named as *P. affinis* by Schischkin, has small fruits with sparse hairs but the styles are divergent exactly as in typical *peregrina*.

S Europe (S France, Italy, Sicily, Malta, Balkans, Crimea), Crete, Cyprus, Syria, Lebanon, Palestine, Turkey, Caucasus, Iran, C Asia (Turkmenistan, Kazakhstan, Tajikistan, Siberia).

8. **Pimpinella kotschyana** *Boiss.*, Ann. Sc. Nat. sér. 3, 1: 133 (1844). — Type: Iraq, Gara Dagh, *Kotschy* 302 (K!, iso.). Boiss., Fl. Orient. 2: 870 (1872); Hand.-Mazz. in Ann. Naturh. Mus. Wien 27: 91 (1913); Nábělek in Publ. Fac. Sci. Univ. Masaryk 35: 123 (1923); Wolff in Pflanzenr. IV. 228, 90: 245 (1927); Zohary, Fl. Palest. ed. 2, 1: 520 (1932); Bornmüller in Beih. Bot. Centralbl. 58B: 277 (1938); Blakelock in Kew Bull. 3: 435 (1948); Zohary in Dep. Agr. Iraq Bull. 31: 110 (1950); Rawi in Dep. Agr. Iraq Tech. Bull. 14: 94 (1964); Mouterde, Nouv. Fl. Lib. et Syr. 2: 608 (1970); Matthews in Fl. Turk. 4: 360 (1972); Engstrand in Fl. Iranica 162: 321 (1987).

P. corymbosa Boiss. var. *kotschyana* (Boiss.) Post, Fl. Syria 349 (1896).
P. haussknechtii Rech.f. & Riedl, Anz. Öst. Akad. Wiss., Math.-Nat. Kl. 13: 4 (1961).

Biennial, (14–)30–50(–60) cm, covered with a short canescent tomentum, considerably branched, branches opposite or more rarely alternate; lower branches very long, upper gradually shorter, secondary branches similar, thus forming a dense corymb. Stem and branches terete, finely striate. First radical leaves entire or obscurely trilobed, ovate-oblong, margins obtusely dentate, the remainder simply pinnate with broad, subsessile, ovate, obtusely dentate segments in 2–4 pairs, all long-petiolate. Stem leaves varying from simply pinnate with broad segments to 1–3-pinnatisect, uppermost trisect with narrow segments or broad, obcuneate and trifid above, sessile. Umbels very numerous, 10–20-rayed, rays 1–5 cm, densely puberulous. Involucre of 3–5 linear bracts, much shorter than rays. Partial umbels ± 20-flowered, pedicels 2–10 mm, densely puberulous, involucel of 6–8 subulate bracts which are shorter than flowering pedicels. Petals not radiate, dorsal surface moderately hairy centrally or sometimes glabrous, flowers ± 2.5 mm in diameter. Fruit subglobose, ± 2 mm, densely lanate with patent white hairs, ribs prominent but concealed amid the hairs. Stylopodia shortly conical. Styles erect flexuose, 2–3 mm, equalling or exceeding the fruit in length, hairy.

HAB. Rocky mountain slopes, under light *Quercus* shade (in denuded Quercetum) on limestone, on red marl banks, in stream bed among *Pinus* woods, on steppic hills; alt. 350–1500(–1750) m; fl. & fr. Jun.–Jul (–Aug).
DISTRIB. Common in the NW and central sectors of the middle and lower forest zones and the NW sector of the moist-steppe zone of Iraq, occasional in the SE forest zone. **MAM**: Jabal Bekher, nr Zakho, *Rawi* 23007?, 23360!; Khantur, NE of Zakho, *Rawi* 23362!; "mons Gara" (Gara Dagh), *Kotschy* 302 (type); Mar Ya'qub, above Simel, *Nábělek* 348; Sandur, *Haines* W1598!; Zawita, *Guest* 3689!, 4633!, 4868!; Sarsang, *Haines* W565!; Atrush, *Guest* 3631!; Pika Ser, *Field & Lazar* 839!; **MRO**: between Harir and Rowanduz, *Bornmüller* 1235; Rowanduz, *Omar, Sahira, Karim & Hamid* 38373!; Handren, *Bornmüller* 1236; between Karoukh and Dargala, *Rawi, Nuri & Alkas* 27726?; Haji Umran, *J.K. Jackson* (3510=)15073!, *Rechinger* 11297!, *Omar Sahira, Karim & Hamid* 38479!; Qurnago, *Rawi & Serhang* 26607!; Pushtashan, 15 km N E of Rania, *id.* 23842!; Sarachawa, 10 km N of Rania, *Omar, Sahira, Karim & Hamid* 36238!; Berrog Mt., between Qandil and Rania, *Rawi & Serhang* 23938!; **MSU**: Darband-i Bazian, *Haussknecht* s.n.!; Tainal, between Darband-i Bazian and Tasluja, *Gillett & Rawi* 11634!; Penjwin, *Rawi* 22523!; ? (or Iran), "Montes Avroman et Schahu", *Haussknecht* s.n.; (or Iran), Hawraman, *Haussknecht* s.n.!; **MJS**: Magharad ravine, Jabal Sinjar, *Handel-Mazzetti* 1388; **FUJ**: between Balad Sinjar and Tal Afar, *Guest* 4176!; Asis Zimmar (Zummar), *H. Mashtuf* 5177!; "Zalan mt." (?Jabal 'Ain Zala), *Rawi, Husham & Nuri* 29402?; **FUJ/FNI**: 81 km NW of Mosul, *Rawi* 22949!; **FNI**: Gerwona, nr Ain Sifni, *Field & Lazar* 724!; Baqasra, nr Ain Sifni, *Salim Effendi* 2586!
NOTE. ZANJAFÎL (Kurd.-Ain Sifni, *Salim* 2586), this Arabic name for ginger, which does not grow in Iraq, seems most dubious.

Turkey, Iran.

9. **Pimpinella olivierioides** *Boiss. et Hausskn.*, Boiss., Fl. Orient. 2: 871 (1872); Wolff in Pflanzenr. IV. 228, 90: 246 (1927); Bornmüller in Beih. Bot. Centralblatt 58B: 227 (1938); Zohary in Dep. Agr. Iraq Bull. 31: 110 (1950); Rawi in Dep. Agr. Iraq Tech. Bull. 14: 94 (1964); Matthews in Fl. Turk. 4: 360 (1972); Engstrand in Fl. Iranica 162: 322 (1987).

Erect biennial, up to 75 cm, whitish-canescent throughout. Stem and branches terete, striate; branches numerous from about the middle, some opposite and some alternate; lower branches long-ascending, upper gradually shorter; secondary branches similarly arranged, whole system thus forming a dense corymb. Radical leaves not seen. Stem leaves pinnatisect with 2 or 3 pairs of lanceolate or ovate, entire or toothed segments, sheath decurrent along the petiole; uppermost trisect or cuneate-spatulate and tridentate above. Peduncles 0.5–3 cm. Umbels very numerous, 6–12-rayed, rays densely hairy, 0.8–4 cm. Involucre of 4–6 conspicuous bracts mostly about half as long as rays, linear and simple or trifid at tip. Partial umbels dense, ± 20–flowered, bracts of involucel numerous, 10–12, broadly linear, equalling at least the length of the 1–5 mm pedicels plus receptacle and often the flower also. Petals very broad, cordate-orbicular, not radiate, densely hairy over the entire dorsal surface, chartaceous with age; flowers 2.5–3 mm in diameter. Fruit subglobose, densely covered with whitish hairs, ribs pronounced but obscured by hairs. Stylopodia shortly conical. Styles long, flexuose, hairy, ± 3 mm.

HAB. In the lower mountain, on barren rocky slopes; alt. 700–1000 m; fl. & fr. Jul -Aug.
DISTRIB. Rare in Iraq; only found in a few localities in the central sector of the lower forest zone. **MRO**: Kuh-i Sefin, above Shaqlawa, *Bornmüller* 1237; Shaqlawa, *Omar, Sahira, Karim & Hamid* 38259!; Chinaruk, *Rawi & Serhang* 25337!; Kew-a Rash, nr Rania, *id.* 23785! **MAM**: nr Shaikhan, *Haussknecht* s.n.; **MSU**: nr Chemchemal, *Rechinger* 10583.

Turkey (N & C Anatolia), Iran.

10. **Pimpinella olivieri** *Boiss.*, Ann. Sc. Nat. sér. 3, 1: 132 (1844); Boiss., Fl. Orient. 2: 870 (1872); Hand.-Mazz. in Ann. Naturh. Mus. Wien 27: 91 (1913); Nábělek in Publ. Fac. Sci. Univ. Masaryk 35: 123 (1923); Wolff in Pflanzenr. IV. 228, 90: 246 (1927); Anth. in Notes Roy. Bot. Gard. Edinb. 18: 289 (1935); Zohary in Dep. Agr. Iraq. Bull. 31: 110 (1950); Riedl in Fl. Lowland Iraq: 466 (1964); Rawi in Dep. Agr. Iraq Tech. Bull. 14: 94 (1964); Mouterde, Nouv. Fl. Lib. et Syr. 2: 608 (1970); Zohary, Fl. Palaest. 2: 425 (1972); Engstrand in Fl. Iranica 162: 323 (1987).

P. bornmuelleri Hausskn., nomen in sched. *Bornmüller* 1257; [*P. corymbosa* (non Boiss.) Blakelock in Kew Bull. 3: 435 (1949).]

Erect biennial, ± 35 cm, furnished with short white-canescent hairs in lower parts, usually subglabrous above. Branches numerous from a little above base, mainly opposite, branching system bushy and corymbose as in the previous two species. Stem and branches terete, finely striate. A few of lowest leaves occasionally simply pinnate with sessile, orbicular-ovate toothed segments in 1–3 pairs, remaining leaves bi- to tri-pinnate with lanceolate segments (only the uppermost trisect), lower ± long-petiolate and the upper becoming sessile. Peduncles 1–3 cm. Umbels very numerous, 3–10-rayed, rays unequal, 0.7–3.5 cm, glabrous or minutely papillose. Involucre of 2–4 narrowly linear bracts, much shorter than rays. Partial umbels 15–20-flowered, pedicels glabrous, unequal. Bracts of involucel narrowly linear, not equalling the longer of the 0.8–5 mm pedicels. Petals oblong, not radiate, glabrous or hairy only along the centre line dorsally, flowers 2.25–3 mm in diameter. Fruit subglobose, ± 2 mm, densely furnished with short granular papillae, ribs evident but not prominent. Stylopodia conspicuous, mammillate, not tapering into the slender, flexuose, persistent, ± 2 mm glabrous styles.

HAB. Steppic hills and plains, fields, alongside an irrigation canal, on silty soil and loam; alt. 50–600 m; fl. & fr. Apr.–May(–Jun.).
DISTRIB. Occasional in the steppe region of Iraq. **MJS**: "M. Singara" (Jabal Sinjar), *Haussknecht* s.n.!; **FUJ/MJS**:?, "Muwasul Tiatan Mukzuk Nuwar" (?nr Jabal Qilat), *Field & Lazar* 467!; **FUJ**: between Sinjar and Ba'aj, *Bharucha & Ani* 8794!; Ilaimid (Tal Julaimid), *Alizzi & Hussain* 33929!; Sairamun, S of Mosul, *Handel-Mazzetti* 1198; between Hammam Ali and Qaiyara, *Handel-Mazzetti* 1164; **FNI**: Mindan, *Rawi* 11363!; **FAR**: Ankawa, nr Arbil, *Bornmüller* 1287!; between Arbil and Altun Kupri, *Nábělek* 360, 492; **FKI**: 10–15 km N of Kirkuk, *Rawi* 21576!; Qara Tepe, *Guest* 1956!; **FNI**: Sa'diya, *Graham* s.n.; 10 km N of Qaraghan (Jalaula), *Gillett & Rawi* 7365!; Koma Sang nr Mandali, *Hadač et al.* 4607; Badra, *Rechinger* 13999; **DWD**: 12 km N of T.1, on road from Husaiba, *Chakr., Rawi, Khatib & Alizzi* 31728!; **DGA**: 4 km E of Samarra, *Rawi* 20326!, *Rechinger* 9552; **LEA**: Ra'aya village, Ba'quba province, *Alizzi & Hussain* 33698!; **LCA**: Tarmiya, *Haines* W121!; Kadhimiya, Baghdad, *Graham* 599.
NOTE. BUIJANA (Kurd.-Jalaula, *Gillett & Rawi* 7365).

Syria, Lebanon?, Turkey.

11. **Pimpinella brachyclada** *Rech.f. et Riedl*, Anz. Österr. Akad. Wiss., Math.-Nat. Kl. 98: 224 (1961). — Type: Iraq, below Rowanduz, *Rechinger* 11247 (W, holo.); Engstrand in Fl. Iranica 162: 332 (1987).

Erect perennial, (30–)60–75(–100) cm, furnished throughout with short, ash-white pubescence, particularly densely so in the region of inflorescence. Stem terete, finely striate, wiry, surrounded at the base by remains of previous years' leaf-sheaths, branching usually from below the middle. Branches long and ascending often flexuose, with many very short side branches. Basal leaves several, long-petiolate, lanceolate-oblong in outline, simply pinnate with (2–)4–5 pairs of leaflets; leaflets up to 4 cm long but usually less, sessile, cuneate-based, coarsely dentate or often subtrilobed, orbicular-ovate in outline, becoming glabrescent with age. Stem leaves rapidly becoming ternate, most of upper linear-spatulate, finally linear and bract-like. Peduncles 3–8(–10) mm. Umbels very small, solitary or somewhat aggregated on short lateral branchlets, fruiting diameter not exceeding 2.5 cm, rays and pedicels shortly patent-hairy; rays usually 3–6, short, 2–8 mm. Involucre absent or of 1 or 2 linear bracts. Partial umbels 10–12-flowered, pedicels 1.5–3 mm; involucel with 2–5 linear short bracteoles. Petals small, pubescent over the entire dorsal surface, flowers 1.5–2 mm in diameter. Fruit ± 3 × 1 mm, rather thinly patent-hairy, ellipsoid. Stylopodia tumid, mammillate, styles long, flexuose-divergent, 1.5–2 mm, glabrous. Fig. 75: 1–7.

HAB. In the mountains, on rocky slopes, on shady north wall of cliff, fissures in limestone and on metamorphic rock in *Quercus* forest, by edges of stream; alt. 600–2400 m; fl. & fr. (Jun.–)Aug.–Sep. (–Nov.).
DISTRIB. Quite common in the central sector of the forest zone of Iraq. **MAM**: Zinta gorge, *E Chapman* 26095!; **MRO**: Rowanduz gorge (Gali Ali Beg), *Guest* 3219!, 13090!, *Guest & Long* 13615!, *Rawi* 24256!; Khalirfan, *Gillett* 9442!, *Haley* 59, *Guest & Husham (Alizzi)* 14539?; below Rowanduz, *Rechinger* 11247 (type); N of Kani Rash, nr Turkish frontier, *Thesiger* 1229; Sairo, *Gillett* 9678!; 47 km N of Rowanduz, *Rawi & Serhang* 24263!; Barsarini gorge, *Guest & Husham (Alizzi)* 15883A!; Pushtashan, *Rawi & Serhang* 24192!; above Pushtashan, nr Iranian frontier, *Rechinger* 11183; Qurnaqo, *id.* 26640; **MSU**: Hawraman mts., *Rawi, Chakr., Nuri & Alizzi* 19827!

Endemic (but near Turkish and Iranian frontiers).

12. **Pimpinella hadacii** *Engstrand* in Fl. Iranica 162: 327 (1987). — Type: Iraq, near Zawiya [Zewiya], *Hadač* 2814 (K!, holo.).

Perennial, ± 35 cm, sparsely pubescent below, ± glabrous above, divaricately branched from near the base, branches very slender; all branches except the uppermost alternate. Radical leaves long-petiolate, to 18 cm, lamina oblong, simply pinnate, leaflets short-stalked, irregularly serrate, cuneate at base; lower cauline leaves similar, upper reduced to narrowly lanceolate to subulate, entire. Peduncles 0.5–1 cm, umbels with 2–3 glabrous or sparsely pilose rays. Partial umbels 2- or 3-flowered. Involucre and involucel absent; pedicels 1–4 mm, sparsely patent-pilose. Flower ± 1.4 mm in diameter. Fruit (immature) ovoid, laterally compressed, 2.5–3 × 1.6 mm, glabrous or very slightly pilose. Stylopodium cup-shaped, ± 0.5 mm long; styles filiform, ± 0.8 mm long, divaricate.

HAB. On limestone rocks; alt. unknown; fl. & fr. Oct.–Nov.
DISTRIB. Very rare; only found once in the forest zone of Iraq. **MSU**: nr Zawiya (Zewiya), *Hadač* 2814 (type).

Endemic. Engstrand (l.c.) notes that his species is close to *P. tragium* but is distinguished by the slender, divaricately branched stems, the few-rayed umbels and the petiolulate basal leaf segments. The description has been taken from the type description; no material other than the type has been seen.

13. **Pimpinella kurdica** *Rech.f. et Riedl*, Anz. Österr. Akad. Wiss., Math.-nat. Kl. 13: 2 (1961). — Type: Jabal Khantur, Sharanish, near Turkish frontier, *Rechinger* 12100 (W!, holo.); Engstrand in Fl. Iranica 162: 327 (1987).

Biennial or perennial, ± 60 cm, densely furnished throughout (particularly in the region of the inflorescence) with short, spreading, ash-white hairs. Stem erect, terete or somewhat angular above, finely striate, branched from near the base. Lowest branches alternate, upper becoming opposite and finely verticillate. Radical leaves very numerous, long-petiolate, 12–20 cm long, lamina oblong-lanceolate in outline, simply pinnate with usually 4 pairs of leaflets; leaflets sessile, ovate, crenate-dentate, broadest at the obliquely cuneate, entire base. Lowest stem leaves similar to the radical, upper rapidly reducing to small and simply pinnate leaves with few entire or toothed leaflets, sessile on the sheaths, or finally lanceolate-spatulate. Peduncles 1–4 cm. Umbels many, with 12–20 rays, 8–22 mm long. Partial umbels ± 15-flowered. Involucre and involucel absent; rays and 2–4 mm pedicels densely furnished with patent hairs longer than their diameter. Petals hairy on midrib dorsally, flowers ± 2 mm diameter. Fruit not seen.

Fig. 75. **Pimpinella brachyclada**. 1, 2, 3, habit × ²/₃; 4, petal × 12; 5, 6, fruit (both faces) × 6; 7, fruit cross section × 14. 1–2, 4 from *Guest & Long* 13615; 3, 5–7 from *Guest* 14539. Drawn by D. Erasmus (1964).

HAB. On a limestone mountain, in rocky fissures; alt. 1200 m; fl. & fr. Sep.
DISTRIB. Very rare; only found once in the forest zone of Iraq. **MAM**: Jabal Khantur, Sharanish, nr Turkish frontier, *Rechinger* 12100 (type).

Endemic.

14. **Pimpinella saxifraga** *L.*, Sp. Pl. ed. 1, 263 (1753); Grossh., Fl. Kavk. ed. 2, 7: 85 (1967).
P. rotundifolia Scop., Fl. Carn. ed. 2, 1: 208 (1772); *P. calvertii* Boiss., Diagn. ser. 3, 2: 73 (1856).

Perennial. Stem to 60 cm tall, puberulous or glabrescent, with ± erect branches; fibrous sheath of persistent petioles of previous year's radical leaves surrounding the base. Basal leaves petiolate, oblong-ovate, 5–20 cm, simply pinnate, lobes ovate, serrate or incised-serrate; cauline leaves smaller, upper undivided and sheathing the stem. Umbels 5–18-rayed, rays 8–40 mm long. Partial umbels with 10–25 flowers. Petals white, obcordate, with inflexed tip, glabrous or shortly pilose. Fruit ovoid, ± 2.5 mm, glabrous. Stylopodia mamillate-depressed; styles slender, 0.–1.5 mm.

HAB. Limestone rocks; alt. ± 2000 m. fl. ?May.
DISTRIB. Rare in Iraq; known from only one record close to the Iranian frontier. **MSU**: Mt. Hawraman nr Tawila, *Rechinger* 10271!

From Europe to W & SW Asia.

15. **Pimpinella tragium** *Vill.*, Prosp. Hist. Pl. Dauph.: 24 (1779) & Hist. Pl. Dauph. 2: 605 (1787); DC., Prodr. 4: 171 (1830); Boiss., Fl. Orient. 2: 871 (1872); Wolff in Pflanzenr. IV. 228, 90: 248 (1927); Hayek in Fl. Penins. Balc. 1: 998 (1927); Zohary, Fl. Palest. ed. 2, 1: 520 (1932); Blakelock in Kew Bull. 3: 435 (1948); Rawi in Dep. Agr. Iraq Tech. Bull. 14: 95 (1964); Mouterde, Nouv. Fl. Lib. et Syr. 2: 608 (1972); Matthews in Fl. Turk. 4: 360 (1972).

Perennial; extremely plastic and variable in habit, from compact and dwarf (5–10 cm) to tall, erect and much-branched (to 1 m), usually ± closely and shortly white-pubescent but sometimes subglabrous or glabrous. Root thick and often woody, often surmounted by few to numerous persistent petioles of previous years' radical leaves. Radical leaves very variable in size, long-petiolate, pinnate with (1–)2–5 pairs of very shortly stalked or subsessile pinnae which may be shallowly or very deeply cut, very rarely 2- or 3-pinnate (not in Iraq). Stem leaves few, similar to the radical leaves but much reduced with fewer, generally more dissected segments, sessile on sheaths; uppermost, or sometimes all cauline leaves, reduced, bract-like. All leaves broadly sheathing. Peduncles 1.8–10.5 cm. Umbels few to many, (2–)6–10(–15)-rayed, rays 5–25 mm, glabrous to densely pubescent. Involucre absent. Partial umbels 10–20-flowered; pedicels 1–5 mm, glabrous to pubescent. Petals not radiate, ± hairy dorsally, often densely so; flowers 2–2.5 mm in diameter. Involucel usually absent, but in most plants one or two partial umbels will be found bearing a few subulate bracteoles. Fruit 2–3 mm, ovoid, shortly tomentose. Stylopodia conical or somewhat depressed; styles slender, flexuose, 2–2.5 mm.

Mediterranean Europe (from Spain to Greece, S Russia and Crimea), Syria, Lebanon, Turkey, Caucasus, Iran, C Asia (Turkmenia), N Africa (Morocco, Algeria).

A veritable hierarchy of infraspecific taxa has been built up around this variable plant, but many of these are now regarded as mere habitat variations which overlap considerably. The treatment here follows that of Tutin in Flora Europaea and Matthews in Flora of Turkey, not out of conviction that such a treatment is the final word but for the sake of uniformity and through inability after much study to offer anything better.

subsp. **tragium** is a slender plant up to ± 30 cm tall with the leaf-lobes ± round and shallowly, rather regularly toothed. It is confined to southern France and possibly occurs in Sicily, but not in Iraq, where the species is represented by:

subsp. **lithophila** (*Schischk.*) *Tutin* in Fedde Rep. 79: 62 (1968) & in Fl. Europ. 2: 331 (1968); Matthews in Fl. Turk. 4: 361 (1972); Engstrand in Fl. Iranica 162: 325 (1987).
[*P. tragium* var. *pseudotragium* (non (DC.) Boiss.) Bornmüller in Beih. Bot. Centralbl. 58B: 277 (1938).]

Plant slender, (8–) 15–60(–75) cm, usually covered with short, white hairs, rarely some leaves glabrous. Leaflets 3–20 mm long, about as wide or slightly wider, mostly broadly cuneate at the base, more rarely feebly cordate, irregularly sharply serrate and frequently incised.

HAB. (of subsp.): Rocky alpine summits and stony mountain slopes, often in shady places; alt. (1200–) 2000–3700 m; fl. & fr. (Jun.–)Jul.–Aug.
DISTRIB. (of subsp.): Quite common in the alpine region, thorn-cushion zone and upper forest zone of Iraq. **MAM**: Mt. Gara (Gara Dagh), *Kotschy* 344!; Sarsang, *Haines* W1324!; Mt. Zawita, NE of Zakho, nr Turkish frontier, *Rawi* 23600!; **MRO**: Mt. ENE of Ser-i Hasan Beg, *Guest* 2923!; Halgurd Dagh, *Guest* 3057, *Gillett* 9594!, 9613!, 12374!, *Guest* 2057!; Chiya-i Mandau, *Guest* 2709!, 2709A!, 2801!; Kuh-i Sefin (Sefin Dagh), *Bornmüller* 1233; Sakri-Sakran, *Bornmüller* 1239; Qandil range, *Rawi* 24114!, 24420!, *Rawi & Serhang* 244665!, 26776!, *Rechinger* 11141;
NOTE. The Iraqi material of this species is all referable to this subspecies, the only differences being in the dimensions of both the entire plants and their parts. Mountain top plants and those from exposed rocky slopes are short, tufted, with usually a mass of persistent petioles of previous years' leaves at the base of stem and very small leaflets. Those from shaded rocks and others probably from among boulders rather than merely stony places are taller, slender and with broader leaflets. Such states occur in practically all plants varying similarly in habitat.
The Bornmüller specimens cited above, and recorded by the collector as *P. tragium* var. *pseudotragium* (DC.) Boiss., have not been seen. However, one of Bornmüller's localities, Sakri-Sakran, is apparently near Chiya-i Mandau [Guest, Fl. Iraq 1: 48 (1966)], where Guest gathered subsp. *lithophila*. In view of this and the otherwise homogeneous nature of the Iraqi material of this species it seems safe enough to include Bornmüller's plants here. In point of fact I doubt if subsp. *lithophila* and subsp. *pseudotragium* can be separated. Matthews in Fl. Turk. states that subsp. *pseudotragium* has crisped-tomentose indumentum and leaf segments (2–)3–6 cm long and subsp. *lithophila* is glabrous or puberulent with leaf segments 0.5–2 cm. I have found the difference between crisped-tomentose and puberulent impossible to interpret from the Iraqi material, and the type of *P. pseudotragium* [Seidkhadzi, *Szovits*, G-DC.] has the *largest* leaf segments less than 2 cm. Even if the two are combined the name subsp. *lithophilum* has priority at subspecific rank, and thus the name of the Iraqi plant will remain unchanged.

S Europe from Spain to the Crimea, Turkey, Caucasus, W & S Iran, C Asia (Turkmenistan), NW Africa.

16. **Pimpinella zagrosica** *Boiss. et Hausskn.*, Boiss., Fl. Orient. 2: 872 (1872); Wolff in Pflanzenr. IV. 228, 90: 254 (1927); Zohary in Dep. Agr. Iraq Bull. 31: 110 (1950) sub "P. zazrovica" sphalm.; Rawi, ibid. Tech. Bull. 14: 95 (1964).

Erect perennial, 30–75 cm, with a stout rootstock; furnished throughout with a ± close, short, white pubescence, somewhat sparingly branched from a little way above the base or about half-way. Stem and branches tough and wiry, terete, finely striate; branches long, ascending, or little further divided. Radical leaves several, long-petiolate, lanceolate-oblong in outline, simply pinnate with 2 or 3 pairs of broadly ovate or suborbicular, deeply and irregularly toothed 8–35 × 9–40 mm leaflets which are sessile, some being decumbent along the rachis. Stem leaves few, pinnae more dissected, uppermost sessile and trifid or reduced and bract-like. Peduncles slender, (3–)5–13 cm. Umbels rather few, 15–20-rayed, rays 5–30 mm, shortly pilose. Partial umbels 10–15-flowered, pedicels 2–5 mm. Involucre and involucel absent. Flowers not radiate, 1.5–2 mm in diameter; petals subglabrous or only hairy dorsally along the centre line. Fruit when ripe 3–3.5 mm, broadest near base, flask-shaped, covered with appressed white hairs, ribs inconspicuous. Stylopodia shortly conical, styles slender and flexuose, or ± recurved, 2–2.5 mm.
It would appear that in this species the young fruiting umbel contracts, with the rays drawing closer together; they spread, however, when the fruit is ripe.

HAB. In the mountains, on deep soil among *Quercus* trees, on a slope in a valley, on rich soil at foot of N-facing cliff; alt. 1100–2000 m; fl. & fr. Jun.-Aug.
DISTRIB. Occasional in the upper forest zone of Iraq, penetrating up into the lower thorn-cushion zone. **MAM**: Sarsang, *Haines* W1327!, *E Chapman* 26420!; **MRO**: Nurobar valley, Qandil range, *Rawi & Serhang* 24153!; **MSU**: Hawara Barza, Hawraman, *Rawi, Husham (Alizzi) & Nuri* 29341!; "Avroman et Schahu", *Haussknecht* s.n.

Section 3. TRAGOSELINUM DC.

Flowers usually white, rarely pink, reddish or purple. Petals always deeply incised above with a well-developed inflexed lobule. Fruit glabrous and smooth.

17. **Pimpinella anthriscoides** *Boiss.*, Fl. Orient. 2: 874 (1872). — Type: Iraq, (or in Iran), Avroman [Hawraman], *Haussknecht* s.n. (K!, syn.); Wolff in Pflanzenr. IV. 228, 90: 306 (1927); Zohary, Fl. Palest. ed. 2, 1: 520 (1932); Zohary in Dep. Agr. Iraq Tech. Bull. 14: 94 (1964); Townsend in Kew Bull. 17: 429 (1964); Grossh., Fl. Kavk. ed. 2, 7: 86 (1967); Mouterde, Nouv. Fl. Lib. et Syr. 2: 609 (1972); Matthews in Fl. Turk. 4: 363 (1972); Engstrand in Fl. Iranica 162: 330 (1987).

Erect perennial, (30–)60–85(–100) cm, quite glabrous throughout, with an incrassate fusiform root. Stem fistular, deeply sulcate, at least below, branches rather few, long and ascending. Basal leaves long-petiolate, lamina triangular in outline, 20–25 × 15 cm, 2–3-pinnate with 6–8 pairs of pinnae, lowest pinnae 3- or 4-pinnulate, segments of pinnae and pinnules often not completely free near the tips of divisions, sharply toothed all round, teeth mucronate or aristate at tips. Stem leaves similar but sessile, smaller, with fewer pairs of pinnae. All leaves ± scabrid or smooth on principal veins beneath. Peduncles long, 5.5–16 cm. Umbels few, large, 15–20-rayed, rays (0.5–)2–4 cm. Involucre of 4–8 extremely slender setaceous bracts. Partial umbels 15–20-flowered, the pedicels unequal, 2–7 mm, flowers 2 mm in diameter; petals falling very early. Bracts of the involucel similar in form to those of the involucre and like them much shorter than subtended rays. Fruit oblong-ovoid, commissural face narrow, ribs pale and obvious but very slender. Stylopodia mammillate, most prominent in young fruit. Styles slender, at first divergent finally reflexed, 1–1.5 mm.

HAB. In mountains in damp places, near springs on muddy soil, by streams; alt. 1650–2200 m; fl. & fr. Jun.
DISTRIB. Rare in Iraq, found in the upper forest zone, perhaps penetrating up into the lower thorn-cushion zone. **MRO**: Haji Umran, *Nuri & Kass* (*Alkas*), 27776!; Halgurd Dagh nr Sarchal, *Hadač* 2707, 2159; **MSU**: Binawa Suta nr Penjwin, *Hadač* 4947; **MSU** (or in Iran), Hawraman, *Haussknecht* s.n. (type)!
NOTE. Very distinct from all the other Iraqi species, apart from being completely glabrous, in the long fruit and *Anthriscus*-like foliage.

Turkey, Syria, Lebanon, Caucasus, Iran.

30. HELLENOCARUM Wolff

Pflanzenr. 90 (IV. 228): 167 (1927)

Glabrous, perennial or biennial herbs with a thickened rootstock. Leaves pinnately decomposite. Umbels compound. Involucre and involucel present, short. Flowers white or yellow. Petals ovate, entire or somewhat retuse, with an inflexed acumen. Calyx teeth very small. Fruit oblong, somewhat laterally compressed. Mericarps convex, the 5 primary ridges filiform, slightly prominent, pale. Vittae cyclic around the endosperm, forming almost a complete ring, equal or the main vallecular and commissural vittae larger. Disk very narrow; stylopodia conical, constricted below, styles slender and reflexed. Endosperm subterete or slightly dorsally compressed, commissural face flat or almost so.

Three species, one in Iraq and Iran, the other two distributed from S Italy to Turkey.
Hellenocarum (from the Gr. ελληνο, *helleno*, Greek and Carum, q.v., the genus as originally conceived being centred on Greece).

1. **Hellenocarum amplifolium** (*Boiss. et Hausskn.*) Kljuykov in Biol. Nauki 8: 62 (1985).

Muretia amplifolia Boiss. et Hausskn., Boiss., Fl. Orient. 2: 859 (1872). — Type: Iraq (or Iran), Hawraman mts., *Haussknecht* s.n. (JE!, iso.); Wolff in Pflanzenr. 90 (IV. 228): 213 1927); Zohary in Dep. Agr. Iraq Bull. 31: 109 (1950); Rawi in Dep. Agr. Iraq Tech. Bull. 14: 93 (1964); Rech.f., Fl. Iranica 162: 257 (1987).

Erect, glabrous, 50 cm or more in height, biennial or perennial herb. Stem slender, terete above, angular and striate below, much-branched from below the middle, branches long and divaricate, with few reduced leaves. Radical leaves petiolate, very broad, deltoid-ovate in outline, bi- to tri-pinnate with oblong, obtuse segments up to ± 12 × 4 mm. Upper leaves abruptly diminishing, ± sessile on the narrowly oblong sheaths with few linear segments, uppermost with 3 small lobes or lamina totally obsolete. Umbels numerous, on 1–11 cm peduncles, 9–16-rayed, rays 1–4 cm. Partial umbels 12–18-flowered; pedicels slender, 2–3 mm. Involucre and involucel of many short, linear, reflexed bracts and bracteoles. Petals

Fig. 76. **Hellenocarum amplifolium**. 1, habit × 2/3; 2, leaf × 2/3; 3, petal × 46; 4, fruits (both faces) × 14; 5, fruit cross section × 28. 1, 3–5 from *Kass* 27745; 2 from photo no. 5743. Drawn by D. Erasmus (1964).

yellow, < 1 mm long, not incised above, trivittate. Fruit oblong-elliptic, 2.5–4 mm, chestnut-brown (not fully ripe, when probably black) with pale ribs. Stylopodia conical; styles slender, reflexed, ± 1.5. mm. Fig. 76: 1–5.

HAB. On rocks and screes of limestone mountains; alt. 800–1500(–2150?) m; fl. & fr. May–Jul.
DISTRIB. Very rare in Iraq; found in the forest zone. **MRO**: nr Dargala in the direction of Karoukh mts, *Nuri & Kass* 27745!; **MSU**: Hawraman mts above Khormal, *Hadač* 5022!; ? (or Iran), Hawraman mts, *Haussknecht* s.n. (type)!

W Iran.

31. SCANDIX L.

Sp. Pl. ed. 1: 256 (1753); Gen. Pl. ed. 5: 124 (1754)

Annual herbs, glabrous or hairy, leaves pinnately decomposite into narrowly linear segments. Umbels compound with few rays or rarely apparently simple (in fact 1-radiate). Involucral bracts none or solitary, bracteoles several, simple or variously divided. Flowers white; petals equal or the outer very radiate. Petals cuneate or obovate, with or without a short inflexed lobule, incised or shallowly emarginate or entire above, the outer sometimes clearly radiant. Calyx teeth minute or obsolete. Fruit subcylindrical or laterally compressed, with a very long beak, constricted at the commissure. Ribs 5, conspicuous, broad or filiform. Vittae in the valleculae solitary, slender or obscure, on commissure 0–2. Endosperm subterete or slightly dorsally compressed, commissural face deeply sulcate, margins involute. Stylopodia small, with a ± well-developed crenulate-edged disk, styles erect. Carpophore undivided or shortly bifid.

Some 15–20 species distributed from Europe and the Mediterranean to SW and C Asia.
Scandix (from the Gr. σκάνδιξ, *skandix*, a name used by various Greek authors probably for chervil, *Anthriscus cerefolium* or according to Liddell & Scott *Scandix pecten-veneris*.)

1. Umbels apparently simple or 2- or 3-radiate with very short rays which are shorter than the fruit (not including the 12–20 mm beak) and scarcely noticeable at the flowering stage; bracts of the involucel pinnatifid . . . . 3. *S. stellata*
   Umbels compound, rays of the main umbel long, exceeding the length of fruit (not including beak); if umbel simple, then beak of fruit exceeding 20 mm and bracts around the base of fruit simple or only 2- or 3-fid at the tip 2
2. Main umbels 5–9-radiate; marginal flowers conspicuously radiate . . . . 2. *S. iberica*
   Main umbels 1–3-radiate (rarely all 1-radiate); marginal flowers only very slightly radiate . . . . 1. *S. pecten-veneris*

1. **Scandix pecten-veneris** *L.*, Sp. Pl. 1: 256 (1753); DC., Prodr. 4: 221 (1830); Boiss., Fl. Orient. 2: 914 (1872); Nábělek in Publ. Fac. Sci. Univ. Masaryk 35: 124 (1923); Hayek, Prodr. Fl. Balc. 1: 1005 (1927); Zohary, Fl. Palest. ed. 2, 1: 531 (1932); Guest in Dep. Agr. Iraq Bull. 27: 89 (1933); Anth. in Notes Roy. Bot. Gard. Edinb. 18: 289 (1935); Bornmüller in Beih. Bot. Centralbl. 58B: 279 (1938); Blakelock in Kew Bull. 3: 436 (1948); Zohary in Dep. Agr. Iraq Bull. 31: 111 (1950); Schischkin in Fl. SSSR 16: 141 (1950); H. Riedl in Fl. Lowland Iraq: 454 (1964); Rawi in Dep. Agr. Iraq Tech. Bull. 14: 95 (1964); Grossh., Fl. Kavk. ed. 2, 7: 37 (1967); Mouterde, Nouv. Fl. Lib. et Syr. 2: 587 (1970); Hedge & Lamond in Fl. Turk. 4: 328 (1972); Zohary, Fl. Palaest. 2: 390 (1972); E. Nasir in Fl. Pak. 20: 94, f. 28 (1972); V. Täckh., Stud. Fl. Egypt ed. 2: 391 (1974); Meikle, Fl. Cyprus 1: 697 (1977); Hedge & Lamond in Fl. Iranica 162: 107 (1987).

S. pecten-veneris var. *brevirostris* Boiss., Fl. Orient. 2: 914 (1872); Zohary, Fl. Palest. ed. 2, 1: 531 (1932); Bornmüller in Beih. Bot. Centralbl. 58B: 280 (1938); Blakelock in Kew Bull. 3: 436 (1948); Rawi in Dep. Agr. Iraq Tech. Bull. 14: 96 (1964); Mouterde, Nouv. Fl. Lib. et Syr. 2: 587 (1970); Zohary, Fl. Palaest. 2: 390 (1972).

Erect, branched, 20–50 cm, almost glabrous to sparingly hairy or with the stem ± densely and shortly spreading-hairy; stem and branches ± deeply sulcate. Leaves all similar but reducing in size upwards, narrowly deltoid or oblong in outline, tripinnate with short, narrowly linear segments, margins scabrid or smooth. Lower leaves long-petiolate, upper sessile, all with broad sheaths furnished with membranous, long-ciliate margins. Peduncles 1.5–8 cm. Umbels with 1–3 stout rays, sometimes true compound umbels with no involucre, sometimes apparently one or more rays being leafless branches since their point of junction is subtended by a true leaf; rays 0.6–2 cm long when in ripe fruit. Partial umbels 3–10-flowered, petals inconspicuously radiate, outer flowers ± 2 mm in diameter. Bracteoles 5, broadly lanceolate, simple or bifid or trifid above, ciliate-margined. Pedicels becoming incrassate and flattened, mostly to 2 mm long in ripe fruit. Fertile part of fruits 7–15 mm long, linear, shortly hispid; beak 28–50 mm, long, flattened and tapering, keel scabrid. Stylopodia very small, disk shortly cylindrical with a crenate edge above; styles short, often less than 1 mm, thick, persistent.

HAB. In the mountains on grassy places among shrubs and trees, shady spots in oakwoods on limestone, in fields in the valleys etc., hillsides, and fields on the upper plain, fields, gardens and ditches on irrigated alluvial plain; up to alt. 700–1100(–1600) m or more; fl. & fr. Mar./Apr.–May/Jun.
DISTRIB. Common in the lower forest zone and steppe region of Iraq, also on the irrigated alluvial plain in the desert region. **MAM**: Amadiya, *Guest* 1254!; **MRO**: Salah ad-Din, *Rawi & Gillett* 10480!, *Helbaek* 737!; Kuh-i Sefin, at Shaqlawa, *Bornmüller* 1252; Rowanduz gorge, *Guest* 2028!; Haji Umran, *Mooney* 427!; Zawita, *Karim, Hamid & M. Jasim* 40851?; Plingan (Balisan?), 17 km NW of Rania, *Rawi, Nuri & Kass* 28627!; between Dera (Harir) and Nura, 16 km NW of Rania *id.* 28595!; Dolako (Dolahraqa?), *id.* 28562?; **MSU**: Jarmo, *Helbaek* 1577!; Sarchinar, *Omar & Karim* 37953?; Sulaimaniya, *Graham* 741!; Kani Sard, between Sulaimaniya and Dukan, *Omar, Karim, Hamza & Hamid* 37352!; Dukan, *Omar & Karim* 37953?; between Sulaimaniya and Arbat, *id.* 38097!; above Sarao (?Kani Sard), 45 km SE of Sulaimaniya, *Haddad & Barkley* 7503?; **FUJ/MJS**: Sinjar, *Omar & Hamid* 36496!; **FUJ/FNI**: Mosul, *Calder* 2058 , *Shahwani* 25173!; **FAR**: 25 km NW of Arbil, *Barkley & Brahim* 4587; **FKI**: Tuz, *Rogers* 0642!; **FPF**: Jabal Hamrin, *Sutherland* 298; Mandali, *Guest* 1802!; ? **FNI/LEA**: Lower Diyala, *P.En.* s.n. (Kew No. 58)?; **DWD**: S bank of R. Euphrates, opposite Rawa, *Gillett & Rawi* 7048!; **LCA**: Khan Bani Sa'd, *Alizzi & Hazim* 33689?; Baghdad, *Paranjpye* s.n.!; comm. Graham, *Rogers* 0108!, *Guest* 1118!, 1118A!, *H. Ahmad* 9825!, *Haines* W595!; Yusufiya, *Polunin, Nábělek, Abdul-Wahhab et al.* 77!; Babylon, *Bornmüller* 372; **LSM**: Amara, in *Anth.*, l.c.; **LBA**: Basra, *Gillett & Rawi* 5954!
NOTE. Venus' Comb, Shepherd's Needle; MINQĀR AL-LAQLAQ ("stork's beak"), Ir., *Guest* 1118, 1802) or HALQ AL-LAQLAQ ("stork's throat"), Ir., *id.*) or sometimes HALQ HĀJI LAQLAQ ("stork, the pilgrim's throat") (*Guest* 1933); KAIZANŪK (Kurd.-Amadiya, *Majid Mustafa* in Guest 1254); ?SŌLAQA (Kurd.-Pers., Mandali, *Guest* 1802). Schischkin (1950) sums up the economic importance of this plant as "a weed, mainly of spring crops" – to which we have nothing to add.

W, C & Mediterranean Europe to W Russia, Crimea, Cyprus, Aegean Isles, Syria, Lebanon, Palestine, Jordan, NE Mediterranean Egypt, Turkey, Caucasus, Iran, Afghanistan, Pakistan, C. Asia (Turkmenistan to Siberia), Kashmir, N Africa (Morocco to Libya). Introduced in many other parts of the world, including S Africa, Tasmania and N America.

2. **Scandix iberica** *M. Bieb.*, Fl. Taur.-Cauc. 1: 425 (1808) et 3: 236 (1819); Boiss., Fl. Orient. 2: 915 (1872); Nábělek in Publ. Fac. Sci. Univ. Masaryk 35: 125 (1923); Hayek, Prodr. Fl. Balc. 1: 1005 (1927); Zohary, Fl. Palest. ed. 2, 1: 531 (1932); Zohary in Dep. Agr. Iraq Bull. 31: 111 (1950); Schischkin in Fl. SSSR 16: 142 (1950); H. Riedl in Fl. Lowland Iraq: 454 (1964); Rawi in Dep. Agr. Iraq Tech. Bull. 14: 95 (1964); Grossh., Fl. Kavk. ed. 2, 7: 38 (1967); Mouterde, Nouv. Fl. Lib. et Syr. 2: 587 (1970); Hedge & Lamond in Fl. Turk. 4: 327 (1972); Zohary, Fl. Palaest. 2: 391 (1972); E. Nasir in Fl. Pak. 20: 92 (1972); Hedge & Lamond in Fl. Iranica 162: 109 (1987).

S. falcata M. Bieb., Fl. Taur.-Cauc. 1: 230 (1808) non Londes, Mém. Soc. Nat. Mosc. 1: 31 (1811).
Wylia iberica Hoffm., Umb. 1: 19 (1814); *Scandix australis* L. subsp. *grandiflora* (non (L.) Thell.) Hedge & Lamond in Fl. Turk. 3: 330 (1964).

Erect, branched, 15–30 cm, almost glabrous or shortly spreading-hairy, particularly around the mid-section, stem and branches more wiry and tough than in the previous species, finely striate. Leaves all similar, the upper small and less-dissected, sessile, the lower petiolate. Outline of larger leaves oblong to broadly deltoid-ovate, lamina tripinnate into short, narrowly linear segments. All leaves with broad ciliate sheaths as in *S. pecten-veneris*. Peduncles ± 1–5 cm. Main umbels 5–9-rayed, rays 1.25–3 cm long when in fruit. Involucre of 1 or 2 narrowly linear or pinnate bracts, or none. Partial umbels 10-flowered, the outer petals very conspicuously radiate, flowers 4–6 mm in diameter. Bracteoles ± 6, lanceolate,

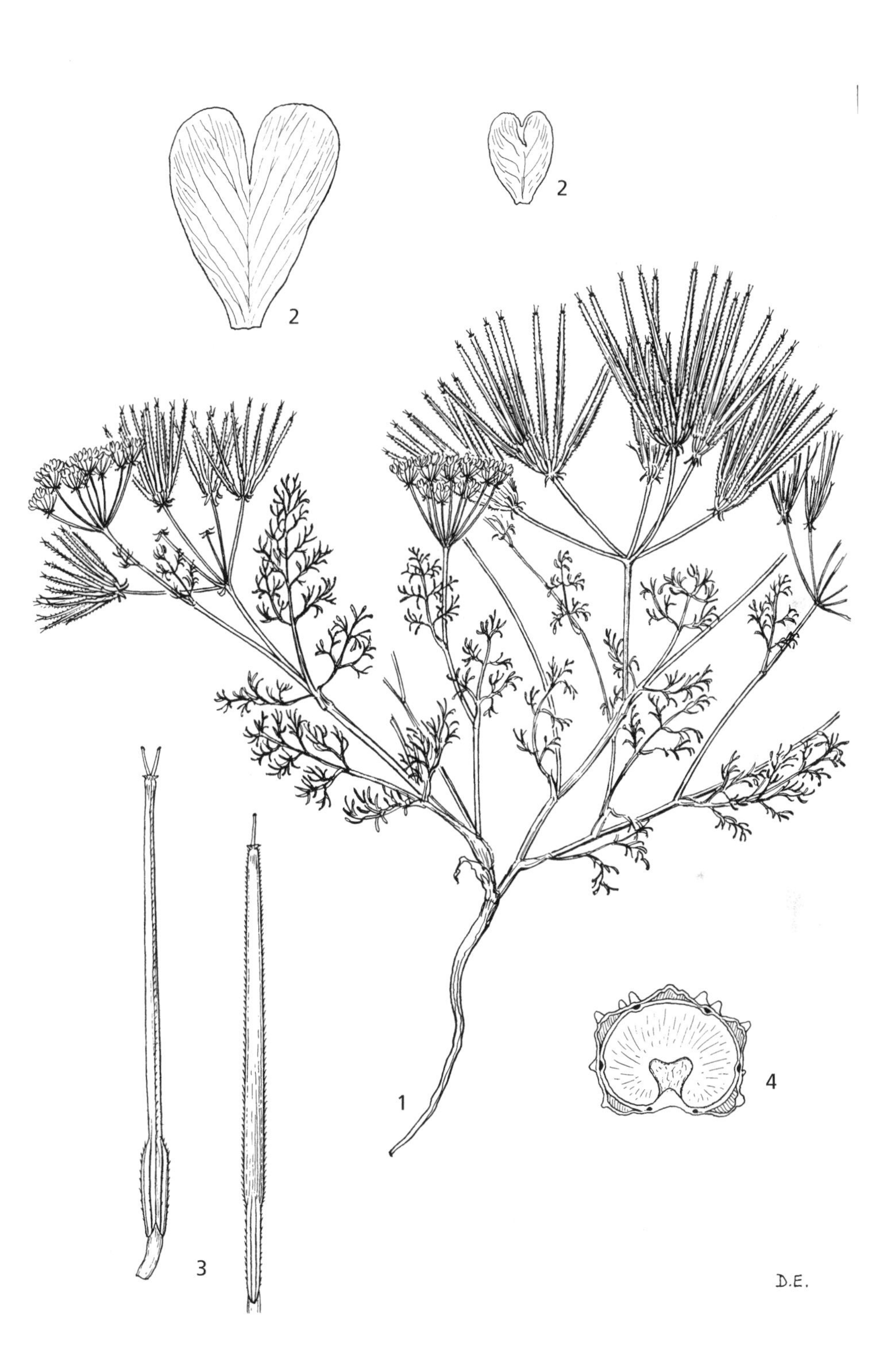

Fig. 77. **Scandix iberica**. 1, habit × ²/₃; 2, petals × 6; 3, fruits (both faces) × 2; 4, fruit cross section × 6. All from *Lazar* 3330. Drawn by D. Erasmus (1964).

entire or 2-lobed, often white-margined, ± whitish-lanate. Fertile part of fruits 8–12 mm long, linear, scabrid, pedicels 2–6(–8) mm, scarcely incrassate and flattened. Beak 20–35 mm, much flattened, keel scabrid-ciliate. Stylopodia very small, disk broader at top than at bottom, crenate-edged. Styles rigid, divergent, 1–1.25 mm, persistent. Fig. 77: 1–4.

HAB. In the mountains, in moist grassy places, among oakwoods in valley and on steppic hillsides; a frequent weed on disturbed ground; alt. 250–1100(–1350) m; fl. & fr. Apr.–Jun.
DISTRIB. Quite common in the lower forest zone of Iraq, occasional in the steppe region. **MAM**: Ain Nuni, *Nábělek* 537; Sarsang, at the foot of Gara Dagh, *Haines* W1001!, *Alkas* 18651!; **MRO**: foot of Jabal Baradost 10–20 km SW of Shanidar, *Erdtman & Goedemans* in *Rechinger* 15603; Rowanduz, *Meade* 194; **MSU/FKI**: between Sulaimaniya and Kirkuk, *Rawi* 21509!; **MSU**: Jarmo, *Helbaek* 423!; between Dukan and Sarsor, *Botany Staff* 43041?; 8 km N of Kani Spi, on Sulaimaniya-Penjwin road, *Rawi* 22410!; Khormal, *Rawi* 8964!; 6 km E of Qara Anjir towards Chemchamal, *Rawi & Gillett* 7564A; Susakan, S of Tawila, *Rawi* 21829!; **FNI**: Mosul, *Lazar* 3330!, 3351!; **FKI**: S of Taktak bridge, between Kirkuk and Koi Sanjaq, *Rawi, Nuri & Kass* 28065!
NOTE. ?KHANAKHLI (Kurd.-Khormal, *Rawi* 8964).

Syria, Lebanon, Palestine, Jordan, Sinai, Turkey, Caucasus, Iran.

3. **Scandix stellata** *Banks & Sol.* in Russell, Nat. Hist. Aleppo, ed. 2, 2: 249 (1794); Blakelock in Kew Bull. 3: 437; Schischkin in Fl. SSSR 16: 146 (1950); Rawi in Dep. Agr. Iraq Tech. Bull. 14: 96 (1964); Mouterde, Nouv. Fl. Lib. et Syr. 2: 588 (1970); Hedge & Lamond in Fl. Turk. 4: 326 (1972); Zohary, Fl. Palaest. 2: 392 (1972); E. Nasir in Fl. Pak. 20: 95, f. 28 (1972); Meikle, Fl. Cyprus 1: 696 (1977); Hedge & Lamond in Fl. Iranica 64: 104 (1987); She Menglan & Watson in Fl. China 14: 29 (2005).

S. pinnatifida Vent., Hort. Cels.: 14 (1800); DC., Prodr. 4: 221 (1830); Boiss., Fl. Orient. 2: 916 (1872); Zohary, Fl. Palest. ed. 2, 1: 531 (1932).
S. hispidula Bertol., Miscell. Bot. 1: 17 (1842).
Scandicium stellatum (Banks et Sol.) Thell., Fedde Rep. 16: 16 (1919); Zohary in Dep. Agr. Iraq Bull. 31: 111 (1950); V. Täckh., Stud. Fl. Egypt ed. 2: 391 (1974).

Erect, branched, 10–30(–50) cm, almost glabrous or shortly spreading-hairy, particularly around the mid-section, stem and branches slender, wiry, striate. Leaves similar throughout, lower petiolate, upper sessile, oblong or deltoid in outline, tripinnate into short, narrowly linear segments, sheaths membranous-margined and ciliate. Umbels with (1–)2–3(–5) rays which are shorter than the ripe fruit, or so nearly absent that the umbels appear simple. Involucre absent. Peduncles 1–5 cm. Partial umbels 3–8-flowered, outer petals not conspicuously radiate, flowers 1.5–2 mm in diameter. Bracteoles of involucel conspicuous, pinnatifid, when rays of umbel are very short forming a false involucre to the apparently simple umbel, pedicels none or to 1 mm. Fertile part of fruits 5–7 mm long, linear, scabrid in upper half, ribs distinct; beak 12–20 mm, somewhat flattened, scabrid over the entire surface but much more strongly so on the keel. Stylopodia minute, disk shortly cylindrical, broader than long, crenate-edged. Styles very short, 0.25 mm. Usually a more delicate and graceful plant than the two preceding species.

HAB. Lower mountain slopes in oakwoods on limestone, on ledges of a limestone cliff, bottom of a shady ravine in light *Quercus* scrub, on a grassy slope near a waterfall, etc.; alt. 500–1350(–1700) m; fl. & fr. Apr.–May.
DISTRIB. Quite common in the lower forest zone of Iraq. **MAM**: Matina (Chiya-i Matin), *Rawi* 8719!; Dohuk gorge, *Emberger, Guest, Long, Schwan & Serkahia* 15415!; Sarsang, *Alkas* 18689!; **MRO**: Salah ad-Din, *E Chapman* 11973!; Sefin Dagh, *Gillett* 8072!, *Bornmüller* 1259; Shaqlawa, *Haines* W756!; Rowanduz gorge, *Guest* 2029!, 2118!; Rowanduz, *Guest* 587!; Pira Magrun, *Haussknecht* s.n., *Gillett* 7780!; Jindian, *Guest* 745!; Gali Warta, 30 km NW by N of Rania, *Rawi, Nuri & Alkas* 28835!; Serkabkhan, 7 km NW by W of Rania, *id.* 28518!; Chewa Rash (Kew-ar Rash), NE of Rania, *id.* 28503!; **MSU**: Jarmo, *Helbaek* 461!; Bakrajo, *Omar & Karim* 38007?; Mt. Hawraman nr Tawila, *Rechinger* 10274; **MJS**: Kursi, *Gillett* 10884!

Mediterranean Europe (from S Spain & France to the Balkans, Greece, SW Russia & Crimea), Cyprus, Aegean Isles, Syria, Lebanon, Palestine, Jordan, Sinai, Egypt, Turkey, Caucasus, Iran, Afghanistan, Pakistan, C Asia (Turkmenistan & Kazakhstan to Siberia), N Africa.

32. **ANTHRISCUS** Pers.

Syn. 1: 320 (1805)

Annual, biennial or more rarely perennial herbs, hirsute or more rarely glabrous. Leaves 2- or 3(–4)-pinnate, the segments often rather broad. Umbels compound. Involucre of 1 or 2 bracts or absent; involucel of several bracteoles. Flowers white, usually polygamous. Calyx teeth obsolete or very small. Petals oblong or cuneate with an inflexed lobule, emarginate or almost entire above, outer sometimes ± radiant and obcordate. Fruit ovate or oblong or linear, shortly attenuate above, beaked, laterally compressed and constricted at the commissure. Ribs 5, somewhat apparent above, obscure below. Vittae solitary in the valleculae, very slender and often inconspicuous, commissural vittae 2. Stylopodia small, conical or depressed. Carpophore entire or bifid. Endosperm subterete or dorsally compressed, commissural face deeply sulcate.

Some 20 species in Europe, temperate Asia and N Africa; two species in Iraq.
Anthriscus (from the Gr. ανφρυσκον, *anfriskon,* the name of an umbelliferous plant in Theophrastus): Chervil, Beakchervil (Am.).

Perennial; main umbels with 10–15 glabrous rays; pedicels with a ring of hairs at the tips; fruit with a very short, indistinct beak . . . . 1. *A. sylvestris*
Annual; main umbels with 3–5 pubescent rays; pedicels lacking a ring of hairs at the tips; fruit with a distinct beak about half the length of the mericarps. . . . . 2. *A. cerefolium*

1. **Anthriscus sylvestris** (*L.*) *Hoffm.*, Gen. Umb.: 40 (1814); DC., Prodr. 4: 223 (1830); Boiss., Fl. Orient. 2: 910 (1872); Zohary, Fl. Palest. ed. 2, 1: 529 (1932); Schischkin in Fl. SSSR 16: 128 (1950); Hedge & Lamond in F. Turk. 4: 323 (1972); Grossh., Fl. Kavk. ed. 2, 7: 34 (1967); Hedge & Lamond in Fl. Iranica 162: 85 (1987); She Menglan, Cannon & Watson in Fl. China 14: 26 (2005).

Chaerophyllum sylvestre L., Sp. Pl. ed. 1: 258 (1753).

Erect perennial, 60–100(–150) cm, much-branched from halfway or below. Stem stout, fistular, sulcate, either glabrous throughout or, with petioles of lower leaves and lower branches, ± hispid or shortly hairy. Upper parts, including the umbel rays, glabrous (excepting bracteole margins and sometimes fruit). Leaves all similar, broadly deltoid in outline tripinnate into ovate to oblong-lanceolate, acute segments, lower long-petiolate, uppermost sessile on sheaths, all glabrous on upper surface, ± hispid on lower surface on veins and along margins. Peduncles 2–4(–7) cm. Umbels many, 4–15(–20)-rayed, fertile rays 2.5–4 cm. Involucre absent. Partial umbels 7–15-flowered, outer flowers distinctly radiate, 4 mm in diameter. Involucel of ± 5 oblong, 2–6 mm, ciliate bracts. Fruit 6–8 × 2–2.5 mm, broadest near the base and attenuate above, glabrous or furnished with short, stiff, curved, tuberculate-based bristles; beak very short, sulcate, less than 1 mm as a rule, not clearly marked off from the body of fruit; primary ribs only faintly visible above, and in the beak; pedicels with a ring of hairs at tip immediately below the fruit. Stylopodia turgidly conical, suddenly attenuate into the persistent, ± 1 mm styles, disk depressed.

Almost throughout Europe; also widely distributed in temperate Asia from Turkey through N & NW Iran, Caucasus, Kashmir, Tibet and the Himalayas to Korea, China and Japan; NW, tropical and S Africa; introduced in N America.

Var. *sylvestris,* with glabrous fruit, occurs throughout most of the range of the species but not in Iraq. In Iraq represented by:

var. **nemorosa** (*M. Bieb.*) *Trautv.* in Act. Hort. Petrop. 5: 437 (1877).

Anthriscus nemorosa (M. Bieb.) Spreng., Umbell. Prodr.: 27 (1813); DC., Prodr. 4: 223 (1830); Boiss., Fl. Orient. 2: 911 (1872); Zohary, Fl. Palest. ed. 2, 1: 530 (1932); Schischkin in Fl. SSSR 16: 127 (1950); Townsend in Kew Bull. 17: 431 (1964); Grossh., Fl. Kavk. ed. 2, 7: 34 (1967); Mouterde, Nouv. Fl. Lib. et Syr. 2: 585 (1970); E. Nasir in Fl. Pak. 20: 96 (1972).
Chaerophyllum nemorosum M. Bieb., Fl. Taur.-Cauc. 1: 232 (1808); Hedge & Lamond in Fl. Turk. 4: 322 (1972).

Fig. 78. **Anthriscus sylvestris** var. **nemorosa**. 1, habit × ²/₃; 2, mature umbel × ²/₃; 3, petals × 6; 5, fruits (both faces) × 3; 5, fruit cross section × 12. All from *Davis* 14822. Drawn by D. Erasmus (1964).

Fruit with short, stiff, curved, tuberculate-based bristles. Fig. 78: 1–5.

HAB. (of var.): Among undergrowth in oak forest on a mountain slope, not common; alt. 1200 m; fl. & fr. not recorded.
DISTRIB. (of var.): Very rare in Iraq. **MAM**: Sarsang above Shaqlawa, *Haines* 1006!; **MRO**: Halgurd Dagh, Tenji Saipaiyan, *Hadač* 2170.
NOTE. As mentioned elsewhere [Kew Bull. 39: 603–604 (1984)] the recently noted occurrence of this plant in Tanzania, again in an *A. sylvestris* region, seems to confirm beyond doubt that *A. nemorosa* is best treated at varietal rank under *A. sylvestris*.

In Europe only in Italy and the Balkan peninsula, otherwise occurring throughout most of the range of the species.

2. **Anthriscus cerefolium** (*L.*) *Hoffm.*, Gen. Umb. ed. 1: 41 (1814); DC., Prodr. 4: 223 (1830); Boiss., Fl. Orient. 2: 913 (1872); Hayek in Prodr. Fl. Balc. 1: 1064 (1927); Schischkin in Fl. SSSR 16: 1336 (1950); Townsend in Kew Bull. 17: 431 (1964); Grossh., Fl. Kavk. ed. 2, 7: 36 (1967); Hedge & Lamond in Fl. Turk. 4: 323 (1972); Hedge & Lamond in Fl. Iranica 162: 85 (1987).

Scandix cerefolium L., Sp. Pl. ed. 1: 257 (1753).

Erect annual, 30–75 cm ± considerably branched from halfway or lower. Stem fistular, striate or somewhat sulcate, mostly glabrous but often pubescent above nodes. Leaves similar throughout, 3–4-pinnate, lamina deltoid-ovate or broadly deltoid, segments shortly oblong or lanceolate, glabrous above, somewhat hairy below; lower leaves long-petiolate, upper sessile. Umbels several, sessile or on peduncles up to 2.5(–3) cm long, often lateral and leaf-opposed; rays 3–5, pubescent, 1.5–3(–3.5) cm; involucre absent. Partial umbels 4–10-flowered, pedicels 2–6 mm; flowers 2 mm in diameter, not obviously radiate; involucel of 2–3 lanceolate, ciliate bracts, 2–4 mm. Fruit 7–10 × ± 1.5 mm, linear with a distinctly demarcated beak about half the length of the body of fruit, glabrous or setose, ribs scarcely apparent save on the beak; pedicels lacking a ring of hairs at apex. Stylopodia conical, tapering so gradually into the styles (1 mm or less) as to be indistinguishable from them; disk shortly cylindrical, longer than broad. A delicate, light green plant with graceful foliage.

DISTRIB. (of species): Probably native in E & SE Europe (Austria to Romania & S Russia), naturalized elsewhere; Turkey, Caucasus, Iran, C Asia (Turkmenistan), N Africa. Introduced in N America.
NOTE. Chervil, Salad chervil (Am.), much cultivated as a herb in France and many other parts of the world.

Var. *cerefolium*, with glabrous fruits throughout the range of the species. In Iraq, the species is represented by:

var. **trichosperma** *Wimm. & Grab.*, Fl. Siles. 1: 291 (1827).
A. longirostris Bertol., Fl. Ital. 3: 197 (1837); Schischkin in Fl. SSSR: 16: 137 (1950).

Fruit (except for the beak) with simple bristles.

HAB. (of var.) In a shady ravine under trees on a N-facing mountain slope; alt. ± 1100 m; fl. & fr. May.
DISTRIB. (of var.): Very rare in Iraq, only found in the forest zone. **MRO**: Shaqlawa, *Haines* W649!; *Haines* s.n.!; Zawita, on lower slopes of Botin, *Haines* s.n.; between Shaikhan and Sakri-Sakran, *Hadač et al.* 5428.
NOTE. As a parallel variant of *A. cerefolium* to var. *nemorosa* of *A. sylvestris*, similarly prevalent in the eastern part of the range of the species, it seems only reasonable to give both similar rank.

SE Europe (to S Russia & Crimea), Turkey, Caucasus, Iran, C Asia (Turkmenistan).

33. **FOENICULUM** Mill.

Gard. Dict. ed. 4 (1754)

Tall, glabrous biennial or perennial herb with 3–5 pinnately decomposite leaves, segments filiform. Umbels compound. Involucre and involucel usually absent. Flowers yellow, petals rather broad, with an obtuse incurved lobule, not emarginate above, not radiate. Calyx teeth obsolete. Fruit oblong or ovoid, slightly laterally compressed, commissure broad. Mericarps subterete, ribs 5, prominent, obtuse, equal or the marginal slightly stouter. Vittae solitary in valleculae, 2 on the commissure. Stylopodia large, conical, entire; styles short. Carpophore bipartite. Endosperm sulcate beneath vittae, dorsally compressed, commissural face flat or slightly concave.

A monotypic genus.
Foeniculum (the diminutive of the Lat. *foenum*, hay); Fennel.

1. **Foeniculum vulgare** Mill., Gard. Dict. ed. 8, no. 1 (1768); DC., Prodr. 4: 142 (1830); Hayek, Prodr. Fl. Balc. 1: 1018 (1927); Zohary, Fl. Palest. ed. 2, 1: 544 (1932); Guest in Dep. Agr. Iraq Bull. 27: 35 (1933); Blakelock in Kew Bull. 3: 434 (1948); Schischkin in Fl. SSSR 16: 542 (1950); H. Riedl in Fl. Lowland Iraq: 466 (1964); Grossh., Fl. Kavk. ed. 2, 7: 97 (1967); Mouterde, Nouv. Fl. Lib. et Syr. 2: 262 (1970); Hedge & Lamond in Fl. Turk. 4: 376 (1972); Zohary, Fl. Palaest. 2: 432 (1972); E. Nasir in Fl. Pak. 20: 61 (1972); Husain & Kasim, Cult. Pl. Iraq 106 (1975); Chakr., Plant Wealth of Iraq 246 (1976); Meikle, Fl. Cyprus 1: 749 (1977); Rech.f., Fl Iranica 162: 346 (1987); She Menglan & Watson in Fl. China 14: 134 (2005).

Anethum foeniculum L., Sp. Pl. ed. 1: 263 (1753).
Foeniculum officinale All., Fl. Ped. 2: 25 (1785); Boiss., Fl. Orient. 2: 975 (1872); Zohary in Dep. Agr. Iraq Bull. 31: 112 (1950); Rawi, ibid. Tech. Bull. 14: 92 (1964).
F. piperitum DC., Prodr. 4: 142 (1830); Boiss., Fl. Orient. 2: 975 (1872).
F. foeniculum (L.) Karst., Fl. Deutschl. 2: 462 (1895).
F. vulgare subsp. *piperitum* (Ucria) Coutinho, Fl. Port.: 450 (1913); V. Täckh., Stud. Fl. Egypt ed. 2: 393 (1794).

Stout, erect, glabrous, glauco-pruinose perennial, mostly 0.6–2 m. Stem solid, tough, round, finely striate, usually much-branched. Leaves 3–5-pinnate into filiform segments of (0.5–)1–5 cm long and in more than one plane, all furnished with lanceolate-oblong, elongate sheaths which in the uppermost leaves often exceed the length of lamina. Outline of lower leaves broadly deltoid or deltoid-ovate. Peduncles 2–11 cm. Umbels many, terminal or leaf-opposed, with (3–)8–30 rays 0.8–5 cm long, unequal. Partial umbels of many (8–30) flowers, pedicels 2–7 mm. Involucre and involucel usually absent, though one or two subulate bracteoles may occasionally be found. Fruits oblong-ovoid, 4–6 mm, ± pruinose, glabrous, ribs prominent. Stylopodia turgidly conic, furrowed. Styles very short (± 0.25 mm.), reflexed. Whole plant with a characteristic and powerful aromatic odour resembling aniseed, especially when bruised. Fig. 79: 1–6.

HAB. in the mountains by the open side of stream near a sheep pool, in rocky places; alt. 950–1350 m; fl. & fr. Jun.–Jul.
DISTRIB. Rare in Iraq; only found native in the middle forest zone. **MAM**: Sarsang, *Haines* W479!; ?**MSU** (or Iran): Hawraman, *Haussknecht* s.n. The remaining specimens are from plants cultivated in the steppe or desert regions: **FUJ/FNI**: Mosul (cult.), *Rawi* 5834!; **DLJ**: Baiji ("among crops on irrigated land nr R. Tigris"), *H.D. Peile* 5!; **LCA**: Baghdad (cult.), *Rawi & Shahwani* 25328!; Rustamiya (cult. "from seed obtained from Mosul") *Lazar* 1371!, *Guest* 3420!; **LBA**: Basra (cult.), *Guest* 309!
NOTE. Common Fennel; GHAZNĀIJ (Ir.-Baghdad, *Guest* 3420). RĀZYĀNIJ, a name then used for fennel in Syria, Egypt, Spain and Morocco, according to Ibn al-Baitar) is the only local name recorded for it which probably refers more specifically to its fruits – commonly on sale in the markets in Iraq where, as elsewhere, they are used in cooking as a condiment – than to the plant itself. No doubt, as mentioned by Guest (1933), fennel is often confused with dill (*Anethum graveolens*, q.v.) the dried leaves of which are also sold as a potherb and used to garnish dishes under the colloquial name SBINT. The fruits of fennel yield an aromatic oil used in medicine as carminative, stomachic and stimulating; as a flavourant they form an ingredient of the well-known children's laxative, Liquorice Powder, according to Wren (1956). Another Arabic name for this plant, SHAMRA BARRÎYA was noted by Handel-Mazzetti in Aleppo (Hakim 26) and the name SHAMRA was confirmed by Sharaf (1928), who also gives SHAMĀR as a variant. The young shoots of the plant are used in Europe as a vegetable. It is widely cultivated in many parts of the world, and often found as an escape.

C & S Europe (Britain, Spain, France, Germany, Switzerland, Italy, Sicily, Balkans to Ukraine), Cyprus, Syria, Lebanon, Palestine, Sinai, Egypt, Arabia, Turkey, Caucasus, Iran, Afghanistan, Pakistan, C Asia (Turkmenistan, Tajikistan), N India, China, Japan, N Africa (Morocco-Libya), Macaronesia (Madeira, Canary Is., Azores); also introduced into many other parts of the world such as Polynesia, tropical & S Africa, S America.

As indicated above, it is usually seen as a cultivated plant in our territory and wild specimens are rare. The garden forms are commonly biennial, with fistular stems; two have been given varietal names:

Fig. 79. **Foeniculum vulgare**. 1, habit × ²/₃; 2, leaf × ²/₃; 3, mature umbel × ²/₃; 4, petal × 8; 5, fruits (both faces) × ²/₃; 6, fruit cross section × 8. 1 from *E. Thurston* 1912; 2, 4 from 22 Jul 1931, *Medlin* s.n.; 3, 5–6 from *Clarke* 47614. Drawn by D. Erasmus (1964).

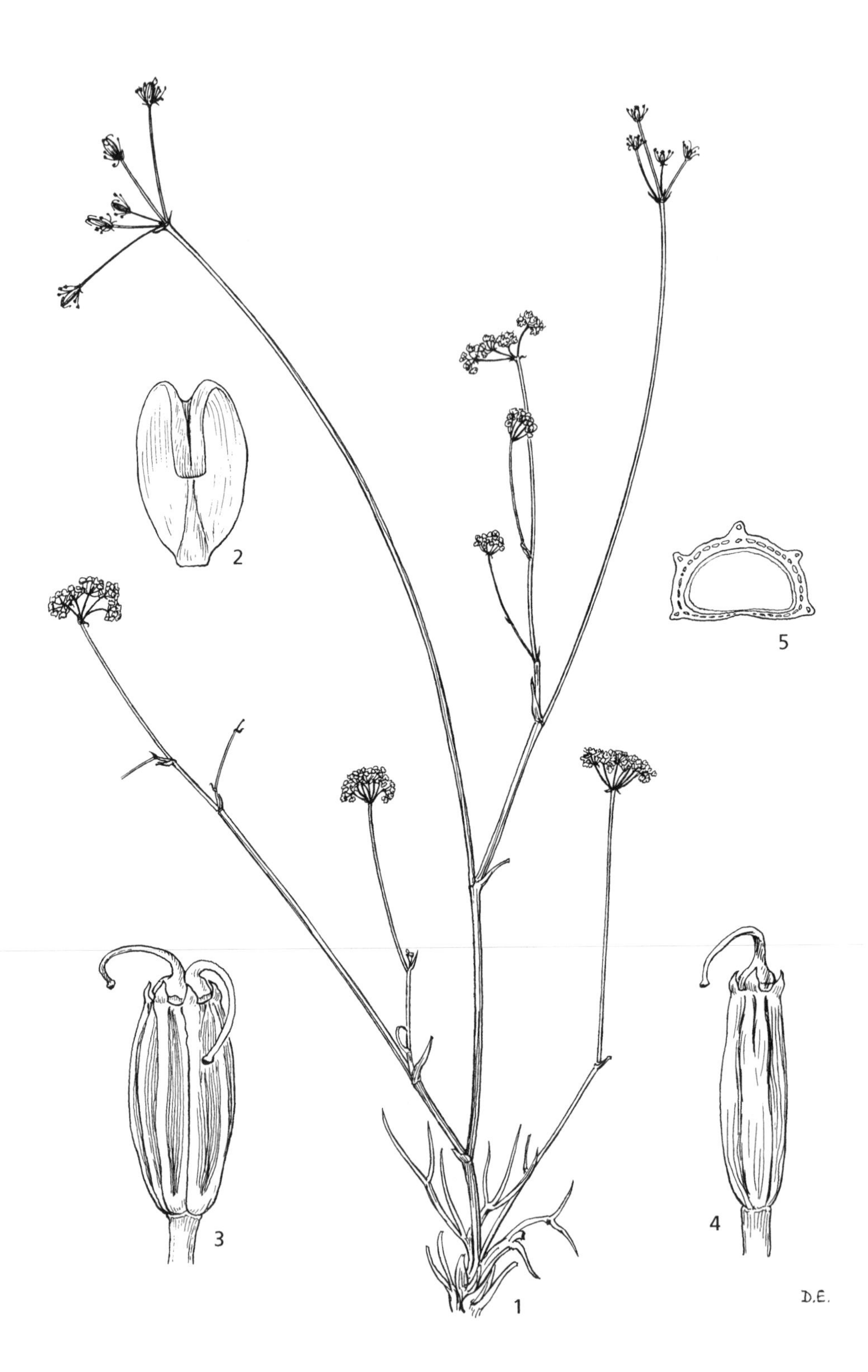

Fig. 80. **Sclerochorton haussknechtii**. 1, habit × 2/3; 2, petals × 22; 3, 4 fruits (both faces) × 6; 5, fruit cross section × 12. All from *Gillett* 9614. Drawn by D. Erasmus (1964).

var. **dulce** (*Mill.*) *Fiori & Paol.*, Fl. Anal. Ital. 2: 173 (1899).

Stems subcompressed, underground surculi inflated and fleshy.

var. **sativum** (*Bertol.*) *Fiori & Paol.*, l.c.

Stems terete, underground surculi not inflated and fleshy.

NOTE. One Iraq specimen, *Haines* 479, has apparently the centre of the stem (unfortunately the main stem is not present) solid with pith and is stated to be perennial. Its habitat also indicates that this may be truly wild. The habitat of *Haussknecht* s.n. (which may be in either Iraq or Iran) is recorded as "in rupestris" (in rocky places) which also indicates a wild plant.

34. **SCLEROCHORTON** Boiss.

Fl. Orient. 2: 968 (1872)

Glabrous, wiry perennial herbs. Leaves pinnately dissected, with thick, rigid grooved segments. Umbels compound. Involucre and involucel present. Flowers white, some of the outer often male, not producing fruit; petals obovate, distinctly emarginate above with a truncate inflexed acumen. Calyx teeth distinct, acute. Fruit oblong-ovoid, somewhat laterally compressed. Mericarps with the 5 primary ridges prominent, laterals marginal; valleculae broad, with 5–7 slender vittae. Commissural face of endosperm concave. Disk small, stylopodia depressed or elongate-conical. Carpophore bipartite.

Two species in SE Europe and SW Asia respectively; one species in Iraq.

Sclerochorton (from the Gr. σκληρός, *skleros*, hard, χόρονός, *khoronos*, crown, presumably from the rigid remains of earlier growth found at the base of the stem).

1. **Sclerochorton haussknechtii** Boiss., Fl. Orient. 2: 968 (1872); Rawi in Dep. Agr. Iraq Tech. Bull. 14: 96 (1964).

Plant 30–40 cm, stiffly erect, alternately branched from near the base, glabrous throughout, glaucous. Stem and branches slender but tough, terete, finely striate, stem base clad with rigid and fibrous remains of previous years' petioles. Leaves irregularly pinnately dissected, or the final divisions furcate, segments thick, grooved, cylindrical or flattened on one side, 2–50 × 0.6–1.5 mm, mostly basal; upper leaves rapidly reducing to sheath-like bracts, branches leafless. Peduncles 2.5–22 cm. Umbels several, rays ± 6, unequal, 0.6–7 cm. Bracteoles and bracts 3–6, subulate, short. Partial umbels 4–12-flowered, pedicels 2–12 mm, unequal. Flowers white, ± 2 mm in diameter. Fruit 5 × 2.5 mm, oblong-ovoid, calyx teeth persistent and obvious. Stylopodia depressed-conic, sulcate. Styles long, reflexed, ± 2.5 mm. Fig. 80: 1–5.

HAB. On a high mountain on metamorphic rock; alt. 2600 m; fl. & fr. Aug.–Sep.
DISTRIB. Very rare in Iraq, only found once over 60 years ago, in the upper thorn-cushion zone. **MRO**: Halgord Dagh, *Gillett* 9614!

Iran (on Mt. Shahu, 3650 m, *Haussknecht* s.n.).

35. **HEPTAPTERA** Marg. & Reut.

Mém. Soc. Phys. Genève 8: 302 (1838)
Colladonia DC., Prodr. 4: 240 (1830), non Sprengel (1824)

Tall, perennial herbs with a papillose indumentum, leaves 1–3-pinnately divided into broad, oblong segments, rarely entire or pinnatipartite. Umbels compound. Involucre and involucel present, of several bracts. Flowers yellow, petals oblong, not or scarcely emarginate above, with an inflexed acumen. Calyx teeth very short or obsolete. Inflorescence usually polygamous with lateral umbels male. Fruit ovoid, oblong or ellipsoid, somewhat dorsally compressed, commissure broad, pericarp spongy; mericarps subterete, usually asymmetrical, one mericarp with one dorsal rib, the other with two, as well as both marginal ribs, furnished with corky or membranous wings, rarely regularly 5–winged. Vittae numerous, adherent to the thin endocarp of the seed. Disk broad, depressed, margins crenulate. Stylopodia usually small, rarely more prominent. Carpophore bipartite. Endosperm with the commissural face deeply sulcate, margins involute.

Eight species in Mediterranean Europe and SW Asia; one species in Iraq.
Heptaptera (from the Gr. επτά, *epta*, seven, πτερόν, *pteron*, wing, from the normally asymmetrically winged fruit, with three wings on one mericarp and four on the other).

1. **Heptaptera anisoptera** (*DC.*) *Tutin* in Fedde Rep. 74: 33 (1966); Mouterde, Nouv. Fl. Lib. et Syr. 2: 620 (1970); Herrnstadt & Heyn in Notes Roy. Bot. Gard. Edinb. 31: 91 (1971) & in Fl. Turk. 4: 389 (1972); Herrnstadt & Heyn in Fl. Iranica 162: 169 (1987).

Prangos anisopetala DC., Prodr. 4: 240 (1830), sphalm. pro *P. anisoptera.*
Anisopleura crenata Fenzl in Flora 26: 459 (1843).
Meliocarpus anisopterus (DC.) Boiss. in Ann. Sci. Nat. sér. 3, 2: 85 (1844).
M. peduncularis Boiss., Diagn. ser. 1, 10: 52 (1849).
Colladonia crenata (Fenzl) Boiss., Fl. Orient. 2: 946 (1872); Zohary, Fl. Palest. ed. 2, 1: 540 (1932); Zohary in Dep. Agr. Iraq Bull. 31: 112 (1950); Rawi, ibid. Tech. Bull. 14: 90 (1964).
Prangos crenata (Fenzl) Drude in Pflanzenfam. 3, 8: 174 (1898); Guest in Dep. Agr. Iraq Bull. 27: 71 (1933); [*Colladonia anatolica* (non Boiss.) Bornmüller in Beih. Bot. Centralbl. 58B: 281 (1938).]
Heptaptera crenata (Fenzl) Tutin in Fedde Rep. 74: 33 (1966); Mouterde, Nouv. Fl. Lib. et Syr. 2: 620 (1970); Zohary, Fl. Palaest. 2: 409 (1972).

Erect, yellowish or glaucous-green, 1–1.5 m. Stem rigid but not very stout, terete, finely striate, often ± scabrous with branched papillate hairs on the striae below, fibrous remains of previous years' leaf-sheaths persistent at base, branched above, branches terete, also finely striate. Basal leaves oblong-ovate in outline, with a long broad sheath attenuate into the petiole, 2 or 3-pinnate into broad, ovate, oblong-elliptic or broadly lanceolate segments up to 15 × 7.5 cm. Middle cauline leaves deltoid to ovate, simply pinnate, upper reduced to ± large sheaths. Lower and middle leaves scabrous with branched papillate hairs on sheaths, primary divisions and underside of main veins, otherwise glabrous, with regular cartilaginous-margined crenulations. Peduncles 6.5–12 cm, those of fertile umbels shortest. Umbels few, very large, 6–20 rays very long, those of fructiferous umbels 6–20 cm, of the laterals shorter. Involucral bracts ± 6, very conspicuous, lanceolate to oblong. Partial umbels many (1–30)-flowered; flowers 5–7 mm in diameter, long-pedicellate, with the fructiferous pedicels to ± 5 cm. Bracteoles similar in form and number to bracts. Fruit elliptic-oblong to cuneiform, (10–)15–22 × 5–7 mm. Marginal ribs broadly (± 2 mm) winged, central dorsal rib merely prominent in one mericarp with intermediate ribs moderately winged, the other mericarp with central winged and intermediates unwinged. Disk broad, depressed, slightly concave, margin undulate. Stylopodia shortly conical. Styles tightly reflexed, ± equalling the diameter of fruit.

HAB. In mountains, on grassland among degraded *Quercus* scrub on a limestone ridge on shady banks by mule track on border of vineyard; alt. 500–1700 m; fl. & fr. May–Aug.
DISTRIB. Occasional in the forest zone of Iraq. **MAM**: Sandur, nr Dohuk, *Haines* W1571!, W 1710!; Baradost Dagh, *Field & Lazar* 875!; on a mountainside near Arbil, *Bornmüller* 1269!; **MRO**: Mergasur, *Hadač et al.* 5820; **MSU**: NE of Darband-i Khan, *Poore* 602!; Malakawa, *Rawi, Husham (Alizzi) & Nuri* 29479!

Syria, Lebanon, Palestine, Turkey, Iran.

36. **PRANGOS** Lindl.

Quart. Journ. Sci. Lond. 19: 7 (1825)

Tall, perennial herbs, leaves pinnately decomposite into linear to filiform segments. Inflorescence polygamous, terminal umbel hermaphrodite and male, laterals mostly male. Umbel compound. Involucre and involucel of several bracts and bracteoles, usually short, persistent or caducous, narrow or rarely broad. Flowers yellow, petals broadly ovate, not or scarcely emarginate above, with an inflexed acumen. Calyx teeth obsolete or sometimes evident. Fruit ovoid, ellipsoid or ± globose, somewhat laterally compressed, the commissure broad, mesocarp corky; mericarps subterete, the 5 ribs furnished with conspicuous flat to extremely undulate wings. Vittae numerous, ± contiguous. Disk broad, explanate, margins usually undulate. Stylopodia usually small, rarely larger, conic. Carpophore bipartite. Endosperm with the face deeply sulcate, margins involute.

Some 30 species distributed from the Mediterranean to C Asia; seven species in Iraq.
Prangos (from a vernacular name reported for *P. pabularia* in northern India).

1. Petals hairy on the dorsal surface .... 7. *P. corymbosa*
   Petals glabrous 2
2. Small mountain plant not exceeding 40 cm; leaves almost entirely radical .... 6. *P. peucedanifolia*
   Taller plants with leafy stems 3
3. Lower part of primary ridges below wings with numerous papillate outgrowths resembling smaller, interrupted secondary wings .... 5. *P. pabularia*
   Lower part of primary ridges smooth or with very few outgrowths 4
4. Wings of primary ridges of fruit straight or slightly undulate above; sepals obsolete or very irregular 5
   Wings of primary ridges of fruit strongly undulate; sepals distinct 6
5. Bracts of involucre and involucel linear-lanceolate rarely as much as 3 mm wide .... 1. *P. ferulacea*
   Bracts of involucre and involucel broad, oblong or oblong-ovate, more than 3 mm wide .... 2. *P. platychlaena*
6. Plant rarely suffused with purplish colouration; fructiferous umbel-rays 4–5(–6) cm; fruit not more than 10 mm wide including the wings; exocarp intrusive between blocks of mericarp tissue .... 4. *P. uloptera*
   Plant commonly ± purplish-suffused; fructiferous umbel-rays (6–) 7–10(–14) cm; fruit more than 10 cm wide including the wings; exocarp not intrusive between blocks of mericarp tissue .... 3. *P. asperula*

1. **Prangos ferulacea** (*L.*) *Lindl.*, Quart. Journ. Sc. 19: 7 (1825); DC., Prodr. 4: 239 (1830); Boiss., Fl. Orient. 2: 937 (1872); Hayek, Prodr. Fl. Balc. 1: 1072 (1927); Guest in Dep. Agr. Iraq Bull. 27: 77 (1933); Bornmüller in Beih. Bot. Centralbl. 58B: 281 (1938); Fedchenko & Schischkin in Fl. SSSR 16: 265 (1950); Rawi in Dep. Agr. Iraq Tech. Bull. 14: 95 (1964); Grossh., Fl. Kavk. ed. 2, 7: 58 (1967); Herrnstadt & Heyn in Fl. Turk. 4: 386 (1972); Zohary, Fl. Palaest. 2: 408 (1972); Herrnstadt & Heyn in Boissiera 26: 39 (1977); Herrnstadt & Heyn in Fl. Iranica 162: 198 (1987).

Laserpitium ferulaceum L., Sp. Pl. ed. 2: 358 (1762).
Cachrys prangoides Boiss., Ann. Sc. Nat. sér. 3, 2: 76 (1844); Rawi in Dep. Agr. Iraq Tech. Bull. 14: 89 (1964).
Prangos stenoptera Boiss. & Buhse, Nouv. Mém. Soc. Nat. Mosc. 12: 104 (1860).
P. carinata Griseb. ex Degen in Term. Közl. 28: 44 (1896).
P. ferulacea var. *carinata* (Griseb.) Fiori, Nouv. Fl. Ital. 2: 98 (1925); Blakelock in Kew Bull. 3: 436 (1948); Rawi in Dep. Agr. Iraq Tech. Bull. 14: 95 (1964).
[*P. pabularia* (non Lindl.) *Guest* in Dep. Agr. Iraq Bull. 27: 77 (1933 p.p.].
Cachrys nematoloba Rech.f. & Riedl, Anz. Österr. Akad. Wiss., Math-Nat. Kl. 1961 (14): 248 (1961); Rawi in Dep. Agr. Iraq Tech. Bull. 14: 89 (1964).

Tall and robust, 0.5–1.5 m, corymbosely branched above, branches opposite or subverticillate; stem sulcate, ± glabrous or papillate, especially at nodes; branches and umbel rays glabrous. Lower leaves petiolate, broadly deltoid-ovate in outline, to 75 cm long, pinnately decomposite into long, mucronate, linear-setacous 5–11 × 0.25–2 mm segments, glabrous or more rarely scabrid; upper leaves sessile, more broadly deltoid, similarly dissected; segments fewer and often longer; sheaths broadly expanded, short, sulcate. Peduncles 5–10 cm. Umbels many, 8–15(–20)-rayed, rays 4–8.5 cm in length in fruiting umbels. Involucre and involucel of 3–8 membranous, linear-lanceolate, caducous bracts and bracteoles. Partial umbels 6–20-flowered, flowers ± 4 mm in diameter, fruiting pedicels 6–12 mm. Sepals obsolute or very irregular. Petals glabrous. Fruit oblong-ovoid to ellipsoid or ± globose, 16–30 × 10–15 mm; wings of primary ridges up to 2 mm wide, straight or only very slightly undulate, sometimes absent or present only as keels, ridges smooth below the wings. Styles 2–3 mm, deflexed; stylopodia flat or angled between the mericarps. Fig. 81: 1–5.

HAB. In the mountains, on high rocky slopes, sometimes also cultivated at lower levels and also found in the vicinity of villages; alt. 950–2200 m; fl. & fr. May–Jun.
DISTRIB. Occasional in the NW and central sectors of the lower thorn-cushion and forest zones of Iraq. **MAM**: Jabal Daimka and Jabal Kashan, nr Guli, *Mudir of Guli* 3147!; Amadiya (cult.), *Guest* 1256!; Jedaida, *Karim, Nuri & K. Hamid* 44804!; Sarsang, *Haines* W1008!; **MRO**: Sakri-Sakran, *Bornmüller* 1262; Kudu, *Rawi* 9227!; Mergasur, *Agnew et al.* 5821; Haji Umran, *Wiltshire* in herb. Haines s.n.!; **MSU**: Hawraman, N of Halabja, *Rawi* 22087!; Kamarspa, between Halabja and Tawila, *Rawi* 22264; above Daramar, *Gillett* 11848!

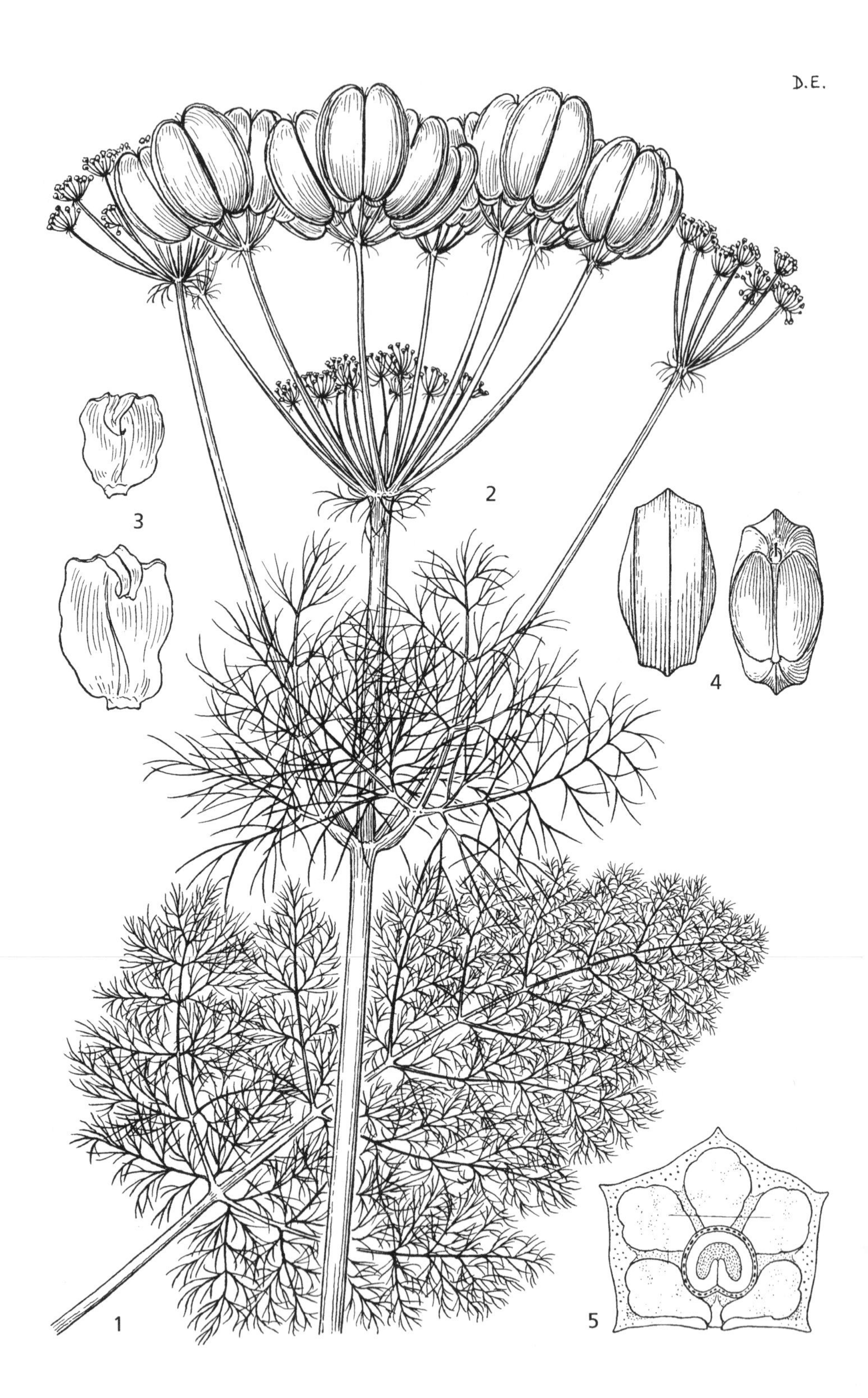

Fig. 81. **Prangos ferulacea**. 1, habit × 2/3; 2, leaf × 1/4; 3, petals × 14; 4, fruits (both faces) × 1; 5, fruit cross section × 2. 1, 3–5 from *Rawi* 22087; 2 from *Rawi* 22264. Drawn by D. Erasmus (1964).

Note. HALIS (Kurd.-Zakho, *Mudir of Guli* 3417 – "an important fodder plant, especially for sheep, larger animals eat it dried as hay. Found in the hills and sometimes cultivated for winter fodder or for seed"), HALIZ (Kurd.-Amadiya, *Guest*, ex *Majid Mustafa*, 1256 – "from a plant cultivated in a village garden but also said to be growing wild in the district. It is eaten by animals"). Haines also notes on the label of his gathering at Sarsang that this plant was "plentiful around village buildings and chaff dumps". Wahby & Edmonds' *Kurdish-English Dictionary* (1966) spells the name HELÎZ and translates the name into English as "hay"; also LOH (Kurd.-Halabja, *Gillett* 11848 – "leaves collected as winter fodder") cf. LU = "the name of a fodder grass" Wahby & Edmonds. There is no doubt that the plant provides a most important winter fodder for livestock in the mountains when the villages are for long under snow.

SE Europe (Italy, Sicily, Balkans), W Syria, Lebanon, Turkey (C & SE Anatolia), Armenia, W, C & S Iran, N Africa (Libya).

2. **Prangos platychlaena** *Boiss.* in Tchihatcheff, Asie Mineure 3 (Bot.), 1: 457 (1860); Boiss., Fl. Orient. 2: 938 (1872); Zohary, Fl. Palest. ed. 2, 1: 539 (1932); Herrnstadt & Heyn in Fl. Turk. 4: 386 (1972); Herrnstadt & Heyn in Fl. Iranica 162: 199 (1987).

Plant tall, 1–1.5 m, glabrous, corymbosely branched above, branches opposite. Stem and branches purplish, terete, striate. Lower leaves ovate-oblong to deltoid-ovate, up to 60 cm, pinnately decomposite into linear-setaceous mucronate segments. Upper leaves reducing, bushy, sessile. Peduncles 7–11 cm. Umbels numerous, 9–20-rayed, rays of fruiting umbels 4–12 cm. Involucre of 4–8 broad oblong to ovate-oblong persistent bracts. Partial umbels 10–20-flowered; flowers ± 4 mm in diameter; fruiting pedicels 9–12 mm; bracteoles similar in number and form to bracts. Sepals obsolete. Petals glabrous. Fruit oblong-ellipsoid, 15–20 × 7–8 mm, wings of primary ridges rather variable, straight to somewhat undulate, especially above, ridges smooth below the base of wings. Styles 2–3 mm, deflexed; stylopodia flat.

Hab. High in the mountains, among *Astragalus* shrubs, on rocky slopes, on metamorphic rocks; alt. 2200–2800 m; fl. & fr. Aug.–Sept.
Distrib. Occasional in the sub-alpine thorn-cushion zone of Iraq. **mro**: Halgurd Dagh, *Gillett* 9616, *Guest & Husham (Alizzi)* 15840; Qandil range, NE of Rania, *Rawi & Serhang* 26824!
Note. KARKUL (Kurd.-Rowanduz, *Gillett* 9616).
I am not completely convinced of the identity of this material with *P. platychlaena*, which in the case of *Guest & Husham* 15840 is of fruits only. Attempts should be made to re-collect it in the above localities.

Turkey.

3. **Prangos asperula** *Boiss.*, Diagn. ser. 1, 10: 54 (1849) & Fl. Orient. 2: 942 (1872); Zohary, Fl. Palest. ed. 2, 1: 539 (1932); Mouterde, Nouv. Fl. Lib. et Syr. 2: 618 (1970); Zohary, Fl. Palaest. 2: 408 (1972); Herrnstadt & Heyn in Boissiera 26: 46 (1977).

Plant 0.75–1.5 m, much-branched above, upper branches opposite. Stem and all vegetative parts of the plant glabrous to ± densely papillate, upper stem and branches, umbel rays and fruit often ± suffused purplish or vinaceous or finely purplish striate. Lower leaves broadly deltoid-ovate, large, to ± 70 cm, petiolate, pinnately decomposite into mucronate, linear-setaceous, 8–30 × 0.2–1.5 mm segments; sheaths long, rather narrowly expanded, firm. Peduncles 5–17 cm. Umbels numerous, (10–)15–25(–30)-rayed, fructiferous rays stout (6–)7–10(–14) cm. Involucre and involucel of several subulate to narrowly oblong-lanceolate bracts and bracteoles, short. Partial umbels 12–20-flowered, flowers 2.5–3 mm in diameter, fruiting pedicels 8–15 mm. Petals glabrous or sometimes papillate, particularly along margins. Fruit broadly oblong-ellipsoid, 15–26 × 10–18 mm with 3–4 mm wide wings, wings strongly undulate but undulations few and large, ridges smooth below the wings. Styles 3–4 mm, deflexed; stylopodia horizontal.

Lebanon, Syria, Iran.

subsp. **asperula**, with the whole plant except for fruit (but sometimes including petals) papillate, and with leaf-lobes 1 mm or more wide, occurs only in Syria and Lebanon. In Iraq the species is represented by:

subsp. **haussknechtii** (*Boiss.*) *Herrnstadt & Heyn* in Boissiera 26: 48 (1977).

P. haussknechtii Boiss., Fl. Orient. 2: 940 (1872); Blakelock in Kew Bull. 3: 436 (1948); Rawi in Dep. Agr. Iraq Tech. Bull. 14: 95 (1964); Herrnstadt & Heyn in Fl Iranica 162: 199 (1987).

Whole plant, including the petals, usually glabrous; leaf lobes 0.2–0.7 mm wide.

HAB. High mountain sides, rocky slopes; alt. 1900–2800 m; fl. & fr. Jul.-Aug.
DISTRIB. Rare in Iraq; only found on two neighbouring mountains in one district of the subalpine thorn-cushion zone. **MRO**: Halgurd (Algurd Dagh), *Rawi* 13748; Goum Tawera, nr Iranian frontier, on Helgurd range, *Rawi & Serhang* 24727!; Chiya-i Mandau, *Guest* 2693!
NOTE. KARKŪL (Kurd.-Rowanduz, *Rawi* 13748), cf. GARGŪL (*Guest*, 1933).

W & C Iran.

4. **Prangos uloptera** *DC.*, Prodr. 4: 239 (1830); Boiss., Fl. Orient. 2: 940 (1872); Bornmüller in Beih. Bot. Centralbl. 58B: 281 (1938); Blakelock in Kew Bull. 3: 436 (1948); Fedchenko & Schischkin in Fl. SSSR 16: 268 (1950); Zohary in Dep. Agr. Iraq Bull. 31: 112 (1950); Rawi, ibid. Tech. Bull. 14: 95 (1964); Herrnstadt & Heyn in Fl. Turk. 4: 383 (1972) & Boissiera 26: 34 (1977); Herrnstadt & Heyn in Fl. Iranica 162: 196 (1987).

Tall, robust, 50–80 cm, corymbosely branched above, upper branches opposite or the terminal verticillate. Stem and branches sulcate, roundish or angled. Indumentum of vegetative parts variable, sometimes absent and stem and leaves quite glabrous, at other times they may be densely scabrid. Lower leaves large, to 45 cm, petiolate, broadly deltoid-ovate in outline, pinnately decomposite into linear-setaceous, mucronate 0.5–1 mm wide segments which are variable in length, in some plants attaining 25 mm, in others at most 8 mm; upper leaves sessile, more broadly deltate, similarly dissected; sheaths broadly expanded, short, firm, striate. Peduncles 3.5–11 cm. Umbels many, rays 6–15(–20), unequal, 4–5(–6) cm long in fruit. Involucre and involucel of 6–8 short, lanceolate, caducous bracts and bracteoles. Partial umbels (4–)6–15-flowered, flowers 3. 5 mm in diameter, fruiting pedicels 4–10 mm. Petals glabrous. Fruit oblong, 8–12 × 4–10 mm including wings; wings very pale, strongly undulate, ridges smooth and devoid of papillae below the base of wings. Styles 2–3 mm, deflexed; stylopodia horizontal, depressed. Fig. 82: 1–5.

HAB. On mountains, on igneous and metamorphic rock, on limestone scree, on dry rocky ridges among oak forest; alt. 1300–3500 m; fl. & fr. (Apr.–)May–Jul.(–Sep.).
DISTRIB. Common, locally dominant, in the central and SE sectors of the alpine region, thorn-cushion and upper forest zones of Iraq. **MRO**: Kuh-i Sefin, *Bornmüller* 1263; **MRO**: Halgurd Dagh, *Gillett* 9577!, 12379!, *Rawi* 13748!, *Guest & Husham (Alizzi)* 15828!; Chiya-i Mandau, above Walza, *Guest* 2694!; Malakh (Gomasur), NE of Rania, *Rawi & Serhang* 24032!; lower southern slopes of Qandil range, *id.* 26701!; **MSU**: Pira Magrun, *Haussknecht* s.n.; *Poore* 628!; Amoret, nr Qaradagh, *Haines* W1146!; Malakawa, nr Penjwin, *Rawi* 22477!; Qopi Qaradagh, *Hadač et al.* 5202; Hawraman, above Daramar, *Gillett* 11882?; ?(or Iran), "Avroman & Schahu", *Haussknecht* s.n.
NOTE. KERMAKH (Kurd.-Rowanduz, *Gillett* 9577 – "used as a winter fodder and locally dominant on steep slopes"). Boissier (op. cit.) also cites the Kurdish name 'Dschinuhr' (CHINŪR?) which Wahby & Edmonds (1966) merely describe as "name of a herb".

Turkey (SE Anatolia), Caucasus, Iran, Afghanistan, C Asia (Turkmenistan to Siberia), N Africa.

5. **Prangos pabularia** *Lindl.* in Quart. Journ. Sci. Lit. Arts 19: 7 (1825); DC., Prodr. 4: 239 (1830); Hook.f., Fl. Brit. Ind. 2: 695 (1879); Boiss., Fl. Orient. Suppl.: 261 (1888); Guest in Dep. Agr. Iraq Bull. 27: 77 (1933), p.p.; Fedchenko & Schischkin in Fl. SSSR 16: 270 (1950); Zohary in Dep. Agr. Iraq Bull. 31: 112 (1950); Rawi, ibid., Tech. Bull. 14: 95 (1964); Herrnstadt & Heyn in Fl. Turk. 4: 383 (1972) & Boissiera 27: 25 (1977); E. Nasir in Fl. Pak. 20: 150 (1972); Herrnstadt & Heyn in Rech. f., Fl. Iranica 162: 193 (1987).

P. lophoptera Boiss., Ann. Sc. Nat. sér. 3, 2: 82 (1844); Boiss., Fl. Orient. 2: 941 (1872); Fedchenko & Schischkin in Fl. SSSR 16: 269 (1950).

Tall, robust, 0.3–1 m, corymbosely branched above, the upper branches opposite or the terminal verticillate. Stem and branches striate, subterete, all vegetative parts glabrous or scabrid. Lower leaves large, to 45 cm, petiolate, broadly deltoid-ovate in outline, pinnately decomposite into linear-setaceous, mucronate, 5–25 × 0.5–1 mm segments. Upper leaves sessile, broadly deltoid, similarly dissected; sheaths short, broadly expanded, striate, firm. Peduncles 4–13 mm. Umbels many, rays 5–15(–20), unequal, length 2–7 cm in fruit. Involucre and involucel of 6–8 lanceolate, caducous bracts and bracteoles, the latter long, occasionally equalling pedicels. Partial umbels 10–15(–20)-flowered, flowers ± 3.5 mm in diameter,

Fig. 82. **Prangos uloptera.** 1, habit × ⅔; 2, leaf × ¼; 3, petal × 20; 4, fruits (both faces) × 2; 5, fruit cross section × 6. 1 from *Gillett* 9577; 2 from *Guest* 2065; 3 from *Poore* 628; 4–5 from *Rawi* 22477. Drawn by D. Erasmus (1964).

fruiting pedicels 5–15 mm. Petals glabrous. Fruit oblong, 8–16(–24) × 4–12 mm, including wings; wings of primary ridges very pale, much undulate-crisped, undulations of adjoining ridges meshing considerably across the valleculae, which may be completely obscured by wings and the large irregular papillae with which the primary ridges are clothed below the base of wings. Styles 3–4 mm, deflexed or flexuose; stylopodia horizontal, depressed.

HAB. On mountain sides, sometimes among *Quercus* woods; alt. 1150–1800 m; fl. & fr. May–Jul.
DISTRIB. Occasional in the upper forest zone of Iraq. **MAM**: Gara Dagh, *Rawi* 9268!, 9298!; Sarsang, *E Chapman* 26371!; **MRO**: Khalana (Kholan), *Rawi* 13835!; **MSU**: Penjwin, *Rawi* 12201!, 12243!; Pira Magrun, *Thesiger* 1126, *Hadač* 2901; Qara Dagh, *Chapman* 26371.
NOTE. BIJAN (Kurd.-Penjwin, Rawi 12243) cf. BÊJÂN (the name of a fodder grass, Wahby & Edmonds, 1966); ?KUWAIRIK (Kurd.-Rowanduz, Rawi 13835).

As mentioned by Guest (1933) *P. pabularia* is recognised as a hay plant in Tibet where it is of great value in the cold arid mountain regions that cannot support better pasture. Watt (1892) examined the records of a number of veterinary surgeons and others who had travelled or resided in Kashmir and Afghanistan and concluded that the plant provides a highly nutritive fodder for sheep and goats. He found that the medicinal properties of its fruit, which was readily obtainable in the local bazaars, were widely recognised and in particular that it had a curative effect on animals affected by the liver fluke. He mentions however that horses which had partaken of the fruit were said to have suffered severely from inflammation of the eyes and even some from temporary blindness. Uphof (1968) lists its medicinal properties as stimulant, carminative, diuretic, stomachic and emmenagogue; it is reputed to promote the expulsion of the foetus in childbirth. According to Fedchenko & Schischkin (1950) there are two forms of this plant one of which cattle will not eat, though the morphological differences between them are so small that it can only be detected in nature. They state that chemical analysis of the two forms (which A. Koroleva names as subsp. *schirin* and subsp. *tez* respectively) revealed that, whereas the former which was sweet and edible yielded only 0.01% of a crude mixture of alkaloids, the latter – the bitter inedible form – yielded ten times that amount of alkaloids ((0.1%).

Turkey (C & S Anatolia), Caucasus, Iran, Afghanistan, Pakistan, C Asia (Tian Shan), N India (Kashmir), W Tibet.

6. **Prangos peucedanifolia** *Fenzl* in Flora 26: 463 (1843); Boiss., Fl. Orient. 2: 938 (1872); Herrnstadt & Heyn in Fl. Turk. 4: 384 (1972) & Boissier 26: 51 (1977); Herrnstadt & Heyn in Fl. Iranica 162: 201 (1987).

P. pumila Boiss. in Ann. Sc. Nat. sér. 3, 2: 77 (1844); Boiss., Fl. Orient. 2: 939 (1872); Nábělek in Publ. Fac. Sci. Univ. Masaryk 35: 125 (1923); Blakelock in Kew Bull. 3: 436 (1948); Zohary in Dep. Agr. Iraq Bull. 31: 112 (1950); Rawi, ibid. Tech. Bull. 14: 95 (1964).

Small herb, 25–35 cm, with few branches from about the middle. Stem terete, tomentellous below, rootstock thickly crowned with the fibrous remains of previous years' leaf-sheaths. Leaves mostly radical, 15–25 cm long, broadly deltoid in outline, pinnate with 3–5 pairs of pinnae, bipinnatisect into long 7–25 × 0.6–2 mm, glabrous, entire, subacute, linear-lanceolate, mucronate segments, pinnae rather remote; sheaths long, inflated, suddenly narrowed to the petiole, closely overlapping one another at the base of stem; upper leaves solitary or very few, sessile, simple or trisect. Peduncles 3–7 cm. Involucre and involucel very small, of few triangular to lanceolate bracts and bracteoles. Partial umbels 7–12-flowered, flowers ± 3 mm in diameter, fruiting pedicels 3–8 mm. Petals glabrous. Fruit glabrous, ovoid-globose, to 12–23 × 10–20 mm including wings of the primary ridges, which are 2 mm wide and straight or slightly undulate. Styles 2.5–3.0 mm, slender and flexuose; stylopodia horizontal or inwardly inclined in fruit, depressed.

HAB. Lower mountain sides, sometimes among *Quercus* trees or open coppiced *Quercus* scrub, steppic conglomerate hills, barren denuded earth; alt. (150)350–950 m; fl. & fr. Mar.–May.
DISTRIB. Quite common in the lower forest zone of Iraq, occasional in the NW sector of the moist-steppe zone. ?**MAM**: (probably, but possibly in Turkey?), "Pl. Mesopot., Kurdistan & Mossul", *Kotschy* 197?; **MRO**: Salah ad-Din, *Gillett* 11283!, *Rawi* 36229!; Shaqlawa, *Gillett* 8067!, 11562!; Baba Chichek (Shaqlawa), *Nábělek* 401; Haibat Sultan Dagh, *Rawi, Nuri & Kass* 28234!; **MSU**: 14 km E of Kirkuk, *Gillett & Rawi* 10675!; Jarmo, *Helbaek* 422!; Jafaran, nr Qaradagh, *Gillett* 7885!; **FUJ**: Jabal Makhul, nr Ain Dibs, *Gillett & Rawi* 7216!; **FKI**: nr Kirkuk, *Guest, Eig & Zohary* 5098!; Tuz, *Rogers* 0229!
NOTE. ?SITUK (Kurd.-Qaradagh, *Gillett* 7885, "fruits edible") cf. SITIK (Kurd., Wahby & Edmonds – "name of a herb"), ?BAUWA (Ir.-Jabal Makhul – "fruit good to eat when young").

Herrnstadt and Heyn under both references given above describe the petals of this species as "whitish", but those Iraqi gatherings bearing colour notes describe the flowers as "yellow" or "bright yellow".

Turkey (SE Anatolia).

7. **Prangos corymbosa** *Boiss.*, in Ann. Sc. Nat. sér. 3, 2: 81 (1844); Boiss., Fl. Orient. 2: 944 (1872); Herrnstadt & Heyn in Fl. Turk. 4: 385 (1972) & Boissiera 26: 56 (1977); Herrnstadt & Heyn in Fl. Iranica 162: 202 (1987).

Plant tall and robust, noted only as 1.2 m, corymbosely branched above, lower branches alternate, uppermost opposite, with reduced sheath-like leaves. Stem finely sulcate, densely crisped-hairy below, decreasingly so above and on the branches. Lower leaves petiolate, broadly deltoid in outline, to 55 cm, ± densely covered, as are the sheaths, with short white hairs; lamina pinnately decomposite into crowded, shortly linear, 2–3 × 0.6–0.8 mm, mucronate segments; upper leaves decreasing to simply pinnate, sessile, and finally becoming sheath-like. Peduncles 3.5–6.5 cm. Umbels many, 6–10-rayed, fruiting rays subglabrous, 3–5.5 cm long. Involucre and involucel very short, deltoid or lanceolate. Partial umbels 7–15-flowered, flowers ± 3.5 mm in diameter, fruiting pedicels unequal, glabrous, 10–15 mm. Petals hairy all over the dorsal surface, long-persistent. Fruit obpyriform to broadly ellipsoid, 15–22 × 12–15 mm, including wings of primary ridges, which are 3 mm wide, straight or slightly undulate. Styles ± 4 mm, slender, deflexed; stylopodia horizontal, depressed.

Hab. In a rocky valley among *Quercus* scrub; alt. ± 1300 m; fl. & fr. June.
Distrib. Very rare in Iraq, only found once in the SE sector of the upper forest zone. **msu**: Qopi Qaradagh, *Haines* W1524.

Turkey.

37. **FERULA** L.

Sp. Pl. ed. 1: 246 (1753); Gen. Pl. ed. 5: 117 (1754)

Perennial herbs, the leaves pinnately decomposite into filiform segments to variously pinnate with segments broad and dentate; sheaths usually large, rigid to flaccid. Umbels compound, polygamous, terminal large, hermaphrodite, usually with two or more long-pedunculate smaller umbels with male and hermaphrodite or only male flowers arising immediately below it. Inflorescence paniculate. Involucre and involucel absent or of a few bracts which are often small and caducous and not to be found at fruiting stage. Flowers yellow, petals oblong or obovate, entire or subentire above with a short inflexed acumen, glabrous or rarely hairy. Calyx teeth small or obsolete. Fruit roundish, oblong or ovoid, dorsally plano-compressed; the three intermediate ribs filiform, scarcely prominent, laterals conspicuously winged and those of each mericarp contiguous, forming a wing to the fruit. Vittae superficial, solitary or 3 or 4 in the valleculae, or more numerous and very slender. Disk broad, explanate, margin ± crenulate. Stylopodia small, depressed or somewhat conical. Carpophore bipartite. Endosperm dorsally flattened, commissural face slightly concave.

Probably some 150 species in temperate Eurasia and the mountains of tropical Africa; six species in Iraq.

Ferula (from the Lat. *ferula*, a rod or cane, used to chastise boys in schools and in former times slaves for minor offences); Giant Fennel (not to be confused with Fennel, *Foeniculum*, q.v.). Asafoetida, the gum resin exuded from the living rhizomes of *F. asa-foetida* (not found in Iraq) is obtained in Iran and W Afghanistan and produces the medicinal oil, oleum Asae Foetidae which is used, according to Wren (1956) as stimulant, antispasmodic and expectorant: also in infantile convulsions and colic. Another use is as a flavourant of sauces. One or two species which occur in mountain regions are much prized, collected and dried for winter animal fodder.

1. Branches thickening at ripe fruit, with a very soft pith; leaf segments broader than 2 mm; sheaths of upper leaves inflated but flaccid 2
   Branches not thickening at ripe fruit, firm; leaf segments not exceeding 1 mm in breadth; sheaths of upper leaves inflated, rigid 3
2. Leaves covered with short, ash-white hairs, the segments oblong, not more than 3 mm wide . . . . 5. *F. rutbaensis*
   Leaves glabrous, segments much more than 3 mm wide, those of the lower cauline leaves flabellate, to 4 cm wide . . . . 6. *F. sphenobasis*
3. Leaf segments 1.5–3 mm broad . . . . 3. *F. oopoda*
   Leaf segments 0.2–1.5 mm broad 4

4. Leaf segments glabrous, filiform-canaliculate . . . . 4. *F. orientalis*
   Leaf segments shortly hairy at least on the lower surface, flat or with narrowly involute margins 5
5. Leaf segments short, the final divisions at most twice as long as broad, often crowded and overlapping, especially at leaf extremities . . . . 1. *F. ovina*
   Leaf segments linear, several times as long as broad, not crowded and overlapping . . . . 2. *F. haussknechtii*

Subgenus 1. PEUCEDANOIDES (Boiss.) Korov.

Branches and peduncles not becoming incrassate in fruiting stage; vittae broad, solitary in the valleculae; mericarps with narrow margin.

1. **Ferula ovina** *Boiss.*, Fl. Orient. 2: 986 (1872); Korovin, Gen. *Ferula* monograph. Illustr. 57, pl. 32 (1947); Rawi in Dep. Agr. Iraq Tech. Bull. 14: 92 (1964); E. Nasir in Fl. Pak. 20: 161, f. 49 (1972); Chakr., Pl. Wealth Iraq 1: 236 (1976), both sub "*F. avina*", sphalm.; She Menglan & Watson in Fl. China 14: 177 (2005).

Peucedanum ovinum Boiss., Diagn. ser. 1, 6: 61 (1846).
[*F. barbeyi* (non Post) H. Riedl in Fl. Lowland Iraq: 468 (1964)].

Plant 60–75 cm tall with numerous branches, glabrous save for the leaves, which are scabrid with short, whitish bristles. Stem and branches terete, finely striate, base of stem covered with fibrous remains of previous years' leaf-sheaths. Lower leaves broadly deltoid-ovate, to 30 cm long, shortly (to 2 cm) petiolate to almost sessile on long, expanded sheaths, petioles and primary divisions glabrous or sparsely hairy below, increasingly hairy upwards, 3- or 4-pinnately decomposite, ultimate segments short, crowded, tips usually cucullate. Upper leaves similarly divided, sessile on ovate to oblong, much inflated, firm sheaths which are amplexicaul, the two edges overlapping at base. Branching paniculate-corymbose, lower branches alternate, upper opposite to verticillate. Umbels many, 4–10-rayed, fruiting rays 1–4 cm. Involucre absent. Partial umbels 6–10-flowered, fruiting pedicels short, 2–4 mm. Flowers ± 4 mm in diameter. Involucel absent or of a few squamiform, caducous bracteoles. Calyx teeth very short. Fruit ovoid-oblong, 8–13 × 6.5–9 mm, vittae broad, solitary in the valleculae. Styles 1.5–2 mm, deflexed, glabrous; stylopodia depressed-conical.

HAB. In desert, mostly on sandy soils, sometimes in rocky places, on gypsaceous limestone hills; alt. 50–650 m; fl. & fr. Apr.–Jun.
DISTRIB. Quite common in the W sector of the desert region of Iraq, particularly in the area round Rutba. **DLJ**: Rawa, *Gillett & Rawi* 7072!; **DWD**: Afaij, Wadi al-Ajrumiya, *Rawi & Nuri* 27177!; Rutba, *Field & Lazar* 135!; 45 km S of Rutba, *Rawi* 21261!; ± 70 km SW of Rutba, *Rawi* 21200!; Faggart al-Ghadaf, *Rawi* 31148!; Wadi al-Ghadaf, 100 km S of Rutba, *Alizzi* 35207!; 260 km NW of Ramadi along Rutba highway, *Rawi* 20970!; W of Falluja, nr southern shore of Lake Habbaniya, *C. Guest* 15191!; **DLJ**: 80 km NW of Rawa, *Chakrawaty, Rawi, Khatib & Alizzi* 31934!
NOTE. ?BAUWA (Ir.-Rawa, *Gillett & Rawi* 7072; "fruits eaten, flowers bright yellow"). On the label of a specimen of this plant collected in Afghanistan a century ago the collector has noted "Height 1 m. Considered excellent grazing for all animals and very fattening for horses" (*Aitchison* 525 in Herb. Kew).
According to Boissier this species attains 2.4 m in height ("5–8 pedalis"), but this does not accord with other authorities, field notes by collectors, or my own field experience of the plant in Jordan.

Iran, Pakistan, Afghanistan, Kazakhstan, Kyrgystan, Tajikstan, W China.

2. **Ferula haussknechtii** *Wolff ex Rech.f.* in Anz. Öst. Akad. Wiss., Math.-Nat. Kl. 89: 175 (1952); Peşmen in Fl. Turk. 4: 448 (1972); Chamberlain & Rech. f. in Fl. Iranica 162: 418 (1987).

Plant to 1.5 m, glabrous except for the leaves; stem finely striate, corymbosely branched above with branches opposite or subverticillate, lower branches alternate. Basal leaves deltoid-ovate in outline, up to 40 cm long, sessile or very shortly petiolate, petioles and primary divisions glabrous or almost so, lamina 4-pinnately decomposite, ultimate segments linear, 3–5 × 1 mm, shortly and remotely petiolate, petioles and primary divisions glabrous or almost so, distinctly scabrid with short white hairs on lower surface, upper leaves with broad, inflated, firm and somewhat shining sheaths, rapidly reducing in size, becoming bipinnatisect with longer (to 12 mm) segments. Umbels many, lower on long, leafless peduncles up to 14 cm, upper shortly pedunculate. Rays of main umbel 7–12, 2–3(–4) cm long in fruit. Partial umbels 6–12-flowered, fruiting pedicels 6–11 mm. Involucre and involucel absent. Fruit 7–15 × 4–11 mm, oblong to ovoid. Styles short, 1.5–2 mm, glabrous; stylopodia depressed-conical.

HAB. On the mountains, upper margin of oak forest, on limestone; alt. 1800 m; fl. & fr. May–Jun.
DISTRIB. Very rare in Iraq, only found once over fifty years ago near the Iranian frontier, on the upper extremity of the SE sector of the forest zone. **MSU**: Hawraman, N of Biyara, *Gillett* 11785!
NOTE. ?KAMAR (Kurd.-Hawraman, *Gillett* 11785, "said to be the best of the fodder species collected for winter use").

Turkey (E Anatolia), W Iran (Luristan).

3. **Ferula oopoda** (*Boiss. & Buhse*) *Boiss.*, Fl. Orient. 2: 984 (1872); Grossh., Fl. Kavk. ed. 2, 7: 108 (1967); Chamberlain & Rech. f. in Fl. Iranica 162: 421 (1987).

Peucedanum oopodum Boiss. & Buhse in Nouv. Mem. Soc. Nat. Mosc. 12: 100 (1860).

Plant 1–2 m high, glabrous. Stem grooved, robust, much-branched. Lower leaves ± 30 cm, 3- or 4-ternate or pinnate, glabrous; ultimate segments filiform, 40–60 × 2–3 mm, 1-veined; sheaths of upper leaves becoming coriaceous, inflated, amplexicaul. Umbels 16–25-radiate, rays ± 4 cm long in fruit. Partial umbels 12–15-flowered. Involucre rudimentary or absent, segments of involucel linear. Petals yellow, glabrous. Fruits oblong-ellipsoid, narrowly winged, 10–12 mm.

HAB. Mountain sides; altitude and flowering time not recorded.
DISTRIB. Very rare in Iraq, only found once close to the Turkish frontier. **MRO**: N side of Mt. Botin, *Hadač et al.* 6149 (figured in Rech. f., Fl. Iranica 162a: t. 377 (1987).

Transcaucasus, Iran.

4. **Ferula orientalis** *L.*, Sp. Pl. ed. 1: 247 (1753); Grossh., Fl. Kavk. ed. 2, 7: 109 (1967); Peşmen in Fl. Turk. 4: 445 (1972); Chamberlain & Rech. f. in Fl. Iranica 162: 418 (1987).

Polycyrtus cachroides Schlecht. in Linnaea 17: 126 (1843).
Peucadenum schlechtendalii Boiss., Ann. Sci. Nat. ser. 2: 311 (1841) & Boiss., Fl. Orient. 2: 985 (1872). — Type: Iraq, Gara Dagh, *Kotschy* 419 (K!, syn.); Zohary in Dep. Agr. Iraq Bull. 31: 112 (1950); *Rawi*, ibid. Tech. Bull. 14: 92 (1964).
Ferula cachroides (Schlecht.) Eug. Korov., l.c.: 56 (1947); Zohary & P.H. Davis in Kew Bull. 1947: 90 (1947), in adnot.

Plant 1.2–1.5 m high, glabrous. Stem terete, finely striate, paniculate-corymbose, upper branches opposite or subverticillate, leafless. Lower leaves very broad, deltoid-ovate in outline, 4–5-pinnately decomposite into setaceous, ± canaliculate segments to 2.5–15 × 0.25–1 mm. Stem leaves reducing to small and very small, sessile on the very large, inflated, firm, very finely striate sheaths, which are completely amplexicaul, overlapping at the basal margins. Umbels many, 8–12(–15)-radiate, rays 2–3.5 cm long at fruit. Partial umbels 8–10-flowered, fruiting pedicels 4–6 mm. Involucre and involucel absent. Flowers ± 4 mm in diameter. Fruit oblong-ovate, 10–14 × 5–7 mm. Styles reflexed, 1.5–2 mm, glabrous; stylopodia depressed-conical.

HAB. On the mountains, among rocks in *Quercus* forest; alt. 1450 m or more; fl. & fr. Jun.–Jul.
DISTRIB. Very rare in Iraq, only found three times in two districts of the upper forest zone. **MAM**: Gara Dagh, *Kotschy* 419 (type of *F. schlechtendalii*)!; Sarsang, *Hadač* 654!; Waziara, *Hadač* 6288; **MRO**: Kani Takht, *Eig et al.* 3864; Sakri-Sakran, *Hadač* 5591; **MSU**: Qopi Qaradagh, *Haines* W1081!

Turkey (E Anatolia), W Iran.

Subgenus 2. MERWIA (B. Fedtch.) Korov.

Branches and peduncles becoming spongy-incrassate at fruiting stage; vittae in valleculae numerous, filiform, sometimes interrupted; mericarps broad, much flattened, with ± broad margins.

4. **Ferula rutbaensis** *C.C. Townsend* in Kew Bull. 20: 77 (1966).

[*F. blanchei* (non Boiss.) Zohary in Dep. Agr. Iraq Bull. 31: 112 (1950); H. Riedl in Fl. Lowland Iraq: 468 (1964); Rawi in Dep. Agr. Iraq Tech. Bull. 14: 92 (1964)].

Plant robust, 75–125 cm, with a sour odour. Stem and branches terete, pithy, striate, glabrous below, whitish pubescent above, with a soft pith, base of stem furnished with fibrous remains of previous years' petioles. Lower branches alternate, upper often subverticillate. Lower leaves broadly deltoid-ovate, to 30 cm, with long sheaths, 2- or 3-pinnate, divisions rather distant; segments flat, ± 10 × 3 mm, oblong-ovate, obtusely toothed, decurrent along the main axes of pinnules, covered, as are primary divisions of the leaf, with a dense whitish pubescence. Upper leaves sessile on inflated but flaccid, not coriaceous, sheaths up to 5 cm long, uppermost leaves reduced to sheath only. Umbels several, rays 12–20, spreading white-hairy when young, 2.5–5.5 cm and becoming glabrescent in ripe fruit. Partial umbels ± 15-flowered, fruiting pedicels short, 2–6 mm. Flowers ± 3.5 mm in diameter. Ovary densely hirsute when young. Ripe fruit large, subrotund, 12 × 12 mm, much flattened, becoming subglabrous with a few short hairs; lateral ribs obscure, remote from the 3 dorsal. Disk very small, styles short, ± 2 mm, glabrous; stylopodia depressed.

HAB. In desert, often on sandy soils, sometimes also on gravelly and silty soils, on sandstone hillside, in wadis among rocks; alt. 200–650 m; fl. & fr. (Apr.–)May–Jun.
DISTRIB. Quite common, locally frequent, in the most westerly sector of the desert region of Iraq. **DLJ/FUJ**: 150 (sub "450" sphalm.) km from Baghdad on the Baghdad-Baiji-Mosul road, *Rawi & Hamadi* 33580!; **DWD**: Wadi al-Walaj, 75 km W of Rutba, *Rawi* 14707!; 10 km N of Rutba, *Rawi & Nuri* 27108!; Afaij, *Alizzi & Hussain* 34059!; 15 km S of Rutba, *Khatib & Hazim* 32436!; 70 km S of Rutba, *Rawi* 31133!; 100 km E of Rutba, *Omar* 34325!; 160 km W of Ramadi, *Chakr., Rawi, Khatib & Alizzi* 32830!

Extreme N Saudi Arabia; may well also occur in SE Syria and E Jordan.

5. **Ferula sphenobasis** *C.C. Townsend*, Kew Bull. 20: 77 (1966); Chamberlain & Rech.f. in Fl. Iranica 162: 395 (1987).

Plant perennial, smelling of onions. Stem erect, to 50 cm, alternately branched above, 1.5 cm thick at base, solid, pithy, terete, finely striate. Rootstock surmounted by fibrous remains of previous years' leaf-sheaths. Branches short, rather crowded, leafless, up to 10 cm long, glabrous with a soft pith. Radical leaves few, pinnate with 2 or 3 pairs of rather narrowly lobed, cuneate-based leaflets; stem leaves few, sessile, lower broadly deltoid in outline, pinnatisect or subbipinnatisect with broader lobes up to 4 cm in length and breadth, which are cuneate at base and narrowly decurrent along the main axis. Sheaths large, inflated, flaccid. Surfaces of leaves and sheaths shortly white-pubescent. Umbels 12–17-rayed, fruiting rays 2–4 cm. Partial umbels about 12-flowered; flowers 3.5–4 mm in diameter; fruiting pedicels 2–6 mm. Involucre and involucel absent. Pedicels and young ovaries shortly hairy. Fruit broadly obovoid, 15–20 × 11–12 mm, sparsely hairy, with broad flat margins. Styles ± 2 mm, deflexed, glabrous; stylopodia depressed.

HAB. In deep gully on soft soil slopes; alt. ± 100 m?; fl. & fr. Mar.–May.
DISTRIB. Very rare in Iraq, only found in one area, not far from the Iranian frontier in the NE sector of the dry steppe zone. **FNI**: 20 km SSW of Khanaqin along the road to Naft Khana, *Haines* W1908! W2008!; Kani Mazi nr Khanaqin, *Hadač et al.* 4650.

W Iran.

38. FERULAGO W.D.J. Koch

in Nov. Act. Acad. Caes. Leop. Carol. 12, 1: 68 (in clav.), 97 (1824)

Perennial herbs with pinnately decomposite leaves. Inflorescence paniculate-corymbose or thyrsoid. Umbels compound, terminals large, hermaphrodite, fructiferous, usually shortly pedunculate, with smaller lateral, male and usually long-pedunculate umbels often arising immediately below them. Involucre and involucel of several bracts, conspicuous and persistent, often reflexed. Flowers yellow or rarely reddish. Petals orbicular to ovate or oblong, entire above with an inflexed acumen. Calyx present, 5-dentate. Fruit roundish to ovate, dorsally plano-compressed; mericarps with 3 dorsal intermediate ribs filiform, prominent or narrowly winged, laterals ± winged, contiguous, forming a margin or wing to the fruit. Vittae numerous and very slender, closely accumbent on the seed, those of commissural face superficial. Disk broad, explanate, margins usually undulate. Stylopodia very small, rarely somewhat conical. Carpophore bipartite. Endosperm dorsally flattened, commissural face concave.

About 40 species distributed in the Mediterranean area and SE Europe to Iran & C Asia; five species in Iraq.

Ferulago, (from *Ferula*, q.v. and the Lat. suffix *-ago*, indicating resemblance).

Ref.: L. Bernardi, "Tentamen revisionis Generis *Ferulago*" in Boissiera 30: 1–182 (1979).

1. Petals hairy on the dorsal surface 2
   Petals glabrous 3
2. Inflorescence branches very abbreviated, furnished with 2 or 3 pairs of large, sheath-like, opposite, striate bracts, general inflorescence appearing as a cylindrical thyrse .... 5. *F. abbreviata*
   Inflorescence branches either not very short and giving a cylindrical thyrse-like general inflorescence, or if so then without 2 or 3 pairs of such bracts .... 4. *F. carduchorum*
3. Leaf segments shortly and rather narrowly lanceolate or oblanceolate, flat; ridges of ripe fruit crisped-laciniate, especially marginal .... 2. *F. macrocarpa*
   Leaf segments not flat; ridges of ripe fruit straight or slightly wavy, not crisped-laciniate 4
4. Vittae on commissural face of fruit very numerous (10+), lateral ridges of fruit scarcely more distant from dorsal ridges than these are from each other .... 1. *F. stellata*
   Vittae on commissural face of fruit 6–8, dorsal ridges of the fruit clearly closer to each other than to marginals .... 3. *F. angulata*

1. **Ferulago stellata** *Boiss.* in Ann. Sc. Nat. sér. 3, 1: 323 (1844); Boiss., Fl. Orient. 2: 1001 (1872); Blakelock in Kew Bull. 3: 434 (1948); Zohary in Dep. Agr. Iraq Bull. 31: 112 (1950); Rawi in Dep. Agr. Iraq Tech. Bull. 14: 92 (1964); Peşmen in Fl. Turk. 4: 462 (1972); Chamberlain in Fl. Iranica 162: 429 (1987).

Plant 60 cm to 1.5 m or more, glabrous. Stem strongly sulcate, ± angular, subterete above, paniculate-corymbose branches also terete, lightly striate. Lower leaves deltoid-oblong to oblong-lanceolate in outline, to ± 40 cm, 3- or 4-pinnately decomposite into fine setaceous segments (2–)5–15(–25) mm long, petiolate; upper leaves sessile, similarly dissected, segments often crowded after the manner of a *Grammosciadium*, only one or two of uppermost merely bifid or trifid or reduced to sheaths. Sheaths narrow, firm, rootstock crowned with fibrous remains of those of previous years. Umbels many, rays (4–)6–12(–18), stout, 30–65 cm long and widely divaricate when fruiting. Involucre of 4–8 linear-lanceolate, flexuose-patent or reflexed bracts. Partial umbels 8–13-flowered, flowers ± 2 mm in diameter. Pedicels 3–15 cm when in fruit, stout. Petals pale yellow, glabrous. Involucel of 4–8 linear-lanceolate bracteoles. Fruit narrowly ellipsoid, 10–16 × 3.5–4.5 mm, tapering at both ends, the 3 dorsal ridges only slightly more distant from the laterals than from one another, carinate or very narrowly winged; vittae numerous all round the endosperm; calyx teeth delicate, small. Styles ± 1 mm, deflexed, slender; stylopodia flat.

HAB. On the mountains, on rocky slopes, in oak forest, on limestone; alt. (500–)1200–1800(–2000) m; fl. & fr. Apr.–Jun.

DISTRIB. Quite common in the forest zone of Iraq. **MAM**: Sarsang, *Haines* W1354!; **MRO**: Rowanduz gorge, Guest 2145; Chiya-i Marmarut, *E.J. Cuckney* 3829; valley above Ari on Ser Kurawa, *Gillett* 9716; **MSU**: Zewiya, *Rawi* 12072!; between Zewiya and Qarachitan, *Gillett* 7784!; Qopi Qaradagh, *Haines* W1561!; Amoret, nr Qaradagh, *id.* W 1154!; Hawraman, N of Biyara, *Gillett* 11756!; Hawraman, *Rawi, Husham (Alizzi) & Nuri* 29356!; Tawila, *Rechinger* 10351; ?**MSU** (and/or only in Iran?), "Montes Avroman et Schahu", *Haussknecht* s.n.; **FNI**: nr Mandali, *Noë* s.n.

NOTE. ?LOH (Kurd.-Hawraman, *Gillett* 11786 – "leaves collected for winter fodder"), ?HALILA (Kurd.-Sulaimaniya, *Gillett* 7784),?ZIA KAWLA (Kurd.-Zewiya, *Rawi* 12072).

Specimens listed above with a query probably belong here but lack either fruit or leaves.

Turkey (E Anatolia), W Iran.

2. **Ferulago macrocarpa** (*Fenzl*) *Boiss.*, Fl. Orient. 2: 1003 (1872); Nábělek in Publ. Fac. Sci. Univ. Masaryk 35: 127 (1923); Bornmüller in Beih. Bot. Centralbl. 58B: 283 (1938); Blakelock in Kew Bull. 3: 434 (1950); Zohary in Dep. Agr. Iraq Bull. 31: 112 (1950); H. Riedl in Fl. Lowland Iraq: 469 (1964); Rawi in Dep. Agr. Iraq Tech. Bull. 14: 92 (1964); Mouterde, Nouv. Fl. Lib. et Syr. 2: 645 (1970); Peşmen in Fl. Turk. 4: 468 (1972).

Uloptera macrocarpa Fenzl in Flora 26: 460 (1843).
Ferulago fieldiana Rech.f. in Anz. Öst. Akad. Wiss., Math.-Nat. Kl. 14: 3 (1952); H. Riedl in Fl. Lowland Iraq: 469 (1964).
F. lophoptera Boiss. in Ann. Sc. Nat. sér. 3, 1: 325 (1844); Rawi in Dep. Agr. Iraq Tech. Bull. 14: 92 (1964).

Plant 45–60 cm, glabrous, glaucous. Stem terete, finely striate, tough, paniculate-corymbosely branched and somewhat flexuose above, branches mostly alternate, uppermost opposite or in a whorl of 3 including the terminal umbel. Basal and lower stem leaves tripinnate, shortly petiolate, broadly deltoid, to 35 cm long, lowest pair of pinnae very long, much longer than the remainder; ultimate segments rather short, lanceolate, flat, 4–7(–12) × 1–3 mm, mucronate. Uppermost leaves of stem and those of branches small, sheath-like, simple or trisect; a few sessile, transitional simply pinnate leaves occur about the middle of stem. Sheaths narrowly expanded, those of radical leaves long, fibrous remains persisting round the base of the stem. Umbels many, rays 5–11, 3–5 cm in length when fruiting. Involucre and involucel of ± 6 lanceolate, patent or reflexed bracts and bracteoles. Partial umbels 6–12-flowered, flowers pale creamy yellow, ± 3 mm in diameter. Petals glabrous. Fruiting pedicels 4–8 mm, slightly exceeding the involucel. Fruit much flattened, obpyriform, 9–16 × 7–11 mm, the 3 dorsal primary ridges with denticulate-erose and somewhat crisped wings, closer to one another than to the laterals, which are ± 1 mm broad, strongly and closely crisped and denticulate-erose; vittae very numerous all round the endosperm, to ± 18 on the commissure; styles 2–3 mm, ± flexuose; stylopodia flat; petals persistent. Fig. 83: 1–6.

HAB. On lower mountain slopes in coppiced *Quercus* scrub, on limestone, steppic hills on sandstone, rocky and stony places; alt. 200–1300 m; fl. & fr. May–Jul.
DISTRIB. Common in the lower forest zone of Iraq, occasional in the steppe region. **MAM**: nr Bauerd (Bawart) S of Zakho, *Kotschy* s.n. (type); Dohuk, *Rawi* 8760!; Sarsang, *Haines* W1010!; Mar Ya'qub, *Nábělek* 355, 503; **MRO**: Salah ad-Din, *Gillett* 11276!; Kuh-i Sefin above Shaqlawa, *Bornmüller* 1264!, 1265!; Shaqlawa, *Gillett* 8063!; Kaiwa Rush (Kew-ar Rash), *Rawi & Serhang* 23770!; **MSU**: 20–25 km E of Qaranjir, *Rawi* 21649!, 21661!; Jarmo, *Helbaek* 1747!, *Haines* W248?, W249!; 20 km NW of Sulaimaniya, *Rawi* 21730!, 22774!; 5 km S of Tawila, *Rawi* 21831!; **FUJ**: Jabal Golat (Qilat), *Field & Lazar* 417!; Meer Khasim (Mir Qasim) between Balad Sinjar and Tal Afar, *id.* 545!; **FAR**: 7 km E of Mahmur, *Gillett* 11219!; nr Arbil, *Haussknecht* s.n.!; **FKI**: banks of R. Zab, nr Kirkuk, *Mrs P. Grigg* 69!; Kani Domlan hills, nr Kirkuk, *Guest* 4287?; Jabal Kumatas, between Qizil Robat (Sa'diya) and Khanaqin, *Nábělek* 299; Koma Sang, *Rawi* 20634!; 10 km E of Mandali, *Rechinger* 9613.

Syria, Turkey (SE Anatolia), W Iran.

3. **Ferulago angulata** (*Schlecht.*) *Boiss.*, Fl. Orient. 2: 1005 (1872). — Type: Iraq, Mt. Gara (Kara Dagh), *Kotschy* 403 (K!, syn.); Nábělek in Publ. Fac. Sci. Univ. Masaryk 35: 121 (1923); Blakelock in Kew Bull. 3: 434 (1950); Zohary in Dep. Agr. Iraq Bull. 31: 113 (1950); Rawi in Dep. Agr. Iraq Tech. Bull. 14: 92 (1964); Peşmen in Fl. Turk. 4: 469 (1972); Chamberlain in Fl. Iranica 162: 431 (1987).

Ferula angulata Schlecht., Linnaea 17: 125 (1843).

Plant tall, 1.5 m or more, glabrous throughout or minutely scabrid, especially on leaves, with sessile or very shortly stipitate glands. Stem sulcate, often strongly so, much-branched above, branches usually alternate below, opposite or verticillate above. Lower leaves broadly deltoid-ovate, petiolate, up to 40 cm long, 4-pinnately decomposite into rigid, setaceous, angular, sulcate segments 15–30 cm long; surface glabrous to densely glandular, where glands are few these are generally to be found at major divisions of leaf. Upper leaves sheath-like, transitional leaves few. Sheaths shortly expanded, firm, striate. Umbels many, 7–12(–15)-rayed, fruiting rays stout, 1.2–3 cm. Involucre and involucel of 6–8 patent or reflexed bracts and bracteoles. Partial umbels 7–17-flowered, petals glabrous, light yellow, flowers ± 2 mm in diameter. Fruiting pedicels unequal, 3–6 mm, equalled by bracteoles. Fruit ellipsoid-ovoid, 8–10 × 4–6 mm, the 3 dorsal ridges much closer to one another than to the laterals, mericarps abruptly constricted below the large, persistent calyx teeth; vittae 6(–8) on the commissure, 8 dorsally. Styles 3 mm, slender and flexuose; stylopodia flat.

HAB. On the mountains, on rocky slopes, limestone crags in *Pinus* forest, in rocky ravine, on cliffs in *Quercus* forest, on metamorphic rock; alt. 800–1850(–2100) m; fl. & fr. Jun.-Aug.

Fig. 83. **Ferulago macrocarpa**. 1, habit base × ²/₃; 2, habit with young fruits and flowers × ²/₃; 3, habit with mature fruits × ²/₃; 4, petal × 10; 5, fruits (both faces) × 2; 6, fruit cross section × 8. 1 from *Helbaek* 1747; 2, 4 from *Rawi* 201831; 3, 5–6 from *Bornmüller* 1265. Drawn by D. Erasmus (1964).

DISTRIB. Common in the NW sector of the forest zone of Iraq, rare in the SE sector, occasionally found above the margin of the thorn-cushion zone. **MAM**: Amadiya, *Sahira* 53791!; Khantur, Rawi 23475!, *Rawi, Nuri & Tikriti* 29002!; nr Turkish frontier on Mt. Zawita, NE of Zakho, *Rawi* 23588!; Mt. Gara (Kara Dagh), *Kotschy* 403 (type)!, 480!, 860!, *Rawi* 9275!; Sarsang, *Haines* W455!, W 1334!, *Hadač* 860!, 861!; Zawita gorge, *Guest* 3711!; Zawita valley, *Guest* 4755!, 4789!, 4815!; Dari, 5 km E of Kani Masi (Ain Nuni), *Omar & Qaisi* 45446!; **MRO**: Shaikh wa Shaikan, *Rawi & Serhang* 24878!; Halgurd Dagh, *Gillett* 9515!; Sula Khal, *Rawi & Serhang* 24695!; Qurnago, *id.* 26623!; Haibat Sultan Dagh, *id.* 25358!; **MSU**: Hawraman mts., *Rawi, Alizzi, Chakr. & Nuri* 19736!

NOTE. ?CHAUR (Kurd., *Gillett* 9515, "collected as fodder for sheep").

Bernardi, l.c., p. 83 cites *Rawi, Nuri & Tikriti* 29002 with some doubt under *F. blancheana* Post. While this specimen is poor, lacking leaves, I see no reason for not referring it to *F. angulata*, which is known from the area. *F. blancheana* is known only from one gathering in the Maraş district of Turkey.

Turkey (SE Anatolia).

4. **Ferulago carduchorum** *Boiss. et Hausskn.* in Boiss., Fl. Orient. 2: 1005 (1872).

F. trifida Boiss. subsp. *carduchorum* (Boiss. et *Hausskn.*) Bornmüller in Beih. Bot. Centralbl. 58B: 283 (1938), incl. var. *kermanensis* Bornmüller

F. angulata (Schlecht.) Boiss. subsp. *carduchorum* (Boiss. et *Hausskn.*) Chamberlain in Fl. Iranica 162: 432 (1987).

Height and habit unknown, but probably at least 1 m tall (see Bernardi, 1.8 m or more). Stem strongly angular-sulcate, with scattered stellate-floccose hairs above and covered, as is the whole plant, with dense minute clavate hairs (older parts often ± glabrescent). Lower leaves deltoid-ovate, to 50 cm, rigid, tripinnately decomposite with distant divisions, ultimate segments remote, linear, 2–12 × 0.6–1 mm, all divisions striate to canaliculate; sheaths very narrow, firm and striate, to 5 cm long. Inflorescence narrowly thyrsoid, opposite or verticillate branches short but slender, pedunculiform. 2.5–4 cm, or longer and more slender with additional lateral umbels, but not short with pairs of large, sheath-like bracts, terminal umbel usually very shortly pedunculate. Umbels with 7–15 rays, fruiting rays 7–25 mm long. Partial umbels 10–12-flowered, flowers ± 2 mm in diameter, not opening wide, fruiting pedicels 1.5–4 mm. Petals greenish yellow to yellow, shortly hairy on the dorsal surface. Involucre and involucel of 4–6 lanceolate, deflexed bracts and bracteoles. Ovary hairy. Fruit (immature) ellipsoid-oblong, 7–8 × 5 mm with primary ribs narrowly winged, distant from the slender laterals, vittae 6 on the commissural face, 8 dorsally. Styles 1–2 mm, deflexed; stylopodia depressed.

HAB. & DISTRIB. No habitat for this plant is recorded in Iraq, the only two gatherings having been made over 150 years ago at 1820–2120 m. ?**MSU** (or in Iran?), Hawraman, *Haussknecht* s.n. (1867); Pira Magrun, *Haussknecht* s.n. (1867).

NOTE. The Pira Magrun specimen consists of leaves only; there is no reasonable doubt as to its identity, though Chamberlain (l.c.) cites it under *F. angulata* subsp. *angulata*.

W & S Iran.

5. **Ferulago abbreviata** *C.C. Townsend* in Kew Bull. 20: 79 (1966). — Type: Iraq, Qopi Qaradagh, *Haines* W 2095 (K!, holo.); Peşmen in Fl. Turk. 4: 471 (1972).

F. bracteata Boiss. et *Hausskn.* in Boiss., Fl. Orient. 2: 1008 (1872); Chamberlain in Fl. Iranica 162: 433 (1987).

Height unknown. Stem strongly sulcate and angular, even above, stellate-floccose and covered, as is the entire plant, with dense minute clavate hairs. Leaves pinnately decomposite (no complete leaf seen), all but the final divisions very elongate, canaliculate with clavate hairs especially dense within the grooves; penultimate segments 2–3 cm, trifid very near the apex into short (1.5–5 mm) ultimate segments, rarely with 1 or 2 lateral divisions below; sheaths not seen. Inflorescence with the appearance of a narrow thyrse owing to the condensed branches and peduncles. Inflorescence branches opposite, or verticillate above, with crowded, numerous, sheath-like, broadly lanceolate, coriaceous, striate bracts 8–12 mm long. Umbels (flowering) with ± 10 short rays. Partial umbels ± 20-flowered, pedicels very short. Petals sulphur-yellow, densely papillose-hairy on the dorsal surface. Involucre of 3 or 4 lanceolate bracts; bracteoles ± 6, lanceolate-acuminate. Ovary hairy. Fruit unknown. Styles very short, ± 0.25 mm, when in flower.

HAB. Open rocky slope near summit of mountain ridge; alt. ± 1750 m?; fl. Jul.
DISTRIB. Very rare, only two gatherings known, from the margin of the upper forest/lower thorn-cushion zone of Iraq. **MJS**: Qopi Qaradagh, *Haines* W 2095 (type)!; Amoret nr Qaradagh, *Haines* s.n.
NOTE. Bernardi, l.c. reduced this species to the synonymy of *F. carduchorum*. This may be correct, for the two are certainly close. However, the broad crowned bracts of the very short inflorescence branches of *F. abbreviata*, very like those of the Turkish *F. bracteata* Boiss. & Hausskn., are unlike anything which I have seen in even the youngest specimens of *F. carduchorum*, such as Haussknecht's Hawraman material; the leaf divisions also appear to be less thick in texture. Thus I prefer to maintain *F. abbreviata* pending population studies in Iraq, since *F. carduchorum* is a rare plant in Iraq which has not been collected by any botanist visiting Qopi Qaradagh.

Endemic.

39. **PEUCEDANUM** L.

Sp. Pl. ed. 1: 245 (1753); Gen. Pl. ed. 5: 116 (1754)

Perennial, rarely biennial or monocarpic herbs, often glabrous. Leaves pinnately decomposite, pinnate or ternately decomposite, segments broad or narrow. Umbels compound. Involucre absent or of few to many bracts, bracteoles usually numerous, rarely absent. Inflorescence polygamous. Flowers white or yellowish, more rarely pink, petals obovate to ovate with an inflexed (unusually long) acumen, emarginate or entire above. Calyx teeth absent or small but distinct. Fruit elliptical to ovate or suborbicular, strongly dorsally plano-compressed. Mericarps flattish or variously convex, intermediate ribs filiform, not or scarcely prominent, laterals flattened, closely contiguous, forming a mostly thin wing to the fruit. Vittae usually solitary in the valleculae (more rarely 2 or 3 and very slender). Disk usually ± flattened, undulate-margined, with small stylopodia. Carpophore bipartite. Endosperm with the commissural face flat or almost so.

Up to 60 species in Eurasia; one species in Iraq.
Peucedanum (πευκέδἄνον, *peukedanon*, the Gr. name of a plant in Theophrastus, perhaps from Gr. πεύκο, pine, referring to the resinous juices of *P. officinale*, the "sulphur root" used in veterinary medicine.

1. **Peucedanum scoparium** *Boiss.*, Diagn. ser. 2, 2: 90 (1856); Boiss., Fl. Orient. 2: 1019 (1872).

Johreniopsis scoparia (Boiss.) M. Pimen. in Fl. Iranica 162: 455 (1987).

Plant stiff, erect, wiry, glabrous and glaucous or yellowish green, attaining up to 1 m but usually less, appearance that of a *Johrenia*, with many ascending branches above. Stem and branches terete, finely striate, stem base clothed with the scaly and fibrous remains of previous years' leaf-sheaths. Lower leaves (often absent or very few at flowering time) lanceolate in outline with long, narrow sheaths, pinnate with 2 or 3 pairs of trifid (more rarely 5-fid) pinnae up to 2.5 cm long, main axis usually with a ± jointed appearance where the pinnae join it and rather fragile there; terminal segment also usually dissected. Upper leaves quickly becoming trisect and finally reducing to white-margined sheaths. Peduncles slender, 1.5–4.5 cm. Umbels many, small, with 3–7 very unequal rays 5–32 mm long. Involucre absent. Partial umbels 3–9-flowered, fruiting pedicels 2–5 mm, flowers yellowish, ± 1.5 mm in diameter. Calyx obsolete. Involucel of 3–6 short, delicate bracteoles. Fruit elliptic, 5–8 mm, disk crenulate-margined, bearing rather prominently conical stylopodia. Vittae solitary in valleculae. Styles deflexed onto the disk, which they scarcely exceed. Fig. 84. 1–7.

HAB. On mountains, on dry open rocky slopes and stony places, on metamorphic rock; alt. 1550–2000 m; fl. & fr. Aug.–Sep.
DISTRIB. Occasional in the upper forest zone of Iraq, penetrating up into the lower thorn-cushion zone. **MAM**: Khantur, NE of Zakho, *Rawi, Tikriti & Nuri* 29008?; Sarsang, *Haines* W1352!; **MRO**: between Ari and Zerwa (Zirva), *Gillett* 9799!; Darband, WSW of Haji Umran, *Rawi* 24268!; Haji Umran, *Guest & Husham* (*Alizzi*) 15899!; **MSU**: Penjwin, *Guest* 12957!

W & S Iran.

Fig. 84. **Peucedanum scoparium**. 1, habit, lower portion × ²/₃; 2, habit with root × ²/₃; 3, habit with flowers and young fruit × ²/₃; 4, habit with mature fruits × ²/₃; 5, petal × 14; 6, fruits (both faces) × 4; 7, fruit cross section × 6. 1, 3–7 from *Guest & Muslim* 15899; 2 from *Wheeler Haines* W1352. Drawn by D. Erasmus (1964).

40. **LEUTEA** M. Pimen.

Fl. Iranica 162: 445 (1987)

Stems solitary, tall, with a fibrous collar; internodes elongate. Stem much-branched, basal leaves shortly petiolate, ternatisect; cauline leaves soon diminishing, upper reduced to scale-like mucronulate appendages or absent. Umbels 3–18-rayed, ultimate branches shorter than fruits. Petals yellowish green. Fruit ellipsoid, up to 6 mm long, margin inconspicuously ribbed.

Until recently regarded as monotypic; now there are three known species, one of which, *L. avicennae* Mozaff., is confined to a single mountain in Hamadan province, Iran and the other, *L. elbursensis* Mozaff., occurs in the Alborz mts.

1. **Leutea rechingeri** (*Leute*) *M. Pimen.* in Fl. Iranica 162: 448 (1987).

Peucedanum rechingeri Leute in Österr. Bot. Zeitschr. 120: 301 (1972). — Type: Iraq, Helgord above Nawanda village, *Rechinger* 11345 (W, holo.).
P. petiolare (non (DC.) Boiss.) Townsend in Kew Bull. 19: 74 (1964).
P. kurdistanica Mozaff. in Bot. Zhurn 88 (4): 113 (2003).

Glabrous, glaucous perennial, 0.3–2 m, branched from base, with a stout rootstock clothed with fibrous remains of petioles. Stem and branches terete, very finely striate, lower leaves pinnately divided into remote, rush-like, cylindrical segments which are bi- or trifurcate at tips, or lowest pinnae again divided with furcate pinnules; all segments 30–100 × 1–1.5 mm. Sheaths of lower leaves expanded, long, firm in texture, very finely striate. Upper leaves quickly reducing, shortly bi- or trifid, and finally narrowing to lanceolate sheath-like scales. Peduncles mostly long and slender, to 15 cm. Umbels 3–5(–7)-rayed, fruiting rays 12–35 mm. Partial umbels 3–6-flowered, fruiting pedicels 2–5 mm. Involucre and involucel absent. Petals greenish white, 1.5–2 mm in diameter. Sepals conspicuous in fruit, acute. Fruit (immature) ellipsoid, 6 × 2.5 mm in diameter, margins not prominent. Vittae solitary in the valleculae. Styles short, about equalling the top of fruit, deflexed; stylopodia depressed.

HAB. On mountain slopes; alt. 2000–3300 m; fl. & fr. Aug.–Sep.
DISTRIB. Very rare and local in the alpine region of Iraq. **MRO**: Helgord range (Halgurd Dagh), *Rawi & Serhang* 24900! Helgord above Nawanda village, *Rechinger* 11345 (type of *P. rechingeri*); N slope of Sarchal (Tenji Saipaiyan), *Hadač* 2217; Hisar-i Sakran, *Hadač et al.* 5631; **MSU**: Pira Magrun, *Prosser* 2027.
NOTE. Type material of *P. kurdistanica* has not been seen; the synonymy is based on an identity of key features. If it proves to be a distinct species, then *L. rechingeri* would be endemic to Iraq.

W Iran.

41. ANETHUM L.

Sp. Pl. ed. 1: 263 (1753); Gen. Pl. ed. 5: 127 (1754)

Tall annual herb with pinnately decomposite leaves. Umbels compound. Involucre and involucral bracts absent. Inflorescence hermaphrodite or commonly polygamous. Flowers yellow. Petals suborbicular, truncate and emarginate above with an incurved acumen. Calyx obsolete. Fruit ellipsoid, much compressed dorsally. Mericarps with intermediate ribs filiform, prominent, laterals winged, contiguous and forming a broad margin to the fruit. Vittae solitary in valleculae, very broad. Disk flattened but small, margins undulate. Stylopodia prominent, conical. Carpophore bipartite. Endosperm with the commissural face flattened.

Anethum (the Latin name for *A. graveolens*).

1. **Anethum graveolens** *L.*, Sp. Pl. ed. 1: 263 (1753); DC., Prodr. 4: 1830); Boiss., Fl. Orient. 2: 1026 (1872); Hayek, Prodr. Fl. Balc. 1: 1039 (1927); Zohary, Fl. Palest. ed. 2, 1: 554 (1932); Guest in Dep. Agr. Iraq Bull. 27: 7 (1933); Blakelock in Kew Bull. 3: 432 (1948); H. Riedl in Fl. Lowland Iraq: 467 (1964); Rawi & Chakr. in Dep. Agr. Iraq Tech. Bull. 15: 13 (1964); Grossh., Fl. Kavk. ed. 2, 7: 97 (1967); Mouterde, Nouv. Fl. Lib. et Syr. 2: 627 (1970); Zohary, Fl. Palaest. 2: 433 (1972); E. Nasir in Fl. Pak. 20: 111 (1972); Husain & Kasim, Cult.

Fig. 85. **Anethum graveolens**. 1, habit × ²/₃; 2, lower portion of stem and root × ²/₃; 3, petal × 10; 4, fruits (both faces) × 6; 5, fruit cross section × 8. 1, 3–5 from *Guest* 2506; 2 from *Field & Lazar* 866. Drawn by D. Erasmus (1964).

Fl. Iraq 105 (1975); Meikle, Fl. Cyprus 1: 751 (1977); Rech.f., Fl. Iranica 162: 345 (1987); She Menglan & Watson in Fl. China 14: 134 (2005).

Peucedanum graveolens (L.) Benth. in Benth. & Hooker, Gen. Pl. 1: 919 (1867).

Erect, glabrous, yellowish or glaucous-green annual herb, corymbosely branched from base or a little above, 0.2–0.9(–1) m. Stem and branches terete, finely striate. Leaves all petiolate, sheathing, 3- or 4(–5)-pinnate into setaceous segments, lower deltoid to oblong-lanceolate in outline, to ± 15 cm long, upper shorter and deltoid-ovate, segments 2–25 × 0.2–0.5 mm, those of upper leaves longest. Peduncles (1.8–)5–22 cm, at least some lateral umbels usually partially male. Umbels many, 5–40-rayed, rays in fruit 2–10 cm. Partial umbels (5–)10–30-flowered, flowers small, ± 2 mm in diameter. Fruiting pedicels 2–8 mm. Fruit ellipsoid, 3–5 × 2–3 mm, tapering to the small crenulate-margined disk. Stylopodia conspicuous, conical, styles persistent, very short, reflexed and scarcely equalling the disk of ripe fruit. Fig. 85: 1–5.

HAB. In gardens, orchards, fields and waste land near villages; widely cultivated and subspontaneous as a weed; alt. up to 700 m or more; fl. & fr. Mar.–May on the plains, Jun.–Jul. in the mountains.
DISTRIB. Occasional in the lower forest zone and steppe region, quite common on alluvial plain in desert region of Iraq. **MAM**: Zawita in gardens below the village (cult.), *Guest* 3740!; **MRO**: Hawdiyan, nr Diana, *Field & Lazar* 866!; **MSU**: Halabja, *Nuri & K. Hamed* 41206!; **DSD**: in irrigated market gardens S of Zubair, *Rawi* 25877!; **LCA**: Bada, N of Baghdad (cult.), *Guest* 2506!; Baghdad, *Paranjpye* s.n.! comm. Graham, *Gillett* 10721!, *Sahira* C. 702? (cult.), *Makki* 336?; Abu Ghraib, *Ani* 6881!; nr Aqurquf (cult.), *Barkley & Hikmat Abbas (al-Ani)* 2518!; Za'faraniya (cult.), *Rawi & Sahwani* 25332!, *H. Ahmed* 9849!; **LEA**: Ba'quba, *Haines* W1003!; **LSM**: 40 km S of Amara, along road to Basra, *Khatib & Alizzi* 32621!; Qurna (cult. in creekside garden), *Guest* 17380!
NOTE. Dill; known in Iraq under several names, most generally, in Lower Iraq, as SBINT and HABBAT HADWA. As noted by Guest (1933) this species is widely cultivated on a small scale throughout our country, principally for its foliage, which is cooked as a vegetable and used to flavour rice. The dried leaves are sold in local markets as SBINT; ripe fruits sold as HABBAT HALWA ("sweet seeds" Arabic) and used as a condiment with cooked meats. In addition to the two above names, dill is often confused with caraway owing to its strong aromatic smell and is hence not infrequently called KARAUYA or (more strictly the name of caraway seed) the fruits of *Carum carvi* q.v. Another name for this plant in the north is SIBAT (Kurd.-Zawita, *Guest* 3740 – "used as a condiment for flavouring").
A medicinal oil (Oleum anethi) can be obtained from the fruit by distillation, its properties being carminative and stomachic. It is recognised locally, as well as elsewhere, as the basis for an excellent remedy for children's complaints such as minor digestive ailments, flatulence etc., and forms the basis of what is known as "dill water" or, alternatively, according to Wren (1956), it can be administered by a few drops of the oil on sugar.

Widely cultivated and often naturalized in Mediterranean and SE Europe, Cyprus, Syria, Lebanon, Palestine, Jordan, Egypt, Arabia, Turkey, Iran etc. to China, N Africa (Libya) and elsewhere.

42. **TRIGONOSCIADIUM** Boiss.

Ann. Sc. Nat. sér. 3, 1: 344 (1844)

Puberulent herbs, biennial, with tuberous rootstocks. Leaves pinnate with few broad, dentate segments, central hermaphrodite, laterals with central flowers male. Umbels compound, bracts of involucre several, or few and caducous; bracteoles several, lanceolate. Flowers white, radiant, outer petals obcordate, remainder notched above and with an incurved acumen, dorsally hairy. Calyx teeth minute or obsolete. Fruit ± rotund, strongly plano-compressed dorsally, pubescent. Mericarps almost flat, the three median ribs slender and not conspicuous, laterals with thickened margins, closely contiguous and forming a wing to the fruit. Vittae solitary in valleculae, shorter than the fruit, somewhat clavate at end. Disk small, crenulate-edged, stylopodia small. Carpophore bipartite. Endosperm with the commissural face flat.

Three species in W Asia; one or possibly two species in Iraq.
Trigonosciadium (from the Gr. τρίγωνος, *trigonos*, triangular and σκιαδειον, *skiadeion*, an umbrella or parasol – from the at least partially 3-angled stem of the species and particularly the first described, *T. tuberosum*).
The character of the small stylopodia seems of little value in separating this genus from *Heracleum*, as the stylopodia in several of the more dwarf species of that genus are equally as small as those of *T. viscidulum*.

Styles hairy; rays more than 20 in number; bracts many . . . . 1. *T. tuberosum*
Styles glabrous; rays 20 or fewer, usually 10–12; bracts few, caducous . . . . . 2. *T. viscidulum*

1. **Trigonosciadium tuberosum** *Boiss.*, Ann. Sc. Nat. sér. 3, 1: 345 (1844). — Type: Iraq, (or in Syria or Turkey), "Mesopotamia", *Aucher* 3723 (K!, syn.); Boiss., Fl. Orient. 2: 1051 (1872); Fl. Turk. 4: 501 (1972); Alava in Fl. Iranica 162: 515 (1987).

Heracleum mesopotamicum Hiroe, Umb. World 1749 (1979).

Root oblong. Stem erect, 30–70 cm, puberulent with short spreading or deflexed hairs, sharply 3-angled below, sparingly dichotomously branched from near base, branches becoming less sharply angled above. Leaves puberulent, short, pinnatisect with 2 pairs of leaflets or ternate, with a broad hirsute sheath which, at least in the upper leaves, decurrent onto the lowest pair of leaflets; leaflets broad, ovate, sharply toothed and sometimes ± trilobed above, cuneate and entire in the lower half. Peduncles 8–16 cm. Umbels few, 14–40-rayed, rays with spreading viscid hairs, 2–8 cm. Involucre of ± 10 reflexed lanceolate bracts. Partial umbels 14–30-flowered, with an involucel of many linear bracteoles. Fruit shortly tomentellous, subrotund, 7–10 × 6–9 mm, deeply emarginate, vittae very short (much less than half the length of the fruit to ²/₃ of its length), inner two shorter than the outer and sometimes very inconspicuous. Disk very small, crenulate-margined, together with the minute stylopodium not exceeding the emargination of fruit. Styles erect or divaricate, 4–5 mm, hairy.

HAB. Not known.
DISTRIB. Very rare, if indeed it occurs in Iraq. Iraq, (or in Syria or Turkey), "Mesopotamia", *Aucher* 3723 (type)!

Turkey (E Anatolia), W Iran.

2. **Trigonosciadium viscidulum** *Boiss. et Hausskn.*, Boiss., Fl. Orient. 2: 1051 (1872). — Type: Iraq, Pir Omar Gudrun (Pira Magrun), *Haussknecht* s.n. (JE, syn.). Blakelock in Kew Bull. 3: 437 (1948); Zohary in Dep. Agr. Iraq Bull. 31: 113 (1950); Rawi, ibid. Tech. Bull. 14: 96 (1964); Fl. Turk. 4: 501 (1972); Alava in Fl. Iranica 162: 515 (1987).

T. undulatum Yıldırımlı in Ot Sistematik Botanik Dergisi 17 (2): 16 (2010).

Root oblong-fusiform, often branched. Stem erect, 30–40 cm, variable as to indumentum, glabrous to shortly puberulent or more long-hairy throughout, the upper internodes viscid-hairy or not. Branched dichotomously from near the base, branches long and somewhat flexuose. Stem often purple-spotted below or purple-suffused, sharply 3- or more- angled, branches less angular. Leaves puberulent, short, lower pinnatisect with 2 or rarely 3 pairs of leaflets, upper ternate, with a broad hairy sheath which in the upper leaves is decurrent on to the lowest pair of leaflets; leaflets broad, ovate, dentate. Peduncles 3–12(–16) cm. Umbels few, 9–20-rayed, rays spreading viscid-hairy, 1.5–7 cm in fruit. Involucre of 1 or 2 setaceous and caducous bracts or absent. Partial umbels 9–30-flowered, with an involucel of 2–5 setaceous bracteoles; fruiting pedicels 3–8 mm. Fruit shortly tomentellous, 6–12 × 6–10 mm, deeply emarginate, subrotund to very broadly obovoid, vittae decidedly longer than half the length of fruit, all ± equal. Disk very small with prominent undulate margins. Stylopodia conical, larger than those of *T. tuberosum* but still very small and not exceeding the emargination of fruit. Styles erect or divergent, 1.5–2 mm, glabrous. Fig. 86: 1–6.

HAB. On mountains; alt. 800–1900 m; fl. & fr. Apr. -Jun.
DISTRIB. Occasional in the forest zone of Iraq. **MAM**: Zakho pass, *Guest* 2284!; Bakarma, *Rawi* 8520A!; Sarsang, *Haines* 1347; **MRO**: Kuh-i Sefin, *Bornmüller* 1274!; Gali Warta, NW of Rania, *Rawi, Nuri & Kass* 28760!; Shaqlawa, *Haines* s.n.; Sakri-Sakran, *Hadač et al.* 5431; Mt. Botin, *Agnew et al.* 6100; **MSU**: Pir Omar Gudrun (Pira Magrun), *Haussknecht* s.n. (type).
NOTE. In view of the variability of the stem indumentum in this species it seems unsafe to rely on the viscidity as a separating character from *T. tuberosum*, particularly since when more material is available the latter may prove equally variable in this respect. The characters by which *T. undulatum* was separated from this species (e.g. undulating margins of mericarps, villous stems and linear bracteoles) appear to fall within the range of variation of *T. viscidulum*.

Turkey (SE Anatolia), W Iran.

Fig. 86. **Trigonosciadium viscidulum**. 1, habit × 2/3; 2, lower portion of stem and root × 2/3; 3, mature umbel × 2/3; 4, petals × 4; 5, fruits (both faces) × 2; 6, fruit cross section × 10. 1–4 from *Rawi et al.* 28760; 5–6 from *Rawi* 8520A. Drawn by D. Erasmus (1964).

43. HERACLEUM L.

Sp. Pl. ed. 1: 249 (1753); Gen. Pl. ed. 5: 118 (1754)

Perennial or biennial herbs, very variable in stature, some gigantic, some dwarf, usually ± pubescent, setose or scabrous. Leaves 1–3-pinnate with broad segments or variably trisect or ternate to entire and ± lobed. Umbels compound, the central hermaphrodite and laterals with central flowers male, or hermaphrodite. Bracts of involucre mostly absent or caducous, bracteoles several. Flowers white, regular, outer usually ± radiant. Petals long-cuneate below, outer obcordate above from the deeply 2-lobed or emarginate tips, inner with an incurved acumen. Calyx teeth small. Fruit ± round, elliptical or obcordate, strongly plano-compressed dorsally, pubescent at least when young. Mericarps almost flat, the 3 median ribs slender and not conspicuous, laterals with thickened margins, closely contiguous and forming a wing to the fruit. Vittae solitary in valleculae, shorter than the fruit, conspicuous and swollen at end and appearing narrowly claviform. Disk and stylopodia variable, conical. Carpophore bipartite. Endosperm with the commissural face flat, commissural vittae 0–2.

About 70 species in the N temperate region and on mountains in the tropics, but specific limits often uncertain; two species in Iraq.

Heracleum (the name of a plant in Theophrastus, from Gr. πάνάκεια χηρακλειον *panakeia herakleion* ("all-heal Hercules"), some species of the genus having been at one time used medicinally); Cow-Parsnip, Cowparsnip (Am.). According to Uphof (1968), the boiled leaves and fruit of the Common Cow-Parsnip (*H. sphondylium*) formed the basis of an alcoholic beverage, known as Bartsch in Slavonic countries, which is still used in some French liqueurs.

Plant stout; leaves entire, lobed and dentate; petals densely hairy; styles hairy . . . . 1. *H. lasiopetalum*
Plant slender, leaves pinnate, segments further deeply divided; petals not or very thinly hairy; styles glabrous . . . . 2. *H. rawianum*

1. **Heracleum lasiopetalum** *Boiss.*, Ann. Sc. Nat. sér. 3, 1: 332 (1844); Boiss., Fl. Orient. 2: 1042 (1872); Rawi in Dep. Agr. Iraq Tech. Bull. 14: 93 (1964); Fl. Turk. 4: 499 (1972);

Tetrataenium lasiopetalum (Boiss.) Mandenova in Rech. f., Fl. Iranica 162: 504 (1987).

Plant erect, 30–60 cm, stoutly built, furnished throughout with broad-based papillate hairs, clothed with remains of petioles at base. Stem somewhat angled, sulcate, with long ascending branches which are sometimes opposite above, alternate in leaf axils below, also sulcate. Leaves simple, toothed, ± trilobed, especially upper; lamina of basal leaves 10–26 cm, basal petioles 7–15 cm; sheaths broad and inflated, strongly ribbed, softly hairy. Peduncles 5–16 cm. Umbels rather few, 11–35-rayed, rays in fruit 2–7.5 cm. Involucre of 1–3 variable bracts, often withering. Partial umbels 10–20-flowered; flowers large, 5–6 mm in diameter. Petals densely hairy, scarcely radiant. Involucel of ± 6 linear-subulate bracteoles. Fruit 9–13 × 7.5–10 mm, densely covered with stiff deflexed hairs of varying length. Vittae scarcely or not visible beneath the indumentum on dorsal face of each mericarp, the two central conspicuous on the commissural face. Disk conspicuous, stylopodia conical, ribbed on the surface and sparsely pilose, margins crenulate. Styles flexuose, very hairy. Petals often persisting for a considerable time on the ripening fruit.

HAB. On high mountain slopes; alt. 1800–2100 m; fl. & fr. Jul.-Sep.
DISTRIB. Rare in Iraq, only found in one district of the central sector of the lower thorn-cushion zone. **MRO**: Halgurd (Algurd Dagh), *Rawi* 13735!; Halgurd Dagh, valley above Nawanda, *Rechinger* 11372; Sula Khal, *Rawi & Serhang* 20212!

Turkey (SE Anatolia), W Iran.

2. **Heracleum rawianum** *C.C. Townsend* in Kew Bull. 27: 429 (1964). — Type: Iraq, Halgurd range, Goum Tawera, ± 2800 m., 1 Sept. 1957, *Rawi & Serhang* 24721 (K!, holo.); Fl. Turk. 4: 498 (1972), sub "H. raweanum" sphalm.; Mandenova in Fl. Iranica 162: 501 (1987).

[*H. humile* (non Sibth. & Sm.) Rawi in Dep. Agr. Iraq Tech. Bull. 14: 93 (1964).]

Slender perennial, 6–35 cm, with a stout rootstock. Stem sparsely hairy, terete, finely sulcate, sparingly branched from base upwards. Leaves mostly radical, glabrous or finely and shortly hairy, oblong in outline, 4–9(–20) cm long, with white-margined expanded sheaths, simply pinnate with 2–4 pairs of sessile, lanceolate-ovate leaflets which are further deeply incised with lanceolate, acute, mucronate segments; upper leaves tripartite or reduced to sheaths. Peduncles 1.5–11 cm. Umbels few, long-pedunculate, 2–5-radiate, fruiting rays 1.5–4.5 cm. Involucre absent or of 2 or 3 very small and caducous bracts. Partial umbels 8–12-flowered. Flowers white, the outer petals shortly radiant, ± 3 mm; all petals glabrous. Fruiting pedicels 3–5 mm. Involucel of 4–8 linear, very short bracteoles, much shorter than pedicels. Fruit ovoid or obovoid, 8–10 × 4–8 mm, thinly hairy when young, mericarps glabrous on both faces when mature. Vittae subequal, the two inner slightly longer and attaining about 1/3 the length of the fruit on the dorsal surface; two only visible on the commissural face very short. Disk small, crenulate-margined, stylopodia conical, tumid, rugose, exceeding the emargination of the fruit. Styles 1–1.5 mm, glabrous, reflexed.

HAB. High in the mountains, on rocky slopes, on a damp scree; alt. 2300–3500 m; fl. & fr. Aug.–Sep.
DISTRIB. Quite common in the alpine region and subalpine upper thorn-cushion zone. **MRO**: Halgurd Dagh, *Gillett* 12350!; Goum Tawera, Halgurd range, *Rawi & Serhang* 24717!; Qandil range, *id.* 26744!, 26823!; NE of Qandil, *id.* 24370!; between summits of Bardanas and Perrish, Qandil range, *id.* 244583!

Turkey (SE Anatolia), W Iran.

44. **OPOPANAX** W.D.J. Koch

Nov. Ac. Acad. Caes. Leop. Carol. 12, 1: 96 (1824)

Perennial herbs, hispid or glabrescent. Leaves pinnate or bipinnate with broad, toothed, sometimes decurrent segments. Umbels compound, polygamous, terminal usually hermaphrodite, laterals with usually at least the inner flowers male. Bracts of involucre and involucel few, small, often caducous. Flowers yellow. Petals broad, subentire above and with a broad, inflexed acumen. Calyx teeth obsolete. Fruit round to broadly obovoid, strongly dorsally plano-compressed, marginal ribs of almost plane mericarps contiguous to form a distinct, tumid border to the fruit; 3 median ribs filiform, not prominent. Vittae 1–3 in the valleculae, percurrent or unequal. Disk almost flat, crenulate-margined, stylopodia very small. Carpophore bipartite. Endosperm with the commissural face flat.

Three species distributed between the Balkans in SE Europe and Iran; two species in Iraq.
Opopanax (from the Gr. οπος, *opos*, a milky juice, πανάξ, *panax*, a remedy, source of the English word panacea). The sap of one species of this genus has been used medicinally. Gum Opopanax, employed in perfumery, is obtained by making incisions in the root of the plant. Campbell Thompson (1946) has found a possible, though perhaps rather dubious, reference to opopanax in the ancient Assyrian medical texts.

At least the lower surface of leaves with papillate hairs which are furcate
or stellate at the tip . . . . 1. *O. hispidus*
Leaf surfaces glabrous. . . . . 2. *O. persicus*

1. **Opopanax hispidus** (*Friv.*) *Griseb.*, Spicil. Fl. Rumel. 1: 378 (1843); Zohary, Fl. Palest. ed. 2, 1: 560 (1932); Rawi in Dep. Agr. Iraq Tech. Bull. 14: 93 (1964); Grossh., Fl. Kavk. ed. 2, 7: 112 (1967); Mouterde, Nouv. Fl. Lib. et Syr. 2: 646 (1970); Chamberlain in Fl. Turk. 4: 472 (1972); Meikle, Fl. Cyprus 1: 757 (1977); Rech.f., Fl. Iranica 162: 438 (1987).

Ferula hispida Friv., Flora 18: 333 (1835).
Pastinaca opopanax L., Sp. Pl. ed. 1: 262 (1753).
P. hispida Fenzl, Flora 26: 462 (1843).
Opopanax orientale Boiss., Ann. Sc. Nat. sér. 3, 1: 330 (1844); Boiss., Fl. Orient. 2: 1059 (1872).

Plant tall, erect, to 3 m. Stem stiff, ± densely retrorse-hispid below, thinly hispid or glabrous above, terete, finely striate, corymbosely branched above. Branches also terete and glabrous, upper commonly opposite or verticillate. Leaves bipinnate, ovate in outline with short but broad and inflated sheaths, to 50 cm long and almost as wide, segments ovate to oblong-lanceolate with regularly serrate, cartilaginous borders, 1.5–10 × 0.7–3 cm; lower pinnae with 2 or 3 or 4 pairs of pinnules; at least the lower pinnules of upper pinnae usually ± decurrent along the main axis. Petioles and main axes of leaves furnished with flattish papillate hairs which are often hirtellous at apex. Leaves with at least lower surface

Fig. 87. **Opopanax hispidus**. 1, habit × 2/3; 2, leaf × 2/3; 3, mature umbel × 2/3; 4, petal × 10; 5, fruits (both faces) × 2; 6, fruit cross section × 9. 1–4 from *Balansa* 1246; 5–6 from *Kennedy* 1388. Drawn by D. Erasmus (1964).

furnished, frequently densely, more rarely sparsely, with hispid hairs which are furcate or stellate at the tips. Leaves diminishing upwards, finally reduced to sheaths. Peduncles 2–13 cm. Umbels several, with 12–20 glabrous rays, fruiting rays 1.6–7 cm. Bracts and bracteoles 1–5, usually small and caducous. Partial umbels 8–12-flowered, flowers ± 2.5 mm in diameter, fruiting pedicels 4–6 mm. Fruit ovoid, 7–12 × 6–8 mm, glabrous, valleculae with 1 or 2 broad, dark vittae, those of the commissural face 6–8 and more conspicuous. Disk much flattened and only slightly convex, shallowly crenulate-edged, stylopodia very small. Styles very short, glabrous, closely deflexed onto the disk and subequalling it. Fig. 87: 1– 6.

HAB. On the lower mountains, in valleys, among denuded *Quercus* forest near water; alt. ± 850 m; fl. & fr. May–Jun.
DISTRIB. Occasional in the SE sector of the lower forest zone of Iraq. **MSU**: between Kirkuk and Sulaimaniya, *Haussknecht* s.n.!; Darband-i Basian, *Haussknecht* s.n.!; Tainan, E of Darband-i Basian, *Gillett & Rawi* 11623!; "Mesopotamia", *Kotschy* s.n.

SE Europe (Italy, Bulgaria, Greece), Aegean Isles, Cyprus, Syria, Turkey, Armenia, Azerbaijan, Georgia, W Iran.

2. **Opopanax persicus** *Boiss.*, Diagn. ser. 1, 10: 36 (1849); Boiss., Fl. Orient. 2: 1059 (1872); Grossh., Fl. Kavk. ed. 2, 7: 112 (1967); Chamberlain in Fl. Turk. 4: 472 (1972); Rech.f., Fl. Iranica 162: 439 (1987).

Plant tall, erect to 2 m. Stem stiff, terete, sulcate or striate, glabrous or sparingly pilose below, corymbosely branched above. Branches terete, glabrous, the lower alternate, the upper opposite or verticillate. Basal leaves bipinnate, deltoid-ovate, to 60 cm long, long-petiolate. Stem leaves similar but smaller, segments ovate-elongate, broadest near base and tapering above with a regularly serrate, cartilaginous border, largest terminal, which may be up to 10 × 4.5 cm, and occasionally lobed on one or both sides near the base; lower pinnae with 1or rarely 2 pairs of pinnules; some or all segments decurrent along the main axis. Petioles of lower leaves with dense papillate hairs, which are very sparse on the rhachis of upper leaves. Leaf surfaces, including veins of lower surface, almost or quite glabrous. Uppermost leaves reducing to sheaths. Umbels several, with 14–20 glabrous rays, rays 1.5–5 cm. Bracts and bracteoles few, small and caducous. Partial umbels 10–12-flowered. Fruiting pedicels 3–6 mm. Fruit ovoid, 9–12 × 6–7 mm, glabrous, valleculae with 1 or 2 broad, dark vittae, those of the commissural face more conspicuous. Disk much flattened, crenulate-edged. Styles very short, glabrous, closely deflexed upon and subequalling the disk.

HAB. By roadside in the mountains, on banks of irrigation canals; alt. 800–850 m; fl. & fr. Jun.
DISTRIB. Occasional in the SE sector of the lower forest zone of Iraq. **MSU**: 20 km WNW of Sulaimaniya, *Rawi* 22775!; Sulaimaniya, *Haines* 1509 (E)! W 1509 (K)!; Tanjaro bridge, nr Sulaimaniya, *Haines* W1509 A!, *Hadač* 5237; Gerdabor (Gerdazubayr?) on Pira Magrun, *Hadač* 4736.
NOTE. Probably not distinct from *O. hispidus*. The Edinburgh sheet of *Haines* 509 has scattered hairs on the lower leaf surfaces and along the veins, while the Kew sheet is quite glabrous. Insufficient fruiting material has been seen to judge the reliability of the fruit characters given by Boissier; population studies in both species are needed.

Turkey (E Anatolia), Armenia, Azerbaijan, Georgia, NW, C & S Iran.

45. **MALABAILA** Hoffm.

Gen. Umb. ed. 1: 125 (1814) non Tausch in Flora 17: 356 (1834)

Perennial herbs with pinnate or 2- or 3-pinnately decomposite leaves, segments often (as in Iraqi species) broad and toothed. Umbels compound, polygamous, central umbel hermaphrodite, outer umbels usually hermaphrodite and (inner flowers) male. Bracts of the involucre few or more, bracteoles several or absent. Flowers yellow. Petals ovate, slightly emarginate or entire above with an inflexed (often obtuse) acumen. Calyx obsolete. Fruit ± round, more rarely elliptical or obovate, strongly dorsally plano-compressed. Mericarps with intermediate ribs filiform, slender, laterals thick, closely contiguous, forming a tumid margin to the fruit. Vittae solitary in the valleculae, 2 on the commissure. Disk crenulate-margined. Stylopodia conical, small or somewhat tumid. Carpophore bipartite. Endosperm with the commissural face flat.

Ten species distributed between from the E Mediterranean to Iran and C Asia; only one polymorphic species in Iraq.

Reference: Pimenov, M.G. & Ostroumova, T.A. (2008). The genus *Malabaila* Hoffmn. (Umbelliferae: Tordylieae): a carpological investigation and taxonomic implications. Feddes Repert. 105(3–4): 141–155.

Malabaila (named in honour of Count Emmanuel Joseph Malabaila, Count of Canale, 1745–1826, Curator of the Prague Botanical Garden).

1. **Malabaila secacul** (*Mill.*) *Boiss.*, Diagn. ser. 1, 10: 42 (1849) sub "sekakul" sphalm.; Boiss., Fl. Orient. 2: 1057 (1872); Nábělek in Publ. Fac. Sci. Univ. Masaryk 35: 128 (1923); Zohary, Fl. Palest. ed. 2, 1: 560 (1932); Bornmüller in Beih. Bot. Centralbl. 58B: 287 (1938); Zohary in Dep. Agr. Iraq Bull. 31: 113 (1950); Rawi in Dep. Agr. Iraq Tech. Bull. 14: 93 (1964); Mouterde, Nouv. Fl. Lib. et Syr. 2: 647 (1970); Chamberlain in Fl. Turk. 4: 487 (1972); Zohary, Fl. Palaest. 2: 446 (1972).

Tordylium secacul Mill., Gard. Dict. ed. 8 no. 5 (1768); DC., Prodr. 4: 198 (1830) sub. "sekakul".
Pastinaca secacul (Mill.) Banks & Sol. in Russell, Nat. Hist. Aleppo ed. 2, 2: 249 (1794).
[*M. loftusii* sensu Guest in Dep. Agr. Iraq Bull. 27: 60 (1933), an unpublished ms. name by Stapf in the Kew herbarium].
Leiotulus secacul (Mill.) Pimenov & T.A. Ostroumova in Feddes Repert. 105 (3–4): 153 (1994).

Plant 30–80 cm, densely puberulent throughout, with long, ascending branches from near the base. Rootstock thick, tuberous, oblong. Stem terete, striate, branches sometimes angled. Radical leaves variable, pinnate with broad, oblong-ovate segments which may be irregularly crenate or lobed, or deeply incised to the midrib, or leaves bipinnate with toothed pinnules; lamina broadly oblong to ovate in outline, to 30 cm long, petiolate. Upper leaves sessile, similarly divided to the lower, but rather few and rapidly reducing in size, uppermost often with 2 or 3 short, linear lobes much shorter than the sheaths. Sheaths large, broadly inflated, tapering above into the petiole, striate. Peduncles 2–14 cm. Umbels several, rays (10–)15–20, unequal, in fruit 2.5–8 cm long, variable in indumentum (as is the upper part of the peduncles) from glabrous to sparsely scabrid with stout-based, often curved bristles or with a dense mixed indumentum including unequal viscidulous hairs. Partial umbels 5–25-flowered. Flowers ± 3 mm in diameter. Petals deeply emarginate, yellow with a brown vitta along the mid-vein. Involucre absent, involucel of 1–6 filiform bracteoles, shorter than the ± 4 mm pedicels. Fruiting pedicels 3–15 mm. Fruit very variable in size and shape, 6–14 × 5–12 mm, oblong to obcordate or almost round, mericarps glabrous or hairy on both faces. Styles 1.5 mm, disk concave, crenulate-margined; stylopodium very shortly conical, not exserted from the apical notch of fruit.

HAB. (of species). Lower mountain slopes, steppic plains & hills, & sub-desert; alt. 150–1350 m; fl. & fr. Apr.–Jun.
DISTRIB. (of species) Quite common in the steppe region of Iraq, occasional in the lower forest zone and in the NW sector of the desert region.

Syria, Lebanon, Palestine, Jordan, Turkey, Iran.

subsp. **secacul**

Commissural face of mericarps glabrous, shiny. Fruit generally smaller, (5–)6–8(–9) mm. Rays of umbel usually subglabrous with scattered stout-based bristles, occasionally viscidulous.

HAB. In the desert on clay and sandy soil, on stony hillsides, weed in a field on the steppe; alt. 200–650 m; fl. & fr. Apr.
DISTRIB. Occasional, mainly in the NW sector of the desert region of Iraq. **MSU**: Pira Magrun, *Haussknecht* s.n.; Qopi Qaradagh, *Hadač et al.* 5205; **FUJ**: 14 km S. of Balad Sinjar, *Chakr., Rawi, Khatib & Alizzi* 33107!; **DWD**: Faggart ash-Shadaf, *Rawi* 31149!; Tal an-Nisr, 48 km NE of Rutba, *Rawi* 31222!; **DLJ**: Tayrat, 60 km NW of Rawa, *Chakr., Rawi, Khatib & Alizzi* 31957!

Syria, Lebanon, Palestine, Jordan, Turkey, NW, C & NE Iran.

subsp. **aucheri** (*Boiss.*) *C.C. Townsend*, Kew Bull. 19 (1): 69 (1964).

M. aucheri Boiss., Ann. Sc. Nat. sér. 3, 1: 336 (1884); Boiss., Fl. Orient. 2: 1057 (1872); Blakelock in Kew Bull. 3: 435 (1948); Zohary in Dep. Agr. Iraq Bull. 31: 113 (1950); H. Riedl in Fl. Lowland Iraq: 469 (1964); Rawi in Dep. Agr. Iraq Tech. Bull. 14: 95 (1964); Chamberlain in Fl. Turk. 4: 483 (1972). *M. erbilensis* Freyn & Bornmüller, Bull. Herb. Boiss. 5: 623 (1897).

Fig. 88. **Malabaila secacul** subsp. **aucheri**. 1, habit × 2/3; 2, mature umbel × 2/3; 3, petal × 12; 4, fruits (both faces) × 3; 5, fruit detail × 8; 6, fruit cross section × 12. 1, 2 from *Guest et al.* 19342; 3–6 from *Guest* 4027. Drawn by D. Erasmus (1964).

Commissural face of mericarps pubescent. Fruit generally larger, (6.5–)8–11(–14) mm. Rays of umbel frequently (in Iraqi material) with abundant viscidulous hairs of diverse lengths. Fig. 88: 1–5.

HAB. On lower mountain slopes, dry rocky open space in *Quercus* forest, stony overgrazed limestone slopes near village, weed in barley and cereal fields on the steppe; alt. 150–1350 m; fl. & fr. Apr.–Jun.
DISTRIB. Occasional in the lower forest zone and steppe region of Iraq. **MRO**: Shaqlawa, *Gillett* 8051!; Baba Chichak, *Nábělek* 402; Koi Sanjaq, *Omar, Karim, Hamza & Hamid* 37164!; **MSU**: Qara Anjir, *Omar & Karim* 37876!; Qopi Qaradagh, *Haines* W1558!; Tawila, *Rawi* 21800!; nr Sulaimaniya, *Thesiger* 342, 346; **FAR**: nr Arbil, *Hadač et al.* 5756; **FNI**: Qizil Robat (Sa'diya), *Guest* 2640A!; Mandali, *Guest* 1728A!
NOTE. ?KIZAR ("carrot", Ir.-Mandali, *Guest* 1728, cf. JAFAR, "carrot", Ar.), JIZIR (Ir.-Mosul, *Loftus* s.n.); GAIZA GIYĀ ("carrot-weed", Kurd.-Arbil, *Guest* 1483) – as Guest (1933) suggests such names are probably used indiscriminately for various umbellifers which bear a superficial resemblance to the carrot; ?DŌLĀB AL-HAWĀ (Ir.-Mosul, *Lazar* 3354).

The subspecies of the following specimens of *M. secacul* are indeterminate: **MAM**: Sarsang, *Haines* W1347!; **MRO**: Kuh-i Sefin, *Bornmüller* 127; Haibat Sultan Dagh, *Rawi, Nuri & Kass* 28169!; **MSU**: Qara Dagh, nr Timasa (or Timara?) village, *Karim, Hamid & Jasim* 40643!; 11 km N of Penjwin, *Rawi* 22901!; **FUJ**: Huqna, between Mosul and Nusaybin, *Loftus* s.n.!; 20 km S of Tal Afar, *Ani & Hasan Hadi* 9798!; **FNI**: Mosul, *Lazar* 3354!; **FAR**: Qosh Tepe, *Guest* 1483!; **FKI**: Tuz, *Guest* 1401!; **FKI/MSU**: 6 km E of Kirkuk, *Botany Staff* 42978?; **FNI**: nr Zurbatiya, *Ani & Bharucha* 7636!; **DWD**: Wadi-al-Ajramiya, *Alizzi & Hussain* 34038?

It is extraordinary that not a single Iraqi gathering of this plant notes its height, the upper limit of which is taken only from material collected from other countries.

M. erbilensis Freyn & Bornmüller in Bull. Herb. Boiss. 5: 623 (1897); Bornmüller in Beih. Bot. Centralbl. 58B: 287 (1938) does not appear to me to be separable from *M. secacul*. It is somewhat intermediate between the two subspecies, with a large fruit with a glabrous or very sparsely hairy commissure and the rays without viscid hairs.

NE Syria, Turkey (very rare), W Iran.

46. **DUCROSIA** Boiss.

in Ann. Sc. Nat. sér. 3, 1: 341 (1844)

Glaucous perennial herb with stout taproots, vegetative parts glabrous. Leaves ternately to pinnately divided. Umbels compound, hermaphrodite. Bracts and bracteoles several, short. Flowers white or greenish. Petals cuneate or obovate, ± entire above with a narrow, inflexed acumen, densely pubescent dorsally. Calyx teeth small. Fruit oblong, ovate or round, hairy, dorsally plano-compressed. Mericarps with the 3 intermediate ribs filiform, slender, laterals swollen and contiguous, forming a tumid margin to the fruit. Dorsal vittae solitary in valleculae; commissural vittae 2, adjacent to the margins. Disk small, crenulate-margined. Stylopodia conic, small to very small. Carpophore bipartite. Endosperm with the commissural face flat.

Five species distributed from Egypt to NW India; two species in Iraq.
Ducrosia (in honour of the Rev. L. Du Cros, a Swiss botanist and friend of Edmond Boissier.)

Flowers yellow; leaves ternately or pinnately decomposite into long, narrowly linear segments, rarely 3 mm wide .... 1. *D. anethifolia*
Flowers yellowish-brown, ± suffused with brown or purple; leaves ternate or trisect, or one of two of the lowest radical leaves with 2 pairs of pinnae, toothed segments broad, 5 mm or more wide in all but the lowest, which are often withered at time of flowering .... 2. *D. flabellifolia*

1. **Ducrosia anethifolia** (*DC.*) *Boiss.*, Ann. Sc. Nat. sér. 3, 1: 342 (1844); Boiss., Fl. Orient. 2: 1036 (1872); Hand.-Mazz. in Ann. Naturh. Mus. Wien 27: 91 (1913); Anth. in Notes Roy. Bot. Gard. Edinb. 18: 289 (1935); Blakelock in Kew Bull. 3: 433 (1948); Zohary in Dep. Agr. Iraq Bull. 31: 113 (1950); H. Riedl in Fl. Lowland Iraq: 467 (1964); Rawi in Dep. Agr. Iraq Tech. Bull. 14: 91 (1964); Mouterde, Nouv. Fl. Lib. et Syr. 2: 654 (1970); E. Nasir in Fl. Pak. 20: 165, f. 50 (1972); Alava in Fl. Iranica 162: 469 (1987); Ghazanfar, Fl. Oman 2: 145 (2007).

Zozinia (sic) *anethifolia* DC., Prodr. 4: 196 (1830).
Ducrosia olivieri Boiss. & Buhse in Nouv. Mém. Soc. Nat. Mosc. 12: 101 (1860).

Erect, 10–40 cm, with a pungent smell; stem glaucous, terete, branched from at or near the base upwards, sulcate below, more finely striate above, hollow or with a very soft pith. Branches many, alternate or the upper occasionally opposite. Lower leaves oblong or ovate in outline, bi- or tri-pinnatisect, with 2 or 3 pairs of leaflets (occasionally one of a pair suppressed) with long, narrowly linear, entire, mucronate, 3–40 × 1–2(–3) mm segments; upper leaves becoming ternatisect and often a few uppermost narrowly linear and simple. Peduncles long, 4–14 cm. Umbels many, large for size of plant, terminal umbel with 10–25 rays, the branch umbels with 6–15 glabrous rays; fruiting rays 1.5–6 cm. Partial umbels 9–25-flowered; flowers yellow, ± 2 mm in diameter. Involucre and involucel similar in form, of about 10 short, reflexed, white-margined bracts and bracteoles. Bracteoles and pedicels glabrous, or the former margined with very short hairs; fruiting pedicels 4–10 mm, glabrous. Fruit round, 5–6 mm, tumid margin broad, pale and conspicuous, central part dark, more pronouncedly hairy, especially on the commissural face. Vittae prominent dorsally. Stylopodia small, not exceeding the emargination of the fruit. Styles short, glabrous, equalling or somewhat exceeding the disk, closely reflexed. Fig. 89: 1–6.

HAB. In the desert, frequently on sandy and sandy gravel soils, occasionally on silty soils in depressions, also on heavily eroded steppic foothills, conglomerate, sandstone etc., very rare in the mountains (only found once over a century ago) on a low montane plain; alt. 50–600(–750) m; fl. & fr. Jun.–Jul.
DISTRIB. Common in the desert region of Iraq, occasional in the steppe region. **MSU**: between Kirkuk and Darband-i Basian, *Haussknecht* s.n.; **FUJ**: Sharqat, *Handel-Mazzetti* 1121; Fatha, *M. al-Khatib & H. Tikriti* 29719!; **FAR/FKI**: nr Altun Kupri, *Guest* 4027A!; **FKI**: Baba Gurgur, *Guest* 4012!; Kirkuk, *Haussknecht* s.n.; **FKI/MSU**: 8 km E of Kirkuk, *Gillett & Rawi* 11596!; **FKI**: 9 km NW of Injana, *Rawi* 22865?; **FNI**: Koma Sang, nr Iranian frontier, *Rawi* 20618!; Makatu, *Guest* 869!; 10 km E of Mandali, *Rechinger* 9623; 4 km E of Badra, *Haddad* 9950?; 78 km NNW of Fakka, *Hazim* 30670!; Jabal Hamrin, *Sutherland* 299; **DWD**: Wadi Hauran, nr Rutba, *Alizzi & Omar* 36191!; 2 km W of Rutba, *Rechinger* 12817!; 4 km SW of Rutba, *Rawi* 21037!; 160–190 km W of Ramadi on road to Rutba, *Rawi* 31325!; 75 km W of Ramadi, *Rawi* 21037!; 30 km W of Ramadi, *Alizzi* 35131?; 20 km W of Ramadi, *Rechinger* 9502; nr Habbaniya, *Rawi & Alizzi* 34464?; **DLJ**: 13 km E of K.3, Chakr., *Rawi, Khatib & Alizzi* 32979!; W of Baiji, Rawi & Hamada 34145?; ± 50 m NW of Ramadi, *Guest, Rawi & Nuri* 13565!; 65 km W of Falluja, *Rawi* 30217!; 55 km NW of Ramadi, *Rawi* 20230!; 34 km NW of Falluja, *Rawi, Shahwani, Kass & Tikriti* 25472!, *Rawi & Shahwani* 25473!; **DGA**: 4 km E of Samarra, *Rechinger* 13497; **DSD**: ± 20 km S.E of Ashuriya, *Guest, Rawi & Rechinger* 19342!, *Rechinger* 9274; 50 km W of Shabicha on track to Ma'niya, *Hazim* 30680?; 45 km WNW of Shabicha, *Rechinger* 9463; 5 km from Ma'niya, *Hazim* 32494; 10 km W of Ma'niya, *Hazim* 32534?; Aidaha, *Guest, Rawi & Rechinger* 19134!; between Aidaha and Ansab, *Fauzi, Hazimi & Hamid* 38827?; ± 30 km E of Salman, near Golaib, *Guest, Rawi & Rechinger* 18764!; *Rechinger* 9274; Saddat, 30 km ESE of Salman, *Guest, Rawi & Rechinger* 18855!, *Rechinger* 9314; 214 km WNW of Busaiya, on road from Salman, *Chakr., Khatib, Rawi & Tikriti* 30002!; 97 km SW of Busaiya, *Khatib & Alizzi* 32764!; Faidhat ach-Chabd, *Rawi, Khatib & Tikriti* 29199!; between Mughaizal and Khadhar al-Ma'i, *Serkahia* 18346!; **LCA**: 25 km E of Falluja, *Ani* 994!; "Falluja desert", *Haines* s.n.; **LEA**: Ba'quba, *Haines* W1004!
NOTE. HAZA (Ir.-Rutba, *Rawi* 21037; Ir.-Salman, *Guest, Rawi & Rechinger* 18855), "smoke from this plant when burned said to be good for eyes affected by bad smell; and it is said to alleviate flatulence, if dried and ground with tobacco for an early morning smoke"; Ir.-Khadr al-Ma'i, *Serkahia* 18346 – "a medicinal plant".

Syria, Arabia, Iran, Afghanistan, Pakistan.

2. **Ducrosia flabellifolia** *Boiss.*, Ann. Sc. Nat. sér. 3, 1: 342 (1844); Boiss., Fl. Orient. 2: 1036 (1872); Anth. in Notes Roy. Bot. Gard. Edinb. 18: 289 (1935); Zohary in Dep. Agr. Iraq Bull. 31: 113 (1950); H. Riedl in Fl. Lowland Iraq: 467 (1964); Rawi in Dep. Agr. Iraq Tech. Bull. 14: 97 (1964); Mouterde, Nouv. Fl. Lib. et Syr. 2: 654 (1970); Alava in Fl. Iranica 162: 471 (1987).

Erect, 8–40 cm, with an unpleasant smell; stems glaucous or yellowish green, finely striate, branched from at or near base, hollow. Branches many, long and ascending, alternate, or opposite above. Lowest leaves pinnate with two pairs of leaflets which are deeply bi- or tri- fid and acutely toothed, often withered at time of flowering. Intermediate leaves ternate with the terminal leaflet cuneate-based, all leaflets ± trisect, segments at least 5 mm wide, acutely toothed. Uppermost leaves trisect, similarly toothed. All leaves very long-stalked, petiole of lower leaves up to 12 cm. Peduncles 3–12 cm. Umbels several to many, terminal umbel with (4–)20–25 glabrous rays, the lateral umbels with (3–)11–18 glabrous rays; fruiting rays 2.5–6 cm. Partial umbels 20–30-flowered; flowers buff, ± suffused with brown or purple, 2–2.5

Fig. 89. **Ducrosia anethifolia.** 1, habit × ²/₃; 2, mature umbel × ²/₃; 3, petal × 12; 4, fruits (both faces) × 3; 5, fruit detail × 8; 6, fruit cross section × 12. 1, 2 from *Guest et al.* 19342; 3–6 from *Guest* 4027. Drawn by D. Erasmus (1964).

mm in diameter. Involucre and involucel similar in form, of 8–10 bracts and bracteoles, the latter considerably downy. Pedicels 2–4 mm in fruit, glabrous or sparingly hairy upwards. Fruit similar to that of *D. anethifolia*, with slightly longer styles.

Hab. Steppic, desert hills and plains, on sandy gravel soil; alt. c. 50–600 m; fl. & fr. Apr.–May(–Jun.).
Distrib. Rather rare in Iraq, in the dry steppe zone and desert region. **FNI**: Jabal Hamrin, *Sutherland* 300; Jabal Hamrin, nr Shahraban (Muqdadiya), *Haines* W1401!; nr Kharbut, 54 km SW of Mandali, *Rechinger* 12744; **DWD**: 15 km S of Rutba, *Khatib & Hazim* 32442; ? **DWD/DLJ**: (or in Syria?) between Aleppo and Baghdad, *G.A. Oliv.* s.n.; **LCA**: 'in desertis Assyriae' (nr Baghdad), *Aucher* 3729!

Syria, Jordan.

47. **ZOSIMA** Hoffm.

Gen. Umb. ed. 1: 145 (1814)
Zozimia auct. mult. sphalm.; *Pichleria* Stapf & Wettst. (1886)

Perennial herbs with stout taproots. Leaves pinnate or pinnately decomposite. Umbels usually polygamous with the inner flowers of partial umbels male. Bracts of the involucre and involucel numerous. Flowers white or greenish. Petals obovate, emarginate above with an inflexed acumen. Calyx teeth obsolete. Fruit roundish, strongly dorsally plano-compressed; mericarps slightly convex, intermediate ribs filiform and very slender, remote from the laterals, which are thick and closely contiguous, forming a tumid border to the fruit. Vittae solitary in valleculae, dark and conspicuous, commissure bivittate, rarely more present. Disk small, depressed, crenulate-margined. Stylopodia very small. Carpophore bipartite. Endosperm with the commissural face flat.

Ten species in W Asia; a single species in Iraq.
Zosima (explained by Hoffmann as dedicated to the three classicist Zosima brothers Anastasii, Nikolai and Zoi, an oblique reference to the triple relationship of the genus with *Pastinaca, Tordylium* and *Heracleum*).

1. **Zosima absinthifolia** (*Vent.*) *Link*, Umb. 145 (1814); DC., Prodr. 4: 195 (1830); Boiss., Fl. Orient. 2: 1037 (1872); Nábělek in Publ. Fac. Sci. Univ. Masaryk 35: 128 (1923); Zohary, Fl. Palest. ed. 2, 1: 558 (1932); Anth. in Notes Roy. Bot. Gard. Edinb. 18: 289 (1935); Blakelock in Kew Bull. 3: 437 (1948); Zohary in Dep. Agr. Iraq Bull. 31: 113 (1950); Rawi in Dep. Agr. Iraq Tech. Bull. 14: 97 (1964); Grossh., Fl. Kavk. ed. 2, 7: 133 (1967); Mouterde, Nouv. Fl. Lib. et Syr. 2: 654(1970); Chamberlain in Fl. Turk. 4: 503 (1972); Zohary, Fl. Palaest. 2: 445 (1972); E. Nasir in Fl. Pak. 20: 168, f. 50 (1972); V. Täckh., Stud. Fl. Egypt ed. 2: 394 (1974); Meikle, Fl. Cyprus 1: 760 (1977); Alava in Fl. Iranica 162: 474 (1987); Boulos, Fl. Egypt 2: 179 (2000).

Heracleum absinthifolium Vent., Choix Pl. Jard. Cels.: 7 (1803).
Zosima orientalis Hoffm., Gen. Umb. ed. 1: 145 (1814).

Plant stout, erect, 20–50 cm, rootstock crowned with fibrous remains of previous years' petioles. Stem deeply sulcate, densely covered, as are all vegetative parts, with a crisped white pubescence, corymbosely branched from near base and occasionally almost to the middle, with a foetid odour when bruised. Branches sulcate or coarsely striate, leafy only near base, greater part of each branch being the 10–35 cm umbel peduncle. Basal leaves oblong-lanceolate in outline, greyish green, 12–30 cm long, bi- to subtripinnate into short, oblanceolate, 1–6 × 0.5–2 mm segments, long-petiolate (4–10 cm), often very similar to those of *Artemisia absinthium*, with somewhat expanded, striate sheaths. Upper leaves becoming more shortly petiolate or sessile, much reducing, becoming pinnate with toothed segments. Umbels several, large for the size of this plant, with 15–30 pubescent unequal rays 2–11 cm long. Partial umbels many- flowered, (12–25), flowers ± 4.5 mm in diameter. Outer petals slightly radiate; pedicels hairy, 4–10 mm in fruit. Peduncle and rays sometimes thickened at the apex in fruit. Bracts and bracteoles similar, 12–15, lanceolate, white-margined. Fruit somewhat variable, normally ± round but sometimes inclined to oblong, 8–13 × 6–9 mm, moderately hairy to densely lanate with long white hairs, commissural face glabrous. Disk small, not exceeding the emargination of the fruit. Styles glabrous, slender and ± flattened, deflexed, 1.5–2 mm. Fig. 90: 1–7.

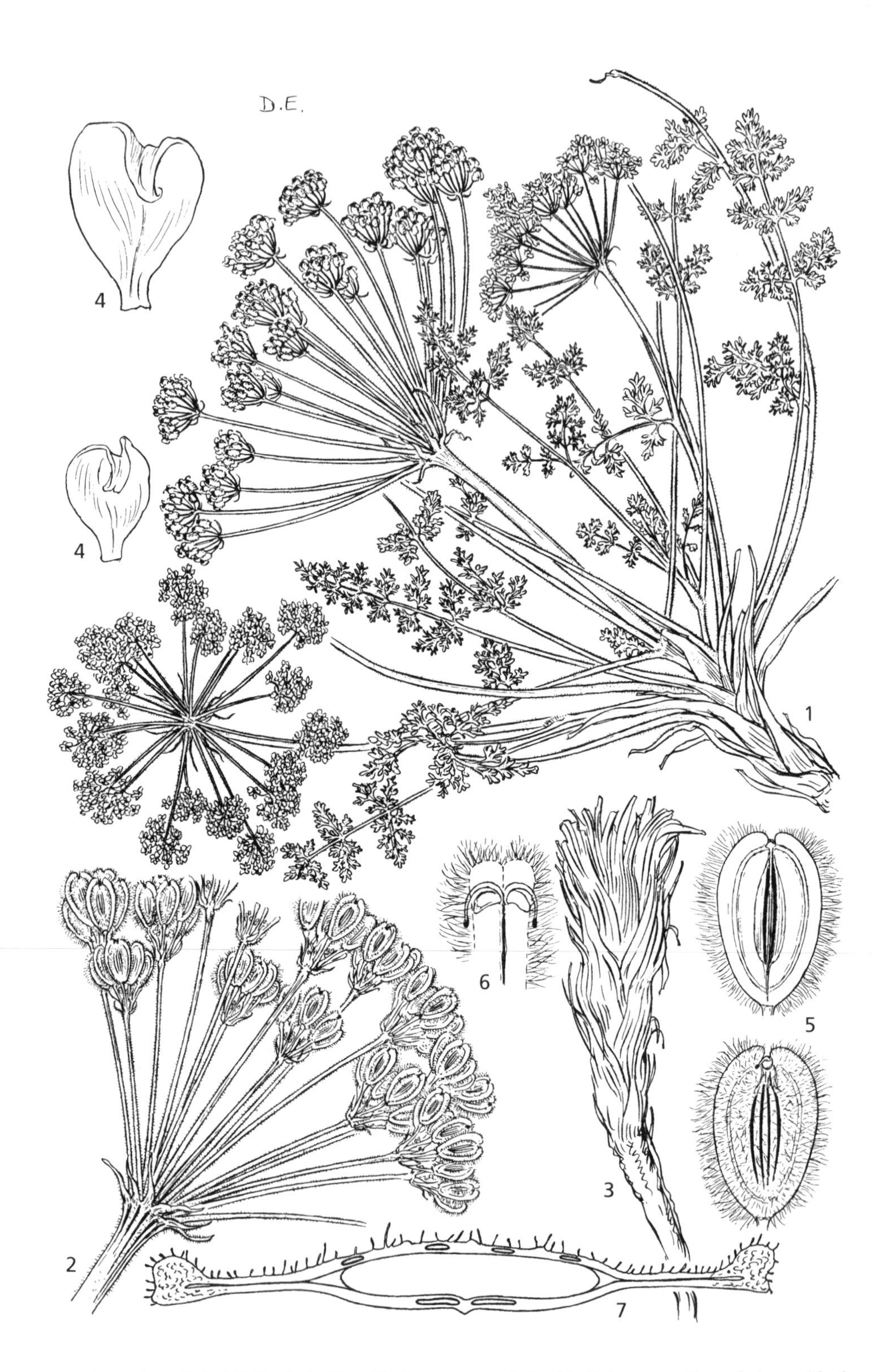

Fig. 90. **Zosima absinthifolia**. 1, habit × ²/₃; 2, mature umbel × ²/₃; 3, lower portion of stem × ²/₃; 4, petals × 12; 5, fruits × 2; 6, fruit detail × 2; 7, fruit cross section × 10. 1–4 from *Helbaek* 694; 5–7 from *Wheeler Haines* W763. Drawn by D. Erasmus (1964).

Hab. On mountains, on high rocky ledges above a gorge, stony overgrazed slopes near village, among coppiced *Quercus* woods, on limestone; alt. (200–) 600–1350 (–2000) m; fl. & fr. (Apr.–)May–Jun.
Distrib. Common in the middle and lower forest zones of Iraq; very rare in the steppe and desert regions. ?**mam**: "Mesopot., Kurdistan & Mossul", *Kotschy* 310!; ? Montafa (?Mantufa), *Field & Lazar* 95!; Khantur, *Rawi* 23369!; Mar Ya'qub, above Simel, *Nábělek* 356; Zawita gorge, *Guest* 2188!; Aqra, *Rawi* 11439!; **mro**: Salah ad-Din, *Gillett* 11278!, *G.W. Chapman* 11979!; W face of mountain, below Salah ad-Din, *Barkley & Haddad* 5727; Kuh-i Sefin, nr Shaqlawa, *Bornmüller* 1273!; Shaqlawa, *Gillett* 8055!, *Haines* W763!, *Hadač* 1451; Saran, Karoukh, *Kass & Nuri* 27270!; Chinarok, *Omar, Karim, Hamza & Hamid* 37228!; Haibat Sultan Dagh, *Rawi, Nuri & Kass* 28190!, 28215!; Gali Dargala, 20–35 km NW by W of Rania, nr Dargala and Berd Agha Gin villages, *id.* 28692!, 28886!, 28909!; **msu**: 12 km E of Chemchamal, *Gillett & Rawi* 7601!; Jarmo, *Helbaek* 694!, *Haines* W254!; Sulaimaniya, *Graham* 621!; Pira Magrun, *Haussknecht* s.n.!; mt. Hawraman nr Tawila, *Rechinger* 10240; between Qarachitan and Zewiya, *Gillett* 7762!; **fki**: Tauq bridge, *Haines* W1389!; Kizil Robat (Sa'diya), in Anth., l.c.; **dwd**: Aujah, 25 km N of Rutba, *Omar* 31499!
Note. ?DAKSI (Kurd.- Shaqlawa, *Gillett* 8055 – "fruit eaten when young"; "goat-grazed on dry slopes", *Haines* W763). Aitchison (1888–94) mentioned several colloquial names for this plant in Afghanistan and noted it as "a very common and well-known herb of which the leaves and roots are eaten, both raw and cooked, by the people" of that country.

One specimen cited above, *Field & Lazar* 95, has two incomplete extra vittae on each side of the usual two on the commissural face of the fruit, but although small seems in no other way remarkable. The normal commissural pair in this species are often very broad, forming a dark area right in to the carpophore.

Cyprus, Syria, Jordan, Sinai, Turkey, Caucasus, Iran, Afghanistan, Pakistan, C Asia (Turkmenistan).

48. **TORDYLIUM** L.

Sp. Pl. ed. 1: 239 (1753); Gen. Pl. ed. 5: 111 (1754)

Annual or rarely biennial herbs, ± pubescent or scabrid-hirsute. Leaves undivided or lobed, dentate, with a ± cordate base, or pinnate with broad, toothed segments, rarely further dissected to subtripinnatisect. Umbels compound, polygamous with outer flowers of partial umbels hermaphrodite and the inner male. Involucre and involucel of short or long, linear bracts and bracteoles, or the former absent or caducous. Flowers white, pinkish or purplish, often conspicuously radiate, smaller petals cuneate with an incurved acumen, emarginate above, larger distinctly 2-lobed, lobes usually unequal. Calyx teeth variable, conspicuous, small, ± unequal. Fruit ± round to ovate-elliptic, all mericarps strongly dorsally plano-compressed or peripheral fruit thus but the inner fruit unicarpellate, smaller and hemispherical, usually ± pubescent, less frequently glabrous; bicarpellate mericarps almost flat, intermediate ribs slender and inconspicuous, laterals thick and closely contiguous, forming a tumid, rugose or moniliform margin to the fruit. Vittae solitary (rarely 3) in the valleculae, 2 to numerous on the commissure. Disk variable, from broad and explanate with a crenulate margin to very small; stylopodia also variable. Carpophore bipartite. Endosperm with the commissural face flat.

About 18 species in Europe, N Africa and W Asia; only one species in Iraq.
Tordylium (from the Gr. τόρδῦλον, *tordylon*, used by Dioscorides for a plant believed to be *T. apulum* L.).

1. **Tordylium aegyptiacum** (*L.*) *Poir.* in Lam., Encycl. 7: 711 (1806); Boiss., Fl. Orient. 2: 1030 (1872); Zohary, Fl. Palest. ed. 2, 1: 556 (1932); Guest in Dep. Agr. Iraq Bull. 27: 100 (1933); Blakelock in Kew Bull. 3: 437 (1948); H. Riedl in Fl. Lowland Iraq: 470 (1964); Rawi in Dep. Agr. Iraq Tech. Bull. 14: 96 (1964); Mouterde, Nouv. Fl. Lib. et Syr. 2: 655 (1970); Alava in Fl. Turk. 4: 511 (1972); Zohary, Fl. Palaest. 2: 442 (1972); Meikle, Fl. Cyprus 1: 763 (1977); Rech.f., Fl. Iranica 162: 466 (1987); Boulos, Fl. Egypt 2: 177 (2000).

Hasselquistia aegyptiaca L., Cent. Pl. 1: 9 (1755).

Plant erect, 12–50(–60) cm, ± hairy or scabrid throughout its vegetative parts. Stem sulcate to angular. Leaves oblong-ovate in outline, lowest long-petiolate, up to 8 cm long, upper rapidly becoming sessile on the expanded sheath, all bipinnate, lower usually with pinnules again deeply divided and thus subtripinnatisect, segments rarely attaining 4 mm in breadth, short and subobtuse. Branches most few, from the base or about the middle of stem, long and ascending. Peduncles 2.5–11 cm. Umbels usually rather few, with 6–15 scabrid or papillate-hairy unequal rays 0.5–8 cm long. Involucre of few, setaceous, recurved bracts, usually short but up to 1 cm in large individuals. Partial umbels with 4–26 flowers, outer flowers very radiant, larger petals to 7 mm. Central partial umbel (umbellule) of many umbels modified to a purplish-black pileiform structure. Involucel of mostly 3 linear bracteoles, reflexed and disposed to one side of the partial umbel, rather long and frequently equalling the scabrid, 3–10 mm fruiting pedicels. Bicarpellate outer fruits round to broadly elliptic, 7–10 × 8–10 mm, frequently glabrous when mature, but a few hairs occasionally persisting centrally, when the mericarps are usually also minutely pruinose-papillose; fruit border ± distinctly crenulate-rugose. Unicarpellate inner fruits much smaller, smooth, swollen outwards from the hollow, contracted commissural face, bead-like pruinose-papillose, ± 5 mm. Disk very small, crenulate-margined, stylopodia small, conic, subequalling the emargination of the fruit. Styles glabrous, reflexed, short, about twice as long as the disk.

HAB. Fields, fallow, abandoned cultivation and waste ground in the lower mountains and on the upper plains; alt. 250–750(–1000) m; fl. & fr. (Mar.–)Apr.–May(–Jun).
DISTRIB. Quite common in the NW and central sectors of the lower forest and moist steppe zones of Iraq. **MJS**: Asi, *Rawi* 8475!, 8476a!; **MAM**: Dohuk, *Makki* (*as-Sidqi ash-Sharbati*) 3293!; **MRO**: Khansad valley, *Guest* 2148!; Shaqlawa, *Haines* W1555!; Armota, nr Koi Sanjaq, *Omar, Karim, Hamza & Hamid* 37246!; Bani Talabini, between Koi Sanjaq and Taktak, *id.* 37158!; **MSU**: Mt. Pira Magrun nr Girdabor, *Hadač et al.* 4728; **MSU/FKI**: between Kirkuk and Sulaimaniya, *Rawi* 21488!; **FUJ**: between Nusaybin and Mosul, *Anon.* (*Loftus*?) s.n.!; Ain Ghazal, *Guest* 4072!; Mosul, *Lazar* 3408!; Hammam Alil, *Ani* 9742!; **FNI**: between Mosul and Zakho, *Christabel & Evan Guest* 13240!; 8 km S of Dohuk, *Anon.* 43355; Tal Kaif, *Mudir of Tal Kaif* 3177!; **FAR**: Arbil, *Shahwani* 25107!, *Haines* W1408!, *Agnew et al.* 5748.
NOTE. ?KHURÎLK (Kurd.-Dohuk, *Makki* 3293); ?KARKAIMA (Ir.-Tal Kaif, *Mudir* 3177 – "useful for sheep grazing"). A weedy plant but apparently of some value for grazing.

E Mediterranean, Cyprus, Syria, Lebanon, Palestine, Jordan, Egypt (rare), Arabia, Turkey.

49. **AINSWORTHIA** Boiss.

in Ann. Sc. Nat. sér. 3, 1: 343 (1844)

Tordylium subgenus *Ainsworthia* (Boiss.) Benth. in Benth. & Hook.f., Genera Plantarum 1, 3: 924 (1867)

Annual herbs, ± hispid or scabrid. Basal leaves simple, cordate, long-petiolate, margins ± regularly crenate; lower cauline leaves simple to pinnate with 1 or 2 pairs of broad leaflets; upper leaves simple or trilobate. Umbels compound, polygamous with outer flowers of partial umbels hermaphrodite and inner male. Involucre and involucel present, filiform. Flowers white, often conspicuously radiant, smaller petals cuneate with an incurved acumen, emarginate above, larger distinctly 2-lobed with lobes unequal. Calyx teeth obsolete. Fruit ± round to ovate-elliptic, all mericarps dorsally plano-compressed or peripheral fruit of partial umbels thus but the inner fruit unicarpellate and convex; mericarps almost flat, intermediate ribs slender and inconspicuous, laterals thick and closely contiguous, forming a tumid but smooth and non-moniliform margin to the fruit, outer face with small white papillae. Vittae solitary in the valleculae, 2 on the commissure. Disk practically obsolete, stylopodia very small with short styles.

Two species in Cyprus, Turkey and the Palestine region; a single species in Iraq.
Ainsworthia (dedicated to the British geologist William Francis Ainsworth, 1807–1896).

1. **Ainsworthia trachycarpa** *Boiss.*, Diagn. ser. 1, 10: 43 (1849); Boiss., Fl. Orient. 2: 1035 (1872); Nábělek in Publ. Fac. Sci. Univ. Masaryk 35: 128 (1923); Zohary, Fl. Palest. ed. 2, 1: 557 (1932); Bornmüller in Beih. Bot. Centralbl. 58B: 286 (1938); Blakelock in Kew Bull. 3: 431 (1948); Zohary in Dep. Agr. Iraq Bull. 31: 113 (1950): Rawi in Dep. Agr. Iraq Tech. Bull. 14: 88 (1964); Mouterde, Nouv. Fl. Lib. et Syr. 2: 657 (1970); Alava in Fl. Turk. 4: 512 (1972); Zohary, Fl. Palaest. 2: 443 (1972); Meikle, Fl. Cyprus 1: 765 (1977); Alava in Fl. Iranica 162: 467 (1987).

A. cordata Boiss. in Ann. Sc. Nat. sér. 3, 1: 343 (1844) et Boiss., Fl. Orient. 2: 1035 (1872), quoad descr., non *A. cordata* (Jacq.) Boiss.
Tordylium cordatum (Jacq.) Poir. subsp. *trachycarpum* (Boiss.) Holmboe, Studies Veg. Cyprus: 141 (1941); *T. trachycarpum* (Boiss.) Al-Eisawi & Jury in Bot. J. Linn. Soc. 97 (4): 395 (1988).

Fig. 91. **Ainsworthia trachycarpa.** 1, habit × 2/3; 2, petals × 8; 3, fruit × 6; 4, fruit cross section × 30. 1, 2 from *Guest* 3097A; 3, 4 from *Rawi* 11333. Drawn by D. Erasmus (1964).

Plant erect, 20–60(–100) cm, slender, hispid to ± scabrid with deflexed hairs in its vegetative parts. Stem and branches coarsely striate, or sulcate in well-grown plants. Basal leaves entire, cordate-oblong or cordate-ovate, 0.7–7(–10) cm long, coarsely crenulate; lower cauline leaves similar or pinnate with one or two pairs of lateral leaflets and a much larger terminal leaflet, crenate or crenate-dentate. Lower cauline leaves long-petiolate, uppermost much reduced, sessile on the broad sheaths. Branches rather few, long, ascending. Peduncles long, 5–25 cm. Umbels few, rays 12–20(–30), unequal, 0.5–3.5 cm, often contracting somewhat in fruit, furnished with straight unicellular bristles and with papillae on adaxial surface. Involucel of many (12 or more) reflexed, scabrid, setaceous bracts 4–15 mm long. Partial umbel with 15–25 flowers, outer petals strongly radiant, 4–6 mm. Involucel of 3–5(–7) setaceous bracteoles, very long (exceeding even the radiant flowers), arranged unilaterally on the outer side of partial umbel. Fruiting pedicels unequal, 2–6 mm. Fruit 3–4.5 mm long, ± 3 mm in diameter, round to broadly elliptical, covered with ± abundant small white papillae. Disk minute, stylopodia very small, not exceeding the emargination of the fruit. Styles reflexed, short, equalling or slightly exceeding the smooth margin of fruit, which is pale and contrasts with the dark centre. Fig. 91. 1–4.

HAB. Lower mountain slopes, weed in fig garden near village; alt. 450–1000 m or more; fl. & fr. Apr.–May.
DISTRIB. Occasional; not very common, found in the lower forest zone of Iraq. **MAM**: Aqra, *Qaimaqam of Aqra* 3097!, 3097A!, *Rawi* 11333!; **MRO**: Kuh-i Sefin, above Shaqlawa, *Bornmüller* 1277!; Shaqlawa, *Haines* W764!; Serderian, between Arbil and Rowanduz, *Nábělek* 430; Rowanduz, *Hadač et al.* 5289; **MSU**: Qarachitan, *Gillett* 7706!; **FNI**: Mindan bridge, *Anders* 1267.
NOTE. ?KARAFSŪK (Kurd.-Aqra, *Qaimaqam* 3097, "fodder plant").

Cyprus, Syria, Lebanon, Palestine, Jordan, Turkey.

50. **JOHRENIA** DC.

Mém. Omb. (Coll. Mem. 5): 54 (1829)

Biennial and perennial herbs with a stout rootstock, glabrous or more rarely pubescent above. Leaves pinnately decomposite into narrow segments, upper reduced to sheaths. Umbels compound. Involucre absent, involucel of few to several short, often caducous bracteoles. Flowers all hermaphrodite, white or yellowish, not radiant. Petals rather broadly oblong, entire or slightly retuse above, with an incurved acumen. Calyx teeth small and caducous, or obsolete. Fruit ellipsoid, ovate or oblong, strongly dorsally plano-compressed. Mericarps only slightly convex. Intermediate ribs narrow, filiform but slightly prominent, laterals thick and closely contiguous, forming a rather narrow tumid border to the fruit. Vittae absent or extremely small and situated at the primary ribs. Disk sometimes somewhat expanded with an undulate margin, sometimes very narrow. Stylopodia small, styles short. Carpophore bipartite. Endosperm with the commissural face almost flat.

Some 20 species distributed in SW & C Asia; a single species in Iraq.
Johrenia (named in honour of the German botanist M.D. Johren of Kolberg, who died in 1718).

1. **Johrenia dichotoma** *DC.*, l.c.; Prodr. 4: 196 (1830); Boiss., Fl. Orient. 2: 1010 (1872); Hand.-Mazz. in Ann. Naturh. Mus. Wien 27: 91 (1913); Zohary, Fl. Palest. ed. 2, 1: 551 (1932); Bornmüller in Beih. Bot. Centralbl. 58B: 285 (1938); Zohary in Dep. Agr. Iraq Bull. 31: 113 (1950); Rawi, ibid. Tech. Bull. 14: 93 (1964); Mouterde, Nouv. Fl. Lib. et Syr. 2: 650 (1970); Chamberlain in Fl. Turk. 4: 331 (1972); Rech.f., Fl. Iranica 162: 378 (1987).

Plant stiff, erect, glabrous and somewhat glaucous, with many ascending, wiry, alternate branches above, (0.3–)0.6–1(–1.5) m. Stem and branches terete, finely striate, base of stem bearing fibrous remains of previous years' leaf-sheaths. Basal and lower cauline leaves narrowly deltoid in outline, to 30 cm long, bi- or subtri-pinnate into narrow (to 10 × 1.5 mm), subacute, entire segments; sheaths narrowly dilated, long and tapering. Such leaves are, however, few, leaves becoming rapidly reduced to white-bordered sheaths. Peduncles 0.9–7 cm. Umbels many, of 4–8(–10) very unequal, 0.5–4 cm rays. Partial umbels of 12 yellowish flowers, ± 2 mm in diameter; petals early caducous; fruiting pedicels 1–3 mm. Bracteoles 1–3, small and linear. Fruit obovoid, 4–6 × 2.5 mm, white margins distinct from the darker centre. Disk obscure, very narrow, crenulate-margined, stylopodia broadly conical. Styles short, deflexed on to the disk, which they scarcely exceed. Fig. 92: 1– 7.

Fig. 92. **Johrenia dichotoma** 1, habit × 2/3; 2, mature umbel × 2/3; 3, petal × 16; 4, 5, 6, fruits (both faces) × 4; 7, fruit cross section × 16. 1 from *Bornmüller* 1266; 2–6 from *Rawi* 9080. Drawn by D. Erasmus (1964).

HAB. On mountains, in stony and rocky places; alt. 700–1300 m; fl. & fr. May–Jun.
DISTRIB. Rare in Iraq, found in three districts in the lower forest zone. **MJS**: above (Balad) Sinjar, *Handel-Mazzetti* 1454; Jabal Sinjar, *Qaisi, Khayat & Karim* 50873!; **MAM**: Bamerne, *Qaisi & Hamad* 45915!; **MRO**: Kuh-i Sefin (Sefin Dagh), above Shaqlawa, *Bornmüller* 1266!, 1268; *Rawi* 9080!

Syria, Lebanon, Turkey (S & Inner Anatolia).

51. **DICHOROPETALUM** Fenzl,

Pugill. Pl. Nov. Syr.: 17 (1842)

Monocarpic or polycarpic perennials with branched foliose stems and a thickened collar of fibrous sheaths clothing the base of stem. Leaves linear to lanceolate or triangular in outline, shortly petiolate; basal leaves with linear, lanceolate or ovate ultimate segments; cauline leaves becoming entire upwards. Umbels 3–9-radiate, usually unequal in length; bracts absent. Petals yellow (in ours). Stylopodium flat or conical. Fruits dorsally compressed, narrowly winged, vallecular vittae absent or weak; dorsal and marginal mericarp ribs inflated (in ours) or keeled.

26 species, many having been transferred from *Peucedanum*; found mainly in the Balkans, SW Asia and N Africa.
Ref.: M. Pimenov & Kljuykov, Critical taxonomic analysis of *Dichoropetalum, Johrenia, Zeravschania* and related genera of Umbelliferae-Apioideae-Peucedaneae in *Willdenowia* 37: 487 (2007).

1. **Dichoropetalum aromaticum** (*Rech. f.*) *Pimenov & Kljuykov* in Willdenowia 37: 487, f. 7e (2007).

Johrenia aromatica Rech. f., Fl. Iranica 162: 376, t. 303 (1987).

Glabrous aromatic perennial with a 15–20 mm thickened collar of fibrous sheaths clothing the base of stem. Stem 60–80 cm, grooved, much branched from base, some branches opposite. Basal leaves unknown, cauline leaves lanceolate, with a sheathing base and scarious margins, apparently 2- or 3-pinnatifid (from limited material) with linear ultimate segments 1(–2) mm broad. Umbel with 5–9 rays, 15–30 mm long in fruit, divaricate; bracts absent. Partial umbels 9–14-flowered; pedicels 3–10 mm long, divaricate. Bracteoles ± 5, linear, ± 2 mm long. Fruits (immature) to 5 × 2.5–3 mm, ellipsoidal, with thickened margins ± 0.5 mm broad; ridges 5, prominent. (Description based on the account by Rechinger in Flora Iranica).

HAB. Recorded only once from Iraq; likely to be found in grassland on foothills.
DISTRIB. Very rare. **MSU**: Nalparaiz, *Hadač et al.* 4851.

W Iran; very local, found only in the Sanandaj region of Iranian Kurdistan.

52. **CORIANDRUM** L.

Sp. Pl. ed. 1: 256 (1753); Gen. Pl. ed. 5: 124 (1754)

Slender glabrous annual herbs with a characteristic odour. Leaves 1–3, bipinnate or pinnatisect. Umbels compound, usually polygamous, with few to many of inner flowers of the partial umbels male. Involucre absent, or of a single linear bract; involucel of few lanceolate to filiform bracteoles. Flowers white or pinkish, radiant, smaller inner petals obovate and emarginate above with an inflexed acumen, outer radiate ones obcordate and 2-lobed above. Calyx distinct, with acute, unequal teeth. Fruit subglobose, mericarps semiterete, separating late or not at all, the 5 primary ridges slender and flexuose, secondary ridges broader and more prominent. Vittae very slender, solitary beneath each secondary ridge, and very obscure, or absent. Commissural vittae absent or solitary and obscure. Disk not obvious, stylopodia conical, entire, smooth. Carpophore entire or bipartite. Endosperm hemispherical, commissural face concave.

Two species in the Mediterranean and SW Asia; a single species in Iraq.
Coriandrum (from κορίαννον, *koriannon*, the Gr. name of the plant in Theophrastus and other authors, perhaps derived from κορις, *koris*, a bug (often with a foul-smelling defence mechanism) and ἀνισάριον, *anisarion*, anise, due to the unpleasant smell of the unripe fruits); coriander.

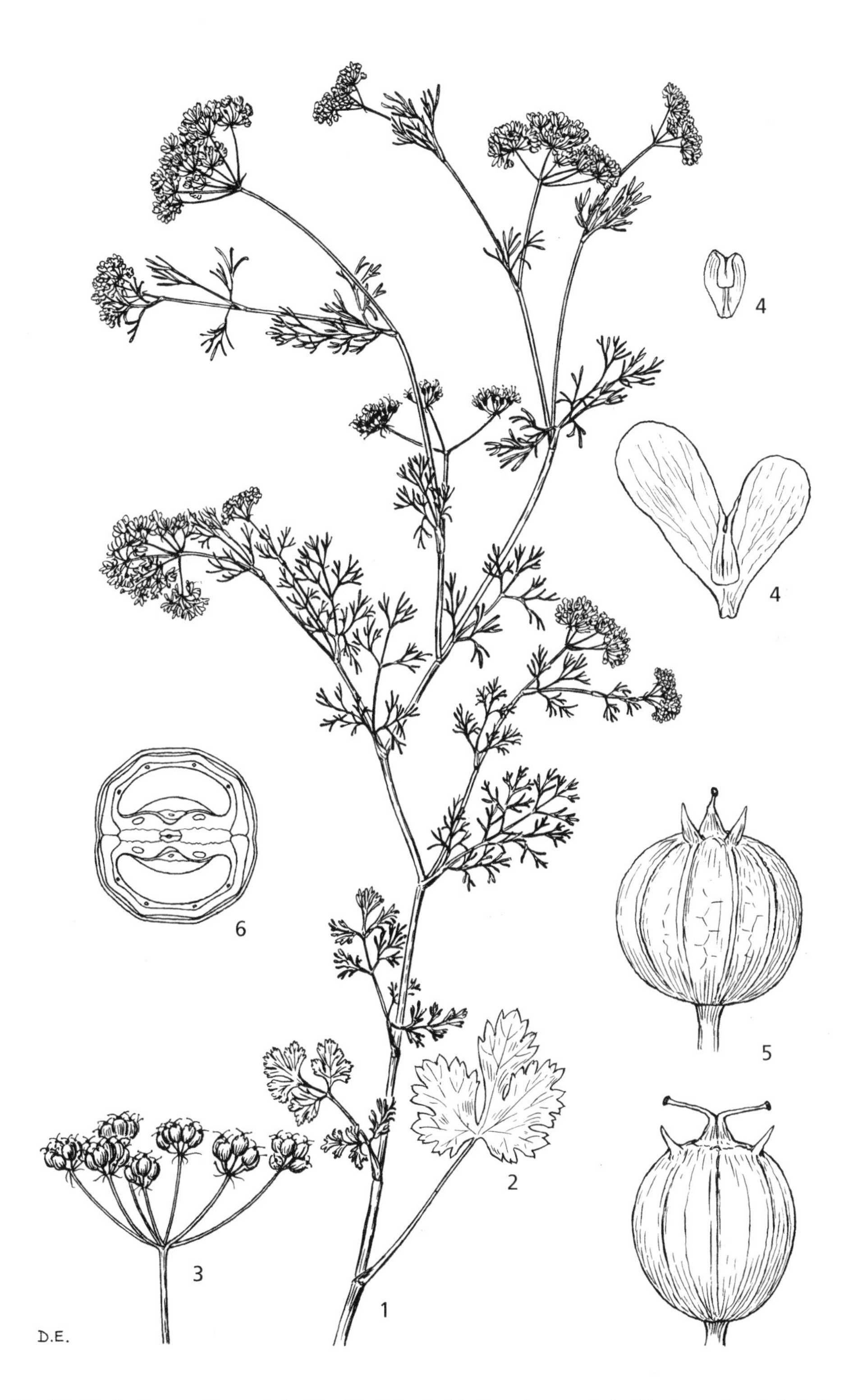

Fig. 93. **Coriandrum sativum**. 1, habit × ²/₃; 2, leaf × ²/₃; 3, mature umbel × ²/₃; 4, petals × 8; 5, fruits (both faces) × 2; 6, fruit cross section × 2. 1–3 from *Turner* 140; 4–5 from *Townsend* 25. Drawn by D. Erasmus (1964).

1. **Coriandrum sativum** *L.*, Sp. Pl. ed. 1: 256 (1753); DC., Prodr. 4: 250 (1830); Boiss., Fl. Orient. 2: 920 (1872); Hayek, Prodr. Fl. Balc. 1: 1074 (1927); Zohary, Fl. Palest. ed. 2, 1: 533 (1932); Guest in Dep. Agr. Iraq Bull. 27: 25 (1933); Anth. in Notes Roy. Bot. Gard. Edinb. 18: 289 (1935); Bornmüller in Beih. Bot. Centralbl. 58B: 280 (1938); Schischkin in Fl. SSSR: 16: 184 (1950); H. Riedl in Fl. Lowland Iraq: 458 (1964); Rawi & Chakr. in Dep. Agr. Iraq Tech. Bull. 15: 30 (1964); Grossh., Fl. Kavk. ed. 2, 7: 48 (1967); Mouterde, Nouv. Fl. Lib. et Syr. 2: 612 (1970); Hedge & Lamond in Fl. Turk. 4: 331 (1972); Zohary, Fl. Palaest. 2: 401 (1972); E. Nasir in Fl. Pak. 20: 26 (1972); V. Täckh., Stud. Fl. Egypt ed. 2: 385 (1974); Husain & Kasim, Cult. Pl. Iraq 105 (1975); Meikle, Fl. Cyprus 1: 717 (1977); Rech.f., Fl. Iranica 162: 161 (1987); Boulos, Fl. Egypt 2: 157 (2000).

Plant with a sweet-musky smell, erect, 20–80 cm, rarely less, with few to many ascending branches from base upwards. Stems solid, ridged. Lowest leaves broadly trilobed to pinnate with 1–4 pairs of broad but lobed and toothed pinnae, occasionally bipinnate. Remaining leaves becoming gradually more dissected upwards, upper irregularly 2- rarely 3-pinnate with rather long, narrowly linear, usually divaricate 2–12 × 0.4–2 mm segments, ± sessile on small, white-margined sheath. Peduncles 1–6 cm. Umbels few to many, with rather few, 3–6(–10) rays, rays 5–22 mm long. Partial umbels 8–20-flowered, with an involucel of 3–5 linear-lanceolate bracteoles; radiate petals 4–5 mm. Fruiting pedicels 2–4 mm. Fruit 4–6 mm, appearing rugose when ripe through the strongly undulate primary ridges, mericarps firmly adherent. Styles slender and flexuose, 1.5–2 mm. Forms with 3 carpels and styles are known. The characteristic odour of the plant is likened to that of bedbugs. Fig. 93: 1–5.

HAB. Widely cultivated as a garden herb on the plains of C & S Iraq (probably also in the lower mountain valleys in the N); a weed in gardens and fields, waste ground, etc., no doubt as an escape from cultivation; up to alt. ± 50 m or more (possibly to 400–500 m or even up to 1000 m?); fl. & fr. Mar.–May (or later?).
DISTRIB. Cultivated and subspontaneous in the desert region of Iraq on irrigated alluvial plains. **DSD**: S of Zubair (cult.), *Rawi* 25874!; **LCA**: Baghdad West, *Gillett* 10722!; Baghdad (cult.), *Bahira* C. 626!; Abu Ghraib, *H. Ahmad* 9869!, *Ani* 6893?; (cult.), *Sahira, Janan & Omar* C. 187!; Za'faraniya, *H. Ahmad* 9850!; Babylon, *Bornmüller* 371; **LSM**: Amara in Anth., l.c.
NOTE. Coriander; KUZBARA (Ar., Ir., Eg., Pal. etc., *Guest* (1935), *Gillett* 10722, *Sahira* C. 626), GESHNÎZ (Pers., Parsa), GIZHNÎZH Kurd., Wahby & Edmonds). As a spice it has been known since the dawn of history. It is used to flavour soups, stews, curries, salads etc. in many countries and in the Hadramaut in S Arabia to flavour bread. According to Schischkin (1950) the young stems are sometimes used as a spice in Georgia while the dried stems provide fuel. He goes on to state that the fruit contains up to 1% of essential coriander oil, of which the main component is coriandrol, and that this is obtained by soaking the fruit in water for about 14 hours and then by distillation. This oil is used in the manufacture of perfumes, soap and liqueurs, and also added to improve the taste of medicines as a substitute for other aromatic substances; the fruit is also sometimes employed as a flavourant and spice in pastries and even in canned meats. Moreover the fruit also contains about 15% of a fatty oil, which can be extracted from the residue left after separation of the essential oil by distillation, while the residual cake, containing some 12% of protein, is fed to cattle. Wren (1956) describes the medicinal properties of the fruit as stimulant and carminative.

Europe, Cyprus, Syria, Lebanon, Palestine, Jordan, Sinai, Egypt, Arabia, Turkey, Caucasus, Iran, Afghanistan, C Asia, N Africa Macaronesia, N & S America, etc. Cultivated and found as an escape almost all over the world, so that its native place of origin remains unknown.

53. **ORMOSCIADIUM** Boiss.

Ann. Sc. Nat. sér. 3, 2: 95 (1844)

Slender, glabrous annual herb. Leaves ± bipinnate, segments narrowly linear. Umbels compound, hermaphrodite or some inner flowers of partial umbels functionally male. Involucre and involucel present, conspicuous, bracts and bracteoles linear and entire or bracts trifid. Flowers white, radiant; smaller petals obcordate with an incurved acumen, larger radiate petals deeply 2-lobed. Calyx teeth obsolete or very small. Fruit oblong-elliptic, dorsally compressed, commissure broad. Mericarps with broad lateral ridges which are strongly moniliform-rugose with conspicuous, and closely contiguous, pale, obtuse projections forming a distinct margin to the fruit; intermediate ridges, both primary and secondary, filiform and not conspicuous; vittae absent. Disk small, with a narrow

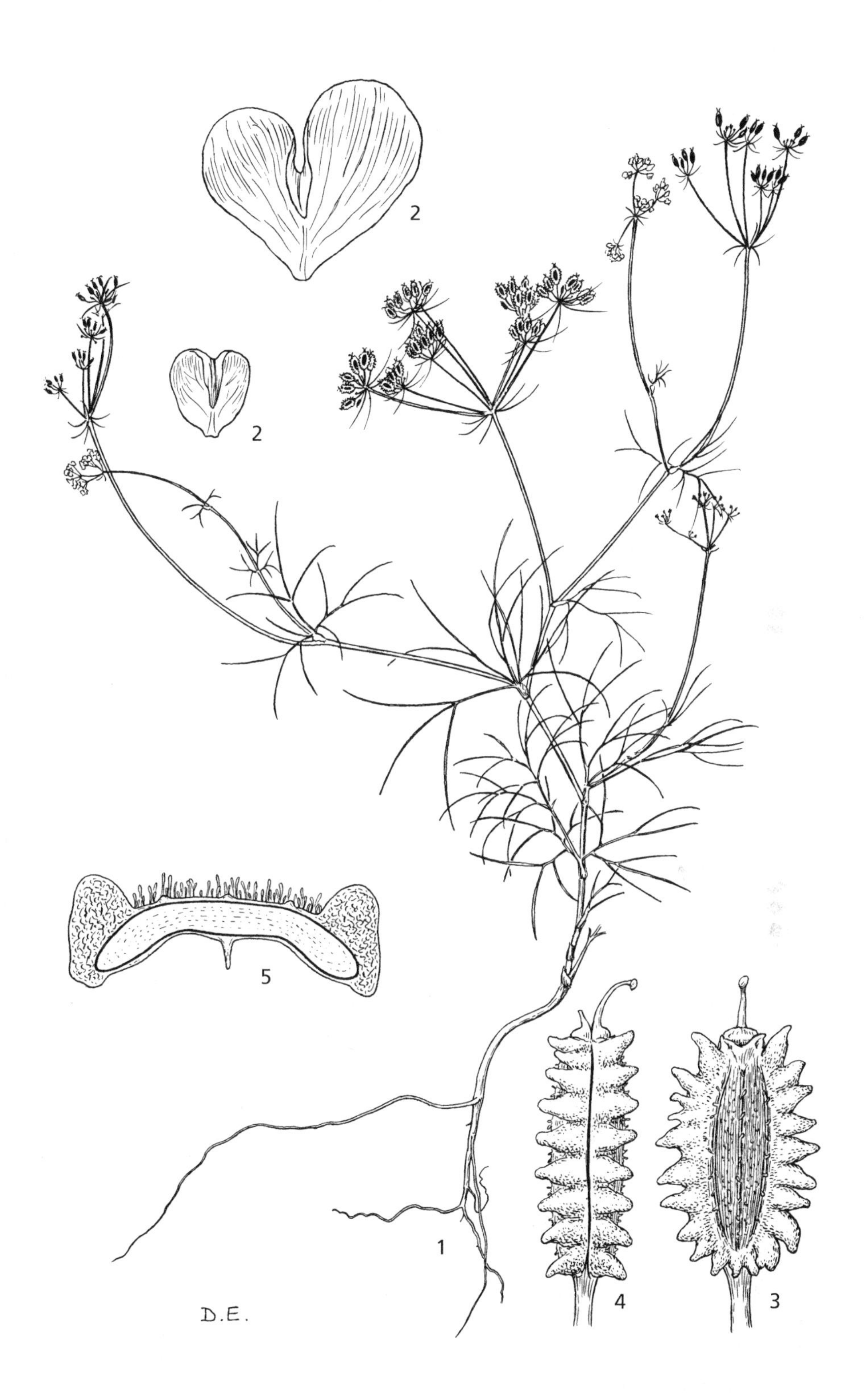

Fig. 94. **Ormosciadium aucheri**. 1, habit × 2/3; 2, petals × 18; 3, fruit front view × 10; 4, fruit side view; 5, fruit cross section × 18. All from *Wheeler Haines* W 1714. Drawn by D. Erasmus (1964).

undulate margin. Stylopodia shortly conical. Carpophore bipartite. Endosperm broad, the commissural face concave.

A monotypic genus.
Ormosciadium (from the Gr. ορμος, *ormos*, a necklace, σκιαδειον, *skiadeion*, an umbrella or parasol, referring to the moniliform edges of the fruit).

1. **Ormosciadium aucheri** *Boiss.*, Ann. Sc. Nat. sér. 3, 2: 95 (1844); Boiss., Fl. Orient. 2: 1029 (1872); Nábělek in Publ. Fac. Sci. Univ. Masaryk 35: 127 (1923); Bornmüller in Beih. Bot. Centralbl. 58B: 286 (1938); Zohary in Dep. Agr. Iraq Bull. 31: 113 (1950); Rawi in Dep. Agr. Iraq Tech. Bull. 14: 94 (1964); Chamberlain in Fl. Turk. 4: 504 (1972); Alava in Fl. Iranica 162: 464 (1987).

O. pulchrum Schischkin in Ref. Nauchn.-Issl. Rabot., Otd. Biol. Nauk, 1945: 7 (1947).

Short, wiry annual 10–20(–30) cm, with a few slender branches from below the middle upwards. Leaves all similar, ± bipinnate, the basal withered at flowering time, lower cauline leaves to 8 cm long, shortly petiolate, all ultimate segments long, setaceous, to 3 cm long in upper leaves, outline of lamina roundish or broadly deltoid; median and upper leaves sessile on short, broadly white-margined sheaths. Peduncles 3–6 cm. Umbels rather few, 4–8-rayed, rays unequal, 6–25 mm. Bracts of involucre conspicuous, setaceous, usually at least half as long as the lowest rays, simple or trisect. Partial umbels ± 12-flowered, radiate; petals 3–6 mm; fruiting pedicels 2–6 mm, much exceeded by the 3 or 4 unilaterally disposed setaceous bracteoles. Fruit 3–4 × 2–2.5 mm, moniliform margin paler than the dark centre when ripe, centre ± sparingly pilose, commissure also dark. Styles loosely deflexed or flexuose, 1 mm, often reddish. Odour similar to that of *Coriandrum*. Fig. 94: 1–5.

HAB. In the mountains, in open *Quercus* wood, in dry valley between eroded slopes with scanty *Quercus* scrub, on sandy conglomerate hills, and on gravel banks; alt. 500–900(–1900) m; fl. & fr. May–Jul.
DISTRIB. Occasional in the forest zone of Iraq. **MAM**: Jabal Bekher, *Rawi* 23156!; Sandur nr Dohuk, *Haines* W 1714!; nr Zakho, *Rechinger* 10715!; between Ain Nuri and Aradin, *Nábělek* 351; "trajectus Nawdust", between Ain Nuni and Amadiya, *Nábělek* 406, 915; **MRO**: Sakri-Sakran, *Bornmüller* 1272!, *Hadač et al.* 5733; **MSU**: Darband-i Khan, *Haines* W1922!; **MSU/FKI**: Tuni Baba hills, 10 km S of Darband-i Khan, *Barkley & Haddad* 7554!; "Assyria" (or in Syria or Turkey?), *Haussknecht* s.n.

Syrian desert, Turkey (E Anatolia), W Iran.

54. **BIFORA** Hoffm.

Gen. Umb. ed. 2: 191 (1816), *nomen cons.*

Slender, glabrous, aromatic annual herbs. Leaves variously pinnate. Umbels compound, rays few. Involucre and involucel absent or of very few subulate bracts. Flowers white, hermaphrodite, usually radiant, non-radiant petals obovate with an incurved acumen, incised above, larger radiant petals deeply 2-lobed. Calyx teeth obsolete. Fruit distinctly didymous, broader than long, commissural faces narrow and excavate. Mericarps subglobose with very obscure, subimpressed primary ribs, finely rugose. Vittae absent. Disk not obvious, stylopodia shortly conical. Carpophore bipartite. Endosperm subspherical, commissural face widely and deeply excavate.

Two or four species distributed from the Mediterranean area to C Asia; a single species in Iraq.
Bifora (from the Latin *biforis*, 2-doored – *bis*, twice, *foris*, a door, referring to two perforations in the pericarp, near the commissure).

1. **Bifora testiculata** (*L.*) *Spreng.* in *Roemer & Schultes*, Syst. Veg. 6: 448(1820); DC., Prodr. 4: 249 (1830); Boiss., Fl. Orient. 2: 921 (1872); Hayek, Prod. Fl. Balc. 1: 1075 (1927); Zohary, Fl. Palest. ed. 2, 1: 534 (1932); Bornmüller in Beih. Bot. Centralbl. 58B: 280 (1938); Schischkin in Fl. SSSR 16: 201 (1950); Zohary in Dep. Agr. Iraq Bull. 31: 111 (1950); Rawi in Dep. Agr. Iraq Tech. Bull. 14: 88 (1964); Mouterde, Nouv. Fl. Lib. et Syr. 2: 613 (1970); Hedge & Lamond in Fl. Turk. 4: 332 (1972); Zohary, Fl. Palaest. 2: 402 (1972); Meikle, Fl. Cyprus 1: 718 (1977); Rech.f., Fl. Iranica 162: 159 (1987).

Fig. 95. **Bifora testiculata**. 1, habit × 2/3; 2, petal × 16; 3, fruit × 4; fruit cross section × 4; 5, fruit longitudnal section × 4. All from *Meikle* 2286. Drawn by D. Erasmus (1964).

Coriandrum testiculatum L., Sp. Pl. ed. 1: 256 (1753).
Bifora dicocca Hoffm., Gen. Umbell. ed. 2: 191 (1816); Grossh., Fl. Kavk. ed. 2, 7: 49 (1967).

Erect annual, 10–40(–50) cm, somewhat sparingly branched from base upwards, branches axillary. Leaves all bi- to subtri-pinnate, lowest petiolate with deeply incised cuneate-ovate segments and oblong in outline, 2–10 cm long, upper progressively more deeply divided into acute, narrowly lanceolate and finally linear 2–10 × 0.5–2 mm segments, upper leaves sessile. Peduncles 0.8–6 cm. Umbels few to many, leaf-opposed, 2- or 3(–5)-rayed, rays 3–6(–10) mm. Involucre absent or of a single setaceous bract. Partial umbels few-flowered, usually 2 or 3 and occasionally reduced to 1 so that the umbel appears to be simple; flowers small (± 1.5 mm in diameter), petals subequal, very early caducous. Involucel of 1or 2 subulate bracteoles. Fruiting pedicels 2–4 mm. Fruit 3–4 × 4.5–7 mm, mericarps hard. Styles minute, reflexed. Fig. 95: 1–5.

HAB. On mountains, under shrubs, on steep slope; alt. 600–1100 m; fl. & fr. May.
DISTRIB. Very rare in Iraq, only found in the Northern sector of the lower forest zone. **MAM**: Dohuk, *Haines* W1007!; **MRO**: on Kuh-i Sefin, at Shaqlawa, *Bornmüller* 1291; **MSU**: Sarchinar, *Hadač* s.n.; Nalparaiz, *Hadač* s.n.
NOTE. Bedevian (1936) calls this plant KUZBARA SAGHIRA ("small coriander"); in other countries it is often found in segetal or ruderal habitats.

Mediterranean Europe (Spain, France, Italy, Corsica, Sardinia, Balkans, Greece), Crete, Aegean Isles, Cyprus, Syria, Lebanon, Palestine, Jordan, Turkey, Caucasus, Iran, C Asia (Turkmenistan), N Africa (Algeria, Libya).

55. CUMINUM L.

Sp. Pl. ed. 1: 254 (1753); Gen. Pl. ed. 5: 121 (1754); Wolff in Engler & Prantl, Pflanzenr. 90 (IV, 228): 22 (1927)

Slender annual (or biennial?) herbs, glabrous with the exception of fruit. Leaves pinnately or ternately dissected, segments very narrow. Umbels compound, rays few to numerous. Involucre and involucel present, bracts linear, conspicuous. Flowers deep magenta or white. Petals oblong or cuneate with an inflexed acumen, incised or deeply lobed above, somewhat unequal but not distinctly radiant. Calyx teeth subulate, unequal, conspicuous. Fruit narrowly ellipsoid-oblong or oblong-ovoid, somewhat compressed laterally. Mericarps with 5 primary ridges prominent, subglabrous or shortly hairy or setose, 4 secondary ridges equally prominent, setulose. Commissure rather narrow. Vittae broad, solitary beneath the secondary ridges, 2 in the commissure. Disk scarcely detectable, stylopodia small and conical. Carpophore bipartite. Endosperm somewhat compressed, commissural face slightly excavate.

Two Mediterranean & SW Asian species extending to C Asia, and one little known species in Sudan; only one species in Iraq.
Cuminum (ancient Roman name of the plant taken from the Gr. κύμῑνον, *cuminon*, cumin, perhaps derived from the old Babylonian name ka-mu-nu); Cumin, sometimes Cummin, KAMMŪN (Ar., Ir., etc.).

1. **Cuminum cyminum** L., Sp. Pl. ed. 1: 254 (1753); DC., Prodr. 4: 201 (1830); Boiss., Fl. Orient. 2: 1080 (1872); Hayek, Prod. Fl. Balc. 1: 1058 (1927); Zohary, Fl. Palest. ed. 2, 1: 568 (1932); Guest in Dep. Agr. Iraq Bull. 27: 26 (1933); Blakelock in Kew Bull. 3: 433 (1948); Schischkin in Fl. SSSR 16: 369 (1950); H. Riedl in Fl. Lowland Iraq: 460 (1964); Rawi in Dep. Agr. Iraq Tech. Bull. 14: 91 (1964); Rawi & Chakr., ibid. Tech. Bull. 15: 33 (1964); Grossh., Fl. Kavk. ed. 2, 7: 73 (1967); E. Nasir in Fl. Pak. 20: 97, f. 29 (1972); Husain & Kasim, Cult. Pl. Iraq 105 (1975); Rech.f., Fl. Iranica 162: 141 (1987); Pu Fading & Watson in Fl. China 14: 75 (2005).

Short erect annual 15–20(–30) cm, rather bushy with ascending branches from near base upwards. Leaves all similar, only the lowest shortly petiolate, most sessile on short, narrow sheaths, ± biternate into long filiform segments (to 4 cm or more in lowest leaves). Peduncles 1.5–5 cm. Umbels many, leaf-opposed, 3–6-rayed, rays 0.5–1 cm, equalled or exceeded by 3–6 bracts of the involucre. Partial umbels few- (2–6) flowered, flowers magenta, ± 2 mm in diameter; fruiting pedicels 2–6 mm, exceeded by the 3 or 4 bracteoles. Fruit 4–6 × 2 mm, styles short, rigid, divergent, often reddish. Fig. 96: 1–4.

HAB. In a silty desert depression, roadsides and waste places on the plains; alt. up to 600 m or more; fl. & fr. Mar.–Jun.
DISTRIB. Occasional in the desert region of Iraq, particularly on the alluvial plain. **DWD**: 35 km W of Rutba, *Rawi* 21154!, *Rechinger* 12535; **LCA**: "Falluja Desert" (on Baghdad-Damascus highway), *Haines* W1621!; Bad'a, nr Baghdad (cult.), *Guest* 2502!
NOTE. Cumin, sometimes also spelled Cummin; KAMMŪN (Ar., Ir., Eg. etc.). The fruits of this plant, generally called Cumin Seed, are a well-known condiment used for flavouring since antiquity. Uphof (1968) sums up its medicinal properties in home remedies as antispasmodic, antihysteric and stomachic. It is a common ingredient in curry powders, cordials and is also used in Leiden cheese. Seeds contain over 7% of essential oil (Schischkin 1950), which is used in perfumery.

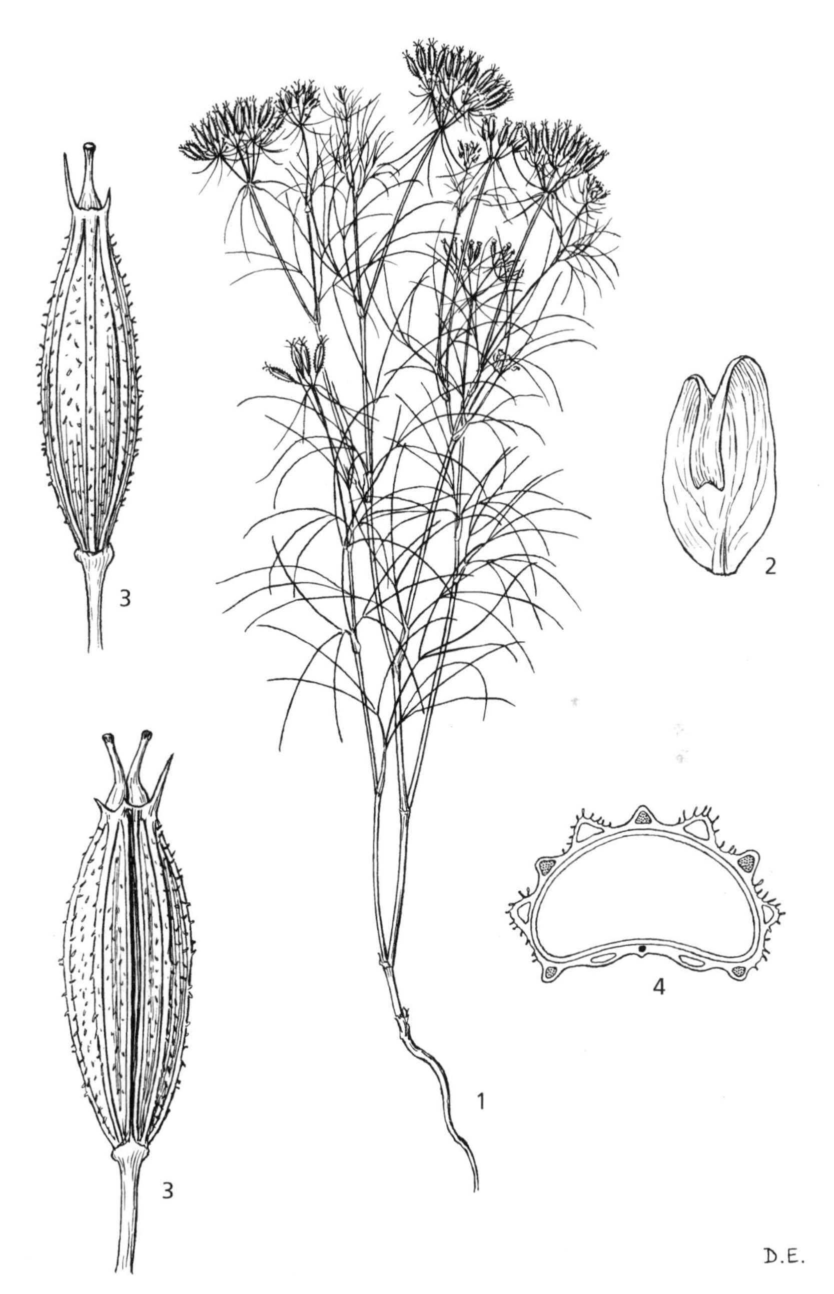

Fig. 96. **Cuminum cyminum**. 1, habit × ²/₃; 2, petal × 20; 3, fruits (both faces) × 8; 4, fruit cross section × 16. All from *Gilliat-Smith* 2304. Drawn by D. Erasmus (1964).

A second species of this genus, *C. setifolium* (Boiss.) K.-Pol. [*Psammogeton setifolius* (Boiss.) Boiss.] has been found at station T.4, on the Syrian section of the Kirkuk-Homs-Banias oil pipeline, and might occur in adjacent Iraq. It can easily be distinguished from *C. cyminum* by the long (5 mm or more) setose hairs on the secondary ridges of the fruit.

SE & Mediterranean Europe, Cyprus, Syria, Arabia, Egypt, Armenia, Azerbaijan, Georgia, Iran, Afghanistan, C Asia (Turkmenistan), N Africa (Morocco, Libya). Widely cultivated also on a small scale within this region as well as in NW India, China (Xinjiang), N America, S America (Chile) and elsewhere.

56. **ARTEDIA** L.

Sp. Pl. ed. 1: 242 (1753); Gen. Pl. ed. 5: 112 (1754)

Glabrous annual herb with tripinnately decomposite leaves with filiform-setaceous segments. Umbels compound, polygamous, with numerous male flowers at the centre of each partial umbel. Involucre and involucel present, very conspicuous, each of numerous pinnate bracts. Flowers white; outer petals very broadly and conspicuously radiant; smaller petals obovate with an incurved lobule and subentire above, larger deeply 2-lobed. Centre of each umbel with a penicillate tuft of blackish-purple hairs which are whitish-hirtellous at the apex. Calyx teeth obsolete. Fruit ovate to ± round, strongly compressed dorsally; mericarps almost flat, primary and 2 secondary ridges slender and filiform; lateral secondary ridges explanate, forming wings which are divided into 5 spatulate, straw-coloured, scaly lobes which taper below into a narrow cuneate base and are crenulate along the outer margin. Commissure broad, floccose, with a prominent central ridge from which transverse rugosities radiate. Vittae solitary beneath secondary ridges, very obscure. Disk not apparent, stylopodia small, conical. Carpophore bipartite. Endosperm flat.

A monotypic genus distributed in W Asia.
Artedia (dedicated to a Swedish friend of Linnaeus, Dr Petrus Artedi 1705–35, who was knowledgeable in several branches of natural science, particularly ichthyology).

1. **Artedia squamata** *L.*, Sp. Pl. ed. 1: 242 (1753); DC., Prodr. 4: 209 (1830); Boiss., Fl. Orient. 2: 1070 (1872); Hand.-Mazz. in Ann. Naturh. Mus. Wien 27: 92 (1913); Nábělek in Publ. Fac. Sci. Univ. Masaryk 35: 128 (1923); Hayek, Prod. Fl. Balc. 1: 1050 (1927); Zohary, Fl. Palest. ed. 2, 1: 562 (1932); Blakelock in Kew Bull. 3: 432 (1948); Zohary in Dep. Agr. Iraq Bull. 31: 113 (1950); H. Riedl in Fl. Lowland Iraq: 471 (1964); Rawi in Dep. Agr. Iraq Tech. Bull. 14: 88 (1964); Mouterde, Nouv. Fl. Lib. et Syr. 2: 602 (1970); Cullen in Fl. Turk. 4: 536 (1972); Zohary, Fl. Palaest. 2: 448 (1972); Meikle, Fl. Cyprus 1: 716 (1977); Leute in Rech. f., Fl. Iranica 162: 120 (1987).

Erect, conspicuous annual, 20–50 cm, with long, ascending branches. Lower leaves petiolate, lanceolate-oblong in outline with a narrow, attenuate sheath, mostly 5–20 × 1.5–4 cm; upper leaves sessile on a shorter, broader sheath with whitish membranous margins and stripes, oblong or ± deltoid in outline; all tripinnate into 2.5–15 mm, setaceous segments. Peduncles slender, 8–20 cm. Umbels few, handsome, contracting somewhat in fruit, tips of rays incurved; radiant petals very large (8–12 mm), very deeply 2-lobed, small non-radiant flowers ± 2 mm in diameter. Rays of umbel (10–)15–25(–30), 5–20 mm long, equalled by the 10–12 white-margined and white-veined bracts of the involucre, which are pinnately divided into setaceous segments and strongly reflexed in fruit. Partial umbels 12–30-flowered, flowers exceeded by bracteoles, which are similar to the involucral bracts but lack the broad white-veined base and are not reflexed in fruit. Fruiting pedicels 2–4 mm. Fruit 9–13 × 8–12 mm, styles short, rigid, divergent. Fig. 97: 1–5.

HAB. On mountain slopes and stony hillsides, by a stream in *Juglans* grove, among coppiced poplar trees (*Populus euphratica*) in valley between limestone boulders, and on steppic plains; alt. (150–)300–1350(–1500) m; fl. & fr. (Apr.–)May–Jun.(–Jul.).
DISTRIB. Common in the middle and lower forest zone of Iraq; occasional in the steppe region. **MJS**: Jabal Sinjar, above Balad Sinjar, *Handel-Mazzetti* 1463; Jabal Sinjar, *Gillett* 11099!; Asi, *Rawi* 8562!; **MAM**: Jabal Bekher, *Rawi* 22979!; **MRO**: Diana, *E.J. Cuckney* 3834!; between Dargala and Karoukh, *Serhang, Nuri & Kass* 27714!; Haibat Sultan Dagh, *Rawi, Nuri & Kass* 28155!; **MSU**: Jarmo, *Helbaek* 1284!, *Haines* W251!; Bakrajo, *Omar, Karim, Hazim & Hamid* 37399!; Pir Omar Gudrun (Pira Magrun), *Haussknecht* s.n.; between Sulaimaniya and Basnaga, G.W. *Chapman* 12023; Dara Tri, *Rawi* 22043!; 5 km S of Tawila,

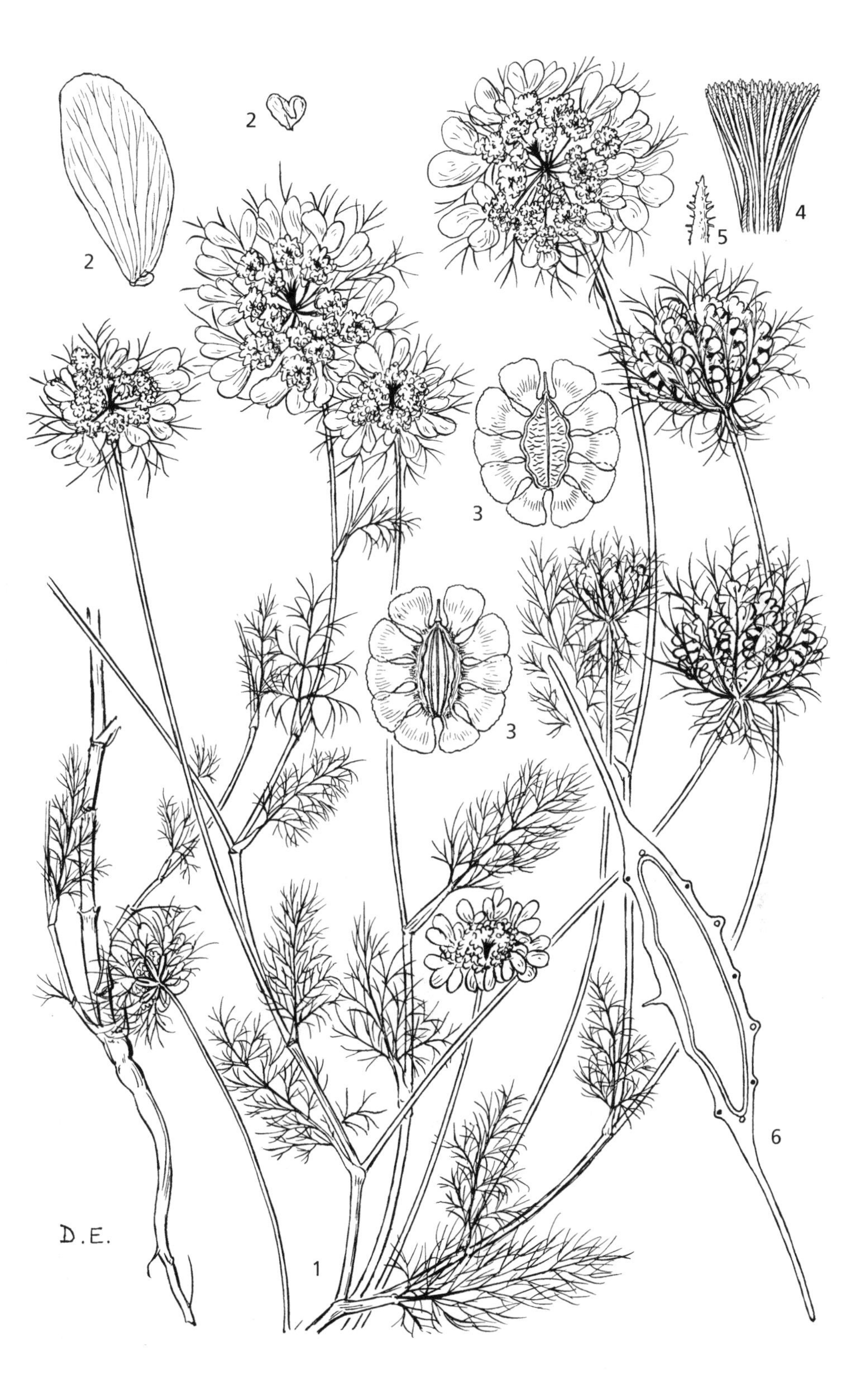

Fig. 97. **Artedia squamata**. 1, habit × ²/₃; 2, petals × 3; 3, fruits (both faces) × 2; 4, perianth tuft × 4; 5, detail of perianth tuft × 16; 6 , fruit cross section × 10. All from *Davis* 19418. Drawn by D. Erasmus (1964).

Rawi 21832!; **MRO/DLJ**: 31 km N of Tikrit on road to Koi Sanjaq, *Rawi, Nuri & Kass* 27977!; **FUJ**: Ain Tallawi, *Rawi* 5664!; Mir Qasim, *Field & Lazar* 547!; **FNI**: fields near Mosul, *Handel-Mazzetti* 1295; **FAR**: E bank of R. Zab near Kellek, *Gillett* 8184!; 7 km E of Mahmur, *Gillett* 11214!; **FKI**: Bani Khelan, *Poore* 365!; **FNI**: Jabal Hamrin, *Sutherland* 302; between Baghdad and Khanaqin, *Prof. E Barnes* s.n.!; between Kasrabad (Qizil Rabat – Sa'diya) and Khanaqin, *Nábělek* 390; Kani Masi, *Morck* 14; 54 km SE of Mandali, *Rechinger* 8673; Naft Khina, *Hadač* 4639.

Greece, Cyprus, Syria, Lebanon, Palestine, Jordan, Turkey, W Iran.

57. DAUCUS L.

Sp. Pl. ed. 1: 242 (1753); Gen. Pl. ed. 5: 113 (1754)

Annual or biennial, generally hispid plants with pinnately decomposite leaves. Umbels compound. Involucre present, bracts simple or more usually pinnately divided. Involucel absent or present, simple or trifid. Flowers white, often radiant; petals obovate or ovate with an inflate acumen, incised above; flowers at the centre of the umbels and partial umbels frequently modified, purplish black and sterile. Calyx teeth small and acute, or absent. Fruit subcylindrical or somewhat dorsally compressed, ellipsoid or oblong-ovoid. Mericarps convex, the 5 primary ridges filiform, divaricate-setose, the 4 secondary more prominent, particularly the laterals, which continue in the same plane as commissural face, lateral primary ridges being inconspicuous; secondary ridges with rows of spines. Commissure usually broad. Vittae solitary beneath secondary ridges. Disk small, entire or with an undulate margin; stylopodia depressed-convex or shortly conical. Carpophore undivided or bipartite. Endosperm subterete or dorsally compressed, commissural face almost flat.

Some 60 species in Europe, Asia, Africa and America; only one (or possibly two) species in Iraq.

Daucus (the Gr. name of the carrot, according to Bailey (1939), Stearn (1972) and other authors; but according to Liddell & Scott (1926) the Gr. δαῦκος, used by Hippocrates, was the name of *Athamanta cretensis*, though *Daucus* was said to have been the name used by Latin authors for the carrot and parsnip); Carrot. JAZAR (Ar., Ir. etc.), GĒZER (Kurd.), GAZAR (Eg.), HŪWĒCH (Pers.).

Tall annual or biennial, up to 1 m; secondary ridges with 6–8
  or by fission 12–16 glochidiate spines in a single row . . . . 1. *D. carota*
Dwarf annual, rarely attaining 30 cm; secondary ridges with 10–12
  spines along each secondary ridge . . . . .2. *D. littoralis*

1. **Daucus carota** *L.*, Sp. Pl. ed. 1: 242 (1753); DC., Prodr. 4: 211 (1830); Boiss., Fl. Orient. 2: 1076 (1872); Hand.-Mazz. in Ann. Naturh. Mus. Wien 27: 92 (1913); Nábělek in Publ. Fac. Sci. Univ. Masaryk 35: 129 (1923); Hayek, Prod. Fl. Balc. 1: 1051 (1927); Zohary, Fl. Palest. ed. 2, 1: 566 (1932); Guest in Dep. Agr. Iraq Bull. 27: 29 (1933); Anth. in Notes Roy. Bot. Gard. Edinb. 18: 289 (1935); Bornmüller in Beih. Bot. Centralbl. 58B: 287 (1938); Blakelock in Kew Bull. 3: 433 (1948); Zohary in Dep. Agr. Iraq Bull. 31: 113 (1950); Schischkin in Fl. SSSR 16: 288 (1951); H. Riedl in Fl. Lowland Iraq: 471 (1964); Rawi in Dep. Agr. Iraq Tech. Bull. 14: 91 (1964); Rawi & Chakr., ibid. 15: 36 (1964); Grossh., Fl. Kavk. ed. 2, 7: 137 (1967); Cullen in Fl. Turk. 4: 531 (1972); Zohary, Fl. Palaest. 2: 452 (1972); E. Nasir in Fl. Pak. 20: 103, f. 31 (1972); Hussain & Kasim, Cult. Pl. Iraq 105 (1975); Meikle, Fl. Cyprus 1: 709 (1977); Rech.f., Fl. Iranica 162: 136 (1987); She Menglan & Watson in Fl. China 14: 205 (2005).

D. guttatus Sm. in Sibth. & Sm., Prodr.; Rech.f., Fl. Iranica 162: 139 (1987).
D. bicolor Sm. in Sibth. & Sm., ibid.
D. broteri Ten. Fl. Nap. 4, Syll. App. 3: 4 (1830) var. *bicolor* (Sibth. & Sm.) Boiss., Fl. Orient. 2: 1074 (1872); Rech.f., Fl. Iranica 162: 136 (1987).
D. microscias Bornmüller & Gauba, Feddes Repert. 49: 262 (1940).

Erect annual or biennial, 30–100 cm, subglabrous to densely hispid, especially in the lower part of the stem. Stem solid, striate or sulcate, hairs in the lower part deflexed. Lower leaves oblong in outline, shortly petiolate to subsessile on the attenuate, expanded sheaths, bi- or tri-pinnatisect into oblong-ovate or broadly lanceolate, incised segments. Upper leaves shorter, sessile on shorter sheaths, 1- or 2-pinnate with narrower, more elongate and divaricate segments. Umbels rather few, to 13 cm in diameter in full flower but frequently

Fig. 98. **Daucus carota**. 1, habit, upper part × ⅔; 2, habit, lower part × ⅔; 3, petals × 8; 4, fruits (both faces) × 4; 5, fruit cross section × 16. 1, 3 from *Rawi* 2276; 2 from *Furse* 3295; 4, 5 from *Rawi* 22958. Drawn by D. Erasmus (1964).

much less, strongly contracted in fruit with the long outer rays curving in over the shorter inner rays; rays very numerous, sparsely hispid to subglabrous. Involucre of 7–12 simply pinnate bracts, bracts glabrous or long-strugose, conspicuous, white-margined below, margin webbing across to the lowest pinnae. Bracteoles ± 8, linear-lanceolate, simple or trifid, membranous-margined, subequalling the largest pedicels. Central flower frequently blackish purple. Outer petals somewhat larger but not conspicuously radiant. Fruit 2–3 mm, secondary ridges with 6–8 or by fission 12–16 slender, glochidiate spines in a single row. Primary ridges filiform, with rows of bristles diverging at 90°. Styles slender, 0.75–1 mm, flexuose-divaricate. Fig. 98: 1–5.

Hab. Mainly segetal and ruderal, on mountains on disturbed ground under light *Quercus* scrub, waste land in gardens, heavily eroded sandstone, by roadside, dry fields, edge of riverine seepage swamps in the steppe, irrigated fields, orchards and gardens in the desert; alt. 50–950(–1150) m; fl. & fr. (Apr.–) Jun.–Jul.(–Aug.).

Distrib. Occasional in the lower forest zone and steppe region of Iraq, rarer on the irrigated alluvial plain in the desert region. **MAM**: (or in Syria/Turkey), "Pl. Mesopot., Kurdistan & Mossul", *Kotschy* 377!; 87 km NW of Mosul, *Rawi* 22958!; Zawita, *Guest* 3752!; Sarsang, *Haines* W530!; Aradin, *Nábělek* 346; Dinarti, *E. Chapman* 26130!; Sarsang, *Haines* W530!; Ba'dhera (Ba'idhra), nr Sheikhan, *Salim* 2598!; **MRO**: Rowanduz district, *Bornmüller* 1281b; Chinaruk, Qandil range, *Rawi & Serhang* 25346; **MSU**: Jarmo, *Helbaek* 1909!, 1907?; 12 km E of Chemchamal, *Rawi & Gillett* 11607!; 20 km WNW of Sulaimaniya, *Rawi* 22776!; Baitua nr Ain Sifni, *Salim* 2580; 12 km E of Chemchamal, *Rawi & Gillett* 11607!; 20 km WNW of Sulaimaniya, *Rawi* 22776!; **FUJ**: between Mosul and Tal Afar, *F. Karim* 37475!; **FUJ/FNI**: below Mosul (along R. Tigris), *Nábělek* 486, *Handel-Mazzetti* 1203; **FNI**: Baitnar, nr Ain Sifni, *Salim* 2580!; Gerwona nr Ain Sifni, *Field & Lazar* 734!; Maltai, nr Tal Kaif, *Nábělek* 349; by Mindan bridge, *E Chapman* 26233; 1 km NE of Mindan, id. 26192!; **FAR**: between Arbil and Ankawa, *Bornmüller* 1281; **LCA**: Baghdad (cult.), *Omar* C. 179!; Abu Ghraib (cult.), *Janan*, C. 584!; Hussainiya nr Kut, *Lazar* 3489!; **LEA**: Ba'quba, *Haines* W996.

Note. Plants with sterile blackish florets at the centre of the umbel have been separated as *D. bicolor*, this is merely a variant, and is found more often in short-lived, annual specimens.

Carrot; JAZAR (Ar.), ?JÎZER (Kurd.), HUWAICH or ZERDEK (Pers., Parsa), HĀWŪCH (Turk.) – some of these names being variously spelled by different authors.? UZAIRAH (Kurd.-Ain Sifni, *Salim* 2580 reported to be presumably the herbage, "poisonous to sheep"; ?GÎYĀ FALLA (Kurd.-Shaikhan, *Salim* 2598 as "a forage plant"). But Wren (1956) states that the plant top is medicinal: diuretic, stimulant and deobstruent, a valuable remedy in folk medicine in cases of dropsy, retention of urine, gravel and bladder afflictions, as also noted by Rawi & Chakravarty (1964) and Chakravarty (1976) who describes its medicinal properties in greater detail. As opposed to the wild forms, the cultivated carrot is widely known as a useful vegetable and grown as such in many countries.

Europe, Aegean Isles, Cyprus, Syria, Lebanon, Palestine, Egypt, Arabia, Turkey, Caucasus, Iran, Pakistan, C Asia (Turkmenistan to Siberia), India, China, Japan, Philippines & Australia, Macaronesia (Madeira, Canary Is.), N Africa (Morocco, Algeria), Ethiopia, S Africa, N, C & S America. Cosmopolitan in temperate and sub-tropical regions of the world.

2. **Daucus littoralis** *Sibth. et Sm.*, Fl. Graeca 3: 65 (1819); Boiss., Fl. Orient. 2: 1074 (1872); Zohary, Fl. Palest. ed. 2, 1: 565 (1932); Zohary, Fl. Palaest. 2: 449 (1972); V. Täckh., Stud. Fl. Egypt ed. 2: 396 (1974); Rech.f., Fl. Iranica 162: 139 (1987); Boulos, Fl. Egypt 2: 183 (2000).

D. muricatus L. var. *littoralis* (Sibth. & Sm.) DC., Prodr. 4: 210 (1830).
D. sp. Blakelock in Kew Bull. 3: 433 (1948).
[*D. leptocarpus* (non Hochst.) Zohary in Dep. Agr. Iraq Bull. 31: 113 (1950).]

Erect or decumbent biennial, subglabrous to rather sparsely hispid, 15–30 cm. Stem slender, subterete, striate, with long, ascending branches from near the base. Lower leaves rather small, petiolate, oblong-lanceolate in outline, bipinnatisect, pinnules further divided into very short, thick, mucronate segments. Upper leaves similarly dissected, sessile. Sheaths short, narrowly expanded. Umbels few, small, rays about 10, unequal, incurved in fruit. Involucre of 3–6 trifid (rarely some simple) bracts, a little shorter than rays. Partial umbels ± 10-flowered, outer petals larger but scarcely conspicuously radiant. Bracteoles ± 6, linear-lanceolate, equalling pedicels. Fruit ± 3 mm long, with 10–12 glochidiate spines along each secondary ridge. Primary ridges filiform, furnished with divergent bristles. Styles 1.0 mm, divergent.

var. **forskahlei** *Boiss.*, Fl. Orient. 2: 1074 (1872).
Caucalis glabra Forssk., Fl. Aegypt.-Arab.: 206 (1775).

Fruit small, 2.25 mm with only 6–9 spines along each secondary ridge.

HAB. on mudflats alongside a tidal river; alt. 1 m; fl. Jun.
DISTRIB. Very rare in Iraq, only found once in the desert region, towards the SE extremity of the alluvial plain. **LBA**: by the Shatt al-'Arab, Siba, *Guest* 25455!
NOTE. HASAICH or KHANNĀQ AD-DIJĀJ (Ir.-Siba, *Guest* 25455).

Aegean Isles, Cyprus, Palestine, Egypt, Turkey, Iran.

58. **ORLAYA** Hoffm.

Gen. Umb. ed. 1, 1: 58 (1814)

Glabrous or pubescent annual herbs with 2- or 3-pinnate leaves. Umbels compound. Involucre and involucel present, conspicuous, bracts and bracteoles frequently scarious-margined. Flowers white; outer petals of outer flowers large, cordate, very strongly radiant, remainder obovate, emarginate with an inflexed lobule. Calyx with 5 small teeth. Fruit ovate, strongly dorsally compressed, commissure broad. Mericarps only slightly convex; primary ridges filiform, setose, secondary ridges more prominent, furnished with stout, biseriate, uncinate prickles, lateral secondary ridges connivent at the commissure. Vittae solitary beneath the secondary ridges. Disk small, entire to somewhat sulcate. Stylopodia conical, styles long and divergent. Carpophore bipartite. Endosperm dorsally compressed, commissural face flat.

Five species in SE Europe and the Mediterranean region to Iran and C Asia; only one species in Iraq.
Orlaya (dedicated to the Muscovite doctor and botanophil Johann Orlay, 1770–1829).

1. **Orlaya daucoides** (*L.*) *Greuter* in Boissiera 13: 92 (1967); Cullen in Fl. Turk. 4: 530 (1972); Zohary, Fl. Palaest. 1: 399 (1972); Meikle, Fl. Cyprus 1: 707 (1977); Rech.f., Fl. Iranica 162: 119 (1987).

Caucalis daucoides L., Sp. Pl. ed. 1: 241 (1753), non Syst. Nat. ed. 12, 2: 205 (1767).
Caucalis platycarpos L., Syst. Nat. ed. 12, 2: 205 (1767), non Sp. Pl. ed. 1: 241 (1753).
Orlaya platycarpos W.D.J. Koch, Nov. Act. Acad. Caes. Leop. Carol. 12 (1): 79 (1824); DC., Prodr. 4: 209 (1830); Boiss., Fl. Orient. 2: 1071 (1872); Hayek, Prodr. Fl. Balc. 1: 1053 (1927); Zohary, Fl. Palest. ed. 2, 1: 563 (1932); Schischkin in Fl. SSSR 16: 183 (1950); Rawi in Dep. Agr. Iraq Tech. Bull. 14: 94 (1964); Grossh., Fl. Kavk. ed. 2, 7: 47 (1967); Mouterde, Nouv. Fl. Lib. et Syr. 2: 603 (1970), non *Caucalis platycarpos* Sp. Pl. ed. 1: 241 (1753) nec ed. 2, 1: 347 (1762)
Daucus platycarpos (L.) Čelak., Bot. Zeit. 31: 44 (1873), nom. illegit..
Orlaya kochii Heywood, Agronomia Lusitanica 22 (1): 13 (1960).

Erect annual, glabrous or more usually somewhat pubescent, especially foliage and lower part of stem, 10–45 cm. Stem slender, wiry, striate, branched from base upwards. Lower leaves long-petiolate, oblong or narrowly deltoid in outline, to 13 × 8 cm, 2- or 3-pinnate, ultimate segments narrowly lanceolate to oblong, to 4 mm long; sheaths short, dilated, with whitish, scarious, ciliate margins. Upper leaves becoming shortly petiolate and sessile, similarly divided. Peduncle (1–)3–7(–12) cm. Umbels several, 2–4-radiate, rays stout, 0.5–2 cm long. Involucre of 2–4 lanceolate bracts with broadly whitish scarious, ciliate margins, simple or rarely somewhat divided at apex. Partial umbels 8–15-flowered, only 2–4 flowers producing ripe fruit; outer petals strongly radiant, ± 3 mm; pedicels very short, ± 2 mm. Involucel of ± 4 bracteoles similar in form to the bracts. Fruit 10–12 × 4–5 mm excluding spines, much flattened dorsally. Primary ribs with two divergent rows of broad-based, large prickles which are uncinate at tips. Calyx teeth small, triangular, persistent. Styles 2 mm in fruit, rigidly divergent. Fig. 99: 1–5.

HAB. On mountains, on steep rock slope under *Quercus* trees, in a shady ravine; alt. 1000–1050 m; fl. & fr. May–Jun.
DISTRIB. Occasional in the lower forest zone of Iraq. **MRO**: Kuh-i Sefin, in fields nr Shaqlawa, *Bornmüller* 1285; Shaqlawa, *Haines* W650A!; **MSU**: Qaradagh, *id.* W 1520!

W, C & S Europe (France, Spain, Germany to Balkans, Greece), Crete, Aegean Isles, Cyprus, Syria, Lebanon, Palestine, Turkey, Caucasus, N Iran, N Africa (Morocco, Algeria).

Fig. 99. **Orlaya daucoides**. 1, habit × 2/3; 2, leaf portion detail × 4; 3, petals × 6; 4, fruit × 3; 5, fruit cross section × 3; 6, styles × 8. 1, 2, 4–6 from *Edmonds* 145; 3 from *Wheeler Haines* 4650A. Drawn by D. Erasmus (1964).

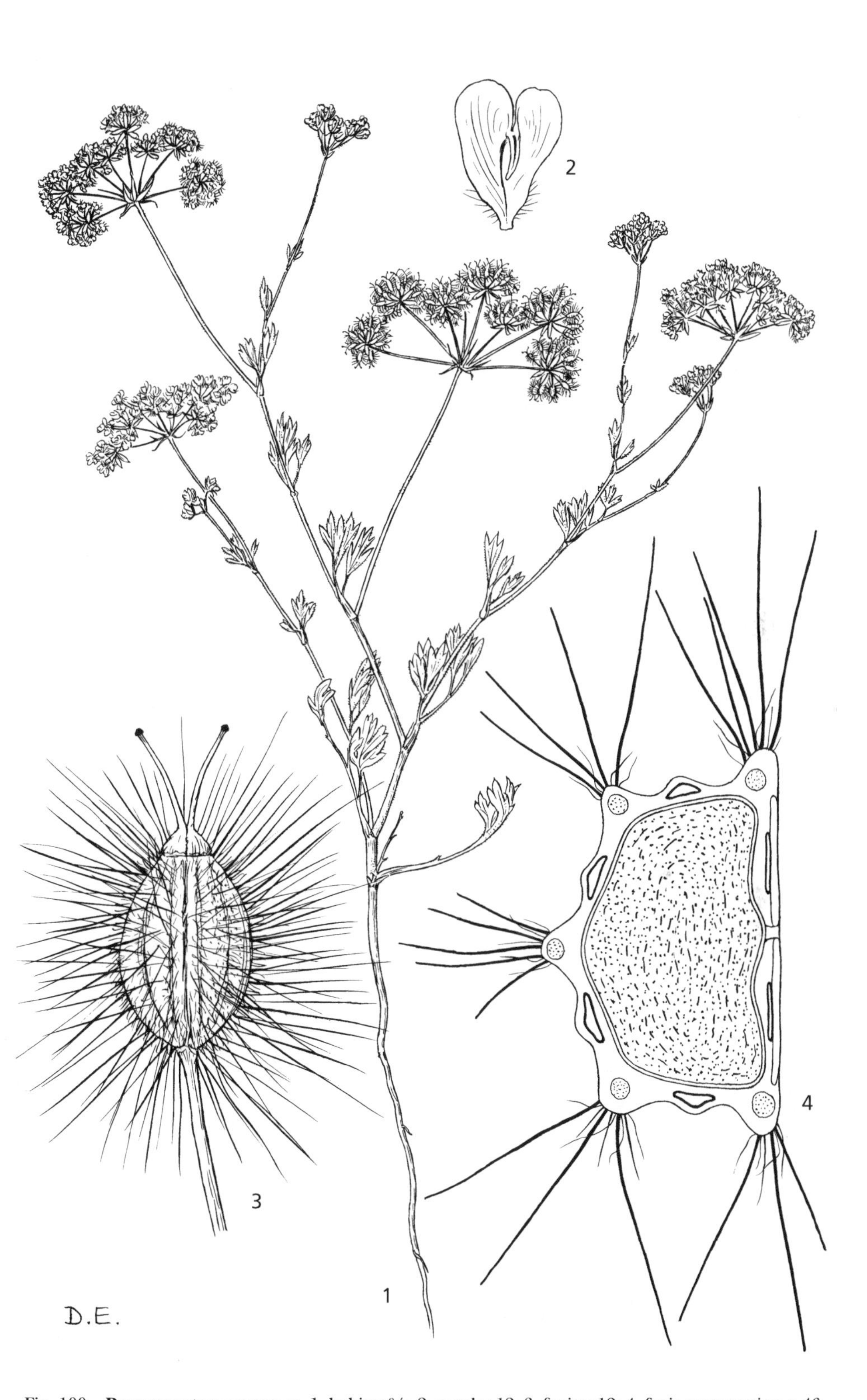

Fig. 100. **Psammogeton canescens**. 1, habit × ²/₃; 2, petal × 12; 3, fruit × 12; 4, fruit cross section × 46. All from *H. Crookshank* 5. Drawn by D. Erasmus (1964).

59. **PSAMMOGETON** Edgew.

in Proc. Linn. Soc. 1 (1845): 253 (1845) et Trans. Linn. Soc. 20: 57 (1846, published 1851)

Annual herbs with 2- or 3-ternatisect pubescent leaves. Umbels compound. Involucre and involucel of several simple or trisect bracts present. Flowers hermaphrodite, white to pink, outer petals somewhat larger but not conspicuously radiant. Petals obovate to ovate, emarginate above with an inflexed lobule or the larger deeply bifid. Calyx teeth obsolete. Fruit ellipsoid-oblong to ellipsoid-ovoid, somewhat laterally compressed; mericarps semiterete; primary ridges filiform, somewhat prominent, secondary ridges similar or more obscure, all ridges furnished with many slender, single-celled, glochidiate bristles or clavate-vesicular hairs. Commissure rather narrow. Vittae solitary beneath secondary ridges, rarely beneath primary ridges only, 2 on the commissure. Disk scarcely obvious, stylopodia short, conical. Carpophore bifid. Endosperm somewhat compressed dorsally, slightly to obviously angular, commissural face slightly concave.

Three to five species, according to taxonomic opinion, distributed from the Mediterranean region to Iran and the W Himalayas; a single species in Iraq.

Psammogeton (from the Gr. ψάμμος, *psammos*, sand and γείτον, *geiton*, neighbour, a reference to the desert habitat of species of this genus.

1. **Psammogeton canescens** (*DC.*) *Vatke*, Ind. Sem. Hort. Berol., App.: 3 (1876); Schischkin in Fl. SSSR 16: 166 (1950); H. Riedl in Fl. Lowland Iraq: 457 (1964); E. Nasir in Fl. Pak. 20: 102, f. 30 (1972); Rech.f., Fl. Iranica 162: 144 (1987).

Athamanta canescens DC., Prodr. 4: 155 (1830).
Pimpinella crinita Boiss., Ann. Sc. Nat. sér. 3, 1: 131 (1844).
Psammogeton biternatum Edgew., Trans. Linn. Soc. 20: 57 (1846).
P. crinitum (Boiss.) Boiss., Fl. Orient. 2: 1078 (1872); Rawi in Dep. Agr. Iraq Tech. Bull. 14: 95 (1964), sub "*crinatum*" sphalm.

Erect annual (5–)8–30 cm, furnished with rather long whitish hairs throughout. Stem much branched dichotomously from near the base in well-grown individuals, pubescent or glabrescent. Leaves all similar (or the uppermost reduced), biternatisect with segments obcuneate, deeply trifid above, these final divisions not more than 1.5 cm in length and ± 5 mm wide, usually much less, basal leaves withered after flowering; lower cauline leaves petiolate; upper sessile on the sheaths, which are membranous-margined basally only, attenuate above and not forming a web onto the primary leaf divisions. Peduncles (1–) 3–9 cm. Umbels few to many, leaf-opposed, 5–12-rayed, rays 0.6–3 cm, lanate. Bracts of the involucre ± 5, lanceolate, lanate, membranous with a coloured midrib, about 1/4 the length of the umbel rays. Involucel of 5(–7) broadly oblong-ovate bracteoles, which are glumaceous, white-lanate, with an excurrent coloured midrib, not equalling the fruit. Fruit (excluding indumentum) 1.5–2 mm, densely furnished with long (to 1.5 mm) hairs which are glochidiate at tip. Stylopodia narrowly conical; styles 1–1.5 mm, rather slender, flexuose-divaricate or divergent. Fig. 100. 1–4.

HAB. In the foothills on gypsaceous sandy soil, and on sandy clay; alt. 50–150 m; fl. & fr. (Apr.–)May–Jun.
DISTRIB. Occasional in the eastern sector of the dry steppe zone of Iraq. **FNI**: 38 km SE of Mandali, *Rawi* 20695!; 16–30 km SE of Badra, *Rechinger* 13961; 22 km SE of Badra, *Rawi* 18429!, *Rechinger* 9237.

W, S & NE Iran, Afghanistan, Pakistan, W Himalayas.

60. **TURGENIOPSIS** Boiss.

in Ann. Sc. Nat. sér. 3, 2: 53 (1844)

Annual herb with pinnately decomposite leaves, glabrous vegetatively. Umbels compound, rays very few. Involucre absent or of 1 or 2 very small bracts, involucel of several setaceous bracteoles. Flowers white, polygamous with inner flowers of partial umbels male, outer petals of hermaphrodite flowers larger but scarcely conspicuously radiant. Petals obovate with an inflexed acumen, emarginate above, larger distinctly 2-lobed. Calyx teeth present, short. Fruit ellipsoid, somewhat laterally compressed, primary ridges slender, shortly ± crisped-setose, secondary ridges very broad with 2 or 3 longitudinal rows of broad-based

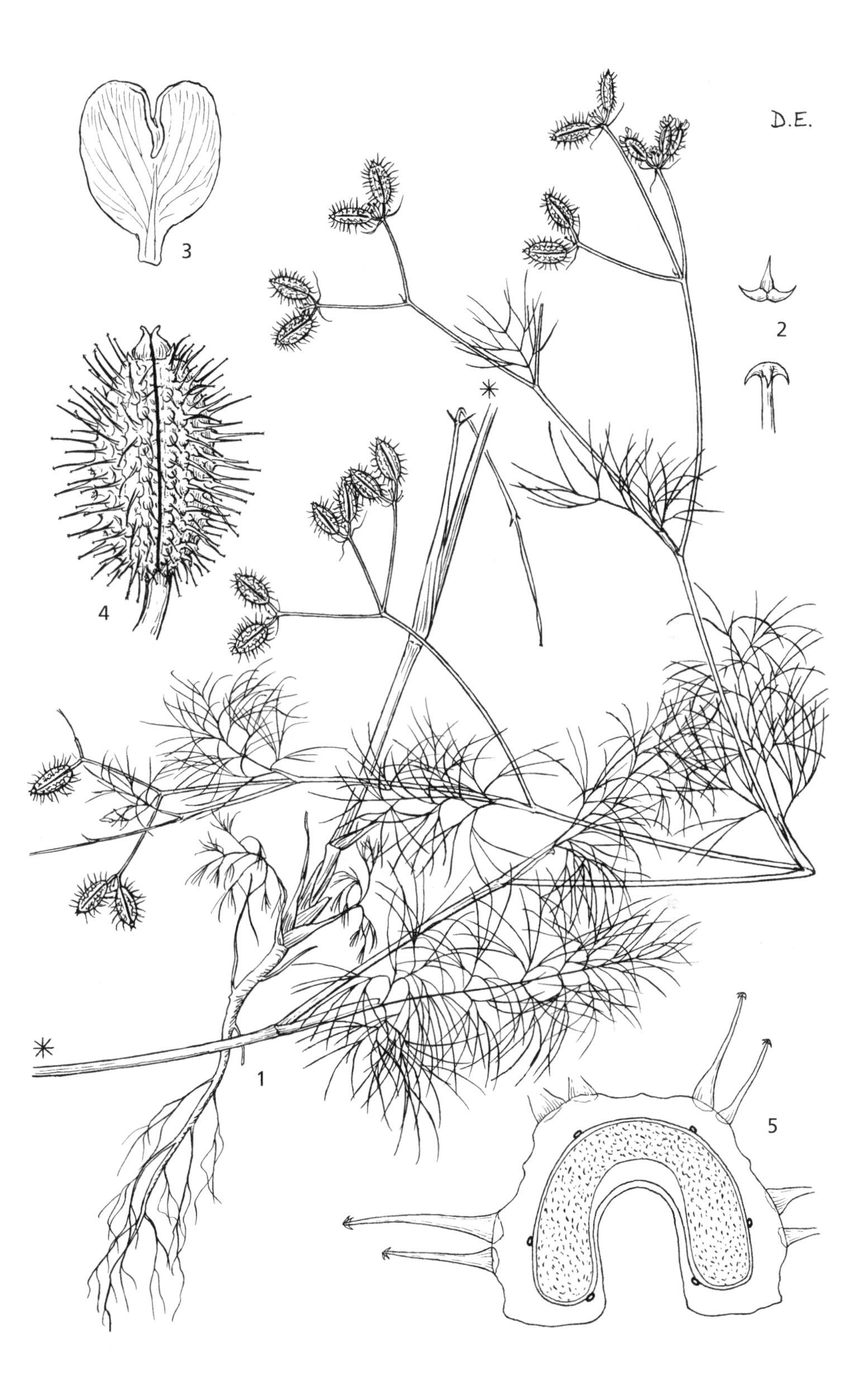

Fig. 101. **Turgeniopsis foeniculacea**. 1, habit × 2/3; 2, detail of bristle × 48; 3, petal × 10; 4, fruit × 4; 5, fruit cross section × 16. All from *Balansa* 574. Drawn by D. Erasmus (1964).

glochidiate bristles. Commissure broad. Vittae solitary beneath the secondary ridges, very slender, 2 similar vittae present within the tips of the endosperm. Disk scarcely obvious, stylopodia stout, conical. Carpophore stout, bifid. Endosperm with the commissural face deeply concave, but not involute at margins.

A monotypic genus.
Turgeniopsis (from *Turgenia*, q.v. and -οψις, *-opsis*, resembling).

1. **Turgeniopsis foeniculacea** Boiss. in Ann. Sc. Nat. sér. 3, 2: 53 (1844); Boiss., Fl. Orient. 2: 1080 (1872); Nábělek in Publ. Fac. Sci. Univ. Masaryk 35: 129 1923); Hayek, Prodr. Fl. Balc. 1: 1058 (1927); Zohary, Fl. Palest. ed. 2, 1: 568 (1932); Zohary in Dep. Agr. Iraq Bull. 31: 113 (1950); Rawi in Dep. Agr. Iraq Tech. Bull. 14: 97 (1964); Mouterde, Nouv. Fl. Lib. et Syr. 2: 601 (1970); Cullen in Fl. Turk. 4: 526 (1972); Rech.f., Fl. Iranica 162: 135 (1987).

Turgenia foeniculacea Fenzl, Pug. pl. nov. Syr.: 18 (1842).
Glochidotheca foeniculacea (*Fenzl*) Fenzl in Russegger, Reise 1 (2): 970 (1843).

Erect annual, 20–50 cm, glabrous save for the fruit and a few sparse bristles on pedicels and involucel. Stem rigid, terete, dichotomously branched from near the base in well-grown plants. Leaves all similar, tripinnatisect, lowest with 2–4 mm filiform segments, deltoid-oblong in outline, to ± 15 cm long, petiolate, membranous-margined sheath attenuate into the petiole above; median and upper leaves becoming rapidly shorter and broader, deltoid-ovate, with longer (to 15 mm) segments, sessile on the sheaths, of which the broad membranous margins are ± auriculate above. Umbels rather few, leaf-opposed and with 1.5–11 cm peduncles, 2–4-rayed, rays 1.5–3.5 cm. Partial umbels 3–8-flowered; outer petals 1.5–2 mm. Involucel of 3 or 4 subulate bracteoles which subequal the 2–5 mm fruiting pedicels, the latter being much incrassate in fruit; 2 or 3 ripe fruit produced per umbel, 7–8 × 3.5–4 mm; stigmas sessile or almost so; stylopodia large, conical. Fig. 101. 1–4.

HAB. On mountains, under *Quercus* trees and in scrub; alt. 1200–1500 m; fl. & fr. (Apr.–)May–Jun.
DISTRIB. Occasional in the upper forest zone of Iraq. **MAM**: Ain Nuni, *Nábělek* 410; Sarsang, *Haines* W1002!; **MRO**: Kani Mazu Shirin, *Hadač et al.* 6028; **MSU**: Jasana, *Omar* 42735!; Benawa nr Penjwin, *Hadač* 4967; Qopi Qaradagh, *Hadač* 5159; Hawraman, *Haussknecht* s.n.

Syria, Lebanon, Turkey, Iran.

61. **LISAEA** Boiss.

in Ann. Sc. Nat. sér. 3, 2: 54 (1844)

Annual herbs, scabrid or setose with glochidiate hairs with short soft hairs in between, leaves simply pinnate, pinnae toothed to pinnatisect. Umbels compound. Involucre and involucel present, conspicuous. Flowers white or pink, polygamous with usually numerous male flowers at the centre of each partial umbel. Outer petals of outer hermaphrodite flowers very conspicuously radiant, deeply bifid, the inner small, emarginate with an inflexed lobule. Calyx teeth distinct, small. Ring of bristles present at tip of fruiting pedicels. Fruit somewhat laterally compressed, ovoid to subrotund, much narrowed to the commissure. Mericarps subterete, the three dorsal primary ridges with rather few very large and broad-based spines at least on one mericarp, laterals and all secondary ridges obsolete or the latter indicated by a row of tubercles or much smaller spines. Vittae obsolete. Disk obvious, stylopodia shortly conical or depressed. Carpophore entire? (not seen in good condition). Endosperm subterete or somewhat compressed, the commissural face sulcate with involute margins.

Three species in SW Asia; two species in Iraq.
Lisaea (named after Domenico Lisa, 1801–1867, an Italian bryologist who studied the flora of the Piedmont area and Sardinia).

One mericarp of fruit with primary ridges with spines, the other without; radiant outer petals 8–11 mm . . . . 1. *L. heterocarpa*
Both mericarps of fruit distinctly spinous, though the spines of one may be smaller especially when young; radiant outer petals 5–7 mm . . . . . 2. *L. strigosa*

1. **Lisaea heterocarpa** (*DC.*) *Boiss.*, Fl. Orient. 2: 1088 (1872). — Type: Iraq (or Iran), Baghdad to Kermanshah, *Olivier & Bruguière* (syn.); Nábělek in Publ. Fac. Sci. Univ. Masaryk 35: 130 (1823); Zohary, Fl. Palest. ed. 2, 1: 573 (1932); Schischkin in Fl. SSSR 16: 178 (1950); Zohary in Dep. Agr. Iraq Bull. 31: 114 (1950); Rawi in Dep. Agr. Iraq Tech. Bull. 14: 93 (1964); Grossh., Fl. Kavk. ed. 2, 7: 46 (1967); Cullen in Fl. Turk. 4: 529 (1972); Rech.f., Fl. Iranica 162: 118 (1987).

Turgenia heterocarpa DC., Prodr. 4: 218 (1830).
Lisaea grandiflora Boiss. in Ann. Sc. Nat. sér. 3, 2: 54 (1844).

Erect annual, 20–50 cm. Stem subterete or finely sulcate, with pale ribs, sparingly branched mostly above the middle, with a mixed indumentum of plentiful, fine, glochidiate hairs on ridges and general short puberulent hairs in between. Leaves rather few, all similar, ovate in outline, simply pinnate with 2–4 pairs of lanceolate, coarsely dentate or more rarely pinnatisect segments; lower leaves shortly petiolate, to 20 cm long, the upper sessile, axes and principal veins with rigid glochidiate bristles beneath, otherwise shortly hairy; all with short, broad, white-margined sheaths which are auriculate above. Umbels on 4–12 cm leaf-opposed peduncles, with 4–13 rays 1–6 cm long and a similar indumentum to that of stem. Involucre of ± 5 lanceolate, entirely membranous bracts, much shorter than rays. Partial umbels many- flowered (15–30), 2–8 bearing fruit. Outer petals of the outer row of flowers very conspicuously radiant, 8–11 mm. Bracteoles ± 8, linear-lanceolate, some a little shorter than fruit or equalling it, membranous-margined below, rigidly setose on midrib dorsally and also along upper margins. Fruiting pedicels 1–3 mm, with a ring of bristles at tip. Fruit ovate, 8–10 mm, the three primary ridges of one mericarp with generally 3 large teeth and scattered smaller ones between, those of the other mericarps edentate; secondary ribs occasionally indicated as tuberculate ridges. Sculpturing and indumentum of the fruit varies considerably; whole surface between the primary ridges may be quite smooth and glabrous or at the other extreme covered with whitish warty tubercles and short stellate down; teeth may also be glabrous or downy. Styles ± 3 mm, stout, flexuose-divaricate.

HAB. In fields and on a rocky clay hillside; alt. (only one recorded) ± 820 m; fl. & fr. May.
DISTRIB. Occasional in the lower forest zone of Iraq. **MAM**: between Faida and Al Baqaq, *Karim, Dabbagh & Hamid* 44924!; Ain Nuni, *Nábělek* 405, 520; **MRO**: Kani Watman, *Nábělek* 433, 411; **MSU**: Tasluja, *Omar, Qaisi & Wedad* 49425!; Tasluja to Sarchinar, *Hadač* 4765; Penjwin, *Hadač et al.* 4866; ?**FNI** (or Iran), Baghdad to Kermanshah, *Olivier & Bruguière* (type).

Turkey (E Anatolia), Armenia, Azerbaijan, Georgia, W, C & S Iran.

2. **Lisaea strigosa** (*Banks & Sol.*) *Eig*, Journ. Bot. 75: 189 (1937); Blakelock in Kew Bull. 3: 434 (1939); Rawi in Dep. Agr. Iraq Tech. Bull. 14: 93 (1964); Mouterde, Nouv. Fl. Lib. et Syr.: 602 (1970); Cullen in Fl. Turk. 4: 529 (1972); Zohary, Fl. Palaest. 2: 401 (1972); Rech.f., Fl. Iranica 162: 119 (1987).

Caucalis strigosa Banks & Sol. in Russ., Nat. Hist. Aleppo ed, 2, 2: 248 (1794).
Lisaea syriaca Boiss. in Ann. Sc. Nat. sér. 3, 2: 55 (1844); Boiss., Fl. Orient. 2: 1088 (1872); Zohary, Fl. Palest. ed. 2, 1: 573 (1932).

Erect annual, 20–50 cm. Stem terete or finely sulcate, especially above, sparingly branched mostly above the middle, with a mixed indumentum of glochidiate hairs on ridges and short puberulent hairs in between. Leaves rather few, all similar, oblong or ovate in outline, simply pinnate with 2–4 pairs of lanceolate, coarsely dentate or more rarely pinnatisect segments; lower leaves petiolate, to 20 cm long, uppermost sessile, all with the axes and principal veins with rigid glochidiate bristles beneath, otherwise shortly hairy; sheaths short, broad, white-margined. Umbels on 5–26 cm, leaf-opposed peduncles, with 4–8 rays 1–5 cm long, indumentum similar to that of stem. Involucre of 5 lanceolate, entirely membranous bracts, much shorter than rays. Partial umbels many (15–30)-flowered, 3–9 bearing fruit; outer petals of outer row of flowers very radiant, 5–7 mm long. Bracteoles ± 8, widely membranous-margined, few with the midrib scabrid dorsally, and not elongate-linear and equalling fruit. Fruiting pedicels (0–)1–4 mm, with an apical ring of bristles. Fruit ovoid, 6–8 mm, the three primary ridges of both mericarps with (4–)5–7 narrower-based glabrous or initially downy prickles which are frequently irregular in size; secondary ridges smooth or sometimes sharply tuberculate, interstices smooth. Styles 3 mm, flexuose-divaricate. Fig. 102: 1–6.

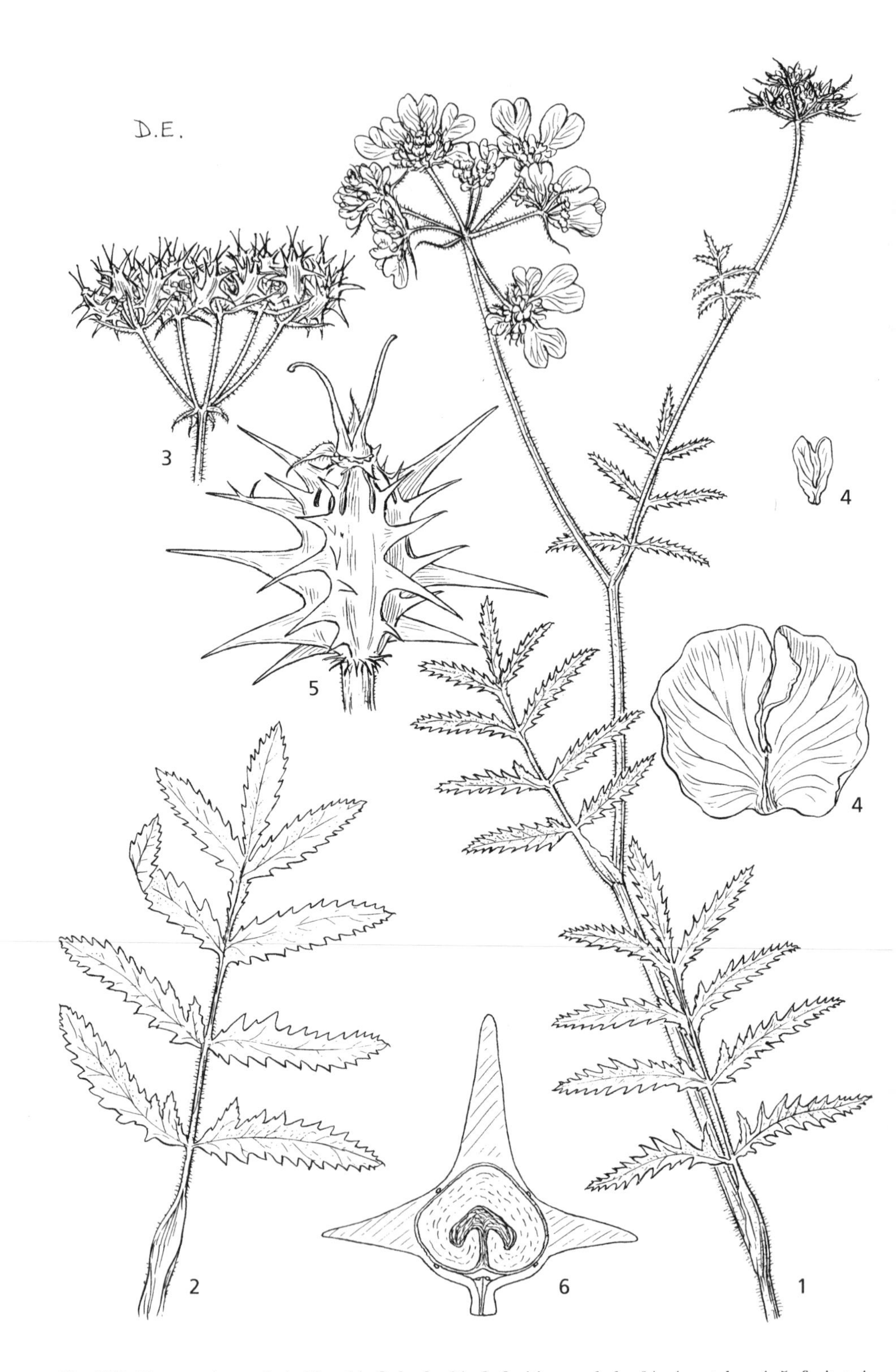

Fig. 102. **Lisaea strigosa**. 1, habit × 2/3; 2, leaf × 2/3; 3, fruiting umbel × 2/3; 4, petals × 4; 5, fruit × 4; 6, fruit cross section × 10. 1 from *Robertson* S/1031; 2–6 from *Gillett* 8300. Drawn by D. Erasmus (1964).

HAB. Recorded only from cereal fields; alt. 380–1000 m; fl. & fr. May–Jun.
DISTRIB. Occasional, but locally common, in the lower forest zone and upper moist steppe zone of Iraq. **MJS**: Asi, *Rawi* 8486A; **MRO**: just E of Rowanduz, *Gillett* 8300!; Shirwan Mazin, *Hadač* 5855; **FUJ**: 10 km N of Mosul, *Hossain* 134!; **FNI**: Baitnar, nr Ain Sifni, *Salim* 2579!
NOTE. LIZAIJ ("clingweed", Ir., Ain Sifni, *Salim* 2579, "not of much feeding value"); a common weed in cereal fields and on disturbed soil near fields.

Palestine and Syria, Turkey (E Anatolia), W Iran.

62. CAUCALIS L.

Sp. Pl. ed. 1: 240 (1753); Gen. Pl. ed. 5: 112 (1754)

Annual herbs, ± hispid or setose, leaves 2- or 3-pinnately decomposite. Umbels compound. Involucre present, small. Involucel present, of several lanceolate bracteoles. Flowers polygamous, centre flowers of each partial umbel male, rarely all hermaphrodite, white or pinkish. Petals obovate, incised above with an incurved lobule, outer petals of outer hermaphrodite flowers somewhat larger. Calyx teeth conspicuous and herbaceous, persistent. Fruit ellipsoid to ovoid, commissure narrow, somewhat laterally compressed. Mericarps subterete, primary ridges filiform, patent-hispid, secondary ridges bearing long (rarely short) stout, rigid spines, which are smooth and uncinate at apex. Vittae small, distinct, solitary beneath secondary ridges; commissure bivittate at points of involution of the endosperm. Disk scarcely obvious, stylopodia conical. Carpophore undivided or shortly bifid. Endosperm subterete or slightly compressed dorsally, deeply sulcate on the commissural face and with margins deeply involute.

A monotypic genus.
Caucalis (Gr. name of an umbelliferous plant in Theophrastus).

1. **Caucalis platycarpos** *L.*, Sp. Pl. ed. 1: 241 (1753); Cullen in Fl. Turk. 4: 526 (1972); Meikle, Fl. Cyprus 1: 704 (1977); Rech.f., Fl. Iranica 162: 116 (1987).

C. daucoides L., Syst. Nat. ed. 12, 2: 205 (1767) non Sp. Pl. ed. 1: 241 (1753); DC., Prodr. 4: 216 (1830); Boiss., Fl. Orient. 2: 1084 (1872); Hayek, Prodr. Fl. Balc. 1: 1054 (1927); Zohary, Fl. Palest. ed. 2, 1: 571 (1932).
Daucus lappula Web. in Wigg., Prim. Fl. Holsat.: 23 (1780).
Caucalis lappula (Web.) Grande, Bull. Ort. Bot. Nap. 5: 194 (1918); Schischkin in Fl. SSSR: 16: 172 (1950); Grossh., Fl. Kavk. ed. 2, 7: 45(1967).

Erect annual, 10–40(–50) cm, with rather sparse whitish papillate hairs on stems and principal axes and veins of leaves. Stem angular-sulcate, branched from near base. Leaves all similar, deltoid-ovate in outline, lower to 7 cm, neatly and regularly bi- or tri-pinnate, segments divided into small, shortly linear, 2–4 × 1–1.5 mm lobes; upper leaves smaller, sessile, lower petiolate, all with rather short, broadly expanded sheaths. Peduncles 2–7 cm. Umbels with 2–5 stout rays 1.5–3 cm long, bracts of involucre (0–)1–3, linear, caducous. Partial umbels 2–12-flowered; bracteoles 2 or 3, linear. Flowers white or pink, outer or (when fewer) all flowers hermaphrodite, inner functionally male; hermaphrodite flowers with petals small, outer to 2.5 mm but not conspicuously radiant. Fruit 8–10 × 4–5 mm long excluding spines, shortly (0.5–3 mm) pedicellate or sessile with a single row of 6–8 stout-based, rigid, uncinate spines (0.5–)2.5–4 mm long disposed along the secondary ridges; primary ridges with patent bristles. Styles short, rigid, less than 1 mm, scarcely divergent.

HAB. Shady places in creekside garden; alt. 2(–1000) m; fl. Apr.
DISTRIB. Very rare in Iraq; recorded in the SE sector of the alluvial plain in the desert region and in the steppe region. **MAM**: Sarsang, *Anders* 2325; **LBA**: Makina Masus, Basra, *Whitehead* 101 in Blatter, l.c.
NOTE. The specimen from Sarsang has not been seen, and the plant was presumably a casual adventive.

Mediterranean Europe, Greece, Cyprus, Turkey, Caucasus, N, C & W Iran.

63. **ASTRODAUCUS**[21] Drude

in Engl. & Prantl, Naturl. Pflanzenfam. 3, 8: 156 (1898)

Biennial, monocarpic, with leaves many times pinnate, resembling those of *Ferula*. Stem up to 1 m tall, branched in upper part. Petals white or cream, the outer ones radiant, unequally lobed, apex inflexed. Fruit subcylindrical, bearing one or two rows of thick pyramidal spines on secondary ridges, primary ridges ciliate or with stellate hairs.

Two species in SE Europe and SW Asia, one restricted to the northern coasts of the Black Sea, the other represented in Iraq.
Astrodaucus (from the Gr. αστερ, *aster* (star) and δαύκος, *daucos*, Gr. (usually interpreted as carrot).

1. **Astrodaucus orientalis** *(L.) Drude* in Engl. & Prantl, Natürl. Pflanzenfam. 3, 8: 271 (1898); Grossh., Fl. Kavk. 3: 135 (1932); Grossh., Fl. Kavk. ed. 2, 7: 44 (1967); Pimenov in Fl. Iranica 162: 114 (1987).

Caucalis orientalis L., Sp. Plant. ed. 1, 241 (1753).
Daucus pulcherrimus (Willd.) W.D. Koch, Nov. Acta Acad. Leop.-Carol. Nat. Cur. 12, 1: 78 (1824); Boiss., Fl. Orient. 2: 1072 (1872).

Tall branched herb with leaves triangular-ovate in outline, many times divided, sparsely hairy; ultimate lobes minute, obtuse. Umbel with 8–12 rays, bracteoles 5, oblong-lanceolate, ciliate. Calyx teeth short. Petals 4 mm, white to pale yellow. Fruit 5–6 mm, with triangular spines on secondary ridges, 2-seriate, longer than the width of fruit, confluent into a wing at the base.

HAB. Clearings in *Quercus* forest, on slate; alt. ± 1400 m. fl. ?May.
DISTRIB. Very rare in Iraq; one record from the moist steppe region. **MSU**: Penjwin, *Rechinger* 12279a.

Turkey, S Russia, Syria, Iran, Caucasus.

64. **TURGENIA** Hoffm.

Gen. Umb. ed. 1: 59 (1814)

Caucalis subgen. *Turgenia* (Hoffm.) Benth. in Benth. et Hook., Genera Plantarum 1: 929 (1867)

Annual herbs, hispid or setose, leaves simply pinnate with broad, toothed leaflets. Umbels compound. Involucre and involucral bracts present, of ovate, broadly scarious-margined bracts and bracteoles. Flowers white or reddish, polygamous as in *Caucalis*. Petals obovate, smaller incised above with an incurved lobule, outer petals of the outer hermaphrodite flowers somewhat larger to much expanded and persistent. No ring of hairs present at top of fruiting pedicels. Fruit ellipsoid, somewhat laterally compressed, commissure narrow. Mericarps subterete, marginal primary ridges filiform, the 3 slender dorsal primary ridges with a single row, and the prominent secondary ridges with 1or 2 rows of straight, scabrid, glochidiate-tipped spines. Vittae distinct, solitary beneath secondary ridges. Disk not obvious, stylopodia conical. Carpophore undivided or shortly bifid. Endosperm subterete or slightly compressed, deeply sulcate on the commissural face and with margins deeply involute.

Two species in W & S Europe, the Mediterranean region, N Africa and SW to C Asia; both represented in Iraq.
Turgenia (named in honor of D. A. Turgenev, Director of the Office of Prince Golitsyn, an old friend and fellow student of Hoffmann at Göttingen University).

Flowers with outer petals of outer hermaphrodite flowers larger than the remainder but not very conspicuously radiant; styles very short, rigid, divergent . . . . 1. *T. latifolia*
Flowers with outer petals very conspicuously radiant; styles long, 3 mm, flexuose-divaricate . . . . 2. *T. lisaeoides*

[21] Description by J R Edmondson.

1. **Turgenia latifolia** (*L.*) *Hoffm.*, Gen. Umb. ed. 1: 59 (1814); DC., Prodr. 4: 218 (1830); Boiss., Fl. Orient. 2: 1087 (1872); Hayek, Prod. Fl. Balc. 1: 1054 (1927); Zohary, Fl. Palest. ed. 2, 1: 572 (1932); Bornmüller in Beih. Bot. Centralbl. 58B: 288 (1938); Blakelock in Kew Bull. 3: 437 (1948); Schischkin in Fl. SSSR 16: 174 (1951); H. Riedl in Fl. Lowland Iraq: 456 (1964); Rawi in Dep. Agr. Iraq Tech. Bull. 14: 96 (1964); Grossh., Fl. Kavk. ed. 2, 7: 46 (1967); Mouterde, Nouv. Fl. Lib. et Syr. 2: 601 (1970); Cullen in Fl. Turk. 4: 527 (1972); Zohary, Fl. Palaest. 2: 398 (1972); E. Nasir in Fl. Pak. 20: 109 (1972); Meikle, Fl. Cyprus 1: 705 (1977); Rech.f., Fl. Iranica 162: 131 (1987); Pan Zehui & Watson in Fl. China 14: 29 (2005).

Tordylium latifolium L., Sp. Pl. ed. 1: 240 (1753).

Fig. 103. **Turgenia latifolia**. 1, habit × ¾; 2, fruit × 7½; 3, fruit corss section × 7½. Reproduced from Fl. Pak. 20: f. 31 (1972), with permission.

Caucalis latifolia (L.) L., Syst. Nat. ed. 12, 2: 205 (1767); Hand.-Mazz. in Ann. Naturh. Mus. Wien 27: 89 (1913); Nábělek in Publ. Fac. Sci. Univ. Masaryk 35: 129 (1923); Guest in Dep. Agr. Iraq Bull. 27: 20 (1933); Anth. in Notes Roy. Bot. Gard. Edinb. 18: 289 (1935); Zohary in Dep. Agr. Iraq Bull. 31: 114 (1950).

Stout annual, 10–40(–60) cm, hispid throughout. Stem sulcate, branched, ribs with long scabrid bristles, both ribs and furrows with short, close pubescence. Leaves all similar, simply pinnate with 3 or 4 pairs of lanceolate or ovate-lanceolate, sharply incised-dentate to pinnatifid, 10–55 × 3–25 mm segments, indumentum of upper surface strigulose and closely appressed, beneath also with long scabrid bristles on principal veins; lower leaves long-petiolate, deltoid-oblong in outline, with broad, white-margined sheaths contracting rather abruptly into the petiole, auriculate; upper leaves shorter, broadly deltoid, sessile on shorter sheaths. Umbels many, leaf-opposed on mostly 1–11 cm peduncles, 2–5-rayed, rays 1.5–8.5 cm, indumentum similar to that of stem. Involucre of 2 or 3(rarely 5) mostly scarious, oblong-lanceolate, ± caducous bracts. Partial umbel of 6–16 deep rose-coloured or white flowers, outer petals somewhat larger but not conspicuously radiant as in the next species, 3–5 fructiferous. Involucel of usually 4–6 ovate or lanceolate-ovate membranous, broadly scarious-margined bracteoles, very conspicuous in bud and early flower. Fruit on short (± 2–5 mm) pedicels, ellipsoid, large and firm, 10–15 × 3 mm excluding the stout-based prickles, of which there are mostly 5–8 on each vertical row; primary ribs with a single row of spines, secondary ribs with (1–)2 rows; all ribs clothed with white wary papillae between spines, as are the faces between ribs. Styles very short, rigid, divergent. Fig. 103: 1–3.

HAB. On mountains, sometimes on stony slopes in denuded *Quercus* wood, more commonly in fields, vineyards, on steppic hills and plains often on wasteland, by roadsides etc.; alt. 200–1400(–1850) m; fl. & fr. (Apr.–)May–Jun.

DISTRIB. Common in the forest zone of Iraq, occasional in the steppe region. **MJS**: Jabal Sinjar, *Handel-Mazzetti* 1495; Kursi, *Gillett* 10900!; **MAM**: above Ain Nuni, *Nábělek* 409; Bekhair (Jabal Bekher), *Rawi* 23012!; Zawita village, *Robertson* 246!; Mar Ya'qub, above Simel, *Nábělek* 353; ? Mosul Liwa, *Mooney* 4321!; **MRO**: Shaqlawa, *Bornmüller* 1280!, *Haines* W760!; Batas, *Rawi* 8466!; 31 km N of Kirkuk, on road to Koi Sanjaq, *Rawi, Nuri & Alkas* 27944!; Chinaruk, *id.* 28279!, 28324!; 5 km SW of Rania, *id.* 28422!; **MRO/FAR**: Arbil province, *Radhi* 3843!; **MSU**: 20–25 km E of Qaranjir, *Rawi* 21654!; Jarmo, *Helbaek* 1222!, *Haines* W291!; 20 km NW of Sulaimaniya on road to Dukan, *Rawi* 21747!; Geqija Dagh, *Gillett & Rawi* 11688!; between Zewiya and Qarachitan, *Gillett* 7831!; Mela Kowa (Malakawa), *Rawi, Husham* (*Alizzi*) *& Nuri* 29576!; Penjwin, *Rawi* 22550!; Hawraman, N of Halabja, *Rawi* 22084!; Khormal, *Rawi* 8840!; Tawila, *Rawi* 21976!; **FUJ**: between Balad Sinjar and Tal Afar, *Guest* 4126A!, 13369!; Mir Qasim, *Field & Lazar* 550!; Qaiyara, *Bayliss* 112!; Sharqat, *Thomas* s.n.! comm. Graham; **FNI**: Tal Kaif, *Mudir of Tal Kaif* 3199!; Baqasra, nr Ain Sifni, *Salim* 2575!; **FAR**: Arbil, *Guest* 3843!; **FKI**: 10–15 km N of Kirkuk, *Rawi* 21588!; **FKI/FNI**: Jabal Hamrin, in Anth., l.c.; **FNI**: Kani Masi and Naft Khana, *Morck* 11, 16 in *Handel-Mazzetti*, l.c.; Jabal Kumatas, *Nábělek* 408 .

NOTE. ?KHUMMĀSH (Ir.-Arbil province, *Radhi* 3843 – "grazed by livestock"); LUZZAIJ ("clingweed", Ir.-Tal Kaif) or KHENJER (dagger-weed", no doubt referring to the prickly fruit, Kurd.-Tal Kaif, Mudir of Tal Kaif 3199 – "useful for sheep grazing"); ?SEPĀN DIRRI (Kurd.-Ain Sifni, *Salim* 2575 – "fodder plant"). Whether or not these unconfirmed colloquial names from the steppe region are valid, the general consensus of opinion indicates that the plant is considered to be of some value for grazing. Schischkin (1950) remarks it is a common weed in crops.

W, C & S Europe, Cyprus, Syria, Lebanon, Palestine, Jordan, Caucasus.

2. **Turgenia lisaeoides** *C.C. Townsend*, Kew Bull. 17: 431 (1964); Rech.f., Fl. Iranica 162: 134 (1987). — Type: Iraq, Gweija Dagh behind Sulaimaniya, *Rawi & Gillett* 11712 (K!, holo.)

Erect annual, 25–40 cm. Stem erect, sulcate, sparsely branched from below the middle, shortly and finely pubescent, ridges with short, white bristles which become longer and more frequent in upper part of stem and the main umbel rays. Leaves all similar, oblong-ovate in outline, simply pinnate with 2 or 3 pairs of lanceolate, 15–50 × 2.5– 10 mm, coarsely dentate to pinnatifid segments, strigose-hairy, lower surfaces with longer strigose hairs on principal veins; lower leaves petiolate, upper sessile, all with rather short, broad, membranous-margined sheaths. Umbels rather few, leaf-opposed, peduncles 3–11 cm, 4–6-rayed, rays 2–4.5 mm, indumentum resembling that of the stem; involucre of 1 or 2 very short bracts, much shorter than rays, membranous and early-caducous. Partial umbels many (7–15)-flowered, 3–5 flowers fructiferous, fruiting pedicels short (3–5 mm); petals white or slightly flushed with pink, the outer very radiant, to 6 mm long. Involucel of 5–8 ovate, pubescent bracteoles, membranous save for the brownish midrib. Fruit ellipsoid, 8–10 ×

3–4 mm (excluding spines), secondary and primary ridges with a single row of 5–6 scabrid spines ± 4 mm long, remainder of surface of fruit shortly whitish papillose. Stylopodia attenuate into the 3 mm, flexuose-divaricate styles.

HAB. Lower mountain sides, on stony slopes, on limestone scree among denuded *Quercus*, in cultivated valley; alt. 1150–1600 m; fl. & fr. Jun.–Jul.
DISTRIB. Found only in the SE section of the upper forest zone, mostly in the furthest SE corner quite close to the Iranian frontier. **MSU**: Gweija Dagh behind Sulaimaniya, *Rawi & Gillett* 11712 (type)!; Tawila, *Rawi* 21867!, *Rechinger* 10294!; Balkha, 7 km W of Tawila, *Rawi* 22340!

Endemic.

65. **TORILIS** Adans.

Fam. Pl. 2: 99 (1763)

Annual herbs, ± hispid or appressed-strigulose, leaves 1–3-pinnate with toothed or rarely entire segments. Umbels compound. Involucre present or absent, involucel of several narrow bracteoles. Flowers white or reddish. Petals obovate, incised above with an incurved lobule, outer often somewhat larger but not conspicuously radiant. Calyx teeth small or very small, triangular or lanceolate, persistent. Fruit linear to ovoid, somewhat laterally compressed, commissure narrow. Mericarps subterete, the 5 primary ridges filiform with longitudinally appressed bristles; secondary ridges with 1–3 rows of long, glochidiate, usually scabrid long or shorter spines, or sometimes mericarps heteromorphic with one spiny and the other verruculose or more shortly spinous. Vittae solitary beneath secondary ridges. Disk scarcely obvious, stylopodia short. Carpophore bifid. Endosperm subterete or slightly compressed, commissural face shallowly sulcate, margins incurved (scarcely so in *T. tenella*), but not strongly involute.

Some 15 species distributed from Macaronesia and the Mediterranean region to E Asia; seven species in Iraq.
Torilis (from the Gr. τορεύω, *toreia*, relief carving or embossing, according to JE Smith (English Flora vol. 2, 1824). As Schischkin (1950), Gilbert-Carter (1964) and others have pointed out, most of Adanson's names appear to have been made up by the author and are meaningless.)

1. Umbels leaf-opposed, very dense, subsessile, very rarely a stout peduncle as long as the diameter of the umbel developing; at least some of the fruits with heteromorphic mericarps, one spinous and the other tuberculate-verrucose . . . . 1. *T. nodosa*
   Umbels not subsessile; if the umbels dense and the peduncle only as long as the diameter of the umbel (*T. stocksiana* forms) then the peduncle slender and mericarps never heteromorphic. 2
2. Fruits with mericarps (in Iraq) heteromorphic, one with longer spines, the other more shortly spinose or hispid-verrucose; terminal leaflet of median leaves elongate, deeply toothed but not completely divided, leaf thus suddenly and longly acuminate. 3
   Fruits with mericarps homomorphic, save frequently for a sessile central fruit in the partial umbel which has uniformly shorter spines; terminal leaflet of median leaves not elongate, completely divided and not forming a long acumen to the leaf 4
3. Rays of umbel (2–)3–12; styles longer, style and stigma together clearly longer than stylopodia. . . . . 2. *T. arvensis*
   Rays of umbel 2–4; stigmas subsessile, style and stigma together not longer than stylopodia. . . . . 3. *T. chrysocarpa*
4. Umbels on peduncles shorter than or subequalling the subtending leaves; umbels with almost always 2 or 3 rays 5
   Umbels on peduncles usually longer than the subtending leaves; umbels usually 5–10-rayed. 6
5. Umbels less dense, rays subequal; pedicels longer, exceeding the shortly lanceolate bracteoles and revealing the central, sessile, more shortly-spined fruit of each partial umbel; styles very short, stigmas ± all sessile . . . . 4. *T. leptophylla*

Umbels dense, rays very unequal and one partial umbel often subsessile; pedicels very short, exceeded by the linear-lanceolate bracteoles; styles obvious, divaricate . . . . 5. *T. stocksiana*

6. Fruit 3.5–4 × 0.7 mm excluding spines; bracteoles equalling pedicels; plant slender . . . . 6. *T. tenella*

Fruit ± 6 × 1 mm excluding spines; bracteoles shorter or subequalling pedicels; plant robust . . . . 7. *T. leptocarpa*

1. **Torilis nodosa** (*L.*) *Gaertn.*, Fruct. 1: 82, t. 20 (1788); DC., Prodr. 4: 219 (1830); Boiss., Fl. Orient. 2: 1083 (1872); Hayek, Prod. Fl. Balc. 1: 1055 (1927); Zohary, Fl. Palest. ed. 2, 1: 570 (1932); Schischkin in Fl. SSSR 16: 162 (1950); H. Riedl in Fl. Lowland Iraq: 455 (1964); Grossh., Fl. Kavk. ed. 2, 7: 43 (1967); Mouterde, Nouv. Fl. Lib. et Syr. 2: 590 (1970); Cullen in Fl. Turk. 4: 519 (1972); Zohary, Fl. Palaest. 2: 397 (1972); E. Nasir in Fl. Pak. 20: 106, f. 32 (1972); V. Täckh., Stud. Fl. Egypt ed. 2: 397 (1974); Meikle, Fl. Cyprus 1: 699 (1977); Peev in Fl. Iranica 162: 123 (1987); Boulos, Fl. Egypt 2: 186 (2000).

Tordylium nodosum L., Sp. Pl. ed. 1: 240 (1753).
Caucalis nodosa (L.) Scop., Fl. Carn ed. 2, 1: 192 (1772).

Erect, decumbent or prostrate annual, 5–35 cm. Stem terete, finely striate, thinly furnished with downwardly-appressed strigulose hairs, with long, ascending branches from the base upwards. Lower leaves petiolate, petioles to 10 cm, lamina broadly deltoid-ovate or deltoid-oblong in outline, bi- to subtri-pinnate, appressed-hairy, ultimate divisions lanceolate, 2–8 × 1–2 mm, acute; upper leaves similar but sessile and more broadly deltoid; sheaths long, white-margined, narrowly expanded. Umbels many, leaf-opposed, usually subsessile but rarely developing a thick peduncle equal in length to the diameter of the umbel (rarely exceeding 20 mm when fruiting); rays 2 or 3, usually very short; pedicels also very short, so that the umbel is very dense and compact. Partial umbels 5–12-flowered. Bracteoles of the involucel 1–3, up to 1/3 the length of fruit. Flowers small, pinkish or white. Fruit ovoid, 2–3 × 2–2.5mm excluding the spines; biseriate glochidiate spines of secondary ridges of mericarps usually dissimilar in outer fruits, those of the outer mericarp long, equalling its diameter, those of the inner shorter, often tuberculiform; inner fruits usually with tuberculiform bristles throughout; rarely all mericarps with long spines; calyx absent or very small and pale. Primary ridges with 1-or 2-seriate longitudinally appressed hairs. Stigmas subsessile. Fig. 104: 1–2.

Hab. In the lower mountains, on sand hills; alt. 900–1000 m; fl. & fr. Apr.–Jun.
Distrib. Rare in Iraq, only found in a few widely spaced localities. **MRO**: Shaqlawa, *Gillett & Rawi* 11572!; **MSU**: Qaranjir, *id.* 7549; Girdabor at foot of mt. Pira Magrun, *Hadač et al.* 4743; mt. Hawraman above Khormal, *Hadač* 5049; **LCA**: Abu Ghraib, *Rechinger* s.n.; **FNI**: nr Badra *Rechinger* 9173b.
Note. Knotted Hedge Parsley (Johns).

W, C & Mediterranean Europe, Cyprus, Syria, Lebanon, Palestine, Jordan, Egypt, Turkey, Caucasus, N & NW Iran, C Asia (Turkmenistan to Tajikistan), N Africa (Morocco to Libya), Macaronesia.

2. **Torilis arvensis** (*Huds.*) *Link*, Enum. Hort. Berol. 1: 265 (1821); Schischkin in Fl. SSSR 16: 159 (1950); H. Riedl in Fl. Lowland Iraq: 455 (1964); Grossh., Fl. Kavk. ed. 2, 7: 42 (1967); Mouterde, Nouv. Fl. Lib. et Syr. 2: 591 (1970); Cullen in Fl. Turk. 4; 520 (1972); Zohary, Fl. Palaest. 2: 395 (1972); Meikle, Fl. Cyprus 1: 700 (1977); Peev in Fl. Iranica 162: 123 (1987).

Caucalis arvensis Huds., Fl. Angl. ed. 1: 98 (1762).

Annual herb with a slender taproot, 0.1–1.25 m, branched from the base and densely bushy with many short, branched stems to tall and simple or with rather few, long, ascending branches. Stems slender, wiry, terete, finely striate, glabrous or with downwardly-appressed strigulose hairs. Basal leaves generally withered by time of fruiting, bi- to subtri-pinnate with pinnatifid segments, deltoid and 3–15 × 2–12 cm in outline, ultimate divisions 2–8 × 1–3 mm, ± furnished with upwardly directed white strigulose hairs especially on lower surface; sheaths 2–5 mm, pale-margined; petiole 1.5–4.5 cm. Median leaves similar or segments becoming narrower and more elongate, more shortly petiolate, sheaths long and narrow with involute margins. Upper leaves similar but reduced, varying to trisect with elongate segments (upper especially long), sessile on narrow sheaths. Umbels very numerous to rather few, on slender 1.5–13 cm peduncles; rays (2–)3–12, appressed-strigulose, 7–30 mm long; involucre usually absent, sometimes one linear bract present. Partial umbels 2–18-flowered, pedicels very unequal, 0–4 mm, indumentum similar to that of rays; involucel of 4–8 bracteoles 1.5–5 mm long. Calyx teeth very small to obsolete. Flowers white to pink or purple, outermost petals of outer partial umbels not to decidedly radiant (to 2 mm), the inner or all petals 0.75–1 mm, strigose-hairy dorsally. Fruit ellipsoid, 3–5 × 2–3 mm, both mericarps densely covered with long, glochidiate-tipped, white to purple spines along the obscure secondary ribs, or one mericarp with spines shorter; stylopodia conical, short; styles very short to slender, 0.2–1.2 mm, style and stigma clearly longer than stylopodia.

DISTRIB. (of species): W, C & S. Europe, SW, S & E Asia, N, tropical and S Africa; introduced in N America and the W Indies.

subsp. **arvensis**

Scandix infesta L., Syst. Nat. ed. 3, 2: 732 (1767).
Torilis divaricata Moench, Meth. Pl. Suppl. 34 (1802).
T. helvetica (Jacq.) Gmel., Fl. Bad. 1: 617 (1805).
T. infesta (L.) Clairv., Man. herb.: 78 (1811); Boiss., Fl. Orient. 2: 1082 (1871); Bornmüller in Beih. Bot. Centralbl. 58B: 288 (1938).
T. infesta subsp. *divaricata* (Moench) Thell. in Hegi, Illustr. Fl. Mitteleur. 5: 1055 (1926); V. Täckh., Stud. Fl. Egypt ed. 1: 206 (1956), pro var.

Plant low and bushy; much-branched from the base upwards. Outer petals of outer flowers of the partial umbels only slightly larger than the remainder, not conspicuously radiant. Fruit with the mericarps commonly homocarpous. Styles very short, rarely twice as long as stylopodia.

HAB. In field in the upper submontane plain; alt. 250 m; fl. & fr. Jul.-Aug.
DISTRIB. Very rare, if indeed it occurs in Iraq; only recorded once as collected nearly a century ago from the moist steppe zone. **FNI**: Nineveh, *Bornmüller* 1284.
NOTE. The above specimen recorded by Bornmüller (1938) has not been seen and may well have been destroyed in World War 2. It may have belonged to the following subspecies, or possibly have been an introduction.

W, C & Mediterranean Europe, Crimea, Syria, Lebanon, Palestine, Turkey, Caucasus, Iran, Afghanistan, C Asia (Turkmenistan to Tian Shan). Introduced into N America and Australia.

subsp. **neglecta** (Roem. & Schult.) Thell., Fl. adv. Montp.: 395 (1912); Hayek, Prodr. Fl. Balc. 1: 1057 (1927); Cullen in Fl. Turk. 4: 520 (1972); Meikle, Fl. Cyprus 1: 701 (1977).

T. radiata Moench, Meth. Pl.: 103 (1794); Schischkin in Fl. SSSR 16: 160 (1950); H. Riedl in Fl. Lowland Iraq: 455 (1964); Mouterde, Nouv. Fl. Lib. et Syr. 2: 591 (1970); Zohary, Fl. Palaest. 2: 396 (1972); V. Täckh., Stud. Fl. Egypt ed. 2: 397 (1974); Peev in Fl. Iranica 162: 124 (1987).
T. neglecta Roem. & Schult., Syst. Veg. ed. 15, 6: 484 (1820); DC., Prodr. 4: 218 (1830); Boiss., Fl. Orient. 2: 1083 (1872); Hand.-Mazz. in Ann. Naturh. Mus. Wien 27: 80 (1913); Zohary, Fl. Palest. ed. 2, 1: 570 (1932); Anth. in Notes Roy. Bot. Gard. Edinb. 18: 437 (1935); Blakelock in Kew Bull. 3: 437 (1948).
T. syriaca Boiss. et Bl. in Boiss., Diagn. ser. 2, 2: 98 (1856).

Plant tall, not bushy, with fewer and longer branches. Outer petals of outer flowers of the partial umbels distinctly larger than the rest, conspicuously radiant. Fruit with the mericarps commonly heterocarpous, one with long bristles, the other hispid-verrucose. Styles slender, usually 3 × as long as the stylopodia.

HAB. (of subsp.) In mountain valleys by shady streams, under *Quercus* and *Juglans* trees, in foothills near seepage pools and rice fields, in shady ditches and beside irrigation canals in gardens and orchards on the alluvial plain; alt. 10–1150(–1180) m; fl. & fr. Jul.–Aug.
DISTRIB. Occasional in the forest zone of Iraq, less common in the steppe region and in irrigated districts in the desert region. **MAM**: 5 km S of Zakho, *Rawi* 23110!; S of Sharanish, Rawi, *Nuri & Tikriti* 29071!; Sarsang, *Haines* W1236!; **MRO**: Shaqlawa, *Omar, Sahira, Karim & Hamid* 38296!; Darband, near Rayat, *Gillett* 9495!; Pushtashan, *Rawi* 23098!, *Rawi & Serhang* 26551!, *Rechinger* 11041; **MSU**: Sulaimaniya, *Haines* W2087!; **FUJ/FNI**: by R. Tigris below Mosul, *Handel-Mazzetti* 1205; **FNI**: Mindan, *E*

Fig. 104. **Torilis nodosa**. 1, habit × 1/2; 2, fruit × 6. **Torilis leptophylla**. 3, habit × ½; 4, fruit × 6; 5, fruit cross section × 15. Reproduced from Fl. Pak. 20: f. 32 (1972), with permission.

Chapman 26223!; **FKI/DGA**: between Kirkuk and Khalis, *Rawi, Chakr., Alizzi & Nuri* 19701!; **LCA**: Daltawa (Khalis), *Guest* 2458!; **LCA**: Abu Ghraib, *Rawi* 10734!, *Haddad* 12867!, *Alizzi* 34322!; Rustamiya, *Lazar* 3414!; **LEA**: Ba'quba, *Omar, Janan al-Mukhtar & Sahira Abdulli* 36998!, *Rechinger* 9745; Ruz (Balad Ruz) in Anth., l.c.; **LSM**: Amara, *id.*
Note. Spreading Hedge Parsley (Eng., Johns); LIZZAIJ ("clingweed", Ir.-Baghdad, Khalis, *Guest* 2458, *Lazar* 3414); GHARAIR (Ir.-Khalis, *Guest* 2458) and ZAND AL-'ARŪS (Ir.-Khalis, *Guest* 2458); both these names being used in Lower Iraq for a number of different species of umbellifers. Handel-Mazzetti noted two Arabic names for this species at Aleppo in Syria: SHAIMRA (*Hakim* 42) and KUZBARA BARRIYA ("wild coriander", *Hakim* 68).

C, S & Mediterranean Europe, Cyprus, Ukraine, Syria, Lebanon, Palestine, Jordan, NE Egypt, Turkey, Caucasus, Iran, Macaronesia, N Africa (Morocco, Algeria, Tunisia).

3. **Torilis chrysocarpa** *Boiss. et Bl.*, Diagn. ser. 2, 2: 98 (1856); Mouterde, Nouv. Fl. Lib. et Syr. 2: 592 (1970); Peev in Rech. f., Fl. Iranica 162: 126 (1987).

Caucalis fallax Boiss. et Bl., Boiss., Fl. Orient. 2: 1086 (1872); Zohary in Dep. Agr. Iraq Bull. 14: 114 (1950); Rawi, ibid. Tech. Bull. 4: 90 (1964).
C. chrysocarpa (Boiss. et Bl.) Bornmüller, Verh. Zool.-bot. Ges. Wien 48: 594 (1898); Zohary, Fl. Palest. ed. 2, 1: 572 (1932).
Torilis leucorhaphis Rech. f. et Riedl, Anz. Öst. Akad. Wiss., Math.-Nat. Kl. 1961 (13): 223 (1961). Type: Iraq, between Dohuk and Amadiya, above Sarsang, *Rechinger* 11646 (W, holo.).

Slender, erect annual, 20–30 cm. Stem terete, finely striate, thinly furnished with downwardly appressed hairs, branched from the middle or below. Leaves bipinnate, deltoid-ovate in outline, segments deeply incised with acute, 3–7 × 1–2 mm lanceolate teeth, lower with a 3–5 cm petiole, uppermost trisect with a long terminal segment and finally much reduced; sheaths narrow, white-margined. Umbels rather few, terminal and leaf-opposed, peduncles slender, 1.5–4 cm; rays 2 or 3(–4), subequal, 6–13 mm, thinly hairy. Involucre absent. Partial umbels 3–6-flowered, pedicels short, 1–3 mm. Involucel of 3–5 bracteoles, somewhat longer than pedicels. Flowers white, small. Fruit ovoid, 3–4 × 1.5 mm; calyx teeth very small, secondary ridges very densely furnished with 1 or 2 rows of subulate, whitish, glochidiate spines; mericarps frequently heterocarpous, one bearing long fine spines longer than the diameter of fruit, the other with the spines shorter. Primary ridges with 1 or 2 rows of longitudinally appressed hairs. Stigmas subsessile, style and stigma not longer than stylopodia.

Hab. In mountains, in thicket and on shady banks between irrigated gardens; alt. 1000–1200 m; fr. Jun.
Distrib. Very rare in Iraq, found in the middle forest zone. **MAM**: between Dohuk and Amadiya, above Sarsang, *Rechinger* 11646 (type of *T. leucoraphis*); Gara Dagh, *Dabbagh & Jasim* 46958!; **MRO**: Shaqlawa, *Gillett* 11569A!; ? **MSU**: (or in Iran?), Hawraman, *Haussknecht* s.n.
Note. This plant requires further investigation. I cannot reduce it totally to the synonymy of *T. arvensis* as was done in the Flora of Turkey, but it is certainly very close to that species. The few-rayed umbels and very short styles (combined styles and stigmas not exceeding the stylopodia) appear to distinguish it. The spines are also longer in proportion to the width of the ripe mericarps.

Syria, Lebanon, Iran.

4. **Torilis leptophylla** (*L.*) *Reichb. f.*, Ic. Fl. Germ. 21: 83, t. 169 (1867); Hayek, Prodr. Fl. Balc. 1: 1056 (1927); Schischkin in Fl. SSSR 16: 163 (1950); H. Riedl in Fl. Lowland Iraq: 456 (1964); Grossh., Fl. Kavk. ed. 2, 7: 43 (1967); Mouterde, Nouv. Fl. Lib. et Syr. 2: 593 (1970); Cullen in Fl. Turk. 4: 522 (1972); Zohary, Fl. Palaest. 2: 396 (1972); E. Nasir in Fl. Pak. 20: 108, f. 32 (1972); V. Täckh., Stud. Fl. Egypt ed. 2, 2: 397 (1974); Meikle, Fl. Cyprus 1: 702 (1977); Peev in Fl. Iranica 162: 127 (1987); Boulos, Fl. Egypt 2: 186 (2000).

Caucalis leptophylla L., Sp. Pl. ed. 1: 242 (1753); DC., Prodr. 4: 216 (1830); Boiss., Fl. Orient. 1: 1084 (1872); Nábělek in Publ. Fac. Sci. Univ. Masaryk 35: 129 (1923); Zohary, Fl. Pal. ed. 2, 1: 571 (1932); Guest in Dep. Agr. Iraq Bull. 27: 20 (1933); Anth. in Notes Roy. Bot. Gard. Edinb. 18: 289 (1935); Bornmüller in Beih. Bot. Centralbl. 58B: 288 (1938); Blakelock in Kew Bull. 3: 433 (1948); Zohary in Dep. Agr. Iraq Bull. 31: 114 (1950); Rawi, ibid. Tech. Bull. 14: 90 (1964).

Erect or decumbent-erect annual, (5–)15–40 cm. Stems very thinly furnished with downwardly appressed strigulose hairs, terete, finely striate, with few to numerous branches from the base upwards. Leaves all similar, bi- to subtri-pinnate, the ultimate segments short and narrow, 1.5–4 × 1 mm, upwardly appressed-hairy; lower leaves with a 2–5 cm petiole, lamina elliptic- or deltoid-oblong in outline, the upper more deltoid and sessile; all with rather long, expanded, white-margined sheaths. Umbels many, leaf-opposed, the peduncle about equalling or shorter than, rarely exceeding, the subtended leaf, 1.5–3(–8) mm; rays 2–3, somewhat unequal, (2–)4–15 mm, thinly appressed-hairy. Involucre absent. Partial umbels 6–8–flowered, the pedicels short, (0–)1–4 mm. Involucel of ± 6 lanceolate-ovate bracteoles, rather shorter than the pedicels. Flowers white or pinkish, small. Fruit 4–6 × 1.5–2 mm, linear-oblong, surmounted by the persistent, rather conspicuous calyx teeth, the secondary ridges furnished with long (longer than the diameter of the mericarps) whitish to purplish glochidiate spines. Stigma sessile or style very short. There is a single sessile fruit at the centre of each partial umbel which usually has considerably shorter spines than the remainder and is clearly visible owing to the length of the pedicels of the outer fruits. Fig. 104: 3–5.

HAB. On the mountains in grassy patches among scrub *Quercus*, on limestone crags in *Pinus* woods, stony hillsides, in valleys, in a shady ravine, on limestone, on stony clay hills, dry steppic hilltops on the plains, also as a weed in a field; alt. (50–)400–700(–1400) m; fl. & fr. Apr.–Jun.
DISTRIB. Very common, locally abundant, in the lower forest zone of Iraq, occasional in the steppe region and on the irrigated alluvial plain in the desert region. **MJS**: Kursi, *Gillett* 10967; Zakho valley, *Guest* 2253!; **MAM**: between Zakho and Sharanish, *Karim, Hamid & M. Jasim* 41132!; Sharifa, nr Amadiya, *O. Polunin* 5138!; Dohuk, *Makki* (*as-Sidqi ash-Sharbati*) 3247!; Zawita, *Guest* 4763!; Sarsang, *Haines* W1012!; Aqra, *Rawi* 11339!, 11462!; Jabal Khantur, *Rechinger* 10837; **MRO**: Khanzad, *Karim, Hamza & M. Jasim* 40864!; Kuh-i Sefin, nr Shaqlawa, *Bornmüller* 1288!; Shaqlawa, *Gillett* 11298!, *Haines* W650!; Rowanduz gorge, Guest 2117!; Serkabkhan, *Rawi, Nuri & Kass* 2824!; Dera, Nura and Plingan, *id.* 28590!, 28688!; Gali Warta, *id.* 28801!; **MSU**: between Qaranjir and Chemchamal, *Gillett & Rawi* 7564!; Shewasur valley, *Omar, Karim, Hamza & Hamid* 37101!; 10 km E of Chemchamal, *Alkas* 18616!; Tainan (Tainal), *Gillett & Rawi* 11645!; 55 km NW of Sulaimaniya along Dukan highway, *Rawi* 21798!; Qara Dagh, nr Timara on Qara Dagh, *Karim, Hamid & M. Jasim* 40576!; Tawila, *Rawi* 22278!; **FUJ**: Tal Kochak, *Gillett* 10815!; Jabal Qilat, *Field & Lazar* 420!; **FUJ/FNI**: Mosul, *Aucher* 3655!, *Rawi* 5665!; **FKI**: Altun Kupri, *Bornmüller* 1289; Kani Domlan hills, *Guest* 4313!, 4314!; **FKI/MSU**: Shuwan, 31 km N of Kirkuk, on road to Koi Sanjaq, *Rawi, Nuri & Kass* 27979!; **DLJ**: Baghdad, *Anon.* 5457; **LEA**: 17 km NE of Ba'quba, *Cowan & Darlington* 175!; **LSM**: Amara, in Anth., l.c.
NOTE. LIZZAIJ ("clingweed", Ir.-Tal Kochak, *Gillett* 10815); ?SIVÎR HŌR (Kurd.-Dohuk, *Makki Beg* 3247); ?NUSAKA (Kurd.-Aqra, *Rawi* 11339).

W & Mediterranean Europe, Crete, Aegean Isles, Cyprus, Syria, Lebanon, Palestine, Jordan, N & NE Egypt, Turkey, Caucasus, Iran, Afghanistan, Pakistan, C Asia (Turkmenistan to Tian Shan), India, China, N Africa (Libya), Macaronesia.

5. **Torilis stocksiana** (*Boiss.*) *Drude*, Pflanzenfam. 3(8): 156 (1898); Schischkin in Fl. SSSR 16: 164 (1950); H. Riedl in Fl. Lowland Iraq: 456 (1964); Peev in Fl. Iranica 162: 129 (1987); Ghazanfar, Fl. Oman Vol. 2.

Caucalis stocksiana Boiss., Diagn. ser. 2, 6: 89 (1859); Boiss., Fl. Orient. 2: 1086 (1872); Bornmüller in Beih. Bot. Centralbl. 58B: 288 (1938); Blakelock in Kew Bull. 3: 433 (1948); Zohary in Dep. Agr. Iraq Bull. 31: 114 (1950); Rawi, ibid. Tech. Bull. 14: 90 (1964).

Erect or decumbent-erect annual, 15–40 cm. Stem terete, finely striate, thinly furnished with downwardly appressed strigulose hairs, branched from the base upwards. Leaves all similar, appressed-hairy, bi- to subtri-pinnate, the ultimate divisions short, subacute, linear-lanceolate, 2.5–6 × 0.75–1.5 mm; lower leaves with a 2–8 cm petiole, lamina elliptic- or deltoid-oblong in outline, the upper short, more pronouncedly deltoid, sessile; all with rather long, expanded, white-margined sheaths. Umbels many, leaf-opposed, the peduncles shorter than the subtending leaves, 0.8–6 cm; rays 2–3, thinly appressed-hairy, short, 0–12 mm, very unequal, the umbel appearing very dense, one partial umbel being generally subsessile. Involucre absent. Partial umbel 6–8-flowered, the pedicels short (to 2 mm, but several fruit often sessile). Involucel of about 6 linear bracteoles, exceeding the pedicels. Flowers white or pinkish, small. Fruit 3.5–4 × 1.5–1.7 mm, oblong, surmounted by the short, pale calyx teeth, the secondary ridges furnished with long (longer than the diameter of the mericarps) whitish or stramineous glochidiate spines. Styles short, divaricate, rather longer than the stylopodia. Endosperm more bluntly angled than in *T. leptophylla*.

HAB. in the lower mountains by the edge of streams and on grassy river banks, in shady irrigated gardens, on the plains in shady hollows in fields, along ditches and irrigation channels, in orchards and date gardens, on damp waste ground; alt. 570–1400 m; fl. & fr. May–Jun. in the mountains, Mar.–Apr. (–May) on the plains.

DISTRIB. Occasional in the lower forest zone and steppe region of Iraq, quite common on the irrigated alluvial plain in the desert region. **MRO**: Shaqlawa, *Gillett & Rawi* 11569B!, 11571!; Koi Sanjaq, *Rawi, Nuri & Kass* 28100!; **MSU**: Jarmo, *Haines* W280!; Pira Magrun, *Haussknecht* s.n.; Malakawa, *Rechinger* 12289; **MSU/FNI**: 30 km S of Darband-i Khan, *Barkley* 7359!; **FUJ**: Makhlat, *Guest* 4281!; **FKI**: Kirkuk, *Rogers* 0181!; **FNI**: Qizil Robat (Sa'diya), *Guest* 1962!; between Naft Khana and Mandali, *Rawi* 12683!; **DWD**: T1, *Omar & Hamid* 36649!; **LCA**: Baghdad, *Noë* s.n. , *H.J. Paranjpye* s.n. comm. *Graham, Bornmüller* 365 , *Rogers* 0421!, *Guest* 1951!; Abu Ghraib, *Alizzi & Omar* 34269!; Rustamiya, *Lazar* 193!, 508!, 3895!; Hilla, *Bornmüller* 366; **LEA**: Abu Saldi Jizera (Abu Jisra), *Barkley & Haddad* 7318!; 6 km N of Ba'quba, *Gillett & Rawi* 7283!; Ba'quba, *Hunting Technical Services* 1A! 1B!; Kut al-Imara, *Gillett* 6579!; **LSM**: Amara, *Alizzi & Omar* 34931!; nr Majarr, *id.* 34439!; 18 km S of Amara, *Botany Staff* 4248!; Qala't Salih, *id.* 42471; nr Azair, *Guest, Rawi & Rechinger* 17669!; **LBA**: Basra, *Bornmüller* 367, *Haines* W140!; Abul Khasib, *Rawi & Haddad* 25956!

NOTE. Quite a common weed in gardens and fields; LIZZAIJ ("clingweed"), Ir.-Baghdad/Ba'quba, *Lazar* 3895/*Gillett & Rawi* 7283).

Although fruit spines' colour in many umbellifer varies from white to purplish, spines in this species appear to be constantly pale.

Arabia, Iran, Afghanistan, Pakistan.

6. **Torilis tenella** (*Del.*) *Reichb.*, Ic. Fl. Germ. 21: 84 (1867); Schischkin in Fl. SSSR 16: 161 (1950); H. Riedl in Fl. Lowland Iraq: 455 (1964); Grossh., Fl. Kavk. ed. 2, 7: 43 (1967); Mouterde, Nouv. Fl. Lib. et Syr. 2: 593 (1970); Cullen in Fl. Turk. 4: 522 (1972); V. Täckh., Stud. Fl. Egypt ed. 2: 397 (1974); Meikle, Fl. Cyprus 1: 703 (1977); Peev in Fl. Iranica 162: 128 (1987); Boulos, Fl. Egypt 2: 186 (2000).

Caucalis tenella Del., Fl. Aeg.: 202, t. 21 (1812); DC., Prodr. 4: 216 (1830); Boiss., Fl. Orient. 2: 1084 (1872); Hand.-Mazz. in Ann. Naturh. Mus. Wien 27: 89 (1913); Nábělek in Publ. Fac. Sci. Univ. Masaryk 35: 129 (1923); Hayek, Prod. Fl. Balc. 1: 1054 (1927); Zohary, Fl. Pal. ed. 2, 1: 571 (1932); Guest in Dep. Agr. Iraq Bull. 27: 20 (1933); Anth. in Notes Roy. Bot. Gard. Edinb. 18: 289 (1935); Bornmüller in Beih. Bot. Centralbl. 58B: 288 (1938); Blakelock in Kew Bull. 3: 433 (1948); Zohary in Dep. Agr. Iraq Bull. 31: 114 (1950); Rawi, ibid. Tech. Bull. 14: 90 (1964).

Slender, erect annual, 10–40 cm. Stem terete, thinly furnished with downwardly-appressed hairs, finely striate, branched from near the base upwards to simple. Leaves deltoid-oblong in outline, 2–3–pinnate with the segments deeply cut into moderately long (to 5 mm), very narrow, acute ultimate divisions; lower leaves with a petiole to 5 cm long, the upper sessile, all thinly appressed-hairy, the sheaths long, narrowly inflated, broadly membranous-margined. Umbels few to several, leaf-opposed, peduncles longer than the subtending leaves, 2–8(–11) cm; rays 5–10, very unequal, 0.4–2.5(–4) cm, sparsely hairy, ± contracted in fruit. Involucre absent. Partial umbels 6–12-flowered; pedicels unequal, short, 1.5–4 mm, at least equalled by the 4–8 subulate, herbaceous bracteoles. Flowers white, small. Fruit linear, 3.5–4 × 0.7 mm, surmounted by the conspicuous, narrowly lanceolate calyx teeth; secondary ridges with a single row of 6–8 slender, greenish or yellowish glochidiate spines 3 × as long as the width of the fruit. Primary ridges with a single row of longitudinally appressed hairs. Styles very short, divergent.

HAB. On a silty plain in the lower mountains, on a rocky limestone slope, mainly in fields on the steppe; alt. (50–)250–900(–1300) m; fl. & fr. (Mar.–)Apr.–May.

DISTRIB. Occasional in the lower forest zone and steppe region of Iraq, rare on the alluvial plain in the desert region. **MJS**: above Balad Sinjar, *Handel-Mazzetti* 1494: **MRO**: Sardariya pass, between Arbil and Rowanduz, *Nábělek* 431; 6 km SW of Rania, *Rawi, Nuri & Kass* 28436!; **FUJ/MJS**: just S of Balad Sinjar, *Barkley* 7947!; **FUJ**: Tal Kochek, *Gillett* 10839!; **FUJ/ FNI**: by R. Tigris below Mosul, *Handel-Mazzetti* 1316; **FNI**: Nineveh, *Guest* 1344!, 1344A!; Nimrud, *Helbaek* 910!; **FAR**: Arbil, *Haines* W1409!; **FKI**: Kirkuk, *Bornmüller* 1287; **FKI/FNI**: Jabal Hamrin, in Anth., l.c.; **FNI**: Badra, *Gillett* 6672!, *Rechinger* 9173a; **DGA**: 13 km E of Samarra, *Rawi* 20366!; **LCA**: nr Daltawa (Khalis), *Rogers* 0286!

SE Europe (Greece), Cyprus, Syria, Lebanon, Palestine, Jordan, Egypt, Turkey (S & E Anatolia), Caucasus, Iran, N Africa (Libya).

7. **Torilis leptocarpa** (*Hochst.*) *C.C. Townsend* in Kew Bull. 17: 434 (1964); Cullen in 4: 523 (1972); Peev in Fl. Iranica 162: 128 (1987).

Daucus leptocarpus Hochst. in Lorent., Wanderungen: 337 (1845); Boiss., Fl. Orient. 2: 1073 (1872); Bornmüller in Beih. Bot. Centralbl. 58B: 287 (1938).

Torilis sintenisii Freyn, Öst. Bot. Zeitschr. 44: 144 (1894.)

Astrodaucus leptocarpus (Hochst.) Drude in Pflanzenfam. 3(8): 157 (1898); Mouterde, Nouv. Fl. Lib. et Syr. 2: 595 (1970).
T. stenocarpa C.C. Townsend, Kew Bull. 17: 435 (1964).

Erect annual, 30–55 cm. Stem terete, finely striate, very thinly furnished with downwardly strigulose hairs, with long ascending branches from near the base or above only. Leaves bi- to subtri-pinnate, deeply cut into short, narrowly lanceolate, 2–4 × 0.6–1 mm subacute segments, thinly appressed-hairy; lower leaves with a petiole to 4 cm long, lamina deltoid-oblong, the upper more pronouncedly deltoid; sheaths narrow, with narrow whitish margins. Umbels few to many, leaf-opposed, on 3–12(–15) cm peduncles longer than the subtending leaf or terminal, with (3–)6–9(–12) thinly appressed-hairy, subequal, 1.5–3(–3.5) cm rays. Involucre absent or of a single linear bract. Partial umbels 8–15-flowered, pedicels unequal, mostly 2–6 mm, some as long as the fruit but central fruit ± sessile with shorter spines than the remainder. Involucel of about 8 subulate bracteoles, shorter than or subequalling the outer pedicels. Flowers white, petals persistent. Fruit linear, ± 6 × 1 mm, calyx teeth small, secondary ridges with ± 12 whitish glochidiate spines 2 × as long as the diameter of mericarps and uniseriate or irregularly biseriate. Primary ridges with 1 or 2(–3) rows of longitudinally appressed hairs. Styles ± 1.5 mm, slender, divaricate.

HAB. On a disturbed hill slope, lately quarried, in *Quercus* woodland, in a low mountain valley and on steppic plain, on irrigated land; alt. (± 5–) 400–900 m; fl. & fr. June.
DISTRIB. Rather rare in Iraq; found in the lower forest zone, moist steppe zone and on alluvial plain in the desert region. **MRO**: Rowanduz, *Hadač & Kader* 5291; Kani Mazu Shirin, *Hadač et al.* 5829; **MSU**: Darband-i Bazian, *Haussknecht* s.n.!; B, *Agnew & Haines* W2170!; Qara Anjir, *Hadač et al.* 5264; **FAR**: Ankawa, nr Arbil, *Bornmüller* 1286!; **LCA**: Baghdad, *Noë* s.n.
NOTE. Although the type of *T. stenocarpa* has more regularly biseriate spines than any other specimen, I no longer feel that it can be maintained as a distinct species.

Syria, Turkey (E Anatolia), N Iran.

66. CHAETOSCIADIUM Boiss.

Boiss., Fl. Orient. 2: 1078 (1872)

Annual herb with 2- or 3-pinnately dissected leaves. Flowers usually hermaphrodite, white or somewhat pink-tinged. Petals small, obovate, emarginate above with an inflexed lobule, outer larger but scarcely conspicuously radiant. Calyx teeth subulate, ciliate, persistently prominent in ripe fruit. Involucre absent, involucel present. Fruit linear-oblong, somewhat dorsally compressed, narrowed above into a short cylindrical column formed by the persistent stylopodia, covered throughout with dense, long, silky hairs; ribs obsolete. Styles long. Dorsal vittae 4, obscure, 2 on the commissural face. Commissural face of endosperm concave.

A monotypic genus.
Chaetosciadium from the Gr. χαίτη, *khaite*, long hair or mane, σκιαδειον, *skiadeion*, an umbrella or parasol, from the silky hairs of the fruit.

1. **Chaetosciadium trichospermum** (*L.*) *Boiss.*, Fl. Orient. 2: 1078 (1872); Zohary, Fl. Palest. ed. 2, 1: 567 (1932); Mouterde, Nouv. Fl. Lib. et Syr. 2: 599 (1970); Peev in Fl. Iranica 162: 142 (1987).

Scandix trichospermum L., Mant.: 57 (1767).
Chaerophyllum trichospermum (L.) Lam., Encycl. 1: 685 (1785).
Torilis trichosperma (L.) Spreng., Sp. Umb.: 142 (1818); DC., Prodr. 4: 220 (1830).
Caucalis trichosperma (L.) Del. in Steud., Nom. Bot. 1: 169 (1821).

Slender annual, 5–50 cm. Stem wiry, slender, terete, finely striate, dichotomously branched from near base, branches long and ascending. Lower leaves deltoid-oblong in outline, long-petiolate with petioles 2.5–6 cm including the long, tapering sheaths, bi-pinnate with the secondary divisions further pinnatifid into short, acute, lanceolate segments 2–4 × 1–2 mm; upper leaves bipinnatisect, sessile on short, wide, broadly scarious-margined sheaths (foliage generally resembling that of a *Torilis*). Umbels several to many, peduncles axillary and terminal, 1.5–6 cm, slender; rays 3–6, slender, short, 3.5–12 mm;

Fig. 105. **Chaetosciadium trichospermum**. 1, habit × 2/3; 2, petals (from the same flower) × 14; 3, fruit × 6; 4, fruit cross section × 60; All fom *Lebanon* 5393. Drawn by D. Erasmus (1964).

partial umbels dense, 6–10-flowered, pedicels 2–3 mm. Involucel of about 8 subulate bracteoles equalling or exceeding pedicels in length. Fruit 3–4 × 0.7 mm, densely covered with white to purplish patent-ascending silky hairs which are often barbellate when young, becoming smooth with age and up to 5 mm long. Styles long and slender, 1–1.5 mm, straight, somewhat divergent. Fig. 105: 1–4.

HAB. On limestone cliffs in dry steppe and stony or rocky hill slopes; alt. 100–160 m; fl. & fr. Mar.–Apr. DISTRIB. Very rare in Iraq; found mainly in one area in the desert region. **MRO**: Diana nr Rowanduz, *Cuckney* 3834; **DLJ**: Rawa, *Rawi & Gillett* 6986!; **DWD**: 8 km E of Ana, *Omar, Qaisi, Hamad & Hamid* 45113!; Jabal Ana, *Khayat & Hamad* 51748!

Syria, Lebanon, Palestine, Jordan.

SPECIES EXPECTED TO OCCUR IN IRAQ

HYDROCOTYLE (now Araliaceae)

Hydrocotyle ranunculoides *L.f.*, which occurs in exclusively aquatic habitats and is thus under-collected, has been recorded from Syria, Palestine, Arabia (Yemen), Iran and Caucasus. It is thus quite likely to occur in Iraq, but no material has so far been collected.

PETROSELINUM (Apiaceae)

According to Parsa (1960), "Qualitas Plantarum (Plant foods for human nutrition)" vol. 7, pp. 65–136, **Petroselinum crispum** (*Mill.*) *Nyman ex Hill* is cultivated in Iraq where its local name is ?MAGHDUNES (p. 88). No specimens have been seen.

SESELI (Apiaceae)

Seseli libanotis *Koch* (syn. *Libanotis montana* Crantz) has been recorded from the extreme SE corner of Turkey, close to the Iraqi border, and can be expected to occur in Iraq; no material has been seen. The distribution of *S. peucedanoides* (Bieb.) Koso-Pol., as shown in Grossheim, Fl. Kavkaza ed. 2, 7: map 100 (1967) also suggests that it might occur in NE Iraq, but no material has been seen.

112. **VALERIANACEAE**

Batsch in Tab. Affin. Regni Veg. 227 (1802)

Emil Hadač[22]

Annual or perennial herbs, rarely suffruticose or shrubby, sometimes with scented rhizomes or tubers, dichasially branched. Leaves exstipulate, opposite or radical, pinnatifid or entire. Flowers in dense cymose inflorescences, hermaphrodite (in ours) or unisexual, zygomorphic or sometimes nearly regular. Calyx annular or toothed (*Valerianella*), often forming in fruit a feathery pappus (*Valeriana, Kentranthus*). Corolla funnel-shaped, mostly with 5 imbricate lobes, sometimes with a long spur. Stamens 1–3, rarely 4, inserted in the corolla tube. Ovary 3-celled, two sterile cells of a different size and shape (smaller or bigger) than the fertile one. Fruit an achene or nut. Seeds without endosperm.

About 16 genera and 400 species, generally distributed except for S Africa, Australia and Polynesia. Three genera in Iraq. *Valeriana* and *Kentranthus* are very rare and found in northern Iraq in the mountains; *Valerianella* is more common.

Under the APG III (2009) classification, Caprifoliaceae is expanded to include Valerianaceae, together with families Diervillaceae, Dipsacaceae, Linnaeaceae and Morinaceae.

1. Annuals; calyx limb of specialised shape or nearly lacking, but never modified into pappus .... 3. **Valerianella**
   Perennials; fruits with a feathery pappus 2
2. Plants woody at base; corolla with a long spur; stamen 1 .... 1. **Kentranthus**
   Rhizomatous herbs; corolla saccate but not spurred; stamens 3 .... 2. **Valeriana**

[22] Institute of Landscape Ecology, Czechoslovakia Academy of Sciences.

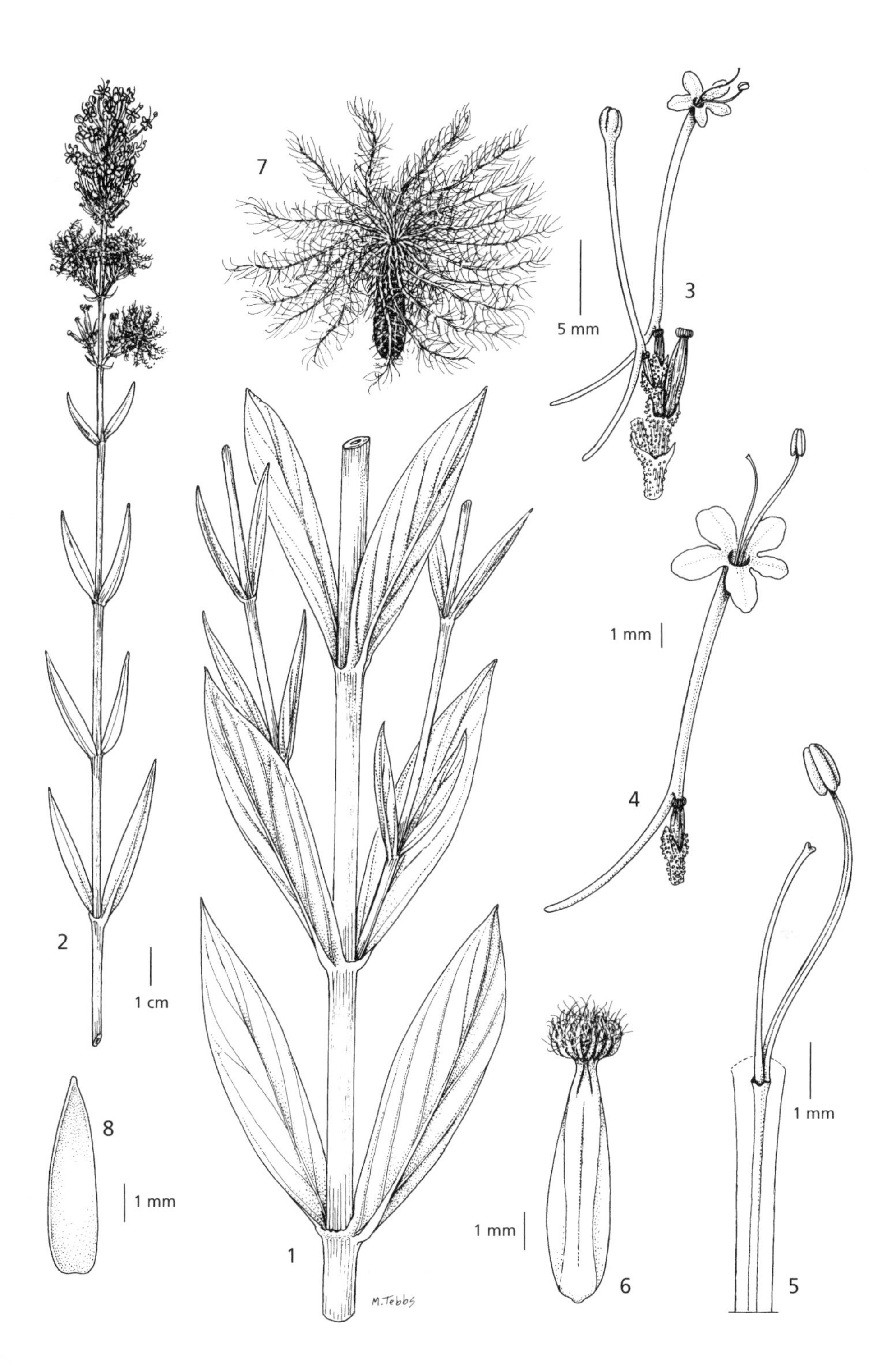

Fig. 106. **Kentranthus longiflorus**. 1, habit; 2, inflorescence; 3, flowers; 4, single flower; 5, corolla tube diswsected to show stamen and style; 6, fruit; 7, fruiting calyx with feathery pappus; 8, seed. All from *Rawi & Serhang* 18308. © Drawn by M. Tebbs (Aug. 2012).

1. **KENTRANTHUS** Neck.

Elem. 1: 122 (1790)
Centranthus DC., Fl. Fr. ed. 3, 4: 238 (1805)

Perennials (or annuals, but not in Iraq), glabrous, shrubby, woody at base. Leaves simple. Flowers in terminal paniculate cymes, pink. Calyx 6–15, segmenst linear, involute during flowering, in fruit forming a feathery pappus. Corolla 5-lobed, tube gibbous near middle, spured at base. Stamen 1. Achene compressed doriventrally, with 3- and 1-veined faces, feathery.

About 12 species distributed in the Mediterranean region; one species in Iraq.
Kentranthus (from Lat. *centrum*, centre and *-anthus*, flower).

1. **Kentranthus longiflorus** (*Stev.*) *Grossh.*, Fl. Kavk. 4: 43 (1934); Fl. SSSR 23: 640 (1958); Hadač in Bull. Coll. Sci. 6:30 (1961); Rech.f., Fl. Iranica 62: 22 (1969).

Centranthus longiflorus Stev., Nouv. Mém. Soc. Nat. Mosc. 1(7): 272 (1829); Boiss., Fl. Orient. 3: 92 (1875); Nábělek in Pub. Fac. Sci. Univ. Masaryk 35: 137 (1928); Zohary, Fl. Palest. ed. 2, 1: 605 (1932); Richardson in Fl. Turk. 4: 558 (1972).

Rhizomatous herb with stems branching, up to 100 cm. Lower leaves oblong, upper lanceolate or linear-lanceolate, ± 7 × 3 cm, entire, obtuse, glaucescent. Corolla pink, tube ± 20 mm; spur of corolla 6–10 mm long. Fig. 106: 1–8.

HAB. On cliffs, near streams; alt.: 1100–2300 m.; fl. Jul., Aug.
DISTRIB. Very rare and local. **MRO**: NE of Rania, *Rawi & Serhang* 18308!; Bisht Ashan mt., *Rechinger* 11189; Halgurd Dagh, *Rawi* 13736!
NOTE. The Iraqi material stands near to the var. *latifolius* in Boiss, Fl. Orient. 3: 92 (1875), with less acute leaves.

Syria, Lebanon, Turkey, Caucasus.

2. **VALERIANA** L.

Sp. Pl. ed. 1: 31 (1753)

Perennial rhizomatous herbs, glabrous or pubescent. Stems simple. Leaves opposite, entire or pinnate. Inflorescence corymbose. Flowers small, pink or whitish. Calyx 6–15, segments linear, involute during flowering, in fruit forming a feathery pappus. Corolla 5-lobed, funnel-shaped, tube gibbous at base. Stamens 3. Achene compressed dorsiventrally, with 3- and 1-veined faces, feathery.

About 250 species in the Northern Hemisphere and in the Cordilleras, also in the S Hemisphere. Two species in Iraq, both rather rare. Zohary (Dep. Agr. Iraq Bull. 31: 140 (1950) names in his list also *V. dioscoridis* Sm. without any locality; no material is available from Iraq, so this is a doubtful record.
Valeriana (based on the personal name *Valerianus* via Old Fr. *valeriane*).

Leaves entire; plant stout, 50–60 cm . . . . 1. *V. alliariifolia*
Leaves pinnate; plant smaller, 20–25 cm. . . . . 2. *V. sisymbriifolia*

1. **Valeriana alliariifolia** *Adams* in Weber & Mohr, Beitr. Naturk. 1: 44 (1805); Kom. in Fl. SSSR 23: 602 (1958); Hadač in Bull. Coll. Sci. 6: 30 (1961); Rech.f., Fl. Iranica 62: 18 (1969); Richardson in Fl. Turk. 4: 552 (1972).

V. alliariaefolia Vahl, Enum. 2: 11 (1805); Boiss., Fl. Orient. 3: 85 (1875); Náb. in Bull. Fac. Sci. Univ. Masaryk 35: 137 (1928); Blakelock in Kew Bull. 3: 441 (1948).

Perennial herb, 50–60 cm tall, stout. Leaves all simple; lower leaves about 12 cm broad and long, cordate at base, acuminate, emarginate-dentate; petioles 5–6 cm; upper leaves smaller and with shorter petioles, uppermost nearly sessile. Bracts ovate to lanceolate, glabrous to pilose. Inflorescence relatively dense. Flowers pink, about 5 mm long. Achene glabrous, 3–4 mm long.

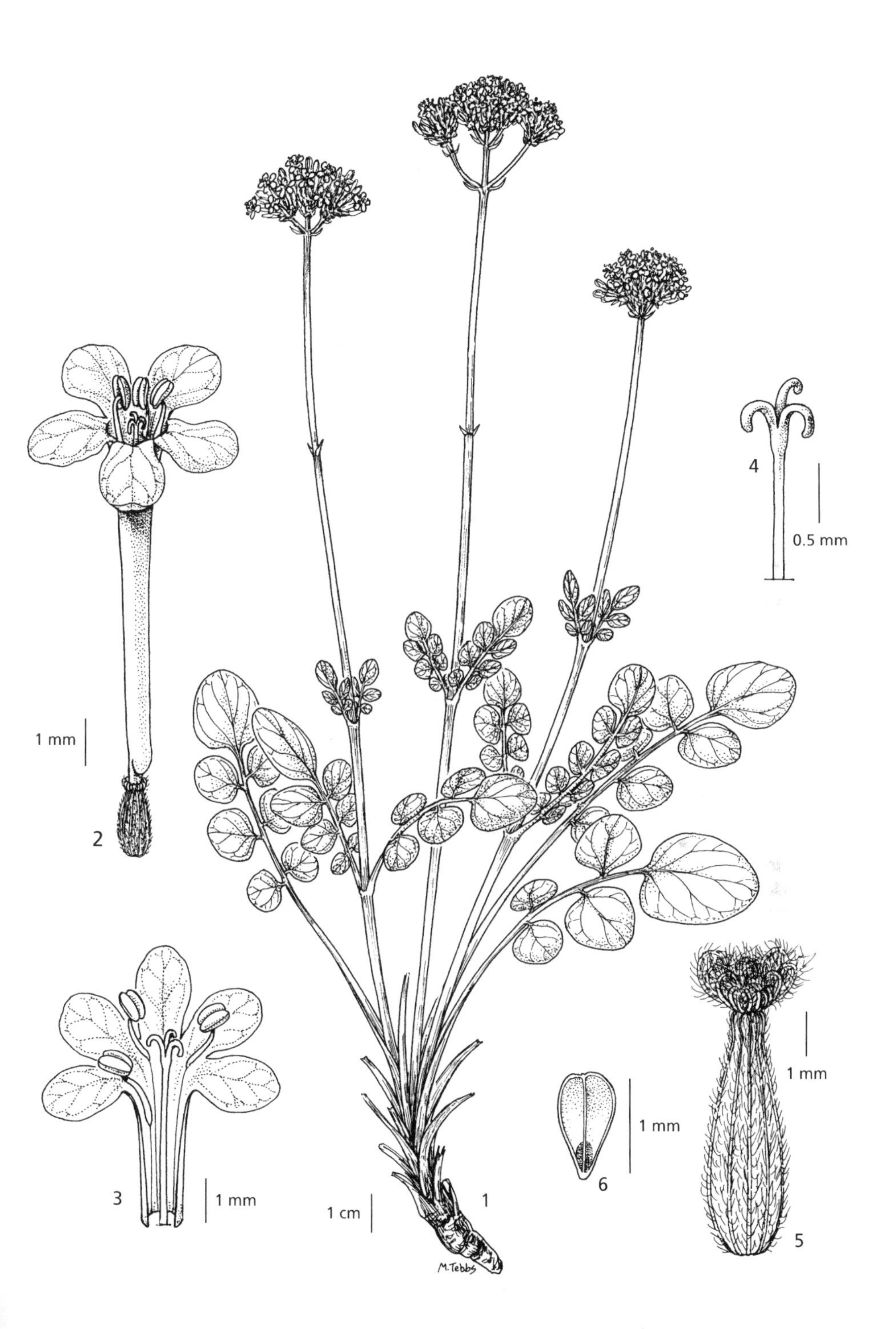

Fig. 107. **Valeriana sisymbriifolia**. 1, habit; 2, flower; 3, corolla opened, top portion; 4, style and stigma; 5, fruit; 6 seed. 1 from *Rawi* 24472; 2–4 from *Guest* 2707; 5, 6 from *Rawi* 18258. © Drawn by M. Tebbs (Aug. 2012).

HAB. In high mountains by springs and brooklets; alt.: 1600–3300 m; fl. Jun.–Jul.
DISTRIB. **MRO**: Helgurd range, *Rawi & Serhang* 24552!; Nawanda, *Agnew & Hadač*; ? Hawara Blinda mt., *Rawi, Nuri & Kass* 27859!; Hassar-i-Sakran, *Hadač & Abd'el Kader* 5535; Ser Kurawa, *Gillett* s.n.!

Turkey, Caucasus, Iran, C Asia.

2. **Valeriana sisymbriifolia** *Vahl*, Enum. Pl. 2: 7 (1805); Boiss., Fl. Orient. 3: 88 (1875); Hand.-Mazz. in Ann. Nat. Hofmus. Wien 27: 429 (1913); Nábělek in Bull. Fac. Sci. Univ. Masaryk 3: 137 (1928); Zohary, Fl. Palest. ed. 2, 1: 604 (1932); Zohary in Dep. Agr. Iraq Bull. 31: 140 (1950); Blakelock in Kew Bull. 3: 441 (1948); Komarov in Fl. SSSR 23: 617 (1958); Hadač in Bull. Coll. Sci. 6: 30 (1961); Rech.f., Fl. Iranica 62: 18 (1969); Richardson in Fl. Turk. 4: 556 (1972).

Perennial herb 20–25 cm tall; rhizome cylindrical; base of stem covered with old leaf bases. Leaves pinnatisect with 2–4 pairs of ovate to ovate-orbicular, entire, shortly petiolate or subsessile leaflets, terminal leaflets larger than laterals, glabrous. Lower stem leaves similar to basal leaves with amplexicaul petioles. Inflorescence dense, ± 2 cm in diameter, spreading in fruit. Corolla pale pink or whitish, 6–8 mm long. Achene ± densely hairy, 3.5–5 mm long. Fig. 107: 1–6.

HAB. On cliffs of northern or western aspect; alt.: 2200–3500 m; fl. Jun.–Aug.
DISTRIB. Occasional in mountainous districts of NE Iraq. **MRO**: Chiya-i Mandau, *Guest* 2707, 2797; Helgurd Dagh, nr. Bermasand lake, *Rawi & Serhang* 24798!; Helgurd Dagh, *Rechinger* 11426, *Ludlow-Hewitt* s.n.; Qandil range, *Rawi* 24472!, *Rechinger* 11173; Sakri-Sakran, *Bornmüller* 1328; Hassar-i (Sakri) Sakran, *Hadač* & *Abd'el Kader* 5651!; Ser Kurawa, *Gillett* s.n.; **MSU**: E of Qala Diza, *Thesiger* 1143; Mt. Hawraman, *Haussknecht* s.n.

Turkey, Caucasus, N, W & E Iran, C Asia.

3. **VALERIANELLA** Mill.

Gard. Dict., abr. ed. 4 (1754)

Annual herbs, erect, glabrous or hairy, dichasially branched. Leaves opposite, basal entire or sinuate-dentate ± sessile, upper leaves with few distinct teeth at base. Flowers small, in head-like cymes, sometimes also solitary in forks of branches, white or pink. Calyx limb sometimes lacking or nearly so, often dentate or crown-like but never with feathery pappus. Corolla funnel-shaped, without spur, 5-lobed. Stamens 3. Fruit an achene, 3-locular, with one fertile and two sterile cells.

For accurate identification, fruiting material is necessary. The most reliable characters (observed through lens) are the shape of the calyx limb and the transverse section of fruits.
About 80 species, mainly in the Mediterranean region. Fifteen species in Iraq, distributed mainly in the northern part of the country.
Valerianella (from *Valeriana*, q.v. and *-ella*, diminutive suffix).

1. Fruits of two kinds: those of forks 1-celled, cylindrical, prismatic, those of cymes ovoid or oblong; calyx limb unequally 3-toothed, two of the teeth small, the third big, thick at base, subulate, resembling a horn .... 1. *V. soyeri*
   All fruits ± similar 2
2. Calyx limb reduced to a small tooth arising from the posterior cell; fruit convex and markedly keeled at back .... 4. *V. carinata*
   Calyx limb subulate to triangular, but not reduced; fruit not keeled at back 3
3. Calyx limb indistinctly 3–14-lobed; fruit tuberculate, prickly at back .... 13. *V. tuberculata*
   Calyx limb 1–6-lobed or truncate; fruit glabrous or pubescent to villous or pruinose, but not tuberculate or prickly at back 4
4. Fruit globose, ± 1.5 mm long, glabrous .... 2. *V. pumila*
   Fruit tubular or slightly conical, but never globose, > 1.5 mm long, variously hairy 5
5. Calyx limb tubular, truncate, nearly entire .... 5. *V. muricata*
   Calyx limb lobed 6

6. Calyx limb 1–3-lobed, lobes ± hooked, entire or denticulate 7
   Calyx limb 6-lobed, each lobe ending in a hook or a mucro 11
7. Calyx limb with 3 lobes, none of which subulate. 8
   Calyx limb with 1–2 subulate lobes. 10
8. Calyx lobes indistinct; fruit 1.5–3 mm, pubescent. . . . . 3. *V. lasiocarpa*
   Calyx lobes distinct; fruit 2.5–6 mm, tomentose to woolly 9
9. Calyx lobes nearly entire or with few shallow acuminate or shortly mucronate teeth; fruit ± 2.5 mm long, tomentose . . . . 11. *V. deserticola*
   Calyx lobes ovate, dentate-aristate; fruit ± 6 mm, densely woolly. . . . . . . 10. *V. dufresnia*
10. Calyx limb with (1–)2 subulate lobes, ± hooked at apex; fruit ± 4 mm long . . . . 14. *V. oxyrhyncha*
   Calyx limb with 1 or 2 lateral lobes, denticulate at apex; fruit ± 3 mm long . . . . 12. *V. szovitsiana*
11. Calyx limb globose, inflated, with 6 small inflexed teeth . . . . 9. *V. vesicaria*
   Calyx limb spreading, campanulate or reduced, not inflated or with inflexed teeth 12
12. Fruit ± 5 mm, dilated at base, with 2 gibbosities, pruinose. . . . . 15. *V. dactylophylla*
   Fruit < 5 mm, not dilated at base, without gibbosities, variously hairy 13
13. Calyx lobes spreading 14
   Calyx campanulate or cup-shaped, lobes not spreading. 15
14. Leaves oblong-lanceolate (4–7 mm wide); calyx lobes ovate, acuminate, ending in a hooked mucro . . . . 8. *V. kotschyi*
   Leaves linear to oblanceolate (1–3 mm wide); calyx lobes acute ending in weak mucros . . . . .3. *V. lasiocarpa*
15. Calyx limb pubescent; fruit 3–3.5 mm . . . . 6. *V. chlorostephana*
   Calyx limb glabrous, sometimes ciliate at margins; fruit ± 2 mm. . . . . .7. *V. coronata*

1. **Valerianella soyeri** *Buchinger* in Boiss., Diagn. ser. 1, 10: 74 (1849); Boiss., Fl. Orient. 3: 102 (1875); Zohary, Fl. Palest. ed. 2, 1: 607 (1932); Zohary in Dep. Agr. Iraq Bull. 31: 140 (1950); Hadač in Bull. Coll. Sci. 6: 32 (1961); Rech.f., Fl. Lowland Iraq: 572 (1964).

Plant glabrous. Upper leaves sessile, incised at base; lower oblong, spatulate. Fruiting branches thickened. Calyx limb unequally 3-toothed, hooked. Fruits irregularly 3-grooved, of two kinds; fertile cell gibbous. One sterile cell as large as the fertile, the other twice as broad.

Hab. Rocky places; fl. Apr.–May.
Distrib. Indicated by Zohary from western desert (dwd), without locality. I (SAG) have not seen any material at K; the record needs confirmation.

E Mediterranean.

2. **Valerianella pumila** (*L.*) *DC.*, Fl. Fr. 4: 242 (1805); Hand.-Mazz. in Ann. Nat. Hofmus. Wien 27: 428 (1913); Náb. in Bull. Fac. Sci. Univ. Masaryk 35: 138 (1928); Lincz. in Fl. SSSR 23: 674 (1958); Hadač in Bull. Coll. Sci. 6: 32 (1961); Rech.f., Fl. Lowland Iraq: 575 (1964); Rech.f., Fl. Iranica 62: 11 (1969); Coode & Matthews in Fl. Turk. 4: 572 (1972).

V. tridentata (Stev.) Krok, Vet. Ak. Handl. Stockh. 5 (1): 73 (1864); Boiss., Fl. Orient. 3: 109 (1875); Zohary, Fl. Palest. ed. 2, 1: 608 (1932); Blakelock in Kew Bull. 3: 442 (1948); Zohary in Dep.Agr. Iraq Bull. 31: 140 (1950).

Plant 5–8 cm tall; stems and leaves scabrous above. Lower leaves linear-spatulate, entire, upper sometimes dentate or incised at base. Flowers white or pinkish. Calyx limb with 3 short lobes. All fruits similar, ± 1.5 mm long, with an umbilical depression in the centre, globose, glabrous, sterile cells broader than the fertile. Fig. 109: 5.

Hab. in cultivated fields, *Poa* steppe; alt.: 50–900 m; fl. Mar.–Apr.
Distrib. Scattered, mostly in the humid steppe zone of northern Iraq. **msu**: Qara Anjir, *Rawi* s.n.; **fuj**: Ain Husan near Sinjar, *Guest* 4236!; **far**: Arbil, *Guest* 1452!; **fki**: Tuz Khurmatu, *Rawi & Gillett s.n.*!; Kirkuk, *Guest* 1377B; Altun Kupri, *Hadač* 13201!; **fni**: Jabal Darawishka, *Guest* 13969!; Jabal Hamrin, *Guest* s.n., *Haines* s.n.; E of Zurbatiya on Iranian border, *Gillett* 6708!; **dlj**: above Rawa, *Rawi & Gillett* s.n.!; **lca**: Baghdad, Jadriya at Tigris, *Haines & Hadač* 4031!

Mediterranean region, Balkans, Syria, Turkey, W Iran, Crimea, Caucasus, C Asia.

3. **Valerianella lasiocarpa** (*Stev.*) *Betcke*, Animadv. Bot. Valer. 26 (1826); Krok in Kongl. Svensk Vet.-Akad. Handl. 5 (1): 34 (1864); Boiss., Fl. Orient. 3: 108 (1875); Náb. in Bull. Fac. Sci. Univ. Masaryk 35: 138 (1928); Blakelock in Kew Bull. 3: 441 (1948); Zohary in Dep. Agr. Iraq Bull. 31: 140 (1950); Lincz. in Fl. SSSR 23: 676 (1958); Hadač in Bull. Coll. Sci. 6: 32 (1961); Zohary, Fl. Lowland Iraq: 574 (1964); Rech.f., Fl. Iranica 62: 15 (1969); Coode & Matthews in Fl. Turk. 4: 580 (1972).

Fedia lasiocarpa Stev. in Mém. Soc. Nat. Mosc. 5: 350 (1817).
Valerianella microstephana Boiss. & Bal. in Boiss., Diagn. ser. 2, 6: 92 (1859).

Plant 10–30 cm tall. Leaves ciliate on margins, dentate, the upper pinnatifid at base. Upper leaves ciliate on margins. Calyx limb with 3 indistinctly isolated lobes or shortly acuminate teeth or sometimes 6-lobed. Fruit tubular, 1.5–3 mm long, pubescent. Fig. 109: 13.

HAB. in fields of the steppe region; alt.: 200–570 m; fl. Apr.–May.
DISTRIB. Occasional in the foothills of middle Iraq. **MSU**: Tanjero valley, *Gowan* 2410!; **FUJ**: Jazira S of Sinjar, near Ba'aj, *Guest* s.n.!; Albayder 78 km S of Sinjar, *Rawi et al.* 32088!; between Sharqat and Hadhr, *Rawi & Gillett* 7261!; Qaiyara, *Bayliss* 47!; **FAR**: 10–15 km from Arbil to Darband, *A. Shahwana* 25284!; **FKI**: Kirkuk, *Guest* 1477!; near Tauq, *Guest, Eig & Zohary* 5094! **FNI**: Jabal Hamrin, *Bornmüller* 1317.

Balkans, Turkey, W & S Iran, Ukraine, Caucasus.

4. **Valerianella carinata** *Loisel.*, Not. Pl. Fr. 149 (1810); Krok in Kongl. Svensk Vet.-Akad. Handl. 5 (1): 61 (1864); Boiss., Fl. Orient. 3: 106 (1875); Zohary, Fl. Pal. ed. 2, 1: 608 (1932); Lincz. in Fl. SSSR 23: 670 (1958); Rech.f., Fl. Iranica 62: 11 (1969); Coode & Matthews in Fl. Turk. 4: 571 (1972).

Plant 10–20 cm tall, scabrid at stem angles. Leaves ciliate; lower leaves spatulate, upper denticulate at base. Flowers in capitate cymes. Calyx limb reduced to a small tooth arising from the posterior cell. Fruit glabrous, ovoid to oblong, convex behind, keeled, grooved in front, deciduous. Sterile cells nearly as broad as the fertile cell. Fig. 109: 6.

HAB. In shady irrigated gardens and fields in lower mountains; very rare; alt. ± 950 m; fl. Apr.–May.
DISTRIB. Rare; only recorded from three somewhat disjunct localities. **MSU**: Shaqlawa, *Haines* W668!; Benava Suta near Penjwin, *Hadač* 4851!; Penjwin, *Hadač* 4868!; **MRO**: Kuh Sefin, *Bornmüller* 1316.

W, C & S Europe, Turkey, N Africa, E Mediterranean, S Iran.

5. **Valerianella muricata** (*Stev.*) *W. Baxt.* in Loud. Hort. Brit. Suppl. 3: 654 (1839); Lincz. in Fl. SSSR 23: 668 (1958); Hadač in Bull. Coll. Sci. 6: 33 (1961); Rech.f., Fl. Lowland Iraq: 574 (1964); Rech.f., Fl. Iranica 62: 16 (1969); Coode & Matthews in Fl. Turk. 4: 581 (1972); Y. Nasir in Fl. Pak. 101: 17 (1976); Collenette, Wildflow. Saudi Ariabia: 742 (1999).

Fedia muricata Stev. apud Roem. & Schultes, Syst. 1: 366 (1817).
Valerianella truncata Reichb., Pl. Crit. 2: 7 (1824); Krok, Kongl. Svensk Vet.-Akad. Handl. 5 (1): 38 (1864); Blakelock in Kew Bull. 3: 442 (1948); Zohary in Dep. Agr. Iraq Bull. 31: 140 (1950).
V. truncata Reichb. var. *muricata* (Stev.) Boiss., Fl. Orient. 3: 105 (1875); Zohary, Fl. Palest. ed. 2, 1: 608 (1932).

Plant 10–15 cm tall, scabrid at leaf veins, leaf edges and stem angles. Lower leaves oblong, upper linear, sometimes dentate at base. Fruiting branches somewhat thickened. Calyx limb tubular, truncate, reticulate-veined, nearly entire, about as broad and long as fruit. Fruit ± 2 mm long. Fig. 109: 11.

HAB. in fields of N & E Iraq; alt. 220–900 m; fl. Apr.
DISTRIB. **MAM**: Shaikh Adi, *Thesiger* 703; **MSU**: Jarmo, *Helbaek* 869! Qaradagh, *Gillett* 7860!; Halabja, *Rawi* 8872!; **FKI**: Kifri, *Rawi & Gillett* s.n.; Kirkuk, *Guest* 1993!; between Tuz and Tauq, *Rawi & Gillett* 7461!; Altun Kupri, *Guest* s.n.; **FNI**: Jabal Hamrin on road to Khanaqin, *Haines* W.912! *Gillett* 7408!; **FNI**: Ba'shiqa, *Hadač* 1540!

Mediterranean region, Balkans, Lebanon, Palestine, Turkey, Iran, Ukraine, Caucasus, C Asia.

6. **Valerianella chlorostephana** *Boiss. & Bal.* in Boiss., Diagn. ser. 2, 6: 92 (1859); Rech.f., Fl. Iranica 62: 14 (1969); Coode & Matthews in Fl. Turk. 4: 577 (1972).

V. boissieri Krok, Kongl. Svensk Vet.-Akad. Handl. 5 (1): 81 (1864); Boiss., Fl. Orient. 3: 111 (1875); Zohary, Fl. Palest. ed. 2, 1: 609 (1932); Hadač in Bull. Coll. Sci. 6: 33 (1961); Rech.f., Fl. Lowland Iraq: 575 (1964); Rech.f., Fl. Iranica 62: 14 (1969).
V. locusta γ *coronata* L., Sp. Pl. 34 (1753).

Plant 5–25 cm tall, pubescent. Leaves oblong, sometimes dentate or pinnatifid. Cymes capitate, globular. Calyx limb cup-shaped, reticulate, with 6 hooked lobes pubescent. Lobes about as long as the fruit. Fruit 3–3.5 mm long, shortly villous. Fig. 109: 10.

HAB. scattered in fields of N & E Iraq; alt.: 50–2000 m; fl. Mar.–May.
DISTRIB. Occasional, mainly in the foothill regions. **MRO**: Rowanduz, *Guest* 2042!; above Sakri-Sakran, *Hadač* 4911!; **MSU**: valley of R. Tanjaro nr Sulaimaniya, *Gowan* 2410; **FUJ**: Jabal Maklub near Ain Dibs, *Rawi & Gillett* 7186!; **FNI**: Jalaula, *Polunin* s.n.; **FNI**: Sleifani near Mosul, *Rawi* 8472!; Ba'shiqa, *Hadač* 153?; **FNI**: Jabal Hamrin on road to Khanaqin, *Haines* QW913!; near Zerbatiya, *Gillett* 6708!; Pilkhana 10 km E of Khanaqin, *Rawi* s.n.; Jabal Darawishka nr Khanaqin, *Guest* s.n.

Syria, Turkey, W Iran.

7. **Valerianella coronata** (*L.*) *DC.*, Fl. Fr. 4: 241 (1805); Krok in Kongl. Svensk Vet.-Akad. Handl. 5 (1): 78 (1864); Zohary, Fl. Palest. ed. 2, 1: 609 (1932); Blakelock in Kew Bull. 3: 441 (1949); Zohary in Dep. Agr. Iraq Bull. .31: 140 (1950); Lincz. in Fl. SSSR 23: 677 (1958); Hadač in Bull. Coll. Sci. 6: 34 (1961); Rech.f., Fl. Lowland Iraq: 575 (1964); Rech.f., Fl. Iranica 62: 13 (1969); Coode & Matthews in Fl. Turk. 4: 576 (1972); Y. Nasir in Fl. Pak. 101: 15 (1976).

Valeriana coronata Willd., Sp. Pl. 1: 184 (1798).
Valeriana locusta L. var. *coronata* L., Sp. Pl. 34 (1753).

var. **coronata**

Plants 5–25 cm tall, pubescent. Leaves oblong, sometimes dentate or pinnatifid. Cymes capitate, globular. Calyx limb cup-shaped, reticulate, glabrous, sometimes ciliate at margins, with 6 hooked lobes; lobes about as long as the fruit. Fruit ± 2 mm long, shortly villous. Fig. 109: 8.

HAB. Unrecorded; alt.: 50–2000 m; fl. Mar.–May.
DISTRIB. Scattered in N & E Iraq. **MJS**: above Kursi, *Gillett* 11036!; **MRO**: Rowanduz, *Guest* 2042!; above Sakri-Sakran, *Hadač* 5512!; **MSU**: Benava Suta, *Hadač* 4923; Penjwin, *Hadač* 4911!; **FUJ**: Jabal Maklub nr Ain Dibs, *Rawi & Gillett* 7186!; Sleifani nr Mosul, *Rawi* 8472!; **FNI**: Ba'shiqa, *Hadač* 153x!; Mindan bridge, *Anders* 1266; **FNI**: Zurbatiya, *Rawi* s.n.
NOTE. A variant with the calyx limb densely hairy on the inside has been recorded from two places in Jabal Sinjar, 500–1000 m: **MJS**: Jabal Sinjar, above Kursi, *Gillett* s.n.!; above Qizil Khan, *Agnew, Hadač & Hashimi* 3635!

C & S Europe, SW & C Asia.

8. **Valerianella kotschyi** *Boiss.*, Diagn. ser. 1, 3: 60 (1843); Boiss., Fl. Orient. 3: 110 (1875); Zohary, Fl. Palest. ed. 2, 1: 609 (1932); Blakelock in Kew Bull. 3: 441 (1948); Zohary in Dep. Agr. Iraq Bull. 31: 141 (1950); Lincz. in Fl. SSSR 23: 678 (1958); Hadač in Bull. Coll. Sci. 6: 34 (1961); Rech.f., Fl. Lowland Iraq: 576 (1964); Rech.f., Fl. Iranica 62: 13 (1969); Coode & Matthews in Fl. Turk. 4: 576 (1972).

Plants 10–30 cm tall, glabrous above, scabrous-pubescent below, branching from the base. Leaves oblong-lanceolate, 4–7 mm wide, denticulate, the upper pinnatifid at base. Calyx limb glabrous, spreading, reticulate, with 6 broadly triangular lobes ending in a hooked mucro; lobes about half as long as the fruit. Fruit woolly, grooved in front, ± 2 mm long. Fig. 109: 9.

HAB. In steppe and fields; alt. 230–1000 m; fl. Apr.–May.
DISTRIB. Scattered, mainly in northern Iraq. **MJS**: above Kursi, *Gillett* 11036!; **MAM**: Dohuk, *Guest* 2232!; **MRO**: Rowanduz, *Gillett* 8324!; Shaqlawa, *Haines* W.921!; **MSU**: Khormal, *Rawi* s.n.; Jarmo, *Helbaek* 1139!; **FAR**: Khanzad valley, *Guest* s.n.; **FNI**: Mosul, *Lazar* 3382!; Tuz, *Hadač* 1115!; Nineveh, *Hadač* s.n.

Syria, Turkey, Ukraine, Caucasus, Pamirs.

9. **Valerianella vesicaria** (*L.*) *Moench*, Meth. Pl. 493 (1794); Krok in Kongl. Svensk Vet.-Akad. Handl. 5 (1): 86 (1864); Boiss., Fl. Orient. 3: 112 (1875); Zohary, Fl. Palest. ed. 2, 1: 609 (1932); Náb. in Bull. Fac. Sci. Univ. Masaryk 35: 138 (1928); Blakelock in Kew Bull. 3: 442 (1948); Zohary in Dep. Agr. Iraq Bull. 31: 141 (1950); Lincz. in Fl. SSSR 23: 680 (1958); Hadač in Bull. Coll. Sci. 6: 35 (1961); Rech.f., Fl. Lowland Iraq: 577 (1964); Rech.f., Fl. Iranica 62: 15 (1969); Coode & Matthews in Fl. Turk. 4: 578 (1972).

Valeriana locusta β *vesicaria* L., Sp. Pl. 33 (1753).
V. vesicaria (L.) Willd., Sp. Pl. 183 (1789).

Plant 5–15 cm tall, puberulent. Leaves dentate, sometimes pinnatifid at base. Calyx limb globose, inflated, reticulate, with 6 small inflexed teeth, ± 2.5 mm long. Fruit woolly. Fig. 109: 12.

HAB. In steppe and fields in the northern half of Iraq; alt. 250–1300(–1500) m; fl. Mar.–Jun.
DISTRIB. Scattered, mainly in the plateau region. **MAM**: Dohuk, *Guest* 2315! Acra, *Kaimaqam of Acra* 3087!; **MRO**: Rowanduz, *Gillett* s.n.!; **MSU**: Dar Mazala, *Hadač* 5125!; Dokkan (Dukan) dam, *Hadač* 1211!; Tawila, *Rawi* s.n.!; **MJS**: Kursi, *Gillett* 10836!; **FUJ**: Mar Jirjis near Mosul, *Rawi & Gillett* 3337!; Ba'shiqa, *Hadač* 1528!; **FKI**: Kirkuk, *Bornmüller* 1824; Kirkuk, Chemchemal road, *Rawi & Gillett* 10564!; Altun Kupri to Kirkuk, *Erdtman & Goedemans* in *Rechinger* 15508; **FNI**: Jabal Hamrin on road to Khanaqin, *Haines* W.920F!; Khanaqin, *Rechinger* 9057a.

Mediterranean area, Balkans, Turkey, Iran, C Asia.

10. **Valerianella dufresnia** *Bunge ex Boiss.*, Fl. Orient. 3: 109 (1875). — Type: Iraq, between Mosul and Baghdad, *Olivier & Bruguière* s.n. (P!, holo.); Hand.-Mazz. in Ann. Nat. Hofmus. Wien 27: 38 (1913); Náb. in Bull. Fac. Sci. Univ. Masaryk 35: 138 (1928); Zohary, Fl. Palest. ed. 2, 1: 609 (1932); Zohary in Dep. Agr. Iraq Bull. 31: 140 (1950); Lincz. in Fl. SSSR 23: 679 (1958); Hadač in Bull. Coll. Sci. 6: 35 (1961); Rech.f., Fl. Lowland Iraq: 576 (1964); Hadač in Kew Bull. 21: 60, fig. 1 (1967); Rech.f., Fl. Iranica 62: 12 (1969); Coode & Matthews in Fl. Turk. 4: 574 (1972); Collenette, Wildflow. Saudi Ariabia: 742 (1999); Breckle, Dittman & Rafiqpoor, Field Guide Afghanistan: Fl. & Veg. 692 (2010).

Dufresnia orientalis DC., Prodr. 4: 625 (1830).
D. conjungens Bunge ex Boiss., Fl. Orient. 3: 109 (1875).
Valerianella orientalis (DC.) Lipsky, non Boiss. et Bal. in Boiss., Fl. Orient. 3: 103 (1875).
V. leiocarpa (C. Koch) O. Ktze, Acta Horti Petrop. 10: 195 (1887); var. *orientalis* (DC.) O. Ktze, l.c.; Blakelock in Kew Bull. 3: 441 (1948).

Plants 5–10 cm tall, glabrous. Leaves oblong-linear, entire. Calyx limb crown-like, 3-lobed to base, lobes reticulate, ovate, dentate-aristate, about as long as the fruit. Sterile cells of carpels broader than the fertile one. Fruit ± 6 mm long, white woolly, grooved in front, with two prominent gibbosities. Fig. 108: 4–5; 109: 7.

HAB. In semi-desert, steppe and fields; alt. 100–500 m; fl. Mar.–Jun.
DISTRIB. Foothills and desert. **FUJ**: between Ain Dibs and Baiji, *Rawi & Gillett* 7170!; **FKI**: 19 km E of Kirkuk, *Rawi & Gillett* 10584!; Tauq, *Polunin* s.n., *Bornmüller* 1325; Altun Kupri, *Hadač* s.n.!; **FNI**: Jabal Hamrin near Khanaqin, *Guest* 1908!; Zurbatiya, *Gillett* 6721!; Jabal al Muwaila, 70 km N of Amara, *Guest, Rawi & Rechinger* 17548!; **LSM/LEA**: Shatt at-Tib, 70 km N of Amara, *Rechinger* 8938; **DLJ**: S of Jazira, 75 km NW from Baghdad, *Agnew & Hadač* 1630!; 60 km NW of Rawa, *Rawi et al.* 31955!; **DGA**: Samarra, *Haines* s.n.; **DWD**: Wadi Muhammadi 63 km W of Ramadi, *Hadač* 4315!; 55 km E of Rutba, *Hadač* 4358!; Khasmat as Sa'abiya, *Agnew, Hadač et al.* 4186!; **DSD**: between Shubeicha and Sha'ib Hisb, *Agnew & Hadač* 4145!; between Sha'ib Hisb and Wadi al-Khirr, *Agnew, Hadač et al.* 4157!

Balkans, Turkey, Syria, Iran, C Asia.

11. **Valerianella deserticola** *Hadač* in Kew Bull. 21: 59, fig. 1 (1967). — Type: Iraq, 35 km E of Nukhaib, *Agnew, Hadač & Hashimi* 4223 (PR!, holo.).

Plants 2–5 cm tall, branched from base, glabrescent. Calyx limb 3-lobed nearly to base, lobes ovate-triangular, reticulate, the middle one about as long as the fruit, the others shorter, ± entire, indistinctly dentate, teeth acuminate or sometimes shortly mucronate, but not aristate. Fruit 2–2.5 mm long, shortly tomentose, antically grooved, sterile cells broader than the fertile one. Fig. 108: 1– 3.

HAB. In deserts and semi-deserts on sand; 50–300 m; fl. Mar.–May.
DISTRIB. Deserts of western and southern Iraq. **DWD**: 35 km E of Nukhaib, Agnew, *Hadač et al.* 4223 (type)!; Nukhaib, *Rawi* 14800A!; 61 km E of Rutba, *Hadač* 4427!; Tal Rafha, *Agnew & Hadač* 4272!; 5 km E of Karbala, *Rawi & Gillett* 6374!; **DLJ**: Jazira, 60 km NNW from Baghdad, *Agnew, Hadač & Haines* 4499!; **DGA**: 23 km S of Samarra, *Rawi & Gillett* 6513!; desert S of Thirthar canal, *Haines* 148!; **DSD**: 5 km E of Karbala, *Rawi & Gillett* 6374!
NOTE. Resembles *V. dufresnia*, but is smaller, with fruits about half the size of *V. dufresnia*, subtomentose (not woolly), calyx teeth never aristate, lateral lobes shorter, not equal as in *V. dufresnia*.
?Endemic; may occur in other parts of the Near East, where it could have been confused with weak specimens of *V. dufresnia*.

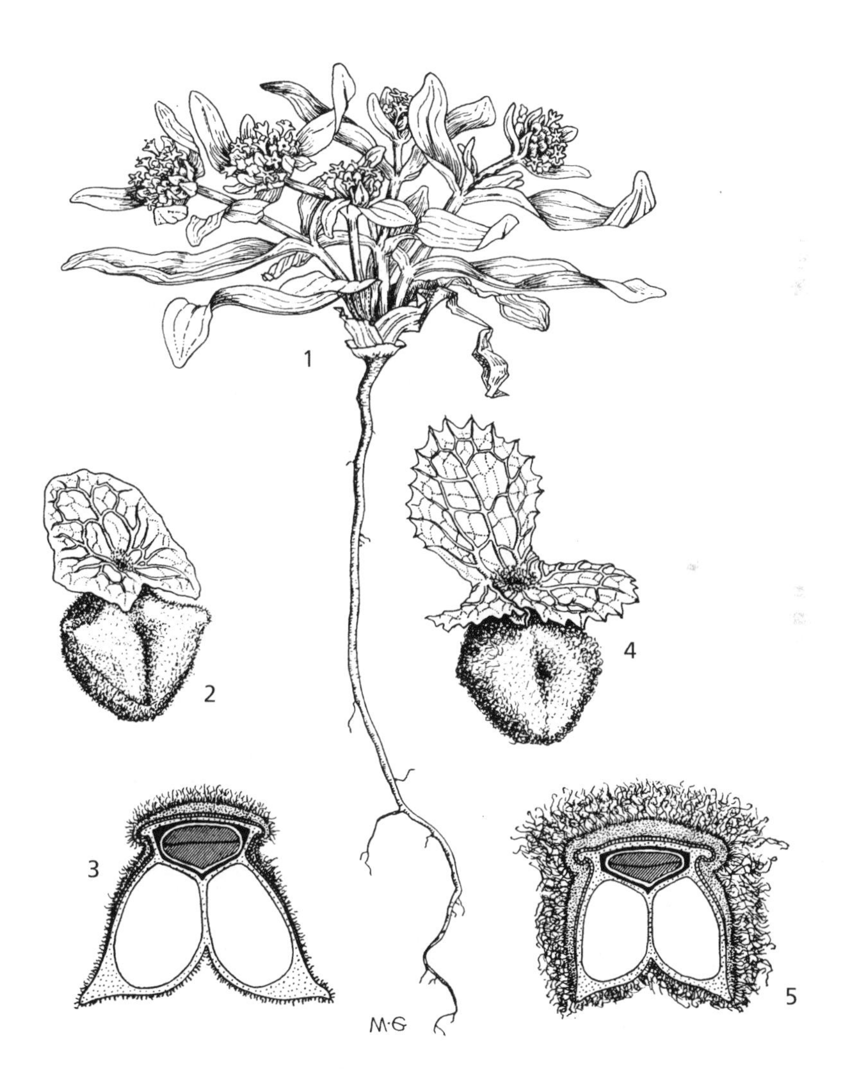

Fig. 108. **Valerianella deserticola**. 1, habit; 2, fruit with enlarged persistent calyx; 3, fruit, cross section. **Valerianella dufresnia**. 4, fruit with enlarged and persistent calyx; 5, fruit, cross section. Taken from Kew Bull. 21: p. 60, fig. 1 (1967).

12. **Valerianella szowitziana** *Fisch. & Mey.*, Ind. Sem. Horti Petrop. 3: 48 (1836); Krok in Kongl. Svensk Vet.-Akad. Handl. 5(1): 58 (1864); Boiss., Fl. Orient. 3: 101 (1875); Zohary, Fl. Palest. ed. 2, 1: 606 (1932); Lincz. in Fl. SSSR 23: 660 (1958); Hadač in Bull. Coll. Sci. 6: 36 (1961); Rech.f., Fl. Lowland Iraq: 574 (1964); Rech.f., Fl. Iranica 62: 9 (1969); Coode & Matthews in Fl. Turk. 4: 563 (1972); Y. Nasir in Fl. Pak. 101: 19 (1976); Collenette, Wildflow. Saudi Ariabia: 743 (1999); Boulos, Fl. Egypt 3: 121 (2002).

V. aucheri Boiss., Diagn. ser. 1, 3: 58 (1843).
V. persica Boiss., l.c. (1843).

Plants hairy, ± 10 cm tall, leaves oblong-linear. Calyx limb reticulate, with 1–2 lateral lobes; horizontal lobe usually denticulate at tip. Fruiting branches not thickened; fruits ± 3 mm long, hispidulous, with an oblong ovoid pit in front. Fig. 109: 3.

HAB. Mostly on irrigated, somewhat sandy and salty fields in lowlands; alt.: 30–100 m; fl. Mar.–Apr.
DISTRIB. Scattered, mainly in the Mesopotamian region. **DLJ**: 4 & 12 km NW of Rawa, *Rawi & Gillett* 7096!; 6 km above Ana, *Rawi & Gillett* 6961!; Sechar near Falluja, *Agnew & Hadač* 3193!; **DGA**: Balad on Samarra to Baghdad road, *Agnew* s.n.; Ashiq near Samarra, *Hadač* 849!; ?**DLJ**: S of Jazira, 75 km from Baghdad, S of Thirthar depression, *Agnew & Hadač* 1614!; **FNI**: Jabal Hamrin on road to Khanaqin, *Hadač* 609!; **FKI**: Tuz, *Agnew, Hadač & Hashimi* 3860!, *Barkley* 7008; 10 km N of Tuz Khurmatu, *Barkley* 7008; **DWD**: Nukhaib, *Rawi* 14800!

Balkans, Egypt, Syria, Palestine, Turkey, Caucasus, Iran, Afghanistan, Pakistan, C Asia.

13. **Valerianella tuberculata** *Boiss.*, Diagn. ser. 1, 3: 59 (1843); Krok in Kongl. Svensk Vet.-Akad. Handl. 5 (1): 47 (1864); Boiss., Fl. Orient. 3: 97 (1875); Zohary, Fl. Palest. ed. 2, 1: 605 (1932); Lincz. in Fl. SSSR 23: 647 (1958); Hadač in Bull. Coll. Sci. 6: 37 (1961); Rech.f., Fl. Lowland Iraq: 573 (1964); Rech.f., Fl. Iranica 62: 4 (1969).

Plants glabrous, to ± 10 cm. Leaves entire. Inflorescence very lax, branches strongly divergent; fruiting branches thickened. Flowers pale purple. Calyx indistinctly 3–14-lobed, lobes with unequal hooks. Fruit about 4 mm long, tuberculate, prickly at the back. Fig. 109: 2.

HAB. Serpentine screes, on mountains; alt. 800–1400 m; fl. Apr.–Jun.
DISTRIB. Rare in northern Iraq. **MJS**: Kursi, *Gillett* 11015!; **MSU**: Malakawa, *Rechinger* 12314, *Rawi* s.n.; Qara Dagh, *Gillett* 7860!; Waziara on Qopi Qaradagh, *Hadač* 6283!

E Mediterranean area, Syria, Turkey, W & C Iran, Afghanistan, C Asia.

14. **Valerianella oxyrhyncha** *Fisch. & Mey.*, Ind. Sem. Horti Petrop. 3: 51 (1837); Rech.f., Fl. Iranica 62: 6 (1969); Coode & Matthews in Fl. Turk. 4: 564 (1972); Y. Nasir in Fl. Pak. 101: 20 (1976); Collenette, Wildflow. Saudi Ariabia: 742 (1999).

V. diodon Boiss., Diagn. ser. 1, 3: 58 (1843); Boiss., Fl. Orient. 3: 99 (1875); Zohary, Fl. Palest. ed. 2, 1: 606 (1932); Zohary in Dep.Agr. Iraq Bull. 31: 140 (1950); Lincz. in Fl. SSSR 23: 653 (1958); Hadač in Bull. Coll. Sci. 6: 37 (1961); Rech.f., Fl. Lowland Iraq: 573 (1964).
V. stocksii Boiss., Fl. Orient. 3: 99 (1875).

Plants 5–10 cm tall. Stem puberulent, especially at base. Leaves glabrous, linear-oblong, dentate. Fruiting branches not thickened. Calyx limb divided into (1–)2 subulate lobes, ± hooked at apex. Fruit ± 4 mm long, grooved in front, with a circular grooved area at base. Fig. 109: 4.

HAB. Lowland steppe; alt. ± 1000 m; fl. Mar.–May.
DISTRIB. Occasional in northern Iraq. **MJS**: Kursi, *Gillett* s.n.!; **FKI**: 6 km S. from Kirkuk, *Agnew, Hadač & Hashimi* 3886!; **DWD**: Nukhaib, *Rawi* s.n.!; Jidd al Fadj, *Agnew & Hadač* 4257!; 6 km above Ana, *Rawi & Gillett* 6961!; **DSD**: 12 km E of Samarra, *Rechinger* 9560; **LEA**: near Deltawa, *Rogers* O285!

Balkans, Palestine, Turkey, Caucasus, NW & NE Iran, Afghanistan, Pakistan, C Asia.

15. **Valerianella dactylophylla** *Boiss. & Hohen.* in Boiss., Diagn. ser. 1, 10: 75 (1849); Krok in Kongl. Svensk. Vet.-Akad. Handl. 5(1): 49 (1864); Boiss., Fl. Orient. 3: 97 (1875); Nábělek in Bull. Fac. Sci. Univ. Masaryk 35: 137 (1928); Zohary, Fl. Palest. ed. 2, 1: 606 (1932); Zohary in Dep.Agr. Iraq Bull. 31: 140 (1950); Lincz. in Fl. SSSR 23: 651 (1958); Hadač in Bull. Coll. Sci. 6: 37 (1961); Rech.f., Fl. Lowland Iraq: 573 (1964); Rech.f., Fl. Iranica 62: 5 (1969).

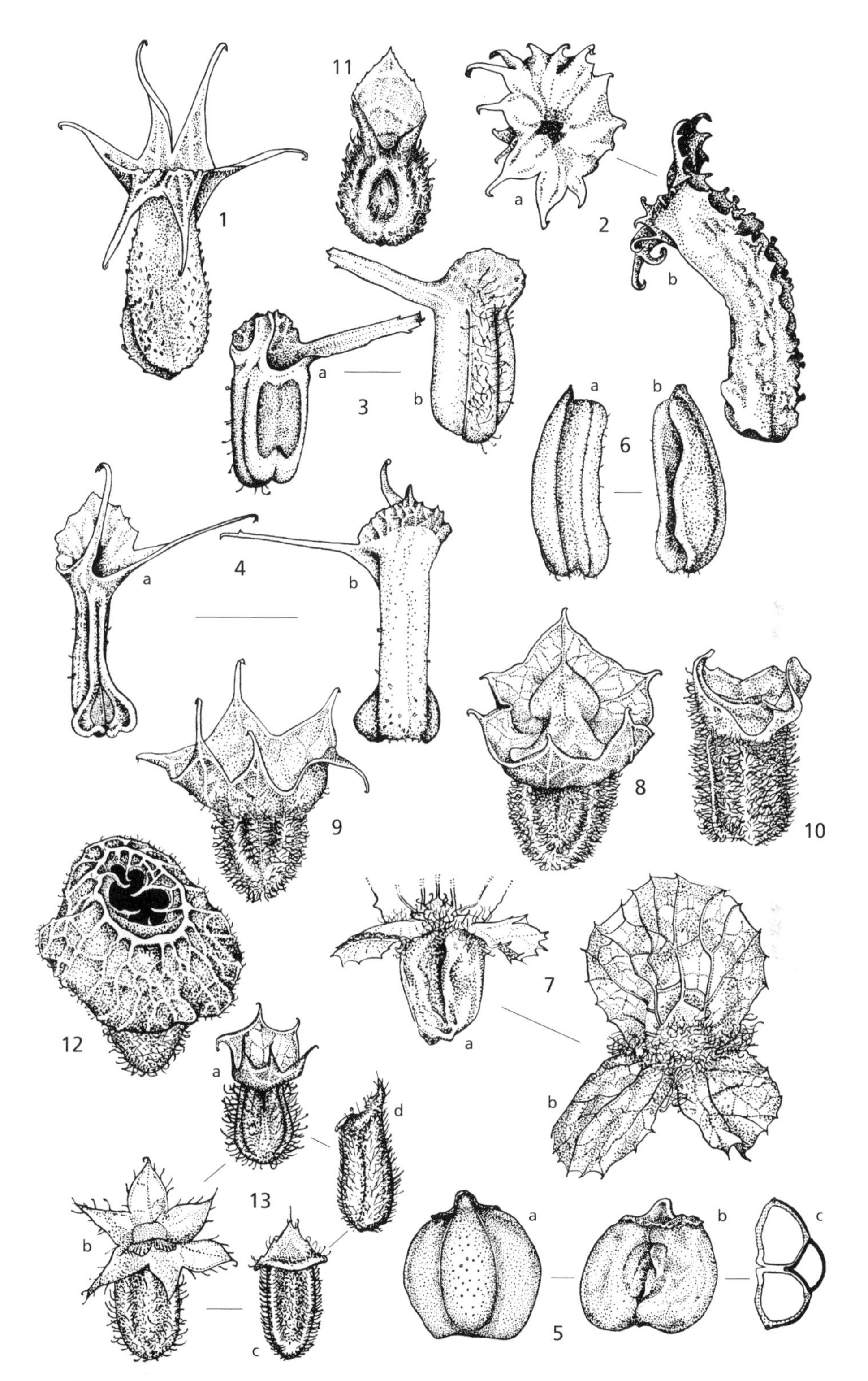

Fig. 109. **Valerianella fruits**. 1, **V. dactylophylla** (15); 2, **V. tuberculata** (13); 3, **V. szovitsiana** (12); 4, **V. oxyrhyncha** (14); 5, **V. pumila** (2); 6, **V. carinata** (4); 7, **V. dufresnia** (10); 8, **V. coronata** (7); 9, **V. kotschyi** (8); 10, **V. chlorostephana** (6); 11, **V. muricata** (5); 12, **V. vesicaria** (9); 13, **V. lasiocarpa** (3). Reproduced from Fl. Turk. 4: 1972, with permission.

Plants up to 30 cm tall, subglabrous. Leaves ciliate at margins, lower leaves oblong, nearly entire, upper digitate. Inflorescence dense, cymes capitate. Calyx limb campanulate, 3–4 mm in diameter, unequally divided into 6 hooked subulate lobes. Fruits ± 5 mm long, pruinose, with a narrow groove in front, dilated at base, with 2 gibbosities. Fig. 109: 1.

Hab. Rocky places in mountains, on serpentine; alt. 800–2400 m; fl. May–Jun.
Distrib. Scattered in the upper forest and thorn-cushion zones of N Iraq. **MJS**: Kursi, *Gillett* 11029!; **MAM**: Matina nr Mosul, *Rawi* s.n.; **MRO**: Botin, *Agnew, Hadač & Haines* 6158!; Nawanda valley, *Hadač* 2669!; Handren Dagh, *Nábělek* s.n.!; between Sheikhan and Sakri-Sakran, *Hadač* 5417!; Kopi Qaradagh, *Haines* W.1180!; Sefin Dagh, *Gillett* 8129!; Mela Kowa, *Rawi* 23498!; Shaqlawa, *Bornmüller* 1326; Kamarspa, *Rawi* 22232!; **MSU**: Penjwin at Malakawa pass, *Rechinger* 12326; Mt. Hawraman above Tawila, *Rechinger* 10317.

Turkey, Syria, Lebanon, Palestine, W & C Iran, C Asia?

113. DIPSACACEAE

Engl. & Prantl., Pflanzenfam. ed. 1, 4: 182–189 (1891).

I. K. Ferguson[23]

Annual or perennial herbs, rarely shrubs. Leaves simple or compound, opposite or rarely whorled, exstipulate. Inflorescence usually a head, or rarely a spike of false whorls (*Morina*) with a calyx-like involucre of bracts. Flowers bisexual, zygomorphic, each with a tubular calyx-like involucel and borne in the axils of imbricate bracteoles. Calyx small, cup-like or tubular, or divided into pappus-like segments. Corolla gamopetalous, 4- or 5-lobed, imbricate. Stamens usually 4, free, alternate with the corolla lobes and inserted towards the top of the tube. Ovary inferior, unilocular, with a solitary ovule pendulous from the apex of the locule. Fruit indehiscent, dry, enclosed within the calyx-like involucel and often crowned with a persistent calyx.

The family consists of nine genera and about 160 species. It occurs primarily in the Mediterranean area and the Near East.
Under the APG III (2009) classification, Caprifoliaceae is expanded to include Dipsacaceae, together with families Diervillaceae, Linnaeaceae, Morinaceae and Valerianaceae.

1. Involucral bracteoles chaffy; involucel 4-angled 2
   Involucral bracteoles hairy or absent; involucel with 8 ridges 3
2. Inflorescence 25–90 mm long, oblong-ovoid .... 1. **Dipsacus**
   Inflorescence 10–20 mm long, ovoid or subglobose .... 2. **Cephalaria**
3. Calyx with 5 bristles .... 3. **Scabiosa**
   Calyx with 12–24 bristles .... 4. **Pterocephalus**

1. DIPSACUS L.

Sp. Pl. ed. 1, 97 (1753); Gen. Pl. ed. 5, 43 (1754)

Stout, erect biennial herbs usually with spiny or prickly stems. Inflorescence usually large, oblong-ovate or rarely globose. Involucral bracteoles chaffy, involucel 4-angled, grooved, truncate at apex with an almost obsolete denticulate margins. Calyx cup-shaped. Corolla 4-fid.

Dipsacus (from Gr. διψα, *dipsa*, thirst).

1. **Dipsacus laciniatus** *L.*, Sp. Pl. ed. 1, 97 (1753); Boiss., Fl. Orient. 3: 116 (1875); Zohary, Fl. Palest., ed. 2, 1: 611 (1932); Zohary in Pal. Journ. Bot., J. ser., 4: 236 (1949); Zohary in Dep. Agr. Iraq Bull., 31: 141 (1950); Lack in Fl. Iranica 168: 3 (1991); Jamzad, Fl. Iran 8: 10 (1993).

D. laciniatus L. var. *comosus* Náb., in Pub. Fac. Sci. Univ. Masaryk 35: 138 (1923).

[23] Royal Botanic Gardens, Kew

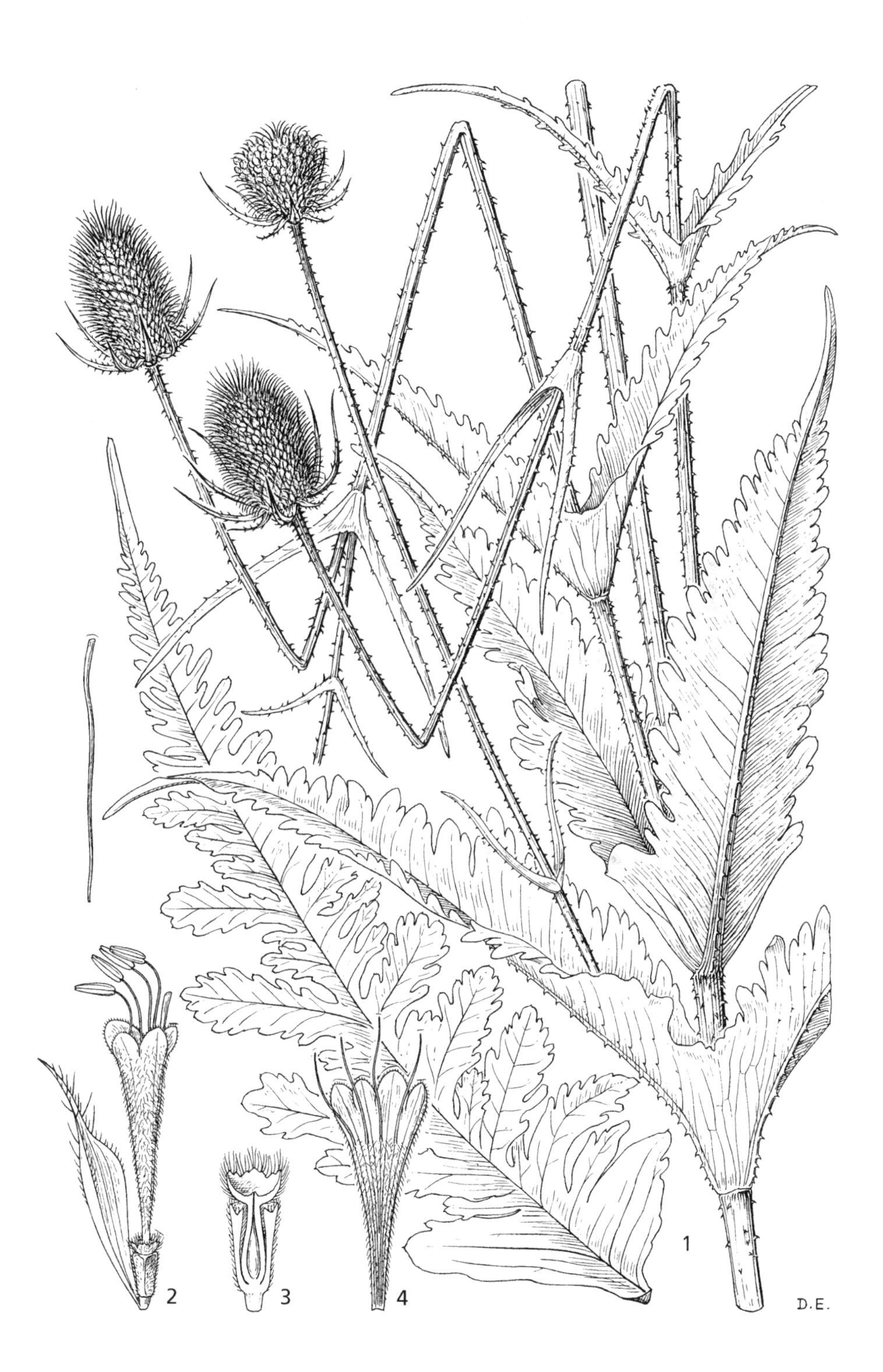

Fig. 110. **Dipsacus laciniatus**. 1, habit × ²/₃; 2, flower × 4; 3, calyx × 13; 4, corolla opened × 4. All from *Davis* 24849 *O. Polunin*. Drawn by D. Erasmus.

Erect biennial reaching to about 1 m tall. Stems stout, ribbed or smooth and usually covered with prickles. Leaves sessile, 10–45 × 3–12 cm, setose-ciliate; the lower oblanceolate, tapering at the base, crenate; the upper connate, pinnatifid. Inflorescence large, 2.5–9 cm long, ovate-oblong. Involucral bracts erect or spreading, linear-lanceolate, subulate, keeled, with spines on the margins and midrib, shorter than or equalling the inflorescence. Involucral bracteoles chaffy, 8–15 × 3–4 mm, oblong, aristate-cuspidate, erect, crowded, equal to or longer than the corolla. Corolla 8–12 mm long, pink or lilac; tube covered with sericeous hairs. Fig. 110: 1–4.

HAB. Damp spots near road; cultivated land in tomato plantation, above tree line, roadside, muddy clay soil, metamorphic rocks; alt. 530–1700 m; fl. Jul.–Aug.
DISTRIB. Very local in northern Iraq. **MRO**: bridge by Ab-e Tanjaro, *Haines* s.n.; Haji Umran, *Gillett* 12438! *Haines* W.1376!; *Rawi & Serhang* 24936! *Rawi* 26829! *Qaisi, Karim & Hamad* 4981!; **MSU**: Qal'a Diza, *Rawi* 9368!; Arbat, *F. Karim* 39301!; Haibat Sultan Dagh 7 km from Quesanjaq (Koi Sanjaq), *Nuri & K. Hamid* 41180!;10 km N of Sayyid Sadiq, *Qaisi & K. Hamad* 43514!; Chemchemal, *Haines* s.n.
NOTE. ?CHAURUKA (Kurd.?), *Rawi* 9368 (Qa'lat Diza).

Europe, N Africa, SW & C Asia; widely introduced.

2. **CEPHALARIA** Schrad.

Cat. Sem. Hort. Gotting. (1814) ex Roem. & Schultes, Syst. 3: 1 (1818)
Z. Szabó, "A *Cephalaria*-Génusz monográfiája" in Mat. Term. Közl. 38 (4) (1940): 1–352 (1940)

Annual, biennial or perennial herbs with glabrous or hairy stems. Levaves opposite, simple or imparipinnate. Inflorescence ovoid or subglobose. Involucral bracteoles chaffy. Involucel 4-angled with 8 ridges and usually with 4 or more long or short bristles. Calyx cup-shaped. Corolla 4-fid, lobes rounded, densely hairy outside. Stamens 4, filaments exceeding corolla. Achenes small, fusiform, glabrous or ciliate.

Cephalaria (from Gr. κεφάλι, *kephali*, head and the possessive Lat. suffix *-aria*).

1. Plant perennial; involucral bracteoles not ending in a bristle .... 1. *C. microcephala*
   Plant annual; involucral bracteoles ending with at least a short bristle 2
2. Involucral bracts (except the outermost) glabrous; corolla lobes about half or more the length of the tube .... 5. *C. stapfii*
   Involucral bracts adpressed-pubescent; corolla lobes about a third the length of the tube 3
3. Leaves weakly lyrate below; flowers blue .... 4. *C. syriaca*
   Leaves distinctly lyrate above and below; flowers cream......................... 4
4. Involucel with 2 bristles longer than the tube and 6 short teeth .....2. *C. dichaetophora*
   Involucel with 4 bristles equal or subequal in length to tube and 4 shorter bristles ....3. *C. setosa*

1. **Cephalaria microcephala** *Boiss.* Diagn. ser. 2, 2: 123 (1856); Boiss., Fl. Orient. 3: 125 (1875); Lack in Fl. Iranica 168: 15 (1991); Jamzad, Fl. Iran 8: 32 (1993).

C. pilosa Boiss. & Huet, Diagn. ser. 2, 2: 122 (1856); Boiss., Fl. Orient. 3: 124 (1875) non *C. pilosa* (L.) Gren. Godr., Fl. Fr. 2: 69 (1850).

Erect perennial herbs up to 50 cm high, pubescent or sericeous. Stems herbaceous, arising from a woody basal stock. Leaves sessile or shortly petiolate, 60–90 × 8–12 mm, oblong-lanceolate, denticulate, or pinnate-lobed below, connate, lyrate above. Inflorescence ovoid, 12–13 mm long, borne on a sparsely pubescent peduncle. Involucral bracts 2–3 × 2–2.5 mm long, ovate, obtuse. Involucral bracteoles 5–7 × 3–5 mm, oblong, acute, ciliate on the upper margins, shorter than the corolla. Calyx with an aristate-denticulate margin. Corolla cream, 7–9 mm long, the lobes about half the length of the tube. Involucel in fruit ± 5 mm long (including bristles), 8-angled with 4 ridges produced apically as bristles equal or subequal in length to the tube and 4 intermediate ridges with shorter bristles ⅓ the length of the tube. Fig. 111: 11.

HAB. *Astragalus* zone on metamorphic rock, serpentinite, red stony soil; alt. 1700–2900(–3180) m; fl. (Jun.–) Aug.–Sep.

Distrib. Occasional in the upper forest and thorn-cushion zones of NE Iraq. **mro**: Hajji Umran, *Rechinger* 11316; Hisar-i Rost, *Haley* 178; Mt. Qandil E of Qala Diza, *Thesiger* 1144; Ser-i Khazni, *Haley* 245; Arl Gird (Helgurd) Dagh, *Gillett* 9512!; Helgord Dagh, E of Bermasand (Berme) lake, *Rawi & Serhang* 24773!; S part of Mt. Karoukh, *Khatib & Nuri* 27479!; Baisar village on Qandil range, *Rawi & Serhang* 24176!; Mt. Sakran 15 km SE of Chuman (Jumani), *Dabbagh & K. Hamad* 46218!; **msu**: Mt. Hawraman, *Rawi, Chakr, Nuri & Alizzi* 19797!
Note. The species is notable for its occurrence on serpentinite in Iraq.

NW, C & NE Iran, Turkmenistan.

2. **Cephalaria dichaetophora** *Boiss.*, Diagn. ser. 1, 6: 71 (1846); Boiss., Fl. Orient. 3: 119 (1875); Zohary in Dep. Agr. Iraq Bull., 31: 141 (1950); Lack in Fl. Iranica 168: 9 (1991); Jamzad, Fl. Iran 8: 19 (1993).

Erect freely branching annual herbs up to 80 cm high. Stems ± tuberculate, puberulous below, ± setose above. Leaves ± setose, ciliate, sessile or shortly petiolate; the lower lyrate-pinnatisect 100–200 × 10–25 mm, with oblong-ovate dentate terminal lobes and linear lateral lobes; the upper 50–70 × 5–15 mm, with oblong-lanceolate, entire terminal lobes and linear lateral lobes. Inflorescence ovoid, 15–20 mm long, borne on a slender sulcate ± setose peduncle. Involucral bracts 3–5 × 1.5–3 mm, oblong-ovate, obtuse, pubescent. Involucral bracteoles 8–15 × 4–7 mm, oblanceolate, the outer acuminate, the inner shortly cuspidate, pubescent, the upper margins ciliate. Calyx aristate with bristles 2.5–3.5 mm long, sericeous. Corolla cream, 12–14 mm long, the lobes about one third the length of the tube. Involucel in fruit 6–8 mm long, sericeous, 8-angled with 2 broad ridges produced into long bristles which are usually longer than the tube, the 6 intermediate ridges with short teeth. Fig. 111: 8.

Hab. Destroyed *Quercus* woods, on limestone, clay slopes, rocky places; alt.: (110–)600–2400 m; fl. Jun.–Jul.
Distrib. Widespread in the forest zone of the mountain region of Iraq. **mam**: Jabal Khantur N of Zakho, *Rechinger* 10778; **mro**: Mt. Qandil between Shahidan and Bisht Ashan, *id.* 11007; between Harir and Baba Chichak, *Bornmüller* 1295; Bisht Ashan, *Rawi & Serhang* 23871!, 24191! **msu**: between Dukan and Mirza Rustam, *Rechinger* 11935; Mt. Hawraman nr Tawila, *Rechinger* 10256; Darband-i Bazian nr. Chemchemal, *id.* 10595; Pira Magrun W of Sulaimaniya, *Thesiger* 1129, *Haussknecht* s.n.; Darband-i Khan, *Sabri & Barkley* 9273, *Poore* 566!; Qara Dagh, *Buthaini Makki* 489; 54 km SE of Mandali, *Rechinger* 12746; Gweija (Goizha) Dagh, *Rawi & Gillett* 11703!; Zalam nr Khormal, *Rawi, Husham (Alizzi) & Nuri* 29418!; **fni**: 55 km SE of Mandali, *Rawi* 20731!

W & S Iran.

3. **Cephalaria setosa** *Boiss. & Hohen.*, Diagn. ser. 1, 2: 107 (1843). — Type: Mt. Gara, *Kotschy* 572 (syn.); Boiss., Fl. Orient. 3: 118 (1875); Nábělek in Pub. Fac. Sci. Univ. Masaryk 35: 138 (1923); Zohary, Fl. Palest. ed. 2, 1: 612 (1932); Blakelock in Kew Bull. 3: 442 (1948); Zohary in Dep. Agr. Iraq Bull., 31: 141 (1950); Lack in Fl. Iranica 168: 10 (1991); Jamzad, Fl. Iran 8: 17 (1993).

Erect freely branched annual herb up to 1 m high. Stems usually tuberculate, ± setose. Leaves ± setose, sessile or shortly petiolate; the lower lyrate-pinnatisect, 100–150 × 25–40 mm, with oblong-ovate, denticulate terminal lobes and oblong-linear lateral lobes; the upper leaves 20–50 × 6–20 mm, lanceolate, entire. Inflorescence ovoid, 10–15 mm long, borne on slender, sulcate ± setose peduncles. Involucral bracts ovate, 2–4 × 1.5–2 mm, acuminate, pubescent. Involucral bracteoles 7–9 × 3–4.5 mm, oblong, aristate, with bristles 1–2 mm, sericeous. Calyx aristate with long and short bristles, sericeous. Corolla cream, ± 9 mm long, lobes 3–4 mm long. Involucel in fruit 4–6 mm long, sericeous, 8-angled with 4 broad ridges 2 of which are produced into bristles equal in length to the tube and 2 with bristles a little longer than the tube and 4 intermediate narrow ridges bearing very short bristles less than one quarter the length of the tube. Fig. 111: 7.

Hab. On limestone, in *Quercus* forest, waste land; 600–1500 m; Fl. Jun.–Jul.
Distrib. Widespread in the forest zone of the mountain region of Iraq. **mro**: Gali Ali Beg, *Omar, Qaisi & Wedad* 49551!; Jabal Baradost N of Rowanduz, *Thesiger* 1201 pp.; Koi Sanjak, *Omar* 37608!; Rowanduz gorge, *Guest* 2977!; **msu**: 34 km W of Sulaimaniya, *Rechinger* 10572; Mt. Hawraman, *Haussknecht* s.n.; Pira Magrun, *Haussknecht* s.n.; **mam**: Mt. Gara, *Kotschy* 372 (type)!; Bamerne 20 km NW of Sarsang,

Qaisi & K. Hamad 45940!; 5 km S of Zakho, *Rechinger* 12131, *Rawi* 23083!; Sarsang, *Haines* 443 p.p.!; Ser Amadiya (Gali Mazurka), *Dabbagh, Qaisi & Hamid* 45969!; Zawita, *Guest* 4600!; Amadiya, Mazurka gorge, *Guest* 3786!; Jabal Bekhair nr. Zakho, *Rawi* 22997!; Dohuk, *Rechinger* 11491, *E. Chapman* 26281!; Zinta gorge, *id.* 26162!; Dinarta village, *id.* 26163!; **MAM /FUJ**: 103 km N of Mosul, *Rechinger* 10643; **FNI**: Hinnis nr Ain Sifni, *Guest* 3613!

Syria, Lebanon, Turkey, W Iran.

4. **Cephalaria syriaca** (*L.*) *Schrad.*, Cat. Sem. Hort. Gott. (1814) ex Roem. & Schult., Syst. Veg. 3: 45 (1818); Boiss., Fl. Orient. 3: 120 (1875); Náb. in Pub. Fac. Sci. Univ. Masaryk 35: 139 (1923); Zohary, Fl. Palest. ed. 2, 1: 612 (1932); Blakelock in Kew Bull. 3: 442 (1948); Zohary in Dep. Agr. Iraq Bull., 31: 141 (1950); V. Täckh., Stud. Fl. Egypt ed. 2: 517 (1974); Y. Nasir in Fl. Pak. 94: 11 (1975); Lack in Fl. Iranica 168: 12 (1991); Jamzad, Fl. Iran 8: 21 (1993).

Scabiosa syriaca L., Sp. Pl. ed. 1, 98 (1753).
Cephalaria boissieri Reuter in Boiss., Diagn. ser. 2, 2: 122 (1856).
C. syriaca subsp. *emigrans* Szabo in Nat. Levm. Kozl. 38: 182 (1940).

Erect usually freely branched annual herbs up to 150 cm high. Stems ± tuberculate, ± setose. Leaves ± setose, ± ciliate, sessile or shortly petiolate, oblong-lanceolate to ovate-elliptic, 30–50 × 15–35 mm, lyrate below, entire or denticulate. Inflorescence ovoid, 10–20 mm long, borne on a long, slender, sulcate ± setose peduncle or sessile in the forks of the branches. Involucral bracts long-ovate or broadly triangular, the outer acuminate, 8–12 × 2–5 mm, oblong, the inner aristate, pubescent. Involucral bracteoles 8–12 × 3–5 mm, oblong, abruptly attenuate, aristate, pubescent, ciliate on the upper margins. Calyx aristate, with bristles up to 1–1.5 mm long, sericeous or rarely pubescent. Corolla blue, exterior sericeous, 8–14 mm long, lobes 3–5 mm long. Involucel in fruit 4–5 mm long, long-sericeous, 8-angled with 4 ridges produced into a bristle equal or subequal in length to the tube and 4 intermediate smaller ridges with bristles of variable length but usually about one quarter the length of the tube. Fig. 111: 9–10.

HAB. Clearings in *Quercus* woods, serpentine screes, rocky places, sandy clay, weed of lettuce patch, cultivated fields (*Triticum, Linum*); alt. (200–)600–1500 m; fl. Apr.–Jun.
DISTRIB. Very common in the forest zone of the mountain region of Iraq, widespread in the steppe zone, rare in the northern desert zone. **MJS**: Asi, *Rawi* 8481!; Jabal Golat between Ain Tellawi & Balad Sinjar, *Field & Lazar* 445!; **MRO**: Handren, at foot of Jabal Baradost, *Thesiger* 824, 1201; Handren, E of Rowanduz, *id.* 1069; Rowanduz, *Bornmüller* 1296; Kuh Sefin above Shaqlawa, *Bornmüller* 1302; **MSU**: Halabja, *Ahmad & Jabar* 50118!; Sulaimaniya, *Haines* s.n.; 40 km SW of Sulaimaniya, *Stutz* 1380; Malakawa pass, *Rechinger* 12280; Jarmo, *Helbaek* 1707;; **MAM**: Dohuk, *Makki Beg* 3246!; Zawita nr Dohuk, *Guest* 3687!, *Chapman* 26297!; Ba'idra nr Shaikhan, *Salim* 2595!; Aqra, *Rawi* 11407!; Zawita gorge, *Chapman* 26136!; Sarsang, *Haines* 443 p.p.; above Sawara Tuka, *Rechinger* 11573; Ispindari saddle, Sawara Tuka, *E. Chapman* 26331!; SW of Aqra, *Anders* 1276; Jabal Bekher nr Zakho, *Field & Lazar* 835!**DLJ**: 50 km SW of Hadhr, *Rawi, Chakr., Khatib & Alizzi* 33076!; **FAR**: Ankawa, *Bornmüller* 1301; **FAR/FKI**: Kirkuk to Arbil, *Thesiger* 459; **FKI**: 37 km E of Kirkuk, *Rechinger* 10041; Kifri, *Rawi & Gillett* 7436!; nr Kirkuk, *Regel* 778; **FUJ**: Tal Kochak, *Gillett* 10818!; S of Sinjar, nr Bi'aj, *Guest* 13397! *Rawi & Hamada* 34165!; 50 km NW of Hatra, *Ani & Brahim* 8781!; **FNI**: Nineveh, *Chapman* 25375! *Noë* s.n.; Nimrud, *Helbaek* 877!; **FNI**: Jabal Hamrin, *Bornmüller* 1300; **DLJ**: 4 km W of K.2, Jezira desert, *Barkley, Palmatier & Juma Brahim* 2060!; **LCA**: Baghdad, *Colvill* s.n.! *Rogers* 0266!; Karrada Mariam, *Haines* W. 176! **LEA**: Ba'quba, *Schläfli* 86!; Nahrwan, *Robertson* s.n.! "Mesopotamia", *Aucher* 781!; **LCA**: Sudur, *Omar, Fawzi, Qaisi & Nuri* 44287!
NOTE. A very variable species. I have examined some of the material cited as subspecies by Szabo but have been unable to recognise his divisions. Some plants from Iraq have narrower entire leaves and a more erect, less spreading inflorescence, smaller capitulum and a less distinctly bristly calyx. However, many intermediate plants showing different combinations and ranges of variation of these characters occur also.
ZÎWÂN (Kurd.), Dohuk (Makki Beg 3246), Jazira S of Sinjar (Guest 13397); "fodder plant", (Salim 2595). MARUR (Kurd.). Worse than *Saponaria* as a taint to wheat. This plant, by contaminating grain in a time of shortage, making the bread dark and smelly, reputedly nearly caused a revolt in Baghdad in 1948–9. The grain is difficult to remove from the wheat; moisten the mixture somewhat before milling, then the weed seed will flatten out and not break up, and may be removed from the flour by sieving (*Rawi & Gillett* 7436).

Mediterranean area, Egypt, Arabia, Palestine, Turkey, Armenia, Azerbaijan, Georgia, Iran, Afghanistan, Pakistan, C Asia.

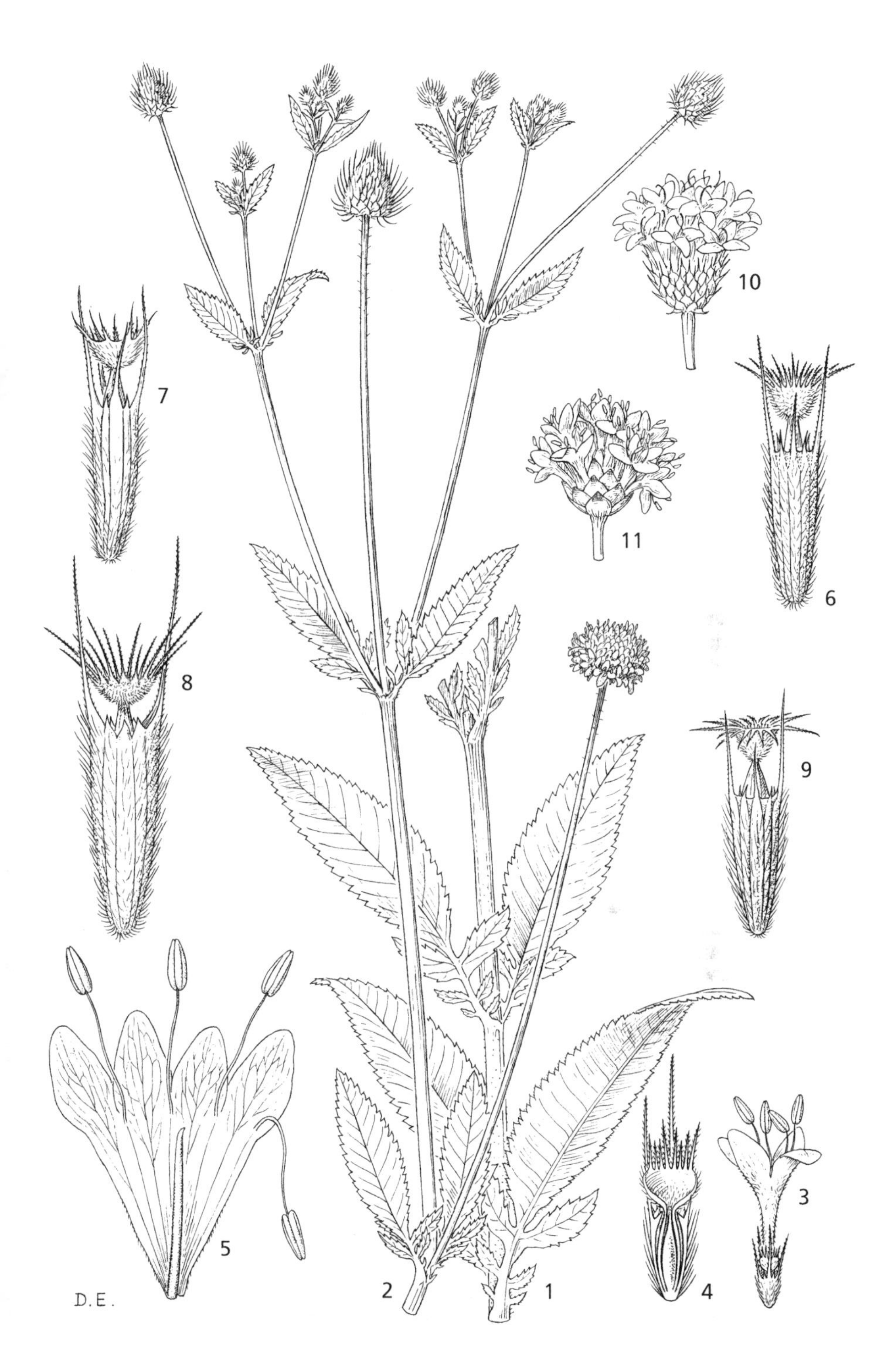

Fig. 111. **Cephalaria stapfii**. 1, section of lower stem with leaves × 1/3; 2, inflorescence × 1/3; 3, flower × 2; 4, involucel of flower × 6; 5, corolla opened × 3; 6, fruiting involucel with persistent calyx × 6. **Cephalaria setosa**. 7, fruiting involucel with persistent calyx × 6. **Cephalaria dichaetophora**. 8, fruiting involucel with persistent calyx × 6. **Cephalaria syriaca**. 9, fruiting involucel with persistent calyx × 6; 9, flowering head × 2. **Cephalaria microcephala**. 11, flowering head × 2. 1 from *Handel-Mazzetti* 1210; 2–5 from *Nabelek* 772; 6, from *Guest* 2977; 7 from *Rawi & Serhang* 23871; 8, 9 from *Chapman* 26173, 10 from *Gardner* s.n. Drawn by D. Erasmus.

5. **Cephalaria stapfii** *Hausskn. ex Bornmüller.* in Beih. Bot. Centralbl., 19 (2): 268 (1906); Hand-Mazz. in Ann. Naturh. Mus. Wien, 27: 429 (1913); Náb. in Pub. Fac. Sci. Univ. Masaryk 35: 139 (1923); Zohary in Dep. Agr. Iraq Bull., 31: 141 (1950); Lack in Fl. Iranica 168: 11 (1991).

Erect, freely branched annual herbs up to 120 cm high. Stems sparsely tuberculate, ± setose below. Leaves glabrous, sessile, 30–100 × 8–25 mm; the lower lyrate-pinnatisect, serrate; the upper oblong-lanceolate, sometimes ciliate. Inflorescence spherical, 8–12 mm across, borne on a sulcate glabrous peduncle. Involucral bracts ovate, obtuse, 3–4 × 2–3 mm; the inner sometimes cuspidate. Involucral bracteoles 5–7 × 2–4 mm, oblong, abruptly attenuate, aristate, with bristles 3.5–5.5 mm long; the inner bracteoles sparsely pubescent, ciliate on the upper margins. Calyx with a few long and numerous short bristles, sericeous. Corolla cream or blue, lobes 4.5–5.5 mm long. Involucel in fruit 4–5 mm long, sericeous, 8-angled with 2 ridges produced into long bristles equal in length to the tube, 2 ridges produced into bristles subequal in length to the tube, and 4 intermediate ridges with short teeth usually less than half the length of the tube. Fig. 111: 1–6.

HAB. Unrecorded.
DISTRIB. Very rare in Iraq. **FUJ**: N of Mosul, *Anders* 1543, 1565. (No material at K).

Turkey (SE Anatolia).

3. SCABIOSA L.

Sp. Pl. ed. 1, 98 (1753); Gen. Pl. ed. 5: 43 (1754)

Annual, biennial or perennial herbs, rarely suffruticose. Stems usually hairy. Inflorescence capitate. Involucral bracteoles minute or absent. Involucel in fruit 8-ridged with a campanulate or rotate crown carrying a variable number of teeth. Calyx limb patelliform, stipitate or nearly sessile with 5 bristles. Corolla 5-fid.

Scabiosa (from Lat. *scabies*, itch, which the rough (scurfy) leaves might have been used to cure.)

1. Plant perennial 2
   Plant annual 3
2. Leaves entire, dentate or rarely pinnatipartite-lyrate; involucral bracts 5–8 mm long and about equalling the inflorescence .... 1. *S. brachycarpa*
   Leaves pinnatipartite into linear-oblong lobes; involucral bracts 8–20 mm long and usually longer than the inflorescence .... 2. *S. argentea*
3. Involucral bracts at least 1.5 × the length of the inflorescence 4
   Involucral bracts shorter than or about equal in length to the inflorescence 5
4. Corolla 4–5 mm long; calyx bristles 6–8 mm long .... 3. *S. divaricata*
   Corolla 7–10 mm long; calyx bristles 12–16 mm long .... 4. *S. macrochaete*
5. Inflorescence radiant 6
   Inflorescence not radiant 8
6. Corolla 18–30 mm long; calyx stalk 3–4 mm long .... 7. *S. leucactis*
   Corolla 10–16(–20) mm long; calyx stalk 1–2 mm long 7
7. Involucel crown 5 mm broad .... 6a. *S. palaestina* var. *palaestina*
   Involucel crown 3–4 mm broad .... 6b. *S. palaestina* var. *persica*
8. Corolla 3–5 mm long; involucral bracts ovate; veins of the involucel crown densely hairy .... 5. *S. olivieri*
   Corolla 6 mm or more long; involucral bracts not ovate; veins of the involucel crown sparsely hairy or glabrous 9
9. Calyx bristles 12–16 mm, long exserted for 10–14 mm from the involucel crown .... 4. *S. macrochaete*
   Calyx bristles 4–6 mm long, not exserted from the involucel crown .... 8. *S. porphyroneura*

1. **Scabiosa brachycarpa** *Boiss. & Hohen.*, Diagn. ser. 1, 2: 113 (1843). — Type: Iraq, Mt. Gara, "Pl. Mesopot., Kurdistan & Mossul", *Kotschy* '255'!; id., anon., [Kotschy] 304 (cited by Boissier in error as 504 in Fl. Orient.; both specimens are types – G, K, P, syn.); Boiss., Fl. Orient. 3: 140 (1875); Blakelock in Kew Bull. 3: 443 (1948); Zohary in Dep. Agr. Iraq Bull. 31: 141 (1950).

Erect perennial herb up to 70 cm high. Stems tomentose-canescent, arising from a woody basal stock. Leaves opposite, tomentose-canescent, 40–70 × 5–12 mm; the lower oblong to narrowly elliptic, entire, dentate or more rarely irregularly pinnatipartite-lyrate; the upper linear-lanceolate, entire. Inflorescence hemispherical, 10–15 mm across, radiant, borne on long slender peduncles 8–14 cm long. Involucral bracts 5–8 × 1–2 mm, linear-lanceolate, acute, tomentose, about equal in length to the inflorescence, often becoming recurved or deflexed in fruit. Corolla white, 8–12 mm long; lobes 4–6 mm long. Involucel in fruit 2.5–4.5 mm long; tube 1.5–2.5 mm long, sulcate for about a third of its length; crown 1–2 mm broad, denticulate, 24-veined, sparsely hairy on the outer surface. Involucral bracteoles a dense tuft of sericeous hairs equal or subequal in length to the involucel tube. Calyx subsessile, bristles 3–4 mm long, scabrid. Fig. 112: 14.

HAB. Rocky plateau, *Quercus* forest; alt. 800–1200 m.; fl. & fr. Jul.-Aug.
DISTRIB. **MAM:** Jabal Bekher, nr Zakho, 90 km N of Mosul, *Rawi* 23059!; Sarsang, *Haines* W. 521!; Badi, nr Dohuk, *Guest* 4426!; Zawita, *id.* 4542! 4577!; Mt. Gara, "Pl. Mesopot., Kurdistan & Mossul", *Kotschy* '255'!; id., anon., [*Kotschy*] 304 (types); Atrush, *Guest* 3641!; **MRO:** Jabal Safin nr Salah ad-Din, *Barkley* 32 Ir. 3381!; Rowanduz, *Omar, Sahira, Karim & Hamid* 38356!, 38359!

Syria, Lebanon. Although this species has been cited for Iran, Jamzad (1993) makes no reference to it.

2. **Scabiosa argentea** *L.*, Sp. Pl. 100 (1753); Zohary, Fl. Palest. ed. 2, 616 (1932); Lack & Rech.f., in Fl. Iranica 168: 41(1991); Jamzad, Fl. Iran 8: 76 (1993).

S. ucranica L., Sp. Pl. ed. 2, 144 (1762); Boiss., Fl. Orient. 3:139 (1875); Náb. in Pub. Fac. Sci. Univ. Masaryk 35: 139 (1923); Zohary in Dep. Agr. Iraq Bull., 31: 141 (1950).

Erect perennial herb, up to 100 cm high. Stems ± adpressed-pilose or more rarely canescent, arising from a woody basal stock. Leaves opposite, pilose or adpressed-canescent, 30–70 × 20–30 mm; the lower pinnatipartite into linear-oblong lobes; the upper pinnatipartite or linear, undivided. Inflorescence hemispherical, 10–15 mm across, radiant, borne on long slender peduncles 8–14 cm long. Involucral bracts 8–20 × 1 mm, oblong–lanceolate, acute, hairy for about half of their length, usually longer than the inflorescence. Corolla white, blue or pink, 8–18 mm long; lobes 6–9 mm long. Involucel in fruit 3.5–5.5 mm long; tube 2.5–3.5 mm long, sulcate for about half of its length; crown 1–2 mm broad, denticulate, 20–24-veined, sparsely hairy on the outer surface. Involucral bracteoles a dense tuft of sericeous hairs equal or subequal in length to the involucel tube. Calyx shortly stipitate, bristles 4–6 mm long, scabrid. Fig. 112: 11.

HAB. *Quercus* forest, mountain slopes, clay hillsides; alt. 1050–1180 m; Fl. Jun.–Jul.
DISTRIB. Occasional in the forest zone of NE Iraq. **MAM:** between Mosul and Zakho, 103 km N of Mosul, *Rechinger* 10651; between Dohuk and Amadiya, 12 km E of Dohuk, *Rechinger* 11498; Mt. Gara, *Kotschy* 551, 302; Diyala: Kani Masi, *Hamad & Fathil* 45628!. **MRO:** Rowanduz, *Omar et al.* 38359; Mt. Sefin, *Karim* 39397!; **MSU:** Amoret, nr. Qaradagh, *Haines* W.1151!; Penjwin, *Rawi* 22572!; nr Penjwin, *Rechinger* 10548.

Mediterranean area from Spain and Algeria to the Balkans, Turkey, Syria, Lebanon, Palestine, Jordan, N & W Iran, Crimea, S Russia, Caucasus.

3. **Scabiosa divaricata** *Jacq.*, Hort. Vindob. 1: 5 (1770); Zohary, Fl. Palest., ed. 2, 1: 616 (1932); Blakelock in Kew Bull. 3: 143 (1948); Jamzad, Fl. Iran 8: 105 (1993).

S. sicula L., Mant. 2: 196 (1771); Boiss., Fl. Orient. 3: 142 (1875); Zohary in Dep. Agr. Iraq Bull 31: 141 (1950).

Erect little-branched annual herb up to 30 cm high, pubescent. Leaves pinnatipartite or lyrate with linear-oblong lobes, 25–40 × 6–12 mm. Inflorescence hemispherical, 8–12 mm across, borne on slender peduncles 6–12 cm long. Involucral bracts 15–25 × 2–4 mm, linear-oblong, 1½ – 2 × the length of the inflorescence, becoming recurved or deflexed in fruit. Corolla pink, mauve or cream, 4–5 mm long; lobes ± 1 mm long. Involucel in fruit 4.5–6.5 mm long; tube 2–3 mm long, sulcate to half of its length; crown 2.5–3.5 mm broad, denticulate, 20–24-veined, sparsely hairy on the outer surface. Involucral bracteoles a dense tuft of sericeous hairs about half the length of the involucel tube. Calyx shortly stipitate, bristles 6–8 mm long, scabrid. Fig. 112: 17.

HAB. Dry river bed, stony open places, limestone mountain with coppiced *Quercus aegilops*; alt. 900–1000 m.; fl. Jun.
DISTRIB. Occasional and scattered in the mountain zone of Iraq. **MAM**: 1 km beyond Ispindari Pass, Sawara Tuka, *E. Chapman* 26351! **MRO**: Shaqlawa, *Gillett* 11294!; Mt. Safin, nr Shaqlawa, *Bornmüller* 1306!; **MSU**: Jarmo, *Helbaek* 1723!. **FKI**: Kani Domlan hills, Kirkuk, *Guest* 4290B!

Balkans, Aegean Is., Cyprus, Syria, Turkey, Lebanon.

4. **Scabiosa macrochaete** *Boiss. & Hausskn.*, Boiss., Fl. Orient. 3: 143 (1875). — Type: Iraq, Mt. Pira Magrun, *Haussknecht* s.n. (K)!; Zohary in Dep. Agr. Bull. 31: 141 (1950); Lack & Rech.f., in Rech. f., Fl. Iranica 168: 44 (1991); Jamzad, Fl. Iran 8: 87 (1993).

Erect annual herbs 20–40 cm high, adpressed-pubescent. Leaves oblong, undivided or sometimes pinnatipartite or lyrate with a large lanceolate terminal lobe and linear lateral lobes, 50–100 × 10–12 mm. Inflorescence hemispherical, 15–30 mm across, borne on short slender peduncles. Involucral bracts 10–20 × 2–5 mm, oblong, acute, densely sericeous at least on the lower half, about $1^1/_2$ × the length of the inflorescence, becoming recurved or deflexed in fruit. Corolla purple, tube 7–9 mm long; lobes ± 1 mm long. Involucel in fruit 8–10 mm long; 5–6 mm long, long-sulcate to about half of its length; crown 3–4 mm broad, denticulate, 25–30-veined, hairy on the outer surface. Involucral bracteoles a dense tuft of sericeous hairs subequal in length to the involucel tube. Calyx stipitate, stalk 2 mm long, bristles 12–16 mm long, scabrid. Fig. 112: 15.

HAB. Vineyards, cultivated area near stream, rocky mountain slopes, dry rocks under *Quercus* scrub; alt. (800–) 1350–1650 m.; fl. Jun.
DISTRIB. Widespread in the mountain region of Iraq. **MRO**: Mt. Kharokh between Saran and Kilki, *Kass & Nuri* 27350!; nr Dergala village, *id.* 27700!; Mt. Sakri Sakran nr Rowanduz, *Bornmüller* 1305!. **MSU**: Mt. Pira Magrun, *Haussknecht* s.n.; Qopi Qara Dagh, *Haines* W.1096!; Penjwin, *Rechinger* 10524, *Rawi* 12231!, 22648!; Pira Magrun, *Haussknecht* s.n. (type)!

NW Iran.

5. **Scabiosa olivieri** *Coult.*, Mem. Dipsac. 36 (1823); Boiss., Fl. Orient. 3: 141 (1875); Náb. in Pub. Fac. Sci. Univ. Masaryk 35: 139 (1923); Zohary, Fl. Palest. ed. 2, 1: 616 (1932); Bobrov in Fl. SSSR 24: 72 (1957); Y. Nasir in Fl. Pak. 94: 6 (1975); Lack & Rech.f. in Fl. Iranica 168: 55 (1991); Jamzad, Fl. Iran 8: 80 (1993).

Erect usually freely branched annual herbs up to 50 cm high, hirsute-tomentose. Leaves oblong-lanceolate, 20–50 × 5–15 mm; the lower undivided or dentate or more rarely pinnatipartite; the upper entire or often with a lobule at the base. Inflorescence hemispherical, 6–18 mm across, borne on short slender peduncles. Involucral bracts 2.5–3.5 × 2–5 mm, ovate, acute, about equal in length to the inflorescence. Corolla purple or white, 3–5 mm long; lobes 1.5–2.5 mm long. Involucel in fruit 3–4.5 mm long; tube 2–2.5 mm long; crown campanulate, 3–4 mm broad, denticulate, 20–24-veined, densely hairy on the outer surface. Involucral bracteoles a dense tuft of sericeous hairs about half the length of the involucel tube. Calyx shortly stipitate, stalk 1 mm long, bristles 4–8 mm long, scabrid. Fig. 112: 16.

HAB. Stony hillside, clay-gypsum hillside, dry steppe, sandy 'haswa' gravel desert, eroded sandstone, gravel soil near river, sandy washes in silty depression; alt. 45–900 m; fl. May–Jul.
DISTRIB. Widespread in the mountains, foothills and plains, extending into the southern deserts of Iraq. **MAM**: Mt. Qaradagh, nr Timara, *Karim, Hamid & Jasim* 40611!; Atrush, *Guest* 3621!, 3641A!; **MJS**: Jabal Qilat, *Field & Lazar* 432!; **FUJ**: Qaiyara, *Bayliss* 135!; 26 km E of Balad Sinjar, *Gillett* 11161!; Sharqat, *Rao* s.n.!; **FAR**: Qal'a Sharqat, *Graham* s.n.; **FAR/FKI**: Altun Kupri, *Guest* 4021!; **FKI**: 36 km from Tuz Khurmatu towards Injana, *Rechinger* 12187; Arbil to Altun Kupri, *Haussknecht* s.n.; nr. Kirkuk, *Rawi & Gillett* 11598!; 8 km E of Kirkuk, *Rawi & Gillett* 11598; **FKI/FNI**: Jabal Hamrin, *Adhaim, Agnew et al.* 1429; **FNI**: Koma Sank police station, nr Mandali, *Rawi* 20677!; 5 km W of Mandali, *Rechinger* 13396 (indumentum approaches *deserticola*); 10 km E of Mandali, *Rechinger* 12758; Mandali, *Haines* s.n.; 20 km N of Badra, *Rechinger* 12730; 5 km N of Zurbatiya, *Qaisi & Khayat* 50623A!; "deserto Mesopotamiae", *Aucher* 779; 25 km E of Badra, *Rawi* 20779!; 5 km SE of Al-Shihabi police station, *Rawi & Haddad* 2625!; **DLJ**: telegraph pole M90 between Baiji and Mosul, *Field & Lazar* 392!; Wadi Thirthar canal nr. Samarra, *Guest & Haidar* 18326!; **DLJ/DGA**: 26 km E of Samarra, *Rawi* 20375!; **DGA**: 50 km SE of Samarra, *Rawi & Shahwani* 25485!; 155 km NE of Sha'bani, *Chakr., Rawi, Khatib & Alizzi* 3292!; **DWD**: 12 km W of Ukhaidhir, *Rawi* 30834!; 16 km W of Falluja, *Chakr. & Rawi* 30289!; Ajrumiya, *Botany staff* 41855!; 15 km NW of Ramadi on road to Hit, *Rawi & Gillett* 6767!; 55 and 31 km W of Ramadi, *Rawi* 20866!,

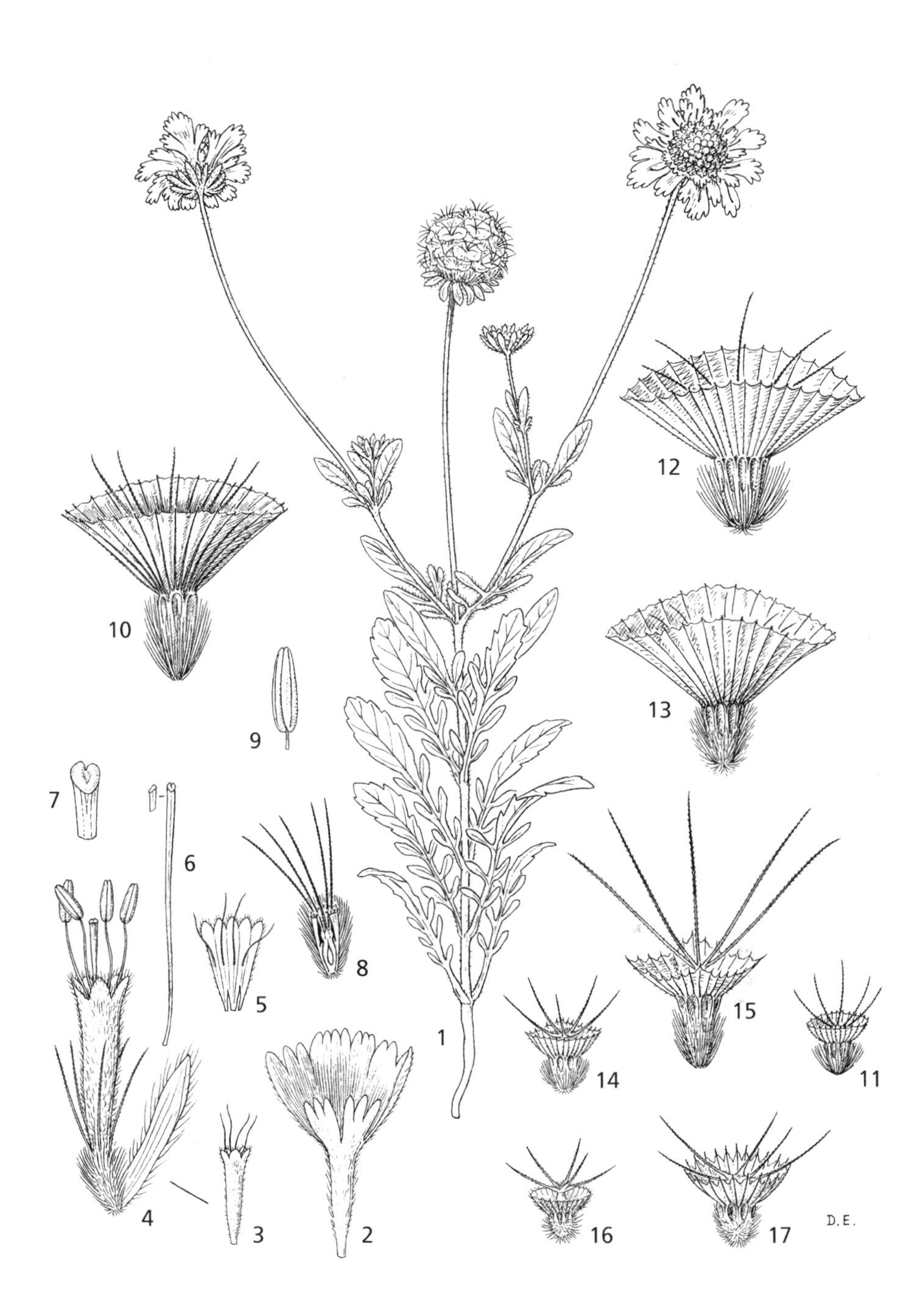

Fig. 112. **Scabiosa leucactis**. 1, habit × 1/2; 2, corolla of outer flower × 2; 3, corolla of inner flower × 2 ; 4, inner flower × 4 ; 5, corolla of inner flower opened × 2 ; 6, style × 4 ; 7, stigma × 12; 8, involucel longitudinal section × 4; 9, anther × 8; 10, fruiting involucel × 2. **Scabiosa argentea**. 11, fruiting involucel × 2. **Scabiosa palaestina** var. **palaestina**. 12, fruiting involucel × 12. **Scabiosa porphyroneura**. 13, fruiting involucel × 2. **Scabiosa brachycarpa**. 14, fruiting involucel × 2. **Scabiosa macrochaete**. 15, fruiting involucel × 2. **Scabiosa olivieri**. 16, fruiting involucel × 2. **Scabiosa divaricata**. 17, fruiting involucel × 2. 1–9 from *Rawi & Haddad* 25601; 10 from *Haines* W.938; 11, from *Schirajevsley* 4184b (Crimean material); 12 from *Rawi* 23098; 13 from *Rawi* 20973; 14 from *Rawi* 23059; 15 from *Rawi* 28648 ; 16 from *Rawi* 20866; 17 from *Gillett* 11294. Drawn by D. Erasmus.

20841!; **DSD:** 40 km WNW of Shabicha, *Guest, Rawi & Rechinger* 19316!; 35 km E by S of Shabicha, *Guest, Rawi & Long* 14074!; 8 km N by W.of Aidaha, *Guest, Rawi & Rechinger* 19154!; 30 km E of Salman, *id.* 18734!; 33 km WNW of Ansab, *id.* 19019!; nr Ansab, *id.* 18931!; 80 km ESE of Buhaira, *Guest & Rawi* 14288!; nr Safai al-Maghif, NE of Ghazlani, *Guest, Rawi & Rechinger* 17252!; Khadhar al-Ma'i, *Khatib & Alizzi* 32700!; 60 km SW of Khadhar al-Ma'i, *id.* 32741!; Umm-Qasr, *Alizzi & Omar* 35005!; 40 km from Khadhar al-Ma'i to Busaiya, *Qaisi, Hamad & Hamid* 48501!; 20 km from Salman to Takhadid, *id.* 48111!
NOTE. ?WARD ES-SA'AT (Ir.), "clock flower" (*Rawi & Gillett* 11598).

Turkey, Syria, Arabia, Palestine, W, C & E Iran, Turkmenistan, Afghanistan, Pakistan, C Asia.

6. **Scabiosa palaestina** *L.*, Mant. 37 (1767); Boiss., Fl. Orient. 3: 144 (1875); Zohary, Fl. Palest. ed. 2, 1: 617 (1932); Zohary in Dep. Agr. Bull. 31: 141 (1950).

Erect annual herbs up to 70 cm high, ± pubescent and often setose. Leaves opposite, oblong-lanceolate or oblong-linear, or linear, dentate or lyrate or pinnatipartite, 30–70 × 5–20 mm. Inflorescence hemispherical, 12–35 mm across, ± radiant. Involucral bracts 8–15 × 2–4 mm, linear to oblong-lanceolate, acute, subequal in length to the inflorescence. Corolla mauve, white or yellow, 10–16(–20) mm long; lobes 5–8 mm long. Involucel in fruit 9–12 mm long; tube 3–4 mm long, sulcate to about half of its length; crown campanulate, 6–8 mm broad, denticulate, 28–36-veined. Involucral bracteoles a dense tuft of sericeous hairs about half the length of the involucel tube. Calyx shortly stipitate, stalk 1–2 mm long, bristles 6–9 mm long, scabrid.

A very variable species. I have examined some of the material cited as varieties by Boissier but I have only been able to recognise two of his divisions.

6a. var. **palaestina**.

S. calocephala Boiss., Diagn. ser. 1, 10: 80 (1849); Patzak in Ann. Naturh. Mus. Wien 65: 23 (1962); Lack & Rech.f., Fl. Iranica 168: 47 (1991).
S. palaestina L. var. *calocephala* Boiss., Fl. Orient. 3: 144 (1875); Náb. in Pub. Fac. Sci. Univ. Masaryk 35: 140 (1923); Zohary, Fl. Palest. ed. 2, 1: 617 (1932); Blakelock in Kew Bull. 3: 143 (1948).
S. palaestina L. var. *genuina* Boiss., Fl. Orient. 3: 144 (1875).
[*S. rotata* (non Bieb.) Zohary in Dep. Agr. Iraq Bull. 31: 142 (1950].

Leaves entire or deltate or lyrate with a large terminal lobe. Inflorescence usually 15–35 mm across. Involucel crown 5 mm, broad with calyx bristles scarcely exserted. Fig. 112: 12.

HAB. Grey sandstone in valley, mountain slope, degraded *Quercus* forest, sandy soil; alt. 105–800 m; fl. (Mar.–)Apr.–Jun.
DISTRIB. Scattered in the mountain and foothill zones of Iraq, rarely in the desert. **MAM:** Bekhair Mt. near Zakho, *Rawi* 22987!; 5 km S of Zakho, *Rawi* 23098!; **MRO**: Handren Dagh, *Gillett* 8304!; Khanzad mt. *Karim, Hamid & Jasim* 40866!; **MSU:** Qaranjir, *Rawi* 21605!; Jarmo, *Helbaek* 1185!, 1230!; **FKI**, banks of R. Zab near Kirkuk, *Grigg* 71!; 2 km S of Sa'diya, *Qaisi* 44213!; **FNI:** Jabal Hamrin (nr Table Mt.), *Guest* 1821!; 25 km E of Badra, *Rawi* 20775!; **DSD:** 15 km SE of Ashuriya, 50 km WNW of Shabicha, *Guest, Rawi & Rechinger* 19353!; Al Samah, *Khatib & Alizzi* 32753!
NOTE. There are a number of poorly branched plants with mauve flowers from Iraq which have been described as *S. palaestina* var. *calocephala* but I have found that these are very variable and I have been unable to find any reliable characters to distinguish them.

Turkey, Syria, Lebanon, Palestine.

6b. var. **persica** *Boiss.*, Fl. Orient. 3: 145 (1875); Zohary, Fl. Palest. ed. 2, 1: 618 (1932); Blakelock in Kew Bull. 65: 23 (1962).

S. persica Boiss., Diagn. ser. 1, 10: 81 (1849); Jamzad, Fl. Iran 8: 97 (1993).

Leaves usually pinnatipartite with linear lobes. Inflorescence 12–15 mm across. Involucel crown 3–4 mm wide with calyx bristles clearly exserted.

HAB. Dry slopes, mountainsides, open grassy slopes, rocky wadi; alt. 500–1900 m; fl. Jun.–Jul.
DISTRIB. Occasional, mainly in the moist steppe zone of Iraq. **MAM:** Tarawanish, 10 km E of Kani Masi, *Qaisi, Hamid & Hamad* 45720!; Dohuk gorge, *E. Chapman* 26266! **MRO**: Arl Gird Dagh (Helgord), *Guest & Ludlow-Hewitt* 2810!; **MSU:** Jarmo, *Haines* W.350!; Chemchemal, *Poore* 328; Qara Anjir 27 km E of Kirkuk, *Rechinger* 15858; **FNI**: Mindan between Aqra and Mosul, *Rawi* 11354; 40 km N of Mosul towards Ain Sifni, *Hossain* 1872!

Iran, Palestine.

7. **Scabiosa leucactis** *Patzak* in Ann. Naturh. Mus. Wien 65; 23 (1962); Lack & Rech.f. in Fl. Iranica 168: 45 (1991); Jamzad, Fl. Iran 8: 91 (1993).

Erect annual herb up to 30 cm high, puberulous-glabrescent. Leaves oblong-spatulate, rarely lyrate-pinnatipartite, 40–80 × 10–15 mm. Inflorescence hemispherical, 25–40 mm across. Involucral bracts oblong-lanceolate, acute, 10–15 × 2–3 mm, shorter than the inflorescence. Corolla white, radiant, 18–30 mm long; lobes 9–14 mm long. Involucel in fruit 12–14 mm long; tube 4–6 mm long, sulcate to about half of its length; crown 8 mm broad with 30–36 veins. Involucral bracteoles a dense tuft of sericeous hairs 5–6 mm long. Calyx stipitate, stalk 3–4 mm long; bristles 7–9 mm long, exserted for 3–4 mm beyond the crown of the involucel. Fig. 112: 1–10.

DISTRIB. **MAM:** 60 km from Mosul to Dohuk, *Karim* 37500; Between Aqra and Mosul, *Rawi* 11370; **MRO:** Kuh Sefin nr Shaqlawa, *Bornmüller* 1309; **MSU:** E of Chemchemal, *Rechinger* 10054; **FUJ:** Mosul, *Anders* 2797!; 20 km NE of Mosul, *Salim Safajy* s.n.; **FNI:** Hinnis 10 km N of Ain Sifni, *Thesiger* 622; **FUJ/FKI**, Fatha NW of Jabal Hamrin, *Calder* 2087; **FKI:** Altun Kupri, *Bornmüller* 1308; **FKI/FNI:** Jabal Hamrin, *Haines* s.n., *Sutherland* S-31; **FNI:** Khanaqin, *Haines* 938; 54 km SE of Mandali towards Badra, *Rechinger* 12740; 20 km N of Badra, *id.* 9691; 22–30 km SE of Badra, *Rechinger* 9133, 9234; Jabal Muwaila 70 km N of Amara, *Rechinger* 8907!, *Guest* et al. 17543; 25 km SW of Badra, *Qaisi et al.* 51346; 17 km SE of Zurbatiya, *Qaisi & Khayat* 50666; 10 km E of Zurbatiya, *Hikmat Abbas & Barucha* 7643; **DGA:** Adhaim, *Agnew* 1430; **LEA:** Shihabi police post, *Rawi & Haddad* 2560; Fakka, 70 km E of Amara, *id.* 25773!

Iran.

8. **Scabiosa porphyroneura** *Blakelock* in Kew Bull. 3: 443 (1948). — Type: Iraq, Jazira Desert, Mar.–Apr. 1933, *C.J. Edmonds* 3803B (K!, holo.). Patzak in Ann. Naturh. Mus. Wien 65: 23 (1962); Lack & Rech.f. in Fl. Iranica 168: 53 (1991); Jamzad, Fl. Iran 8: 102 (1993).

S. aucheri Boiss. sensu Fl. Palest. ed. 2, 1: 618 (1932).

Small erect or spreading annual herb up to 20 cm high, pubescent. Stems branching from the base. Leaves 30–60 × 10–15 mm, lanceolate or linear-lanceolate, undivided or pinnatipartite-lyrate with a large lanceolate terminal lobe and linear lateral lobes. Inflorescence hemispherical, 15–20 mm across. Involucral bracts 10–18 × 2–5 mm, linear-lanceolate, acute, longer than the inflorescence. Corolla yellow, 8–10 mm long, scarcely radiant; lobes 2–4 mm long. Involucel in fruit 10–12 mm long; the tube ± 4 mm. long, sulcate to about half of its length; crown 6–8 mm broad with 24–28 purple veins. Involucral bracteoles a dense tuft of sericeous hairs 5 mm long. Calyx shortly stipitate, the stalk 1–2 mm long; bristles 4–6 mm long, scabrid, included within the crown of the involucel. Fig. 112: 13.

HAB. Limestone cliffs in dry steppe, rocky clay valley, sandy-clay soil, gravel, sandy washes, stony plain; alt. 150–500 m.; fl. Mar.–Apr.
DISTRIB. Widespread in the Desert region of Iraq. **DLJ:** Jazira desert, *C.J. Edmonds* 3803B (type)!; Um al-Matin 74 km from Baiji to Haditha, *Chakr., Rawi, Khatib & Alizzi* 31890!; midway between Rawa and Haditha in Jazira, *Omar, Qaisi, Hamad & Hamid* 451152!**DWD:** S bank of Euphrates opposite Rawa, *Rawi & Gillett* 7050!; Wadi Hauran nr Baghdad, *Omar, Qaisi, Hamad & Hamid* 44576!; Tal an-Nisr 48 km NE of Rutba, *Rawi* 31219!; 35 km W of Nukhaib, Rawi 14759!; 60 km N of Rutba, *Rawi, Chakr., Khatib & Alizzi* 32918!; 260 km NW of Ramadi, *Rawi* 20973!; bottom of Wadi Hawran, 12 km SE of H.1, *Chakr., Rawi, Khatib & Alizzi* 31621!; **DSD:** 40 km WNW of Shabicha, *Guest, Rawi & Rechinger* 19314!; 50 km N of As-Salman, *Rawi* 15869!; 33 km W of Ansab, on Saudi frontier, *Guest, Rawi & Rechinger* 19020!; Jabal Ana (Anaisa?), *Khayat & Hamad* 51721!; **FNI:** Koma Sank police station nr. Mandali on Iranian frontier, *Rawi* 20667!

Syria, Palestine, N & C Iran, Afghanistan, Pakistan.

4. PTEROCEPHALUS

Vaill. ex Adanson, Fam. 2: 152 (1763)

Herbs or shrubs with upright or spreading, woody or herbaceous, usually hairy stems. Inflorescence capitate. Involucral bracteoles hairy or absent. Involucel in fruit 8-ridged, sulcate, ending with minute teeth or in a short crown. Calyx limb short, stipitate with 12–24 plumose bristles. Corolla 5-fid.

Pterocephalus (from Gr. πτερο, *ptero*, feather and κεφάλι, *kephali*, head).

1. Plant annual 2
   Plant perennial 3
2. Fruiting involucel with a short membranous crown and an internal collar surrounding the neck of the ovary . . . . 1. *P. brevis*
   Fruiting involucel toothed at the margin; crown absent; internal collar lacking . . . . 2. *P. plumosus*
3. Dwarf shrub to 15 cm high; stems procumbent or creeping 4
   Erect perennial, exceeding 15 cm in height 5
4. Plant usually densely caespitose, tomentose-canescent; leaves clearly divided, 20–45 × 6–8 mm . . . . 3. *P. pyrethrifolius*
   Plant usually lax and spreading in habit, glandular-pubescent; leaves rarely divided, 30–60 × 10–18 mm . . . . 4. *P. laxus*
5. Leaves entire . . . . 5. *P. canus*
   Leaves dentate or more deeply divided 6
6. Flowers yellow . . . . 6. *P. pulverulentus*
   Flowers white or pink 7
7. Corolla 16–20 mm; calyx bristles 10–14 mm . . . . 7. *P. strictus*
   Corolla 10–14 mm; calyx bristles 7–11 mm 8
8. Plant glabrous or sparsely hairy . . . . 8a. *P. kurdicus* var. *kurdicus*
   Plant glandular-hairy . . . . 8b. *P. kurdicus* var. *viscosissimus*

1. **Pterocephalus brevis** *Coult.*, Mem. Dipsac. 32 (1923); Burtt in Notes R. Bot. Gard. Edinb. 22: 280 (1957); Rech.f., Fl. Iranica 168: 31 (1991); Jamzad, Fl. Iran 8: 59 (1993).

P. involucratus (Sibth. & Sm.) Sprengel, Syst. Veg. 1: 384 (1825); Boiss., Fl. Orient. 3: 148 (1875); Náb. in Pub. Fac. Sci. Univ. Masaryk 35: 140 (1923); Zohary, Fl. Palest. ed. 2, 1: 619 (1932); Blakelock in Kew Bull. 3: 442 (1948); Zohary in Dep. Agr. Iraq Bull., 31: 142 (1950).

Erect annual herb up to 40 cm high, tomentose-canescent with glandular and eglandular hairs. Stems ± freely branched from the base. Leaves 30–100 × 15–25 mm, pinnatipartite, with acute, linear-oblong, pinnatifid, decurrent lobes. Inflorescence hemispherical, 10–20 mm across. Involucral bracts 10–20 × 2–3 mm, linear-lanceolate, acute, equal to or a little longer than the inflorescence. Corolla pink to dark purple, 10–13 mm long; lobes 5–6 mm long. Involucel in fruit 3–4 mm long, with a very narrow membranous crown and an internal collar surrounding the neck of the ovary. Involucral bracteoles absent. Calyx sessile to the neck of the involucel crown, with 12 awns 6–8 mm long and free almost to the base. Fig. 113: 9–10.

HAB. Stony soil on hillside, dry silt along sandstone outcrops, dry stony places; alt. 180–750 m.; fl. May–Jun.
DISTRIB. Widespread in the foothills, extending into the semi-desert zone. **MAM**: Dohuk, *Makki Beg* 3266!; **MRO**: Ziraga (Zirva?) S of Arbil, *Uvarov;* MSU, 4 km E of Qaranjir, *Rawi* 21626!; **MSU**: Jarmo, *Helbaek* 1580!, 1766!; Jarmo nr Chemchamal, *Haines*; 10-18 km from Kirkuk towards Sulaimaniya, *Rechinger* 10007b; **FUJ**: S of Mosul, *Anders* 1908, 2796; Qaiyara, *Bayliss* 106!; **FKI**: 103 km N of Mosul, *Rechinger* 12176!; **FKI/FNI**: Jabal Hamrin, *Sutherland* 31B, *Bornmüller* s.n.; **FNI**: 10 km E of Mandali, *Rechinger* 9645!; Koma Sang police station nr Mandali, *Rawi* 26670!; **FAR**: Makhmur, *Hunting Aerosurveys* 122 (R.B. 63)!; **DWD**: Gara to Rutba, *Rawi* 19930!
NOTE. ?GIYĀ BÎZHINK (Kurd.) (Dohuk, *Mekki Beg ar-Sidqi ash-Sharbati*).

Greece, Cyprus, Turkey, S. Iran, Syria, Palestine, Jordan, Egypt, Libya.

2. **Pterocephalus plumosus** (*L.*) *Coult.*, Mem. Dipsac. 31 (1823); Boiss., Fl. Orient. 3: 147 (1875); Nábělek in Pub. Fac. Sci. Univ. Masaryk 35: 140 (1923); Zohary, Fl. Palest. ed. 2, 1: 618 (1932); Zohary in Dep. Agr. Iraq Bull., 31: 142 (1950); Bobrov in Fl. SSSR 24: 51 (1957); Rech.f., Fl. Iranica 168: 32 (1991); Jamzad, Fl. Iran 8: 61 (1993).

Scabiosa papposa L., Sp. Pl. ed. 1, 101 (1753) – nec L., Syst. Nat. ed. 12, 112 (1767).
Knautia plumosa L., Mant. Alt. 197 (1771).
Pterocephalus plumosus (L.) Coult. var. *macrochaetus* Bornmüller in Beih. Bot. Centralbl. 58: 300 (1938).

Erect annual herb up to 40 cm high, tomentose with long and short glandular and eglandular hairs. Stems ± branched above. Leaves 30–150 × 6–50 mm, the lower oblong, entire, crenate-dentate or lyrate; the upper entire or pinnatipartite or lyrate with a large lanceolate terminal lobe and small linear lateral lobes. Inflorescence hemispherical, 15–30 mm across. Involucral bracts 12–24 × 2.5–4 mm, linear-lanceolate, acute, equal to or longer than the inflorescence. Corolla pink or purplish, 9–12 mm long; lobes 4–6 mm long. Involucel in fruit 4–5 mm long, with a toothed margin; crown and internal collar absent. Involucral bracteoles absent. Calyx shortly stipitate, with 11–12 awns, 9–11 mm long and free almost to the base. Fig. 113: 11–12.

HAB. Wet ground nr. wadi, rocky clay, dry river bed, mountain slope, on limestone at upper limit of *Quercus aegilops* forest; alt. 680–1450 (–2000) m.; fl. May–Jul.
DISTRIB. **MAM**: Sarsang, *Anders* 2320, *Haines* W.437!; **FKI**: between Mosul and Zakho, *Rechinger* 12165; Jabal Khantur nr Sharanish, *Rechinger* 12043; Amadiya, *Leatherdale* 92; Gali Zawita, *Qaisi, Khayat & Karim* 51124!; Khantur mt., NE of Zakho, *Rawi* 23307!; 5 km S of Zakho, *Rawi* 23119!; Mt. Bekher nr Zakho, *Rawi* 23020!; 1 km beyond Ispindari pass, Sawara Tuka, *E. Chapman* 26352!; Ser Amadiya (Gali Mazurka), *Dabbagh, Qaisi & Hamid* 45965!; **MAM/FNI**: Mindan between Aqra and Mosul, *Rawi* 11372!; **MRO**: Kuh Sefin nr Shaqlawa, *Bornmüller* 1312 (type of *P. plumosus* var. *macrochaetus*)!; **MSU**: Mt. Avroman (Hawraman), Susakan nr Tawila, *Rechinger* 10138; Mt. Avroman (Hawraman), above Darimar, *Gillett* 11847!; Kirkuk to Sulaimaniya, *Rechinger* 10584; Jarmo, *Helbaek* 1423, 1761!, *Haines* W. 351!; Darband-i Basian, *Rawi* 22820!; 10 km W of Tawila, *Rawi* 22147!; **MJS**: Jabal Sinjar, *Shrief & Hamad* 50253!
NOTE. ?SHALEMA (Kurd.), *Gillett* 11847.

Balkans, Greece, Turkey, Ukraine, Armenia, Syria, Lebanon, Palestine, Jordan, Caucasus, N, C & W Iran, Turkmenistan, Libya.

3. **Pterocephalus pyrethrifolius** *Boiss. & Hohen.*, Diagn. ser. 1, 2: 109 (1843). — Type: Iraq, Amadiya District, Chiya-e Gara, *Kotschy* 352 (K!, P, W, syn.); Boiss., Fl. Orient. 3: 149 (1875); Náb. in Pub. Fac. Sci. Univ. Masaryk 35: 140 (1923); Blakelock in Kew Bull. 3: 442 (1948); Zohary in Dep. Agr. Iraq Bull., 31: 142 (1950); Rech.f, Fl. Iranica 168: 25 (1991); Jamzad, Fl. Iran 8: 47 (1993).

Dwarf suffruticose shrub up to 10 cm high, usually densely caespitose, tomentose-canescent, ± glandular. Stems freely branched, procumbent or creeping, up to 25 cm long, often tortuous. Leaves 20–45 × 6–8 mm; the lower sometimes scarcely divided, oblong-ovate, crenate-dentate, cuneate and tapering to a short petiole with one or two small linear lateral lobes or pinnatipartite-lyrate with oblong-ovate, pinnatifid, terminal lobes and linear-oblong lateral lobes; the upper pinnatipartite-lyrate with oblong-cuneate, pinnatifid terminal lobes and linear or linear-oblong lateral lobes. Inflorescence hemispherical, 15–30 mm across. Involucral bracts 6–12 × 1.5–2.5 mm, oblong-lanceolate, acute. Corolla pink or lilac, scarcely radiant, 14–18 mm long; lobes 5–6 mm long. Involucel in fruit 2.5–3.5 mm long with a membranous crown 1–2 mm broad. Involucral bracteoles absent. Calyx shortly stipitate with 10–13 awns 14–16 mm long free to the base.

HAB. Rocky places, open gentle slopes, mountain summit; alt. 1880–2100 m; fl. Jun.-Aug.
DISTRIB. Occasional in the mountain zone of Iraq. **MAM**: Chiya-e Gara, *Kotschy* 352 (type)!; by the Turkish frontier at Sharanish N of Zakho, *Rechinger* 10983; Ser Amadiya, *Haines* W. 2066!, *Guest* 4983!; Zawita N of Zakho, *Rawi* 23601!; **MSU**: Mt. Hawraman, *Rawi, Chakr., Nuri & Alizzi* 19790!

Turkey, W Iran.

4. **Pterocephalus laxus** *I.K. Ferguson* in Kew Bull. 20: 249 (1966). — Type: Iraq, Amadiya District, Sevara Gaurik valley NE of Zakho, between Sharanish and Zawita, *Rawi* 23653 (K!, holo.); Rech.f., Fl. Iranica 168: 26 (1991).

Dwarf suffruticose shrub up to 15 cm high, usually lax and spreading in habit, glandular-pubescent. Stem freely branched, procumbent or creeping up to 35 cm long. Leaves 30–60 × 10–18 mm, ovate-elliptic, crenate-dentate, cuneate and tapering to a short petiole or occasionally divided below with a large ovate-elliptic terminal lobe and one or two small linear lateral branches. Inflorescence hemispherical, 25–30 mm across. Involucral bracts 14–21 × 2–4 mm, lanceolate, acute, shorter than the corolla. Corolla lilac, scarcely radiant, 15–20 mm long; lobes 4–5 mm long. Involucel in fruit 2–4 mm long with a membranous crown 1–2 mm broad. Involucral bracteoles absent. Calyx shortly stipitate with 10–11 awns 12–15 mm long, free to the base.

HAB. Mountain slopes, near water; alt. 1200–1500 m; fl. Jul.
DISTRIB. Locally common in the upper forest zone of N Iraq. **MAM**: Sevara Gaurik valley NE of Zakho,

between Sharanish and Zawita, *Rawi* 23653 (type); by Turkish frontier, pass between Sharanish and Masis N of Zakho, *Rechinger* 10872; above Basingera, *Rechinger* 11528; 25 km NE of Zakho, Sawara Hina, *Rawi* 23454!; Bamerne 20 km NW of Sarsang, *Qaisi & Hamad* 45918!; Dori (Dudi) village 5 km E of Kani Masi, *Omar & Qaisi* 45426!

Endemic.

5. **Pterocephalus canus** *Coult. ex DC.*, Prodr. 4: 653 (1830); Boiss., Fl. Orient. 3: 151 (1875); Zohary in Dep. Agr. Iraq Bull., 31: 142 (1950); Rech.f., Fl. Iranica 168: 28 (1991); Jamzad, Fl. Iran 8: 52 (1993).

Erect perennial up to 30 cm high, tomentose-canescent with long and short glandular and eglandular hairs. Stem freely branched, arising from a basal stock, woody below, herbaceous above. Leaves 45–75 × 10–15 mm, ovate-oblong, entire, tapering to a short petiole. Inflorescence hemispherical, 20–30 mm across. Involucral bracts 8–14 × 1–2.5 mm, linear-lanceolate, acute, subequal in length to the inflorescence. Corolla yellow, 10–12 mm long; lobes 3–4 mm long. Involucel in fruit 5–6 mm long with a short tomentose crown. Involucral bracteoles absent. Calyx shortly stipitate, stalk c. 1 mm, with 16–20 awns 6–10 mm long free almost to the base. Fig. 113: 1–8.

HAB. Rocky limestone slopes, *Rhus-Quercus* forest on hillside, by stream in *Juglans* grove; alt. 1200–1800 m; fl. Jun.
DISTRIB. Locally common in the mountain region of Iraq. **MSU**: Pira Magrun, *Haussknecht* s.n.!; nr. Penjwin, *Rechinger* 10475, *Rawi* 12251!; Malakawa pass, *Rechinger* 10433, *Rawi* 22502!, *Rawi, Husham (Alizzi) & Nuri* 29466!; Mt. Hawraman by Tawila, *Rechinger* 10249; Qopi Qaradagh, *Poore* 585!; Dara Tri, between Halabja and Tawila, *Rawi* 22011A!
NOTE. ?SARKHIRRA (Kurd.?), Penjwin (*Rawi* 12251).

SE Turkey, Iran.

6. **Pterocephalus pulverulentus** *Boiss. et Bal. ex Boiss.*, Fl. Orient. 3: 152 (1875); Zohary, Fl. Palest. ed. 2, 1: 619 (1932); Zohary in Dep. Agr. Iraq Bull. 31: 142 (1950).

Erect perennial up to 30 cm high, with crisped, white, tomentulose, glandular or eglandular hairs. Stems freely branched, woody below, herbaceous above. Leaves 20–60 × 5–10 mm, lanceolate, pinnatipartite with short ovate-oblong, denticulate or entire lobes. Inflorescence hemispherical, 8–10 mm across. Involucral bracts 9–11 × 1–2 mm, oblong-lanceolate, acute, subequal in length to the inflorescence. Corolla yellow. Involucel in fruit with a short hairy crown. Involucral bracteoles absent. Calyx with 16–20 awns.

NOTE. I have not seen any material of this species from Iraq and the record is based on a specimen cited by Boissier in *Fl. Orientalis* from "Sindjar Mesopotamiae", (**MJS**).

Jordan, Syria, Saudi Arabia, Palestine, Yemen.

7. **Pterocephalus strictus** *Boiss. & Hohen.*, Diagn. ser. 1, 2: 110 (1843); Hand-Mazz. in Ann. Naturh. Mus. Wien 27: 429 (1913); Náb. in Pub. Fac. Sci. Univ. Masaryk 35: 141 (1923); Blakelock in Kew Bull. 3: 442 (1948); Zohary in Dep. Agr. Iraq Bull., 31: 142 (1950); Rech.f., Fl. Iranica 168: 21 (1991).

Erect perennial up to 30 cm high, tomentose-pubescent with long eglandular and short glandular hairs, viscid. Stems freely branched, woody below, herbaceous above. Leaves 45–70 × 10–15 mm; the lower often cuneate, tapering to a short petiole, crenate-dentate, or occasionally weakly lyrate; the upper oblong-lanceolate, entire or denticulate or pinnatipartite into acute, linear-oblong lobes, sessile or shortly petiolate. Inflorescence hemispherical, 20–25 mm across. Involucral bracts 6–12 × 1.5–3 mm, oblong-lanceolate, acute, about half the length of the inflorescence. Corolla white or pink, 16–20 mm long; lobes 4–6 mm long. Involucel in fruit 3–4 mm long with a narrow membranous tomentose crown ± 0.5 mm broad. Involucral bracteoles absent. Calyx with 16–20 awns 10–14 mm long free almost to the base.

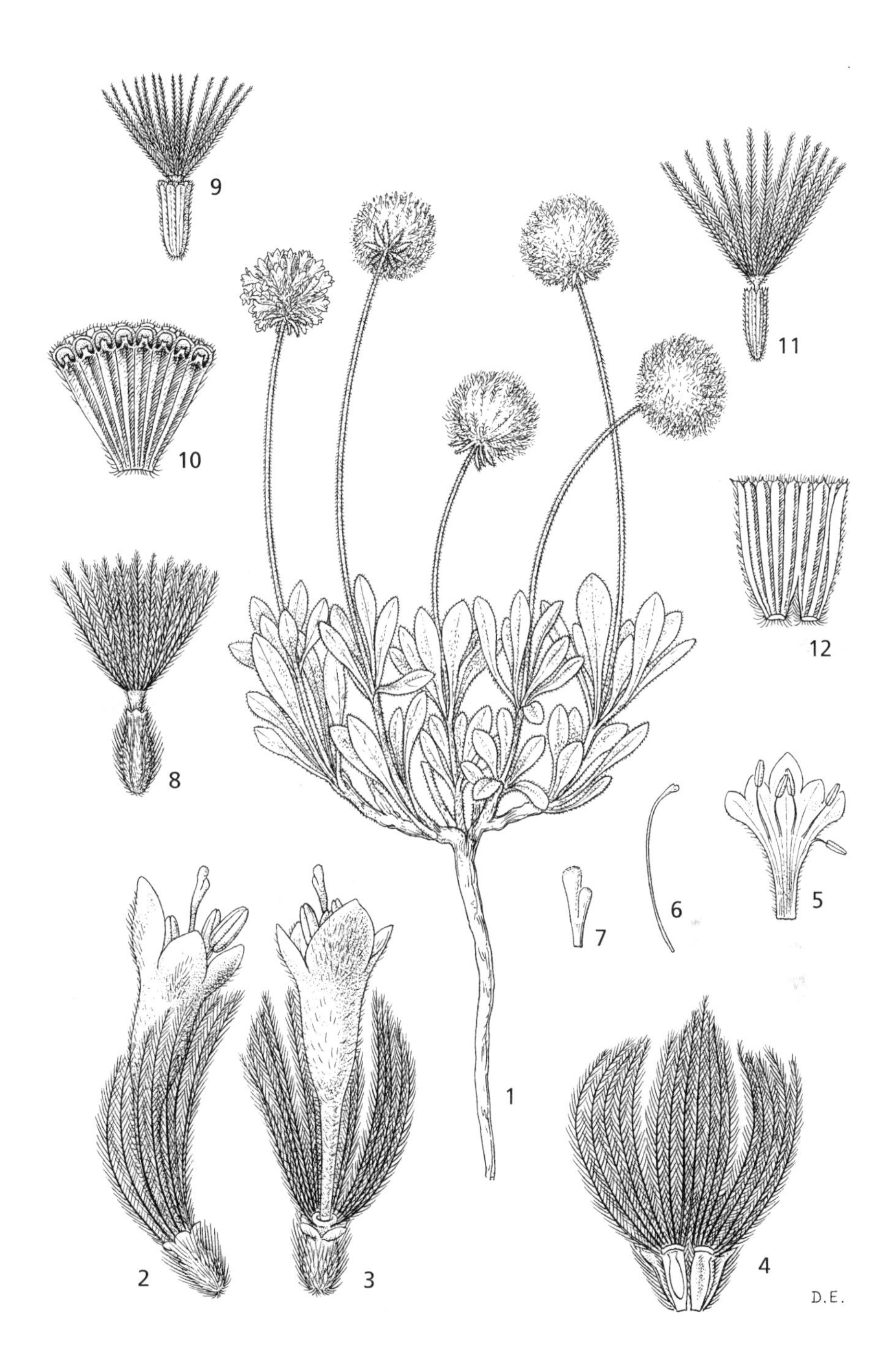

Fig. 113. **Pterocephalus canus**. 1, habit × 1/2; 2, flower side view × 6; 3, flower × 6 ; 4, involucel longitudinal section × 6 ; 5, corolla opened × 3 ; 6, style × 3 ; 7, stigma × 12; 8, fruiting involucel × 3. **Pterocephalus brevis**. 9, fruiting involucel × 3; 10, involucel opened × 6. **Pterocephalus plumosus**. 11, fruiting involucel × 3; 12, involucel opened × 6. 1–7 from *Rawi* 22502; 8 from *Rawi* 29466; 9, 10 from *Kotschy* 100; 11, 12 from *Rawi* 22147. Drawn by D. Erasmus

HAB. *Quercus* forest, grey soil with sandstone, limestone rocks, rocky mountain slopes; alt. 450–1400 m; fl. May–Jul.
DISTRIB. Quite local but abundant in the northern sector of the mountain zone. **MAM**: 23 km NE of Zakho, 2 km S of Sharanish, *Rawi* 23183!; Bikhair Mt. nr Zakho, *id.* 23005!; Dohuk gorge, *Guest* 3705!, 1586!; Zawita gorge, *id.* 3717! 4456!; Sawara Tuka, *Rawi* 9237!; Mt. Gara, *Kotschy* 307!; Atrush, N of Mosul, *Guest* 3624!, *Chapman* 9355!; 20 km NW of Sarsang, *Qaisi & Hamad* 45911!; 10 km S of Sarsang, *id.* 45812!; Sara Amadiya (Gali Mazurka), *Dabbagh, Qaisi & Hamid* 45987!; Yosifka (Aysifku) village 18 km NW of Sarsang, *Dabbagh & Hamid* 45767!; Zawita, *Dabbagh & Qaisi* 45320!; 10 km from Zawita to Atrush, *Qaisi, Hamad & Hamid* 46093!; Dohuk 12 km NE of Sarsang, *Dabbagh & Hamad* 46144!; **MRO**: Rowanduz gorge, *Guest* 2986!
NOTE. ?NISAK (Kurd.), Sawara Tuka, *Rawi* 9237.

Turkey.

8. **Pterocephalus kurdicus** *Vatke* in Linnaea 38: 734 (1874); Rech.f., Fl. Iranica 168: 22 (1991); Jamzad, Fl. Iran 8: 41 (1993).

Erect perennial up to 50 cm high, glabrous or hairy, ± glandular. Stems freely branched, woody below, herbaceous above. Leaves 30–65 × 8–16 mm; the lower cuneate, crenate-dentate, tapering to a short petiole; the intermediate lanceolate, entire or weakly lyrate with a lanceolate terminal lobe and linear lateral lobes; the upper leaves linear. Inflorescence hemispherical, 14–25 mm across. Involucral bracts 5–14 × 1.5–2 mm, oblong-ovate, acute. Corolla pink or lilac, 10–14 mm long; lobes 3–4 mm long. Involucral bracteoles absent. Calyx shortly stipitate with 14–24 awns 7–9 mm long free almost to the base.

A very variable species. Although I have recognised two varieties, the occurrence of glands on the indumentum varies a great deal in plants from Iraq that I have seen. *P. szovitsii* Boiss. is recorded from Iraq but I have not seen any authentic specimens. However, I have examined some of Boissier's type material from Iran and I have been unable to distinguish this from Iraq plants with glandular hairs described as *P. nestorianus* Nábělek. I have found that the number of calyx bristles in the type material of *P. szovitsii* at Kew varies from 1–18 and does not fit Boissier's description in *Fl. Orientalis*.

The leaves are broader and the inflorescence larger in this specimen compared with the type material of *P. nestorianus*, but these characters do not seem to be constant as intermediate plants occur.

8a. var. **kurdicus**

P. putkianus Boiss. & Kotschy in Boiss., Fl. Orient. 3: 150 (1875); Nábělek in Pub. Fac. Sci. Univ. Masaryk 35: 141 (1923); Zohary in Dep. Agr. Iraq Bull., 31: 142 (1950); Jamzad, Fl. Iran 8: 43 (1993).

Plant glabrous or sparsely hairy, eglandular. Involucral bracts 5–7 mm long.

HAB. Rocky clay mountainside; alt. 700–1550 m; fl. Jun.–Aug.
DISTRIB. Rare in the Mountain region of NE Iraq. **MAM**: ?Mt. Zalen (Zalam?), *Rawi, Husham (Alizzi) & Nuri* 29384!; **MSU**: Avroman (Hawraman), *Haussknecht* s.n.!

8b. var. **viscosissimus** *Bornmüller* in Beih. Bot. Centralbl. 58: 302 (1938); Blakelock in Kew Bull. 3: 442 (1948); Jamzad, Fl. Iran 8: 43 (1993).

P. nestorianus Nábělek in Pub. Fac. Sci. Univ. Masaryk 35: 141 (1923); Zohary in Dep. Agr. Iraq Bull., 31: 142 (1950).

Plant glandular-hairy. Involucral bracts 7–9 mm long.

HAB. on rocks; alt. 800–1500 m; fl. Aug.
DISTRIB. Mainly confined to northernmost Iraq. **MAM**: N of Atrush, ± 70 km N of Mosul, *M. Hossain* s.n.!; Amadiya, *Guest* 3778!; Amadiya, Mazurka gorge, *Guest* 2778!; **MJS**: Karsi, Jabal Sinjar, *Qaisi* 49753!, *Sharif & K. Hamad* 50263!, *Qaisi, Khayat & Karim* 50861!; Jabal Sinjar, N of the town, *Gillett* 11090!; **MJS/FUJ**: Jabal Khatchra nr. Balad Sinjar, *Field & Lazar* 657!

Distribution (of species): W Iran, Turkey.

114. CAMPANULACEAE

Juss., Gen. Pl. 163 (1789); Lammers in Novon 8(1): 36 (1998) & *World Checklist and Bibliography of Campanulaceae* (2007)

H. A. Alizzi[24]

Herbs or shrubs, rarely small trees, usually with milky sap. Leaves alternate or rarely opposite, simple, entire or toothed, petiolate or sessile, exstipulate. Flowers actinomorphic, hermaphrodite. Calyx-tube usually adnate to ovary; limb 3–10-lobed; lobes equal or slightly unequal, valvate or imbricate. Corolla 3–10-lobed, usually gamopetalous, campanulate or tubular, rarely rotate, valvate in bud, sometimes irregular and two-lipped. Stamens free or connate, inserted at the base of corolla or on disk, as many as corolla-lobes and alternate with them; anthers 2-locular, introrse. Ovary inferior or half-inferior, very rarely superior, usually 2–5-locular with axile placentas or rarely 1-locular with parietal placentas; ovules numerous in each loculus. Style simple, usually with pollen-collecting hairs; stigmas as many as loculi of ovary. Fruit a capsule or berry. Seeds usually small, sometimes winged, with an oily or starchy endosperm and a straight embryo.

A cosmopolitan family of 79 genera and 1900 species mainly distributed in temperate regions, the subtropics and on tropical mountains. Four genera in Iraq; *Sphenoclea* previously included in Campanulaceae is treated in a separate family Sphenocleaceae (Vol. 5, part 1).

Economically, the principal value of this family is horticultural, particularly the many species of the genus *Campanula* commonly cultivated as ornamental herbs in the gardens of Europe, America and elsewhere.

1. Calyx and corolla 8–10-lobed; stamens 8–10 . . . . 1. **Michauxia**
   Calyx and corolla 5-lobed; stamens 5 2
2. Corolla lobes free to the base or very shortly united . . . . .3. **Asyneuma**
   Corolla lobes distinctly united; corolla tubular, campanulate or rotate; flowers not in corymbose or umbellate cymes 3
3. Corolla rotate; plants annual . . . . .4. **Legousia**
   Corolla tubular or campanulate; plants often perennial . . . . .2. **Campanula**

1. MICHAUXIA L'Hérit., *nom. cons.*

L'Hérit., Monogr. (1788); Aiton, Hort. Kew. 2: 8 (1789); Lam., Illustr. 2: 433, t. 295 (1792).
Mindium Adans., *nom. rejic.*

Robust biennial herbs. Stems or branches usually erect or ascending, hollow. Leaves simple, scabrid, petiolate or sessile. Flowers sessile or pedicellate. Calyx 8–10-lobed with reflexed appendages between lobes. Corolla 8–10-lobed. Stamens 8–10, free; filaments dilated at base. Ovary inferior, 8–10-locular. Style hairy. Stigmas 8–10. Capsule hemispherical, dehiscent by 8 valves near the base. Seeds numerous, small.

Seven species distributed from the E Mediterranean to Caucasus and Iran.

Michauxia (in honour of the French botanist, André Michaux, 1746–1803); Dart Bellflower (Am.). Most of the species of *Michauxia* have been tried in cultivation and one or two are still sometimes grown for their flowers in European and American gardens, among them our native species *M. laevigata* Vent., q.v. Another species, *M. campanuloides* L'Hérit. ex Aiton, which Bailey (1953) terms "a conspicuous summer bloomer"... ... "though rather a coarse plant, of stately aspect when well-grown" ..., is popular in American gardens and makes a notable addition to the flower borders.

Plant up to 70 cm tall; inflorescence narrowly conical; appendages of calyx small . . . . 1. *M. nuda*
Plant more than 100 cm tall; inflorescence spicate; appendages of calyx large and covering calyx tube . . . . .2. *M. laevigata*

[24] † Formerly National Herbarium of Iraq, Baghdad. *Michauxia* & *Campanula* (in part) revised and updated by J. Osborne; *Asyeuma, Legousia* & *Campanula* (in part) updated by Shahina A. Ghazanfar.

1. **Michauxia nuda** *A.DC.* in DC., Prodr. 7: 457 (1839). — Type: "ad Ninevan" (Nineveh), *Aucher-Eloy* s.n. (K!, P, G, syn.); Boiss., Fl. Orient. 3: 891 (1875); Hand.-Mazz. in Ann. Naturh. Mus. Wien 27: 431 (1913); Nábělek in Publ. Fac. Sci. Univ. Masaryk 70: 3 (1926); Blakelock in Kew Bull. 4: 518 (1949); Zohary in Dep. Agr. Iraq Bull. 31: 148 (1950); Rech.f., Fl. Lowland Iraq: 585 (1964); Rawi in Dep. Agr. Iraq Tech. Bull. 14: 128 (1964).

Mindium nudum (DC.) Rech. f. & Schiman-Czeika, Fl. Iranica 13: 48 (1965).

Biennial herb. Stem to ± 70 cm high, glabrous or slightly scabrid with scattered small prickles, or pilose towards base. Leaves scabrid or sometimes pilose; radical leaves petiolate, oblong in outline, decurrent, 8–16 × 4–7 cm, irregularly lobed or pinnatipartite; petiole 5–8 cm; cauline leaves sessile, triangular-lanceolate, 5–8 ×1–2 cm, serrate. Inflorescence narrowly conical; lower flowers terminal on erect branches; lateral flowers pedicellate or subsessile; pedicel stout, glabrous, 15–20 mm. Calyx appendiculate, 8-lobed; appendages small, 1–1.5 × 1 mm. Calyx lobes triangular-lanceolate, scabrid along midrib and margin, 4–7 × 1–1.5 mm wide; calyx tube obconical, 4–5 × 3–4 mm wide, distinctly veined, glabrous or scabrid along veins. Corolla 8-lobed, 10–15 mm long, lobes shorter than tube, scabrid along veins, and hairy between the veins. Stamens 8, filaments slightly hairy, 5 mm with 1.4 mm wide base; anthers acute or linear-oblong, 8–10 × 1 mm. Ovary 8-locular. Style hairy, 15–20 mm; stigmas 8. Capsule obconical, 5 × 3–4 mm, opening by 8 valves near the base. Seeds numerous, small, glabrous, brown, 1.5 × 0.6–0.8 mm. Fig. 114: 1–8.

HAB. In the lower mountains on dry rocky slopes, limestone crags, crevices in cliffs; alt. 500–1400 m; fl. & fr. May–Jul.
DISTRIB. Quite common in the NE sector of the lower forest zone of Iraq. **MAM**: 85 km NW of Mosul, *Rawi* 23126!; Jabal Bekher, *Rawi* 22971!; Khantur, *Rawi* 23303!; Mt. Gara (Gara Dagh), *Kotschy* 489; Dohuk, *Rechinger* 11981; Zawita, *Guest* 4520!, *Omar* 37671!; Zawita gorge, *Guest* 3719!; Amadiya, *Haines* 1718!; Sulaf, *Omar* 37721!; 3 km W of Amadiya, *Rechinger* 11623?; Aqra, *Rawi* 11489!; **MAM/MRO**: Bekme gorge, *Gillett* 8224!; Arl Gird Dagh, *Kass & Nuri* 27200!; Gali Ali Beg, Karim, Hamide and Jasim 40907!; **MJS**: N of Balad Sinjar, *Gillett* 11092!; Asi, *Rawi* 8411!; **MJS/FUJ**: Jabal Khatchra, *Field & Lazar* 615! 656!; **FUJ**: Dubardan mountain (40 km from Mosul to Aqra), *Al Dabbagh & Jasim* 46820!; **FNI**: "ad Ninevan" (Nineveh), *Aucher-Eloy* s.n. [type].
NOTE. The precise location of Aucher-Eloy's type specimen labelled on the Kew Herbarium sheet as above and cited by Boissier in Boiss., Fl. Orient. as "in rupestribus aridis Mesopotamiae ad Ninivam" appears questionable: the ruins of Niniveh are situated across the Tigris from Mosul in a cultivated rolling plain in the moist steppe zone some 250 m below the lowest altitude recorded by any of the later collectors. Moreover the historic interest and accessibility of Niniveh ensures that it is visited by nearly all botanists who pass by Mosul and the species has never been found there again, nor so far from the mountains during almost the last two centuries, i.e. since 1826.

E Syria (Jabal Abdul-Aziz, Hand.-Mazz.).

2. **Michauxia laevigata** *Vent.*, Descr. Pl. Nouv.: 81 (1802); Graham in Curtis Bot. Mag. 59, t. 3128 (1832); Boiss., Fl. Orient. 3: 891 (1875); Bornmüller in Beih. Bot. Centralbl. 61B: 72 (1941) sub "levigata" sphalm.; Zohary in Dep. Agr. Iraq Bull. 31: 143 (1950); Fedorov in Fl. URSS 24: 386 (1957); Rawi in Dep. Agr. Iraq Tech. Bull. 14: 128 (1964); Damboldt in Fl. Turk. 6: 82 (1978).

Mindium laevigatum (Vent.) Rech. f. & Schiman-Czeika in Fl. Iranica 13: 47 (1965).

Biennial herb. Stem hollow, more than 100 cm high, glabrous or with few short prickles or slightly pilose towards base. Leaves scabrid, dentate, lower and radical leaves ovate-oblong, petiolate, 10–17 × 4–8 cm, decurrent; cauline leaves sessile, oblong, 3–7 × 1.5–2 cm. Inflorescence spicate; flowers pedicellate or sessile; pedicel stout, usually glabrous or with few prickles near the apex, 5–15 mm. Calyx appendiculate; appendages large, triangular-lanceolate, 5–6 × 3–4 mm, usually covering calyx-tube; calyx lobes 8, triangular-lanceolate, ± 10 × 3–4 mm, ciliate along margin and prickly along midrib; calyx tube cup-shaped, distinctly veined, scabrid along veins, 6 × 5 mm. Corolla 8-lobed, usually scabrid along veins, 15–20 mm long. Stamens 8; filaments hairy-papillose, 4–5 mm; anthers 2-locular, linear to linear-oblong, 12 × 1 mm, usually glabrous. Ovary 8-locular. Capsule 8-locular, obconical, dehiscent by 8 valves near base. Seeds numerous, small, ovate, glabrous, brown, ± 1.5 × 0.5 mm. Fig. frontispiece.

HAB. Rocky slopes and stony mountainsides, in oak forest, under walnut shade by a stream; alt. 1000–1600(–2000) m; fl. & fr. Jun.-Aug.

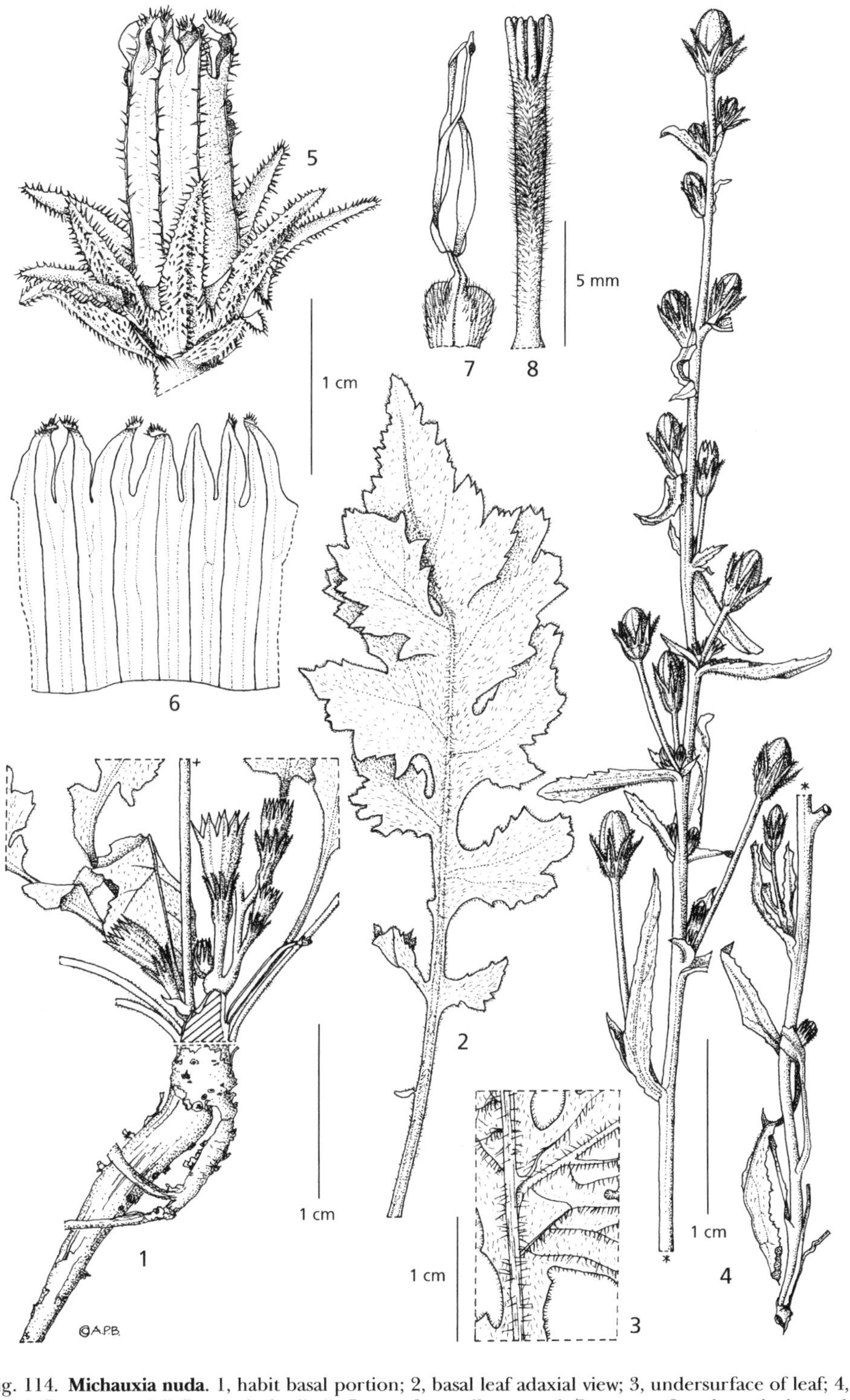

Fig. 114. **Michauxia nuda**. 1, habit basal portion; 2, basal leaf adaxial view; 3, undersurface of leaf; 4, inflorescence (all flowers in bud); 5, flower; 6, corolla opened; 7, stamen; 8, style and stigma. 1 from *Gillett* 8224 & *Rawi* 22971; 2 from *Wheeler Haines* W1718; 3, 4 from *Zamita*, 4 July 1976; 5–8 from *Omar et al.* s.n. © Drawn by A.P. Brown (Sep. 2012).

DISTRIB. Occasional in the SE sector of the upper forest zone of Iraq. **MRO**: Halgurd (Algurd Dagh), above Nawanda, *Bornmüller* 1558; Pushtashan, 15 km NE of Rania, *Rawi & Serhang* 23864!; Qerna Qaw valley (Qurnaqo), N of Pushtashan, *id.* 26596!; **MSU**: Pira Magrun, *Haussknecht* s.n.; Amoret, nr Qaradagh, *Haines* 1178!; Qopi Qaradagh, *id.* 1517!; Mela Kawa (Malakawa), on Sulaimaniya-Penjwin road, *Rawi* 22482!; Tawila, *Rawi* 21908!, *Rechinger* 10266.
NOTE. No local name has been noted for this species, which Chittenden (1951) terms an effective garden border plant. It was described as a horticultural subject by Graham nearly 180 years ago [Curtis Bot. Mag. 59: t. 3218 (1832)], having been cultivated in the Botanic Garden at Edinburgh from seeds brought from Iran in 1829, and would certainly appear to be worthy of trial in our gardens, particularly in Upper Iraq.

Caucasus, N & W Iran.

Michauxia tchihatcheffii Fisch. & C.A. Mey. is reported from Gali Balinda, northeast of Dohuk City near the Turkish border (Nature Iraq (2009), Key Biodiversity Survey of Kurdistan, N1-1209-03, unpublished report), but I have not seen any material of this species.

2. CAMPANULA L.

Sp. Pl. ed. 1: 163 (1753); Gen. Pl. ed. 5: 77 (1754).

Annual, biennial or perennial herbs. Stems erect or decumbent, usually hairy. Leaves alternate, petiolate or sessile, serrate or entire. Flowers hermaphrodite, solitary or in terminal corymbs or spikes, pedicellate or sessile. Calyx 5-lobed, with or without reflexed appendages in between the lobes; calyx tube usually adnate to the ovary. Corolla campanulate or tubular, rarely rotate, 5-lobed. Stamens 5, free; filaments usually dilated at the base; anthers 2-locular, linear or linear-oblong. Ovary inferior, 3–5-locular; ovules numerous. Stigmas 3–5. Capsule 3–5-locular, usually crowned with the persistent calyx-lobes, opening by 3–5 lateral valves at base or at the middle or beneath apex. Seeds small, numerous, yellowish to brown.

A genus of some 300 species, mostly perennial herbs, widely distributed over the northern hemisphere, particularly in the Mediterranean region and W Asia: 16 native species in Iraq.

Campanula (diminutive of the Lat. word *campana*, bell, referring to the bell-shaped form of corolla); bellflower. It is also sometimes loosely called Harebell, though this name is better reserved for the common harebell (*C. rotundifolia*); another source of confusion is that the latter plant is usually called bluebell in Scotland and the U.S.A. whereas in England the name bluebell is used for another plant (*Hyacynthoides non-scripta*) belonging to a different family (Liliaceae).

Originally, perhaps on account of their striking appearance rather than due to any intrinsic merit, some species of *Campanula* and of other genera were prized as medicinal plants. Linnaeus perpetuated an old herbalist name in his species *C. erinus* (p. 169), and the old generic names *Trachelium* and *Tracheliopsis* also remind us of this ancient use of these herbs – but, as medicinal plants, the Campanulaceae have long fallen into disfavour.

Though certain species contain alkaloids, none now provides medicinal products or essential oils for commerce; possibly for this reason they are generally palatable and some species play a minor role as constituents of our upland and mountain pastures, being sometimes named as such by the Kurdish shepherds. One edible species of *Campanula*, the rampion (*C. rapunculus*), though widely distributed throughout C & S Europe, N Africa and W Asia, has not been found in Iraq; this biennial plant was formerly much cultivated as a vegetable in European gardens for its tuberous roots which are eaten boiled or (with the lower leaves) raw as a salad.

However, it is as colourful ornamental garden herbs in borders and rockeries, generally with bell-shaped flowers of striking bluish-mauve hue, that the Campanulaceae are best known. The most popular cultivated species of bellflower is the Canterbury bell (*C. medium*), a favourite garden flower in Europe for over 400 years, of which there are many cultivars with large bell-shaped flowers ranging from blue and pink to mauve, purple and white. Bali (1946) lists this biennial among common garden flowers in Iraq: it is usually propagated from seed sown in autumn and flowers in late spring and early summer in the following year. There are several double forms, among them the well-known cup-and-saucer flower, known horticulturally as 'Calycanthema'; this is merely a garden form and not a true variety. Hussain & Kasim (1975) also list *C. medium* among garden plants now cultivated in Lower Iraq; but no specimens of this species from Iraq have yet been seen.

1. Calyx appendiculate (with appendages in between lobes) 2
   Calyx exappendiculate (without appendages in between lobes) 7
2. Flowers sessile 3
   Flowers pedicellate 6

3. Flowers surrounded by large involucral bracts .... 6. *C. involucrata*
   Flowers not surrounded by involucral bracts.................................. 4
4. Appendages in between calyx lobes inflated and covering the capsule; plant perennial .... 9. *C. stricta*
   Appendages in between calyx lobes not inflated; plant annual or biennial (*fastigiata*) 5
5. Stems fastigiated, somewhat succulent; leaves glabrous or pubescent, usually 3–5-toothed at apex; flowers axillary towards apex........... 5. *C. fastigiata*
   Stems slender, not succulent; leaves hispid, entire; flowers terminal ... 1. *C. propinqua*
6. Plants hispid all over; cauline leaves very small; flowers in panicles; calyx tube obconical, 1.5 mm × 1 mm, hirsute....................... 7. *C. radula*
   Plant glabrous or slightly scabrid near the apex; cauline leaves narrowly ovate, 3–8 × 1–4.5 cm; flowers in long spikes; calyx tube globose, 3–4 × 2–2.5 mm, hispid along veins .... 8. *C. sclerotricha*
7. Flowers sessile.. 8
   Flowers pedicellate 9
8. Flowers in dense clusters, surrounded by involucral bracts 1.5–2.5 cm long .... 10. *C. glomerata*
   Flowers 1–2, not surrounded by involucral bracts..................... 2. *C. erinus*
9. Leaves reniform to reniform-cordate.. 10
   Leaves linear to lanceolate, spatulate, elliptic, ovate to oblong 12
10. Leaves almost sessile to very shortly petiolate, < 4 mm; stems and leaves rather densely pubescent .... 13. *C. mardinensis*
    Leaves distinctly petiolate, > 4 mm; stems and leaves very shortly pubescent or glabrous 11
11. Leaves shortly pubescent, conspicuously and sharply toothed or lobed .. 12. *C. acutiloba*
    Leaves glabrous or subglabrous, shortly denticulate................ 14. *C. perpusilla*
12. Flowers terminal, usually solitary, ebracteolate; stigma with 3 long branches .... 11. *C. stevenii*
    Flowers axillary, in racemes or capitate in terminal corymbs, bracteolate; stigma shortly lobed.. 13
13. Flowers in a capitate terminal corymb or umbel; plants densely soft hairy .... 16. *C. postii*
    Flowers in racemes; plant indumentum not as above........................ 14
14. Racemes densely leafy; bracts conspicuous; calyx lobes ovate .... 15. *C. conferta*
    Racemes not densely leafy; calyx lobes linear to lanceolate 15
15. Stems flexuous, filiform, ± scabrid with short reflexed hairs or prickles; calyx lobes triangular-lanceolate, glabrous................. 4. *C. flaccidula*
    Stems flaccid and trailing, distinctly angled, scabrid with small reflexed hairs or prickles; calyx lobes linear or linear-lanceolate, scabrid, at least along the margins and veins, ± recurved............. 3. *C. retrorsa*

1. **Campanula propinqua** *Fisch. & C.A.Mey.*, Ind. Sem. Horti Petrop. 2: 32 (1836); Boiss., Fl. Orient. 3: 930 (1875); Nábělek in Publ. Fac. Sci. Univ. Masaryk 70: 6 (1926); Zohary in Dep. Agr. Iraq Bull. 31: 143 (1950); Fedorov in Fl. SSSR 24: 190 (1957); Rawi in Dep. Agr. Iraq Tech. Bull. 14: 128 (1964); Damboldt in Fl. Turk. 6: 48 (1978).

C. reuteriana Boiss. & Balansa in Boiss. Diagn. Pl. Orient. 9 (ser. 2) 3: 108. (1856); Rech.f., Fl. Lowland Iraq: 585 (1964)

C. propinqua Fisch. & C.A.Mey. var. vel forma *grandiflora* Milne-Redhead in Curtis Bot. Mag. 157: f. 9349 (1934); Blakelock in Kew Bull. 4: 517 (1949)

C. cecilii Rech.f. & Schiman-Czeika in Fl. Iranica 13: 12 (1965)

Annual herb, strigose or hispid all over. Stems dichotomous, slender, 7–35 cm high, erect or decumbent. Leaves narrowly oblong, sessile or nearly so, entire, 5–35 × 2–10 mm. Flowers terminal, sessile. Calyx appendiculate; calyx lobes linear-lanceolate, 10–25 × 1–1.5 mm, longer than calyx tube and shorter than corolla; calyx tube ± 4.5 × 3 mm. Corolla funnel-shaped, 15–25(–30) mm long, usually hirsute along veins and margins of lobes; lobes acute. Stamens subglabrous; filaments short. Ovary 3-locular. Style hairy, ± 10 mm long; stigma 3-fid. Capsule 3-locular, usually nodding, ± 7 × 5 mm. Seeds numerous, ovate, 0.8 × 0.5 mm, shiny, light yellow. Fig. 115: 1–9.

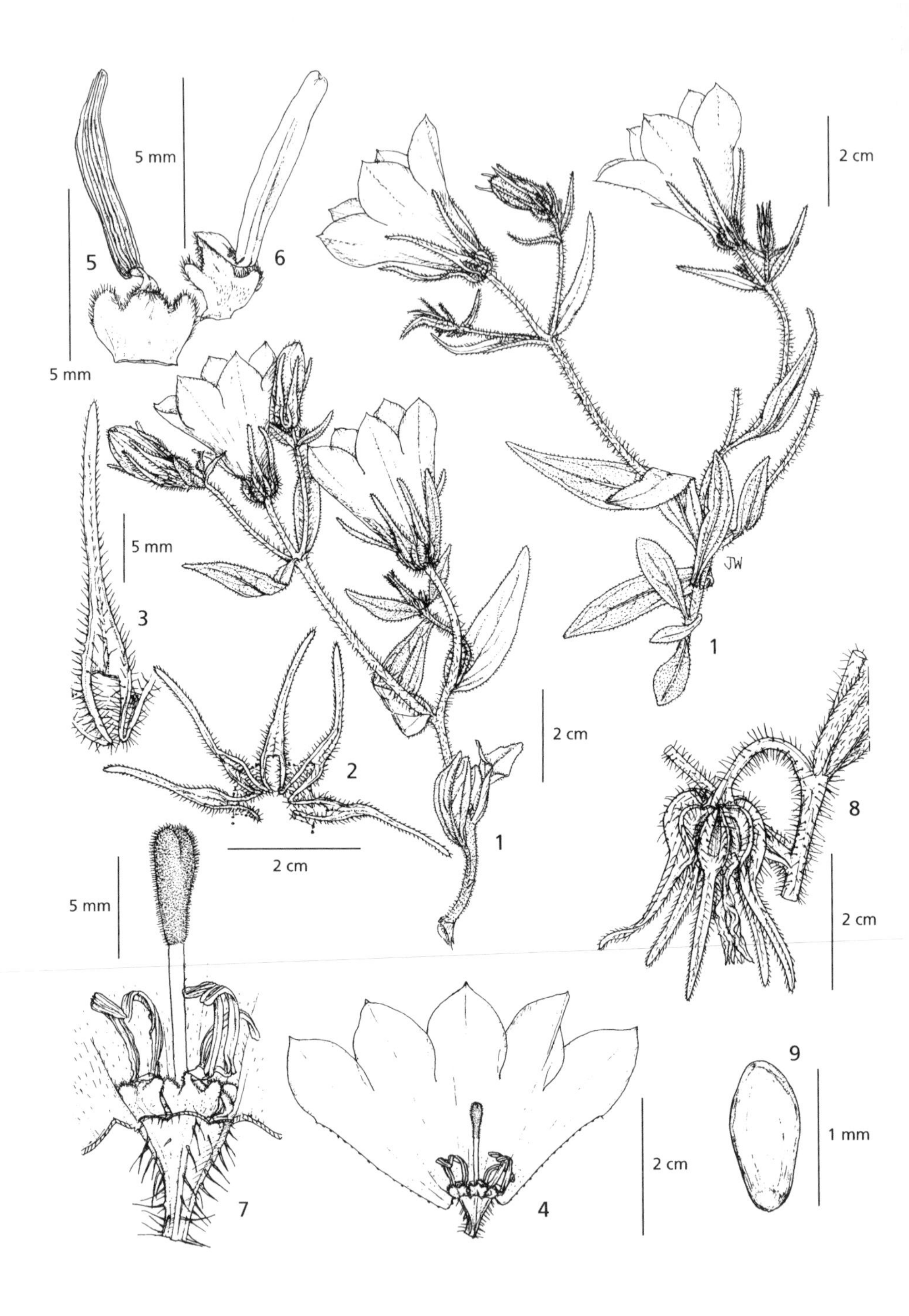

Fig. 115. **Campanula propinqua**. 1, habit; 2, calyx flattened, abaxial view; 3, calyx lobe, abaxial view; 4, corolla opened; 5, 6, stamens abaxial view (5), adaxial view (6); 7, corolla removed to show ovary, stamens and style; 8, fruit; 9, seed. 1–7 from *Barkley & Barkley* 5571; 8 from *Rawi* 28470; 9 from *Nuri & Kass* 27310. © Drawn by J. Beentje (Sep. 2012).

HAB. In the mountains on steep rocky slopes, sometimes in oak forest or scrub, by a stream in walnut grove, in terraced fig gardens on limestone, rarely in fields, on stony hillsides, on a steppic gypseous plain, in dry gravelly wadi bed in the plains; alt. (200–)500–1400(–1800) m; fl. & fr. (Apr.–) May–Jun. (–Jul.).
DISTRIB. Common in the forest zone of Iraq, occasional in the moist-steppe zone. **MAM**: Khantur, NE of Zakho, *Rawi* 23323!; Matina (Chiya-i Matin), Rawi 8722!; Between Amadiya and Sarsang, *Karim, Hamid & Jasim* 40934!; Dohuk, *Makki Beg Sidqi ash-Sharbati* 3262!; Zawita, *Hunting Aerosurveys* 67!; Gara Dagh nr Sarsang, *Al Kas* 18697!; **MRO**: Aqra, *Rawi* 11442!, 11482!; Bekme gorge, *Helbaek* 763!; Rowanduz gorge, *Ludlow-Hewitt* 2360!; Jabal Baradost, nr Diana, *Field & Lazar* 919!; Chiya-i Marmarut, *Cuckney* 3827!; Saran, nr Kani Kawan spring, *Karoukh, Nuri, Kass & Serhang* 27310!; Haibat Sultan Dagh, N of Koi Sanjaq, *Rawi, Nuri & Kass* 28146!; 2 km N of Chinaruk, *id.* 28340!; Chewa (Kew-a) Rash, *id.* 28470!; Gali Warta, 30 km NW by N of Rania, *id.* 28838!; **MSU**: Shewa Sur, NW by N of Chemchamal, *Omar, Karim, Hamza & Hamid* 37106!, 37110!; Jarmo, *Helbaek* 1133!, 1870!, Haines W. 324!; Bakrajo, *Omar & Karim* 37990!; Jasana, N of Surdash, *Omar* 42792!; Qarachitan, on Pira Magrun, *Gillett* 7832!; Mela Kowa (Malakawa), *Rawi* 22446!; 10 km W of Tawila, *Rawi* 22121!; Tawila, *Rawi* 21919!; Dara Tri, on Tawila-Halabja road, *Rawi* 21990!; Khormal, *Rawi* 8442!; Hawraman, on Iranian frontier N of Halabja, *Rawi* 22071!; **MJS**: Kursi, *Gillett* 11023!; above Kursi [Karsi], *id.* 11045A!; Asi (Dair Asi), *Rawi* 8501!; **FUJ/FNI**: nr Mosul, *Loftus* 17!; **FNI**: Mahad, nr Shaikhan, *Salim* 2619!; **FAR**: 10 km north of Erbil [Arbil], *Barkley & Barkley* 5571!; **FNI**: Dar-i Diwan, *Poore* 302!; **FNI**: Khanaqin, *Anon.* 5727!
NOTE. Near East bellflower (Am.); KŪZARAK (Kurd.-Shaikhan, *Salim* 2619 – fodder plant"), NAFALI GUR (Kurd.-Dohuk, *Makki Beg* 3262 – "fodder plant").
This species has been in cultivation for over a century and is well known to horticulturists in Europe. It was grown in the Botanic Garden in St. Petersburg from seed collected in Iranian Azerbaijan in 1835.
Milne-Redhead (op. cit.) described a handsome large-flowered form or variety, *C. propinqua grandiflora* Hort., grown in England from seed brought from the Rowanduz district of Iraq in 1930. It is a much-branched half-hardy annual with large violet-mauve flowers of variable hue, easy to grow on light sandy soils. Several of the specimens cited above are of this form, which differs only in flower-size from typical material of this species.

Turkey (S & E Anatolia), Caucasus, W Iran.

2. **Campanula erinus** *L.*, Sp. Pl. ed. 1: 169 (1753); Boiss., Fl. Orient. 3: 932 (1875); Nábělek in Publ. Fac. Sci. Univ. Masaryk 70: 7 (1926); Zohary, Fl. Palest. ed. 2, 2: 167 (1932); Bornmüller in Beih. Bot. Centralbl. 61B: 74 (1941); Blakelock in Kew Bull. 4: 517 (1949); Zohary in Dep. Agr. Iraq Bull. 31: 143 (1950); Fedorov in Fl. URSS 24: 189 (1957); Rawi in Dep. Agr. Iraq Tech. Bull. 14: 128 (1964); Rech. f. & Schiman-Czeika in Fl. Iranica 13: 14 (1965); Fedorov in Fl. Europaea 4: 88 (1976); Zohary, Fl. Palaest. 3: 281 (1978); Damboldt in Fl. Turk. 6: 52 (1978); Boulos, Fl. Egypt 3: 131 (2002); Ghazanfar, Fl. Oman 3: 178 (in press).

Annual herb, hirsute all over, 3–20 cm high. Stems single or dichotomous, slender. Leaves sessile or rarely shortly petiolate, ovate to elliptic, 3–15 × 2–6 mm, margin shallowly serrate to deeply lobed. Flowers 1–2, small, sessile. Calyx exappendiculate; calyx lobes triangular-lanceolate, longer than the tube, 2–5 × 1.5–2 mm; calyx tube broadly obconical. Corolla longer than calyx-lobes, blue or purple to white or yellow, usually slightly hairy along margins of lobes. Stamens nearly glabrous; anthers linear-oblong. Ovary small, 3-locular. Style hairy, short, ± 3 mm long; stigma 3-fid. Capsule 3-locular, turbinate, nodding. Seeds light brown, elliptic, glossy, 0.6 × 0.2 mm. Fig. 116: 1–8.

HAB. Crevices in limestone rocks in mountains; among rocks at the foot of riverside cliffs; on dry, steppic gypseous hills; alt. (150–)200–1300 m; fl. & fr. (Mar.–)Apr.–May.
DISTRIB. Occasional in the lower forest zone and steppe region of Iraq. **MAM**: Aqra, *Rawi* 11310!, 11329B!; **MRO**: Sefin Dagh, *Gillett* 8157!; Kuh- Sefin (Sefin Dagh), nr Shaqlawa, *Bornmüller* 1556; Shaqlawa, *Haines* 431; Baradost, nr Shan-i Dar, *Erdtman & Goedemans* in *Rechinger* 15692; **MSU**: Jarmo, *Helbaek* 558!, 714!, 1830!, 1848!; **FUJ**: Makhlat, *Guest* 4268!; **FNI**: Eski Kelek, *Gillett* & *Rawi* 10411!; **FNI**: Koma Sang, nr Mandali, *Rawi* 20682!
NOTE. Forked Bellflower. Though rather a weedy plant this species is sometimes cultivated for its reddish-violet to white flowers, particularly in S Europe. Chittenden (1951) mentions it as a weedy but hardy garden annual; but both Clifford Crook (1951) and Bailey (1953) consider it hardly worthy of space in the garden.

Mediterranean Europe (Spain to Greece), Aegean Isles, Cyprus, Syria, Lebanon, Palestine, Jordan, Mediterranean Egypt, Arabia, Turkey, Caucasus, W, C & S Iran, N Africa.

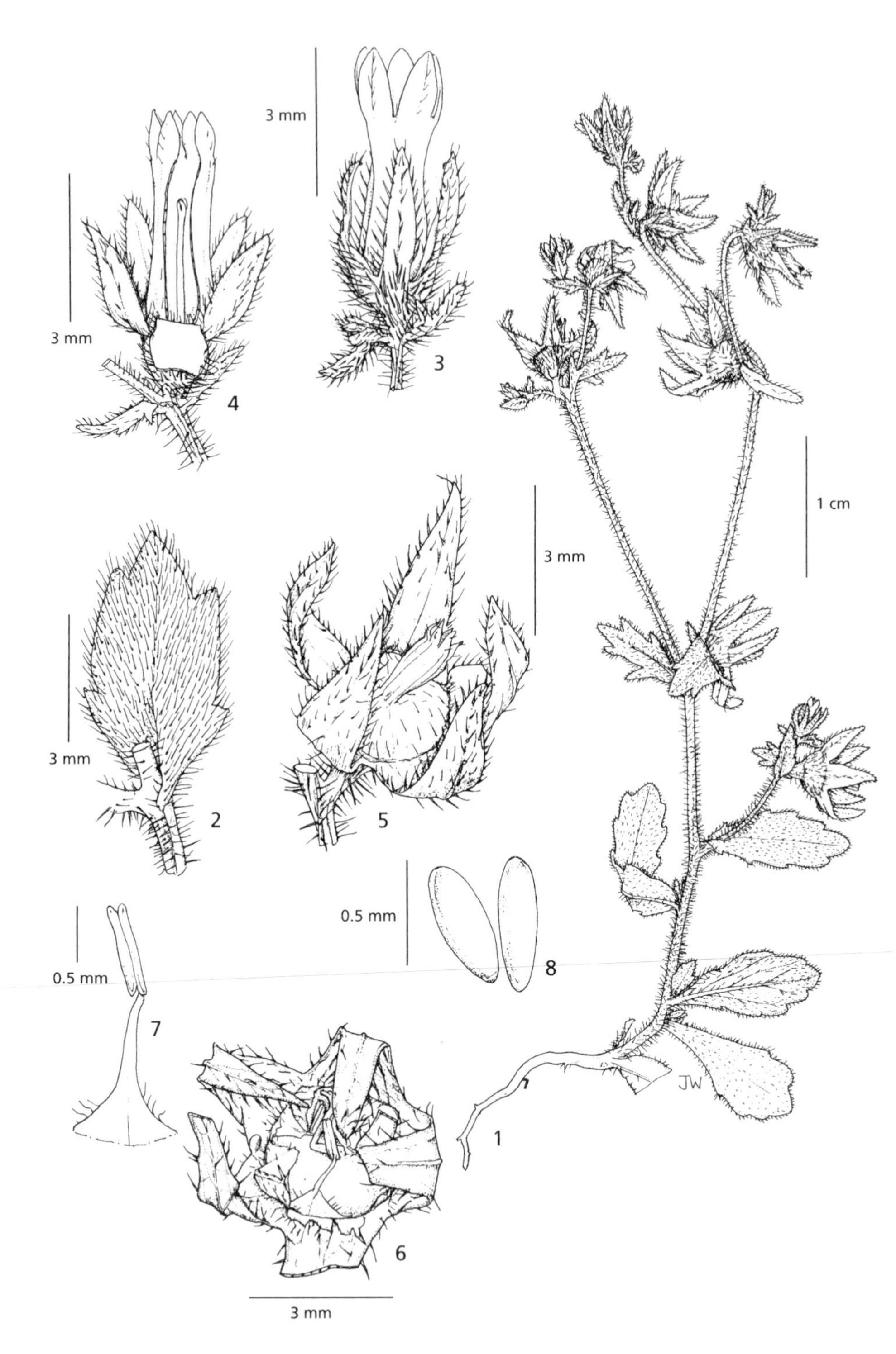

Fig. 116. **Campanula erinus**. 1, habit; 2, leaf upper surface; 3, flower side view; 4, flower with corolla opened; 5, mature flower from above; 6, mature flower with capsule; 7, stamen; 8, seed. 1–5, 8 from *Gillett* 8157; 6–7 from *Rawi* 20682. © Drawn by J. Beentje (Sep. 2012).

3. **Campanula retrorsa** *Labill.*, Ic. Pl. Syr. 5: 5, t. 3 (1812); Boiss., Fl. Orient. 3: 942 (1875); Nábělek in Publ. Fac. Sci. Univ. Masaryk 70: 7 (1926); Zohary, Fl. Palest. ed. 2, 2: 168 (1932); Zohary in Dep. Agr. Iraq Bull. 31: 144 (1950); Rawi in Dep. Agr. Iraq Tech. Bull. 14: 128 (1964); Rech. f. & Schiman-Czeika in Fl. Iranica 13: 37 (1965); Zohary, Fl. Palaest. 3: 282 (1978); Damboldt in Fl. Turk. 6: 61 (1978).

Annual herb, with flaccid, trailing stems or branches. Stem distinctly angled, scabrid with small reflexed hairs or prickles, 15–40 cm long. Leaves minutely toothed and shallowly crenate, the lower leaves ovate-spatulate, petiolate, 10–40 × 5–10 mm; upper leaves lanceolate or linear-lanceolate, sessile, 10–20 × 2–4 mm. Flowers long-pedicellate. Calyx exappendiculate; calyx lobes linear or linear-lanceolate, 1-veined, scabrid, at least along the margins and veins, ± recurved, longer than the calyx tube, 7–25 × 1–2 mm; calyx tube narrowly obconical, glabrous or nearly so, distinctly veined, 3–5 × 1.5–2 mm. Corolla glabrous, as long as or slightly shorter than the calyx lobes, infundibuliform. Stamens villous near the base; filaments short; anthers linear. Ovary 3–locular. Style hairy, 10 mm long; stigma 3-fid. Capsule 3–locular. Seeds numerous, 0.7 × 0.3 mm, glossy, light brown.

HAB. On lower mountain slopes, in oak forest or coppiced oak, on limestone; alt. 600–900(–1800) m; fl. & fr. Apr.–June.
DISTRIB. Occasional in one district only of the central sector of the lower forest zone of Iraq. **MRO**: Kan-i Watman, *Nábělek* 2130; Between Rowanduz and Agoyan, *Kass, Nuri & Serhang* 27261!; Handren Dagh, above Rowanduz, *Gillett* 8301!; Haji Umran, *Chapman* 11933!; **FNI**: Hinnis, 9 km N of Ain Si, *Thesiger* 627; **LCA**: Karrada Mariam, Baghdad, *Wheeler Haines* W.178!
NOTE. *Wheeler Haines* W.178 was a fragment of a single weed found in a lettuce patch on the bank of R. Tigris. This annual species has occasionally been cultivated as a garden plant in Europe.

S Turkey, Syria, Lebanon, Palestine, Jordan.

4. **Campanula flaccidula** *Vatke* in Linnaea 38: 712 (1874); Blakelock in Kew Bull. 4: 517 (1949); Rawi in Dep. Agr. Iraq Tech. Bull. 14: 128 (1964); Rech. f. & Schiman-Czeika in Fl. Iranica 13: 37 (1965); Damboldt in Fl. Turk. 6: 61 (1978).

C. singarensis Boiss. et Hausskn. in Boiss., Fl. Orient. 3: 943 (1875); Nábělek in Publ. Fac. Sci. Univ. Masaryk 70: 7 (1926); Bornmüller in Beih. Bot. Centralbl. 61B: 74 (1941); Zohary in Dep. Agr. Iraq Bull. 31: 144 (1950); Rawi in Dep. Agr. Iraq Tech. Bull. 14: 128 (1964).

Annual herb with stems (5–)7–15(–20) cm long. Stems slender, flexuous, filiform, ± scabrid with short reflexed hairs or prickles, decumbent. Leaves petiolate, or subsessile near apex of stem, broadly ovate to elliptic, (4–)7–25(–35) × 4–15(–25) mm, entire to minutely toothed and shallowly crenate, glabrous or with few scattered stiff hairs along the margin and midrib, apex acute to obtuse, base cuneate or abruptly attenuate; petiole slightly strigose, 3–5 mm long. Flowers small, pedicellate; pedicel filiform, slender, glabrous, 5–15(–20) mm long. Calyx exappendiculate, glabrous; calyx lobes 5, triangular-lanceolate, longer than calyx tube and usually longer than corolla, 5–10 × 1–1.5 mm; calyx tube cup-shaped, 1–4 × 1–3 mm. Corolla campanulate, blue or violet, glabrous. Stamens glabrous; anthers linear-oblong. Style hairy, ± 4 mm long; stigma 3-fid. Capsule 3-locular. Seeds numerous, elliptic, 0.6 × 0.2 mm, glossy, light brown.

HAB. Rocky slopes in the mountains, on moist shady limestone cliffs, in rock clefts and crannies; alt. 600–1600 m; fl. & fr. Apr.–Jun.
DISTRIB. Common and locally abundant, in the lower forest zone of Iraq. **MJS**: Jabal Sinjar, *Haussknecht* s.n.!, Omar, Al-Khayat & Al-Kaisi 52532!; Kursi, *Gillett* 10881!; **MAM**: Aqra, *Rawi* 11453!; Gali Mazurka (2 km NW of Amadiya), Al-Dabbagh & Jasim 46868!; **MRO**: Kuh-i Sefin, above Shaqlawa, *Bornmüller* 1557!; Sefin Dagh, *Gillett* 8158!; Shaqlawa, *Wheeler Haines* W670!; Shakh-i Harir, *Emberger, Guest, Long, Schwan & (Sa'dun) Jusuf (Serkahia)* 15435!; Rowanduz gorge, *id.* 15524!, *Cuckney* 3832!; Hamdiyan on Baradost, *Thesiger* 832 ?; Zeita, nr Shirwan Mazin, *Haines* s.n.; Serkapkan, 7 km NW of Rania, *Rawi, Nuri & Kass* 28514!; **MSU**: on Iranian frontier, nr Tawila, *Rechinger* 10257!; Dara Tri, between Tawila and Halabja, *Rawi* 22026!

SE Turkey, W Iran.

5. **Campanula fastigiata** *Dufour* ex *Schult.* in Roem. & Schult., Syst. Veg. 5: 136 (1819); Fedorov in Fl. Europaea 4: 78 (1976); Damboldt in Fl. Turk. 6: 55 (1978); Aghabeigi & Assadi in Edinb. J. Bot. 65(3): 381 (2008).

Brachycodon fastigiatus (Dufour ex Schult.) Fedorov in Fl. SSSR 24: 342 (1957); *B. fastigiata* (Dufour ex Schult.) Fedorov in Fl. Azerb. 8: 164 (1961).

Small, succulent, erect, annual or biennial herb 2–5(–8) cm tall. Stems simple or with fastigiate branches. Leaves sessile to shortly petiolate, glabrous or pubescent, broadly obovate to linear oblong (becoming narrower on the upper part of the stem), cuneate at base, usually 3–5-toothed at apex, sometimes entire. Flowers erect, axillary towards the apex of branches, sessile. Calyx appendiculate; calyx lobes erect, linear-lanceolate, obtuse to acute, occasionally dentate, 2–5 mm, usually longer than corolla. Corolla pale blue or mauve, obconical, 1.5–3 mm, lobed to $^1/_3$. Ovary obconical, 1–2 mm. Style included; stigmas 3. Capsule obconical, 5-veined, 1.5–5 mm. Seeds ellipsoid, 0.3–0.4 × 0.2 mm, glossy, light brown.

HAB. Dry, saline soils; dry steppe on gypsaceous hills; alt. 200 m; fl. & fr. March (from a single specimen only).
DISTRIB. Occasional. **FUJ**: Qaiyarah,Mosul, *Bayliss* 87!; **FNI**: near Badra, *Gillett* 6713!

Spain, N Africa, Caucasus, SW & C Asia.

6. **Campanula involucrata** *Aucher ex A.DC.* in DC., Prodr. 7: 467 (1839); Boiss., Fl. Orient. 3: 926 (1875); Nábělek in Publ. Fac. Sci. Univ. Masaryk 70: 5 (1926); Zohary in Dep. Agr. Iraq Bull. 31: 143 (1950); Rawi in Dep. Agr. Iraq Tech. Bull. 14: 128 (1964); Rech. f. & Schiman-Czeika in Fl. Iranica 13: 30 (1965); Damboldt in Fl. Turk. 6: 20 (1978).

Perennial herb, branching from base, hirsute or pubescent. Branches of stem erect or ascending, ± 20–50 cm long. Leaves sessile, serrulate, oblong-lanceolate, rather acute, 1–4 × 0.5–2 cm wide. Flowers sessile, capitate, hidden amongst or surrounded by involucral bracts; bracts ovate, reticulate. Calyx appendiculate, hirsute all over, 5-lobed; calyx lobes linear-lanceolate, longer than the tube, 5–7 × 0.5–1 mm; calyx tube obconical, 4 × 2 mm. Corolla 5-lobed, longer than calyx tube, hirsute all over, dark lilac. Stamens nearly glabrous, or with few scattered hairs; filaments short, 2 mm long; anthers linear. Ovary 2-locular. Style hirsute, 10–12 mm; stigma bifid. Capsule 2-locular. Seeds numerous, yellowish, 2–2.5 × 1 mm.

HAB. Rocky mountain slopes, in oak forest and coppiced oak, on limestone; alt. 1150–1600(–2000) m; fl. & fr. May–Jun.(–Jul.).
DISTRIB. Occasional in the upper forest zone of Iraq, sometimes extending up into the lower thorn-cushion zone. **MAM**: Sarsang, *E. Chapman* 26403!, *Haines* W. 1156!; **MRO**: Kurruk Dar (Kurek Dagh), nr Rowanduz, *Nábělek* 2118; Seiwaka (Sewok), *Kass, Nuri & Serhang* 27973!, 27588!; Kawriesh (Quraish), on E spur of Karoukh, *id.* 27166!; Gali Warta, 30 km NW by N of Rania, *Rawi, Nuri & Kass* 28736!; Mts. S of Rayat, *Thesiger* 1053; **MSU**: Penjwin, *Rawi* 12183!, 12207!, *Rechinger* 10520; Hawraman, *Gillett* 11883!; Spur of Hawraman, N of Biyara, *id.* 11823!; Hawara Birza, *Rawi, Husham Alizzi & Nuri* 29360!; ?**MSU** (or only in Iran?), "Avroman et Schahu", *Haussknecht* s.n.

Turkey, NW & W Iran.

7. **Campanula radula** *Fisch.* ex *Fenzl* in P.de Tchihatcheff, Asie Min., Bot. 2: 395 (1860); Boiss., Fl. Orient. 3: 909 (1875); Nábělek in Publ. Fac. Sci. Univ. Masaryk 70: 4 (1926); Guest in Dep. Agr. Iraq Bull. 27: 17 (1933); Blakelock in Kew Bull. 4: 517 (1949); Zohary in Dep. Agr. Iraq Bull. 31: 148 (1950); Fedorov in Fl. URSS 24: 248 (1957); Rawi in Dep. Agr. Iraq Tech. Bull. 14: 128 (1964); Rech. f. & Schiman-Czeika in Fl. Iranica 13: 14 (1965).

Perennial herb, hispid all over. Stems rigid, erect, branching near the top. 20–40 cm tall. Stem leaves ovate, sometimes broadly so, biserrate, coriaceous, shortly petiolate or sessile, 1.5–10 × 1–8 cm; cauline leaves and inflorescences very small. Flowers in panicles, pedicellate. Calyx hirsute all over, appendiculate; calyx lobes triangular-lanceolate, usually longer than calyx tube and shorter than corolla, 2–4.5 × 1 mm; calyx tube obconical, nearly 1.5 mm × 1 mm. Corolla hispid all over, 1.5–2 cm long; lobes shorter than corolla tube. Stamens glabrous or nearly so; filaments short, 2–3 mm long; anthers 2-locular, linear or linear-oblong. Ovary 3-locular. Style hairy, 10–12 mm long; stigma 3-fid. Capsule 3-locular. Seeds small, 1.5 × 0.1 mm, usually not straight.

HAB. & DISTRIB. (of species). As for var. *radula*.

var. **radula**. Type: Iraq, Gara Dagh, S of Amadiya, *Kotschy* 423 (K!, syn.).

HAB. (of var.) In the mountains along cracks and deep fissures in limestone cliffs, in damp shady places, N-facing rock faces etc.; alt. 550–1000(–1800) m; fl. & fr. Aug.–Sep.(–Oct.).
DISTRIB. (of var.) Occasional in the forest zone of Iraq. **MAM**: Khantur, NE of Zakho, *Rawi, Nuri & Tikriti* 29028; Gara Dagh, S of Amadiya, *Kotschy* 423 [type]!; Sarsang, *Haines* W. 1224!; **MRO**: Rowanduz gorge (Gali Ali Beg), *Guest* 1592!, 13012!, 13089!, *Gillett* 9447!, *J.K. Jackson* 3521!, *Rechinger* 11255 ?, *Rawi* 26798!, *M. ash-Sharif* 41773 ?

Turkey (SE Anatolia).

var. **minor** *Boiss.*, Boiss., Fl. Orient. 3: 909 (1875); Grossh., Fl. Kavk. 4: 63 (1950).

C. *coriacea* Davis in Notes Roy. Bot. Gard. Edinb. 24: 29, t. 4 (1962); Damboldt in Fl. Turk. 6: 29 (1978).

Plant smaller than var. *radula*. Leaves ovate, 6–10 × 5–8 mm. Calyx lobes triangular. Corolla smaller, ± 10 mm long.

HAB. (of var.). Rocky mountain; alt. ± 2000 m; fl. June.
DISTRIB. (of var.) Very rare in the forest zone of Iraq. "Kurdistan", *A.H. Layard* s.n.!; **MSU:** Hawraman mts., *Rawi, Chakr., Nuri & Alizzi* 19740!
NOTE. Layard's gathering used to be the only known record of this variety which was not re-found until some 80 years later, but as the precise location of Layard's specimen is not stated, it may have been from Turkey or Iran.

E Turkey, ?Caucasus.

8. **Campanula sclerotricha** *Boiss.* in Diagn. ser. 1, 11: 66 (1849); Boiss., Fl. Orient. 3: 901 (1875); Nábělek in Publ. Fac. Sci. Univ. Masaryk 70: 3 (1926); Blakelock in Kew Bull. 4: 517 (1949); Zohary in Dep. Agr. Iraq Bull. 31: 143 (1950); Fedorov in Fl. URSS 24: 213 (1957); Rawi in Dep. Agr. Iraq Tech. Bull. 14: 128 (1964); Rech. f. & Schiman-Czeika in Fl. Iranica 13: 14 (1965); Damboldt in Fl. Turk. 6: 23 (1978).

Perennial herb with few branches. Stem reddish-tinged, glabrous or slightly scabrid usually near the apex, 30–120 cm long. Leaves simple, radical leaves large, denticulate, to 16 × 12 cm, petiole to 25 cm long; cauline leaves narrowly ovate, 3–8 ×1–4.5 cm, with 2–6 cm long petiole, hispid, serrate or biserrate, truncate or cordate at base. Flowers pedicellate; pedicel thick, 6–15 mm, usually glabrous. Calyx appendiculate, hirsute or scabrid along the margins and veins; calyx lobes triangular-lanceolate, much longer than calyx tube, 10–20 × 2–3.5 mm; calyx tube globose, 3–4 × 2–2.5 mm, hispid along veins. Corolla campanulate, hispid along veins, ± 20 mm; corolla lobes triangular, shorter than the tube. Corolla tube usually shorter than calyx lobes. Stamens 5, free; filaments short, 2–3 mm; anthers 2-locular, hairy, linear. Ovary 3-locular, style elongate, 10–20(–25) mm, hairy; stigma 3-fid. Capsule bell-shaped, 4–8 × 3–5 mm, 3-locular, hispid along the veins. Seeds small, 1.5 × 0.8 mm, broadly elliptic. Fig. 117: 1–8.

HAB. On the mountains in damp shady places, by streams, on wet mossy rocks in a stream bed, in a hedgerow near a stream, shady places in oak forest, also under walnut or plane shade by stream; alt. 1050–1500(–2000) m; fl. & fr. Jul.-Aug.
DISTRIB. Common in the upper forest zone of Iraq. **MAM**: Sharanish, NE of Zakho, *Rawi et al.* 29034!; Chalki Nasara nr Wazi on banks of R. Khabur, *Thesiger* 1247?; Amadiya, *Guest* 3766!; Sarsang, *Haines* W. 410!, W. 1600!; **MRO**: Halgurd Dagh, nr Nawanda, *Guest & Ludlow-Hewitt* 2779!; Hisar-i Rost, *Haley* 103 ?; S of Ser Kurawa, *Gillett* 9801!; between Darband and Gunda Zhor, *Gillett* 12391! (12311?); Qandil range, above Pushtashan, *Rechinger* 11202; N of Pushtashan, *Rawi & Serhang* 18311?; Pushtashan, *id.* 26484!; NE of Rania, *id.* 18237!, 23876?; Qurnago valley, *Rawi* 26620!; between Qurnago and Kani Rash, *Rawi* 26679!; **MRO/MSU**: (or probably in Iran), "Return from Oroomah (Urumiya) through the Kurdish mountains", *Captain Garden* s.n. (1857)!

Turkey (E Anatolia), Caucasia, W & C Iran, Afghanistan.

9. **Campanula stricta** *L.*, Sp. Pl. ed. 2: 238 (1762); Boiss., Fl. Orient. 3: 923 (1875); Nábělek in Publ. Fac. Sci. Univ. Masaryk 70: 4 (1926); Zohary, Fl. Palest. ed. 2, 2: 165 (1932); Fedorov in Fl. URSS 24: 212 (1957); Rawi in Dep. Agr. Iraq Tech. Bull. 14: 128 (1964); Rech. f. & Schiman-Czeika in Fl. Iranica 13: 28 (1965); Damboldt in Fl. Turk. 6: 42 (1978).

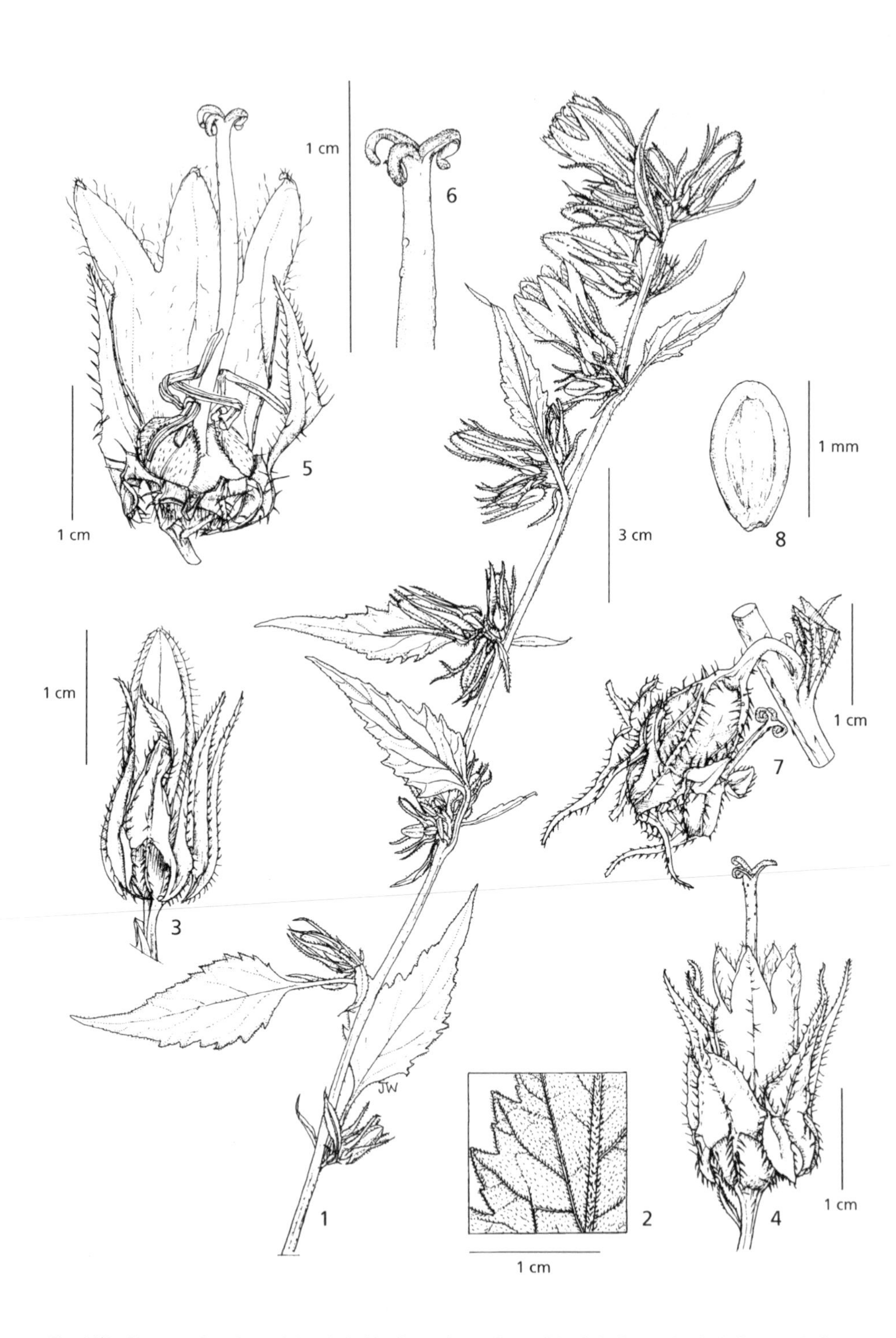

Fig. 117. **Campanula sclerotricha**. 1, habit; 2, undersurface of leaf; 3, flower bud; 4, flower; 5, flower opened; 6, style and stigma; 7, capsule; 8, seed. 1 from *Rawi & Serhang* 18311; 2–6 from *Rawi & Serhang* 18237; 7–8 from *Rawi & Serhang* 26484. © Drawn by J. Beentje (Sep. 2012).

var. **stricta**

Perennial herb, scabrid all over, branching from base. Stems ascending, rigid, 20–50 cm long. Leaves usually stiff; the lower oblong or ovate-oblong, 2–5 × 0.5–0.8 cm, entire or denticulate; cauline leaves sessile, linear-lanceolate or lanceolate, 1.5–2.5 × 0.5–0.9 cm. Flowers sessile, bracteolate. Calyx appendiculate; calyx lobes triangular-lanceolate, 4–8 × 2–3 mm; appendages inflated and covering the capsule. Corolla longer than calyx-lobes, hairy or sometimes nearly glabrous; corolla-tube funnel-shaped. Stamens glabrous or nearly so; filaments short, 3–4 mm long, usually slightly hairy near base. Ovary 3-locular, style hairy, 12–15 mm; stigma 3-fid. Capsule 3-locular, 5–7 mm × 5–6 mm. Seeds numerous, 2.5–3 × ± 1.5 mm, light yellow.

HAB. On dry stony ground on a mountain; alt. ± 1800 m; fl. & fr. Aug.
DISTRIB. Very rare in Iraq: only found once or twice in the lower thorn-cushion zone of Iraq. **MRO**: mountain N E of Sairo, *Gillett* 8687!; **MRO** (or probably in Iran?), "return from Oroomah (Urmiya) through Kurdish mountains", *Captain Gordon* (1857) s.n.

Syria, Lebanon, Turkey, Caucasus, NW Iran.

10. **Campanula glomerata** *L.*, Sp. Pl. ed. 1: 166 (1753); Hooker, Fl. Londin. t.146 (1828); Boiss., Fl. Orient. 3: 927 (1875); Zohary, Fl. Palest. ed. 2, 2: 165 (1933); Bornmüller in Beih. Bot. Centralbl. 61: 73 (1941); Blakelock in Kew Bull. 4: 517 (1949); Zohary in Dep. Agr. Iraq Bull. 31: 143 (1950); Fedorov in Fl. URSS 24: 202 (1957); Rawi in Dep. Agr. Iraq Tech. Bull. 14: 128 (1964); Rech. f. & Schiman-Czeika in Fl. Iranica 13: 29 (1965); Damboldt in Fl. Turk. 6: 18 (1978).

subsp. **glomerata**

Perennial herb; stems erect, pubescent or glabrescent or with scattered stiff short hairs, 20–60 cm long. Leaves scabrid, serrulate, ovate or oblong-lanceolate, 5–7 × 1–2 cm; lower and radical leaves petiolate, petiole 2–6 cm; upper leaves sessile. Flowers sessile, in dense clusters surrounded by involucral bracts; bracts oblong-lanceolate, 1.5–2.5 cm long. Calyx exappendiculate; calyx lobes ciliate with short stiff hairs or strigose all over, lanceolate, 5–7 × 1–2 mm, longer than calyx tube and shorter than corolla; calyx tube obconical, glabrescent or scabrid, 4 × 2 mm. Corolla scabrid or glabrous, funnel-shaped; lobes shorter than tube. Stamens 3, glabrous or nearly so; filaments short, 2–3 mm long; anthers linear or linear-oblong. Ovary 3-locular. Style hairy, 10–12 mm long; stigma 3-fid. Capsule 3-locular. Seeds numerous, small, yellowish.

HAB. High in the mountains, in damp places, among *Astragalus* bushes by a stream; alt. (600–) 1800–3000 m; fl. & fr. Jun.–Jul.
DISTRIB. Occasional in the central sector of the thorn-cushion zone of Iraq. **MRO**: Kuh-i Sefin, *Bornmueller* 1554; Halgurd Dagh, *Guest & Ludlow-Hewitt* 2842!, 2944!; Sakri-Sakran, *Bornmueller* s.n.; Serin, NE of Rania, *Rawi & Serhang* 18240A!

Most of Europe (except the extreme north), Syria, Lebanon, Turkey, Caucasus, NW & C Iran, C Asia, Siberia.

11. **Campanula stevenii** *M. Bieb.* in Fl. Taur.-Cauc. 3: 138 (1819); Boiss., Fl. Orient. 3: 936 (1875); Fedorov in Fl. URSS 24: 316 (1957); Rawi in Dep. Agr. Iraq Tech. Bull. 14: 128 (1964); Rech. f. & Schiman-Czeika in Rech. f., Fl. Iranica 13: 35 (1965); Damboldt in Fl. Turk. 6: 56 (1978).

subsp. **stevenii**

Perennial herb sparingly branched from the base. Stems slender, erect or decumbent, slightly hairy, 20–40 cm long. Radical leaves oblong-spatulate, petiolate, crenate, 15–40 × 5–10 mm, slightly hairy; petiole 15–35(–40) mm long; cauline leaves small, 10–25 × 1–3 mm, sessile, crenate. Flowers usually terminal, pedicellate. Calyx exappendiculate, glabrous; calyx lobes lanceolate, shorter than calyx tube, 4–5 × 1–1.3 mm; calyx tube obconical, 6 × 2 mm. Corolla glabrous, funnel-shaped. Stamens nearly glabrous; filaments short, 2 mm; anthers linear. Ovary 3-locular. Style hairy, 10–14 mm long; stigma distinctly 3-branched; branches of stigma 7–9 mm long. Capsule 3-locular. Seeds numerous, small, brown, 1 × 0.5 mm.

HAB. By the wayside on a high mountain track; alt. 2500 m; fl. & fr. Aug.–Sep.
DISTRIB. Very rare in Iraq – only found once (or possibly twice) in the central sector of the middle thorn-cushion zone. **MRO**: by the road near Bardanas on Qandil range, *Rawi & Serhang* 24618!; (Turkey or Iran), "Kurdistan mountains", *A.H. Layard* 292 ?

Turkey, NW & N Iran.

12. **Campanula acutiloba** *Vatke* in Linnaea 38: 709 (1874). — Type: Iraq, Pira Magrun, *Haussknecht* s.n. (K!, syn.); Bornmüller in Beih. Bot. Centralbl. 61: 73 (1941); Blakelock in Kew Bull. 4: 517 (1949); Rawi in Dep. Agr. Iraq Tech. Bull. 14: 127 (1964); Rech. f. & Schiman-Czeika in Fl. Iranica 13: 22 (1965); Damboldt in Fl. Turk. 6: 38 (1978).

C. cissophylla Boiss. et Hausskn., Boiss., Fl. Orient. 3: 920 (1875) & Pl. Nov. Or. 1: 2 (1875); Nábělek in Publ. Fac. Sci. Univ. Masaryk 70: 4 (1926); Zohary in Dep. Agr. Iraq Bull. 31: 143 (1950); Rawi, ibid. Tech. Bull. 14: 127 (1964).

Perennial herb with a shrubby base, hirsute all over. Stems or branches slender, filiform, 7–20 cm long. Leaves reniform-cordate, denticulate, petiolate, 4–6 × 5–12 mm, with 5–7 lobes. Petiole filiform, hirsute, 4–20 mm. Flowers small, pedicellate; pedicel filiform, hirsute, 8–20(–30) mm. Calyx exappendiculate, calyx lobes triangular-lanceolate, 2–4 × < 1 mm, longer than the calyx tube; calyx tube cup-shaped, 1.5–2 × 1–2 mm. Corolla funnel-shaped, hirsute all over. Stamens glabrescent; filaments short; anthers 2-locular, oblong-linear. Ovary small, 3-locular. Style hairy, 4–4.5 mm long; stigma 3–fid. Capsule 3-locular, 1 mm. Seeds very small, ± 1 mm long or less, yellowish to light brown. Fig. 118: 1–8.

HAB. On lower mountain slopes, in crannies on limestone rock, on a shady rock face, sometimes in oak forest or among coppiced oak; alt. 600–1200(–1400) m; fl. & fr. (Apr.–)May–Jun.
DISTRIB. Occasional throughout the lower forest zone of Iraq. **MAM**: Gara Dagh, *Kotschy* 377?, *Rawi* 9296!; Sarsang, *Haines* W. 431; Zinta gorge, *Chapman* 26097!; **MRO**: Serderian, between Arbil and Rowanduz, *Nábělek* 2115; Kuh-i Sefin (Sefin Dagh), *Bornmueller* 1551, *Gillett* 8160!; Gali Ali Beg (Rowanduz gorge), *Bornmueller* 1552!, *Guest* 2971!, *Gillett* 8319!; Rowanduz, *Nábělek* 2116, *Rechinger* 11248; **MSU**: Pira Magrun, *Haussknecht* s.n. (type)!; *Rawi* 12117! *Thesiger* 1133; in the pass between Qarachitan and Zewiya, *Gillett* 7753!; Qara Dagh, *Haines* s.n.; Amoret, nr Qaradagh, *Haines* W. 1133!
NOTE. Though this charming and delicate little species appears never to have been brought into cultivation, Clifford Crook (1951) terms it the Asian counterpart of the "garganica group", namely the variety or form of the Adria Bellflower (*C. elatine* var. *garganica*), which is popular in European and N American rock gardens.

Turkey (SE Anatolia), NW Iran.

13. **Campanula mardinensis** *Bornmüller & Sint.* in Mitth. Thüring. Bot. Vereins, n.f. 20: 31 (1905); Nábělek in Publ. Fac. Sci. Univ. Masaryk 70: 4 (1926); Zohary in Dep. Agr. Iraq Bull. 31: 143 (1950); Rech. f. & Schiman-Czeika in Fl. Iranica 13: 23 (1965); Damboldt in Fl. Turk. 6: 38 (1978).

Perennial herb with filiform decumbent stems, densely pubescent all over. Leaves reniform or cordate, denticulate, 4–6 × 5–12 mm, shortly petiolate or subsessile. Flowers small, pedicellate. Calyx exappendiculate, calyx lobes linear-lanceolate, longer than calyx tube, 3 × 0.5 mm; calyx tube cup-shaped, 2 × 1.5 mm. Corolla longer than calyx, hirsute all over, 5–7 mm long. Stamens nearly glabrous, 3 mm. Ovary small, 3-locular. Style hairy, 4 mm; stigma 3-fid. Capsule small, 3-locular. Seeds small, numerous, yellowish.

HAB. Lower mountain slopes, fissures in rocky limestone; alt. 700–1200 m; fl. & fr. Jun.–Jul.
DISTRIB. Rare in Iraq; only found twice, not far from the Turkish frontier, in the NW sector of the lower forest zone. **MAM**: Mar Ya'qub, above Simel, *Nábělek* 2117; Jabal Khantur, N of Zakho, *Rechinger* 10792.
NOTE. The plant is very similar to *C. acutiloba*, but the leaves are subsessile or very shortly petiolate. Nábělek's specimen differs from the type ("Mesopotamia Kurdica prope Mardin", *Sintenis*) in being altogether less hairy.

Turkey (E Anatolia).

14. **Campanula perpusilla** *A.DC.* in DC., Prodr. 7: 474 (1839); Boiss., Fl. Orient. 3: 920 (1875); Nábělek in Publ. Fac. Sci. Univ. Masaryk 70: 4 (1926); Rawi in Dep. Agr. Iraq Tech. Bull. 14: 128 (1964); Rech. f. & Schiman-Czeika in Fl. Iranica 13: 21 (1965); Damboldt in Fl. Turk. 6: 21 (1978).

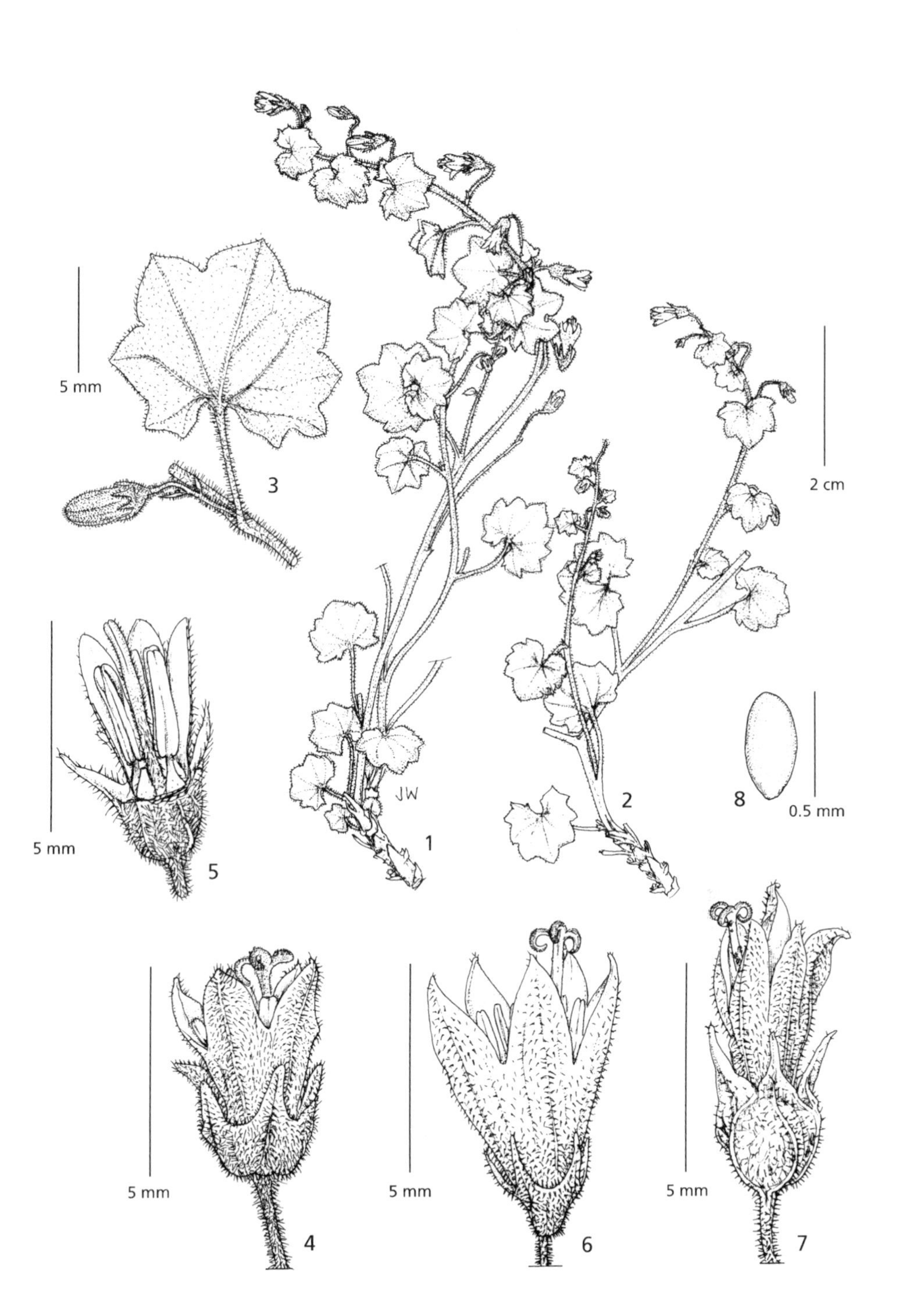

Fig. 118. **Campanula acutiloba**. 1, 2, habit; 3, leaf with bud; 4, young flower side view; 5, flower with front petals removed; 6, flower side view; 7, mature flower at fruiting stage; 8, seed. 1–5 from *Gillett* 8160; 6–8 from *Wheeler Haines* W1133. © Drawn by J. Beentje (Sep. 2012).

Perennial herb, glabrous all over or sparsely hairy, branched from base. Stems decumbent, thin or filiform, 3–4 cm. Leaves petiolate, reniform-cordate or orbicular-cordate, denticulate, 4–15 × 5–20 mm; petiole very long, 3–6 cm, filiform. Flowers small, pedicellate. Calyx exappendiculate; calyx lobes 5, linear, 2.5–3 × 0.3–0.5 mm; calyx tube narrowly oblong, shorter than lobes, 1–1.3 × 0.3–0.5 mm. Corolla 5-lobed, narrowly funnel-shaped; tube of corolla much longer than calyx tube, 4–5 mm long. Stamens 5, free, filaments dilated at base, ± 1.4 mm. Ovary small, 3-locular; style straight, ± 4.5 mm; stigma small, slightly 3-fid. Capsule small, 3-locular; seeds numerous, very small.

HAB. In the mountains, rock fissures, in shade; alt. 800–2600 m; fl. Apr., Jun.–Jul.
DISTRIB. Rare in Iraq – only found twice in the forest zone. **MSU**: Pira Magrun, *Haussknecht* s.n.; Darband-i Khan, *Haines* 2165.

W Iran, Turkey?

15. **Campanula conferta** *A.DC.* in DC., Prodr. 7: 468 (1839); Boiss., Fl. Orient. 3: 916 (1875); Rawi in Dep. Agr. Iraq Tech. Bull. 14: 127 (1964); Rech. f. & Schiman-Czeika in Rech. f., Fl. Iranica 13: 27 (1965); Damboldt in Fl. Turk. 6: 36 (1978); Aghabeigi & Assadi in Edinb. J. Bot. 65(3): 381 (2008).

Perennial herb, branching from base. Stems decumbent or ascending, slightly hairy, 10–30 cm long. Leaves usually pubescent or sometimes glabrous with ciliate margins; radical leaves spatulate, 3.5–7.5 × 6–10 mm; cauline leaves ovate-oblong, sessile, serrate, acute, 10–22 × 4–10 mm. Racemes densely leafy. Flowers mostly axillary, bracteolate, shortly pedicellate; bracts ovate-elliptic, 6–8 × ± 4.5 mm. Calyx exappendiculate; calyx 5-lobed, pubescent; calyx tube obconical, 3–5 × ± 1.5 mm; calyx lobes ovate, longer than the tube, 5–7 × 2–3 mm. Corolla campanulate to funnel-shaped, hairy all over or rarely glabrous, 5-lobed, 5–8 mm; lobes shorter than the tube. Stamens 5; filaments short, ± 1.5 mm, dilated at base. Ovary 3-locular. Style hairy, erect, protruding from the corolla, ± 9 mm long. Stigma slightly 3-fid. Capsule nodding. Seeds numerous.

HAB. Stony mountainsides; alt. 1500–2000 m; fl. & fr. May–Jun.
DISTRIB. Rare in Iraq; only found in four locations in the upper forest zone. **MAM**: Mt. Zawita, NE of Zakho, *Rawi* 23590!, *Rechinger* 10963!; **MRO**: Zeita, nr Shirwan Mazna, *Haines* s.n.; E of Karoukh Dagh, between Seiwaka (Sewok) and Dargala, Alkas, *Nuri & Serhang* 27595!; between Kujar and Kani Gawra, *id.* 27640!

N Syria, Iran, Turkey (E Anatolia).

16. **Campanula postii** (*Boiss.*) *Harms*, Naturl. Pflanzenfam., Nachtr. 1: 318 (1897); Damboldt in Fl. Turk. 6: 42 (1978).

Tracheliopsis postii (Boiss.) Buser in Bull. Herb. Boiss. 2: 527 (1894); Zohary, Fl. Palest. ed. 2, 2: 171 (1933); Rawi in Dep. Agr. Iraq Tech. Bull. 14: 129 (1964).
Trachelium postii Boiss., Fl. Orient. Suppl.: 336 (1888).

Perennial herb. Stems arising from fleshy rootstock, densely covered with fine, soft hairs, 20–40 cm long. Leaves sessile, elliptical, 2.5–6 × 1–3 cm, upper surface hispidulous, lower surface tomentellous, serrate. Flowers in a capitate terminal corymb or umbel, pedicellate; pedicel hispid, 5–10 mm. Calyx exappendiculate, hispid along veins and margins or hispid all over; calyx lobes 5, triangular-lanceolate, 3–3.5 × 1.8–2 mm; calyx tube turbinate, shorter than lobes. Corolla tubular to funnel-shaped, glabrous or nearly so or hispid all over; lobes of corolla 5, shorter than the tube. Stamens 5, free; filaments slightly hairy, 6–8 mm; anthers glabrous or nearly so. Ovary 3–locular, 3 × 2.5 mm. Style long, 2–3 × as long as the corolla, 14–16 mm, filiform, usually glabrous or with a few hairs near the apex; stigmas 3. Capsule erect, cup-shaped, 3.5 × 3 mm, dehiscent by 3 valves near the base. Seeds numerous, oblong, glabrous, yellowish.

HAB. In the mountains on cliffs in rock crevices; alt. 1000–1200 m; fl. & fr. Jul.-Aug.
DISTRIB. Rare in Iraq – only found in two neighbouring places near the Turkish frontier in the NW sector of the middle forest zone. **MAM**: *Sharanish*, ± 25 km NE of Zakho, *Rawi & Rechinger* 16689!, *Rawi* 23692!; Khantur mt., NE of Zakho, *Rawi* 23328!

Syria, Turkey (S & E Anatolia).

3. **ASYNEUMA** Griseb. et Schenk

in Wiegm. Archiv. 18(1): 335 (1852)
Podanthum (G. Don) Boiss., Fl. Orient. 3: 945 (1875)

Perennial or biennial herbs. Stems erect or ascending. Leaves simple, alternate, petiolate or sessile. Flowers solitary or clustered in spikes or racemes, sessile or pedicellate. Calyx 5-lobed. Corolla divided to base into 5 narrow lobes. Stamens 5, free. Ovary inferior, 3-locular. Style simple, usually hairy. Stigma 3-fid. Capsule 3-locular, dehiscent by 3 lateral valves near apex, middle or near base. Seeds numerous, small.

Some 50 species in the Mediterranean region to Caucasus and Iran; five species in Iraq.
Asyneuma (name of uncertain origin. Fedorov (1957) suggests that the Gr. word ασυνευμα might be translated as 'not together', meaning in this case the character typical of the genus: separation of the corolla lobes, as opposed to the connate lobes in *Phyteuma*.

1. Capsule nodding, dehiscent by 3 lateral valves to near base .... 5. *A. persicum*
   Capsule erect, dehiscent by 3 lateral valves to the middle
   or to near the apex 2
2. Upper leaves cordate-amplexicaul; lowers leaves petiolate.......................
   .... 1. *A. amplexicaule* subsp. *aucheri*
   Upper leaves elliptic to lanceolate to linear-lanceolate 3
3. Flowers pedicellate in lax branched or unbranched spikes .... 2. *A. lobelioides*
   Flowers sessile or shortly pedicellate 4
4. Lower leaves oblong, 1.5–2.5 × 0.8–1.4 cm; calyx scabrid;
   capsule erect, oblong, crowned with the persistent erect
   calyx lobes.......................................3. *A. rigidum* subsp. *rigidum*
   Lower leaves oblong-lanceolate, 8–10 × 1.5–2 cm; calyx
   glabrous; capsule not crowned with persistent erect calyx lobes .....4. *A. pulchellum*

1. **Asyneuma amplexicaule** (*Willd.*) *Hand.-Mazz.* in Ann. Naturh. Mus. Wien 27: 431 (1913); Bornmüller in Beih. Bot. Centralbl. 38(2): 338 (1921); Fedorov in Fl. URSS 24: 406 (1957); Rawi in Dep. Agr. Iraq Tech. Bull. 14: 127 (1964); Rech. f. & Schiman-Czeika in Rech. f., Fl. Iranica 13: 43 (1965); Damboldt in Fl. Turk. 6: 68 (1978).

Phyteuma amplexicaule Willd., Sp. Pl. 1: 925 (1798), as 'amplexicaulis'.
Campanula amplexicaulis (Willd.) Boiss., Diagn. Pl. Or. Ser. 1, 11: 77 (1849), non Michx., (1803).
Podanthum amplexicaule (Willd.) Boiss., Fl. Orient. 3: 948 (1875).

subsp. **aucheri** (*A.DC.*) *Bornmüller* in Beih. Bot. Centralbl. 38: 339 (1921); Rech. f. & Schiman-Czeika in Fl. Iranica 13: 43 (1965).

Phyteuma aucheri A.DC., Prodr., 7: 456 (1839); *Podanthum amplexicaule* var. *aucheri* (A.DC.) Hausskn. ex Bornmüller in Beih. Bot. Centralbl. 20(2): 177 (1906).

Perennial herb. Stem erect, glabrous or pubescent particularly towards the base, 30–70 cm tall. Leaves serrate, usually glabrous or glabrescent on upper surface and glabrous or hairy on lower surface; lower leaves shortly petiolate, ovate-oblong, 3.5–7 × 1.5–2.5 cm; upper leaves sessile, lanceolate, cordate-amplexicaul, 2–4 × 0.4–0.5 cm, uppermost leaves small, 5–10 × 1–2 mm. Flowers small, numerous, shortly pedicellate, crowded in sessile clusters. Calyx glabrous; calyx lobes triangular-lanceolate, 1 × 0.4–0.5 mm; calyx tube ovoid, as long as or slightly longer than its lobes, 1–1.3 × 0.5 mm. Corolla glabrous, longer than calyx-lobes, 6–8 mm. Stamens glabrous; filaments short, 1.5–2 mm long, dilated at base, usually slightly ciliate with short hyaline hairs. Style hairy, 8–12(–14) mm. Capsule erect, ovoid, 1 × 0.5 mm, dehiscent by 3 lateral valves near the middle. Seeds small, brown.

HAB. High mountain slopes, in damp places; alt. 1800–2300(–2600) m; fl. & fr. Jul.-Aug.
DISTRIB. Occasional in the subalpine thorn-cushion zone of Iraq. **MAM**: nr Zawita, *Rechinger* 12000; **MRO**: on the NE slopes of Halgurd Dagh, *Gillett* 9551!; by lake Kermasur on Qandil range, *Rawi & Serhang* 18290!; lower slopes of Qandil range, NE of Rania, *id.* 26708!; Serin, NE of Rania, *id.* 18239!
NOTE. subsp. *amplexicaule* differs in its sessile leaves, and is found from Turkey to Caucasus. All Iraq material is referable to subsp. *aucheri*.

Turkey, Caucasus, N Iran.

2. **Asyneuma lobelioides** (*Willd.*) *Hand.-Mazz.* in Ann. Naturh. Mus. Wien 27: 431 (1913); Blakelock in Kew Bull. 4: 517 (1949); Fedorov in Fl. URSS 24: 417 (1957); Rawi in Dep. Agr. Iraq Tech. Bull. 14: 127 (1964); Rech. f. & Schiman-Czeika in Fl. Iranica 13: 45 (1965); Damboldt in Fl. Turk. 6: 72 (1978).

Phyteuma lobelioides Willd., Phytogeogr. 1: 20, t. 4, f. 2 (1794).
Podanthum lobelioides (Willd.) Boiss., Fl. Orient. 3: 953 (1875).
Asyneuma lobelioides var. *filipes* Nábělek in Publ. Fac. Sci. Univ. Masaryk 70: 8 (1926); Blakelock in Kew Bull. 4: 517 (1949); Rawi in Dep. Agr. Iraq Tech. Bull. 14: 127 (1964).

Perennial herb. Stems erect or ascending, sometimes decumbent, arising from a woody base, slightly hairy especially towards the base and glabrescent or glabrous towards the apex, 15–70 cm tall, simple or branched. Leaves simple, denticulate, undulate or entire, hairy; radical leaves petiolate, oblong-lanceolate, 4–10 × 0.5–1.5 cm, petiole 3–4 cm; cauline leaves few to many, sessile or sometimes tapering to a short petiole, elliptic to linear-lanceolate, 2.5–5 × 0.2–1.5 mm. Flowers pedicellate, in lax bracteolate branched or unbranched spikes; bracts small, triangular-lanceolate, 1.5 × 0.5 mm. Calyx glabrous, lobes oblong-lanceolate, 2–3.5 mm wide; calyx tube ovoid-oblong, 1.5–3 × 0.5–1.2 mm. Corolla glabrous, longer than calyx-lobes. Filaments 2 mm long, dilated and ciliate at the base. Style hairy, 8–10 mm. Capsule oblong-ovate, erect, glabrous, 3–5 × 1.5–2 mm, crowned with the persistent erect calyx lobes, dehiscent by 3 lateral valves near apex. Seeds small, rotund, 1 × 0.8 mm, brown.

HAB. Stony mountain sides and rocky slopes, among *Astragalus* bushes; alt. (2000–)2600–3000(–3600) m; fl. & fr. Jul.-Aug.
DISTRIB. Occasional, but locally common, in the central sector of the alpine and thorn-cushion zones of Iraq. **MRO**: Halgurd Dagh, *Guest & Ludlow-Hewitt* 2966!, *Gillett* 9565!, 12334!, 12358!; Hisar-i Rost, *Guest & Alizzi* 15819!; Ser Kurawa, *Gillett* 9763!; **MSU**: Beribadan, *Ludlow-Hewitt* 1515!

Turkey (SE Anatolia), Iran, Armenia, Caucasus.

3. **Asyneuma rigidum** (*Willd.*) *Grossh.*, Opred. Rast. Kavk. 426 (1949); Fedorov in Fl. URSS 24: 410 (1957); Damboldt in Fl. Turk. 6: 77 (1978).

subsp. **rigidum** *Damboldt* in Fl. Turk. 6: 7 (1978).

Phyteuma rigidum Willd., Sp. Pl. 1: 925 (1798).
Phyteuma lanceolata Willd., Sp. Pl. 1: 924 (1798).
Podanthum lanceolatum (Willd.) Boiss., Fl. Orient. 3: 951 (1875).
Asyneuma lanceolata var. *rigidum* (Willd.) Boiss., Fl. Orient. 3: 952 (1875).
Asyneuma lanceolata (Willd.) Hand.-Mazz. in Ann. Naturh. Mus. Wien 27: 431 (1913); Nábělek in Publ. Fac. Sci. Univ. Masaryk 70: 7 (1926); Fedorov in Fl. URSS 24: 409 (1957); Rawi in Dep. Agr. Iraq Tech. Bull. 14: 127 (1964); Rech. f. & Schiman-Czeika in Fl. Iranica 13: 45 (1965).

Perennial herb. Stems slender, pubescent or scabrid, erect or ascending, usually simple or branched towards base, 25–60 cm tall. Leaves denticulate, scabrid or glabrous, the lower oblong, petiolate, 1.5–2.5 × 0.8–1.4 cm, the upper linear-lanceolate, sessile, 15–40 × 2–6(–10) mm. Flowers small, in loose spikes, usually sessile or shortly pedicellate. Calyx scabrid; calyx lobes lanceolate, 2–3 × 0.7–1 mm; calyx tube obconical, nearly 1.5–2 × 1 mm. Corolla 5–7 mm. Filaments dilated and ciliate at base, 2 mm. Style hairy, 8–10 mm. Capsule erect, oblong, 5–7 × 2–3 mm, dehiscent by 3 lateral valves near apex, crowned with the persistent erect calyx lobes. Seeds small, oblong, 2 × 0.8 mm, brown.

DISTRIB. (of species): W Syria, Sinai, Turkey, Caucasus, Iran, N Africa (Algeria).
HAB. High mountain slopes, by a stream, along a roadside; alt. 2200–2850 m; fl. & fr. Aug.-Sep.
DISTRIB. Occasional in the central sector of the thorn-cushion zone of Iraq. **MRO**: Warshanka – Magar mt., *Rawi & Serhang* 24335!; between Bardanas and the Qandil range, *id.* 24634!; NE of Qandil, *id.* 24389!

Turkey (Anatolia), Caucasus, NW Iran.

4. **Asyneuma pulchellum** (*Fisch. et Mey.*) *Bornmüller* in Beih. Bot. Centralbl. 38(2): 348 (1921); Fedorov in Fl. URSS 24: 414 (1957); Rawi in Dep. Agr. Iraq Tech. Bull. 14: 127 (1964); Rech. f. & Schiman-Czeika in Fl. Iranica 13: 46 (1965); Damboldt in Fl. Turk. 6: 79 (1978).

Phyteuma pulchellum Fisch. & Mey., Ind. Sem. Petrop. 1: 35 (1835).

Podanthum pulchellum (Fisch. & Mey.) Boiss., Fl. Orient. 3: 947 (1875).

Biennial herb. Stems branched, slightly hairy or glabrescent, 50–80 cm tall. Leaves hirsute on both sides; radical leaves oblong-lanceolate, 8–10 × 1.5–2 cm, petiolate, denticulate; petiole 6–10 cm; cauline leaves few, sessile, 3–5 × 1 cm. Flowers sessile, bracteolate, forming a spike consisting of distant clusters each of 2–5 flowers; bracts small, reniform, denticulate, 2–3.5 × 2–5 mm, hirsute or glabrous. Calyx glabrous or nearly so, calyx lobes lanceolate, 3–5 × 1–1.5 mm; calyx tube turbinate, 2–3 × 1–1.5 mm. Corolla usually glabrous, 4–7 mm. Filaments hairy, 2 mm. Style hairy, 6–8 mm. Capsule erect, glabrous, ovoid, 3-locular, 5 × 3 mm, dehiscent by 3 lateral valves near apex. Seeds numerous, small, 1.5 × 1 mm, yellow-brownish.

Hab. Stony mountain sides and steep dry rocky slopes, in oak forest and oak-juniper scrub; alt. 1300–1700 m; fl. & fr. Jun.-Aug.
Distrib. Rather rare in the NW and C sectors of the upper forest zone of Iraq. **MAM**: 27 km NNE of Zakho, nr Sharanish, *Rawi* 23270!; Sarsang, *Haines* W. 1201!, 1717!; **MRO**: Haji Umran, *Omar, Sahira, Karim & Hamid* 38473.

Turkey (SE Anatolia), Caucasus, W, NW & N Iran.

5. **Asyneuma persicum** (*A.DC.*) *Bornmüller* in Beih. Bot. Centralbl. 38(2): 347 (1921); Rech. f. & Schiman-Czeika in Fl. Iranica 13: 41 (1965).

Campanula persica A.DC. in DC., Prodr. 7: 483 (1839); Damboldt in Fl. Turk. 6: 39 (1978), *nom. illegit.*
Phyteuma asperum Boiss., Diagn. Pl. Orient. Ser.1, 11: 72 (1849).
Campanula syspirensis C. Koch in Linnaea 23: 639 (1850); Bornmüller in Beih. Bot. Centralbl. 61: 73 (1941); Blakelock in Kew Bull. 4: 517 (1949); Rawi in Dep. Agr. Iraq Tech. Bull. 14: 128 (1964).
Podanthum persicum (A.DC.) Boiss., Fl. Orient. 3: 956 (1875); Zohary in Dep. Agr. Iraq Bull. 31: 144 (1950); Rawi, ibid. Tech. Bull. 14: 129 (1964).
Podanthum asperum (Boiss.) Boiss., Fl. Orient. 3: 955 (1875); Zohary in Dep. Agr. Iraq Bull. 31: 144 (1950); Rawi, ibid. Tech. Bull. 14: 129 (1964).
Asyneuma asperum (Boiss.) Rech. f. & Schiman-Czeika in Fl. Iranica 13: 41 (1965).

Perennial herb, much branched from base. Stems and branches erect or ascending, slightly scabrid or hairy, greenish, 20–40 cm long. Leaves scabrid or hairy; radical leaves petiolate, decurrent, 10–12 × 1–4 mm, the uppermost leaves small, lanceolate. Flowers pedicellate or subsessile, bracteolate, in spikes. Calyx hairy all over or sometimes glabrous or subglabrous; calyx lobes triangular-lanceolate, 1–4 × 0.5–1 mm, usually reflexed; calyx tube ovate-turbinate, 1.5–2.5 × 1–1.5 mm. Corolla hirsute all over, or sometimes almost glabrous, longer than calyx lobes, corolla tube 1/3–1/4 the length of the lobes. Stamens 5, free, filaments dilated at base, 1.5–2 mm. Style hairy, 8–10 mm. Capsule nodding, ovoid, 3–5 × 2–4 mm, dehiscing by 3 lateral valves near base. Seeds small, yellow-brown, 2–3 × 1–1.3 mm. Fig. 119: 1–8.

Hab. Dry rocky mountain slopes, on barren metamorphic rock, on serpentine and limestone, on a cliff face with S aspect, on scree, occasionally near streams; alt. (1500–)1800–3000(–3300) m; fl. & fr. Jul.–Aug.(–Sep.).
Distrib. Common in the central sector of the alpine and subalpine thorn-cushion zones of Iraq, occasional in the upper forest area. **MAM**: Gali Zawita, NW of Zakho, *Rawi* 23626 ?; Mt. Gara, *Kotschy* 377 ?; Sarsang, *Haines* W. 431 ?; **MRO**: Halgurd Dagh, *Guest* 2968!, *Gillett* 9560!, 9575!, 12322!, 12344!, 12449!; Bermasand lake on N side of Halgurd, *Rawi & Serhang* 24771!; Musaiyif Nawanda, *Guest & Alizzi* 15808!, *Rechinger* 11395 ?; Chiya-i Mandau, *Guest & Ludlow-Hewitt* 2796!; Sakri Sakran, *Bornmüller* 1550!; Rayat, *Guest* 13048 ?, 13080!, *Guest & Alizzi* 15887!; in valley between Rayat and Haji Umran, *Rechinger* 11281; Sula Khal, *Rawi & Serhang* 24689!; Sia Kuh, *Ludlow-Hewitt* 1534!; Haji Umran, Guest 13052, *Gillett* 946!, *Rawi* 24277!; Warshanka-Magar, *Rawi & Serhang* 24327!; Qandil range, *id.* 24061!, 26729!; Gom-i Kermasur (lake) on Qandil range, *id.* 24143!; NE of Qandil, *id.* 24375!, 24425!; Serin, on the way from Rania to Qandil, *id.* 24012!; **MSU**: Pira Magrun, *Haines* W. 1857 ?; Zalam, on Hawraman range, *Rawi, Husham (Alizzi) & Nuri* 29376!; Hawraman,*Rawi, Chakr., Nuri & Alizzi* 19724!, 19744!; ? **MSU**: (or in Iran?), "Avroman & Schahu", *Haussknecht* s.n.; **FKI/DGA**: between Kirkuk and Khalis, *Rawi, Chakr., Nuri & Alizzi* 19711!
Note. This shrubby herb is said to be locally known in Halgurd district as BURRUK (Kurd. *Guest & Alizzi*). So far as we are aware it has never been brought into cultivation (not listed in the RHS "Plant Finder").

Turkey (SE Anatolia), W & S Iran.

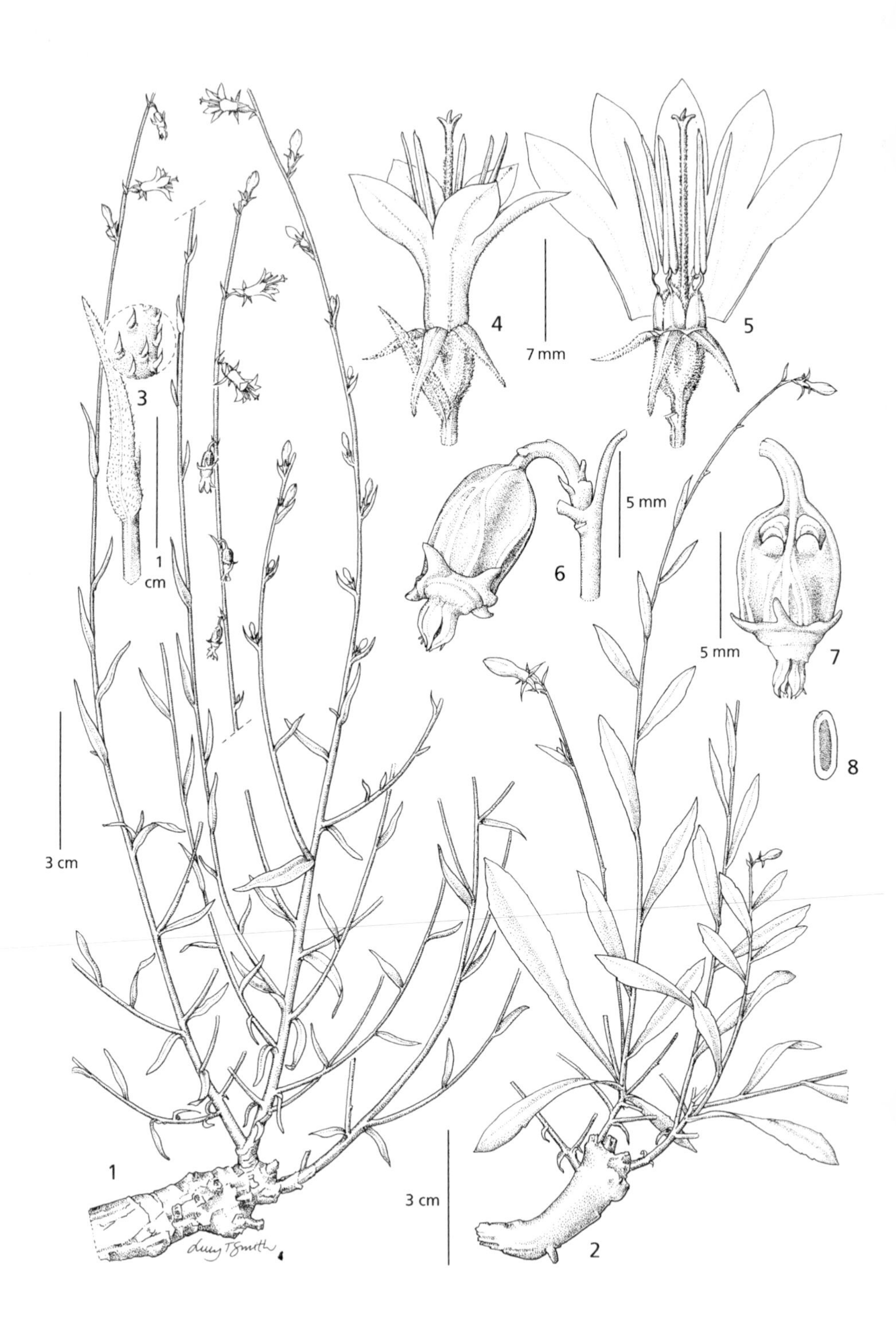

Fig. 119. **Asyneuma persicum**. 1, 2, habit; 3, leaf with hair detail; 4, flower; 5, flower with corolla opened; 6, 7, fruit, different views; 8, seed. © Drawn by L. Smith (Sep. 2012).

4. LEGOUSIA Durand

Fl. Bourg. 1: 37 (1782)
Specularia A. DC., Monogr. Campan.: 44 (1830)

Annuals. Stem erect or ascending, simple or branching usually from base. Leaves simple, alternate, sessile, exstipulate. Flowers solitary or in cymes. Calyx 5-lobed. Corolla rotate, 5-lobed. Stamens 5, free; filaments dilated at base. Ovary 3-locular. Styles hairy; stigma 3-fid. Capsules linear-oblong, prismatic, dehiscent by small lateral valves beneath apex. Seeds small.

Some 15 species distributed in the Northern Hemisphere, one or two extending south into S America; three species in Iraq.

Legousia (named after the 17th century French traveller, Le Gouz – according to Fedorov). The name of its synonym, *Specularia*, is said to be from the Latin, speculum, mirror, the common European species (*L.* (*S.*) *veneris*) having been widely known as Venus' looking-glass.

1. Flowers solitary or in cymes; calyx lobes reflexed or curved downwards; corolla lobes acute .... 1. *L. falcata*
   Flowers solitary or in terminal cymes; calyx lobes erect or ascending; corolla lobes mucronate 2
2. Calyx lobes as long as the ovary at flowering; filaments glabrous; capsule constricted at apex .... 2. *L. speculum-veneris*
   Calyx lobes about half as long as the ovary at flowering; filaments ciliate at base; capsule not constricted at apex .... 3. *L. pentagonia*

1. **Legousia falcata** (*Ten.*) *Fritsch* ex *Janch.* in Mitt. Naturwiss. Vereins. Univ. Wien n.s. 5: 100 (1907); Bornmüller in Beih. Bot. Centralbl. 61: 74 (1941); Fedorov in Fl. URSS 24: 431 (1957); Rech. f. & Schiman-Czeika in Fl. Iranica 13: 5 (1965); Zohary, Fl. Palaest. 3: 284 (1978); Damboldt in Fl. Turk. 6: 84 (1978).

Prismatocarpus falcatus Ten., Fl. Nap. Prodr.: 16 (1811–1815); Fl. Nap. 1: 77, t. 20 (1811–1815).
Campanula falcata Roem. & Schult., Syst. 5: 154 (1819).
Specularia falcata A. DC., Monogr. Campan.: 345 (1830); Boiss., Fl. Orient. 3: 960 (1875); Nábělek in Publ. Fac. Sci. Univ. Masaryk 70: 8 (1926); Zohary in Dep. Agr. Iraq Bull. 31: 144 (1950); Rawi, ibid. Tech. Bull. 14: 129 (1964).

Annual herb. Stems ascending or decumbent, slightly scabrid with small stiff hairs, 15–40 cm. Leaves slightly crenate, scabrid with few, small hairs and sometimes nearly glabrescent; lower leaves ovate, 10–20 × 10–15 mm, petiolate or nearly so; upper leaves ovate-oblong to lanceolate, 20–25(–30) × 10–14 mm. Flowers solitary or in cymes. Calyx lobes oblong-linear, 10–15 × 1–1.5 mm, reflexed or curved downwards, usually glabrous or sometimes scabrid; calyx tube oblong, 10 × 2 mm, usually scabrid with sparse stiff hairs. Corolla small, 6–8 mm, glabrous or nearly so; lobes acute. Filaments short. Ovary oblong; style hairy, 6–8 mm. Capsule oblong, 8–10(–12) × 1.5–2 mm, erect, usually scabrid or sometimes glabrescent, dehiscent by small valves near the apex. Seeds small, orbicular or rotund, ± 1 mm in diameter, glabrous, dark brown. Fig. 120: 9–11.

HAB. Lower mountain slopes; alt. 900–1600 m; fl. & fr. May–Jun.
DISTRIB. Occasional in the middle forest zone of Iraq. **MAM**: Aqra, *Rawi* 11447A!; **MRO**: Shaqlawa, *Gillett* 11570!; Kuh-i Sefin, nr Shaqlawa, *Bornmüller* 1565; Serkupkhan, N of Rania, *Rawi, Nuri & Kass* 28529!; **MSU**: Pira Magrun and Hawraman, *Haussknecht* s.n. in *Bornmüller*, l.c.; Pira Magrun, *Rawi* 12120!
NOTE. GÎYA PANJI (Kurd.-Sulaimaniya, *Rawi* 12120).

S Europe (Spain to Greece), Aegean Isles, Cyprus, Syria, Lebanon, Palestine, Jordan, Turkey, Caucasus, W & S Iran, N Africa (Morocco, Algeria), Macaronesia (Madeira, Canary Is.).

2. **Legousia speculum-veneris** (*L.*) *Chaix* in Vill., Pl. Dauph. 1: 338 (1786); Zohary, Fl. Palaest. 3: 283 (1978); Fedorov in Fl. URSS 24: 428 (1957); Rech. f. & Schiman-Czeika in Fl. Iranica 13: 6 (1965); Damboldt in Fl. Turk. 6: 85 (1978).

Campanula speculum-veneris L., Sp. Pl. ed. 1: 168 (1753).
Specularia speculum (L.) A.DC., Monogr. Campan.: 346 (1830); Boiss., Fl. Orient. 3: 958 (1875).
S. speculum-veneris (L.) Caruel in Parl., Fl. Ital. 8: 139 (1888).

Annual herb. Stem erect or ascending, usually branching from base, scabrid with short, reflexed, stiff hairs, 10–30 cm long. Leaves sessile, serrulate or shallowly serrate, usually scabrid, lower leaves ovate, upper leaves oblong-lanceolate. Flowers solitary or in terminal cymes, sessile. Calyx lobes linear or linear-lanceolate, about as long as the ovary at flowering. Corolla rotate, scabrid or sometimes nearly smooth, ± 10 mm; lobes mucronate. Stamens free; filaments short, glabrous. Style viscid, hairy. Capsule oblong, constricted at apex.

HAB. Mountains, and fallow fields; alt. 900–1400 m; fl. & fr. Mar.-May.
DISTRIB. Apparently rare in the mountains of northern Iraq; citations from Fl. Iranica. **MAM**: Amidiya [Amadiyah], Shaikh Adi, *Thesiger* 704, 718. **MRO**: Baradost Dagh, Handren, *Thesiger* 829 p.p.

C & S Europe, Cyprus, Crete, Syria, Caucasus, Turkey, W Iran.

3. **Legousia pentagonia** (*L.*) *Thell.* in Vierteljahrsschr. Naturf. Ges. Zürich 46: 465 (1908); Bornmüller in Beih. Bot. Centralbl. 61: 74 (1941); Zohary in Fl. Palest. 3d. 2, 2: 170 (1933); Fedorov in Fl. URSS 24: 429 (1957); Rech. f. & Schiman-Czeika in Fl. Iranica 13: 6 (1965); Zohary, Fl. Palaest. 3: 284 (1978); Damboldt in Fl. Turk. 6: 85 (1978).

Campanula pentagonia L., Sp. Pl. ed. 1: 169 (1753).
Specularia pentagonia (L.) A. DC., Monogr. Campan.: 344 (1830); Boiss., Fl. Orient. 3: 959 (1875); Nábělek in Publ. Fac. Sci. Univ. Masaryk 70: 8 (1926); Zohary, Fl. Palest. ed. 2, 2: 170 (1933); Blakelock in Kew Bull. 4: 518 (1949); Zohary in Dep. Agr. Iraq Bull. 31: 144 (1950); Rawi, ibid. Tech. Bull. 14: 29 (1964).

Annual herb. Stem erect or ascending, usually branching from base, scabrid with short, reflexed, stiff hairs, 7–25 cm long. Leaves sessile, 5–20(–25) × 3–8 mm, serrulate or shallowly serrate, usually scabrid, lower leaves ovate, upper leaves oblong-lanceolate. Flowers solitary or in terminal cymes, sessile. Calyx scabrid or sometimes nearly smooth; lobes linear or linear-lanceolate, longer than the corolla tube, 8–15 × 1 mm, erect; calyx tube oblong, 8–20 × 1–2 mm. Corolla purple, rotate, scabrid or sometimes nearly smooth, 10–15 mm; lobes mucronate. Stamens free; filaments short, ciliate at base. Style viscid, hairy, about 8–12 mm long. Capsule oblong, 4–8 × 0.5–2 mm. Seeds small, 1 × 0.8 mm, brown, shining. Fig. 120: 1–8.

HAB. Mountain pastures, fields; alt. (500–)900–1400 m; fl. & fr. Apr.-May(–Jun.).
DISTRIB. Occasional in the middle forest zone of Iraq. **MAM**: Dohuk, *Guest* 2321; Betas, on Jabal Bekher, *Rawi* 8626 ?; Aqra district, *Anon.* 26069!; **MRO**: on Kuh-i Sefin at Shaqlawa, *Bornmüller* 1564, *Haines* W. 669; Sefin Dagh, *Gillett* 8128; Baradost, nr Shan-i Dar, *Erdtman & Goedemans* in *Rechinger* 15581, 15620; Zeita, nr Shirwan Mazin, *Haines* 2136; Kurek Dagh, nr Rowanduz, *Nábělek* 2107; Saran, by Kani Kawan spring, *Nuri, Kass & Serhang* 27314; Gali Warta, 30 km NW by N of Rania, *Rawi, Nuri & Kass* 28834; **MSU**: Sulaimaniya, *Haussknecht* s.n.!; Gweija Dagh, nr. Sulaimaniya, *Mooney* 4331; "Kurdistan", *Mrs A.M. Low* 180.
NOTE. Blakelock (1949) suggested that it was doubtful whether this species was distinct from *L. speculum-veneris* as the size of the corolla and the degree of constriction at apex of immature capsules varies.

Mediterranean Europe (Balkans, Greece), Crete, Aegean Isles, Syria, Lebanon, Palestine, Jordan, Turkey, Caucasus, W Iran.

Species of doubtful occurrence

Legousia meiklei (in ms.) was described by Alizzi from apparently two collections from the lower moist-steppe zone of Iraq, but unfortunately this material cannot be located either at the National Herbarium, Baghdad or at K. The description as in the original manuscript is reproduced here, with the hope that the collection may turn up if it has been misplaced and not destroyed. It appears to be a good species, differing in its small stature and in being glabrous.

Legousia meiklei Alizzi (in ms.). — Type: Iraq, Upper hills and foothill region, Kifri, *Gillett & Rawi* 7393! (BAG, holo. – not located).

Small annual herb, 3–5 cm long, glabrous. Stem erect, filiform, branching from base. Leaves small, sessile, crenate at or toward the apex and tapering toward the base. Flowers small, in terminal cymes. Calyx 5-lobed; lobes narrowly oblong, 3–4 × 0.5–0.6 mm, longer than calyx tube; calyx tube funnel-shaped or obconical, 1.5–3 × 1–1.6 mm. Corolla glabrous, mauve, lobes shorter than corolla-tube, acute. Stamens glabrous; filaments membranous, ±

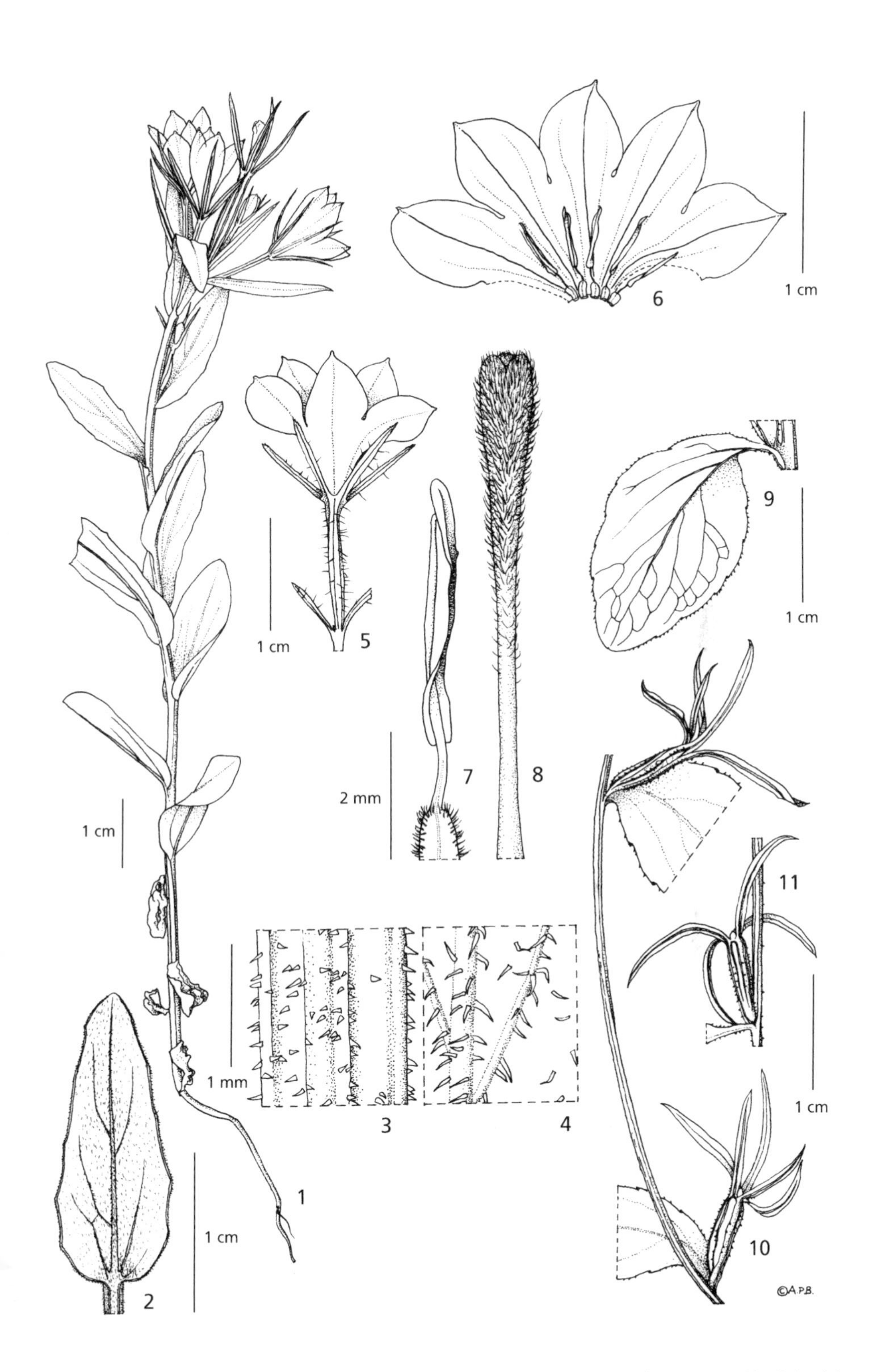

Fig. 120. **Legousia pentagonia**. 1, habit; 2, leaf; 3, indumentum of stem; 4, indumentum of leaf abaxial surface; 5, flower; 6, corolla opened; 7, stamen; 8, style and stigma. **Legousia falcata**. 9, leaf; 10, habit with fruit; 11, fruit with calyx. 1–4 from *Rawi et al.* 21834; 5–8 from *Rustam* 2321; 9–11 from *Rawi* 12120. © Drawn by A.P. Brown (Sep. 2012).

1 mm long. Style hairy, 1–1.5 mm long. Capsule obconical, 1 × 0.7 mm, dehiscent by 3 valves near apex. Seeds numerous, very minute, 0.5 × 0.3 mm, glabrous, brown.

HAB. Dry soil on rather saline steppic plain; alt. 200–250 m; fl. & fr. Apr.
DISTRIB. Rare; only found twice in the lower moist-steppe zone of Iraq. **FUJ:** Qaiyara, *Bayliss* 87; **FKI:** Kifri, *Gillett & Rawi* 7393 (type).

INDEX TO FAMILIES, GENERA AND SPECIES

Scientific names of accepted species and infraspecific taxa are in **bold**, synonyms and misapplied names are in *italics*. Names of valid taxa which are not recorded from Iraq are in normal font, and subgeneric and sectional names are in small capitals. Where a name appears on more than one page, the principal page reference is given in **bold**.

INDEX TO VERNACULAR NAMES

These are cited in the text along with their source, and in some cases are translated into English. Arabic and Kurdish names are capitalized in this list; English names are given in lower case letters. Kurdish names prefixed by GÎYĀ (weed or plant) can be found under this name rather than in their normal alphabetical position.